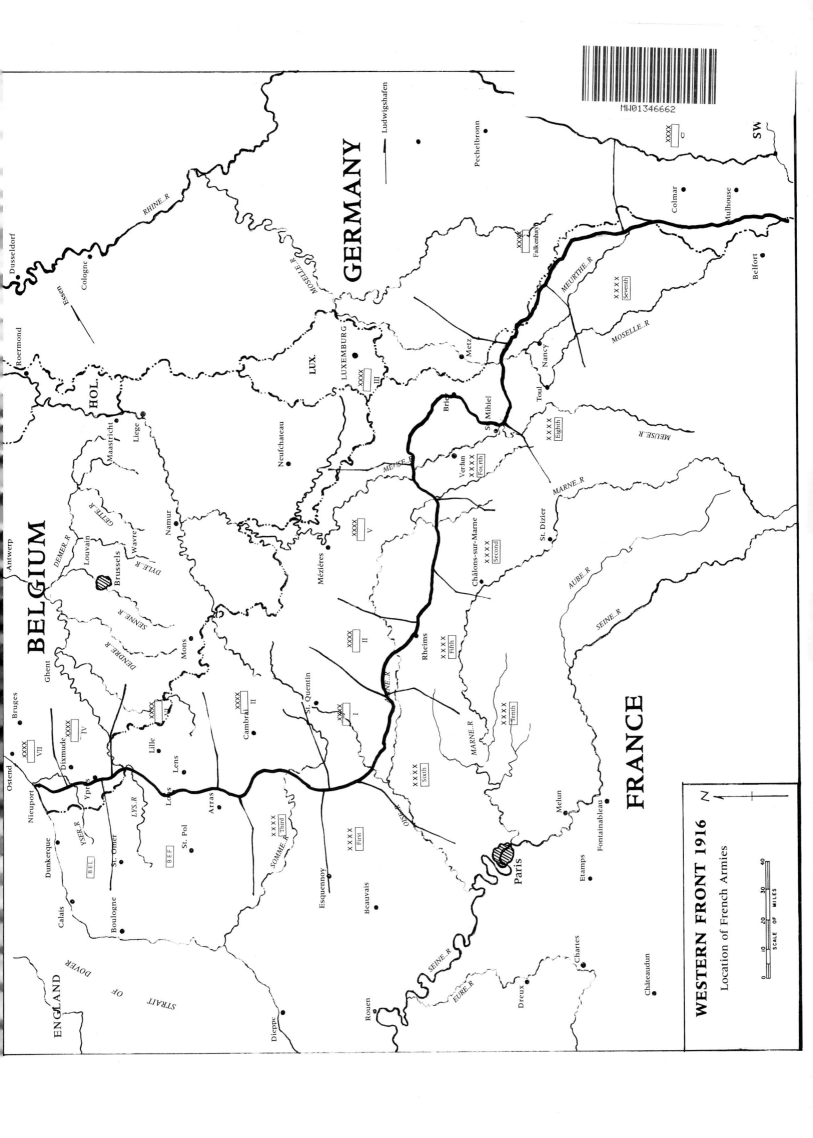

French Aircraft
of the
First World War

Dr. James J. Davilla
Arthur M. Soltan

Dedication

This work is dedicated to my parents, Ray and June Davilla. Dr. James J. Davilla

This work is dedicated to Henri Phély, my father-in-law and Salmson 2 A2 pilot. Arthur M. Soltan

© 1997 by Dr. James J. Davilla and Flying Machines Press

Published in the United States by Flying Machines Press, 35 Chelsea Street, Stratford, Connecticut, 06497

Book and cover design, layout, and typesetting by John W. Herris.

Color aircraft illustrations, color section design, and color section text by Alan E. Durkota.

Scale drawings by Carl Ahremark, Martin Digmayer, Colin Owers, Dennis Punnett, and Ian Stair.

Digital image editing by Aaron Weaver and John W. Herris.

Text edited by R.D. Layman.

Printed and Bound in the United States by Walsworth Publishing Company, Inc., Marceline, MO

All rights reserved. No part of this publication may be reproduced, stored in a retrieval system, or transmitted in any form by any electronic or mechanical copying system without the written permission of the publishers.

Library of Congress Cataloging-in-Publication Data

Davilla, James J., 1951–
French Aircraft of the First World War / James J. Davilla, Arthur M. Soltan.
    p.   cm.
Includes bibliographical references and index.
ISBN 0-9637110-4-0 (alk. paper)
1. Airplanes, Military—France—History.
2. World War, 1914–1918—Aerial operations, French.
3. World War, 1914–1918—Equipment.
I. Soltan, Arthur M., 1930–   . II. Title.
UG1245.F8D38   1997
623.7'461'094409041—dc21   96-52092
                      CIP

Flying Machines Press

For the latest information on books from Flying Machines Press, visit our web site at www.flying-machines.com

Contents: French Aircraft of the First World War

Preface	v	Janoir	284
The Aviation Militaire 1914–1918	1	Laboratory Eiffel	285
The Aviation Maritime 1914–1918	21	L.A.F.	287
The French Aviation Industry 1914–1918	33	Labourdette-Halbronn	287
A.R.1 and A.R.2	37	Larnaudi	287
Astoux-Vedrines	46	Latecoere	288
Astra	47	Latham	288
Audenis	48	L.D.	289
B.A.J.	49	Letord	290
Bassan-Gue	50	Levasseur	299
Bernard	51	Levy-Besson and Georges Levy	300
Bille SACANA	53	Liore et Olivier	305
Blériot	54	Moineau	308
Borel	76	Moncassin	308
Breguet	82	Morane-Saulnier	309
Brun Cottan	132	Nieuport	349
Canton	132	Noel	421
Caproni/R.E.P.	133	Papin-Rouilly	421
Carroll	139	Ponnier	422
Caudron	141	Rausser	428
Clément-Bayard	176	Renault	428
Coanda-Delaunay-Belleville	177	R.E.P.	430
Courtois-Suffit	178	Salmson	435
Coutant	179	Paul Schmitt	451
De Bruyère	180	Schneider *Henri Paul*	458
Deconde	181	SEA	459
De Marcay	181	Semenaud	462
De Monge	182	Short	463
Delattre	184	S.I.A.	463
Deperdussin	184	Sikorsky	465
Descamps	188	Sopwith	465
Donnet-Denhaut	189	SPAD	474
Dorand	198	Tellier	529
Dormay	201	Van Den Born	539
Doutre	201	Vendôme	539
Dufaux	202	Vickers	540
D.N.F.	203	Voisin	541
E.G.A.	205	Weymann	571
Farman	206	Wibault	572
FLO	257	Appendix: Gliders and Glisseurs	573
F.B.A.\Schrek	257	Colors and Markings	575
Galvin	265	Addendum	611
Goupy	266	Photo Collection	612
Gourdou-Lesseurre	266	Glossary	614
Hanriot	270	Bibliography	614
Hochart	284	Index	616

Scale Drawings

Notes: Drawings are 1/72 scale unless otherwise indicated; * indicates 1/144 scale
CA = Carl Ahremark; MD = Martin Digmayer; CO = Colin Owers, DP = Dennis Punnett; IS = Ian Stair

Aircraft	Illustrator	Page	Aircraft	Illustrator	Page	Aircraft	Illustrator	Page
A.R.1 and A.R.2	CO	41–43	FBA Type C	IS	250B	Nieuport 24	IS	393
Astra Bomber *	CA	47	FBA Type H	CA	274A	Nieuport 25	IS	400
B.A.J. C2 Fighter	DP	50	FBA Type S	CA	274B	Nieuport 27	IS	401
Bernard AB1 Bomber *	CA	52	FBA Type D Cannon Fighter *	CA	264	Nieuport 28	IS	406
Bernard SAB1	DP	53	Galvin Floatplane Fighter (Hc)	DP	265	Nieuport 275-hp Lorraine	IS	409
Bleriot 11	IS	59	Gourdou-Lesseure 2 C1 (Glb)	CA	267	Nieuport 29	IS	413
Bleriot 53 *	CA	64	Gourdou-Lesseure C1 (Gla)	CA	268	Nieuport 31	DP	420
Bleriot Four-Engine Bomber *	CA	66	Hanriot HD.1	CA	271	Ponnier L.1	IS	422
Bleriot 67 *	CA	68	Hanriot HD.2	CA	274	Ponnier M.1	DP	424
Bleriot 71 (not to scale)	CA	69	Hanriot HD.3	IS	276	Ponnier M.2	CA	425
Bleriot 73 *	CA	72	Hanriot HD.5	DP	279	Ponnier P.I	CA	427
Bleriot 74 *	CA	73	Hanriot HD.6	IS	280	Renault O1 Bomber *	CA	429
Borel Floatplane	CA	77	Hanriot HD.7	DP	282	R.E.P. Parasol	CA	430
Borel B.O.2 *	CA	78	Hanriot HD.9	IS	283	R.E.P Type N	CA	432
Borel-Boccaccio Type 3000	CA	80	Janoir J-1 Flying Boat *	CA	285	R.E.P. C1 Fighter	CA	433
Breguet AG 4	CA	87	Laboratory Eiffel Fighter	DP	286	Salmson S.M.1	CA	378B
Breguet SN3 *	CA	90	Latham CH.1	CA	289	Salmson 2	IS	445
Breguet BLC	CA	90A	Letord 1	CA	282A	Salmson 3	DP	447
Breguet-Michelin BM4	CA	90B	Letord 2	CA	282B	Salmson 4	CA	538A
Breguet 5	CA	114A	Letord 3 *	CA	295	Salmson 5	CA	449
Breguet 11 *	CA	100	Letord 5	CA	314A	Salmson 7	CA	450
Breguet 12	CA	114B	Letord 6 *	CA	296	Paul Schmitt PS 6	CA	466A
Breguet AV 1 *	CA	101	Letord 9 *	CA	298	Paul Schmitt PS 7	CA	466B
Breguet AV 2 *	CA	102	Levy-Besson "Alerte"			Schneider *Henri-Paul* Bomber *	CA	459
Breguet 14 A2/B2	CO	119–123	Triplane Flying Boat *	CA	301	SEA 4 C2	CA	461
Breguet 16	CA	146A	Levy-Besson			S.I.A.-Coanda Bomber *	CA	464
Breguet 17	IS	131	Georges-Levy 40 HB 2	CA	306B	Sopwith 1 A2 Strutter	CA	468
Caproni CEP 2 *	CA	136	Liore Et Olivier Leo 4	CA	306	SPAD SA.2 and SA.4	IS	476
Carrol A2	CA	139	Liore Et Olivier Leo 5	CA	307	SPAD SD *	CA	480
Caudron G.3	IS	148	Morane-Saulnier Type G	IS	310	SPAD SE *	CA	481
Caudron G.4	IS	154	Morane-Saulnier Type H	IS	312	SPAD SII *	CA	484
Caudron G.6	CA	146B	Morane-Saulnier Type I	IS	314	SPAD 7	IS	491
Caudron R.4	CA	154A	Morane-Saulnier Type L	IS	319	SPAD 11	MD	496
Caudron R.11	CO	154B	Morane-Saulnier Type N	IS	322	SPAD 12	CO	500
Caudron C.22 *	CA	172	Morane-Saulnier Type P	CO	325	SPAD 13	CO	507
Caudron C.23 *	CA	174	Morane-Saulnier Type S *	CA	327	SPAD 14	CO	513
Clement Bayard Bomber *	CA	176	Morane-Saulnier Type T	CA	314B	SPAD 15	DP	515
CSL C1	CA	178	Morane-Saulnier Type TRK *	CA	330	SPAD 16	MD	517
De Bruyère C1	MD	180	Morane-Saulnier Type U	DP	331	SPAD 17	CO	522
De Marcay C1	CA	182	Morane-Saulnier Type V	DP	332	SPAD 20	DP	525
Deperdussin TT	IS	187	Morane-Saulnier Type AC	DP	333	SPAD 22	DP	527
Descamps 27	IS	189	Morane-Saulnier Type AF	IS	336	Tellier T.2 (Side View)	CA	529
Donnet-Denhaut Leveque			Morane-Saulnier Type AFH	IS	337	Tellier T.3	CA	530A
Flying Boat *	CA	191	Morane-Saulnier Type AI	DP	338	Tellier T.4 *	CA	532
Donnet-Denhaut D.D.2 *	CA	193	Morane-Saulnier Type ANL	CA	342	Tellier T.5	CA	534
Donnet-Denhaut D.D.8	CA	306A	Morane-Saulnier Type ANR	CA	343	Tellier T.6 (Side View)	CA	535
Donnet-Denhaut D.D.10 *	CA	197	Morane-Saulnier Type ANS	IS	344	Tellier T.7 *	CA	530B
Dorand Do.1	CA	199	Morane-Saulnier Type BB	CO	347	Tellier *Vonna* (not to scale)	CA	538
Dorand Armed Interceptor	CA	200	Nieuport 6M	IS	350	Tellier-Nieuport S (not to scale)	CA	538
Dufaux Fighter	DP	202	Nieuport 6H Floatplane	CA	353	Vendôme Twin-Engine		
DNF Bomber *	CA	204	Nieuport 10 and Nine	IS	356	Reconnaissance Airplane *	CA	540
Farman H.F.20 *	CA	210	Nieuport 11 and 16	IS	362	Voisin L	CA	543
Farman H.F.27 *	CA	216	Nieuport 12	CA	366	Voisin 3 (LAS)	IS	548
Farman H.F.30 *	CA	218	Nieuport Triplane #1	IS	370	Voisin 1915 Triplane Bomber *	CA	570A
Farman M.F.7	CA	218A	Nieuport Triplane #2	IS	371	Voisin 4	CA	538B
Farman M.F.11	IS	218B	Nieuport Triplane #3	IS	372	Voisin 5	CA	555
Farman F.30 *	CA	245	Nieuport 14	CA	378A	Voisin 8	IS	562A
Farman F.31 *	CA	250	Nieuport 15 *	CA	377	Voisin 9	CA	562
Farman F.40	CA	242A	Nieuport 17	IS	381	Voisin 10	CA	562B
Farman F.50	CA	242B	Nieuport 18/19 Bomber Project *	CA	386	Voisin 12 *	CA	568
Farman F.51 (not to scale)	CA	255	Nieuport 20	CA	388	Voisin E.28 *	CA	570B
Farman Unknown (not to scale)	CA	256	Nieuport 21	IS	389	Weymann W-1	DP	571
FBA Type B	CA	250A	Nieuport 23	IS	391	Wibault Wib.1	DP	573

Preface

Despite the voluminous literature on military aviation during the First World War, surprisingly little has been written about French aircraft during this period. Detailed works have appeared on British, German, Russian, and even Austro-Hungarian aircraft from 1914 to 1918, while French aircraft remain cloaked in mystery.

The reasons for the lack of information on French aviation remain unclear. A large part of the explanation lies with the French defeat in 1940. In the interwar period the French produced a series of 150 books organized into 12 tomes covering their military forces during the Great War. The 13th tome was to have been devoted to French military aviation, but the impending outbreak of war prevented this section from ever appearing. If this final tome was to have been as comprehensive as the preceding volumes, it is likely that it would have rivaled Britain's *The War in the Air* in detail and accuracy.

As German forces approached Paris many of the major French aircraft manufacturers elected to destroy their records rather then let them fall into the hands of the enemy. It has been reported that the records for Blériot-SPAD were destroyed in a huge bonfire held on the company grounds. Other companies are believed to have tossed their records into the Seine river. The French military archives were seized by the Germans and were probably removed to Germany. While portions have been recovered, the French records for this period remain sadly incomplete.

Since the aircraft manufacturers and military failed to record their activities during the First World War, much of the information that is available comes from secondary sources. For example, *L'Aeronautique Pendant La Guerra Mondiale,* written in 1919, covers all facets of French aviation but is incomplete when it comes to aircraft types and manufacturers. Another frequently cited work, *La Guerre Aerienne,* was written at the end of the war and contains little information on aircraft and is often vague concerning operations.

Another factor to be remembered is that the War resulted in huge losses and tremendous suffering for the French people. It is not surprising that the French had little enthusiasm for studying the recent conflict, and this may have contributed to the paucity of literature devoted to military aviation during this period.

Because of the lack of French histories, many historians and writers have depended on the foreign sources that were available. This meant that British and German works were heavily consulted. The British histories, quite appropriately, dealt with the activities of the RFC, RAF, and RNAS with only a brief mention of the Aviation Militaire. The German histories, which tended to be anecdotal and frequently propagandistic, often denigrated the French air service. Some memoirs imply that German aviators held the French in contempt, particularly in comparison with the RFC. It would seem that these comments were usually due to the hostility engendered by the harsh surrender terms imposed by the French on the German people and appear to have little basis in fact. Unfortunately, some writers have elected to take these works at face value.

In an attempt to avoid these pitfalls, the authors have attempted to make extensive use of archival information. Approximately 60,000 pages of information were obtained on over 100 rolls of microfilm. This information consists of the GQG and Armée daily reports on aerial operations, decisions, aircraft losses, escadrille records, and contacts with foreign air services. Although this information was primarily operational in nature, there was also a significant amount of material on the aircraft themselves. This has enabled the writers to uncover previously unknown material on the various types of aircraft serving with operational units. Furthermore, it permitted us to provide a uniquely French perspective on their airplanes rather than having to depend exclusively on foreign sources such as attaché reports. The microfilm rolls have since been donated to the Hoover Institute at Stanford University in Palo Alto, California. Here they will be available to anyone wishing to do further research on French aviation.

Operational Service

Because so little has been written concerning the history of the Aviation Militaire, the authors felt that it was essential that information on operational service be included in this book. All units known to have used a specific aircraft type are included, as are details of the assignments of these units down to the level of the Corps d'Armée. As much detail as possible is given concerning the French tactical and strategic bombing campaigns. This is of particular importance as it was the operational requirements of the French bombing campaign that determined the type and numbers of bombers the Aviation Militaire obtained. The structure of the French fighter escadrilles developed during the war in response to both operational requirements and the quality and quantity of aircraft types available. In order to show the evolution of the French fighter force, the authors have recorded details on the fighter units assigned to the individual Armées and the Groupes de Combat. Although it was not possible within the confines of this book to give a complete history of the Aviation Militaire, it is hoped that the information included will provide the reader with the background against which French military aircraft evolved.

French aircraft equipped a significant part of virtually every Allied air service during the war. The American, Belgium, British, Greek, Italian, Russian, Romanian, and Serbian air forces all relied on aircraft produced by the French aviation industry. In order to show the magnitude of this contribution, the authors have included brief coverage of all French airplanes known to have served with foreign air services.

Designation System

One of the most arcane and complicated areas of French aviation history is the subject of aircraft designations. Virtually every book written on French World War One aviation uses a different designation system for the aircraft and escadrilles. Furthermore, the official documents use a bewildering number of different designations for the same escadrille or aircraft. For example, The Maurice Farman 11 is shown in official records as an M.F.11, MF 11, or Army type IX, X, XI, XII, XIII, XIV, XV, XVI, XVII, XVIII, XIX, XX, XXI, XXII, XXIII, XXIV, XXV, XXVII, XXVIII, XXIX, XXX, XXXI, XXXII, XXXIII, XXXIV, XXXV,XXXVI, XXXVII, XXXVIII, XXXIX, C, CI, CII, CIII, or CIV.

Given these difficulties, the authors have attempted to simplify and clarify the French designation system. First, it was decided to avoid the use of Roman numerals. These were used only intermittently in official documents and can quickly become cumbersome for the reader.

Escadrille designations are presented in capital letters without periods, while periods are used for aircraft. Thus, escadrille Breguet 127 is presented as BR 127 instead of Br 127, BR-127, or BR.127, while a Maurice Farman 11 airplane is shown as M.F.11. This convention should help readers distinguish whether the authors are referring to an aircraft or escadrille. For example, A.R.1 refers to the aircraft, while AR 1 refers to the escadrille.

The manufacturers usually had their own designations for the airplanes they produced. These designations are included in the text but the most commonly used designation is chosen when referring to a specific aircraft type. For example, the Morane-Saulnier Type P had the STAé designation MoS.21. Since this designation was rarely used, the writers ignore the STAé appellation in favor of Morane-Saulnier's type designation. On the other hand, the Salmson-Moineau A92H is referred to by its better known STAé designation: S.M.1.

The authors have attempted to include all known manufacturer, SFA, and STAé designations for each aircraft. However, as mentioned above, the most common designation is used in preference over the less frequently utilized ones.

Even the identity of the manufacturers, themselves, is open to different interpretations. The Deperdussin firm was reformed as the Société Anonyme pour l'Aviation et ses Derives, or SPAD. The firm's airplanes have appeared to in published works as S.P.A.D., SPAD, and Spad. The STAé designations for 1917 used the abbreviation Spa. The authors have chosen to use the SPAD designation for these aircraft, although any of the above appellations are equally valid.

The definitive work on French military aircraft designations has been produced by Jean Devaux in Pegase number 51; the reader can be confidently referred to this article for further information on this often confusing subject.

Three-Views

Due to time constraints it was necessary to obtain the 180 three-views used in this book from several different sources. Our sincere thanks go to Carl Ahremark in Sweden who produced most of the three-views for this book; these drawings constitute a unique historical document in and of themselves. The authors and artist would like to note that some of these three-views were based on photographs (in some cases only two or three were available) due to the lack of official plans. Those drawings which were based on severely limited photographic material are listed as provisional. Our gratitude also goes to Malcolm English at *Air International* who arranged for us to reproduce some of the late Dennis Punnett's three-view drawings; these were used for some of the aircraft which did not develop beyond the prototype stage. Finally, our thanks to Colin Owers, Ian Stair, and Martin Digmayer for the magnificent drawings they produced for our book. Drawings are generally to 1/72 scale except for the larger aircraft which are to 1/144. For binding strength the foldout drawings had to be spaced individually throughout the book.

Color Plates

The color plates are primarily the work of Alan Durkota.

Acknowledgments

A book such as this could not have been written without the help of many others who generously shared their knowledge.

We offer our sincere thanks to Jack Bruce in England. Jack not only reviewed and critiqued the completed manuscript, he also lent us a number of photographs for use in this volume. We should also mention Mr. Bruce's book *Airplanes of the Royal Flying Corps* which is rightly regarded as a landmark in World War One aviation history. The authors made frequent reference to this work while researching the history of French aircraft in British service. Jack's superlative books and magazine articles on French fighters were also extremely helpful.

Peter Grosz supplied us with British intelligence reports concerning aircraft projects. These documents provided the authors with information on previously unrecorded aircraft types. Allan Toelle, whose forthcoming book on French camouflage and markings will likely prove to be the definitive work on the subject, not only lent us photographs from his collection, but also reviewed all the photographs for our book. His advice and suggestions have been much appreciated.

During two visit to Paris in 1994 and 1996 the authors received considerable help from the staffs at the SHAA, SHAT, and SHM. At the SHAA, General Robineau, M. Hoder, and his delightful secretary Mme. Tarin need to be thanked for their help in securing several key documents and several hundred photographs. Mme. Marie-Annick Hepp at the SHAT obtained over 100 rolls of microfilm for our use. Also, our thanks to the staff at the SHM for their support. At the Musée de l'Air we wish to thank M. Lorant and M. Nicolaou for their assistance during a difficult time for the museum.

Others who assisted the authors include Leo Opdycke, who was a constant source of support, Frank Portier, Frank Bailey, and Malcolm Passingham. M. Le Roi shared with us his extensive knowledge of French naval aviation during the war. Our thanks also to the French naval aviation historical society (ARDHAN) for reviewing the portion of our manuscript dealing with French naval aviation operations 1914–1918. Noel Shirley supplied the authors with key information on French airplanes in service with the U.S. Navy. Our thanks also to the staff of the U.S. Air Force Historical Research Agency for providing us with the entire Gorrell History on microfilm, as well as aircraft record cards. Our gratitude also goes to the Smithsonian Air and Space Museum and the Naval Historical center for providing us with U.S. Naval aircraft record cards. Nigel Eastway of the Russian Aviation Research Group of Air Britain supplied us with translations of Shavrov's and Duz's books on Russian aircraft and aviation during the First World War. Thanks also to Augie Blume for sharing his extensive knowledge of Serbian, Russian, and Italian aviation with us. Former Spanish Republican fighter pilot Jose Falco supplied us with information on French aircraft in Spanish service. Ellic Sommer assisted us with the section on the SPAD SA series of airplanes. Gregory Alegi and Roberto Gentilli reviewed and critiqued the section on the license-built Caproni bombers. Thanks are also due to Capt. Efthimiadis of the Hellenic Air Force General Staff who supplied us with the Official History of the Greek Air Service during and after the First World War. Tom Nilsson contributed much valuable information and advice on the French Groupe's de Bombardement. We wish to acknowledge the scholarship of Lennart Andersson whose work on Swedish aircraft, Soviet aviation, small air forces in the 1920s (which appeared in the Small Air Forces Observer) was of great help. The works of the late Raymond Danel, Jean Cuny, and Jean Liron were also consulted. Harry Woodman offered us advice on French aircraft armament. Also, thanks to our mail carriers Susan and Albert who has faithfully delivered every parcel, letter, and magazine used in writing this book.

Special thanks to Jack Herris for his courage and enthusiasm in supporting this project.

Finally, we must give special thanks to our families who tolerated the extended periods of time we spent working on this book. Special thanks to Nicole Soltan who aided us in translating key French documents and who made numerous calls to France on our behalf. To Jayne and Sarah Davilla our thanks for their support. Thanks to Phillip Soltan for his kind assistance in helping us obtain key photographs from the archives. Thanks also to Jack Tyson for producing artwork for this book. Thanks to Doctors Gopi Ayer, Monica Delzeit, M. Goretsky, and Bassam Saffouri for their support during our trips to France.

To all who have helped in producing this book whom we may have failed to include in this acknowledgment, our sincere thanks. Finally, the authors wish to clearly state that any factual errors or omissions are solely their responsibility.

James J. Davilla
Arthur M. Soltan
San Jose, California, 1997

Aviation Militaire 1909–1918

The Beginning: 1909–1914

The introduction of the heavier-than-air machine into French military aviation can be traced back to 1909. At that time General Roques, who was director of engineering for the Ministry of War, purchased five aircraft for evaluation. These were two 50-hp Gnome Farmans, two 30-hp Wright-Barquand Wrights, and one 25-hp La-Manche Anzani. These airplanes were purchased in order to evaluate their military potential, and were not intended to enter operational service.

After the air meet at Rheims the Artillery Service decided to purchase several two-seat airplanes to assess their value in the role of artillery spotting for 75-mm cannon. Major Estienne, who was head of the Aviation Establishment for the Artillery, selected three Farmans, two Wrights, and two Antoinettes. Tests were conducted with these machines at Vincennes (which would later become a major testing center for French military aircraft). Bombing and army co-operation missions were also flown. The artillery service later concluded that its needs could be best met by a single-seat, armored, and armed airplane.

Schools for military aviators were opened at Châlons and Vincennes. These would provide the aviation service with many of its first pilots.

The division of military aviation between the engineering and artillery branches was resolved on 22 October, 1910, when all airplanes were placed under a Permanent Inspector of Military Aeronautics: General Roques.

The Farmans and Blériots were found to be superior to other types because of their sturdiness and reliability. The Aviation Militaire would, therefore, be formed primarily with aircraft from these two manufacturers.

Airplanes were allowed to participate for the first time in French military maneuvers held at Picardy in September 1910. Their distribution was:
2nd Armée: two Farmans, one Sommer, and one Blériot
9th Armée: two Farmans, one Blériot, and one Wright

The airplanes performed successfully, resulting in additional orders being placed for 20 Farmans and 20 Blériots.

Until 1911 the military had contented itself with adopting civil designs for military missions. However, it was decided that a competition should be held to select a machine intended from the outset for reconnaissance. This competition was held that same year; the requirements were for a three-seat airplane (pilot, mechanic, and observer) with a range of 300 km, able to operate from poorly prepared airfields, carrying a payload of 300 kg with a maximum speed of 50 km/h, and which could be rapidly disassembled for easy transport. No fewer than 43 manufacturers submitted 140 designs to meet these demanding requirements. However, when the competition began on 8 October 1911 only 31 airplanes were entered. These are shown in Table 1.

After the competition was completed on 28 November 1911, the winners, in order of preference, were the Nieuport 4M, the Breguet G 3, and the 100-hp Gnome Deperdussin Type B 1911 Military Airplane. However, the competition had exhausted the military's 1911 funds for airplanes and only a small number of each of these airplanes was purchased. All three types were formed into a single unit.

In 1912 it was decided that it would be simpler to maintain a unit if all the airplanes were of a single type. Therefore on 29 March 1912 the French Aeronautic Forces were formed under Inspector General Hirshauer with the following units (now termed escadrilles, or squadrons) each with a single type of aircraft: HF 1 at Châlons, MF 2 at Buc, BL 3 at Pau, D 4 at Saint Cyr, and MF 5 at Saint Cyr.

The Artillery Service had a single escadrille with Blériot single-seaters and the cavalry had a BLC (Blériot Cavalry) unit with three Blériots. Later a few Borel-Blériot and Hanriot single-seaters were also obtained.

During the 1912 military maneuvers the Artillery Service found that airplanes would be more effective if they could be used for low-altitude reconnaissance missions. This, of course, exposed them to ground fire, so these machines would have to be armor-plated.

On 16 April 1913 the Aviation Militaire was placed under the command of C.A. (corps d'armée) commanders or local administrators, thus diminishing the authority of the permanent inspector for aviation. Four months later Colonel Hirschauer resigned and General Bernard was appointed chief of the Aeronautic Services. Bernard concurred with the Artillery Service that armored airplanes would be of the most value. The various airplane manufacturers were asked to produce four different types of armored aircraft:
1. A single-seater for the high-speed scout mission.

Pre-war lineup of Blériot 11s at Belfort. SHAA B83.932.

Table 1: Airplanes Participating in the 1911 Military Concours
100-hp Gnome Breguet biplane
110-hp Dansette-Gillet Breguet biplane
70-hp Renault Maurice Farman biplane with staggered wings
75-hp Voisin canard
70-hp Maurice Farman
70-hp Gnome Coanda (twin-engine)
140-hp Gnome Breguet G 3 biplane
100-hp Gnome Goupy
70- or 80-hp Chenu Goupy
60-hp Antoinette monoplane
120-hp Canton-Unné Breguet
100-hp Gnome Blériot
140-hp Gnome Voisin canard
75-hp Renault Astra triplane
80-hp Chenu Breguet
75-hp Renault Paulhan triplane
70-hp Gnome Henry Farman
100-hp Gnome Henry Farman
70-hp Renault Henri Farman biplane with staggered wings
80- or 85-hp Canton-Unné Breguet
70-hp Labor Savary
140-hp Gnome Blériot
100-hp Gnome Borel
100-hp Clerget Hanriot
100-hp Gnome Deperdussin Type B 1911 Military Airplane
100-hp Clerget Deperdussin Type B 1911 Military Airplane
80-hp Anzani Deperdussin Type B 1911 Military Airplane
75-hp Chenu Astra-Wright
60-hp Renault Astra-Wright
100-hp Gnome Voisin
100-hp Gnome Nieuport 4M.

2. A two-seat airplane for long-range reconnaissance.
3. A two-seat interceptor to destroy enemy airplanes.
4. A three-seat bomber.

Several manufacturers submitted armored airplanes from 1913 through 1914, and during 1914 some were assigned to operational units (in particular, the C.R.P., or Camp Retranche Paris, formed in 1914). However, the performance of these was quite poor, the engines being insufficiently powerful to give a heavily-armored aircraft adequate performance.

The 1913 military maneuvers also employed airplanes. It was found that the reconnaissance information supplied by the Aviation Militaire was extremely useful to the C.A. commanders.

In 1914 it was concluded that armored airplanes were impractical, and it was decided to standardize production on the Morane-Saulnier Type G, the Blériot 11 and Blériot-Gouin monoplanes, and the Voisin 1913 with the 130/140-hp Canton-Unné engine.

On 4 April 1914 the Aviation Militaire became a separate department of the Ministry of War; it was now an independent service. In that same month there were 126 airplanes in service with front-line escadrilles, 126 in reserve, and large numbers in the training units. Also by August 1914, 657 pilots had been trained, of whom 220 were still on active duty.

As France entered the maelstrom of the First World War two main conclusions had been made concerning aircraft procurement. The first was that standardization on a few select types was desirable. While this was a worthwhile goal, during the war this concentration on large production runs of a small number of types would eventually lead to the Aviation Militaire being equipped with aircraft that were inferior to those of the Germans and Austro-Hungarians.

The second decision reached by the Aviation Militaire and its manufacturing service, the SFA (Service Fabrication Aéronatique, or Aeronautics Production Service), was that armored airplanes were impractical for the time being. Only at the very end of the war would the service request that the manufacturers develop an armored aircraft for ground-attack missions. The airplanes designed in 1918 to meet the S 2 and S 3 specifications were all failures because their engines would still not permit an armored aircraft to have high performance.

1914

Political

The outbreak of war was accompanied by an outpouring of patriotism in France. The public and military were convinced of their superiority over the Germans and were confident that the enemy would be defeated in short order. Perhaps influenced by

General Joffre inspects a Dorand biplane at Villacoublay in June 1914. Dorand's design was armored and had a laterally-mounted engine. The caption states that the aircraft was equipped with a T.S.F. unit. Renaud.

the climate in France at this time, General Bernard stopped all further aircraft production, ceased pilot training, demobilized the factories, and sent many of France's best aircraft designers and manufacturers, as well as their employees, to the front lines.

The SFA was subordinated to the Director of Aeronautics in September 1914, and the STAé (the Aeronautics Technical Section) was formed in 1914. The SFA was to guide the manufacturers in the production of aircraft for the Aviation Militaire, while the STAé was to evaluate the airplanes produced by the factories before they entered service with the front-line escadrilles.

On 8 October a plan was formulated to increase the size of the Aviation Militaire to 384 aircraft in 65 escadrilles. As the replacement rate for aircraft at the front (due to combat damage, operational mishaps, and training accidents) was 50 percent a month, this would require 100 aircraft to be produced every 30 days.

Because General Bernard had created havoc in the French aircraft industry by his demobilization of the factories, on 25 October he was replaced as Director of Aeronautics by General Hirschauer. Colonel Barès was made commander of the Air Services in the field. Hirschauer immediately discharged the desperately needed aircraft designers and skilled tradesman from operational units. In addition, orders were placed with the industry for 2,300 aircraft and 4,400 engines. However, as 1914 came to a close, only 192 aircraft were in service.

Operational

At the outbreak of war the order of battle for the Aviation Militaire was:

1st Armée: MF 5 at Epinal, BL 9 at Epinal, BR 17 at Dijon, BL 18 at Dijon, BL 3 at Belfort, BL 10 at Belfort.
2nd Armée: HF 1 at Toul, MF 8 at Nancy, HF 19 at Lyons, MF 20 at Lyons.
3rd Armée: MF 2 at Verdun, HF 7 at Verdun, HF 13 at Châlons, MF 16 at Châlons.
4th Armée: V 14 at Camp Châlons, V 21 at Camp Châlons.
5th Armée: D 4 at Maubeuge, D 6 at Rheims, C 11 at Douai, N 12 at Rheims, REP 15 at Rheims.
Cavalry Flights: BLC 2 at Nancy (2nd Cavalry Division), BLC 4 at Rheims (4th Cavalry Division).

On 30 August the C.R.P. for the protection of Paris was established. Several airplanes, some of them armored, were assigned to this new base.

On 2 September 1914 a Breguet AG 4, piloted by Louis Breguet himself, spotted the German army turning to the east of Paris. This information was confirmed by aircraft of REP 15 and MF 16 and the 6th Armée was quickly despatched to attack the German flank. The resulting Battle of the Marne was a major victory for the Allies, a victory made possible in large part by the Aviation Militaire.

As the front line was consolidated in 1914, the escadrilles now turned to the missions they would perform for the remainder of the war: reconnaissance, bombing, and pursuit.

The reconnaissance units began to experiment with artillery spotting and reconnaissance missions. A combination of flares, flags, and TSF (Transmission Sans Fibre—wireless transmission) were used to convey information to the ground forces. The first experiments in aerial photography were also made.

The first bomb group, GB 1 (Groupe de Bombardement 1), was formed on 27 September 1914. Equipped with Voisin 3s, the unit made attacks on enemy supply and communication lines. However, poor weather inhibited operations for most of 1914.

R.E.P. Parasol serial number 8460 flown by Flt. Lt. J.E.D. Erroll Boyd of the RNAS after capture and internment in Holland on 3 October 1915. F. Gerdessen.

Some of the Voisin 3s of V 24 were armed experimentally with 8-mm Hotchkiss guns. An airplane of that unit piloted by J. Frantz and his mechanic, Quenault, scored the first aerial victory on 5 October 1914.

A GQG memo dated 16 December 1914, proposed the following structure for the Aviation Militaire:
1. One Groupe de Bombardement (GB) assigned to the GQG.
2. Two strategic reconnaissance escadrilles assigned to each Armée, equipped with either M.F.11s or Morane-Saulnier Ls.
3. One army co-operation escadrille to be assigned to each Corps d'Armée. This instruction was later expanded to include an escadrille to be assigned to the Artillery Regiments. Caudron G.3s and, later, G.4s would equip these units.

1915
Political

In September the Aviation Militaire underwent several changes. Hirschauer was forced out of his position as Director of Aeronautics as a result of political in-fighting. René Besnard became the Secretary of State for Aviation Militaire. Besnard intended to create a system which would enable him to determine the requirements of the Aviation Militaire for new aircraft, while at the same time overseeing production. There were four major departments:
1. Inspector General for Aircraft Production.
2. Technical Section (STAé).
3. Production Service (SFA).
4. Repair Workshops.

Technical

The leaders of the Aviation Militaire decided to concentrate aircraft production on a few dependable types. The Caudron G.4s and M.F.11s would be used for reconnaissance and army co-operation duties, the Voisin 3s would be used for bombing, and the Morane-Saulnier Ls would undertake fighter escort and patrol missions. The reconnaissance escadrilles would be assigned to both armées and corps d'armées, while the fighters would be assigned only to the armées. The bombers would be formed into their own autonomous groups.

By March, the number of aircraft produced per month had risen to 432, of which 15 percent were supplied to the RFC and Imperial Russian Air Service. At this time there were 53 escadrilles (most on the Western Front), with 11 replacement units and 11 in reserve. There were two escadrilles with the A.F.O. (Army Forces East—French army units in eastern and southern Europe).

At this time most of the fighter units had Morane-Saulnier Ls, but as 1915 progressed the two-seat Nieuport 10s would be modified into single-seat fighters. The pursuit aircraft of this period were armed with machine guns placed to fire over or around the propeller. Roland Garros tested a Morane-Saulnier G fitted with deflectors on its propeller. This eventually

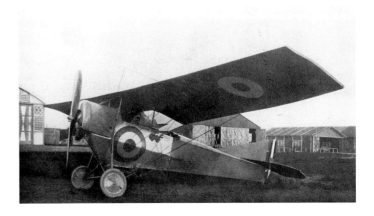

Morane-Saulnier L. The crew of de Bernis and Jacotet had forced the German airplane in the background to land and are posing with the captured plane's crew on 28 April 1915. SHAA B81.1514.

Caudron G.4 of C 66 at Fretoy in 1917. A 1915 type, the Caudron G.4 remained in service long after it was obsolete. SHAA B88.1218.

led to the production of a small number of Morane-Saulnier Ns with deflector blades. However, by the time the Type Ns had entered service, a dedicated fighter derivative of the Nieuport 10, the Nieuport 11, had become the standard fighter of the Aviation Militaire.

Another major innovation was introduced by the French in August 1915. Several Caudron G.4s and Nieuport 11s were formed into the Groupe de Chasse de Malzéville. The purpose of this ad hoc unit was to defend GB 1 and GB 2 from German attacks on the bombers' base at Malzéville. This marked one of the first occasions that fighters were put into a single unit to achieve local air superiority. As the war progressed French fighter units would grow from temporary units with a few airplanes to mass formations with hundreds of fighters.

During 1915 the reconnaissance escadrilles retained their now rapidly aging M.F.11s and G.4s. Both types were vulnerable to attacks from the rear and, as the performance of the German fighters improved, the French escadrilles began to suffer heavy losses. The French procurement system would ensure that deliveries of these vulnerable aircraft would continue well into 1916, with tragic results for their aircrews.

The French wanted a heavy bomber which could damage industrial targets deep in Germany. In June the Aviation Militaire was under political pressure to develop a heavy bomber which could carry 300 kg of bombs, have a range of 600 km, and cruise at 120 km/h at an altitude of 2,000 m. The 1915 concours puissant (competition for a powerful aircraft) specifically requested a bomber capable of meeting these requirements. Two aircraft were successful: the Breguet SN 3 and the Paul Schmitt S.B.R. The SN 3 was further developed into the B.M.4 and the S.B.R. became the Paul Schmitt 7. In service both aircraft were found to be easy prey for enemy fighters, and both had unsatisfactory range and payload. The Paul Schmitt 7 did not even enter service until 1917, by which time it was found to be worthless and had to be withdrawn after only four months.

In 1915 the emphasis was on quantity, not quality, and this meant continued production of the obsolescent Caudron G.4s and M.F.11s. A new requirement for 800 three-seat multi-purpose aircraft was formulated. It was hoped that this new airplane, which would have two engines and would carry a front and rear gunner, would prove less susceptible to the newer German fighters. It would be used for long-range reconnaissance and bombing missions as well as to provide escort for army co-operation aircraft attached to the individual armées. However, the industry was less interested in developing new aircraft then it was in fulfilling the lucrative contracts for older types. Therefore, the Caudron R.4 and G.6 reconnaissance machines did not enter service until well into 1916.

The Aviation Militaire issued a specification for a C3 fighter. This was intended to carry a crew of three: one pilot and two gunners. The C3 fighter would escort the vulnerable reconnaissance aircraft and bombers and could have been used to fly barrier patrols. The H.F.35 and the Caudron G and R series were probably intended to meet this specification. In a similar vein, the SFA issued a specification for the "F" series of airplanes. These were to be literal flying fortresses intended to destroy enemy formations by the sheer intensity of their firepower. The Breguet 11 and Van den Born F.5 were intended to meet this specification. Neither type was selected for production.

The strength of the Aviation Militaire stood at 756 aircraft on 15 August, 1915. These were divided as follows:

Aircraft Type	Front-Line Units	Reserves	Training Units
M.F.7/11	193	29	12
Nieuport 10/11	42	32	7
Morane-Saulnier L/LA	57	6	7
Caudron G.3/G.4	64	35	10
Voisin 3/4	129	20	10
Breguet 4/5	2		
Vendôme	1		

Nieuport 10 serial number 1739 of the Imperial Russian Air Service. SHAA B85.2585.

Operational

The Aviation Militaire had 48 operational escadrilles in March 1915. These were:

Army Cooperation Escadrilles:

MF 1 assigned to 33rd C.A. (based at Heramville)
MF 2 assigned to 5th C.A. (Clermont en Argonne)
BL 3 assigned to the 6th Armée (Monteaubert)
C 4 assigned to the 1st C.A. (Bouvencourt)
MF 5 assigned to the 1st C.C. (Commercy and Toul)
C 6 assigned to the 5th Armée (Muizon)
MF 7 assigned to the 6th C.A. (Verdun)
MF 8 assigned to the 2nd Armée (Vaux)
C 9 assigned to the 1st Armée (Villers les Nancy)
C 10 assigned to the 6th Armée (Hautefontaine)
C 11 assigned to the 2nd C.A. (Châlons)
MS 12 assigned to the 5th Armée (Châlons)
C 13 assigned to the 3rd Armée (Verdun)
MF 14 assigned to the 7th Armée (Épinal)
REP 15 assigned to the 10th Armée (Béthune)
MF 16 assigned to the 6th Armée (Vauciennes)
C 17 assigned to the 1st Armée (Toul)
C 18 assigned to the 1st Armée (Toul)
MF 19 assigned to the 13th C.A. (Amiens)
MF 20 assigned to the 2nd Armée (Villers les Bretonneux)
V 21 assigned to the 4th Armée (Châlons)
MF 22 assigned to the 4th Armée (Châlons)
MS 23 assigned to the 2nd Armée (Treux)
V 24 assigned to the 5th Armée (Muizon)
MF 25 assigned to the 3rd Armée, but was an independent bomber unit (Ste Menehould)
MS 26 (Dunkerque)
C 27 assigned to the 21st C.A. (Bethune)
C 28 assigned to the 11th C.A. (Lavillers)
V 29 assigned to the 9th Armée (Herlin le Sec). Later in March the escadrille re-equipped with M.F.11s and became an independent bombing unit based at Belfort.
C 30 assigned to the 6th Armée (Cravancon)
MS 31 assigned to the 1st Armée (Villers les Nancy)
HF 32 assigned to the 3rd Armée (Hermaville)
MF 33 assigned to the 9th C.A. (Poperinghe)
C 34 (Belfort)
MF 35 assigned to 20th C.A. (Poperinghe)
MF 36 (Porthem)
MS 37 assigned to the 3rd Armée (Ste Menehould)
MS 38 assigned to the 4th Armée (Châlons)
C 39 assigned to the 5th Armée (Villers Allerand)

Groupes de Bombardement:

GB 1 with escadrilles VB 101, VB 102, and VB 103 (Châlons)
GB 2 with escadrilles VB 104, VB 105, and VB 106 (Dunkerque)
GB 3 with escadrilles VB 107, VB 108, and VB 109 (Belfort)

In April 1915 the GQG issued new instructions on the organization of the escadrilles.
1. Each armée was to have an M.F.11 escadrille for reconnaissance and a Morane-Saulnier L escadrille for aerial combat and long-range reconnaissance.
2. Each corps d'armée was to have either a M.F.11 or a Caudron G.4 escadrille for army co-operation duties.
3. The Groupes de Bombardement remained autonomous.

The Battle of Artois in May and June 1915 gave the Aviation Militaire the chance to evaluate this new organization. The corps d'armée escadrilles were the most active and gave considerable assistance to the C.A. units by providing photographs, tactical reconnaissance information, and regulating artillery fire by TSF. The Groupes de Bombardement were assigned to attack enemy bivouacs, columns marching to the front, artillery batteries, and balloons. Unfortunately, the GB units were incapable of hitting such relatively small targets from their attacking altitude of 2,000 meters. Furthermore, the bombs they carried were usually not powerful enough to do significant damage even when a target was hit.

The French drew the following conclusions from the Battle of Artois about the utilization of the Aviation Militaire:
1. Military aviation was most useful to the army commanders when it provided tactical information such as the location of enemy troops and the regulation of artillery fire. It was also extremely important in tracking the advance of friendly troops during major ground offensives.
2. The basic unit of the Aviation Militaire should be the army co-operation escadrille assigned to the corps d'armée.
3. The S.A.L. (Section Artillery Lourde—heavy artillery section) escadrilles were to closely co-ordinate their activities with, and be subordinate to, the C.A. escadrilles.

It was decided that each corps d'armée would require the following airplanes:
1. One airplane for photo reconnaissance.
2. Two airplanes for army cooperation duties.
3. One airplane for artillery regulation.

It was estimated that the C.A. escadrilles would be required to fly between 24 and 32 sorties per day to accomplish these missions. This, in turn, meant that a minimum of 15 aircraft would be needed for each escadrille. Thus, the standard size for an escadrille was set at 15 airplanes.

The S.A.L. units would perform similar duties for the artillery units. The reason the artillery units were given their own air force was based on the fact that the artillery considered itself to be independent of the corps d'armée and even the groupe d'armée. An artillery officer stated this quite clearly when he said "In the same fashion that we possess our own mobile observation posts, we must have our own balloons and aircraft." In theory, at least, it was intended that the S.A.L. units would be subordinate to, and cooperate with, the C.A. aviation commander.

The Groupes de Bombardement underwent several changes

in 1915. The bomber units were still using the Voisin 3s and cannon-armed 4s with limited success. The Voisins could not carry a heavy payload and their limited range put the more enticing industrial targets in the Essen region out of reach. One of the first mass bombing raids of the war was made on 27 May, 1915, when 18 Voisins of GB 1 bombed Ludwigshafen. Other attacks followed and the initial results were encouraging. MF 25 and MF 29 were independent bombing units equipped with M.F.11s converted to bombers. GB 2, 3, and 4 were formed later in 1915. However, the results of day and night bombing raids proved to be disappointing, with little damage being inflicted on the enemy. The Germans reacted quickly to these attacks by intensifying anti-aircraft artillery and concentrating their fighters at likely sites of attack. This resulted in the bombers suffering heavy losses. The decision was made in September to concentrate on tactical targets, at least until a suitable heavy bomber became available. The GB units based at Malzéville was dispersed. GB 1, 2, and 4 moved to Champagne, while GB 3 was based at Artois. They were to now provide direct support for the troops at the front by bombing German supply lines and also by attacking German artillery units. Strategic bombing fell into such disfavor that GB 4 was disbanded in November and VB 102 and VB 103 were detached from GB 1. These later two escadrilles were to re-equip with Nieuport 11s and become fighter units.

By October there were 50 army cooperation units, four Groupes de Bombardement, and a small number of dedicated fighter units. A contemporary French report stated that during the initial stages of the Battle of Artois "the Boche aviation had been timid and fearful, but little-by-little adopted a more aggressive defense, with increasing attacks on our reconnaissance machines." Indeed, in the autumn of 1915 the Fokker E.I had appeared. This nimble fighter with a synchronized machine gun took a heavy toll of the French Farmans, Caudrons, and Voisins. The machine also outclassed the Morane-Saulnier L and LA fighters. The result was heavy losses for the army cooperation machines, resulting in hostility between the fighter and C.A. escadrilles. This animosity would disappear over time, but would reappear with the Battle of Chemin des Dames in 1917. The Fokker threat would be removed only when the Nieuport 11 appeared in large numbers as 1916 progressed, and when the fighters were grouped into formations large enough to achieve control of the air over the battlefield.

In October 1915 new plans for the structure of the Aviation Militaire in 1916 were announced. It was hoped to have 45 army cooperation and 34 artillery spotting escadrilles. These units were to operate two-seat reconnaissance machines for short-range missions. Several escadrilles of the previously mentioned twin-engined aircraft were to be formed for both long-range reconnaissance missions and to provide fighter escort for the more vulnerable two-seaters.

The structure of the Aviation Militaire in December 1915 was:
D.A.V.: C 11, C 18, MF 63, N 67
D.A.L.: C 9, C 42, C 66, C 69, MF 45, MF 58, MF 70, N 48, N 65, N 68
1st Armée: C 17, MF 5, MF 44, N 31
2nd Armée: C 13, C 51, C 64, MF 8, MF 35, MF 40, MF 50, MF 60, N 23
3rd Armée: MF 2, MF 25, MF 32, N 37
4th Armée: C 28, C 46, MF 7, MF 41, MF 51, N 38, V 21
5th Armée: C 6, C 30, C 39, C 53, N 12, V 24
6th Armée: C 4, C 10, C 43, C 47, MF 19, MF 62, N 3
7th Armée: C 34, F 14, MF 20, MF 59, N 49
10th Armée: C 27, C 56, C 57, MF 1, MF 16, MF 22, MF 33, MF 35, N 15, N 57, N 59
GB 1: VB 101, VB 114
GB 2: VB 104, VB 105, VB 106

GB 3: VB 107, VB 108, VB 109
C.R.P.: Escadrilles 94, 95, 96, 97

Foreign Service

In May 1915, MF 98 Tenedos arrived with six M.F.11s, with three more following later. The unit flew numerous reconnaissance missions until being withdrawn in February 1916.

V 85, V 86, V 88, V 89, N 90, and N 91 were assigned to the A.F.O. and provided support for the French army units on the Romanian and Greek fronts.

V 84, MF 97, and MF 99 were assigned to the Serbian front.

1916
Political

Several political changes occurred in the first quarter of 1916. Undersecretary Réne Besnard resigned after the airplane manufacturers complained that his attempts to reorganize the aviation industry were too radical. Colonel Régnier assumed Besnard's post. Colonel Stammler was named head of the SFA. Finally, Commandant Dorand separated the STAé from the SFA. Dorand's first goal was to replace the obsolete M.F.11 and F.40 pushers. When the French manufacturers in general, and the Farman firm in particular, expressed a hesitancy to abandon the pusher layout, Dorand had Capitaine Lepére design a new short-range reconnaissance machine. This aircraft would later become known as the A.R.1 (although some sources referred to them, incorrectly, as the Dorand). Another machine, a three-seater for long-range reconnaissance to replace the unsuccessful Caudron R.4s and delayed G.6s, was also designed by the STAé. These airplanes would later become known by the firm that would manufacture them under license—Letord. The SFA concentrated on the output of aircraft types which were already in production.

Technical

In February 1916, the Aviation Militaire had the following aircraft on strength:

Caudron G.3	141
Caudron G.4	167
Maurice Farman 7	71
Maurice Farman 11	101
Nieuport 10	120
Nieuport 11	90
Morane-Saulnier L/LA	18
Caudron R.4	1
Ponnier N	5
SPAD SA.2	4
Breguet 4	22
Breguet 5	11
Voisin 4	20
Voisin 3/5	159
Caproni (R.E.P.-built)	7

The most surprising feature of this list is that it shows that the bulk of the Aviation Militaire's equipment still consisted of obsolescent M.F.7s, G.3s, and Nieuport 10s. The French were to encounter heavy losses when these types were pitted against the Fokker E and Pfalz E fighters during the Battle of Verdun.

The failure of the bomber campaign was to directly affect the outcome of the 1916 concours puissant. Astra, Breguet, Caudron, D.N.F., Morane-Saulnier, and SPAD all entered aircraft in the competition. The Morane-Saulnier S and SPAD SE were both able to fulfill the requirements of the competition and a request for 300 Morane-Saulnier S bombers was placed by Colonel Barès. However, the French parliament refused to

Nieuport 11 N571. The Nieuport 11 was instrumental in gaining air superiority from the German Fokker and Pfalz monoplanes and was widely used by Allied air services. SHAA B83.3506.

provide this much money for a bomber and cut the order to 100. Barès responded by canceling the entire order. This meant that the French strategic bombing force would be forced to rely on the antiquated Voisins as well as a handful of license-built Caproni bombers up until the last few months of the war. During the first part of 1916, GB 1 and GB 2 limited their attacks to the train stations at Metz, the factories in the Saar Valley, and the blast furnaces at Lorraine; all these targets were within 300 km of the bombers' airfields at Malzéville.

The situation for the day bombing units was not much better. The Breguet-Michelin 4s were of limited value and the M.F.11s and F.40s were regarded as dangerous by their crews. It had been hoped that Sopwith 1½ Strutters could be built under license and supplied quickly to the army cooperation and day bomber escadrilles. However, delays in production prevented the first example from reaching the front lines until the spring of 1917.

The reconnaissance units were also forced to retain their Farmans and Caudrons. The STAé had completed development of both the A.R.1 short-range and the Letord long-range machines, but they, too, would not reach front-line escadrilles until early 1917. The Caudron G.6 became available in late 1916.

Operational

The German plan in the air was to achieve complete superiority over the Aviation Militaire in the skies above Verdun. This would prevent the Allies from assessing the German troop movements, while at the same time enabling the German reconnaissance machines to observe the Allied troops with immunity. Initially, the Germans were able to destroy the slow Caudrons and Farmans. C 4, C 6, C 27, C 33, C 46, C 53, C 66, N 3, N 48, N 67, MF 1, MF 25, MF 35, MF 55, F 218, VB 101, V 110, and GBM 5 all participated in the battle. The French fighter units were divided along the front: N 67 and N 23 (R.F.V.—Retranche Fortress Verdun), N 12 (Armée de l'Aisne); N 37 (Armée d'Argonne), N 3 (Armée de Picardie); N 15, N 57, N 69 (10th Armée), N 65 (GB 1), and N 26 (Armée de Belgique—Belgian Army). However, they were so dispersed that they could not achieve complete control of the skies over the Verdun battlefield. This resulted in the Germans being able to conduct successful reconnaissance of the French lines, while the army co-operation aircraft suffered heavily at the hands of the German fighters.

Colonel Barès assigned Major Trincornot de Rose the task of re-establishing control of the skies over Verdun. De Rose responded by concentrating Nieuport 11s and Morane-Saulnier Ns in the Verdun-Bar-le-Duc area. This ad hoc formation was designated a provisional Groupe de Chasse. By 15 March the following fighter escadrilles had been formed into the provisional Groupe de Chasse:

N 65 and N 67 formed the nucleus of the Groupe de Chasse and were based at Bar-le-Duc.
N 23 was based at Vadelaincourt and was kept in reserve.
N 37 based at Brocourt in the vicinity of Argonne.
N 15 and N 69 also at Bar le Duc in the 10th Armée region.
N 57 arrived later and was based at Lemmes.

The escadrilles were to perform offensive patrols and were to seek out and destroy the enemy. This was a major change in the French policy of using fighters, which had previously called for fighters to fly barrage patrols over specific sectors of the front and destroy any enemy aircraft entering that sector. The French were pleased with the results of concentrating their fighters into compact, mobile units which could establish local air superiority over a chosen sector of the front. The provisional units gave way to the permanent Groupes de Chasse (GC) and, by the end of the war 13 such GC units were in existence. By late March the French had once again attained air superiority over Verdun.

While Verdun had established the importance of the fighter, the opposite was true for the bomber units. Once again, the French bombers had achieved little despite heavy losses. Soon after the Battle of Verdun, GB 2 was disbanded and its escadrilles were re-equipped and re-trained for army co-operation duties. This left GB 1, which had a handful of obsolescent Voisin 5s and some Caproni 2s, and GB 3 with a mixture of Breguet-Michelin 4s and M.F.11s. However, the Breguet-Michelin bombers, which were soon formed into their own bomb group, GBM 5, had serious engine difficulties and were found to be of little use until fitted with new engines. In October GB 4 was re-formed.

The army cooperation escadrilles continued to fly the Caudron G.4, M.F.11, and Farman F.40. In an attempt to improve the crew's chances for survival, some of these were now fitted with cameras of longer focal length. This permitted the aircraft to fly higher and thus avoid the enemy defenses. Also, TSF units were now coming into more widespread use with the reconnaissance units.

The Battle of the Somme, which began in July, required the French fighter units of the G.A.N. (Groupe d'Armée North—Army Group North) to establish air superiority over the front. The French were fortunate that the Nieuport 16 and, later in the battle, the Nieuport 17 were just entering escadrille service. The Aviation Militaire again formed their fighter units into two larger organizations in order to coordinate their operations, as well as to simplify logistics. The first unit was Groupement de Combat de la Somme, which had been created in April in anticipation of the forthcoming battle. GC Somme eventually consisted of N 3, N 26, N 73, and N 103. A second unit, Groupement de Chasse Cachy, had N 5, N 23, N 37, N 62, N 65,

N 67, and N 69. The fighter escadrilles were equipped with Nieuport 11s, 16s, and 17s and had a significantly better performance than the German monoplane fighters. Once aerial superiority had been established the Nieuports were used to strafe enemy troops. The German ground troops were so demoralized by the French aerial attacks that it has been said that they cursed the Allies and their own air service with equal fervor. Approximately three or four fighters were assigned to the C.A. escadrilles to escort the reconnaissance machines. The bomber units, particularly GB 3, VB 101, and GBM 5, became directly involved with the battle and concentrated on tactical missions against transport targets and airfields from April through July. GB 3, GB 4, and GBM 5 confined their operations primarily to night attacks as attacks during the day usually resulted in heavy losses.

The Oberndorf raid on 12 October 1916 showed the futility of attempting bombing missions during the day without adequate fighter escort. A total of 17 M.F.11s, B.M.4s and Farman 40s from GB 4 attacked the Mauser factory at Oberndorf. They were escorted by seven Breguet 5 escort fighters and 18 RFC Sopwith 1½ Strutters. Twelve of the French airplanes were destroyed by the enemy defenses. As a result of these appalling losses, the French GQG decided to abandon strategic bombing operations during the day and, instead, switched the day bombing units to tactical targets near the front lines where adequate protection could be provided by fighters.

By the fall of 1916, the German fighter units were receiving the Albatros D.I and Halberstadt D.II fighters. These began to take a heavy toll of not only the French reconnaissance aircraft and bombers but also of the Nieuport fighters. To help counter the new German fighters, the French began to create permanent fighter groups to replace the previous provisional Groupes de Chasse, which were temporary organizations. GC 11, 12, and 13 were formed in October. Of equal importance was the availability of the superior SPAD 7, of which 143 had been delivered by the end of 1916.

In October it was decided to redesignate the escadrilles as follows:
Fighter units: 1–99
Bomber units: 100–199
S.A.L./C.A. units: 201–299
This renumbering was haphazardly applied to the escadrilles.
On 15 November, 1916, the order of battle for the Aviation Militaire was:

1st Armée:
Assigned to 1st Armée: N 102 and VB 101 (based at Sacy-le-Grand)
Assigned to C.A.: F 209 (Ferme Porte) and F 19 (Pierrefonds)
Reserves: Parc Aero No.3 (Port Salut)

2nd Armée:
Assigned to 2nd Armée: N 23 (Vadelaincourt) and C 105 (Julvécourt)
Assigned to 16th C.A.: F 50 (Autrécourt)
Assigned to 15th C.A.: C 13 (Autrécourt)
Assigned to 31st C.A.: F 44 (Froidos)
Assigned to 2nd C.A.: F 5, F 55, F 216, F 221 (Lemmes), F 8, F 25, C 104 (Vadelaincourt)
Assigned to 14th C.A.: F 20 (Senoncourt)
Assigned to 3rd C.A.: C 4 (Souilly)
S.A.L. units: C 220, C 224 (Lemmes), C 228 (Sulvácourt)
Ecole de tir d'artillerie de Coole (Camp de Mailly)
Reserve units: Parc Aero No.5 (Bar-le-Duc), Iere Reserve No.5 (Saint-Dizier)

3rd Armée:
Assigned to 30th C.A.: C 18 and C 28 (Grivesnes)

4th Armée:
Assigned to Armée: N 38 (Cuperly-la-Cheppe and Châlons-Suippes), C 64 (La Cheppe)
Assigned to S.A.L. : C 212 (Le Tilloy)
Assigned to 17th C.A.: C 56 and C 4 (Ferme d'Alger)
Assigned to 1st C.A.: C 53 (Auve)
Assigned to 7th C.A.: F 72 (St.-Menehould)
Assigned to 18th C.A.: C 6 (Mailly)
Reserve units: Parc Aero No.4 (Châlons), Iere Reserve No.4 (Troyes)

5th Armée:
Assigned to Armée; N 76 (Magneux-les-Fismes)
Assigned to S.A.L.: F 210 (Arcy-St.-Restitute)
Assigned to 4th C.A.: F 40 (Rosnay)
Assigned to 37th C.A.: C 30 (Baslieux-les-Fismes)
Assigned to 38th C.A.: C 39 (Rosnay)
Reserve units: Parc No.1 (Rosnay), Iere Reserve No.1 (Noisy-le-Sec)

6th Armée:
Assigned to Armée: N 62, N 112, C 21, and C 46 (Chipilly)
1st C.A.C.: C 51 (Bougainville)
Reserve units: Aero Park No. 2 and No.102 (St.-Fuscien)
Iere reserve Aero No. 2 (Liancourt-Rantigny)
Assigned to 9th C.A.: F 33, F 24, and F 208 (Croix-Comtesse)
Assigned to 20th C.A.: F 204, C 207, and F 35 (Fricourt)
Assigned to 6th C.A.: F 7, C 17, and C 43 (Bois des Tailles)
Assigned to S.A.L. G.A.N: F 215 (Morlancourt)

7th Armée:
Assigned to Armée: N 49 (Fontaine), F 14 (Griecourt with 76th D.I.), F 59 (Corcieux with 161st D.I.), C 34 (Romagny with 52nd D.I.)
Assigned to 34th C.A.: C 61 (Fontaine)
Reserve units: Parc No.9 (Belfort), Reserve No.9 (Épinal)

D.A.L.:
Assigned to Armée: N 68 (Madeline, N 75 (Lunéville), N 77 (Toul), C 42 (Villers-les-Nancy)
Assigned to C.A.: C 9 (Villers-les-Nancy), C 222 and F 58 (Luneville), F 45 (Art-sur-Meurthe), F 70 (Toul), F 71 (Manoncourt), F 223 (Saizerais)
Reserve units: Parc Aero No.1 (Art-sur-Meurthe), Parc Aero No.6 (Toul), Iere Reserve No.6 and No.10 (Charmes)

10th Armée:
Assigned to Armée: N 69 and N 15 (Grivesnes)
Assigned to A.L.G.P.: C 106 (Grivesnes)
GB 3: V 107, V 108, and V 109, VC 111, VC 113 (Esquenoy)
Assigned to 12th C.A.: F 22, F 211, and F 206 (Le Pilier)
Assigned to 2nd C.A.C.: F 16, C 47, and F 203 (Le Hamel)
Assigned to 2nd C.A.: C 11, F 52, and C 202 (Villers-Bretonneaux)
Assigned to 21st C.A.: F 60, C 27, F 201, and F 205 (Marcelcave)
Assigned to 10th C.A.: F 32 and F 54 (Moreuil)
Reserve units: F 218 (Cours d'artillerie de Beauvais), C 10 (Service du Camp d'Instruction), Parc Aero No. 7 (Berny St. Noye), Parc Aero No.103 (Esquenoy), Iere Reserve No.7 (Esquenoy)
Assigned to 36th C.A.: V 116 (St.-Pol), MF 36 (Furnes), C 74 (Hondschoote)
Reserve units: Aero Parc No.8 and Reserve No.8 (St.-Pol)
GB 1: CEP 115, VC 110, VB 114
GB 2: N 65, VB 104, VB 105, VB 106
GB 4: VB 110, VB 111, VB 112
GBM 5: BM 117, BM 118, BM 119, BM 121
GBM 6: BM 120

Sopwith 1½ Strutter (French designation Sopwith 1A2). By the time French-built examples reached the front in 1917 (nearly a year after British Strutters reached operations), the type was obsolescent. Significant numbers of the type were still operational in French service into the spring of 1918. This example demonstrates the five-color camouflage scheme applied to later production aircraft. MA photo.

Independent bombing escadrilles: MF 25, MF 29, MF 123
D.C.A.: Escadrille d'Antilly, Escadrille d'Meyzieux
C.R.P.: Escadrilles 393, 394, 395, 396, 397

In November the Aviation Militaire consisted of 1,418 front-line aircraft: 328 fighters, 834 reconnaissance machines, and 253 bombers. No less than 95 percent of the reconnaissance machines were Farmans and Caudrons, while the bombers were Voisins and Capronis. The fighters were mostly obsolescent Nieuport 11s and 17s. There were also 49 aircraft with the A.F.O., 196 trainers, and 409 machines in the General Reserves.

As 1916 ended the Aviation Militaire had made several decisions that would determine its efficacy for the remainder of the war. Most fighter units would be placed into groups known as Groupes de Combat. This would allow the fighters to achieve control of the air over selected areas of the front. The fighters would patrol their lines aggressively but would not make deep penetrations behind the German lines (unless they were accompanying reconnaissance or bomber machines). The Royal Flying Corps, on the other hand, felt that the battle should be carried directly to the enemy. For this policy it earned the admiration of the Germans as well as enormous losses in both aircraft and personnel. Individual fighter escadrilles were also assigned to each of the Army Groups primarily to furnish air cover for the reconnaissance units assigned at the armée and corps levels.

The bombers were relegated to tactical support missions, primarily against railway stations, camps, and supply depots. Strategic bombing against German cities was restricted to raids made in retaliation for German attacks on French cities. From this point until well after the Second World War, the strategic bomber would have a limited role in French aviation policy.

Finally, the reconnaissance escadrilles continued to perform the mundane but vital duties of artillery spotting, reconnaissance, spy dropping, and photography. Army co-operation escadrilles were now assigned to virtually every corps d'armée as well as the armées.

However well thought-out the French plans were, they would be of little use until the Aviation Militaire could be supplied with modern airplanes. As the year ended virtually all the French aircraft were obsolescent and, in many cases, obsolete. All this would change as 1917 progressed.

Foreign Service

The following French aviation units were active with the A.F.O.:
A.F.O.: (Greek and Romanian fronts): V 83, V 85, V 86, V 88, V 89, N 90, N 91
Serbia: MF 82, V 84, H 387, MF 98, MF 99

1917
Political

A number of political changes took place in early 1917. In February the Directorate General of Aeronautics was established. It was created to coordinate the activities of the Aviation Militaire with the navy and the Allies, as well as controlling the flow of supplies from the manufacturers and reserves to the front-line units. Colonel Barès was forced from office during that same month; he was another victim of the Byzantine French political system. Major de Peuty replaced Barès as the director of the Aeronautical Service. In March a new prime minister came to power and, under this new regime, the undersecretary of aeronautics was established under Daniel Vincent. Vincent had one major goal: modernize the Aviation Militaire. Under his leadership production of new aircraft types increased markedly and the monthly production of aero engines expanded to twice the previous rate.

On 2 August, 1917, Colonel (later General) Duval was appointed commander of the Aviation Militaire by Pétain. Guiffart was named director of the Aviation Militaire and Major Guignard became head of the SFA. Jacques Louis Dumesnil was named the undersecretary for aeronautics in September. He had served as an officer with C 11.

Vincent forced the STAé to stop developing new airplanes. These STAé-initiated designs had placed it in competition with the very manufacturers it was supposed to be assisting. The STAé was now instructed to confine its activities to research, evaluation of new designs to meet the specifications Vincent had set out, and to provide technical support to the aircraft manufacturers.

Technical

Unfortunately, by the beginning of 1917 the aircraft being delivered from the factories were types which had originally been intended for service in 1916: Sopwith 1½ Strutter, A.R.1, P.S.7, Morane-Saulnier T, Caudron G.6, and Voisin 8. By the time these reached front-line units they were no match for the German fighters. Partially as a result of these inferior machines and partly due to bad weather, the Aviation Militaire played only a limited role during the Battle of Artois (9 March).

The three-seat, long-range reconnaissance airplane had proved to be troublesome. The Caudron G.6s, R.4s, Salmson-Moineau S.M.1s, and Morane-Saulnier Ts had all had significant technical problems. However, the utility of these airplanes was obvious. To simplify the use of three-seat airplanes, the GQG issued a memorandum on 16 March calling for the elimination of many of the Caudron, Morane-Saulnier, and S.M. escadrilles because of the "serious inconveniences" they imposed on the escadrille's personnel. It is probable the maintenance problems, as detailed in the main text, that these airplanes experienced had resulted in a low sortie rate. Furthermore, it was noted that having these large machines all in one escadrille resulted in overcrowding of the airfields. The memo went on to specify that these airplanes would be dispersed to the various escadrilles for long-range reconnaissance, "special" artillery regulation missions (meaning very long-range artillery bombardments), and barrage flights (barrier patrols). Presumably, the Aviation Militaire's mechanics would find it easier to deal with a few three-seaters in each squadron rather than having to service an entire escadrille of these complex machines.

The C.A. and base escadrilles were to have 15 airplanes of which four would be three-seaters. The A.L.G.P. and heavy artillery escadrilles were each to have a complement of 15, of which five were three-seaters. In 1917 the STAé-designed Letord series entered service and provided a much higher level of performance and dependability than their predecessors. Although largely overlooked by historians, the three-seat airplanes and in particular the Letord series played a major role in enabling the army co-operation escadrilles to carry out their missions.

By August the following aircraft types were in service with the G.A.N. and G.A.E.:

Aircraft	Escadrilles	RGA	Damaged	Total
A.R.1/2	216	95	5	316
Breguet				
B.M.4/5	67	36		103
Br.12	1			1
Br.14	24	3		27
Caproni Bombers	16			16
Caudron				
G.3	11	7		18
G.4	139	75	1	215
G.6	133	169		302
R 4	53	10		63
Farman				
M.F.11	1	3		4
F.40/44	109	66	5	180
F.43	1	3		4
F.41 bis	1			1
F.40 ter	41	16	2	59
Letord 1, 2, 4, and 5	89	9		98
Morane-Saulnier				
Type P	61	55	2	118
Type T		13		13
Nieuport				
Nieuport 11	15	13		28
Nieuport 16/17	317	143	9	469
Nieuport 24	4	10	1	15
Nieuport 14	16	5	1	22
Salmson S.M.1	26	6		32
Schmitt P.S.7	43	14	1	58
Sopwith				
1½ Strutter: recon.	234	142	1	377
1½ Strutter: bomber	142	175	1	320
SPAD				
SPAD 7 (180-hp)	445	47	3	495
SPAD 7 (200-hp)	2			2
SPAD 12	1			1
SPAD 13	17	6		23
Voisin				
Voisin 5	2	14	2	18
Voisin 8	51	24	6	81
Voisin 8 Ca	33	30	1	64
Other Types:	1	9	3	13

This gives a total of 2,312 aircraft with escadrilles on the

Western Front, with 1,198 in reserve and 46 unavailable due to damage.

As 1917 came to an end the obsolete G.4s, M.F.11s, and F.40s with which the Aviation Militaire had begun the year were at last being replaced by more modern types. The following aircraft were assigned to escadrilles serving on the Western Front on 1 November 1917:

Fighters:

Nieuport 16/17/24	310
SPAD 7/13	444

Reconnaissance:

F.40 series	51
G.4/G.6	153
Nieuport 12	4
Morane-Saulnier P/T	33
A.R.1/A.R.2	313
Breguet 14 A2	77
Sopwith 1½ Strutter	351
Letord 1/2/4/5	121
Salmson S.M.1	26
SPAD 11	78

Bombers:

F.40 series	15
Voisin 8	102
B.M.4/5	47
Breguet 14 B2	77
Caproni (license built)	12
Sopwith 1½ Strutter	157
Schmitt P.S.7	31

This gives a total of 2,402 aircraft with the escadrilles assigned to the G.A.N. Plans now called for a force of 4,000 airplanes by July 1918. There were to be 1,500 aircraft with the reconnaissance units, 1,500 fighters, and 1,000 bombers.

Operational

The GQG plan for 1917 called for the Aviation Militaire to expand to 187 escadrilles, each with ten aircraft. The emphasis would be on reconnaissance units and the fighters needed to protect them. The bombers in general, and the strategic bombers in particular, were assigned low priority. In March the estimate for the size of the escadrilles was increased to 15 airplanes per unit; a total of 2,665 aircraft would be needed to support these escadrilles.

The order of battle for the Aviation Militaire in February 1917 was:

1st Armée: N 102 (based at Sacy-le-Grand and assigned to the 1st Armée), VB 101 (Sacy-le-Grand), F 1 (Pierrefonds, 33 C.A.), F 63 (Pierrefonds, 1st C.C.), R 213 (Pierrefonds, 1st Armée), F 19 (Remy, 13 C.A.), F 209 (Ferme Porte, 35 C.A.), F 216 (Gournay), C 10 (Corbeaulieu, 35 C.A.)

2nd Armée: N 23 (Souilly, 2nd Armée), F 5 (Lemmes, 15 C.A.), C 13 (Lemmes, 15 C.A.), C 220 (Lemmes, 15 C.A.), C 224 (Lemmes, 15 C.A.), F 25 (Vadelaincourt), F 40 (Couroure, 4th C.A.), C 4 (Couroure, 3rd C.A.), F 44 (Froidos, 31 C.A.), F 50 (Autrecourt, 16 C.A.), F 221 (Julvecourt, 16 C.A.), C 105 (Autrecourt), C 18 (Senoncourt, 30 C.A.), C 104 (Behonne)

3rd Armée: N 79 (Mesnil-St.-Georges, 3rd Armée), R 217 (Mesnil-St.-Georges, 3rd Armée), N 112 (Grivesnes, 3rd Armée), F 16 (Grivesnes, 3rd Armée), C 66 (Grivesnes, 3rd Armée), C 122 (Grivesnes, 3rd Armée), C 28 (Grivesnes, 14th C.A.)

4th Armée: N 38 (La Noblette, 4th Armée), N 78 (La Noblette, 4th Armée), F 22 (Suippes, 12th C.A), F 33 (Mailly, 9th C.A.), F 41 (St. Menehould, 32nd C.A.), F 71 (St. Menehould, 8th C.A.), F 35 (Matougues, Center d'Instruction), C 216 (Matougues, Center d'Instruction), C 228 (Matougues, Center d'Instruction), C 56 (Ferme d'Alger, 17th C.A.), C 214 (Ferme d'Alger, 4th Armée), C 64 (La Cheppe, 4th Armée), C 212 (Le Tilloy)

The unusual Salmson S.M.1 three-seat reconnaissance plane of 1917. A single engine drove two propellers via a complex gear train. SHAA B86.4203.

SPAD 7s of the Imperial Russian Air Service. First introduced in 1916, the fast, robust SPAD 7 replaced the obsolescent Nieuport fighters in most French fighter squadrons by the end of 1917. The SPAD 7 was also widely used by Allied air services. Via Colin Owers.

5th Armée: N 31 (Rosnay, 5th Armée), N 76 (Rosnay, 5th Armée), N 83 (Rosnay, 5th Armée), F 2 (Rosnay, 5th C.A.), F 24 (Rosnay, 38th C.A.), F 72 (Bouleuse, 7th C.A.), R 210 (Bouleuse, 5th Armée), C 17 (Hourges, 1st C.A.), C 53 (Hourges, 1st C.A.), C 39 (Nogent-Sermiers, 38th C.A.)

6th Armée: N 48 (La Bonne Maison, 6th Armée), N 80 (La Bonne Maison, 6th Armée), N 62 (La Cense, 6th Armée), F 35 (La Cense, 20th C.A.), F 7 (La Cense, 2nd C.C.), F 7 (Mont-St.-Martin, 6th C.A.), C 30 (37th C.A.)

7th Armée: N 49 (Fontaine, 7th Armée), N 81 (Fontaine, 7th Armée), N 82 (Fontaine, 7th Armée), C 61 (Fontaine, 34th C.A.), F 14 (Girecourt, 129th D.I.), F 59 (Corcieux, 161st D.I.), C 34 (Romagny, 52nd D.I.), C 27 (Luxeuil, 21st C.A.)

8th Armée: N 68 (Madeleine, 8th Armée), N 75 (Lunnevie, 8th Armée), F 58 (Lunnevie, 8th Armée), C 22 (Lunnevie, 8th Armée), N 77 (Toul, 8th Armée), C 11 (Toul, 2nd C.A.), F 70 (Toul, 8th Armée), F 45 (Saint Meurthe), F 223 (Saizerais), C 9 (Villers-les-Nancy, 8th Armée), C 42 (Villers-les-Nancy, 8th Armée)

10th Armée: N 15 (Cachy, 10th Armée), N 69 (Moreuil, 10th Armée), F 60 (Marcelcave, 18th C.A.), C 6 (Marcelcave, 18th C.A.), C 202 (Marcelcave, 18th C.A.), F 206 (Bougainville), C 227.

Groupes de Combat:

GC 11: N 12 (Vadelaincourt, 2nd Armée), N 57 (Vadelaincourt, 2nd Armée)

GC 12: N 3 (Mannoncourt, 8th Armée), N 26 (Mannon-court, 8th Armée), N 73 (Mannoncourt, 8th Armée), N 103 (Mannoncourt, 8th Armée)

GC 13: N 37 (Ravenel, 3rd Armée), N 65 (Ravenel, 3rd Armée), N 67 (Ravenel, 3rd Armée), N 124 (Ravenel, 3rd Armée)

Groupes de Bombardement:

GB 1: CEP 115 (Malzéville), VC 110 (Malzéville), VC 114 (Malzéville)

GB 3: VB 107 (Esquennoy), VB 108 (Esquennoy), VB 109 (Esquennoy), VC 111 (Esquennoy), VC 113 (Esquennoy)

GB 4: SOP 29 (Ochey), F 123 (Ochey), BM 120 (Ochey)

GBM 5: BM 117 (Malzéville), BM 118 (Malzéville), BM 119 (Malzéville), BM 121 (Malzéville)

R.F.D.-Rosendael: C 74 (Hundschoote, Belgian Army), F 36 (Furness, G.A.N.), VC 116 (St. Pol-sur-Mer)

Escadrilles of the Interior: N 311

Escadrille de Tarcenay: 304th D.C.A. (Cresnot), 307th D.C.A. (Lyon)

C.R.P. (Paris): Escadrilles 350, 351, 352, 353, 354, 393, 394, 395, 396, 397

T.O.E.:

A.F.O.: V 383, V 385, V 386, V 388, V 389, N 390, N 391
Serbia: MF 382, MF 384, H 387, MF 398, MF 399
Russia: N 581, N 582
North Africa: VR 551, VR 552

A.R.1 of F 55; the left side of the unit insignia was white, the right side was red. During 1917 the A.R.1 replaced the obsolete pusher designs for reconnaissance. SHAA B89.3359.

The Battle of Chemin des Dames began on 17 April. It was anticipated that French forces would be able to break through the German lines, and the Aviation Militaire was assigned the task of spearheading the advance. It was to destroy German reconnaissance and bomber aircraft above the sectors in which the French armies would be advancing. The French fighter forces were to play a key role in the battle. GC 11 (N 12, N 31, N 48, N57), GC 12 (N 3, N 26, N 46, N 73, N 103), and GC 14 (N 75, N 80, N 83, N 87) were placed under a unique organization, the Groupement de Combat of the G.A.R., which was to coordinate the activities of all the fighter units. Independent fighter escadrilles included N 15 (10th Armée), N 62 (6th Armée), N 65 (G.A.N.), N 69 (10th Armée), and N 76 (5th Armée). Approximately 100 fighters were assigned temporarily to the 4th Armée to provide protection of its reconnaissance units. These fighter units were N 37, N 38, N 81, N 102, and N 112. Fighters were also drawn from F 72. GB 1 was the bomber unit assigned to the 4th Armée.

Overall, the G.A.R. (composed of the 4th, 5th, 6th, and 10th Armées for the battle) had 400 fighters (of which approximately 50 percent were SPAD 7s) and 600 bomber and reconnaissance airplanes (from 47 reconnaissance units and GB 1 and GB 3). The Aviation Militaire was facing 210 fighters of Jagdstaffeln 1, 2, 4, 9, 10, 13, 14, 17, 19, 21, 22, and 24 and 40 bombers of KG 2 and KG 4.

Unfortunately, the weather on 9 April 1917 inhibited the army cooperation units from accomplishing much over the battlefield. The fighter units had great success as they hunted German airplanes over the front lines. However, this meant that there was inadequate protection for the army cooperation and bomber units, which suffered appalling losses. Although units such as GC 14 experienced difficulties as they were just beginning the transition to SPAD 7s, the fighter escadrilles destroyed large numbers of observation balloons and aircraft. In fact, the appearance of the SPAD 7 in large numbers was one of the key elements enabling the Groupes de Combat to establish and maintain aerial superiority. French fighters claimed 72 German aircraft while losing only 17.

However, the success of the French fighters came at an enormous cost to the army cooperation escadrilles, which suffered 108 crew members killed, wounded, or missing. The commanders of the various corps d'armée were infuriated by what they saw as the abandonment of their airplanes by the Groupes de Chasse. The arrival of General Pétain at the GQG resulted in the decentralization of the French fighter forces. This was intended to assure that a massacre of the C.A.'s escadrilles would never happen again. This policy would stay in effect until 1918, meaning that the Aviation Militaire had abandoned the attempt to maintain aerial supremacy over the front for the remainder of 1917.

By June the goal of 2,665 airplanes had been achieved. These were distributed as follows: 2,172 on the Western Front, 209 with the C.R.P., and 284 with the A.F.O.

The Battle of Flanders began on 7 June and ended on 6 November. The air units of the 1st and 5th Armées were active, and MF 25, MF 36, VC 116, VB 109, and SOP 108 concentrated their attentions on German aerodromes. For the most part, the Germans did not attack the French aerodromes; when they did, the French losses were initially small. For example, MF 25's aerodrome was bombed in August and September but the unit was able to continue operations by the simple expedient of moving to a new base. GB 1 joined in the campaign against the aerodromes in September. On 26 September the Germans struck back at aerodromes at Vadelaincourt, Senard, Lemmes, and Osche. This time the Germans achieved complete surprise and attacked with large numbers of aircraft; the result was heavier French losses than on previous raids. GB 1, GB 4, and MF 25 continued their attacks with some success, but they were never able to equal the damage that had been done by the Germans on 26 September. Yet again, the weakness of the French bomber force had been exposed.

By the end of 1917, French aircraft production was steadily increasing. More important than numbers was the quality of the airplanes the Aviation Militaire would be receiving: the Breguet 14, Salmson 2, and SPAD 13. Furthermore, a return to a policy of concentrating the air force over selected areas of the front would enable the Aviation Militaire to control the skies above the battlefields for the remainder of the war.

Foreign Service
Italy

The French sent the 10th Armée to Italy in late 1917 to assist the Italians in their struggle against Austro-Hungary. The

Paul Schmitt 7. This type reached the front a year later than expected and served only a short time before being withdrawn. It was vulnerable to fighter attack due to its modest performance and lack of maneuverability. SHAA B88.1361.

Voisin 8. Despite its obsolete pusher configuration, the reliable Voisin 8 was more successful than its contemporary, the Paul Schmitt 7. SHAA B77.1199.

following escadrilles were assigned to the expeditionary force on 21 November 1917:

10th Armée: N 69, N 62 at Porto Nuova de Verona
31st C.A.: AR 44, SOP 221, SOP 36 at Ghedi; SOP 206 and AR 14 at Porto Nuova

SOP 206 was used primarily over the Isonzo front. Other units were engaged primarily in army co-operation duties that included photomapping the front near Bassano and Piave.

Additional units followed the next month and the order of battle for 16 December was:

10th Armée: SPA 69 at Porto Nuova de Verona and N 82 at San Pietro di Godego
31st C.A.: AR 44: SOP 36, 206, 221 at Castello di Godego
12th C.A.: AR 22 and 254 at Castello di Godego; AR 14 at San Pietro in Gu

The army cooperation units continued to map the Piave Valley, Brenta, Astice, and Adige. SPA 69 and N 82 provided fighter escort during these missions.

The escadrilles assigned to the 31st C.A. located enemy troop concentrations and artillery sites before the attack on Monte Tomba. The 12th C.A.'s units initially performed reconnaissance east of Asiago. Subsequently they participated in the siege of Brenta and were the only source for up-to-date information for the army commanders. This intelligence was obtained by making dangerous low-level passes over the battlefield.

During the Tomba offensive the aviation units bombed and strafed enemy artillery and machine gun positions.

Escadrille 92/392/561 had been sent to Italy in 1915. It provided fighter escort for French and Italian flying boats operating from Venice.

Russia

The Imperial Russian Air Service (IRAS) had depended heavily on the output of the French aviation industry. In addition to airplanes, the Aviation Militaire sent a number of advisers and volunteer pilots to aid the IRAS in its fight against the Central Powers. However, the integration of French pilots into Russian units had not gone smoothly and at least two Russian airplanes were inadvertently shot down by French pilots. This, plus the frustrating inefficiency of the IRAS units, resulted in the decision to form two all-French escadrilles.

N 581, with eight Nieuport 17s and seven SPAD 7s, was formed in February 1917. SOP 582, with Sopwith 1½ Strutters, was formed at the same time. Both escadrilles were assigned to

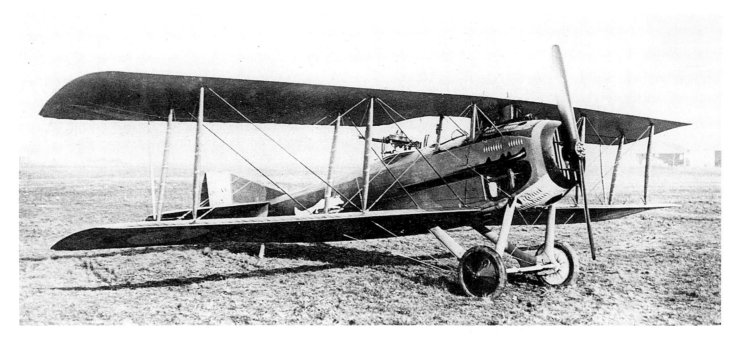

SPAD 11. Designed as a two-seat fighter, it was used for short-range reconnaissance when its handling qualities and maneuverability proved insufficient for its intended role. Photo via Colin Owers.

the 7th Russian Army and began operations in May 1917. Based at Boutchatch, Galicia, they served alongside Captain Kozakov's Nieuport fighter group. In response to the German offensive of 19 July on the 6th Army front, both escadrilles were transferred to Buzcacz. In August both moved to the Romanian front, where they were assigned to the 3rd Corps. In February 1918, both units, N 581 (now re-designated SPA 581 because it was re-equipped entirely with SPAD 7s) and SOP 582 were ordered back to Moscow. SPA 581 passed its SPAD 7s to the Czechoslovakian air service, while SOP 582's planes were given to the Russians. The French aircrew returned to France in 1918. In 1918 both escadrilles were reformed as SAL 581 and SAL 582 and sent to Poland as part of the French military mission to that country.

Other Countries

French military missions provided aircraft and aircrews to assist the fledgling Serbian, Greek, and Romanian air services. Further details can be found in the sections on Nieuport, SPAD, and Breguet aircraft.

1918
Political

On 11 January Dorand was replaced as the director of the STAé. His replacement was Albert Caquot who set up the following departments:
1. Service General d'Avions
2. Service General d'Moteurs
3. Service Essais en Vol

The first two departments would control the acquisition of aircraft and engines, respectively, while the third was responsible for flight-testing prototypes.

General Pétain insisted that the key to defeating the German air service was to concentrate the Aviation Militaire's forces over selected areas of the front. This resulted in the formation of combat wings containing both GC and GB units.

Technical

In 1917 and early 1918, specifications for 12 types of aircraft were released to the industry. These aircraft were evaluated during 1918:

C1 (single-seat fighter)—payload 220–270 kg, ceiling 9,000 m and 240 km/h speed: SPAD 15, Nieuport 28, Morane-Saulnier 27, Morane-Saulnier 29, C.S.L.1, Laboratory Eiffel Fighter, G.L.1, Wibault 1, SPAD 17, HD.9, SPAD 18, SPAD 21, SPAD 22, Nieuport 29, SPAD 20, Descamps 27, HD.7, De Marcay 2, SAB 1, Moineau fighter, and Semenaud fighter.

C2 (two-seat fighter or protégé)—payload 375 kg, ceiling 8,000 m, speed 220 km/h: Borel Boccacio 1, BAJ fighter, Vickers F.B.24G, F.31, SEA 4, HD.3, Breguet 17, F.30, HD.5, Latecoere C2, Morane Saulnier 32, Morane-Saulnier 33, Morane-Saulnier 34, HD.6.

C3 (three-seat escort fighter)—payload 520 kg, ceiling 5,000 m, speed 190 km/h: Caudron R.12.

Ap2 (short-range reconnaissance)—380 kg payload, ceiling 8,500 m, speed 195 km/h: SEA 4, Breguet 14 with Liberty engine.

A2 (two-seat army co-operation)—payload 450 kg, ceiling 7,000 m, speed 200 km/h: SPAD 11/16, Salmson 2, Breguet 14 with 265 hp Fiat engine, Salmson 7, Breguet 14 A2 with 300 hp Renault, SEA 4.

Ab2 (two-seater with light armament)—350 kg payload, ceiling 8,500 m, speed 175 km/h: Salmson 4.

A3 (three-seat long-range reconnaissance aircraft)—620 kg payload, ceiling 8,500 m, speed 195 km/h: Letord 4 and 5, Caudron R.11 and R.12.

B2 (two-seat bomber)—740 kg payload, 7,500 m ceiling, 190 km/h speed: Breguet 14 B2 with 300 hp Renault with Rateau turbocompressor, Nieuport 30.

Bn2 (medium bomber)—900 kg payload, ceiling 5,000 m, speed 150 km/h: Bernard 1, Breguet 16, Voisin 11, Farman F.50.

BN2/3 (two or three-seat heavy night bomber)—payload 1,800 kg, ceiling 5,000 m, speed 150 km/h: Caudron C.23, Caproni 3, Letord 9, Delattre BN3, Bassan-Gue BN 4, Voisin 12, Renault O1, Sikorsky BN3, S.I.A-Coanda BN2.

S2 (two-seat ground attack)—ceiling 4,500 m, speed 190 km/h: LeO 5 S2, Canton S2, Hochart S2.

S3 (heavy ground attack)—ceiling 3,000 m, speed 160 km/h: Henri-Paul Schneider S3.

The aircraft assigned to the Aviation Militaire on 11 November 1918 reflected the extensive modernization that had taken place in late 1917 and early 1918. The following airplanes were in front-line service:

Letord 1 of BR 210. Carrying serial number 7, it is one of the earliest aircraft. A long-range, three-seat reconnaissance plane, it was more conventional, reliable, and successful than the Salmson S.M.1 designed for the same role. SHAA B79.3011.

Fighters:
SPAD 13	1152
Caudron R.11	60
Total:	1212

Bombers:
Breguet 14 B2	225
Voisin 10	135
Farman F.50	45
Caproni 3	20
Total:	425

Reconnaissance:
Breguet 14 A2	645
Salmson 2 A2	530
SPAD 11/16	305
Caudron R.11	30
Voisin 10	75
Total:	1585

The total number of aircraft in service with the Aviation Militaire on the Western Front (including those with the R.G.A. and undergoing repair) was 3,222: 1,152 fighters, 1,585 reconnaissance airplanes, 285 day bombers, and 200 night bombers.

Operational

Based on the earlier successes achieved by concentrating the GC units at the Battle of Chemin-des-Dames, the GQG decided to adopt even larger formations that would include both fighter and bombing aircraft. In March the Groupements Ménard and Féquant were adopted; each combined at least one fighter and one bomber group. Later multiple GC and GB units would be combined into Division Aériennes (Aerial Divisions) which could establish local air superiority and then use it to strafe and bomb enemy troops and lines of supply. These later formations proved to be too large to base at a single aerodrome or even set of aerodromes, but the Aerial Divisions were retained until the end of the war. In fact, two aerial brigades were formed in mid-June; the first under Major de Göys, the other under Féquant. The brigades were to establish local air superiority so that the day bombers could perform ground-attack missions, while the night bombers concentrated on transportation targets. The improvement in both quantity and quality of the French fighters enabled the fighter groups to be concentrated over selected areas of the front, while still leaving enough independent fighter escadrilles to protect the army co-operation units.

The night bomber units were still equipped with Voisin 10s and Caproni 2s at the beginning of 1918. Because of this inferior force, the GB units limited their attacks to the German rail network in the Saarbrucken and Lorraine basins in an attempt to block the transportation of iron ore.

A special school, the Center d'Instructions de Chasse et de Bombardement, was established to provide advanced training for fighter and bomber pilots. This aided the front-line escadrilles by relieving them of the responsibility of combat training.

In an attempt to win the war before the United States could become actively involved in the conflict, the Germans planned as series of devastating attacks in early 1918. They launched their attack on 21 March in what became known as the Battle of Picardie. Using their fighter groups, the French quickly achieved air superiority and, within two days, the bomber escadrilles were free to attack the bridges and other crossing points over the Somme and Crozat canal. The Féquant and Ménard Groups were both sent to the front and were able to achieve and hold control of the skies.

The order of battle for the Aviation Militaire on 15 April, as the final phase of the war began, was:

G.A.R.
Groupment Ménard
1st Escadre:
GC 15: SPA 37, SPA 81, SPA 93, SPA 97, R 46 based at Montagne
GC 18: SPA 48, SPA 94, SPA 153, SPA 155 based at Montagne
GC 19: SPA 73, SPA 85, SPA 95, SPA 96 based at Airaisnes
12th Escadre:
GB 5: BR 117, BR 120, BR 127 based at Plessis-Belleville
GB 6: BR 66, BR 108, BR 111 based at Plessis-Belleville
GB 9: BR 29, BR 123, BR 129 based at Plessis-Belleville

Breguet 14 B2s of BR 134. Produced in large numbers during 1918, the Breguet 14 B2 was one of the best day bombers of the war, and was also used by the US Air Service. A reconnaissance version, the Breguet 14 A2, was also successful. SHAA B89.4658.

Groupement Féquant
2nd Escadre:
GC 11: SPA 12, SPA 31, SPA 57, SPA 154 based at Tille
GC 13: SPA 15, SPA 65, SPA 84, SPA 88 based at Fouquerolles
GC 17: SPA 77, SPA 89, SPA 91, SPA 100 based at Fouquerolles
13th Escadre:
GB 3: BR 107. BR 126, BR 128 based at Maisonneuve
GB 4: BR 131, BR 132, BR 134 based at Bonne-Maison
11th Escadre:
GB 1: VB 110, VB 114, VB 116 based at Passy-en-Valois
GB 7: VR 118, VR 119, BM 121 based at Cramaille
GB 8: VB 109, VB 113, VB 125 based at Chateau-Thierey
1st Armée:
Assigned to 1st Armée: GC 12: SPA 3, SPA 26, SPA 67, SPA 103 based at Hetomesnil
Assigned to 1st Armée: BR 45, SPA 102, SAL 230, SPA 228, SAL 232 based at Esquennoy, BR 208 based at Bovelles
5th C.A.: SPA-Bi 2, SOP 105, SOP 250 based at Poix
9th C.A.: AR 256, AR 257, SOP 237, SOP 217 based at Grandvillers, SAL 33 based at Esquennoy
6th C.A.: BR 7, SAL 204 based at Esquennoy, SOP 5, SOP 141, AR 288 based at Viefvillers
31st C.A.: SPA-Bi 286, SAL 16, SOP 285, SOP 104 based at Montagne, AR 44, AR 275 at Etampes
3rd Armée:
Assigned to 3rd Armée: GC 16: SPA 78, SPA 112, SPA 150, SPA 151 based at Besles
Assigned to 3rd Armée: SPA 79, MS 158, BR 210 based at Sacy
Artillerie Lourde de l'Armée: SOP 234 based at Remy
88th Artillery Lourde: SOP 231 based at Sacy
18th C.A.: SAL 6, AR 268, BR 209 based at Remy; SAL 52 based at Corbeaulieu
34th C.A.: SAL 61, SOP 271 SOP 285, BR 201 based at Remy
35th C.A.: SAL 10, SOP 282, SAL 225 based at Sacy-le-Grand
62nd D.I.: SAL 1, AR 272 based at Corbeaulieu

G.A.N.
GC 21: N 98, SPA 124, N 157, SPA 163 based at La Noblette
GB 3: VB 101, CEP 115, CEP 130 based at Villeneuve-les-Vertus
GB 18 (Italian): Squadriglias 3, 4, and 15 based at Villeneuve
6th Armée:
Assigned to 6th Armée: SPA 62, BR 213 SOP 216 based at Saint-Amand, Lafayette Escadrille based at Bonnemaison
1st C.A.: SPA-Bi 53, SOP 17, SOP 255, SPA-Bi 212 based at La Cense
30th C.A.: SAL 18, SOP 278 based at Mont-Saint-Martin, SOP 9 based at Ambrief, SPA-Bi 284, SOP 55 based at Mont-de-Soissons
11th C.A.: SAL 8, SOP 251, BR 205 based at Mont-de-Soissons
38th C.A.: SAL 39, SOP 273 based at Rosnay
2nd C.A.: BR 11, SOP 269 based at Ambrief
37th D.I.: AR 21 based at Ambrief, SAL 52 based at Shery
4th Armée:
Assigned to 4th Armée: SPA 38, MS 156 at Melette, BR 227 based at Montagne
1st C.A.C.: SAL 51, SOP 260 based at Bouzy
3rd C.A.: SAL 4, SOP 280 based at La Cheppe
4th C.A.: SPA-Bi 140, AR 40, AR 267 based at Ferme d'Alger
8th C.A.: SPA 54, AR 262 base at Dancourt, AR 71 based at Auve, SAL 203, SPA-Bi 266 based at Bouy

G.A.E.
GC 20: SPA 78, N 99, N 159, N 162 based at Manoncourt
2nd Armée:
Assigned to 2nd Armée: SPA 23, N 160 based at Souilly
10th C.A.: SAL 32, AR 59, AR 261 based at Issoncourt
17th C.A.: C 56, SOP 281 based at Vadelaincourt
13th C.A.: SPA 64 based at Vadelaincourt, AR 19 based at Foucaucourt, AR 259 based at Froidos
68th D.I.: SOP 43 based at Bulainville, SOP 287 based at Beauzee
311th A.L.: SPA 215 based at Lemmes

7th Armée:
Assigned to the 7th Armée: SPA 49, N 152 based at Chaux
21st C.A.: SAL 27, SOP 252 based at Dognerville, SOP 106 based at Corcieux, SOP 223 based at Fontaine
40th C.A.: AR 58 based at Bessoncourt
Assigned to the City of Belfort: N 315 based at Chaux
8th Armée:
Assigned to 8th Armée: N 90 based at Malzéville, N 87 based at Luneville, N 92 based at Belrain
7th C.A.: SPA 34, AR 72 based at Luneville, AR 259 based at Moyen
15th C.A.: SAL 13, SOP 270 based at Villers-les-Nancy
32nd C.A.: SAL 122 based at Saizerais, AR 41 based at Gonderville, AR 258 based at Toul
2nd C.A.C.: SAL 47, SOP 277 based at Belrain
Assigned to the City of Nancy: N 314 based at Malzéville

Flanders

36th C.A.: SAL 74, SOP 276, BR 218 based near Bray-Dunes
2nd C.C.: SAL 24, SOP 279 based near Bray-Dunes

Italian Army

12th C.A.: AR 22 based at Nove, AR 254 based at Verone
Escadrille de Venise: N 561 at Le Lido

GQG Reserves

10th Armée:
Assigned to the 10th Armée: GC 14: SPA 75, SPA 80, SPA 83, SPA 86 based at Fienvillers
Assigned to the 10th Armée: SPA 69, SOP 22 based at Fienvillers, N 82 based at Ambrief,
Assigned to the 10th Armée: SAL 28 based at Sacy-le-Grand, BR 207 based at Frenvillers, SOP 222 based at May-en-Multien
14th C.A.: SPA-Bi 20, SPA 60, SOP 236 based at Fienvillers
5th Armée:
Assigned to the 5th Armée: SPA 76, MS 161, BR 220 based at Le Bourget, SOP 223 based at Cernon, BR 211 based at La Cheppe
16th C.A.: SAL 50, AR 274 based at Verrines
20th C.A.: BR 35, AR 70, AR 253 based at Suzay
1st C.C.: SPA 63, SAL 30 based at Plessis-Beleville
Groupement De L'Oise:
66th D.I.: AR 289 based at May-en-Multien
Artillery Escadrilles:
AR 14 (46th D.I.), SOP 214, SOP 206, SOP 221 based at Etampes
BR 218, BR 233, SOP 226, SOP 229 based at Mairy
BR 244 based at Augny (assigned to the GQC)
BR 219 based at Dugny
SOP 36 (47th D.I.) based at Etampes, SOP 238 (47th D.I.) based at Rozoy-en-Brie
SPA 265 (48th D.I.) based at Cernon

T.O.E. Escadrilles

Assigned to the A.F.O.: BR 501, BR 503, BR 504, BR 505, SPA 506, SPA 507, SPA 508, BR 509, BR 510
Serbia: AR 521,, BR 522, N/SPA 523, BR 524, BR 525
Greece: N 531, BR 532
North Africa: VR 541, VR 542, C 543, C 544, VR 551, VR 552, F 553, F 554, VR 555, AR 556, VR 558
Indochina: V 571

The structure of the Aviation Militaire on 15 April shows a force in transition between the obsolescent Sopwith 1½ Strutters, A.R.1s, and Nieuport 17s and the new SPAD 7s, SPAD 13s, Salmson 2s, and Breguet 14s.

On 14 May Groupements Ménard and Féquant, as well as the night bomber groups Villomé and Chabert, were formed into the 1st Aerial Division. This enabled the units to coordinate their attacks on German positions. General Duval had direct control of the Aerial Division. In June the groupements were redesignated brigades.

The Germans launched their next offensive on 27 May at Chemin des Dames. In order to protect Tardenois, which was threatened by the Germans, Groupement Ménard was sent to the area the same day and within three hours was attacking German troops. Aerial superiority was once again established in only two days. Groupement Féquant arrived four days later.

During May the superlative Fokker D.VII became available to the Germans. This new fighter was able to inflict significant damage on the French bombers and reconnaissance aircraft. However, the twin-gun SPAD 13 was now available to the French and by sheer force of numbers these new fighters soon overwhelmed the weakened Luftstreitkräfte.

The final German offensive was in July 1918 when further attempts were made to cross the Marne. With the aid of the bomber escadrilles, the French were able to stop the German advance.

Now the Allies began a series of advances that would ultimately defeat Germany. The Battle of Ile de France lasted from 18 July to 4 August. The 1st Division's fighter units were able to protect the Breguet 14 bombers as they attacked the retreating German troops. Cover for the Breguets was provided by the SPAD 13s and Caudron R.11 escort fighters. Attacks concentrated on the Vesle and Ardre Valleys in support of the 5th Armée.

The Battle of Santerre lasted from 8 to 30 August; the Aviation Militaire was particularly active over the 2nd and 3rd Armée fronts. At this time the fighters were often being used to strafe and bomb enemy troops, some of the SPAD 7s and 13s having been modified to carry a small bombload.

In preparation for the Battle of Saint Mihiel (12 to 30 September) the 1st Air Division moved to Lorraine. When the battle began the French had no trouble defeating the weakened German fighter force and the bombers were usually able to roam the front-lines with complete impunity. The French supplied 900 aircraft, the British 160, and an additional 440 aircraft were supplied by the United States, Belgium, and Italy.

During the war's final weeks, massive formations of up to 200 bombers were being assembled, supported by SPAD 13s in group strength, as well as several escadrilles of Caudron R.11 escort fighters. In October an entire German division was destroyed by aerial attacks. The Air Division ended the war at Lorraine, where it was being prepared for one final assault. However, on 11 November the Germans capitulated. The 1st Aerial Division was credited with the destruction of 637 airplanes and had dropped 1,360 tons of bombs from May until 11 November 1918.

Foreign Service

During the first three months of 1918, the 10th Armée remained in Italy. Its aviation units on 1 January were:
10th Armée: SPA 69 and N 82 at San Pietro in Gu
31st C.A.: AR 4, SOP 36, SOP 206, SOP 221 at Castello di Godego; AR 14 (46th D.I.) at San Pietro in Gu
12th C.A.: AR 22, AR 254, SOP 214 at Castelgomberto

SPA 69 and N 82 conducted long-range reconnaissance missions over Adige, Brenta, and Piave, and provided fighter escort for the army cooperation units of the 31st C.A. The 31st C.A. attacked anti-aircraft batteries and photographed the front-lines, especially at Bretagne and Bassano, which were near the 46th D.I. In January the 12th C.A. was kept in reserve. It flew training missions and participated in maneuvers with the Italian and French armies.

A month later the units assigned were:
10th Armée: SPA 69 and N 82 at San Pietro
31st C.A.: AR 44 and AR 275 at Castelgomberto; SOP 214

Levasseur-built SPAD 13 of the US Air Service. During 1918 the fast, robust SPAD 13 gradually replaced all other single-seat fighters in French service and became the mainstay of the US Air Service. MA 1417.

Below: Caudron R.11 of the 12eme Escadrille of 12eme RAB. The heavily-armed, three-seat Caudron R.11 was the other fighter in French service at the end of the war. Used as an escort for day bombers, it was also successful as a reconnaissance aircraft. SHAA B87.4948.

(Artillery Breganze) at Castelgomberto; SOP 221 (Artillery Basane) at Nove; SOP 36 (47th D.I.) at San Pietro in Gu
12th C.A.: AR 22, SOP 206, AR 14 at Castello di Godego; AR 254 at Verone; N 561 at Venice

All the units continued their support of the French armies but several problems were hampering their activities. The SPAD 7s were found to be difficult to fly and were given only to the best pilots in the escadrille. The Nieuport 17s were inferior to both the SPAD and enemy aircraft. No fewer than 14 Sopwith 1½ Strutters were lost during the first three months of action, secondary to engine failures and crashes. Despite these accidents only three men were killed: one in an aircraft that broke apart in the air and two in an airplane that caught fire in flight. The weather was poor and the bitter cold imposed severe hardship on the crews during long-range reconnaissance missions. The A.R.1s gave good service and some were modified for use as night bombers. The airfields were in poor condition and surrounded by trees; the latter accounted for many of the Sopwith 1½ losses. Finally, the aviation commanders complained that the front was too quiet and the fighters, in particular, saw little action.

March was to be the last month the 10th Armée was in Italy. The last order of battle for these units was given on 15 March:
10th Armée: SPA 69, N 82 at San Pietro in Gu
31st C.A.: AR 44 and AR 275 at Castelgomberto
12th C.A.: AR 22, SOP 36, SOP 221 (46th D.I), and SOP 206 at Nove, AR 254 at Verone, SOP 214 at Castelgomberto, and AR 14 at Fontanelle, N 561 at Venice

Shortly after this, the 10th Armée and its aviation units returned to France.

By November 1918 the following units were still based in Italy: SAL 22, SAL 254 (12th C.A.), SPA 201 (Protection of Venice).

Other Countries

French aviation units assigned to the A.F.O. at the end of the war were:
Serbia: SPA 523, BR 521, BR 522, BR 254, BR 255
Greece: SPA 531, BR 532, BR 533, BR 534

At the time of the Armistice the Aviation Militaire consisted of the following units:

Salmson 2 A2 serial number 479 of SAL 58. By war's end the robust and reliable Salmson 2 A2 was the backbone of the short-range reconnaissance units and was also the standard reconnaissance aircraft of the US Air Service. SHAA B86.4194.

1st Brigade d'Aviation (Groupement Ménard)

Escadre de Combat 1: GC 15 : SPA 37, SPA 81, SPA 93, SPA 97 at Melette; GC 18: SPA 48, SPA 94, SPA 153, SPA 155 at La Noblette; GC 19: SPA 73, SPA 85, SPA 95, SPA 96 at Ferme d'Alcer

Escadre de Bombardement 12: GB 5: BR 117, BR 120, BR 127 at Matouges; GB 6: BR 66, BR 108, BR 111 at Tantonville; GB 9: BR 29, BR 123, BR 129 at Plivot

2nd Brigade d'Aviation (Groupement Féquant)

Escadre de Combat 2: GC 13: SPA 15, SPA 65, SPA 84, SPA 88 at Marios, C 17: SPA 77, SPA 89, SPA 91, SPA 100 at Auve: GC 20: SPA 68, SPA 99, SPA 158, SPA 159, SPA 162 at Tilley

Escadre de Bombardement 13: GB 3: BR 107, BR 126, BR 128 at Couyville; GB 4: BR 131, BR 132, BR 134 at Saint-Dizier; C 46 and R 246 at Saint-Dizier

G.A.R.

GC 14: SPA 75, SPA 80, SPA 83, SPA 86, SPA 166 at Clastres; Escadre 14

GB 8: VB 109, BR 113, VB 125 at Clastres; GB 10: VB 101, VB 116, VB 133 at Clatres

1st Armée: SPA 102 (assigned to 1st Armée), BR 201 at Ercheu; VR 293, BR 234, BR 202 (Libermont), SAL 203 (Flavy-le-Meldeux), SAL 204 (Manicourt), SAL 71 (8th C.A., Fe Du Rendez-Vous), SPA-Bi 54 (8th C.A., Fe Du Rendez-Vous), SAL 262 (8th C.A., Fe Du Rendez-Vous), BR 35 (20th C.A., Avricourt), SAL 70 (20th C.A., Avricourt), SAL 253 (20th C.A., Avricourt), BR 44 (31st C.A., Libermont), BR 275 (31st C.A., Libermont), SAL 74 (36th C.A., Manicourt), SPA-Bi 276 (36th C.A., Manicourt), SPA-Bi 21 (37th D.I.,Manicourt; SPA 14 (46th D.I., Flavy-le-Meldeux; SPA-Bi 36 (47th D.I.,Libermont), BR 283 (67th D.I., Fe Du Rendez-Vous

3rd Armée: SPA 79 (3rd Armée, Cerny-lès-Bucy), BR 209 (3rd Armée, Croutoy) BR 226 (3rd Armée, Vaucastille), BR 217 (3rd Armée, Tigny), SAL 16 (2nd C.A., Coucy-les-Eppes, SAL 50 (16th C.A., Cerby-lès-Bucy), BR 274 (16th C.A., Cerby-lès-Bucy) SAL 6 (18th C.A., Berny-Loisy), SPA-Bi 268 (18th C.A., Berny-Loisy), SAL 19 (35th C.A., Lavergny), BR 282 (35th C.A., Lavergny)

G.A.F. (Flanders)

SAL 24 (22nd C.C., Bray-Dunes), BR 279 (22nd C.C., Bray-Dunes), SPA 313 (St. Pol Naval Base)

6th Armée (Belgium): GC 23: SPA 82, SPA 150, SPA 160, SPA 161, SPA 169, SPA 170 at Rumbecke; SPA 62 (6th Armée, Rumbecke), BR 239 (6th Armée, Rumbecke), SAL 225 (6th Armée), Rumbecke), SAL 61 (34th C.A., Wyngène), BR 271 (34th C.A., Wyngène), SAL 18 (30th C.A., Thielt), SPA 278 (30th C.A., Thielt), SAL 34 (7th C.A., Ingelmunster), SAL 72 (7th C.A., Ingelmunster), SAL 259 (7th C.A., Ingelmunster), SAL 25 (2nd C.C.), BR 279 (2nd C.C.)

G.A.C.

Escadre 11: GB 1: F 25, F 110, F 114 at Villenieuve; GB 7: VB 118, VB 119, VB 121 at Villeseneux; GB 51: VB 135, VB 136, VB 137 at La Cheppe; GB 18 (Italian) CAP 3, CAP 14, CAP 15

5th Armée: GC 16: SPA 78, SPA 112, SPA 151 SPA 156 at Rosney, SPA 76 (5th Armée, Montbre), BR 216 (5th Armée, Montbre), BR 222 (5th Armée, Montbre); BR 231 (5th Armée, Anthienay), SAL 230 (5th Armée, Menil-Lepinois), BR 223 (5th Armée, Saint Étienne), SPA-Bi 140 (4th C.A., Menil-Leoinois), SAL 40 (4th C.A., Menil-Leoinois), BR 267 (4th C.A., Menil-Leoinois), SPA-Bi 2 (5th C.A., Roucy), SAL 105 (5th C.A., Roucy), BR 250 (5th C.A., Roucy), SPA-Bi 64 (13th C.A., St. Étienne-Saint Suippe), SAL 19 (13th C.A., St. Étienne-Saint Suippe), SAL 264 (13th C.A., St. Étienne-Saint Suippe), SAL 27 (21st C.A.), SAL 108 (21st C.A.), SAL 252 (21st C.A.), SAL 288 (45th D.I., St. Étienne-Saint Suippe), BR 287 (52nd D.I., Roucy), SPA 285 (62nd D.I., Coulognes-en-Tardenn)

4th Armée: GC 12: SPA 3, SPA 26, SPA 67, SPA 103, SPA 167 at Hauvine GC 21: SPA 98, SPA 124, SPA 157, SPA 163, SPA 164 at Machault GC 22: SPA 38, SPA 87, SPA 92, SPA 152

Farman F.50 of the 207eme of the 2eme RAB (ex-VB 109). Arriving late in the war, the Farman F.50 finally gave the French a twin-engine night bomber comparable to the AEG, Friedrichshafen, and Gotha bombers long used by Germany. SHAA B93.2619.

at Pomacle SPA 156 (4th Armée, Cauroy), BR 207 (4th Armée, Cauroy), BR 213 (4th Armée, Cauroy), BR 232 (4th Armée, Fe Scay), BR 237 (4th Armée, Fe Scay), VR 291 (4th Armée, Aussonce), BR 210 (4th Armée, Herbisse), BR 223 (4th Armée, Waegemoulin), SPA-Bi 212 (4th Armée, La Nuville-en-Tourne-A-Puy), SAL 33 (9th C.A., Fe Scay), SAL 256 (9th C.A., Fe Scay), BR 257 (9th C.A., Fe Scay), SAL 8 (11th C.A., Le Neuville-on-Tourne-A-Puy), SPA-Bi 55 (11th C.A., Le Neuville-on-Tourne-A-Puy), SAL 251 (11th C.A., Le Neuville-on-Tourne-A-Puy), SPA-Bi 60 (14th C.A., Saint Étienne-a-Arnes), SAL 263 (14th C.A., Saint Étienne-a-Arnes), SAL 39 (38th C.A., Wagemoulin), SAL 273 (38th C.A., Wagemoulin), SPA-Bi 265 (48th D.I., Fe Scay), SPA-Bi 266 (87th D.I., Waegemoulin)

1st U.S. Army: BR 225 (Gourcelles), BR 228 (Prete), BR 234 (Prete), BR 236 (Prete), BR 219 (Courcelles-Saint Aire), SPA 229 (Courcelles-Saint Aire), BR 205 (Courcelles-Saint Aire), BR 214 (Foucaucourt), BR 215 (Foucau-court), BR 208 (Vadelain Court), BR 211 (Clermont), SPA-Bi 284 (3rd C.A., Beauzée), SAL 56 (17th C.A., Beauzée), BR 281 (17th C.A., Beauzée)

2nd Armée: SPA 23 (Brabant-le-Roi), BR 243 (Brabant-le-Roi), VR 209 (Loupy-le-Château)

2nd U.S. Army: SAL 47 (Rumont), SAL 28 (Rumont), SAL 227 (Rumont)

G.A.E.

GB 2: CAP 115, CAP 130 at Epiez
8th Armée: GC 11: SPA 12, SPA 31, SPA 57, SPA 154, SPA 165 at Pont-Saint-Vincent SPA 90 (8th Armée, Manoncourt), BR 244 (8th Armée, Manoncourt), VR 292 (8th Armée, Manoncourt), R 241 (8th Armée, Hutrey), BR 206 (8th Armée, Burthecourt), BR 269 (2nd C.A., Burthecourt), SAL 5 (6th C.A., La Neuville-Dev,-Bayon), BR 7 (6th C.A., La Neuville-Dev,-Bayon), BR 141 (6th C.A., La Neuville-Dev,-Bayon), SAL 58 (73rd D.I., Ochey)

Protection of Nancy: SPA 314 at Pont Saint-Vincent
7th Armée: SAL 49 (7th Armée, Chaux), R 242 (7th Armée, Chaux), BR 245 (7th Armée, Chaux), BR 221 (7th Armée, Chaux), SPA-Bi 53 (1st C.A., Chaux-Nord), SAL 17 (1st C.A., Lure), SPA 253 (1st C.A., Besoncourt), SAL 32 (10th C.A., Dogneville), SAL 59 at Dogneville, SPA-Bi 261 (10th C.A., Corcieux), SPA-Bi 286 (40th C.A., Mezine), SPA-Bi 42 (40th C.A., Lure), SAL 52 (38th D.I., Bour-guignon), BR 43 (68th D.I., Luneville) BR 9 (167th D.I., Besoncourt)

Protection of Belfort: SPA 315 at Belfort
10th Armée: SAL 69 (10th Armée), BR 224 (10th Armée), BR 294 (10th Armée), SAL 4 (3rd C.A), SAL 280 (3rd C.A), SAL 41 (32nd C.A.), SAL 122 (32nd C.A.), SPA-Bi 258 (32nd C.A.), SAL 1 (33rd C.A.), BR 272 (33rd C.A.), BR 218 (33rd C.A.) SAL 51 (1st Armée Artilliere), BR 260 (1st Armée Artilliere), SAL 30 (1st C.C.), SPA-Bi 63 (1st C.C.), BR 104 (1st Armée Customs)

D.C.A.

Protection of Besançon: SPA 412 at Tarcenay
Protection of Creusot: SPA 441 at Autully
Protection of Lyon: SPA 442
Protection of Paris: SPA 57, SPA 124, SPA 461, SPA 462, SPA 463
Coastal Defense: LET 481, LET 482, LET 483, LET 484, LET 485, LET 486, LET 487, LET 491

At the time of the Armistice the Aviation Militaire comprised 3,222 aircraft in front-line service, virtually all of which were modern types. The French possessed the largest, and possibly best organized, air force in the world. However, by the Second World War the Armée de l'Air would be obsolete in both equipment and doctrine. But that is another story...

Aviation Maritime 1914–1918

Introduction

The use of airplanes by the French Navy had been under study since 1910, when a section of aviation was created by recommendation of a commission of the general staff. The aviation section comprised two elements: lighter-than-air machines and airplanes. These two elements were to develop together, but airplanes were to become dominant in the later stages of the war.

Under the pressure of war, the Aviation Maritime was to develop numerous coastal installations for anti-submarine and reconnaissance operations. The operations in the North Sea-Channel-Brittany area were a struggle against submarines, while those in the Mediterranean and Middle East were mainly patrolling and reconnaissance.

The command structure of the naval air service was constantly changed during the war and in the process became cumbersome. The various installations were not given complete autonomy, which led to delays in implementing new equipment and patrolling techniques.

Naval aviation was further hampered by equipment originally designed for land operations. When adopted by the naval service this equipment had to be modified for maritime use; for example, the unreliability of aircraft engines meant that floatation devices had to be fitted to land-based aircraft.

1914

At the outbreak of war, the aviation section consisted of some 25 machines of various types located at Fréjus and Saint-Raphaël. The navy also had a modified cruiser, the *Foudre*, converted to a seaplane carrier.[1]

The machines were: three Caudrons of various types, 13 Nieuport 6H/M monoplanes, one Voisin "canard," seven Voisins of various types, and one Breguet. At least one escadrille of Nieuport monoplanes was based at St. Raphaël; at the outbreak of war these aircraft had been commandeered from a Turkish order.

Pilots were trained in Saint Raphaël, Bizerte, and Hourtin, after first obtaining their land pilot certificates in an army flying school. Observers were trained in Saint Raphaël and gunners at Cazeaux under army supervision.

The first two centers were established in December of 1914, at Dunkerque and Boulogne; their primary mission was anti-submarine activities. The airplanes from these bases were able to perform coastal patrols and stage raids against the German submarine bases at Zeebrugge and Ostende. It was initially recommended that two airplanes be stationed at Boulogne and six at Dunkerque. On 24 August 1914 a directive requested a study of the installation of bombs on airplanes to be based in the Adriatic. An escadrille was later formed there for patrol operations.

1915

Organization

The Aviation Maritime developed land-based centers and support bases known as Poste de Combat (P.C.) and Poste de Relache (P.R.). The last was a rest station. The centers were located at strategic points along the coast with the support bases covering areas between centers.

Coastal operations developed on the western shores of France and in the Mediterranean. The submarine war in the west was fought with airplanes, airships, and surface ships. The Mediterranean and Middle East required co-ordination with ground activity. The Adriatic involved anti-submarine patrols, but not on the scale of the west.

---

(1) The *Foudre*, although officially rated as a cruiser, had been built as a carrier for small torpedo boats, a function soon abandoned. She was converted to a seaplane carrier in early 1912, and the first takeoff of an airplane from a French ship was made on 8 May, 1914, from a temporary platform on her foredeck. After unsuccessful operations in the Adriatic early in the war she was dispatched to Port Saïd in late 1914. She operated briefly at the Dardanelles during March 1915.

Nieuport 6H being loaded on board the SS *Rabenfels* (a captured German merchant vessel) in late April 1915. SHAA B88.3305.

On 31 December 1915 the Centers de Aviation were located as follows:

France	Italy	Greece
Dunkerque	Brindisi	Salonika
Boulogne		
Le Harve		

Operations
France

By January 1915 the installations at Dunkerque and Boulogne were completed. Boulogne had five bombers: one each of Breguet, Farman, and F.B.A. and two Voisin 3s. One F.B.A. was added in late January. These airplanes were then divided between Dunkerque and Boulogne.

The first action against Ostende and Zeebrugge was on 4 February 1915. The airplanes were sent out separately, which proved to be ineffective. After this raid they were sent in groups. The bombs carried had only two to four kilograms of high explosive. Some 10-kg bombs were carried with a delayed fuse to allow a six-meter immersion in water before detonation.

At the end of February Dunkerque was reinforced by another escadrille of F.B.A. Type B flying boats and land-based Voisins. Some of the F.B.A. boats were later armed with cannon.

The naval high command ordered naval personnel, trained by VB 102, to increase bombing operations. Requests to replace the airplanes with Morane-Saulniers and Nieuports were refused by the Aviation Militaire. Therefore, the Aviation Maritime continued to use the outdated Voisins for bombing raids. During April, May, and June the Voisins completed 35 sorties and dropped 552 bombs on Ostende and Zeebrugge.

The performance of the F.B.A. Type B flying boats was disappointing. Their biggest problem was the inability to climb above 2,000 meters, which exposed them to anti-aircraft fire; also their submarine attacks were ineffective. This situation persisted through the balance of 1915. The obvious need for better equipment was clear, as well as the need for airplanes capable of long-range reconnaissance.

The Aviation Maritime's airplanes also performed reconnaissance, regulated naval fire, and located enemy submarines. The flying boats attempted to sink the submarines with bombs if they came upon them on the surface. However, the primitive equipment combined with the small target presented by a submarine made this type of attack difficult and usually unproductive. Airplanes were later fitted with cannon in an attempt to rectify this situation. It was hoped that cannon fire would force the submarine to dive, enabling airplanes to drop bombs while the submarine was temporarily blinded. While some success was experienced by the use of these tactical innovations, real success was achieved only when adequate co-ordination developed between airplanes and ships. In these instances the flying boats would locate the submarine for surface vessels which would then attack the enemy submarine.

Overseas

Naval activities were limited during the first years of the war.

Bases at Venice and Brindisi were at the disposal of the French. They served as patrol and reconnaissance facilities to detect submarines in the Adriatic and Mediterranean. Some bombing raids were conducted against enemy bases in the Trieste area and the arsenal at Pola.

In December 1915 Nieuport 6Hs, during reconnaissance of the Syrian coast, discovered Turkish troops using the road from the Hedjaz to Maan. The Nieuports also participated in attacks on railways and encampments, as well as raids on Ber-Saba and Adana. The escadrille was disbanded in April 1916 after the British organized their own air service.

Subsequently, the Nieuport escadrille was sent to Egypt and remained operational at Port Saïd from late 1915 to early 1916. The French Nieuports cooperated with the British to protect the Suez Canal, and performed reconnaissance in the Syrian sector.

1916
Organization

As the submarine warfare intensified more centers and P.C.s were established to patrol the affected coast lines and, eventually, led to continuous patrols of all the French and Mediterranean areas.

On 31 December, 1916, the organization of the Aviation Maritime was as follows:

Donnet-Denhaut D.D.2. The D.D.2 was a successful 1916 design; about 400 were built for anti-submarine patrol. B83.3779.

Centers de Aviation:

France	North Africa	Greece
Dunkerque	Bizerte	Salonika
Boulogne		Corfu
Le Harve		Argostoli
La Pallice		
Toulon		

Poste d'Combat:

North Africa

Bône

Operations

France

During 1916 much the same situation existed as in the preceding year. The number of bomber escadrilles was increased and a fighter escadrille was added in April. Heavier bombs were developed by the naval artillery works at Toulon which included incendiary weapons.

Weather was a major concern during winter operations in the Atlantic. Weather in the channel area was bad for many months during the year, which hampered bombing and patrol flights. It was difficult for the airplanes to fly from a rough sea, and fog reduced the chances of spotting submarines.

Communications between ships and airplanes and their bases was accomplished by a combination of wireless, flares, carrier pigeon, and signal flags. The flags were used to signal ships and were strung on a line trailing from the airplanes. The pigeons were a particular help for downed pilots, and doubtless saved many lives.

The weather in March and April allowed the aging Voisins to bomb Ostende and Westende. They continued to attack various land targets. For example, in September they bombed the forest of Houthulst, the station at Engel, factories at Bruges, and the aerodrome at Mariakerker. In October F.B.A. flying boats hit the Ostende dock installations without much success.

The raids did provide some secondary results. The German submarines and other vessels using Ostende and Zeebrugge were required to undergo repair work at night, and special shelters were built to protect the submarines. In retaliation, the Germans staged a raid on Dunkerque on 23 December 1916 that caused a great deal of damage from incendiary bombs. Three hangers containing 12 F.B.A.s were burned.

At least seven Sopwith Babies were supplied to the Aviation Maritime; serial numbers were 8128, 8129, 8185, N1106, N1121, N1430, and N1431. Although some were used at Dunkerque, none were built in France.

Overseas

Airplanes in the Mediterranean had better weather, but were faced with problems of maintenance and supply. The Mediterranean patrols did have an advantage in spotting submarines and mines because of shallow, clearer water. Again, the lack of adequate equipment hampered chances of successful attacks.

In addition to the *Foudre,* some converted merchant vessels served as carriers for French seaplanes. These included two German ships seized at Port Saïd in 1914, the *Aenne Rickmers* and *Rabenfels;* the *Campinas,* a French cargo-liner converted at Port Saïd in 1915; the *Rouen,* a former English Channel passenger vessel, and the *Dorade,* a former German ship, probably a trawler. All these served in the Mediterranean. There were also two former English Channel packets, the *Nord* and *Pas-de-Calais,* sidewheel paddle vessels, based at Dunkerque and Cherbourg, respectively.[2]

An escadrille of F.B.A.s was stationed in Venice, under a plan conceived by the French, British, and Italians, to patrol and bomb the Austrian coast. For about two years this understaffed and ill-equipped escadrille undertook its tasks but suffered heavy losses. Although the F.B.A.s were updated using more powerful engines, these planes were never equal to the Austrian flying boats. The main task of the escadrille was to spot submarines in the northern Adriatic. They were also used to bomb Trieste, Miramare, Citta Nuova, and Parenzo. A base was established at Gardo to permit the F.B.A.s to conduct several raids on Pola. However, the Austrians staged a raid on Gardo and forced the F.B.A.s to withdraw. The escadrille inflicted heavy damage on an Austrian submarine when an F.B.A. dropped its bombs from only 15 meters. This attack resulted in submarine commanders becoming cautious when in an area covered by the F.B.A.s. When disbanded in April of 1917 not a single officer of the original escadrille survived.

---

(2) *Aenne Rickmers* and *Rabenfels* were in British navy service and in 1915 were renamed *Anne* and *Raven II,* respectively. They operated Nieuport floatplanes from the *Foudre;* the pilots were French and the observers were British. Nieuports also operated at various times aboard British cruisers in the Mediterranean and Red Sea during 1915.

Levy-Besson flying boat at Saint Raphael. More than 100 Levy Besson flying boats served with the Aviation Maritime. SHAA B84.2534.

1917

Organization

The year 1917 was the turning point for the Aviation Maritime. On the first of April the high command decided that the defeat of the German submarines was essential. A buildup of the Aviation Maritime, originally proposed at the end of 1915, was now put into effect. The result was a massive increase in its size and strength.

The Aviation Maritime was divided into the following geographic districts:

Patrol of the English Channel and North Sea

1st division covering the North Sea - P.C. based at Dunkerque

a) Aviation Maritime in the G.A.N. Zone

There were two escadrilles under the command of Lieutenant de Vaisseau Lofevre; one at Dunkerque and one at Saint-Pol; together they had 32 aircraft of which 24 were in service and eight were in reserve.

Center de Saint-Pol - equipped with bombers and fighters intended for use against land-based airfields. However, a German bombing attack destroyed the bombing escadrille on 1 October, 1917. Another attack on 18 December 1917 badly damaged the base at Dunkerque and forced the dispersal of its aircraft among Dunkerque, Calais, and Gravelines. This seriously diminished the Dunkerque's base efficacy, as by 10 January 1918, only 12 aircraft remained.

During 1917 two additional units were created:

P.C. Calais

P.C. Ostende

b) Aviation Maritime - English Channel

2nd division covering the English Channel - Center de Boulogne. In July there were 16 aircraft based at Boulogne under the command of Lieutenant de Vaisseau Serre.

P.C. Dieppe - six aircraft.

Escadrille côtière (coastal) de Eu - formed in July.

Center de dirigeables de Marquise-Rinxent - commanded by Lieutenant de Vaisseau Larrouy.

Patrol of the Atlantic Ocean and the English Channel

a) 1st Division for Normandy

P.C. Cherbourg - created on 3 April 1917, it had 16 seaplanes. At Chantereyne construction of hangars began in 1917; by 1918 it had 24 aircraft. Poste de Combat Port-en-Bessin had six seaplanes.

Center at Harve - commanded by Lieutenant de Vaisseau Flamanc, it had 16 seaplanes on strength. These were primarily F.B.A.s. P.C. Fecamp was opened in August 1917 and received its first aircraft at the end of the year.

Escadrille côtière de Lion-sur-Mer - a côtière escadrille of the Aviation Militaire was based here. Under the command of Capitaine Lafay, this unit patrolled Ouistreham for submarines and mines.

Center de Guernesey - crewmen from the Venice escadrille were based here. It was under the command of Lieutenant de Vaisseau Le Cours Grand-Maison and had 12 seaplanes. Unfortunately, bad weather at the base severely limited their activity from 1917 through 1918.

Centers de dirigeables at Havre

Centers de dirigeables at Montebourg.

b) 2nd Division for Bretagne

P.C. Brest

Center de Tréguier - created on 17 June with 16 seaplanes. It had a P.C. unit based at the Morlaix river. The base was later turned over to the Americans. P.C. La Penze was created at the end of 1917.

Center Camaret - opened on 16 January 1917 and had 18 aircraft by July. P.C. Ouessant functioned as a rest station.

Escadrille côtière de Plomeur - An Aviation Militaire unit was based here but did not see action until 1918.

Center de dirigeables de Guipavas - two airships (by the end of the war eight airships were based here).

Patrouilles de Loire - created 4 June 1917.

Center de Lorient - commanded by Lieutenant de Vaisseau Destrem, this center began operations on 10 May 1917. It initially had 12 aircraft, but by November there were 16 seaplanes in service. A section was established to study techniques to find and attack submarines and to escort convoys. P.C. Croisic entered service 10 July 1917. On 27 November the P.C. was handed over to the Americans.

P.C. Ile Tudy - created 27 June 1917 under the command of Enseigne de Vaisseau Plurien. By October it had 16 aircraft; that same month the base was taken over by the Americans.

Escadrille côtière de Quiberon - commanded by Capitaine Bourdes, this unit entered action on 25 June 1917.

Escadrille côtière de La Baule - commenced activities on 1 June 1917 under the command of Capitaine Lallemand. The unit escorted convoys from this base until transferred to Croisic.

Center de dirigeables de Paimbouef - one airship; in 1918 the base was turned over to the Americans.

c) 3rd Division for Gascogne and Biscaye

P.C. La Rochelle

Center de La Pallice - commanded by Lieutenant de Vaisseau Truzy, this unit had 12 seaplanes and patrolled between Arcachon and le Croisic. In July it had 16 aircraft. P.C. Sables d'Olonne had six aircraft.

Center de Bayonne - commanded by Lieutenant de Vaisseau Vielhomme, the unit patrolled from le Pays Basque to Arcachon. Component units were P.C. Socoa, Escadrille cotiere des Sables-d'Olonne (commanded by Capitaine Thouvenin), Escadrille côtière du Verdon (commanded by Capitaine Fontaine), P.C. Cazeaux (opened on 12 July 1917 with three seaplanes), and P.C. Hourtin.

Center de dirigeables at Rochefort - commanded by Lieutenant de Vaisseau Sable; it had seven airships.

Patrol of the Mediterranean

a) 1st Division for Algeria

P.C. Bone

Center de Bizerte - had 32 aircraft on strength in 1917.

Center de Oran - Had a total of 12 Donnet-Denhaut D.D. 2s and D.D. 8s.

Center de Arzew - created in late 1917 with 12 seaplanes.

P.C. Petit-Port

P.C. Mostaganem - installed June 1917 with three seaplanes; by October there were six.

P.C. Beni-Saf - Created August 1917.

Center de Alger - created in March 1917 with 12 aircraft under the command of Enseigne de Vaisseau Le Vayer. By 6 June the number of aircraft had doubled to 24. These aircraft were D.D.2s, D.D.8s, Georges-Levy HB 2s, and Tellier T.2s. They escorted convoys from Gibraltar and Bizerte.

P.C. Bougie - created 9 February 1917 with three seaplanes.

P.C. Tenès - entered service 20 June 1917, but no aircraft were ever assigned to the base.

P.C. Cherchell - installed in June 1917 and had some aircraft based there. Postes de relache at de Dellys and Port Queydon.

P.C. Djidjelli - six seaplanes.

Center de Bône - construction completed in fall 1917. By

September 12 seaplanes were on strength and were used to patrol convoy routes. However, operations were limited due to frequent engine breakdowns.

P.C. Collo - created 1 May 1917. Bad weather prevented operations; it was so bad that seaplanes could not be hoisted onto the slip. For this reason the unit became a poste de relache.

P.C. Philippeville - created 19 February 1917. Functioned as a poste de relache.

Center Bizerte - Created in 1916, this base served as a staging area for aircraft being assigned to Bone, Kelibia, and Narsala. The center had a complement of 36 aircraft. It also acted as a training base for pilots being sent to other centers in Algeria and Tunisia. On 12 January 1917 a hangar big enough to hold 16 seaplanes was built at the base. During the first months of 1917 the base had between 35 and 40 aircraft; due to a shortage of pilots only a dozen were operational at any given time.

This situation was rectified in October 1917 when the following changes were introduced:
1. The base complement was reduced to 32 seaplanes, and the postes de Tabarka and Kelibia were to have six aircraft; this raised the total to the requested level of 44.
2. A center was opened at Marsala and 16 seaplanes were assigned to it.
3. The training of pilots was discontinued at Bizerte, in part due to the bad weather at the base.
4. Aircraft with folding wings were to be transported on the aviation transport *Normandie*. They had been found to be of limited usefulness because they carried an inadequate payload.(3)
5. Some of the duties and installations at Bizerte were transferred to the Center de Karouba.

However, none of these changes could improve the problems with the aero engines and, as a result, the center had only 14 airplanes in service at any given time.

P.C. Kelibia - Poor weather also inhibited operations from this site; during August and September only 19 sorties could be flown.

Poste de relache San Antiocco - plans to establish a base here were abandoned when the Italians established another seaplane base on Sardinia.

P.C. Tabarka - opened in March 1917.

Center Sousse - established on 19 February 1917 under the command of Professur d'Hydrographie Bertin, this base had 12 aircraft on strength by August. The seaplanes from this base escorted convoys between Bizerte and Sicily and from Bizerte to southern Tunisia. By 4 October there were 16 aircraft in service. Unlike some of the other bases, the weather at Sousse was conducive to flying operations. A poste de relache was established at Shabba in November. A plan to open a P.C. at Djerba was abandoned because of the relatively light coastal traffic in the area.

P.C. Lampedusa - created on 30 March 1917 with four aircraft used to escort Bizerte-Malta convoys until the station was closed in July.

Center de Marsala - poste de relache established in June 1916. It was subsequently decided to open a seaplane base here to augment the radius of action for aircraft based at Bizerte. This was accomplished in January 1918 with a planned complement of 16 seaplanes.

L'escadrille côtière de Sfax - created on 24 March 1917, reportedly with nine Caudron G.4s (but more likely these were G.6s) and four additional aircraft in reserve. It began operations in 1918.

Center de dirigeables de la Senia - Entered service in 1918.

Center de dirigeables d'Alger-Baraki - Airship *A.T.6* was based here beginning in November 1917.

Center de dirigeables de Bizerte - in June 1917 this base had the airship *Lorraine* and the balloon *La Tunisie*.

b) 2nd Division for Provence

P.C. Toulon

Center de Perpignan - commanded by Lieutenant de Vaisseau Le Villain and became operational on 18 June 1917. It had 12 seaplanes and patrolled from Barcelona to Saintes-Maries-de-la-Mer. Poor weather limited operations.

Center de Sète - created 23 February 1917 as P.C. Perpignan, it became a center on 20 September under the command of Lieutenant de Vaisseau Roux.

Center de Toulon - formed in 1916. In 1917 it was under the command of Capitaine de Fregate Richard. The unit had 16 seaplanes. The seaplanes (F.B.A.s and Donnet-Denhauts) escorted convoys and searched for mines from Marseille to Italy. An escadrille was later detached to Saint-Georges. Postes de

---

(3) The identity of the *Normandie* has not been confirmed. It is not known if the aircraft with folding wings were to have been the F.B.A. Type S or Short 184.

F.B.A. Type H based at l'Ecole de Tir at Cazaux. Developed from the F.B.A. Type C, the Type H was the most extensively produced flying boat of the war and served with Great Britain and Italy in addition to France. SHAA B84.610.

Relache were established at Ciotat, Porquerolles, and Saint-Tropez.

Center de Antibes - created 17 July as a P.C., it became a center on 16 November 1916 and was commanded by Lieutenant de Vaisseau Val.

Center d'Ajaccio - created 26 September 1917 with 16 seaplanes.

P.C. Calvi - created in December with six seaplanes.

Center de Bastia - created 17 December 1917 with 16 seaplanes.

Escadrille cotiere de Marsaille - commanded by Capitaine Picheral.

Center de dirigeables de Aubagne - began operations in June 1917.

Center de dirigeables de Cuers-Pierrefeu - created June 1917, but did not begin operations until the end of the year.

Center de dirigeables de Mezzana - created 4 September 1917 under Lieutenant de Vaisseau Jouglard.

c) 3rd Division for the central Mediterranean
P.C. at Milos

Center de Corfu - created in 1916, it had 36 seaplanes: 24 Donnet-Denhauts and 12 Telliers. Four Sopwith 1-1/2 Strutter fighters were assigned in August and September 1917. Because of engine difficulties only about seven aircraft were in operation at any given time. Lieutenant de Vaisseau Hautefeuille assumed command of the base on 13 August 1917. The Sopwith 1½ Strutter fighters were of little value as they required meticulous servicing that severely limited the number of sorties that could be flown.

P.C. Mourtzo - located on the south of Corfu. The decision to establish this unit was made in December 1917. It did not begin operations, however, until 1918.

Escadrille d'avions de chasse de Potamos - This was a unit of the Aviation Militaire and was equipped with Nieuport fighters. It was placed under the control of Sous-Lieutenant Philbert in August 1917 and began operations in October.

Center de montage de Brindisi - Aircraft intended for bases in the eastern Mediterranean were sent here in cases for assembly. Once assembled, they were flown to their intended bases.

Center de Salonika - initially proposed in 1915; the decision to build this base was made on 18 October 1916. The base was constructed during 1917 and became operational during 1917. P.C. Kassandra, P.C. Panomi, P.C. Skiatho, and P.C. Sikia were under the control of this center. The aircraft at this center were exclusively F.B.A. Type Hs. A lack of qualified personnel and poor sanitary conditions severely limited the center's efficiency.

P.C. Kassandra - This base had six seaplanes used to patrol Salonika. It was inactivated on 16 April 1917.

Escadrille du Campinas - The seaplane carrier *Campinas* was used in the Aegean Sea and the Ionian islands. It was sent to Corfu in April 1917 and used its F.B.A. Type Cs for barrage flights. On 3 June it was sent to Salamines and used its aircraft to track troop movements north of the Corinth Canal. On 17 June its aircraft patrolled Patras; strong winds, however, made flights over the bay difficult. The *Campinas* was subsequently sent to Piraeus as this station had better facilities to protect the fragile F.B.A. Type Cs from the elements. On 29 November the *Campinas* was sent to Milos, where it remained until the end of hostilities.

Center de Piraeus - created in 1917, this base had eight F.B.A. flying boats on strength. Initially based at Argostoli, it was subsequently sent to Mitylene, and finally to Piraeus in December 1917.

P.C. Vathi - based on the northwest coast of Ithica, this base was created in July 1917 to escort convoys along the Tarento-Itea route.

Center de Milos - created in 1917 with 16 seaplanes.

P.C. Navarin - created in December 1917. It began operations with six seaplanes in 1918.

Poste de Skiatho - base proposed in 1916 but not established as a base de relache until 1918.

Poste de Sikia - poste de relache; did not become operational until 1918.

Escadrille côtière de Panormos - Six Farmans (probably F.40s) were lent to the Aviation Maritime by the Aviation Militaire in October 1917. They were intended to perform coastal patrol missions from the base. Patrols did not begin until 1918.

Escadrille côtière de Kourtesis - Escadrille G 488 was sent from Bizerte to Patras on 30 September 1917. It was placed under the control of the Aviation Maritime. The aircraft, reportedly 12 G.4s (but it is much more likely that they were G.6s), were assigned to the field at Kourtesis in December 1917. Reconnaissance sorties did not begin until May 1918.

Center de Milos - It was decided to create a new base at Milos in December 1917. It began operations in February 1918 and by August had 20 F.B.A. flying boats and two Tellier T.6s with cannon.

Operations
Anti-Submarine

During the later part of the war naval airplanes and airships took part in convoy escort. These were intended to attack surfaced submarines or at least force then to stay hidden. Table 3 notes the dramatic increase in operations during 1917.

The tactics for convoy escort developed in 1917 called for a pair of flying boats to fly 15 miles ahead of the convoy to search for submarines and mines. Airships were probably more effective than aircraft in preventing submarine attacks (no convoy escorted by a French airship was attacked in 1917), and

Tellier T.3 at Corfu in June 1918. More than 100 Tellier flying boats served with the Aviation Maritime and the U.S. Navy. SHAA B78.1782.

were more effective than airplanes in searching for mines. However, the airships could be seen from a long distance and, while this forced the submarine to submerge, it also meant that the enemy boat would escape detection. However, airplanes had the dual advantage of being smaller and faster, so it was much easier for them to find and attack submarines.

Notable operations during 1917 included:

5 January - seaplanes from La Pallice rescued an English steamer from a submarine attack.

12 February - a section from La Pallice saved a sailing ship from a submarine attack.

19 February - seaplanes from Bône defended an English cargo ship from German attacks.

25 February - a seaplane from Le Harve forced a German submarine that was patrolling the shipping lanes to dive.

16 March - a section from La Pallice saved a cargo ship that was being attacked by enemy submarines.

13 July - A section from Brest attacked a submarine that was stalking a convoy and forced the German boat to submerge.

15 July - seaplanes from Brest attacked a submarine and forced it to dive before it could attack an allied ship.

23 July - seaplanes from Brest made numerous attacks on a German submarine stalking a convoy.

14 August - a section of seaplanes from Brest bombed a submarine as it was attacking a convoy. Another seaplane from Brest forced an enemy submarine to dive.

20 September - seaplanes from Bône, while escorting a convoy, discovered a German submarine preparing to attack.

9 October - aircraft from Bône saved an Italian schooner from a submarine attack.

10 October - seaplanes from Bône discovered a submarine waiting for a convoy and forced the German boat to dive.

21 October - a section of airplanes from Camaret attacked a submarine before it could engage a convoy of sailing ships.

4 December - seaplanes attacked a submarine stalking a convoy.

8 December - seaplanes from Port-en-Bessin discovered a submarine waiting three miles ahead of a convoy.

14 December - a section from Tréguier saved a sailing ship from submarine attack. On the same day two seaplanes attacked a submarine in the Gulf of Corinth. This was probably the German *UC 38*, which was sunk later that day by one or more French destroyers.

15 December - seaplanes from Oran attacked a submarine in the vicinity of a convoy.

As can be seen from the above incidents, contacts between French seaplanes and submarines were rare, and it was unusual for French aerial attacks to inflict significant damage on submarines. However, simply by forcing the German boats to dive, the French aircraft saved innumerable ships, cargoes, and lives.

Bombing

The airplanes based at Dunkerque launched a major raid in January 1917, attacking the Bruges-Zeebrugge canal. They succeeded in blocking 26 ships at Zeebrugge and Bruges. The French and British attacks on these German-controlled seaports were beginning to take their toll. For this reason the Germans began a systematic attack against Dunkerque and St. Pol. The French retaliated by regular bombing of all targets along the German-occupied coast. Heavy raids were made in July against Ostende, Zeebrugge, Ghistelles, and Middelkerke. Most of the raids were at night and only small groups of airplanes were committed. Even though the importance of the naval targets was well understood, the headquarters in Paris could not allocate sufficient resources to the naval center at Dunkerque to enable the bombers to inflict serious damage.

In September and October the Germans again staged devastating raids on St. Pol and Dunkerque. Dunkerque received

7,514 bombs, which killed 233 and wounded 336. The German attacks continued through December 1917, effectively shutting the center down.

1918
Organization

By the end of the war the number of centers grew from two to 33, the number of Poste de Combat to 14. (Some centers were planned for 1918, but were not completed, and some P.C.s became centers.) Table 1 provides a detailed order of battle for the Aviation Maritime in 1918. The number of airplanes assigned in 1918 to these bases reached a total of 1,264 of all types.

The French negotiated with the Americans to establish naval bases along the French coast to assist in the battle against German submarines. Equipped with French flying boats, the Americans, on 10 March 1918, established a presence at Dunkerque. Before the end of the war they were able to make several raids in conjunction with French squadrons.

Equipment

Patrol: Types included F.B.A., Donnet-Denhat flying boats, and the Tellier with a 200-hp Hispano-Suiza engine. Patrol seaplanes were armed with two 35-kg bombs and a machine gun and usually carried wireless.

Bombardment: This class comprised landplanes placed into naval service. The navy received Voisin 3s and 5s, M.F.7s, M.F.11s, Farman F.40s, and Breguet-Michelin 4s. Most of these were used for bombing around the Dunkerque area.

Alert: This role was to be filled by the F.B.A. Type C. There were a number of other types submitted by various manufacturers to fulfill this role.

Fighting Seaplanes: These were floatplanes adapted from land versions. The SPAD 14 and the HD.2 were used in this role, but only at the Dunkerque and Corfu centers. Sopwith 1½ Strutters mounted on floats were also used.

High Seas: This was an ambitious building program that did not come to fruition. It was probably inspired by the British success with multi-engine planes. The specifications called for four-seat machines armed with a machine gun and a cannon with 35 rounds. Wireless was also to be carried. A large number of manufacturers submitted airplanes to meet this category but the end of the war prevented the development of most of these.

Shipboard Airplanes: An F.B.A. Type C was designed to operate from a naval ship but launch and recovery were very slow and, added to the changes necessary to accommodate an airplane, caused the navy to lose interest in the idea. Some experiments were made in late 1918 on development of a carrier and of airplanes suitable for carrier operations. The turret of the battleship *Paris* was covered with a platform from which airplanes were launched. However, serious development of carrier aviation had to wait until after the Armistice.

Operations

The balance of 1918 saw the navy engaged in escort duties and the continued fight against submarines. With United States assistance, the French were able to maintain constant patrols in the areas were the German submarines were operating.

Important operations during 1918 included:

31 January - a Tellier and an F.B.A. flying boat from Guernesy attacked a submarine. The F.B.A. carried two "E" bombs, the Tellier only one. The submarine was probably damaged.

20 February - a section of seaplanes from Corfu attacked a submarine with two "F" bombs (52 kg), and two "D" bombs (22 kg). An oil slick was spotted suggesting that the submarine was damaged.

21 February - a section of seaplanes from Marseille attacked a submarine in the vicinity of two convoys.

12 March - in a mistaken-identity attack, the French airship *A.T.O.*, flying from Le Havre, bombed and sank the British submarine *D-3* in the English Channel.[4]

12 April - A section of aircraft from Piraeus bombed a submarine that was stalking a convoy.

23 April - a section of seaplanes from Ile Tudy attacked a submarine. Two airplanes from Guernesey attacked a submarine with bombs and notified their base of the discovery by releasing two homing pigeons. The base subsequently despatched a third seaplane which spotted a periscope and bombed the submarine.

29 April - a sailing ship was saved from a submarine attack by seaplanes from Sousse which bombed the submarine.

9 May - three seaplanes from Gamaret dropped three bombs on a large submarine. An oil slick was seen, suggesting that the ship had been damaged.

13 May - seaplanes from *Campinas* located a mine field near the Aegean island of Milos and guided to it the minesweeper *Rateau* and trawler *Morse*, which removed the mines.

18 May - seaplanes from Oran and Arzew twice attacked the German submarine *U-39*, damaging it so severely that it had to put into Cartagena, Spain, where it was interned for the rest of the war.[5]

24 May - seaplanes from Camaret spotted a submarine and dropped "F" (52 kg) bombs on it at 8:55 a.m. At 10:50 a.m. another submarine was seen and attacked with four "F" bombs. An oil slick was seen. At 4:00 p.m. four bombs were dropped on a submarine by other aircraft from the base. Again, an oil slick was seen. At 5:15 p.m. a periscope was seen and bombs were dropped. Finally, at 5:45 p.m., seaplanes from Camaret dropped five "F" bombs on a submarine.

31 May - at 7:45 a.m. two seaplanes from Guernesey attacked a submarine with two bombs, saving an English sailing ship from attack. At 9:55 a.m. other aircraft from Guernesey attacked a submarine with two bombs.

6 June - two G.L. 40 HB.2 seaplanes (L-B 18 and L-B 24) from Le Havre bombed a submarine that had been spotted by aircraft from Seine. The submarine was forced to dive but it is not known if it was damaged.

6 July - two aircraft from Marsala attacked a submarine near the straits of Sicily. Three bombs were dropped and the submarine was forced to dive.

20 July - seaplanes from Bayonne bombed a submarine.

23 July - seaplanes from Bayonne dropped two bombs at a submarine.

28 July - at 11:30 a.m. aircraft from Camaret dropped two bombs at a submarine. At 2:00 p.m. two more bombs were dropped on another submarine.

10 August - At 7:30 a.m. two seaplanes from La Pallice dropped two bombs at a submarine. At 9:25 a.m. other aircraft from the base attacked two submarines with bombs. At 4:15 p.m. a third attack was made on a submarine. Although no damage was confirmed, the submarine or submarines were forced to break off their attack on a convoy.

5 September - seaplanes from Aveiro attacked a submarine with four bombs, forcing it to dive.

13 September - two aircraft of LET 487 based at Lion-sur-Mer escorted a convoy from Boulogne to Le Havre. During the mission a submarine was attacked with two bombs.

Summary

At the beginning of the war naval airplanes were used primarily for reconnaissance. The machines of the 1914–15 period were fragile and difficult to maintain with engines that were subject to frequent failure. This situation was to continue until the beginning of 1917, when the submarine threat had to be addressed.

With the buildup of anti-submarine forces in the channel area and North Sea, the importance of aircraft for patrol and attack was realized. Although resources for naval aviation were much more limited than those given to the Aviation Militaire, the Aviation Maritime's flying boats helped turned the course of the submarine war in favor of the Allies, and provided the framework for the future Aéronavale.

---

(4) The airship's commander was absolved of blame for this accident, which was caused by a mix-up in recognition signals.
(5) The French, unaware that both attacks were on the same target, believed at the time that two different submarines had been attacked.

Donnet-Denhaut D.D.10. Developed in response to the Aviation Maritime's requirement for a high seas flying boat, the DD.10 went into production too late to see operational service before the end of the war.

Table 1: Organization of Naval Centers in 1918

Code	Center	Poste de Combat	Coastal Squadrons & Locations
North Atlantic and Channel			
North Sea			
D	Dunkerque	Calais	SPA 313 St. Pol
	Ostende	High Sea (project)	
Channel			
B	Boulogne (was PC)	Dieppe	LET 485 Eu
Normandy Lower Seine			
H	Le Havre (was PC)	Fécamp	LET 487 Lion-sur-Mer
CH	Cherbourg	Port-en-Bessin	
G	Guernesey (was PC)	Also PC	High Sea (project)
	Montebourg	Not put in service	Former Venice Squadron
Atlantic			
Brittany			
T	Tréguier (was PC)	Ile Bréhat	Transferred to U.S. Navy
	(Transferred to La Penzé)		
B			
P	La Penzé (was CP)		
C	Camaret	Ouessant	LET 491 Plomeur High Sea (Project)
Lower Loire			
L	Lorient	Croisic	LET 484 Le Croisic (F)
	Ile Trudy	High Sea (C)	
	Ile d'Yeu	LET 483 Quiberon	
		High Sea Project	
Gascone			
V	La Pallice		LET 482 Sables (S)
		LET 481 Verdon (V)	
		High Sea Project	
Biscaye			
B	Bayonne	Hourtin	High Sea Hourtin (H)
	Socoa	High Sea Cazaux (C)	
	Aviro (Portugal)		
Morocco			
C	Casablanca	Safi (Proposed)	G 489 Mazagan
	Dakar (Senegal)		
Mediterranean			
Eastern Mediterranean and Corsica			
P	Perignan		
I	Cette	Marseille-Cap Janet	LET 486 Marseille (M)
H	Toulon	St. Raphael	Fighters St. Rapfael (C)
		High Sea (project)	
R	Antibes	Nice	Mail Nice (N)
J	Ajaccio	Calvi	High Sea (project)
Algeria-Tunisia			
W	Arzew	Cherchell	
D	Djidjelli (was CP)		
O	Oran	Nemours	
	Mostaganem		
A	Algiers	Tenés	High Sea (project)
	Bougie		
Q	Bône	Collo	
P	Bizerte	Tabarka	High Sea (project)
	Kelibia		
S	Sousse	Lampedusa	G 490 Sfax
L	Marsala (was CP)		
Western Mediterranean			
O	Corfu	Morietzo	N 561 Protection Venice
		N 562 Fighter Corfu	
		Campinas (shipboard)	
G	Pireaus	Vathi	High Sea (project)
		Fighter-Seaplane (project)	
N	Navarin (was CP)		G 488 Kourtésis
M	Milos		
U	Salonique	Cassandre	High Sea (project)
		Skhiato	Panomi (squadron?)

Table 2: Coastal Squadrons

Coastal squadrons were army units attached to the navy, and given a letter designation that corresponds to the center attached.

Coastal Squadron	Letter Code	Original Coastal Name	Established	Dissolved
LET 481	V	Le Veron 551	July 1917	Nov. 1918
LET 482	S	Sables-d'Olonne 552	July 1917	Dec. 1918
LET 483	Q	Quiberon 553	March 1917	Dec. 1918
LET 484	F	Le Croisic 554	April 1917	Dec. 1918
LET 485	E	Eu 555	July 1917	Dec. 1918
LET 486	M	Marseille	July 1917	Dec. 1918
LET 487	L	Lion-sur-Mer	June 1917	Dec. 1918
G 488	?	Tunisia, Transferred to Greece as Kourtésis Patras G 488	Aug. 1917	Dec. 1918
G 489	?	Mazagan (Morocco) G 489	March 1917	Dec. 1918
G 490	S	Sfax (Tunisia)	March 1918	Dec. 1918
LET 491	N	Plomeur	June 1918	Dec. 1918

Most coastal escadrilles used Voisin bombers or Caudron G.4s and G.6s. Despite the LET designation, it appears that only a small number of Letord 1, 2, 4, and 5s were assigned to these units. The Letord series was not adopted by the Aviation Maritime.

Hanriot HD.2s on beaching gear. SHAA B87.5020.

Table 3: French Naval Aircraft Forces and Activities, 1914–1918

Date	Center	CP	Sqdn.	Aircraft	Personnel	Missions	Hours	Sub Attacks	Probables	Mines	Escort
1914											
Aug.			2	8	208						
Sep.			2								
Oct.			2								
Nov.			2								
Dec.	1	1	2								
1915											
Jan.	1	1	2	20	291						
Feb.	1	1	2								
Mar.	2	0	2								
Apr.	2	0	2								
May	2	0	3								
June	2	0	3								
July	2	1	3	54	500			2	1		
Aug.	2	1	4								
Sep.	2	1	4								
Oct.	2	1	4						1		
Nov.	2	1	5								
Dec.	3	1	4								
1916											
Jan.	3	1	3	64	805	360		1		2	
Feb.	3	1	3			400		1	1		
Mar.	4	0	3			500				1	
Apr.	5	2	3			560				1	
May	5	2	4			600					
June	5	3	4			440					
July	6	3	4	96	1942	480					
Aug.	6	4	3			760		2	2		
Sep.	6	4	3			830		3	1		
Oct.	6	5	4			650		1		2	
Nov.	6	5	3			470		3			
Dec.	6	5	3			470		3			
1917											
Jan.	9	3	3	159	2194	1000	1000	2			
Feb.	10	4	4			1200	1000	8	1	1	
Mar.	11	3	4			1200	1200	7			
Apr.	12	4	4			1900	2200	16			
May	13	5	4			2630	3300	22	7	2	
June	17	7	6			3150	3450	15	1	6	
July	18	10	9	277	4140	3500	4650	18	3	1	
Aug.	21	11	10			4100	4500	15	3	12	
Sep.	22	12	11			4900	5900	14	4	9	
Oct.	23	14	13			3374	3700	31	6	7	
Nov.	23	14	14			3040	4000	9	3	1	
Dec.	23	14	15			1900	2900	14	1	4	402
1918											
Jan.	25	17	15	691	7772	2160	4100	7	2	1	360
Feb.	25	17	15			2125	3900	12	1	1	540
Mar.	25	18	15			2200	4400	10	1	4	645
Apr.	27	17	15			2820	4970	8	3	7	690
May	27	19	15			3500	7280	41	5	9	740
June	29	22	15			3700	7560	16	2	11	990
July	33	20	15	1119	10583	3950	7850	24	4	15	1080
Aug.	32	19	16			4200	9050	31	8	1	1350
Sep.	32	20	15			2850	6300	15	4	1	910
Oct.	32	20	14			2800	5800	12	2	1	800
Nov.*	32	21	12	1264	11059	800	1530	1		2	205

* 11 Nov. 1918: 702 pilots, 693 observers, aircraft 6,970 various. (Includes all aircraft, even non-operational and army co-operation aircraft.)

Hanriot HD.2 being launched off the Paris during trials, 1918. SHAA B75.459.

Farman F.40 floatplane trainer at the Ecole de Tir at Cazaux during the war. SHAA B84.583.

The French Aviation Industry

In 1914 French aviation manufacturing was a cottage industry which produced 60 aircraft a month. By 1918 it had become a military-industrial complex of 62 firms with 102 factories. During the war the French produced more than 50,000 aircraft and 90,000 aero engines. However, the creation of this giant industry was accomplished only after several years of struggle.

1914

The German invasion resulted in the destruction or forced evacuation of large parts of the aviation industry. In addition, the manufacture of carburetors, steel, and textiles on which the aviation industry depended was also disrupted. The factories in the north, northeast, and the Aisne valley were primarily affected. As a result, in August 1914 only 50 aircraft were produced for the military, and production rose slowly during 1914. From August through December only 541 military aircraft were produced, along with 860 aircraft engines.

The SFA (Service Fabrication Aeronautique or Aeronautics Production Service) was based at Chalais-Meudon at the beginning of the war. It, too, was displaced by the German advance, finally being installed at Lyons in September. The SFA oversaw the stabilization of the aircraft industry and encouraged the relocation of Caudron, Farman, R.E.P., and Voisin to Lyon; Gnome, Le Rhône, and Salmson also moved to Lyon. When the front stabilized in December, the SFA returned to Paris, where approximately 90% of French airframe and engine production would eventually be centered.

The forced evacuation of their factories and a shortage of raw materials were not the only difficulties facing the manufacturers. General Benard's belief that there would be a quick French victory resulted in the drafting of large numbers of pilots, designers, and workers from the industry.

The SFA decided to concentrate on the mass production of a small number of aircraft types rather than allowing each manufacturer to produce its own design. The main types selected were the Morane-Saulnier L (to be built by Morane-Saulnier and Breguet), the Caudron G.3 (Blériot, Caudron, and Deperdussin), and the Voisin 3 (Breguet, Nieuport, R.E.P., and Voisin). It was hoped that this would result in a rapid increase in aircraft production. Le Rhône engines would be built by Le Rhône and Hispano-Suiza; Renaults by Darracq, Peugeot, and Renault; and Salmsons by Darracq, Peugeot, and Salmson.

As can be seen from Table 1, the changes implemented by the SFA resulted in a slow but steady improvement in the rate of aircraft production:

Table 1: 1914 Production (War Months)		
Month	Aircraft	Engines
August	50	40
September	62	100
October	100	137
November	137	209
December	192	374
Total	541	860

1915

In 1915 there was continued improvement in aircraft and engine production. The monthly output of airframes nearly doubled while the rate of engine construction increased by nearly 80%. Approximately 15% of the aeroplanes produced in French factories were supplied to the British and Russians.

The decision to concentrate on a few basic designs enabled the French to carefully supervise their development and production. Capitaine Albert Etévé of the Aeroplane service despatched officers to the Caudron, Farman, Morane-Saulnier, Nieuport, and Voisin factories.

Production was hampered by the lack of skilled personnel and raw materials. The former problem was at least partially alleviated when the military consented to having some of the conscripted workers returned to the aviation industry.

Standardization on a few types enabled the industry to expand production, but it also resulted in stagnation in the development of new types. This situation was exacerbated by the government's decision to launch a heavy bomber program. Manufacturers such as Voisin, Morane-Saulnier, Blériot, and Breguet spent a large amount of time and money hoping to gain lucrative contracts for heavy bombers. However, success proved to be elusive as all the types entered in the 1915 concours puissant proved to be unsuitable for the heavy bomber role. Only Breguet's design was close to being acceptable, although Paul Schmitt's bomber was also ordered in small numbers for use as an interim type. The French parliament (more so than the military) still demanded a heavy bomber, which resulted in the 1916 concours puissant. This meant that during 1915 virtually all the major manufacturers squandered additional resources developing aeroplanes which would never see service. However, there was one benefit to this program—it sped development of the 150-hp Hispano-Suiza 8A engine which would prove to be of great value in fighters.

The GQG wanted more fighters, two-seat reconnaissance machines, and twin-engine multi-purpose aircraft designed for fighter escort, long-range reconnaissance, and bombing. This was to be accomplished, in part, by reducing the production of bombers. The Undersecretary of State for Aviation, René Besnard, supported the GQG's request despite parliament's call for more bombers; his refusal to build a heavy bomber fleet would be a major factor in his dismissal in late 1915.

Besnard also reorganized the SFA and abandoned the system whereby a single group of officers monitored the production of a single type of aeroplane. Now several types would be supervised

Right: Caudron G.4s under construction.

by a single officer, which resulted in less familiarity with the unique problems or qualities of a specific design.

Five firms initiated the production of aeroplanes in 1915—Borel, Burlat, Panhard, Peugeot, and Michelin—the latter building Breguet's BM4 bomber. The increase in manufacturers was driven at least in part by the tremendous profits being made in the aircraft industry. Caudron's income rose almost 30-fold from 1914 to 1915; Nieuport's multiplied 40-fold. The large profits were important because they enabled the firms to expand and increase their production rates. Emmanuel Chadeau has estimated that 48% of the profits generated during 1914–15 were used to pay overtime and to buy expensive raw materials. Salmson, for example, tripled its plant size and hired 1,200 additional workers. By the end 1915 it was building its own magnetos and aircraft. However, the government became increasingly obsessed with what might today be termed "excessive" profits and took punitive action in 1916.

Table 2: 1915 Production

Month	Aircraft	Engines
January	262	307
February	280	370
March	431	696
April	421	584
May	355	652
June	378	603
July	340	538
August	383	571
September	454	533
October	402	648
November	367	687
December	416	897
Total	4,489	7,086

1916

1916 saw a period of stagnation in the French aircraft industry. Mediocre aircraft remained in large scale production, while introduction of new designs was hampered by technical problems and shortages of workers and raw materials. For the first trimester production remained stagnant at 500 aircraft per month, and slowly climbed to 745 per month by the end of the year. The fact that most of these aircraft were obsolete types underscores the serious problems facing the Aviation Militaire. The situation for engine production was somewhat better with a production increase from 1,000 engines per month at the beginning of the year to 1,800 per month at the end. The number of engines produced in 1916 was 2.3 times that in 1915.

In 1916 12 more manufacturers began to produce aeroplanes: Ariés, Ballot, Brasier, Clément-Bayard, Darracq, De La Fresnay, Delaunay-Belleville, Despujols, Hanriot, Niepce (later DNF), S.E.C.M., and Sarrasin-Fréres.

Engine manufacturers also began to build aircraft under license. Salmson, as mentioned above, began aircraft production in late 1915 and in 1916 received a contract for 1,000 Sopwith 1½ Strutters. Renault also entered the field of aircraft production, later building the STAé's A.R.1 under license.

As mentioned earlier, the government had become agitated by the high profits being generated by the aircraft manufacturers. It was decided to impose a "war tax" on the industry. This tax only angered the firms and does not seem to have significantly reduced the prices of aircraft or engines. A typical aircraft with motor cost approximately $5,000 in 1914; in 1916 the cost was $7,000. Furthermore, the tax inhibited investment in new equipment and overtime expenses.

Despite the "war tax" the major companies continued to thrive. Caudron's sales in 1916 were 220% higher than 1915, Nieuport's were 216% higher, and Voisin's were ten times greater than 1915. By 1916 the work force was 68,920 (compared to 30,960 in 1915). Gross profits of the 13 major manufactures were nearly 12 million dollars in 1915; in 1916 the figure was 54 million dollars. Furthermore, the government had difficulty in controlling prices; for example, the large engine manufacturers, whose products were critical to the war effort, simply refused to substantially discount their prices.

Indeed, the aero engine manufacturers were so successful that the most important and efficient companies were able to dominate their competitors and expand their business beyond engine production. In 1916 Renault produced 22% of the engines in France and Salmson produced more than 17%. As mentioned already, Renault and Salmson began to build aircraft under license in 1916. During 1917 Salmson began to build its own designs and Renault followed suit a year later.

New aircraft, particularly a new two-seat reconnaissance aircraft and a light bomber, were desperately needed by the Aviation Militaire. However, the parliament continued to insist on a heavy bomber. During the 1916 concours puissant Astra, Breguet, Caudron, D.N.F., Morane-Saulnier, and SPAD all submitted designs. Both the SPAD SE and Morane-Saulnier Type S were found to be adequate. Ironically, now that a suitable bomber had been found, the declining fortunes of the Aviation Militaire resulted in neither aircraft being built. The air war over Verdun emphasized the need for fighter aircraft in large numbers. It was decided that SPAD's production of fighters might suffer if it were to start building bombers so the SPAD SE was rejected (an equivalent design proposed by Nieuport was rejected for the same reason). Morane-Saulnier, however, could produce its new bomber without endangering France's supply of fighters; the Morane-Saulnier Type S was selected for production and 300 were ordered. However, the French government now reversed itself, deciding that bomber production had lower priority than fighters and that only 100 of the Morane-Saulnier Type S would be needed. The order was subsequently canceled; thus a substantial portion of the French aviation industry had, once again, wasted enormous resources on a project which fell victim to the capriciousness of the French government.

Colonel Henri-Jacques Régnier was appointed head of the Twelfth Directorate of Aeronautics in February 1916. One of his first actions was to separate the STAé (Technical Section of Aeronautics) and the SI (Industrial Section) from the SFA. Commandant Dorand was appointed head of the STAé in 1916. His mission was to test new aircraft being built by the aircraft manufacturers and to assist in their development. Dorand, however, had built aircraft prewar and had his own ideas about what the aviation industry should be producing. In short order he had Capitaine G. Lepére design a two-seat tractor for army co-operation duties; a new, multipurpose, twin-engine aircraft was also developed. Although well intentioned, Dorand's actions

The French armaments industry labor force. MA.

Table 3: 1916 Production

Month	Aircraft	Engines
January	501	1,001
February	422	965
March	505	1,178
April	526	1,249
May	603	1,262
June	541	1,295
July	664	1,552
August	745	1,561
September	814	1,579
October	731	1,727
November	752	1,624
December	745	1,792
Total	7,549	16,785

Table 4: 1917 Production

Month	Aircraft	Engines
January	846	1,579
February	832	1,204
March	1,227	1,552
April	1,107	1,721
May	1,258	1,986
June	1,143	1,885
July	1,330	1,960
August	1,302	1,965
September	1,276	1,899
October	1,470	2,089
November	1,550	2,537
December	1,576	2,715
Total	14,915	23,092

only served to antagonize the very industry he was supposed to be supporting. Both the A.R.1 and the twin-engine aircraft (known by its primary manufacturer, Letord) entered service, but they were limited successes. The following year Dorand was removed, and the STAé was ordered to cease the design of new aircraft.

Some firms, such as Blériot, Farman, and Voisin, did not adopt to the need for developing new designs. Blériot and Voisin concentrated enormous effort in building the heavy bomber the French parliament insisted it wanted; none of the companies' designs were successful. As a result Blériot spent most of the war building SPAD fighters. Voisin continued to build his light bombers which were little changed from the prewar models; he ceased to produce aircraft after the war ended. Commandant Dorand approached Farman about building a new two-seat reconnaissance aeroplane to replace the M.F.11s and F.40s then in service. The firm initially rejected the offer; as a consequence, the Farman brothers spent the remainder of the war building A.R.1s under license. It is unclear whether these firms' refusal to adopt new designs was because of a desire to continue mass production of proven designs (which was quite lucrative for the company) or came out of a genuine belief that their designs were the best solution to the Aviation Militaire's needs.

The aircraft manufacturers were, on their own, solving the problem of the Aviation Militaire's qualitative inferiority. The SPAD fighters, Salmson 2 A2, Breguet 14 A2 and B2, and Farman F.50 were all beginning development. Although developmental and production problems delayed the entry of these aircraft into service, once they arrived the French air arm was rejuvenated.

1917

In 1917 the production of aircraft would increase from 7,549 in 1916 to 14,915—nearly 200% higher. Similarly the manufacture of aero engines would jump from 16,785 in 1916 to 23,092 in 1917, a 138% increase. More importantly, the qualitative improvement in aircraft and engines more than matched the numerical increase. By late 1917 factories were stopping production of obsolete A.R.1s and 2s, F.40s, and Sopwith 1½ Strutters and switching to SPAD fighters, Breguet 14s, and Salmson 2s. Not only were these aircraft more complicated and, therefore, time-consuming to build, but they relied on special materials such as, in the case of the Breguet 14, duralumin.

Such an improvement in quantity and quality would not have been possible without a significant change in the aviation industry—the increasing use of subcontractors. In fact, during 1917 many of the larger firms became concerned as they saw their subcontractors taking an increasingly larger share of the market. The major manufacturers saw a decrease in their capital of 38%, primarily due to the increased use of subcontractors to produce aircraft. These smaller firms had accounted for less than 10% of production at the start of the war. During 1917 subcontractors accounted for 45% of the aircraft and 85% of the Hispano-Suiza engines produced.

There were 22 new manufacturers who started aircraft and engine in 1917; these were: Ateliers de Colombes, Bathiat, Bernard, Bessoneau, Caffort, Chenard and Walker, Doriot-Flandin, Fives-Lille, Gregoire, Janoir, Kellner, Latécoère, Leflaive, de Lesseps, Levasseur, de Marcay, Mayen, Massenet-Maille, S.A.C.A., S.C.A.F., SEA, and S.E.C.A. At least in part due to the war tax, the major manufacturers concentrated on the development of new types, leaving the subcontractors to build the majority of aircraft.

Engine production was similarly affected by the use of subcontractors. During 1917 the seven major manufacturers of aircraft engines saw their share of the market reduced to approximately 15%; the remaining 85% was produced by the subcontractors.

In March 1917 the GQG proposed a goal of 2,665 frontline planes by the end of 1917; and a 4,000-plane force for April 1918 was suggested to the War Committee. Daniel Vincent became undersecretary of the Twelfth Directorate for aviation in March 1917. One of his first actions was to halt the STAé's development of its own aircraft. He also attempted to standardize production on a few modern types and encouraged (some might say forced) a large number of companies to accept license production of aircraft designed by other manufacturers by promising adequate supplies of raw materials and workers if they complied.

The government attempted to increase aircraft production by converting some automobile factories to aircraft manufacturing. There was also an attempt to increase the number of skilled workers available to the industry. However, even in factories

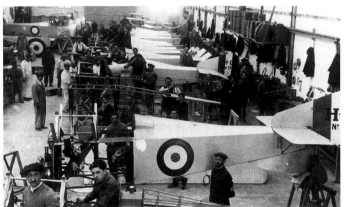

Hanriot HD.1 and HD.2 production line.

where there was an adequate number of workers, strikes served to further hinder production. Metal, automotive, and aircraft manufacturers went on strike in May and September, affecting 57,000 workers and 32 aircraft factories.

Despite these difficulties the French aviation industry was recovering. In 1918 it would manufacture large numbers of high quality aircraft, primarily due to the productivity of the subcontractors.

1918

During 1918 the French aircraft industry built 24,652 aircraft, a 165% increase from 1917. This number is even more impressive when the post-Armistice slow down in production is taken into account. Engine production also climbed to 44,563 in 1918, nearly a 200% increase over the preceding year. By July 1918 the April 1918 goal of 2,870 new aircraft had been nearly reached with 2,827 aeroplanes in service. All of these aeroplanes were modern types.

In August 1918 approximately 40% of the aircraft produced were fighters (SPAD 13s), 33% reconnaissance planes (Salmson 2s, SPAD 16s, Breguet 14 A2s), 15% bombers (Breguet 14 B2s, Farman F.50s, Voisin 10s); and 12% trainers. Production figures for the most important French aircraft in 1918 are as follows:

| SPAD 13 | 7,300 | Salmson 2 | 3,200 |
| Breguet 14 | 5,500 | Voisin 10 | 1,850 |

The aircraft were more sophisticated and, hence, more expensive. The average price of an airframe (without engine or armament) in 1918 was $6,000—20% higher than in 1917 and twice as high as at the start of the war.

In 1918 there were 36 major manufacturers: Bellanger, Bernard, Blériot, Borel, Breguet, Caudron, Compagnie Générale des Omnibus, Darracq, Esnault-Pelterie (REP), Farman, Grémont, Hanriot, Janoir, Kellner, Latécoère, de Lesseps, Letord, Levasseur, Lioré et Olivier, De Marcay, Morane-Saulnier, Michelin, Nieuport, Niepce et Fetterer (DNF), Nhup, Régy, Renault, Salmson, SAIB, Savary et Lafresnaye, Scaf, Schmitt, SEA, Sidam, SPAD, and Voisin-Lafresnaye.

Subcontractors did most of the airframe manufacturing in 1918, capturing nearly two-thirds of the total business that year. Subcontractors for engine manufacturers had an even larger share of the market than the airframe builders—62.7%. Original manufacturers accounted for only 39% of airframes and 37.3% of engines. Blériot-SPAD serves as an example. In 1916 the company built 98.2% of the SPAD 7s built that year. In 1917 it was producing 57% of the SPAD fighters. However, by 1918 that number had reversed: Blériot-SPAD was building only 43% of the total, while subcontractors accounted for the remaining 57%.

Postwar, many of the firms that had started as subcontractors became major airframe producers in their own right. Such firms included Amiot, Bernard, Latécoère, SEA (later Bloch and Potez), Levasseur, and Lioré et Olivier.

It should be remembered that, even at this stage of the war, approximately 20% of French output was going to the allies—primarily the United States, Belgium, and Romania. Of the 51,700 French aircraft built during the war, 9,460 were supplied to the allies. The breakdown of foreign shipments is as follows:

United States	3,300	Belgium	400
United Kingdom	2,000	Romania	300
Russia	2,000	Other countries	160
Italy	1,300		

In 1918 there were 23 major engine manufacturers. These were: Anzani, Aries, Ballot, Brasier, Caffort, Chenard et walcker, Clément-Bayard, Clerget, Darracq, De Dion. Delaunay-Belleville, Fives-Lille, Gnome Rhône, Grégoire, Hispano-Suiza, Lorraine-Dietrich, Mayen, Peugeot, Renault, Salmson, S.C.A.F.,

Compagnie Général des Omnibus, and Voisin-Lefebrve.

The engines built in 1918, like the airframes, were considerably more sophisticated than those built the previous year. In 1917 the number of rotary engines built (10,757) almost equaled the number of stationary engines (11,395); there were also 1,223 radials built. By 1918 stationary engines vastly outnumbered rotary engines (29,461 versus 6,349). French engines were supplemented by some imported engines; for example, some Breguet 14s were equipped with Fiat A12bis engines.

The production increases of 1917–18 enabled the French to form large combat groups. There were enough SPAD fighters to protect the reconnaissance aircraft and enough remaining to form fighter groups capable of establishing aerial superiority over selected areas of the front. Production of the Breguet 14 bombers resulted in the formation of several new Groupe's de Bombardement. These would be used effectively in ground attacks against the German troops and their supply lines. Finally, the new reconnaissance aircraft (Salmson 2s and Breguet 14 A2s) would give their crews a chance to survive even determined fighter opposition.

Pétain proposed an even more ambitious plan for a 6,000 aeroplane force. This was to be composed of 300-hp, single-seat fighters, two-seat fighters, twin-engine multi-role aircraft, armored ground attack aircraft, medium bombers, and multi-engine strategic bombers. These requirements led to the specifications for new aircraft listed on pages 14 and 15.

While the plan for a 6,000 plane air force was certainly far reaching, it was also unrealistic. By November there were 3,222 aircraft in service: 1,152 fighters, 1,585 reconnaissance aeroplanes, 285 day bombers, and 200 night bombers. The aviation reserves had an additional 2,600 aircraft. The reasons for France's inability to reach the goal of 4,000 frontline aeroplanes are twofold: first, a substantial portion of its production was diverted to the United States; second, combat attrition approached 50% for some types. By the time of the Armistice a completely armed and equipped airframe was being completed every 15 minutes around the clock; a new engine was being delivered to the military every 10 minutes. Therefore, the inability to meet the goal of 4,000 aeroplanes cannot be attributed to a failure of the aircraft industry.

Table 5: 1918 Production

Month	Aircraft	Engines
January	1,714	2,567
February	1,668	3,117
March	1,647	3,139
April	2,100	4,029
May	2,071	3,847
June	2,459	4,274
July	2,622	4,490
August	2,912	4,320
September	2,322	3,934
October	2,362	4,196
November	1,392	3,502
December	1,383	3,148
Total	24,652	44,563

During the First World War the French aviation industry produced 51,700 airframes; this was just behind the United Kingdom, which built 52,027 aircraft. France was the clear leader in engine production with 92,386 being manufactured during the war years. The French industry had not only made the Aviation Militaire the world's largest air service at the time of the armistice, but had produced an additional 9,460 aircraft to supply the air services of Italy, Russia, the United States, the United Kingdom, and other countries.

A.R.1 and A.R.2

A.R.1 serial number 3015. This view emphasizes the pronounced negative stagger of the upper wings. SHAA B75.511.

By 1916 it was increasingly obvious that the Farman F.40 was no longer capable of performing daytime reconnaissance. The F.40's pusher configuration prevented defense against attacks from behind, a fact which the German fighters fully exploited. Therefore Colonel Dorand, who headed the STAé, formulated a requirement for a two-seat reconnaissance aircraft with tractor configuration that would improve its ability to defend against rearward attacks. Dorand approached Farman about producing the new plane; however, Farman was reluctant to abandon the pusher configuration and the STAé had to find another engineer. Captain G. Lepére (who later designed the Packard LUSAC-11 and 21 in 1918) was selected to develop the new aircraft, which was designated A.R.

The A.R.1 was later known as the Dorand A.R.1 because of Commandant Dorand's association with the type. However, Dorand had no direct involvement with either the design or production of the aircraft beyond drawing up the initial specification. The designation A.R. has at least three alternative meanings. The Renault firm, which built the engine, believed that it stood for "Avion Renault" or "Avant Renault." Renault insisted that any aircraft using Renault engines should have the letter "R" for "Renault" included in its designation. The Aviation Militaire, however, had no interest in acknowledging Renault's contribution to the A.R., and insisted the designation stood for "Avion de Reconnaissance." In fact, the STAé designation for the airplane was A.R.1 and A.R.2, regardless of whether a Renault or Lorraine engine was fitted.

The A.R.1 was a tractor biplane with negatively staggered wings and powered initially by a 160-hp Renault engine. Later versions had a 190-hp Renault 8Gd or a 240-hp Lorraine 8A. The crew was seated in tandem with the pilot just below the top wing and a gunner located beneath a cutout in the upper wing. The pilot's forward vision was hampered by the rhino exhaust of the 160-hp Renault. Later versions had the 240-hp Renault 8Gd, in which the exhausts passed along either side of the engine cowl, giving the pilot a clearer view ahead. The pilot's upward view was quite limited because of his position underneath the top wing; he had two small windows in the floor to enhance downward vision. There was also a window and camera opening in the floor of the observer's cockpit. Four bomb cells were situated between the pilot and the observer and permitted vertical storage and release of four 120-mm bombs. The pilot fired a synchronized 7.7-mm Vickers machine gun fixed on the starboard side of the fuselage, while the observer had one or two Lewis guns on a movable mount. The radiator was initially located in the nose with an auxiliary radiator under the nose. Subsequent aircraft had the radiator under the fuselage above the lower wing. The two-bay wings were of equal span. The fuselage was suspended between the upper and lower wing by struts made of ash. Only the lower wing had dihedral. Ailerons were located on only the upper wing. The rudder and elevator were rectangular.

Variants

There were two other versions of the A.R. The first, and most important, was the A.R.2 which had a tightly cowled 190-hp Renault 8Gd/Gdx, or 240-hp Lorraine 8Bb engine. There were airfoil wing radiators, and the surface area of the wings was reduced. The other version was the A.R.1 D2 trainer, which had a Renault 12d.

Production

An order for 750 A.R.1s was placed on 24 September, 1916. Numerous firms including Farman and Letord built the aircraft under license. However, as with virtually all French aircraft of the time, production was seriously delayed. It had been

A.R.1 assigned to AR 33.

anticipated that by February 1917 delivery of 645 would have been completed; however, not a single aircraft had been delivered by that date. Slowly, the production rate began to increase and by 1 August, 1917, there were 216 A.R.s in service with the escadrilles, with an additional 80 in the general reserves. A total of 1,435 Dorands of all types was eventually built.

Operational Service

The A.R.1 entered service in early 1917. AR 1 was one of the first escadrilles to receive the new A.R.1s. The new plane had been eagerly awaited by the members of F 1, who had been so disgusted with the poor performance of their F.40s that they had adopted a winged snail as the escadrille insignia. Unfortunately, the A.R.1 would not prove a significant improvement over the F.40. In April 1917 F 1 became AR 1; at that time AR 1 was assigned to the 33rd C.A. (Corps d'Armée) and participated in the Battle of Chemin des Dames. The crews of AR 1 provided the crucial, if mundane, task of artillery spotting. They enabled the French artillery to strike effectively at the German troops and artillery units and were active over the Vauclerc plateau. In August 1917 the crews were allowed to recuperate from the intense activities during the Battle of Chemin des Dames by moving to a quiet sector where the front lines were stable. Based at Lure, along with the 33rd C.A., AR 1 saw little activity. On 11 September AR 1 moved to Fontaine, where the aggressiveness of German fighters resulted in a large number of combats. The unit returned to Lure in late January 1918 and re-equipped with Salmson 2s to become SAL 1.

No fewer than 49 escadrilles were equipped with A.R.1s; 44 of these units served on the Western front. Escadrilles 14, 22, 44, 254, and 275 (which operated a mix of A.R.s and Sopwith 1½ Strutters) also saw service on the Italian front. Other escadrilles which received the A.R.1s were:

AR 2, formed from F 2 during 1917. It was equipped with A.R.1s and a few Letords and was assigned to the 5th C.A. It re-equipped with SPAD 16s in November 1917.

AR 8, formed from MF 8 in October 1917. It was assigned to the 11th C.A., and re-equipped with Salmson 2 A2s in February 1918.

AR 14, formed from F 14 in September 1917. It was assigned to the 31st C.A. and was based at Porto Nuove in Italy as part of the 10th Armée. In December it was based at San Pietro in Gu and in February 1918 at Castello di Godego. In March AR 14 became part of the 46th Division Alpins. It returned to France in April and was active in the Third Battle of Flanders from May to June, during which time it was re-equipped with Salmson 2s.

AR 16, formed from F 16 in July 1917. It was assigned to the 18th C.A. in the 10th Armée sector. AR 16 was subsequently active at Vosges (July) and Champagne (October). In January 1918 it re-equipped with Salmson 2s.

AR 19, formed from F 19 in early 1917 and assigned to the 13th C.A. AR 19 was active in the battles at Noyon and Saint-Quentin. In July it moved to Verdun in the 2nd Armée sector where it was particularly active in ground attacks on enemy positions. In April 1918 it was based at Foucaucourt; shortly after it became SAL 19.

AR 20, formed from F 20 in August 1917. It was assigned to the 14th C.A. and was active in the 6th Armée sector. It became SPA-Bi 20 in early 1918.

A.R.2 utilized as an ambulance aircraft as indicated by the prominent red cross on the fuselage. SHAA B85.199.

A.R.1 serial number 1073. SHAA B88.91.

Below: A.R.1 at Rosnay (near the Marne) in June 1917.

AR 21, formed from C 21 in March 1918. It was assigned to the 37th D.I. and was based at Ambrief. It re-equipped with SPAD 16s in June 1918.

AR 22, formed from F 22 in July 1917. In December 1917 it was assigned to the 12th C.A. and was based at Castello di Godego. In April 1918 AR 22 was based at Nove. In May the escadrille re-equipped with Salmson 2s.

AR 32, created from F 32 in 1917. It was assigned to the 10th C.A. but it became SAL 32 in December 1917.

AR 33, formed from MF 33 in June 1917. It was assigned to the 9th C.A. and re-equipped with Salmson 2s only six months later.

AR 35, formed from MF 35 in June 1917. It was assigned to the 20th C.A. and based at Ville-Savoye. It was active during the Battle of Chemin des Dames. During the six months the unit used A.R.1s (along with several Caudron R.4s) it flew 173 reconnaissance missions, 268 artillery regulations, 65 photographic missions, and 54 infantry co-operation missions. AR 35 was subsequently assigned to the 35th C.A. and moved to Lorraine, where it was based at Villers-les-Nancy. In November 1917 it re-equipped with Breguet 14 A2s.

AR 40, formed from MF 40 in 1917. In April 1918 AR 40 was assigned to the 4th C.A. and was based at Ferme d'Alger. It became SAL 40 in mid-1918.

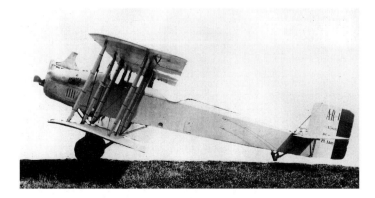

A.R.1. The pilot's vision was hampered by the rhino exhaust of the 160-hp Renault. SHAA B83.5644.

A.R.2 with the 240-hp Renault 8Gd. The exhausts passed along either side of the cowl, giving the pilot a better view. SHAA B76.1968.

A.R.1. MA 6807.

A.R.2 serial number 787. MA 76933.

A.R.2 serial number 169. MA.

A.R.1 serial number 1012. MA.

A.R.1 and A.R.2

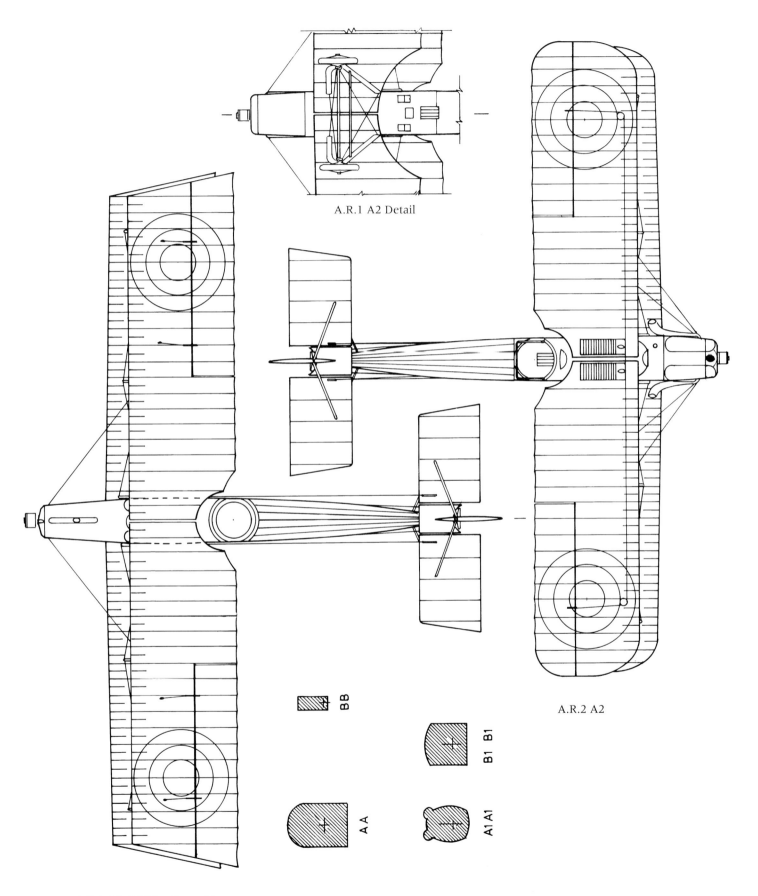

A.R.1 A2 Detail

A.R.2 A2

A.R.1 A2 (200-hp Renault)

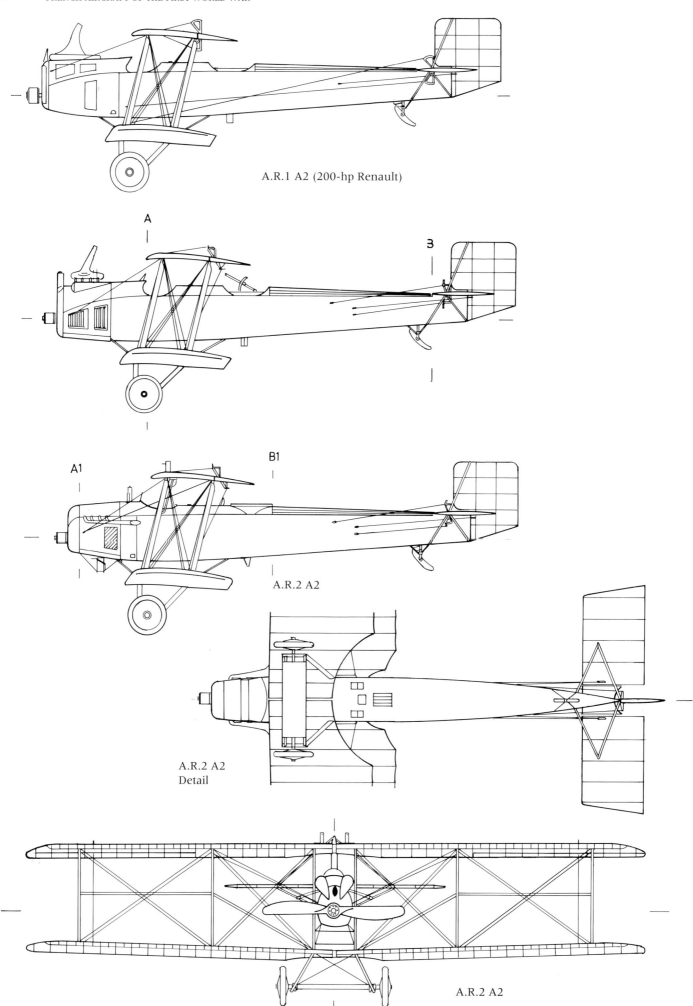

A.R.1 and A.R.2

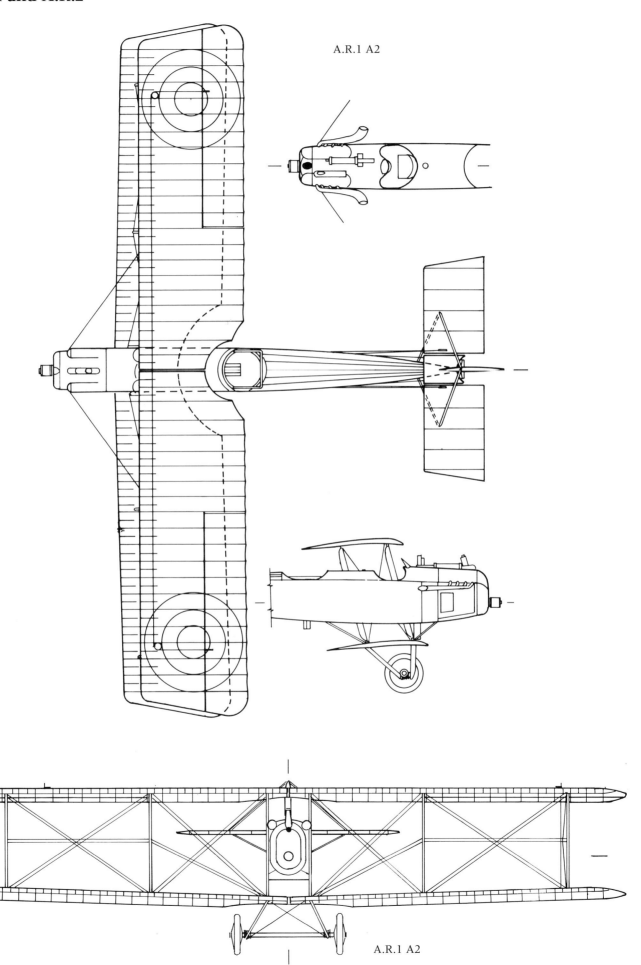

AR 41, formed from F 41 in 1917. Commanded by Capitaine Delaitre, it was initially based at Lorraine. In April 1918 it was assigned to the 32nd C.A. and was based at Gonderville. It re-equipped with Salmson 2s in June 1918.

AR 44, formed from F 44 in June 1917. It was assigned to the 31th C.A. which, along with the 10th Armée, was sent to Italy in late 1917. AR 44 was initially based at Ghedi, then Verone and, finally, Castello di Godego. It re-equipped with Breguet 14 A2s in January 1918.

AR 45, formed from F 45 in late 1917. It was assigned to the 1st Armée and converted to Breguet 14 A2s in March 1918.

AR 50, formed from F 50 in 1917. It was assigned to the 16th C.A. and became SAL 50 in February 1918.

AR 52, formed from F 52 in late 1917. It was assigned to the 38th D.I. AR 52 was active over Montdidier and Noyon during the German offensives of 1918. It re-equipped with Salmson 2s in March 1918.

AR 58, formed from F 58. It was assigned to the 40th C.A. and was based at Bessoncourt. It became SAL 58 in February 1918.

AR 59, formed from F 59 in October 1917. It was assigned to the 10th C.A. and was based at Issoncourt. It became SAL 59 in May 1918.

AR 70, formed from F 70 in late 1917. It was assigned to the 20th C.A. and was based at Suzay. AR 70 re-equipped with Salmson 2s in April 1918.

AR 71, formed from F 71 in late 1917. Assigned to the 8th C.A., it was based at Auve. AR 71 became SAL 71 in April 1918.

AR 72, formed from F 72 in September 1917, was assigned to the 7th C.A. and based at Moyen. AR 72 re-equipped with Salmson 2s in March 1918.

AR 201, formed from F 201 in July 1917. It was assigned to the 13th C.A. and was based along the front near Verdun. It moved to Mont-de-Soissons during the battle of La Malmaison. Later it transferred to the 8th Armée sector and was based at Meurthe-et-Moselle. AR 201 re-equipped with Breguet 14 A2s in April 1918.

AR 203, formed from F 203 in October 1917. It was attached to the 4th Armée and became SAL 203 in February 1918.

AR 205, formed from F 205 in late 1917. It re-equipped with Breguet 14 A2s in February 1918.

AR 211, formed from F 211 during 1917. It was initially commanded by Lieutenant Chamouton. AR 211 moved to Tilloy and was active during the Battle of La Malmaison. At the beginning of 1918 it moved to La Cheppe in the 4th Armée sector. In February AR 211 transitioned to Breguet 14 A2s.

AR 230, formed in August 1917 and probably based in the 1st Armée sector. It became SAL 230 in March 1918.

AR 233, formed in January 1918 from elements of AR 211. It re-equipped with Breguet 14 A2s in April 1918.

AR 253, formed from elements of AR 70 in November 1917. It was assigned to the 20th C.A. and transitioned to Salmson 2s in May 1918.

AR 254, formed in November 1917. It was assigned to the 12th C.A. and the unit accompanied the 10th Armée to Italy. AR 254 was based at Castello di Godego initially and performed reconnaissance east of Asiago. Subsequently it participated in the siege of Brenta. During the Tomba offensive the escadrille bombed and strafed enemy artillery and machine gun positions. In April 1918 AR 254 was based at Verone. In August the unit re-equipped with Salmson 2s.

AR 256, formed from elements of AR 33 in January 1918. It was assigned to the 9th C.A. AR 256 was commanded by Lieutenant Hervé and based at Esquennoy, where it operated from during the Battle of the Somme. It became SAL 256 in May 1918.

AR 257, created from elements of AR 45 in January 1918. Commanded by Lieutenant Rotival, it was assigned to the 152nd D.I. which was part of the 9th C.A. In April it moved to Courcelles, where it participated in the Battle of the Somme. AR 257 received Breguet 14 A2s in July 1918.

AR 258, formed in January 1918. It was initially based at Saizerais and commanded by Capitaine Houssay. As part of the 32nd C.A. it flew artillery regulation and ground strafing missions over Chenicourt-Limey. The unit re-equipped with Breguet 14 A2s in July 1918.

AR 259, formed from elements of AR 22 in February 1918. It was initially assigned to the 13th C.A. and later to the 2nd Armée. It used A.R.1s until these were replaced by Salmson 2s in May 1918.

AR 261, formed from elements of AR 32 and AR 59 in February 1918. It was assigned to the 10th C.A. and based with the 2nd Armée at Issoncourt. It became SPA-Bi 261 in July 1918.

AR 262, formed in January 1918. It was assigned to the 8th C.A. and was commanded by Lieutenant Guérin. It was based at Dancourt and was re-equipped with Salmson 2s in July 1918.

AR 264, formed from elements of AR 19 in January 1918. It was initially assigned to the 13th C.A. and, commanded by Capitaine Daum, was active over the Argonne and the Avocourt forest. AR 264 also participated in the attack on Cheppy in March. It re-equipped with Salmson 2s in August 1918.

AR 267, formed from elements of AR 40 in February 1918. It was assigned to the 4th C.A. and based at Ferme d'Alger. AR 267 re-equipped with Breguet 14 A2s in July 1918.

AR 268, formed in February 1918. It was assigned to the 18th C.A. and based at Remy. AR 268 became SPA-Bi 268 in August 1918.

AR 272, formed from elements of AR 1 in February 1918. It was assigned to the 33rd C.A. In April AR 272 was assigned to the 62nd D.I. and was based at Corbeaulieu. It re-equipped with Breguet 14 A2s in July 1918.

AR 274, formed from elements of AR 50 in February 1918. It was assigned to the 16th C.A. and was based at Verrines. AR 274 became BR 274 in July 1918.

AR 275, formed from elements of AR 44 in February 1918. It was assigned to the 31st C.A. and was based at Étampes. It re-equipped with Breguet 14 A2s two months later.

AR 288, formed from elements of AR 52 in February 1918. It was assigned to the 45th D.I. in the 6th C.A. sector. It was based at Viefvillers and re-equipped with Salmson 2s three months later.

AR 289, formed from AR 58 in March 1918. It was assigned to the 66th D.I. and was based at May-en-Multien. It became SPA-Bi 289 in June.

T.O.E. Units

Five other escadrilles were T.O.E. units assigned to French army corps based in foreign countries:

AR 251 was based on the Serbian front.

AR 382 was based on the Serbian front.

AR 521 was based on the Serbian front.

AR 533 was a mixed French/Greek escadrille and saw action on the Greek front.

AR 556 was formed in September 1917 and designated as the Escadrille de la subdivision Rabat.

Most of the escadrilles supplied with A.R.1s had previously used M.F.11s and F.40s. The tractor layout of the A.R.1s was seen as presenting a marked improvement over the pusher configuration of the Farman aircraft. The gunner's Lewis guns had a clear field of fire, which helped to discourage attacks from the rear. Also, the performance of the A.R.1 was marginally superior to the M.F.11 and the F.40. The AR escadrilles flew reconnaissance missions over the front and were particularly useful spotting artillery. As the A.R.1 had a limited range, many units were equipped with a few Letord aircraft for long-range reconnaissance. It was clear, however, by the summer of 1917 that the A.R.s were rapidly becoming obsolete. The aircraft

Above and above right: Two views of an A.R.2 with an experimental engine. MA 369 and 370.

Another view of A.R.2 serial number 169. SHAA B84.994.

became increasingly vulnerable to German fighters, and many AR escadrilles were assigned to less active sections of the front during late 1917. Despite the fact that the A.R.s were now of limited usefulness, the Aviation Militaire had little choice but to keep them in service. The superior Salmson 2 A2s and Breguet 14 A2s were not yet available in sufficient numbers to completely equip the reconnaissance units.

In escadrille service the A.R.1s had some minor problems. The throttle attachment to the fuselage was found to be defective and had to be replaced. On the first 20 airplanes the radiators were found to be ineffective and the engines were frequently overheating. The problem was corrected in subsequent machines. The 10th Armée complained that the A.R.1s built by Farman had more defects than those built by other manufacturers.

A.R.1s were used as night bombers while serving with the 10th Armée in Italy. They were equipped with cockpit lights for the crew, and a row of spotlights were placed on the landing gear to enable targets on the ground to be spotted. These A.R.1s carried up to 120 Type P lance bombs.

The A.R.1s and A.R.2s were replaced by Salmson 2 A2s, Breguet 14 A2s, and SPAD 11s in late 1917.

Although the A.R.1s were superior to the Farmans they replaced, the "Dorands" were obsolescent by the time they entered service. It has been noted that Colonel Dorand believed the type had performed as well as could have been expected considering the mediocre engines with which they had been equipped.

Foreign Service
Greece

The French supplied the Greek air service with 37 A.R.1s in 1917. They were assigned to the 532 Mira Vomvarthismou ke Anagnorisseos (532 Bombing and Reconnaissance Squadron) which had been formed on 10 December, 1917. This unit was based at Gorgop and operated a mixture of A.R.1s and Breguet 14s. 532 Mira saw action bombing enemy positions in the Vardar valley, the fortifications of Shar-di-Legen, and the Axos river. The 533 Mira Dioxeos (533 Fighter Squadron), formed in June 1918, also operated a mixture of A.R.1s and Breguet 14s. Twenty-two A.R.1s remained in Greek service at the end of the war. In early 1919 elements of both 532 and 533 Miras saw service in the Greco-Turkish War. However, it seems that only the Breguet 14s were actually used in combat, as the A.R.1s were now considered to be obsolete. The A.R.1s were retired in 1923.

Serbia

On 18 May, 1917, F 382 replaced its F.40s with A.R.1s and was designated AR 382. Another A.R.1 unit, AR 521, subsequently became the 1st Serbian Escadrille on 17 January, 1918. It was initially based at Ostrovo and moved to Vertekop in July 1918. This unit participated in attacks on Bulgarian troops and reinforcements along the front. In early autumn the escadrille was sent to Uskub and operations were conducted against the retreating Germans and Austro-Hungarians. It appears that the

A.R.1s had been replaced by the Breguet 14s before the war ended.

United Kingdom

The RFC borrowed a single example of an A.R.1 from French Escadrille SPA-Bi 2 in January 1918. It had been planned to use it to attack the German battlecruiser *Goeben*, but no such attack was ever made.

United States

Desperate to get its pilots into combat, the United States Air Service accepted aircraft types purchased from the French which were clearly obsolete. Along with Sopwith 1½ Strutters and SPAD 11s, a total of 22 A.R.1s and 120 A.R.2s were purchased. The 1st Observation Squadron had a dozen A.R.1s on strength by May 1918. The 12th Aero Squadron was also equipped with A.R.1s and A.R.2s, the 89th Aero Squadron had five A.R.1s in May 1918, and the 91st Observation Squadron also had a few A.R. aircraft in 1918. The remaining A.R.1s and A.R.2s were based at Gondrecourt, Meuse.

Fortunately for the Americans, the A.R.1s and A.R.2s were used primarily for training and most of the squadrons replaced them with more modern types before entering combat. In April the 1st Aero Squadron had re-equipped with SPAD 11s (although a few A.R.1s remained with the unit). By the first week in June, the 12th Aero Squadron (in the Baccarat sector) had re-equipped with Salmson 2 A2s. The 91st Aero Squadron replaced their "Dorands" with Salmson 2 A2s in April 1918. The A.R.1s and A.R.2s were greatly disliked by the American pilots, who contended that the A.R. designation actually stood for 'Antique Rattletraps.'

A.R.1 Two-Seat Reconnaissance Aircraft with 190-hp Renault 8Gd

Span 13.27 m; length 9.30 m; wing area 50.17 sq. m
Empty weight 810 kg; loaded weight 1,250 kg

Maximum speed:	2,000 m	152 km/h
	3,000 m	147 km/h
	4,000 m	141 km/h
Climb:	2,000 m	11 minutes
	3,000 m	22 minutes 20 seconds
	4,000 m	39 minutes

Ceiling 5,500 m; range 375 km
Armament: four 120-mm bombs, a synchronized 7.7-mm Vickers machine gun, and one or two Lewis guns on a movable mount.

A.R.2 Two-Seat Reconnaissance Aircraft with 240-hp Lorraine 8Bb

Dimensions same as A.R.1
Empty weight 825 kg; maximum weight 1,250 kg

Maximum speed:	2,000 m	159 km/h
Climb:	2,000 m	14.5 minutes
	3,000 m	28.2 minutes

Service ceiling 3,000 m; absolute ceiling 4,500 m
Range 375 km; endurance 3 hours at 3,000 m
Armament: same as A.R.1.
Total production of the A.R.1 and A.R.2 was 1,435 aircraft.

Astoux-Vedrines

The Astoux-Vedrines triplane was tested at the Belgian aerodrome of Étampes during the First World War. The Étampes field, which had previously been used by the Farman and Blériot firms to evaluate aircraft, was used by the French to test fly a number of unusual designs. The plane was designed by Astoux and test-flown by Jules Vedrines and so has become known as the Astoux-Vedrines triplane.

The aircraft had three wings of narrow chord with a single interplane strut on each side connecting all three wings. The wings were staggered so that the top wing was foremost and the bottom was at the extreme rear. An unusual aspect of the design was that the incidence of the wings could be varied in flight. A large spinner helped to give the fuselage a streamlined silhouette. A small fixed fin extended above and below the tail and a large rudder was attached to the fin. A tailskid extended from the lower fin. The engine was a 130-hp Clerget.

Jules Vedrines decided to adopt a careful test program and would set the wing incidence on the ground before flying the aircraft. Another pilot, named Simon, destroyed the aircraft in a crash, possibly due to failure of the variable incidence wing. It does not seem that the triplane was rebuilt or that the design was developed further.

Astoux-Vedrines Triplane Experimental Aircraft with 130-hp Clerget

One built

A crowd of onlookers examine the ill-fated Astoux-Vedrines triplane. MA 11.790.

Astra

Astra Bomber

The Astra Société de Constructions Aéronautique produced a series of airships before the war. In 1909 it obtained a license from the Wright Brothers to build modified Wright Flyers. The company later built several military biplanes, including the Type C of 1912 and the Type CM of 1913. The main differences between the two designs was that the Type C was constructed of wood while the CM was made of metal (hence the M designation). Three of the CMs were ordered by the Royal Naval Air Service but were never delivered.

The Astra firm produced a heavy bomber which, it was hoped, would meet the requirements of the 1916 concours puissant. Little is known about this plane and it does not seem that any photographs of it have survived. It is known that the Astra bomber was powered by three 220-hp Renault engines. Its configuration was much like the Caproni bomber, with a twin fuselage and an abbreviated central fuselage. A pair of wheels was located under all three fuselage sections. There was a machine-gun position in the nose and one on each boom.

The Astra bomber did not meet the specifications of the concours and was not selected for production. It appears that further development of the type was abandoned.

The Astra firm continued to produce aircraft under license during the First World War and in 1921 merged with Nieuport to become Nieuport-Astra.

Astra Heavy Bomber with Three 220-hp Renault Engines
Span: 25.5m; length 14.2 m; wing area 140 sq. m
Empty weight 2,300 kg; loaded weight 3,500 kg
Maximum speed: 125 km/h; ceiling 3,800 m; climb to 2,000 m in 24 minutes
One built

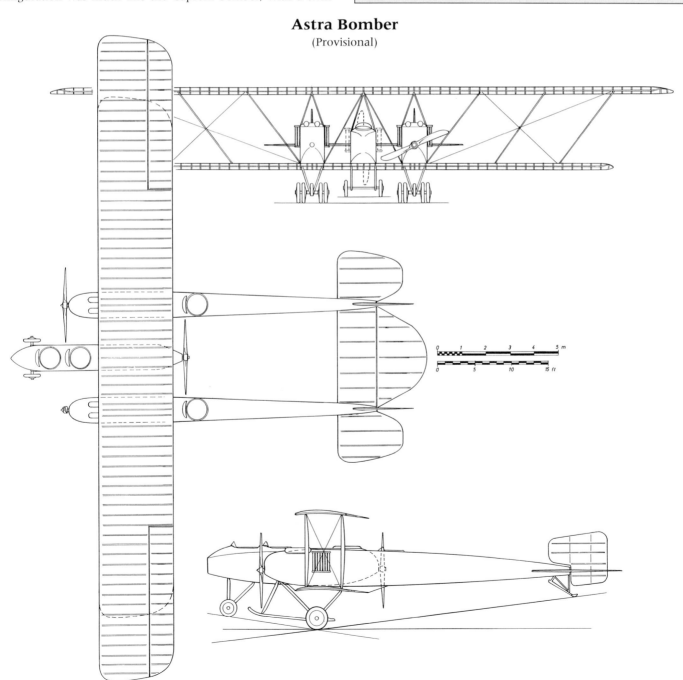

Astra Bomber
(Provisional)

Astra-Paulhan Flying Boat

The Astra firm produced a flying boat which had twin-engines and twin-fuselages. In many respects it was similar to the preceding Astra bomber. The aircraft was tested at the Saint Raphaël naval base in 1919. No further details, aside from the two photographs shown, are available.

Right: The Astra-Paulhan flying boat at Saint Raphaël in 1919. 06603.
Below: The Astra-Paulhan flying boat. B87.5023.

Audenis E.P.2 and C2

Charles Audenis was a pioneer French aviator who, with the collaboration of his friend, Jean Jacob, designed and built two training aircraft before the First World War. Audenis subsequently worked for the Voisin firm and was eventually accepted as a pilot by the Aviation Militaire. He served as a flying instructor for about a year and then was posted to MF 1. Sometime during the course of the war Audenis and Jacob built a two-seat plane with a single-bay wing. It was designated the type E.P.2 and was intended for use as a trainer. It was powered by a single 80-hp Le Rhône engine, although an 80-hp Clerget, a 120-hp Le Rhône, or a 130-hp Clerget engine could be fitted.

The single bay biplane wings were of equal span. The wings were constructed in three sections with the center section supported by N-shaped cabane struts. The fuselage was in two parts to facilitate disassembly and transportation. The front of the fuselage was fabric covered, while the rear section was of monocoque construction. The vertical tail was triangular with elliptical elevators and rudders. The landing gear was made of poplar with a duralumin axle. There were dual controls with the student seated in the front seat.

Later in 1916 Audenis designed, built, and tested another aircraft. It was intended to meet the C2 category and was a two-seat biplane with equal-span wings. The lower wing was set well below the fuselage by a series of struts. There was a single bay of I-shaped interplane struts. The engine was probably a 130-hp Clerget. The aeroplane appears to have had a single synchronized machine gun fired by the

Audenis E.P.2 trainer. MA 31986.

pilot, while the observer had a ring mount. The rear fuselage was sharply tapered. The maximum speed was 180 km/h. The type was not accepted by the Aviation Militaire and Audenis returned to his regular duties with MF 1 (now AR 1). He subsequently helped design the B.A.J. C2 fighter of 1918. (See below.)

Below: Audenis C2 fighter. MA 31989.

E.P.2 Primary Trainer with 80-hp Le Rhône
Span 10.20 m; length 7.65 m; height 2.80 m; wing area 21 sq. m
Empty weight 460 kg; loaded weight 714 kg
One built

Audenis C2 Two-Seat Fighter
Maximum speed: 180 km/h
One built

B.A.J. C2

The B.A.J. was a two-seat fighter designed to meet the C2 specifications of 1918. The C2 category was for an aircraft armed with a 7.7- or 11-mm machine gun synchronized to fire through the propeller arc, two machine guns (or a single cannon) in a turret controlled by the gunner, and a machine gun firing through the underside. The aircraft was to have a payload of 375 kg, a maximum ceiling of 8000 m, a cruising altitude of 5000 m, and a maximum speed of 220 km/h.

The aircraft submitted to meet the C2 specification included a number of types which used the Hispano-Suiza 8Fb of 300-hp. These were the Hanriot-Dupont HD.5, the Borel-Boccacio type 3000, and the B.A.J. C2.

In 1916 Charles Audenis had designed two military aircraft that were not accepted for production (see above). His next design appeared in 1918 and was designated B.A.J. C2, the initials B.A.J. standing for Boncourt-Audenis-Jacob. Jacob was a lieutenant with AR 1 who had been killed during the war; the inclusion of his name in the designation implies that before his death he collaborated with Audenis on the design.

The aircraft was powered by a 300-hp Hispano-Suiza 8Fb and had a two-bay wing. The upper wing had a slight dihedral; the lower wing had none. Single interplane struts braced the wings. The interplane bracing wires were attached to the spars. The wing bracing proved satisfactory during static testing; the wings did not fail until a load factor of nine had been reached. The ailerons were horn-balanced (as were the rudder and elevators) and were on the lower wings only. Armament was a synchronized 7.7-mm Vickers machine gun fired by the pilot and two 7.7-mm Lewis guns on a T.O.3 ring mount for the observer.

The B.A.J. C2 was completed by 1 May 1918 and arrived for testing at Villacoublay in late November of that year. However, it was not flown until 28 January 1919. A second prototype was also subsequently tested at Villacoublay. The type was not adopted by the Aviation Militaire.

B.A.J. C2 Two-Seat Fighter with 300-hp Hispano-Suiza 8Fb
No known specifications
Armament: a synchronized 7.7-mm Vickers machine gun and two 7.7-mm Lewis guns on a T.O.3 ring
Two built

B.A.J. C2 fighter. The aircraft was powered by a 300-hp Hispano-Suiza 8Fb and had a two-bay wing. MA31989.

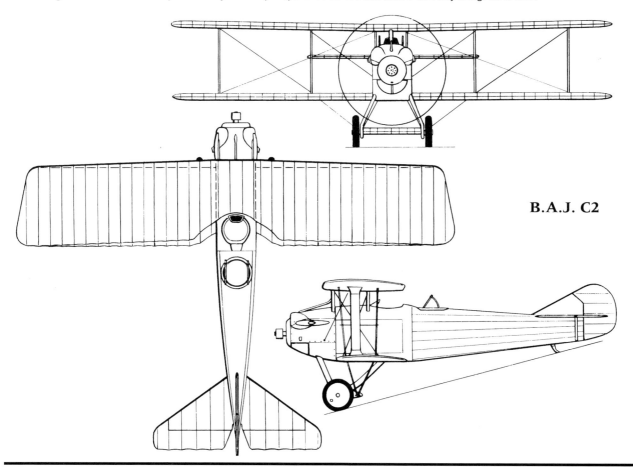

B.A.J. C2

Bassan-Gué BN4 Bomber

The Aviation Militaire, interested in obtaining a strategic night bomber in the same class as the RAF's Handley Pages, issued the BN4 specification in August 1918.

The BN3/4 requirement was for a heavy bomber to be used for night attacks, which meant that it would need little, if any, defensive armament. It was to carry a crew of three or four and have a 1,200-kg bomb load. Projected date for entry into service was 1919. The Bassan-Gué bomber was one of the aircraft designed to meet the specification.

The Bassan-Gué was to have been a triplane powered by three 450-hp Renault engines. However, the STAé had decided to reserve these engines for use on fighters; strategic bombers were not being given top priority for engine production. Instead it was asked that the design accept Hispano engines of only 300 hp. This meant the Bassan-Gué bomber would have been underpowered by 33 percent or would have had to been modified to accept four engines. It is not surprising, therefore, that the design was abandoned before it ever left the drawing board.

Bernard

The twin-engine Bernard A.B.1. The nose gunner was the bomber's only protection. B91.2712.

Bernard A.B.1, A.B.2, A.B.3, and A.B.4

The Bernard firm had produced SPAD 13s and 16s under license and in 1918 responded to the BN2 specification for a medium night bomber. The aircraft, designated A.B.1, was in competition with the Breguet 16, Farman F.50, and Voisin 11.

The A.B.1 was a biplane powered by two 180-hp Hispano-Suiza engines. The fuselage was made of wood and had a rectangular cross-section. The engines were suspended between the upper and lower wings with the two propellers positioned only inches from the pilot's cockpit. A set of M-shaped struts supported the engines from the top wing and a set of W-shaped struts supported them from the bottom wing. The undercarriage consisted of a pair of wheels below each of the engines. The tail was of conventional configuration with a single rudder and horizontal tailplane. The wing had two bays of struts outboard of each engine. The top wing was longer than the lower and there were ailerons on both the upper and lower wings. Protection was provided by a gunner/bombardier in the nose firing a single 7.7-mm machine gun. As all raids were to take place at night, this was considered an adequate defense.

The Farman F.50 and Breguet 16 were selected for production. However, the Bernard design had compared favorably in terms of speed and bomb load, and an order was placed for ten aircraft. None of the Bernard A.B.1s was ever used operationally, and it appears that none was ever assigned to an escadrille.

In an attempt to improve the aircraft's performance, Bernard substituted two Hispano-Suiza engines of 200-hp. Despite the improvement in performance created by the new engines, the Bernard A.B.2, as it was designated, was not chosen for production.

Postwar, the company attempted to sell the machine as a postal aircraft designated A.B.3 or a postal/passenger carrier designated A.B.4. Both types could use either two 180-hp or 200-hp Hispano-Suiza engines. The 1919 company brochure assured prospective customers that the all-wood construction insured a sturdy airframe which would withstand great stresses in any climate. Furthermore, the Hispano-Suiza engines were touted as having the distinct advantage of being available in large numbers at low prices.

The A.B.3 was to be used as a mailplane and could carry 905 kg of cargo. Only a single A.B.3 was built.

The A.B.4 was an airliner version of the projected A.B.2 bomber and could carry up to seven passengers. Only a single example was built.

Bernard A.B.1 Bn2 Night Bomber with Two 180-hp Hispano-Suiza Engines

Span 18.95 m; length 11.30 m; height 3.65 m; wing area 83 sq. m
Empty weight 1,660 kg; loaded weight 2,895 kg
Maximum speed: 161 km/h at 2000 m; ceiling 4,900 m; climb to 4,000 m in 15 minutes 45 seconds
Armament: one 7.7-mm machine gun
Ten built

Bernard A.B.2 Bn2 Night Bomber with Two 200-hp Hispano-Suiza Engines

Data as A.B.1 except: Empty weight 1,640 kg; loaded weight 3,084 kg; bomb load 640 kg
Maximum speed: 180 km/h
One built

Bernard A.B.3 Mail Plane with Two 180-hp or 200-hp Hispano-Suiza Engines

Data as A.B.1 except: Span 19.47 m; wing area 83 sq. m
Empty weight (180-hp HS) 1,570 kg; empty weight (200-hp HS) 1,660 kg; loaded weight (180-hp HS) 2,923 kg; loaded weight (200-hp HS) 3,163 kg
Maximum speed: 160 km/h at 2,000 m; climb to 2,000 m in 14 minutes; climb to 4,000 m in 40 minutes; ceiling 4,900 m
One built

Bernard A.B.4 Airliner with Two 200-hp Hispano-Suiza Engines

Dimensions identical to A.B.3
Empty weight (180-hp HS) 1,600 kg, empty weight (200-hp HS) 1,690 kg; loaded weight (180-hp HS) 2,953kg; loaded weight (200-hp HS) 3,193 kg
Maximum speed: 165 km/h at 2,000 m; climb to 2,000 m in 16 minutes, climb to 4,000 m in 49 minutes; ceiling 5,300 m
One built

The Bernard A.B.1 was powered by two 180-hp Hispano-Suiza engines. This aircraft has the standard four-bladed propellers. Renaud.

Bernard A.B.1

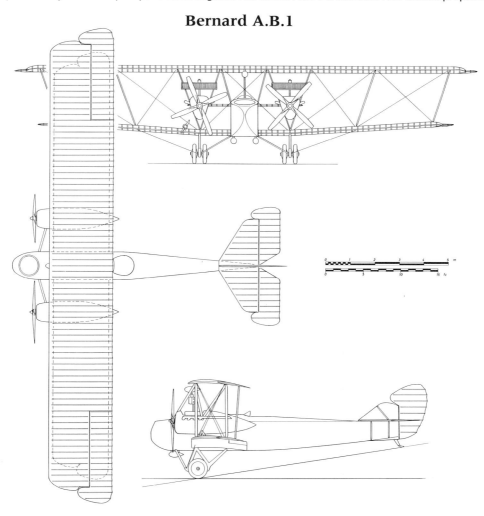

Bernard S.A.B. C1 Fighter

To meet the C1 specification calling for a fighter powered by a 300-hp engine, Adolphe Bernard, Louis Bechereau, Marc Birkigt, and Louis Blériot designed a two-bay biplane. Designated the Quatre B, or Fourth B (after the fact that its four creators' last names all began with B) it later received the more prosaic appellation S.A.B. (Société Avions Bernard) C1. The new fighter had a 300-hp Hispano-Suiza liquid-cooled engine and was a two-bay biplane with a corpulent but streamlined fuselage.

The S.A.B. C1 did not fly until 11 November, 1918, the day

The second of five S.A.B. C1s built by the Levasseur firm. 31989.

of the Armistice. Five examples were built by the Levasseur firm. These featured an annular radiator just behind the propeller cone. However, the mediocre performance of the S.A.B. C1, combined with the Armistice, meant that no further examples beyond the original five were produced.

Bernard S.A.B. C1 Fighter with 300-hp Hispano-Suiza
Span 9.30 m; length 6.90 m; height 2.55 m; wing area 28.90 sq. m
Empty weight 783 kg; loaded weight 1,122 kg
Maximum speed 210 km/h; climb to 2,000 m in 5.78 min.
Five built.

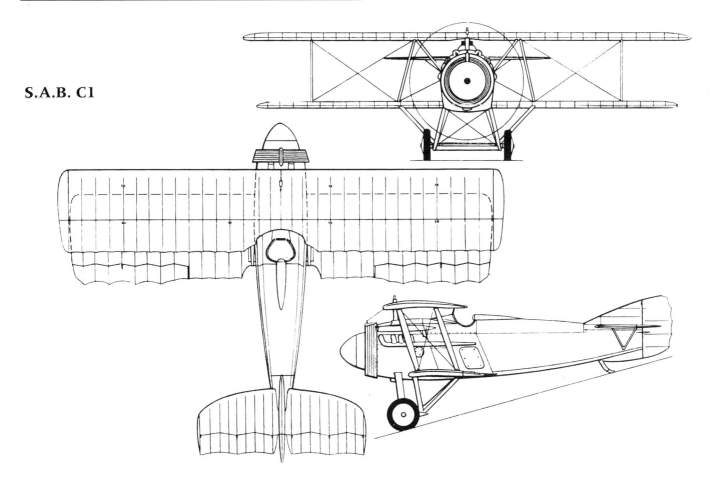

S.A.B. C1

Bille S.A.C.A.N.A. Triplane Bombers

Among the diverse firms that submitted designs to meet the BN 3/4 specifications of 1918, the Bille (S.A.C.A.N.A.) firm submitted three triplanes in early 1918. Two seem to have been identical except that one was to have six 400-hp Lorraine engines and the other to have six 400-hp Liberty engines. This dual design was probably necessitated by the difficult engine availability situation in 1918; many of the better engines were being reserved for fighters. Both aircraft had a combined wing surface area of 500 sq. m, fuel for 12 hours endurance, a maximum speed of 160 km/h at 4,000 m, and could climb to 4,000 m in one hour.

The third design was also a huge triplane with a combined wing area of 600 sq. m. Power was to have been supplied by six S.A.N.A. engines of 700-hp each. The estimated weights were equally impressive and included an empty weight of 16,032 kg, loaded weight of 25,054 kg, and a payload of 5,000 kg. Enough fuel was to be carried for 12 hours endurance. The estimated maximum speed at 4,000 m was 160 km/h, and climb to 4,000 m was to take 60 minutes. This design also remained an unbuilt project.

Blériot Aéronautique

Louis Blériot had an inauspicious start to his aviation career when, in 1908, he constructed an ornithopter that never flew. In 1904 he built a floatplane glider that crashed and sank on its first flight. The next aircraft was built by Gabriel Voisin for Blériot. It too was a floatplane but was powered by two Antoinette 25-hp engines and had ring-shaped wings. Despite numerous modifications (including fitting a single 50-hp Antoinette in place of the two smaller engines) the aircraft never flew. It was one of the earliest to be fitted with ailerons.

Blériot designed his next aircraft (the fifth) himself—a canard monoplane with swept-back wings. It was covered with paper and powered by a 24-hp Antoinette engine. It, too, proved incapable of flight. Undeterred, Blériot built still another craft. Named *Dragonfly*, it was a monoplane with tandem wings and also used the 24-hp Antoinette engine. Its longest flight was 80 meters.

The Blériot 7 was built in 1907 and was a tractor monoplane. It had a hinged rudder and elevators that could move in opposite directions for roll control. The longest flight was 180 meters.

The Blériot 8 had a lattice-like fuselage and used ailerons for control. Modifications were made by Blériot when he had the opportunity to see the Wrights fly in 1908. Partly due to these changes, Blériot was able to complete a 28-km cross-country flight.

The Blériot 9 was a monoplane with an extraordinarily slender fuselage and radiators fixed to the undercarriage. Built in 1908, it never flew.

The Blériot 10 was a biplane quite similar to Gabriel Voisin's canard designs. It apparently was unsuccessful and never flew.

Blériot returned to the monoplane layout for his next, and far more successful, design—the Blériot 11 in which he made his famous flight across the English Channel. This resulted in his being swamped with orders for his aircraft. Small numbers of the Blériot 11s were used during the first few months of the war. However, Blériot's subsequent designs, especially his heavy bombers, were far less successful. During the war Blériot produced aircraft under license for other manufacturers, particularly the SPAD firm.

Prewar Blériot Planes 1910–1914

1. Type 12—50-hp Gnome G or 35-hp or 60-hp ENV engine or a 40-hp ANV engine: two-seat trainer. Five built in 1909–1910.
2. Type 13—100-hp Gnome double Omega engine; four-seater built in December 1910. On one flight the type 13 carried ten persons.
3. Type 14—50-hp Omega Gnome monoplane. Two built in late 1910.
4. Type 15—100-hp Gnome double Omega. Project only.
5. Type 20—50-hp Gnome Omega; experimental landplane. One built January 1911.
6. Type 21—70-hp Gnome Lambda; two-seat reconnaissance machine. Twelve were built in 1911. Blériot produced a floatplane version of the Type 21 with an 80-hp Gnome Lambda in 1913.
7. Type 22—two-seat plane with an 80-hp Gnome Lambda. One built in 1911.
8. Type 23—100-hp Gnome double Omega racing aircraft. Three built in May 1911. Two participated in the Gordon Bennet Cup races.
9. Type 24—100-hp Gnome double Omega or 140-hp double Gamma, five-seater. Built September 1911.
10. Type 25—50-hp Gnome Omega, canard design. One built in 1911. The Type 26 was a triplane development.
11. Type 27—50-hp Gnome Omega or 70-hp Gnome Gamma; racing plane. Two built August 1911.
12. Type 28—30-hp Anzani 3 A2; two-seater. Built in 1911.
13. Type 29—70-hp Gnome Gamma; reconnaissance plane. 1911 project; never built.
14. Type 30—80-hp Anzani; four-seater; never built.
15. Type 32—Blériot engine 1912 project; never built.
16. Type 33—70-hp Gnome Gamma or 80-hp Gnome Lambda; two-seater with a canard configuration. One built in July 1912.
17. Type 36—80-hp Gnome Lambda; armored two-seater built in October 1912. Named *La Torpille*.
18. Type 37—80-hp Gnome Lambda or 100-hp Gnome Delta side-by-side two-seater with the engine in the rear of the fuselage. One built in 1913.
19. Type 40—80-hp Gnome 7A. One built in 1913 and presented to the French army at Buc for evaluation.
20. Type 41—80-hp Gnome Lambda; canard layout. One built in 1913.
21. Type 42—80-hp Gnome Lambda; canard design. Built in 1913.
22. Type 46—80-hp Gnome Lambda; single-seater designed in 1914. Never built.

Blériot 11

At the beginning of the First World War no fewer than eight escadrilles were using the Blériot 11—an aircraft little changed from the machine in which Blériot crossed the English Channel.

It was a shoulder-wing monoplane with a small, balanced rudder. Construction was of wood and fabric. The fuselage was assembled as a box girder of rectangular cross-section; its forward portion was covered with plywood or fabric. The wings had two ash spars with ribs made of poplar. The wing tips were curved, and roll control was via wing warping. The one-piece rudder was completely moveable, with no fixed fin, and the stabilizers had variable incidence. All versions of the Blériot 11 featured a raised, spindly undercarriage. The split-axle landing gear was made of wood reinforced by steel cable. The landing gear wheels were 1.60 meters apart.

Cutouts in the rear part of the wings were intended to improve the pilot's downward vision—poor downward vision was a major problem with other aircraft with shoulder mounted wings. Unfortunately, the observer was seated in front of the pilot and over the center of the wing; in this location, his view of the ground was very poor. Although they were initially unarmed, some Blériot 11s were later modified to carry a modest bomb load.

Blériot 11 BL 237 of BL 10 flown by pilot Zarapoff at Belfort in December 1913. B90.2389.

Blériot 11 with Anzani engine which was used as a trainer by the Imperial Russian Air Service. B86.248.

Below: Blériot 11 BL 181 at the Blériot school at Pau in October 1915. B79.3200.

Variants (Military)

Type 11 1912—50-hp Gnome engine. Reinforced wings and landing gear struts. Movable tail.

Type 11 1913—60-hp Clerget, otherwise same as Blériot 11 1912, except for the deletion of the landing gear reinforcement.

Type 11 Ecole Militaire—Powered by a 30-hp Anzani engine, this version was designed for training and was used by both civilian schools and the military aviation school at Étampes. It featured double tail skids.

Type 11 Artillerie—Blériot 11 with modifications to facilitate disassembly and assembly for ease of transport.

Type 11 Penguin—Powered by a 35-hp Anzani; used for ground instruction only.

Type 11-1 Artillerie—50-hp Gnome, single-seater with modified rudder and elevator. This version was designed with a collapsible fuselage easily disassembled for transport.

Type 11-2 Artillerie—70-hp Gnome engine, two-seater parasol with a modified undercarriage and rudder.

Type 11bis-side-by-side seating; equipped with a "pigeon" tail and oval rudder. Appeared in February 1910. Powered by 50-hp or 70-hp Gnome engines. Exported to the Netherlands, Russia, and Japan.

Type 11-2 Genie—70-hp Gnome, two-seater with modified landing gear. This variant participated in the military maneuvers in September 1912. One was later displayed at the Salon de Paris equipped with a Hotchkiss machine gun. For further details see the section on operational use.

Type 11-2 Hydro—Floatplane version of the Blériot 11-2 with a 80-hp Gnome or 80-hp Le Rhône engine. Tandem two-seater with three floats built in October 1913. Another version, with only two floats, had a rudder that was enlarged and extended below the aircraft. The floats could be replaced with wheels.

Type 11-3—A 100-hp Gnome double Omega, three-seater with a balanced elevator; this aircraft carried serial number 14. A second example of this type (serial number 26) was fitted with a 140-hp Gnome double Gamma engine. Both were entered in the military concours of 1911 and both were damaged during the trials. Further development of the Type 11-3 was abandoned because it was unstable and the landing gear was too fragile.

Type 11-Brevet Gouin—a parasol design powered by a 60-hp or 80-hp Gnome engine. It was designed by a Lieutenant Gouin in February 1914. The parasol configuration was intended to enhance the crew's downward vision. It participated in the 1914 concours securite. The rudder was split into two parts and could be used as an air brake. Twenty examples were ordered by the War Ministry on 15 October 1914. Many more were built by Blériot Aircraft Limited in England and in Italy by S.I.T.

A tandem two-seater with an 80-hp Le Rhône or a 100-hp Gnome was built in July 1914, but none was ordered.

Operational Service

Blériot 11s played a significant part in the development of the Aviation Militaire. In 1909 there were two in military service: a single-seater and a two-seater. Four additional Blériot 11s were offered to the army by the journal *Le Temps* in 1910.

Two Blériot 11s participated in the maneuvers at Picardie in September 1910. Pleased by the results of these early trials, the artillery ordered 20 Blériot 11s: 17 two-seaters and three single-seaters. BL 3, based at Pau, was formed in 1911 and was equipped with Blériot 11-2s.

The Blériot 11s were being constantly developed in response to military needs. The Commission du Genie (Commission of Engineers) at Vincennes recommended many modifications to the Blériot 11-2. Named the Blériot 11-2 Genie, the new design had cutouts in the wing roots to facilitate downward view, a modified tail wheel to improve landing characteristics, and streamlining of the upper fuselage. Another Blériot 11-2 Genie was fitted with a rifle and carried grenades. In 1910, a Blériot 11 single-seater was equipped with a 37-mm cannon with five shells that could be fired through the propeller hub. The shock of the cannon fire ruptured the motor shaft and further trails were abandoned. In 1912 a Hotchkiss machine gun was installed on a Blériot 11-2, fitted on a tripod to fire over the propeller. Lieutenant Bellenger successfully conducted trials with this weapon in December 1912.

In February 1912 four Blériot 11s were sent to Morocco to assist the army. They flew reconnaissance missions.

A modified Blériot 11 with a 50-hp Gnome engine and owned by Roland Garros was offered to the government on 12 September 1912. Flown by a Lieutenant Rose, it participated in military maneuvers on 16 September 1912. His information proved to be extremely valuable and enabled the cavalry to surprise the opposing division.

Blériot 11 at Vesoul during the September 1911 maneuvers. The pilot is Lieutenant Malherbe. B91.0101.

Blériot 11 a l'Ecole d'Avord during the war. B92.1567.

At the start of the war the following escadrilles used the standard Blériot 11 with the shoulder-mounted wing:

BL 3, formed in July 1912 under the command of Lieutenant Bellenger. The unit participated in the 1912 maneuvers and was subsequently based at Belfort. At the beginning of the war it was assigned to l'Armée d'Alsace. BL 3 was active in the actions around Mulhouse and its missions included bombing German troops near the Vosges. In August 1914 BL 3 moved to the 6th Armée sector in defense of Paris. During the Battle of the Marne BL 3 was assigned to General Foch's 9th Armée. In September BL 3 moved to the 6th Armée sector. Early in 1915 Morane-Saulnier Ls replaced the Blériot 11s.

BL 9, formed in 1912, participated in the 1913 maneuvers at Toulouse. At the beginning of the war BL 9 was assigned to the 1st Armée at Épinal. Operating a mix of Blériot 11s and M.F.7s, the unit was active in the Battles at Alsace, Sarrebourg, Mortagne, Flirey, and Woëvre. In August 1915 BL 9 re-equipped with Caudron G.4s.

BL 10, formed in 1912 and active during the 1913 maneuvers. At the beginning of the war, it was based at Belfort under the command of Capitaine Zaparoff. It was assigned to the 1st Armée. In September 1915 BL 10 was assigned to the 9th Armée and participated in the battle of the Marne and at Alsace. It re-equipped with Caudron G.3s in April 1915.

BL 18, formed in 1913 at Dijon and at the beginning of the war assigned to the 1st Armée, based at Épinal. The unit was subsequently based at Nancy, Toul, and Verdun. At the end of 1914 BL 18 experimented with night bombing of rail stations at Metz and Arnaville. In February the unit re-equipped with Caudron G.3s.

BL 30, formed in September 1914 and initially under the command of Lieutenant Illac. Later Lieutenant Van der Vaero assumed command, and BL 30 was assigned to the 6th Armée on the Aisne front. In January 1915 it re-equipped with Caudron G.3s.

The Blériot 11 Artillerie were assigned to the following escadrilles:

BLC 2, attached to the 2nd Cavalry Division and subsequently the 2nd Armée.

BLC 4, formed before the outbreak of war. It was assigned to the 4th Cavalry Division. The Blériot 11s were unsatisfactory because they were prone to engine failure, and the unit was disbanded in January 1915.

BLC 5, formed in June 1914. It was assigned to the 5th Cavalry Division and was based at Reims. A short time later it was assigned to the Corps de Cavalerie Sordert. The unit was active over Belgium and then moved to Maubeuge, where it supported the French troops at the Somme. The escadrille was also active over the Marne near de Péronne and Saint-Quentin. BLC 5 was based at Lys in October, Dunkerque in December, and Arras in January 1915. It was assigned to the 1st Cavalry Corps. It moved to Champagne in February, and later to Hauts de Meuse, Saint-Dizier, and Picardie. BLC 5 was disbanded in August 1915.

The Blériot 11s were most frequently used for reconnaissance missions. The early escadrilles were often shifted around the battlefield as situations changed. For example, from early September to 4 November 1914 BL 3 moved to seven locations along the front.

On occasion the Blériot 11s of BL 3 were used as makeshift bombers. Usually steel darts, called flechettes, were dropped in groups of 500 from boxes attached to the side of the aircraft. Bombs, usually modified 75-mm shells, were also dropped. The first aerial night bombardment of the war was undertaken by Blériot 11s from BL 18 when Captain Maz Boucher attacked German factories at Metz.

Aerial combat was rare and usually inconclusive. BL 3's first combat occurred on 4 November when a Blériot 11 was fired upon by a German aircraft armed with a carbine. Neither plane was damaged.

The Blériot 11s were soon eclipsed by the more modern M.F.11s, Caudron G.3s and Morane-Saulnier Type Ls arriving at the front. For the rest of 1914 and into early 1915, the Blériot 11 escadrilles were relegated to artillery spotting. In 1915 the Blériot 11 was recognized as being decidedly obsolete and its lack of downward visibility for the observer was recognized as being particularly troublesome. The 1st Armée commander complained of the Blériot 11's poor climb and gliding characteristics. For all these reasons, Blériot 11 escadrilles were rapidly converted to other aircraft. BL 3 converted to Morane-Saulnier L's to become MS 3 in March 1915, while BL 9, BL 10, BL 18, and BL 30 re-equipped with Caudron G.3's and G.4s during the winter of 1915.

The pilots who flew the Blériot 11 Artillerie for artillery spotting liked it because it was easy to disassemble for transport. However, the BLC units were disbanded in 1915 and their Blériots were turned over the aviation schools for use as trainers. The Blériot 11s were used at the schools at Amberieu, Chartres, Chateauroux, Istres, Crotoy, and Pau.

The Blériot 11 was the only Blériot design to see service in any significant numbers during the war. After the orders for Blériot 11s were completed the Blériot firm manufactured Caudron G.3s and G.4s under license.

Foreign Service
Argentina
A single Blériot 11 with a 50-hp Omega engine was obtained for the Escuela de Aviacion Militar in 1922.

Australia
A single Blériot 11 was used by the Australian Central Flying School in 1914. It was given serial CFS 6 and was used for ground instruction.

Blériot 11 escadrille in 1914. All the aircraft are two-seaters; serial numbers are BL 217, BL 221, and BL 222. Renaud.

Belgium
The Belgium air service obtained a single Blériot 11 at the beginning of the war when two private citizens, Jan Olieslagers and a pilot named Tyck, placed themselves and their machine at the disposal of the army. It was assigned to the 5th Escadrille. On 20 August 1914 the French sent two Blériot 11 trainers to Antwerp. These served until February 1915.

Bulgaria
Bulgaria purchased a number of Blériot 11s. They were used against the Romanian air service (also equipped with Blériot 11s) during the Second Balkan War, but they were unarmed and overmatched by Romanian Blériot 11s fitted with machine guns.

Chile
Chile acquired seven Blériot 11s in 1913. These had a variety of different engines: two of 35-hp, three of 50-hp, and two of 80-hp.

Denmark
The Danish air service purchased one Blériot 11 in 1915. It was probably used by the Haerens Flyveskole but was no longer in use by the end of the year.

Greece
A single Turkish Blériot 11 was captured in October 1912. It was used for reconnaissance flights over Thessalonica until it crashed on 4 April 1913.

Guatemala
The Escuela de Aviación has a single Bleriot 11-2 on strength in July 1914. It was destroyed by an American instructor shortly after it arrived.

Italy
The Italian Air Service used Blériot 11s early in the war, having purchased five in 1910. These were:
1. 50-hp Gnome—one two-seater and one single-seater.
2. 35-hp Gnome—one single-seater.
3. 25-hp Gnome—two single-seat trainers.

Other examples were later purchased and were supplied to the training units at Aviano and Cascina Malpensa.

The Italians used four main types of Blériot 11s, all built under license by S.I.T. (the Societa Italiana Transaerea). They were:
1. Blériot 11 Monoposto—a single-seater with a 50-hp Gnome engine. it was soon discovered to be of little use in wartime.
2. Blériot 11 "Parasol"—70-hp Gnome; this saw widespread use with the squadriglias.
3. Blériot 11 "Idro"—seaplane version of the 11 with a 90-hp Le Rhône engine, twin floats and a tail float. Only one example produced.
4. Blériot 11-2—two-seat trainer with an 80-hp Gnome engine. Beginning in April 1914 47 examples were purchased

In 1911 two Blériot 11s served alongside Nieuport 4s in Libya as part of the 1st Flottiglia di Aeroplani de Tripolia, which consisted of nine machines that undertook reconnaissance, bombing, and even leaflet-dropping operations during the Turko-Italian War. What was probably history's first wartime reconnaissance flight by a heavier-than-air machine was performed on 22 October 1911 by Capitano Carlo Piazza in a Blériot 11.

When Italy entered the First World War, Blériot 11s served with Squadriglias 1ª, 2ª, 3ª, 4ª, 13ª, and 14ª. They were assigned as follows:

1st Gruppo: 1ª, 2ª, 3ª, 13ª, and 14ª Squadriglias.
3rd Gruppo: 4ª Squadriglia.

In 1915 the 1st Gruppo was assigned to the 3rd Armata, while the 4ª Squadriglia was attached to the Venice Fortified Harbor Headquarters.

Thirty Blériot 11s were assigned to the front but had been withdrawn as being unserviceable by 1 December 1915.

Japan
Japan purchased a single Blériot 11-2 trainer in 1911. Powered by a 50-hp Gnome N 1 engine. it was delivered to the Tokorosawa army airfield in March 1911, but on 13 March it disintegrated in flight, killing the pilot and his passenger—the first men in Japan to die in an aviation accident.

New Zealand
A Blériot 11-2 became New Zealand's first military plane in 1913. Purchased by the Imperial Air Fleet Committee in Britain and arriving in New Zealand in September 1913, it was named *Britannia*. At the start of the First World War it was donated to the RFC and sent to Brooklands in late 1914.

Romania
In 1911 George Bibescu purchased four Blériot trainers for use at his flying school at the Cotorceni aerodrome. These machines were subsequently given to the War Ministry, which had re-established a flying school in April 1912. The syllabus called for students to begin training on Farmans and then advance to Blériot 11s. Four additional trainers were purchased from France in 1912. By the autumn of that year the military school had seven Blériot two-seat trainers with 80-hp engines and two single-seaters with 50-hp engines.

In 1912, six Romanian Blériot 11s equipped with machine guns established aerial superiority over the unarmed Bulgarian air force during the Second Balkan War.

Blériot 11. The presence of the mounted officers suggests that this aircraft may have been assigned to a Bleriot escadrille de cavalerie: BLC 2, BLC 4, or BLC 5. Renaud.

Blériot 11 two-seater in Egypt. Photo via Colin Owers.

The Romanian air service still had six Blériot 11s on strength in the autumn of 1915 when it entered the war. All were assigned to Grupul 3. However, it appears they were used for only a limited time in the reconnaissance role for by 1916 the unit had re-equipped with other types.

Russia

The Imperial Russian Air Service purchased and later built a large number of Blériot 11s. Twenty were purchased for the flying schools in 1910 and remained in service until 1915. Engines fitted to these planes included 25-hp Anzanis, 40-hp ENV ("Labor"), and 50-hp Gnomes.

The Blériot 11 was built under license by the Dux plant from 1911 through 1912; these planes were initially used for reconnaissance and later assigned to flying schools. They were also built under license by RBVZ and Shchetinin beginning in 1911. These were withdrawn from front-line service by 1916.

A few Blériot 11-2bis trainers with side-by-side seating were also used by the flying schools (as were a few of the Blériot 21s) from 1911 through 1912. The Russians also had a single Blériot 11-3bis, a three-seater with a 100-hp Gnome engine, a single Blériot 12, a few Blériot 21s, and a single Blériot 27.

Serbia

The first Serbian military pilots were trained in France, many at the Blériot schools. When these pilots returned to Serbia in late 1912 they took with them a Blériot 11 powered by a 50-hp Gnome (which was dubbed *Orlic* or *Eagle*) and two 70-hp Blériot 11 two-seaters. One of the Blériot 11s was supplied to the escadrille assigned to support the Serbian army near Skadar during the Second Balkan War. While flying a Blériot 11 on a reconnaissance mission, Sgt. Tomic encountered a Bulgarian machine. As neither aircraft was armed the pilots saluted each other—both were alumni at the Blériot School. On 13 July 1914 Tomic took his Blériot 11 to Valjevo. A few days later a second Blériot 11 (also from the Blériot school) was assembled. These with other machines of the air service, were assigned to the 3rd Serbian Army when the First World War began. However, by mid-August there had been significant attrition due to crashes and the Serbian air service then had only three planes—one Blériot 11 and two Deperdussins. On 17 August the lone Blériot 11 made a reconnaissance flight over the Veliki-Bosnak area. Another Blériot 11 was added to the force when the repair shop at Nish was able to assemble a second machine; it was flown by a Greek volunteer. Around this time the Serbian air service decided to buy two Blériot two-seaters (probably Blériot 11-2s or 2bis) and arm them with machine guns. The French readily agreed to the purchase in August 1914. However, when French units arrived on 24 March 1915 they brought with them M.F.11s, and it was this type that was supplied to the Serbians.

Sweden

The Swedish Aviation Company purchased five Blériot 11s. Four were assigned serials 7, 11, 13, and 17. The fifth was used as a "penguin" and did not receive a serial number.

Switzerland

The Swiss air service purchased two Blériot 11s in August 1914 from their private owners, Lieutenant Lurgin and Oskar Bider. The Blériot 11s were initially used for reconnaissance and later as trainers. The machines were given numbers 22 and 23. One was struck off charge in 1917.

Turkey

An Ottoman air service began to form in 1911 when Yüzbasi (Captain) Fesa and Üstegmen (Lieutenant) Kenan were sent to the Blériot flight school at Pau. Some Blériot 11-2s were acquired and these flew numerous reconnaissance missions during the First Balkan War, which began on 6 November 1912. By 1913 there were three 11-2 Genies with 70/80-hp Gnome engines on strength. The Turkish Army subsequently received two new Blériot 11s, one being named *Osmanli*. The air service also functioned after Turkey entered the Second Balkan War (30 June–10 August 1913).

In July 1914 the Turks received three 35-hp Blériot Penguins for use as ground trainers. Long-range flights were attempted in a Blériot 11 named *Ertugrul*. It crashed in Palestine but was recovered and repaired. After Turkey entered World War One this aircraft was sent to Canakkale on the Dardanelles for reconnaissance missions. It was retired on 22 March 1915 and is believed to have been scrapped at Constantinople.

A U.S. War Department report dated 29 October 1914 (the date that Turkey entered the war) stated that the Turkish inventory at the Yesilköy flying school included two Blériot monoplanes and three Penguins. Turkish records listed two Blériot 11-2s named *Edremit* and *Tarik Ibn Ziyad*.

The Turks tried to send two Blériot 11s to the Caucus front, but on 6 November 1914 the three-ship convoy including the vessel carrying the aircraft was intercepted by a Russian naval squadron and sunk. The two Turkish pilots were plucked from the water and became prisoners.

The surviving Blériot 11s continued in the reconnaissance role, but were soon replaced by more up-to-date German aircraft. One Blériot was reportedly shot down by a Russian warship, but this cannot be confirmed.

United Kingdom

A Blériot 11-2 was entered in the 1912 military trials. It was adopted by the RFC and given the serial 221. On 28 January 1913 Blériot 11-2 No.219 was presented to the War Office by the International Correspondence School. The aircraft remained in service until August 1914 when it was transferred to the

Central Flying School. It was destroyed four months later.

Another Blériot that participated in the trials was a Type 21. Unlike the 11-2, which had tandem seating, the 21 had side-by-side seating. It seems that this aircraft was the property of a Lt. R.A. Cammell, who operated it from the camp at Larkhill on Salisbury Plain. It was subsequently used by the air battalion, which designated it the B2. By 12 October 1912 it was in service with No.3 Squadron and was renumbered 251. It was planned to equip the aircraft with a rotary engine, and it was sent to the Royal Aircraft Factory. However, these plans were abandoned in October 1913 and the aircraft did not re-enter service.

An additional Blériot 11 was obtained in September 1912 and

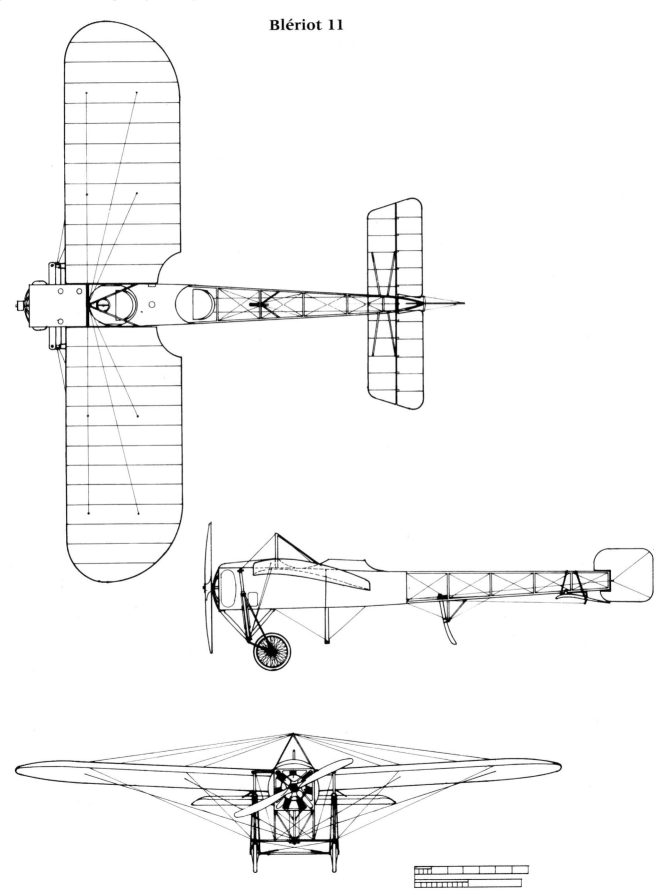

Blériot 11

Blériot parasol flown by Capitaine De Marancourt in 1915. The parasol design was intended to enhance the crew's downward view. B87.3513.

served with No.3 Squadron in 1912 and 1913. It crashed in October 1913 and was struck off charge two months later.

More of the Blériot 11s were ordered in 1913 and it appears that nine of them with 80-hp Gnomes and four with 50-hp Gnomes were in service before the war. Aircraft with 50-hp Gnomes included 293, 297, 298, (all with No.3 Squadron), and 323 (No 6 Squadron), 573, 574, 621 (impressed), and 673. Those with 70-hp or 80-hp engines included 271, 292 (No.3 Squadron), 296 (No.3 Squadron), 374 (No.3 Squadron), 374 (No.3 Squadron), 375 (No.3 Squadron), 388, 389, 473, 570, 571, 572, 606, 619 (impressed), 626 (impressed), 647, 662, 681, and 706.

Of the Blériot 11s on strength only three (serial numbers 608, 619, and 626) were sent to the continent with the BEF at the start of war. Additional Blériot 11s were ordered by the RFC. Blériot Aeronautics, which built Blériot aircraft in Britain under license, produced three Type 11s and nine were acquired from Blériot in France. These were supplied to No.3 Squadron. An additional 18 Blériot 11s were obtained later; all had 70-hp or 80-hp Gnome engines. Ten more Blériot 11s, with various engines, were obtained by the RFC during the first few months of the war.

The Blériots served with Nos.3, 6, 9, and 16 Squadrons; all were active on the Western Front. The aircraft were used for reconnaissance but their relatively light construction resulted in numerous mechanical failures and the last one (1811 of No.3 Squadron) was struck off charge on 10 June 1915.

The remaining Blériot 11s were retired to the training units and, for this purpose, additional aircraft were obtained in May 1916. Surprisingly, the Blériot 11 was not declared obsolete until the autumn of 1918.

The RNAS also ordered Blériot 11s. They were assigned serials 39, 908, 3214–3238, 3890–3893, and 3947–3952. All but the 3947–3952 had 80-hp Gnome engines, the latter having 70-hp Gnomes. They were assigned primarily to the Eastchurch station and No.1 Wing RNAS.

A single example of a Blériot Parasol (in England) was impressed at the outbreak of the war. It was flown to Farnbrough and given the serial 616. The machine was assigned to No.3 Squadron and was one of the few Blériots sent to France. It was powered by an 80-hp Gnome, but had the tail surfaces of a 50-hp machine. After 616 was damaged it was fitted with the tail of an 80-hp machine. The aircraft was used for reconnaissance and even bombing raids carrying 16 hand grenades, two bombs on the fuselage racks, and a Melinite bomb. It subsequently saw service with No.5 Squadron and was struck off charge in May 1915.

Apparently the Blériot parasol gave satisfactory service and additional examples were obtained from Blériot Aeronautics in England. These were given serials 575 (Aircraft Park at St. Omer), 576–579 (No.9 Squadron), 580–586, 616, 2861, and 2862. The last to be struck off charge was 576, which served until 1 June 1915.

After retirement from front-line squadrons, the parasols found their way to the training units, including Reserve Plane Squadrons 2, 4, and 13 and the 4th Wing at Netheravon. The parasols served with these units until late 1915.

Fifteen Blériot 11 Brevet Gouin Parasols were ordered by the RNAS. Equipped with a variety of Gnome engines (70-hp, 80-hp, or 90-hp), they served with Eastchurch Station and Nos.1 and 3 Wings.

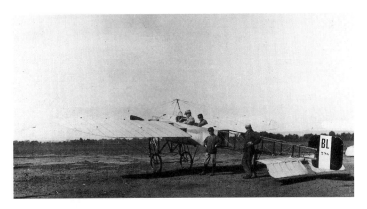

Blériot 11 serial 296. B87.4998. Courtesy Alan Durkota.

Blériot 11 1912 Reconnaissance Plane with 50-hp Gnome
Span 8.90 m; length 7.65 m; wing area 15 sq. m
Empty weight 240 kg; loaded weight 370 kg
Max speed: 90 km/h

Blériot 11 Artillerie Reconnaissance Plane with 50-hp Gnome
As above but with length increased to 8.00 m

Blériot 11 1913 Reconnaissance Plane with 60-hp Clerget
Span 8.90 m; length 7.75 m; wing area 16 sq. m
Empty weight 265 kg; loaded weight 415 kg
Max speed: 100 km/h

Blériot 11-2 (French) Two-Seat Reconnaissance Plane with 70-hp Gnome
Span 9.70 m; length 8.45 m; height 2.50 m; wing area 18 sq. m
Empty weight 335 kg; loaded weight 585 kg
Maximum speed: 100 km/h; climb to 1,000 m took 12 minutes; range 330 km
Armament: Winchester carbines, flechettes, and bombs

Blériot 11 Two-Seat Reconnaissance Plane with 60-hp Gnome
Span 8.94 m, length 7.12 m, height 3.10 m, wing area 15 sq. m
Empty weight 280 kg; loaded weight 450 kg
Maximum speed: 110 km/h, climb to 1,000 m took 7 minutes; range 330 km
Armament: Winchester carbines, flechettes, and bombs

Blériot 11—Artillerie Two-Seat Reconnaissance Plane with 70-hp Gnome
Span 8.90 m, length 7.80 m, height 2.10 m, wing area 15 sq. m
Empty weight 280 kg; loaded weight 400 kg
Maximum speed 90 km/h
Armament: Winchester carbines, flechettes, and bombs

Blériot 11—Brevet Gouin (Parasol) Two-Seat Reconnaissance Plane with 60-hp Gnome
Span 9.20 m, length 7.80 m, wing area 18 sq. m
Empty weight 310 kg; loaded weight 420 kg
Maximum speed: 115 km/hr
Armament: Winchester carbines, flechettes, and bombs
100 built

Blériot 11bis Two-Seat Reconnaissance Plane or Trainer with 50 or 70-hp Gnome
Span 11.00 m: length 8.50 m; wing area 25 sq. m
Empty weight 350 kg

Blériot 11-2 Hydro Seaplane with Three Floats and 80-hp Gnome or Le Rhône (Specifications for 11-2 Hydro with Two Floats in Parenthesis)
Span 11.10 m (11.05 m); length 8.875 m (9.00 m); wing area 19.00 sq. m (21 sq. m)
Empty weight 360 kg (500 kg); loaded weight 560 kg (750 kg)

Blériot 11-2 Type Genie Reconnaissance Plane with 70-hp Gnome
Span 9.70 m; length 8.30 m; wing area 18 sq. m
Empty weight 320 kg; loaded weight 550 kg
Maximum speed: 120 km/h

Blériot 11-3 Type Reconnaissance Plane with 100-hp Gnome Double Gamma
Span 11.70 m; length 8.70 m; wing area 25 sq. m
Empty weight 475 kg; loaded weight 975 kg
Max speed: 100 km/h
Approximately 700 Blériot 11s were built, plus an additional 100 Blériot Gouins.

S.I.T.-built Blériot 11 Monoposto Single-Seater with 50-hp Gnome
Span 8.90 m; length 7.80 m; height 2.10 m; wing area 15 sq. m
Empty weight 280 kg; loaded weight 400 kg
Maximum speed: 90km/h

S.I.T.-built Blériot 11 Parasol with 70-hp Gnome
Span 9.20 m; length 7.80 m; height 2.95 m; wing area 18 sq. m
Empty weight 310 kg; loaded weight 480 kg
Maximum speed: 110 km/h

S.I.T.-built Blériot 11 Idro Floatplane with 90-hp Le Rhône
Span 11.05 m; length 9.00 m; height 3.0 m; wing area 24 sq. m
Empty weight 500 kg; loaded weight 740 kg
Maximum speed: 95 km/h
One built

S.I.T.-built Blériot 11-2 (Italian) Two-Seat Reconnaissance Plane with 70-hp Gnome
Span 10.30 m, length 8.40 m, height 2.45 m, wing area 20.33 sq. m
Empty weight 345 kg; loaded weight 585 kg
Maximum speed: 95 km/h; endurance 3 hr 30 min
Armament Winchester carbines for the crew, flechettes, and bombs
47 built

Blériot 11bis Two-Seat Reconnaissance Plane with 50-hp Gnome built in Russia by RBVZ, Shchetinin, and Dux
Span 8.9 m, length 7.75 m, wing area 14.5 sq. m
Empty weight 240 kg; loaded weight 370 kg
Maximum speed: 95 km/h

Blériot 11 Two-Seat Reconnaissance Plane with 50-hp Gnome built in Russia by Dux
Span 8.9 m, length 7.20 m, wing area 20.9 sq. m
Empty weight 295 kg; loaded weight 440 kg
Maximum speed: 90 km/h

Blériot 36 and *La Vache*

In 1912 Blériot produced the Type 36 to meet the specification for a two-seat, armored long-range reconnaissance machine. It was powered by an 80-hp Lambda engine but was not adopted because the armor restricted the performance.

In 1913 Cmdt. Lucas-Gerardville ordered a modified Blériot 36 powered by a considerably more powerful 160-hp Double Lambda; it was probably hoped that the new engine would provide a more acceptable performance. Originally intended to destroy enemy airships, the new airplane first flew in July 1914. It was assigned to the famous French aviator Jules Vedrines, who was serving with N 3 at that time. It appears that Vedrines was the only pilot who flew the aircraft, and that he performed a number of reconnaissance missions, usually with volunteer observers. Vedrines named the craft *La Vache* (*The Cow*) and had the name painted in large red letters on the fuselage. The mediocre performance of the type despite its powerful engine and cooling problems ensured that only the single example was built.

> **Blériot *La Vache* Two-Seat Armored Reconnaissance Plane with 160-hp Double Lambda**
> Wing span 10.22 m; length 8.33 m
> One built

Blériot 36. This lightly-armored airplane could withstand ground-fire from an altitude as low as 500 meters. The crew of two was seated side by side. B87.1670.

Blériot with 160-hp Gnome Double Lambda of Jules Vedrines named *LA VACHE* (*The Cow*). This version was also armored. It carried a crew of two in tandem, and its landing gear was extensively modified to handle the greater weight. Renaud.

Blériot 39 Armored Monoplane

In 1913 General Bernard formulated a requirement for two types of armored aircraft: a two-seater with a maximum speed of 100 km/h to be used for longer-range reconnaissance and a single-seater for high-speed (120 km/h) reconnaissance. Blériot, along with Breguet, Dorand, Ponnier, Deperdussin, Clement-Bayard, Morane-Saulnier, and Voisin, built a machine to meet these specifications. While the Blériot 36 had been intended to fill the long-range reconnaissance role, the Blériot 39 was a single-seater designed to meet the latter specification. It had a shorter fuselage than the standard Type 11 but had a larger wing with increased span, presumably to allow it to carry the heavy armor. The engine was an 80-hp Gnome. However, the armored aircraft program was a complete failure because the aircraft engines of the time were not powerful enough to overcome the weight of the armor. The Blériot 39 was not used operationally by the Aviation Militaire.

> **Blériot 39 Single-Seat Armored Reconnaissance Aircraft with 80-hp Gnome**
> Wing span 10.10 m, length 6.15 m, wing area 19.00 sq. m
> Empty weight 440 kg, loaded weight 615 kg
> Maximum speed 120 km/h, climb rate 100 meters/minute
> One built

The Blériot 39 single-seat armored reconnaissance aircraft showing its armored cowling. B87.1770.

Blériot 43

The Blériot 43 was a derivative of the Type 36. The main differences were that the Type 43 had a monocoque fuselage and an 80-hp Gnome 7A engine. The crew was seated in tandem with the observer located behind the wing; a large porthole provided him with an excellent view of the ground. A speaking tube facilitated communication between pilot and observer. The fuselage was constructed of wood and cloth; it does not appear that it was armored. The Blériot 43 was

The Blériot 43 had a monocoque fuselage and an 80-hp Gnome 7A engine. Reairsche.

Blériot 43 Two-Seat Reconnaissance Aircraft with 80-hp Gnome 7A
Span 10.10 m, length 6.12 m; height 3.10 m, wing area 19.3 sq. m
Empty weight 350 kg; loaded weight 625 kg
Maximum speed 120 km/h
One built

probably intended to meet the same long-range reconnaissance specification as the Type 36, but without the excessive weight of the armor. However, its performance was not significantly

Blériot 44

Built in 1914, the Blériot 44 was a single-seat artillery cooperation aircraft powered by an 80-hp Gnome 7A engine. The pilot's position was behind the wing to improve his field of vision. The engine and fuel tank were situated in the nose to counteract the weight of the pilot who was in the rear. While the new arrangement provided a better view for the pilot than

Blériot 44 Single-Seat Artillery Cooperation Aircraft with 80-hp Gnome 7A
No specifications known
One built

the standard Blériot 11, the Type 44 was not selected by the military, and only the single prototype was built.

The cockpit of the Blériot 44 was located behind the wing to facilitate the pilot's downward view. #411.

Blériot 45

The Blériot 45 was an unorthodox single-seat aircraft intended for reconnaissance. The attempt to solve the problem of pilot vision took an approach directly opposite that tried in the Type 44 by putting the pilot in the nose of the plane well ahead of the wing. The engine was located within the fuselage, the propeller shaft passing between the pilot's legs. The propeller was placed at the extreme nose. The six cylinders of the engine were accommodated in cutouts in the rear of the fuselage. the tail and rudder were identical to that used on the Blériot 44. The Blériot 45 was not selected for use.

Blériot 45 Single-Seat Artillery Cooperation Aircraft with 80-hp Gnome 7A
No specifications known
One built

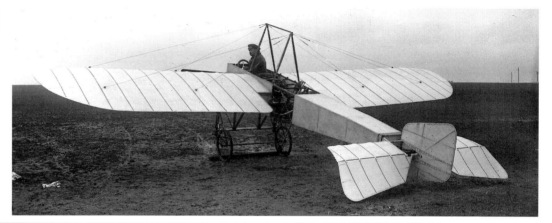

The Blériot 45's cockpit was located in the nose while the engine was located mid-fuselage. #12425.

Blériot Twin-Engine Aircraft (Blériot 53)

Built for reconnaissance missions in 1915, this unusual plane had two 80-hp Le Rhône engines in nacelles mounted between the wings. The fuel tanks were located directly behind the engines. The forward fuselage was conventional, while the rear portion appears to have been left partially without fabric covering. There was a rectangular rudder and biplane horizontal stabilizers. The three-bay wings appear to have been of equal span. It is believed that tests were later conducted with two 100-hp Anzani engines. The type, which may have been designated the Blériot 53, was not selected for production.

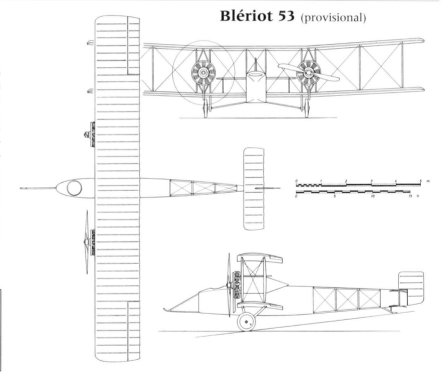

Blériot 53 (provisional)

Blériot Army Cooperation Aircraft with Two 80-hp Le Rhône engines
Span 13.50 m; length 9 m; wing area 27 sq. m
Maximum speed 95 km/h
One built

Blériot 53. It is believed that the twin-engine aircraft was intended to meet the same specification as the Caudron G.4. The twin engines were intended to improve the aircraft's performance compared to the Caudron G.3 and M.F.11, which could not be flown in poor weather. Louis Blériot.

Blériot Monoplane

In 1915 Blériot proposed a military monoplane with a large, shoulder mounted wing. However, Bleriot proposed having an additional set of wings of shorter span and much reduced chord; one wing would be mounted above the fuselage, the other suspended beneath the fuselage. This gave the appearance of a triplane, but Bleriot termed the new design as a monoplane with auxiliary surfaces. The fuselage terminated in a conventional rudder with no fixed fin. As with the preceding Type 53, there were biplane stabilizers. However the ends of the stabilizers were covered by large fins. Blériot's proposed design was given brevet 79988 in November 1915, but apparently it was never built.

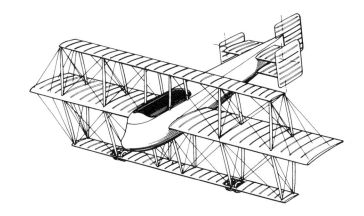

Blériot Twin-Fuselage Monoplane

In 1916 Blériot proposed to the Ministère de la Guerre a project for a twin-fuselage monoplane. This study would later form the basis for the Blériot 77 (see below). The design did not pass beyond the wind tunnel model stage. Based on the model it is known that this was to have been a pusher with engine behind the crew. The crew of two was seated side by side in the central nacelle. Estimated wing area was 40 sq. m.

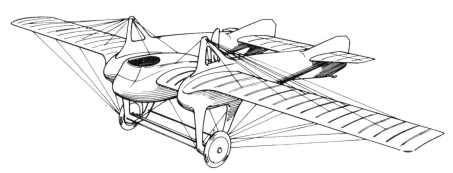

Blériot Four-Engine Bomber

The next Blériot design was an enlarged version of the Type 53 army cooperation plane. The four-engine bomber carried a crew of four and its engines, mounted between the four-bay upper and lower wings, were 120-hp Anzani 10A4s. The aircraft flew at Buc in 1915 or 1916, and it may have been intended to enter it in the 1916 concours puissant. In any event, the Blériot bomber was not selected for service use.

Blériot Heavy Bomber with Four 120-hp Anzani 10A4 Engines
No specifications known
One built

Below: Blériot four-engine bomber. The men in this view emphasize the large size of the aircraft. Louis Blériot.

Below: Blériot four-engine bomber. This aircraft may have been a scale-up of the preceding Blériot 53. Louis Blériot.

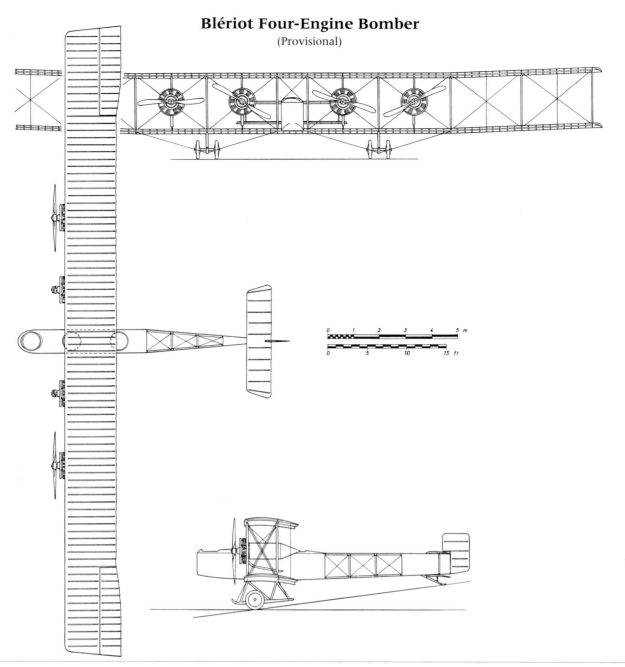

Blériot Four-Engine Bomber
(Provisional)

Blériot 65

The Blériot 65 was a proposal for a two-seat biplane fighter. The aircraft was to have been a biplane powered by a 200-hp Anzani engine Armament was to have consisted of two machine guns; one mounted fore the other aft. Historian Jean Devaux has suggested that this plane would have been in the same category as the SPAD SA.3 two-seat fighter and the Ponnier P.I. It is not known if the aircraft was ever built.

Blériot Two-Seat Fighter with 200-hp Anzani (all data provisional)
Length 7.845 m; height 3 m; wing area 32 sq. m
Loaded weight 1100 kg
Armament: two machine guns

Blériot 67

Many companies responded to the request for a well-protected strategic bomber capable of raiding German cities. Along with seven other manufactures, Blériot entered such an aircraft in the 1916 concours puissant.

The competition emphasized that the aircraft should not only have adequate range and bomb load, but it also had to be able to defend itself against enemy fighters. It had been discovered in 1915 that the Caudron, Farman, and Voisin bombers carried inadequate defensive armament and were suffering heavy losses to German fighters.

Louis Blériot had been interested in producing a heavy bomber for the concours puissant and in February 1916 had created three designs for such a machine, two were quadraplanes, the third was a biplane. All featured two engines above the fuselage and two below. They were closely grouped near the centerline where, should an engine fail, the effects of asymmetric thrust would be less pronounced. None of these projects advanced beyond the drawing board, but the unique engine arrangement was used in Blériot's next heavy bomber, the type 67.

The Blériot 67 was designed with the aid of an engineer named Touillet. In addition to being heavily armed with

Side view of the Blériot 67 showing reinforcement of the trailing edges of the upper and lower wings. MA36379.

The Blériot 67 was powered by four 100-hp Gnome 9B engines. B88.454.

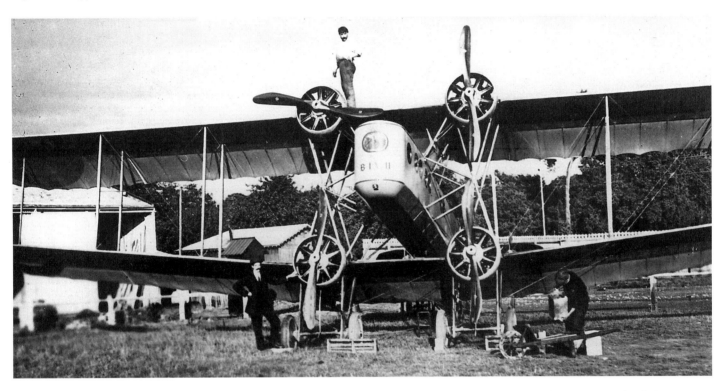
Blériot 67 at Buc. The engines were grouped near the centerline so asymmetric thrust effects would be reduced should an engine fail. B88.452.

Blériot Bomber Projects

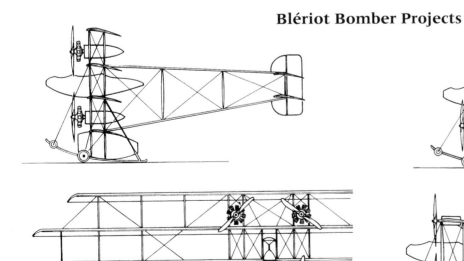
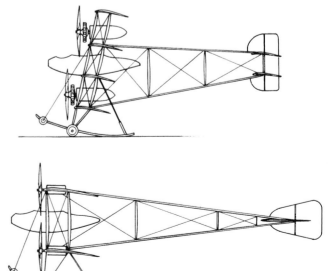

machine guns fore and aft, the Type 67 had to fill the other requirements, including a range of 600 km and a speed of 140 km/h at 2,000 meters. The narrow fuselage was positioned between the two wings. The engines were closely grouped around the centerline, with one each on the port upper, starboard upper, port lower, and starboard lower wings. Unfortunately, the large number of struts and wires needed to brace the fuselage and wings created excessive drag, and this eroded the aircraft's performance. The engines were four Gnome 9Bs of 100-hp. The undercarriage consisted of two pairs of wheels located beneath each of the lower engines. There was a biplane tail with three fins and rudders. A crew of three was carried.

Like most of the other entries, the Blériot 67 was unable to meet the specifications of the competition and no production orders were placed. Only one was built, and it crashed in September 1916 while undergoing flight testing at Buc.

Blériot 67 Heavy Bomber with Four 100-hp Gnome 9B Engines
Wing span 19.40 m; length 11.80 m; wing area 89 sq m
Empty weight 1,800 kg; loaded weight 3,500 kg
One built

Blériot 67

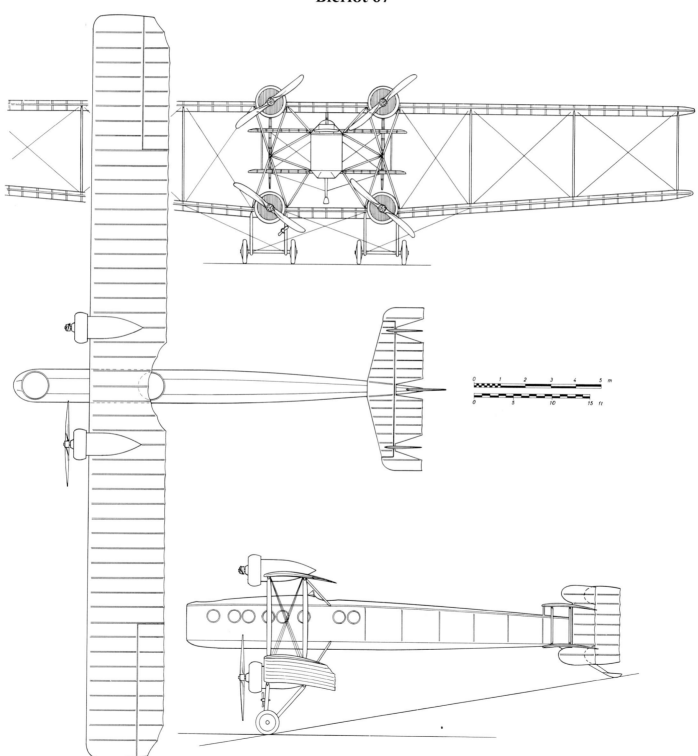

Blériot 71

The failure of the Blériot 67 to fulfill the requirements of the 1916 concours forced Blériot to redesign his strategic bomber. The resulting aircraft was intended to meet the BN 3 specification for a heavy bomber. It also was designed by engineer Touillet. Initially, it was intended to place the fuselage on the lower wing, but at some point before construction the type was redesigned so that the fuselage was suspended between the upper and lower wings. The craft was similar to the Blériot 67, but much larger and powered by four 220-hp Hispano-Suiza 8B engines generating considerably more power. It featured a similar tail with a biplane stabilizer. The original design had three sets of rudders but this was changed to two sets. Only one aircraft was built. The Blériot 71 was damaged on 15 May 1918 at Villacoublay when a Breguet 14 B1 sustained an engine failure and while landing dead stick, caused the Blériot 71 to crash into a ditch in the center of the airfield. The fuselage was fractured. It appears that the plane was not rebuilt, probably because it had not been selected by the Aviation Militaire for series production. The Farman F.50 and Caudron C.23 were instead chosen for the BN 3 specification.

Blériot 71 Heavy Bomber with Four 220-hp Hispano-Suiza 8B Engines
Wing span 26.30; length 14 m; height 6 m; wing area 140 sq. m
Empty weight 3,200 kg; gross weight 6,530 kg
Maximum speed at ground level: 140 km/h
Endurance 6 hours 40 minutes
One built

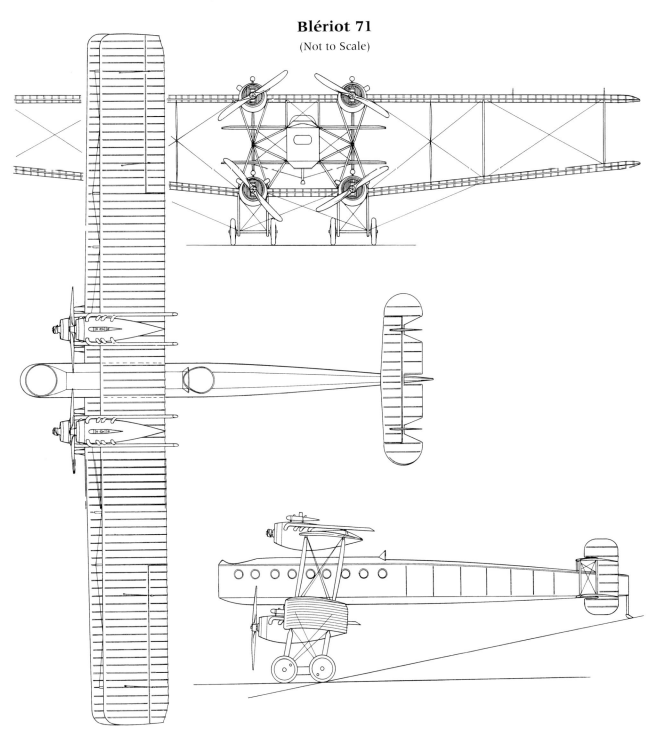

Blériot 71
(Not to Scale)

Top: The Blériot 71 was similar to the Blériot 67, but much larger and was powered by four 220-hp Hispano-Suiza 8B engines generating considerably more power than the four 100-hp engines of the Blériot 67. Louis Blériot.

Above: Only one Blériot 71 was built. Louis Blériot.

Left: The Blériot 71 was damaged in an accident on 15 May 1918 at Villacoublay when a Breguet 14 B1 suffered an engine failure and crashed into the Blériot 71. Louis Blériot.

The Blériot 71. Louis Blériot.

Below: The ground crew give an impression of the size of the Blériot 71. MA1784.

Blériot 73/74/75/76

The failure of the French industry to produce an adequate bomber had forced the Aviation Militaire to have Italian Caproni bombers built under license by the R.E.P. firm. In the meantime, the French manufacturers started work on a whole set of new designs to produce an acceptable night bomber of the BN 3 (Bombardment Nuit 3, night bomber with a crew of three) category. The follow up to the Blériot 71 day bomber was the Blériot 73, designed from the outset as a night bomber.

The Blériot 73, again designed by Touillet, was a large biplane with four 300-hp Hispano-Suiza 8Fb engines in a similar layout

This view of the Blériot 73 shows the four-wheel bogie landing gear and the upswept tail. Louis Blériot.

The Blériot 73 was powered by four 300-hp Hispano-Suiza 8Fb engines in a layout similar to that used on the Blériot 67. Louis Blériot.

Blériot 73

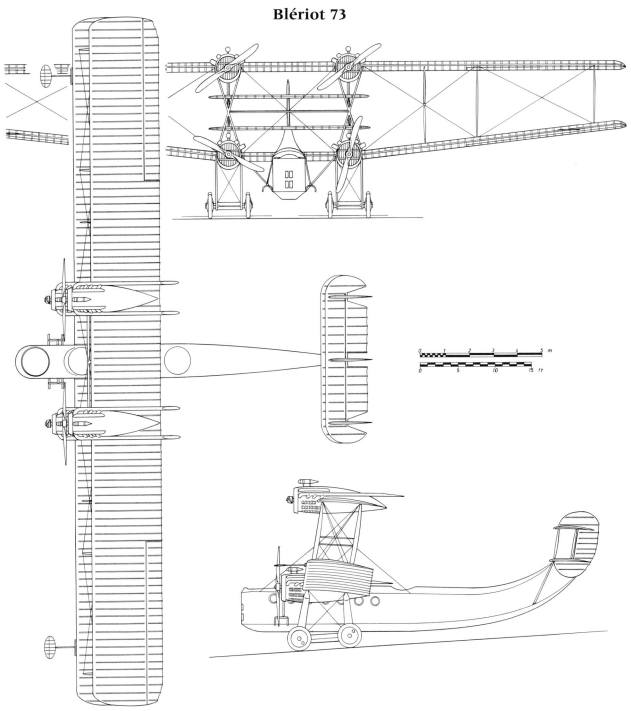

Blériot 73. CC.

Blériot 74

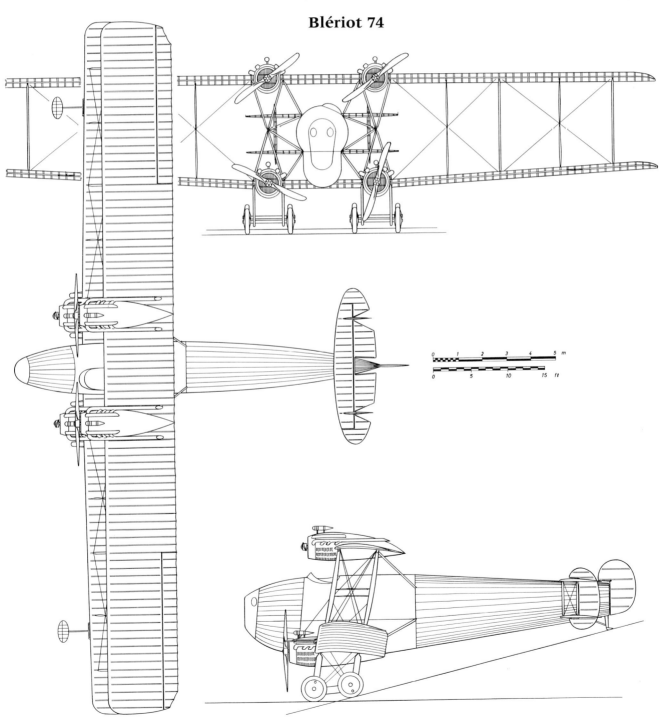

The Blériot 74. B82.921.

to that of the Blériot 67. The bottom wing was attached directly to the top of the fuselage, which had an upswept tail and was constructed like that of a flying boat hull. This aircraft also had a biplane tail with three fins and rudders. The undercarriage consisted of a four-wheel bogie under each of the lower engines; this arrangement allowed for efficient weight distribution. Another feature of the undercarriage was a nose wheel to prevent the aircraft from nosing over (a common problem when landing at night). Aileron balances were of the Constatin vane style and greatly eased the work of the pilot in controlling the aircraft. Bombs were carried in racks mounted on the lower wing and fuselage sides.

Flight testing was conducted at Buc in July 1918. On the first flight the test pilot (named Poullet) was killed when the Blériot 73 crashed on landing when a gust of wind blew the aircraft off its landing path. Testing was completed in January of 1920; but the war's end ensured further development was abandoned.

The Blériot 74 was initially planned as yet another attempt to meet the requirements for a night bomber. Other aircraft produced to meet this requirement included the Voisin 12, Letord 9, Farman F.50 and F.60, Caudron C.23, Breguet 16 Bn2, S.I.A. Bn2, and the Sikorsky Bn2.

Like its predecessors, the Type 74 was powered by four 300-hp Hispano Suiza 8Fb engines closely grouped around the fuselage to reduce asymmetric thrust problems with one engine out. The fuselage was mounted on the lower wing. The landing gear of the types 71 and 73 was retained. The bombardier was to be seated in the bottom of the aircraft with a trapdoor underneath the aircraft to permit maximum visibility. A machine gun was also to be located in this position. A machine gun turret was to be located beneath the rear of the aircraft. There was extensive use of duralumin in the support structure which reduced the weight of the aircraft. The end of the war

Blériot 73. Louis Blériot.

Blériot 73 Heavy Bomber with Four 300-hp Hispano-Suiza 8Fb Engines
Span 28 m; length 14.50 m; height 6.15 m; wing area 148 sq. m
Empty weight 3,200 kg; loaded weight 6,880 kg
Maximum speed: 130 km/h, endurance 6 hours 40 minutes
One built

Blériot 74 Heavy Bomber/Airliner with Four 300-hp Hispano-Suiza 8Fb Engines
Span 27 m; length 15.40 m; height 6.40 m; wing area 148.6 sq. m
Empty weight 3,800 kg, payload 2,250 kg, loaded weight 7,550 kg, including 1,600 liters of fuel and 120 liters of oil
Maximum speed: 140 km/h, endurance 6 hours 20 minutes
One built

Blériot 75 Airliner with Four 300-hp Hispano-Suiza 8Fb Engines
Span 27 m,, length 18.3 m, height 6.4 m, wing area 144 sq. m
Empty weight 3,800 kg, loaded weight 7,500 kg
Maximum speed: 155 km/h, ceiling 4,000 m
One built

Blériot 76 Heavy Bomber Project with Four 300-hp Hispano-Suiza 8Fb Engines
Estimated wing span 27.0 m, length 20.85 m, height 6.75 m
Project only

Blériot 73. Bombs were to be carried in racks mounted the lower wing and fuselage sides. Louis Blériot.

resulted in the Blériot firm's decision to redesign it as an airliner. Andre Herbemont took the wings of the proposed Blériot 74 and designed an entirely new fuselage partly of biconvex shape and accommodating 56 passengers.

During testing at Villacoublay on 22 January 1920 oscillations ruptured the tail and the Blériot 74 crashed, killing its pilot, Armand Berthelot, an ace with 11 victories.

The Type 75, also called the Aerobus, was the same plane as the Blériot 74 but had a new wing that had a pronounced dihedral. The fuselage was lengthened by 3 meters over the Type 74 and the center vertical tail surface was enlarged. Tested by Jean Casale, it had a good performance. However, no orders were forthcoming from the airlines because of the availability of far cheaper war surplus planes.

The Blériot 76, although under study in 1920, warrants mention here because it was designed to meet the wartime BN 4 specification and was a development of the Blériot 75. The bomber had the same engine arrangement as the Types 67, 71, 73, and 74, but the lower engines were located farther outboard than the upper engines. The Type 76 had slightly swept-back wings and a much thicker monocoque fuselage. The engines were the same four 300-hp Hispano Suiza 8Fbs. The undercarriage consisted of a pair of massive wheels; apparently Herbemont had planned later to modify the undercarriage so that it would be retractable. As with so many of the BN 3/4 bomber designs, the Blériot 76 design was not selected for production and remained a project only.

Blériot 77

Although this aircraft was developed in the early 1920s it warrants mention here because it stems from a 1916 design by Louis Bechereau. Proposed by Bechereau to the Ministry of War in 1916, it was an unusual monoplane with two separate fuselages. It was only at the end of the war, however, that detailed study of the design (under the Type 77 Bn4 designation) was begun. The graceful fuselage was similar to the later Blériot 125. The monoplane wing had no external bracing and the internal structure was to be made of wood and lined with cloth. In 1921 a model of the Blériot 77 was tested by the Laboratory Eiffel, but the aircraft did not progress beyond the project stage.

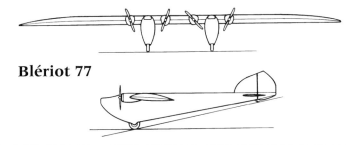

Blériot 77

Blériot 7L "High Seas" Flying Boat

The "high seas" specification issued by the Aviation Maritime in 1918 was intended to provide the navy with a flying boat comparable to the British Felixstowe series. The specification called for an aircraft with a crew of four, a T.S.F. wireless, a 75-mm cannon with 35 rounds, and two machine guns. A large number of firms submitted designs.

The Blériot 7L was to have been a biplane with a single-step hull powered by four 220-hp Hispano-Suiza engines. The two wings had a combined surface area of 135 sq m. Estimated performance data include a payload of 2,000 kg; maximum speed of 150 km/h at 2,000 meters, and ability to climb to 2,000 m in 30 minutes. The aircraft was still being designed in April 1918. It does not appear that it was ever built, remaining a project only.

Société Anonyme des Établissements Borel

Gabriel Borel opened a flying school at Mourmelon in 1910 and produced a series of aircraft before 1914. Numbered 1 through 19, they were for the most part single-engined monoplanes intended for private use. However, in 1912 Borel turned his attention to military aviation and developed a fighter for use against airships. It was a single-engine pusher monoplane with a 80-hp Gnome-Rhône engine. The forward nacelle had side-by-side seating for the two crew members, and two windows were placed in the nose to improve the downward view. The fuel tank was located behind the cockpit and the engine behind the fuel tank. The rudder and horizontal stabilizer were supported by twin booms made of wood. The landing gear was exceptionally tall and the lower portion of the twin booms attached directly to the undercarriage skids. The aircraft was test-flown by a pilot named Daucourt in 1913, and although it displayed excellent flight characteristics, it was not selected for service with the Aviation Militaire.

Borel's factory at Mourmelon was closed shortly after the war began because its workers were drafted into the army. However, with the realization that the war would not be over quickly, the army released workers with aviation skills and Établissements Borel was able to reopen in 1915. Although Borel eventually opened four factories, most of the planes they produced were license-built versions of Caudron G.3s and G.4s, Nieuports, and SPAD fighters. In 1916 Paul Boccaccio designed a bomber with two coupled Hispano-Suiza engines which remained unbuilt. However, a limited number of Borel's designs did see operational service. In 1918 the Établissements Borel became SGCIM (Société Generale des Constructions Industrielles et Mecaniques). Shortly after the war ended SGCIM ceased aircraft production.

Borel Floatplane

The Borel floatplane was the only prewar Borel design to see operational service. It was a small monoplane with twin floats attached to the lower fuselage and a third float on a fin that extended beneath the rear fuselage. Control was by wing warping, and there was a large angular rudder. The engine was a 70-hp Gnome rotary, and a crew of two was carried. The aircraft were built in 1911 and most were used by the Aviation Maritime's training centers at La Vidamee and Buc. They also participated in maneuvers in western France in 1912. While it does not appear that the aircraft were operational with the French navy at the outbreak of the war, some may still have been in use as trainers.

Foreign Service
Brazil
Brazil purchased a single Borel floatplane in 1917. It was given Brazilian naval air service serial number 4 and remained operational until 1919.

United Kingdom
The RNAS acquired eight examples in 1912; these were given serial numbers 37, 48, 83, 84, 85, 86, 87, and 88. Some were in service with the RNAS when the First World War began. These were:
37—Acquired in 1913, Isle of Grain on 11 March 1913; crashed

Borel Floatplane with 70-hp Gnome Rotary
Span 9.6 m, length 6.7 m, height 2.5 m
Max speed: 90 km/h

18 March 1913.
48—From March to July 1913 aboard cruiser *HMS Hermes*. Wrecked by bad weather in July. Repaired and used for training. Fitted with a wheel undercarriage and stationed at the Isle of Grain. SOC December 1914.
83—On strength at Calshot in late July 1913. New wings fitted early December 1913. Remained at Calshot until September 1913 by which time it was converted to a landplane and was based at Eastchurch. SOC December 1914.
84—Based at the Isle of Grain in August 1914.
85—Based at Cromarty in July 1913. To Fort George in November 1913. In September it was at Eastchurch and had been converted to a landplane. It was SOC in December 1914.
86—Based at Leven in July 1913. Flew in the naval maneuvers of September 1913. It subsequently was stationed at Port Laing. In February 1914 it was based at Dundee and was at Leven on 25 March. The next month it was returned to Dundee. In September it was at Eastchurch and had been converted to a landplane. SOC in December 1914.
87—Delivered to Isle of Grain 25 August 1913; deleted June 1914.
88—Delivered to the Isle of Grain in August 1913. Crashed in late 1913 and was SOC.

The Borel floatplane was built in 1911, and most were used by the French Aviation Maritime's training centers at La Vidamee and Buc. B87.1760.

The Borel floatplane which flew the Paris to Deauville course on 24 August 1914. B87.1705.

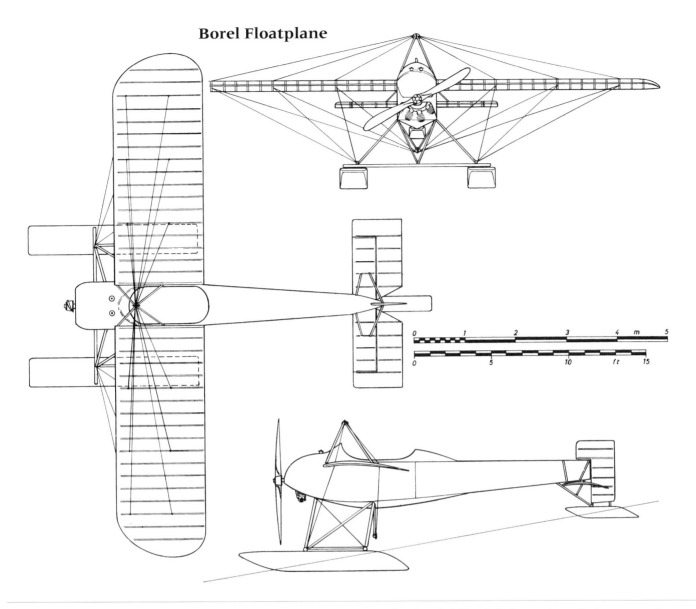

Borel Floatplane

Borel-Odier B.O.2

The Eiffel Laboratories made a number of experiments using models of floatplanes in 1915 and 1916. One result was a large twin-float aircraft intended for use as a torpedo carrier. The aircraft was designed by the Établissements Borel and built by Antoine Odier; it incorporated much of the knowledge accumulated by the Eiffel studies.

The Borel-Odier B.O.2 was intended to meet the naval requirement for a patrol plane and torpedo bomber. This called for a maximum speed of 140 km/h, which was slightly above the Borel-Odier's maximum. The Coutant flying boat, G.L.40 HB 2, Tellier T.4, and Donnet-Denhaut D.D.10 were also designed to meet this specification and all were selected for production.

Borel's aircraft was a large three-bay biplane with the upper wing longer than the lower. The engines used on the production

Borel-Odier B.O.2 in postal service. Renaud.

machines where 220-hp Hispano-Suiza 8Bs. The aircraft had a triple tail unit that may have been intended to make it easier to maneuver on the water. The torpedo was housed in a bay beneath the wing close to the wing/fuselage joint. Armament was two 7.7-mm Lewis machine guns on flexible mounts and one 650-kg torpedo.

The aircraft was tested at St. Raphaël but was destroyed during its first flight in August 1916. It appears that it had two 160-hp Salmson engines but subsequent machines had the Hispano-Suizas mentioned above. Tests with another Borel-Odier torpedo plane subsequently showed that the aircraft could carry a useful load of more than 1,200 kg in rough sea conditions. It could take off after a run of about 18 meters and could fly for 114 minutes with one engine shut down.

The performance was apparently enough to satisfy the navy and 90–92 aircraft were ordered. Deliveries did not begin until August 1917 and only a small number were in service at the time of the Armistice. Some B.O.2s were used at P.C. Nice for postal service between France and Corsica. Because of poor performance and numerous accidents few B.O.2s saw service. Postwar it was used to carry mail. A transport version designed to carry ten passengers was developed in 1919; it was destroyed during testing.

Borel-Odier B.O.2 Torpedo Floatplane with Two 220-hp Hispano-Suiza 8Ba Engines

Span 20.0 m; length 11.23 m; height 3.93 m; wing area 80 sq. m
Empty weight 1,200 kg, loaded weight 2,400 kg
Maximum speed at sea level: 124 km/h; range 520 km
Armament: two 7.7 mm Lewis guns on flexible mounts and one 650 kg torpedo.
90–92 built

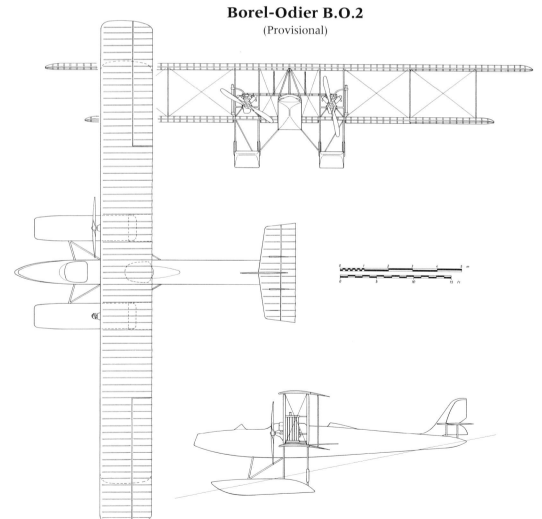

Above: Borel-Odier B.O.2 taxiing. The aircraft had a triple tail unit which may have been intended to make the aircraft easier to maneuver while taxiing. Reairche collection.

Above left: Borel-Odier B.O.2. The engines used on the production machines were 220-hp Hispano-Suiza 8Bs. Reairche collection.

Borel C1

The Borel C1 was designed to meet the C1 specification of 1918, which called for a single-seat fighter to be armed with two machine guns (either 7.7-mm or 11-mm) synchronized to fire though the propeller. An alternative armament of a machine gun and a 37-mm cannon was acceptable. Two additional machine guns (which were to be mounted in a turret and manned by a second crewman) were considered desirable. This later version was called a single-seat protégé. A photographic reconnaissance version was planned.

Two aircraft were designed to meet this specification: the SPAD 20 and the Borel C1 *Flandre.* The Borel design was to have been powered by a 300-hp Hispano-Suiza 8F. The aircraft was under development at the time of the Armistice. Shortly thereafter a decision was made to delay work on the type until the Borel-Boccaccio C2 (Type 3000) had been flight tested. However, it seems that development of the Borel C1 was never completed.

Borel-Boccaccio Type 3000 (C2)

The C2 specification of 1918 called for a two-seat fighter carrying a single synchronized machine gun or cannon, two machine guns in a turret behind the pilot, and a machine gun fitted to fire downward. The payload as to be 375 kg, maximum ceiling was to be 8,000 meters with a service ceiling of 5,000 meters, and a maximum speed of 220 km/h was required. There were many aircraft submitted to meet this specification, three of which used the 300-hp Hispano-Suiza 8Fb engine: the Hanriot-Dupont HD.5, the B.A.J. C1, and the Borel-Boccaccio C2.

Work on the Borel design was completed by March 1918, wind-tunnel models having been used to help develop the basic shape. The aircraft was not completed until 1919.

The Borel-Bocaccio Type 3000 was a two-bay biplane with dihedral on the upper wing only. The interior structure included a latticework arrangement for strength. The leading edges of the wings were covered in plywood and the trailing edge was made of wire. Ailerons were on the lower wing only. The engine mount consisted of two supports made of walnut and were padded with leather to decrease vibrations. The motor supports were reinforced by sheet metal. An Odier starter was mounted on the engine. A large access panel allowed the carburetor, magnetos, starter, and plugs to be easily serviced.

The fuselage was comprised of four longerons of ash and spruce interconnected by a latticework of spruce and piano wire. There were two fuel tanks which could be jettisoned in the event of a fire. The oil tank held 15 liters and was exposed to the air. The rear of the fuselage was made entirely of wood and the tailskid was made of ash. There was a ventral radiator between the landing gear struts. The aircraft had space for two cameras under the pilot's seat. When the aircraft was flown as a single seater the guns of the rear turret would have two machine guns fitted and could be fixed to fire to the front of the aircraft, above the wing. The fuselage was streamlined, and the Hispano-Suiza 8Fb and the fixed Vickers 7.7-mm machine gun were closely faired. A large spinner was fitted. There were prominent cutouts on the trailing edges of both the upper and lower wings to facilitate the crew's view. Armament consisted of one Vickers 7.7-mm machine gun (with the option for a second). There was

Above: Borel C1. This version was called a monoplace protégé; it could be flown as a single-seater with a second crewman for rear protection. B93.2497.

Above right: The Borel-Boccaccio Type 3000 appears to be almost identical to the C1 except for the cabane struts, which have been modified. This suggests that the wing was moved to the rear, perhaps to compensate for an altered center of gravity. B93.12620.

Right: Borel-Boccaccio Type 3000. The two Lamblin radiators were mounted on the undercarriage. MA36389.

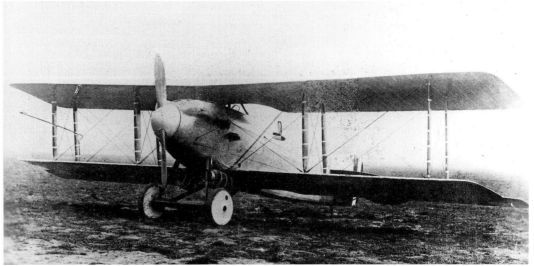

Borel-Boccaccio Type 3000 C2; only the upper wings had dihedral. B91.2663.

Borel-Bocaccio Type 3000 Two-Seat Fighter with 300-hp Hispano-Suiza 8Fb		
Span 11.4 m; length 7.095 m; height 2.65 m; wing area 33 sq. m		
Empty weight 897 kg; loaded weight 1,315 kg		
Maximum speed:	sea level	242 km/h
	1,000 m	260 km/h
	2,000 m	237 km/h
	4,000 m	230 km/h
Climb to:	1,000 m	2 min. 47 sec.
	2,000 m	6 min. 29 sec.
	4,000 m	15 min.
	6,000 m	25 min. 53 sec.
Ceiling 7,500 m; range 500 km; endurance 3 hours		
Armament: two fixed 7.7-mm Vickers machine guns; one or two 7.7-mm Lewis guns on a T.O.3 gun mount.		
One built		

a T.O.3 ring mount to which two 7.7-mm Lewis guns could be fitted. A third Lewis could be fitted to fire through the floor. Thus the Borel-Boccaccio Type 3000 met the armament requirement for the C2 class. Static testing revealed that the airframe was quite sturdy with a load factor of seven.

Modifications after initial flight testing included replacing the ventral radiator with two Lamblin radiators mounted on the undercarriage. The exhaust manifolds were shortened, and the tailplane bracing was strengthened by the addition of two struts.

The type performed well and easily exceeded the C2 requirements. However, the SPAD 20 had been completed earlier (having been flown in August 1918) and was chosen for series production.

Borel-Boccaccio Type 3000

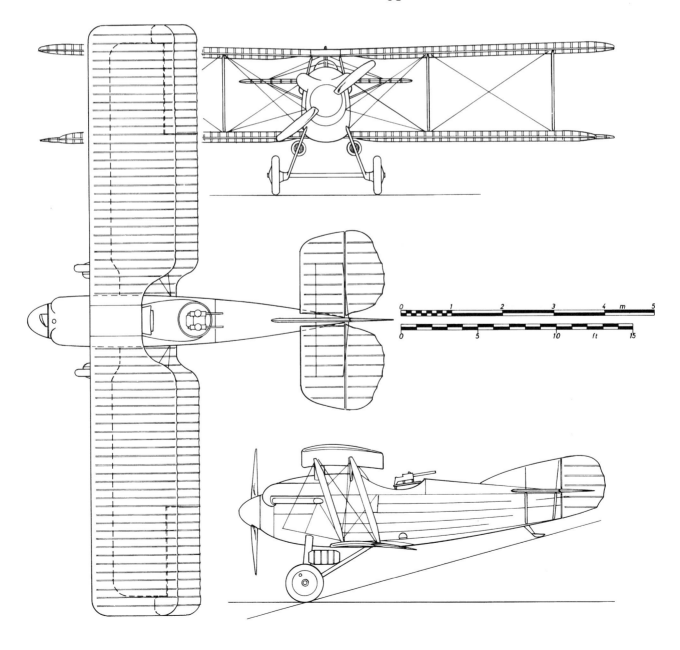

Borel-Boccaccio Type 3000 C2. The fuselage was well streamlined with a 300-hp Hispano-Suiza 8Fb engine. B76.115.

The rear gunner's position of the Borel-Boccaccio Type 3000 C2 could be fitted with either one or two Lewis machine-guns. MA36384.

Borel High Seas Flying Boat

The decision of the French navy in 1918 to issue a request for a large, multi-engine flying boat comparable to the British Felixstowe series resulted in an extraordinarily large number of submissions from the aviation industry. The specification called for a flying boat equipped with a T.S.F. wireless, a 75-mm cannon with 35 rounds, and two machine guns.

The Borel firm submitted a design for a biplane with twin floats, three engines, and a crew of three. The engines were to have been 400-hp Lorraines or Libertys. No other projected specifications are known; it seems likely that the aircraft remained an unbuilt project.

Borel Twin-Engine Floatplane

In mid-1918 Borel had a twin-engine floatplane under construction. It was to have been powered by two 400-hp Liberty engines and to have carried a crew of three. It is not known if this was to have been an entirely new design, or if it was a development of the Borel-Odier floatplane. It is not known if construction was ever completed.

Borel Twin-Engine, Three-Seat Floatplane with Two 400-hp Liberty Engines (all data estimated)
Wing area 140 sq. m
Payload 2,000 kg
Maximum speed: 150 km/h at 2,000 m

Société des Avions Louis Breguet

The first Breguet design was a helicopter. Louis and Jacques Breguet, in collaboration with Professor Charles Richet, built the Breguet-Richet Gyroplane in 1906. In September 1907 it made what may have been the first vertical take off and landing of a heavier-than-air machine. Although it flew, the helicopter was unstable in flight and development was abandoned. A second, more sophisticated, machine was built in 1908 but it, too, was unstable. It was destroyed in a hurricane in 1909.

The first Breguet fixed-wing plane was built in 1908. The Breguet 1 flew in June of that year. It was a biplane which employed wing-warping for control and had a biplane tail with twin rudders. Power was initially supplied by a 50-hp Renault engine, and in August 1910 the Breguet 1 set a record by carrying six persons. The Société des Avions Louis Breguet was formed the next year.

Another machine was built in 1912—a tractor biplane with wing warping for control and a large cockpit seating three. The aircraft had a tricycle undercarriage, and the entire tail could be rotated for pitch control. The engine was either a 90-hp Salmson radial (which was water-cooled) or a 100-hp Gnome.

Another variant of the basic Breguet tractor biplane was fitted with floats and won the Grand Prix at Monaco in March 1913. It had a single float under the fuselage and a single stabilizing float under each wing and the tail.

The sequence for allotting a designation to an aircraft was as shown in the following table:

Engine Make	Type Number				
	1	2	2bis	3	4
G: Gnome	50/60-hp	70-hp	80-hp	100-hp	140/160-hp
R: R.E.P.	50/60-hp				
L: Renault	50/60-hp	70-hp	100-hp		
C: Chenu	40-hp	80-hp			
D: Dansette	55-hp	110-hp			
U: Canton-Unné	70/80-hp	110-hp/130-hp			200-hp
O: Le Rhône	50-hp	80-hp			

Thus the Breguet designation Type G3 No.24 indicates the 24th example of a Breguet design powered by a 100-hp Gnome.

Other prewar Breguet aircraft include:

1. Number 1 (1909) with 50-hp Levavasseur engine. Many modifications to basic airframe; later a Gobron-Brillie engine was fitted.
2. Type L1 (1909) with Renault 50-hp engine
3. Type G1 (1910) with 50-hp Gnome. Set a world speed record with one passenger (86.3 km/h on 19 January 1911) and two passengers (79.7 km/on 6 March 1911). In September 1911 an example was fitted with an 80-hp Canton-Unné engine and flew with the pilot and one passenger from Casablanca to Fez in Morocco.
4. Type L1 and Type G1 (June 1910) were built in response to a request by Colonel Hirschauer for an aircraft for artillery spotting. Both types were similar to the Breguet G1 of 1910 and both were built in the winter of 1910. The Types L1 and G1 were tractor biplanes with four-wheeled undercarriages. The Type L1 had a 50-hp Renault engine while the Type G1 had a 60-hp Gnome engine. Using the Type L1, Louis Breguet participated in the 1910 maneuvers in Picardie carrying Captain Madiot as an observer. Both types were so successful that Breguet was nominated for the Legion d'Honneur. Two each of the Type L1 and G1 were built for the Aviation Militaire.

In the autumn of 1911 the British War Office purchased a single Breguet L1; by October 1911 the aircraft had arrived in England. It was initially given the serial B3, later changed to 202. The aircraft was described as being difficult to fly and requiring enormous strength to steer. By June 1912 it was based at Farnborough. It was destroyed in a crash landing and formally struck off charge in December 1913.

5. Type L2 bis (1910) with 90-hp Renault; built for Leon Bathiat.
6. Type R1 (1910) with 70-hp R.E.P. engine; entered in Coupe Michelin; carried three persons.
7. Type G (1911) with 90-hp Gnome engine; two examples built and given company numbers 8 and 10. Later modified into an R1 with a 50/60-hp R.E.P. engine.
8. Type U1 (1911) 80-hp Canton-Unné; company number 14. A single example was ordered by the RFC in 1913. It served with No.4 Squadron until being deleted in January 1914.
9. Type L1 (1911) 55-hp Renault ; company number 23.
10. Type G3 (1911) 90-hp or 100-hp Gnome. Set a record by carrying 11 persons. Company number 24. Two examples built.
11. Type L2 (1911) 115-hp Renault. Participated in military concours at Reims. The RFC acquired two:
    a) No.212—Breguet L2 with 70-hp Renault. Assigned to No.2 Squadron; in April 1913 the Renault was replaced by an 85-hp Salmson. In December 1912 assigned to No.4 Squadron. SOC December 1913.

In 1913 the Breguet firm acquired the rights to produce the Bristol-Coanda two-seat aircraft under license. Aircraft No.228, shown here, was delivered to Douai in May 1914. However, no Bristol-Coanda two-seaters were built by Breguet. B93.2698.

Breguet Militaire Type 1912 with 80-hp Gnome engine at the Étampes training center in 1913. B86.610.

Breguet Type 1912 with BR 17 at Mortcerf airfield in February 1914. B86.614.

b) No.213—Breguet L2 with a 70-hp Renault. Assigned to No.2 Squadron. Assigned to No.4 Squadron in December 1912. SOC in January 1914.

12. Type L1 (1912), 50/60-hp Renault; purchased by British, French, and Swedish air services.
13. Type HU3 (1912), 200-hp Canton-Unné; seaplane version of G3 with a large central float, a tail float, and an additional float under each wingtip. Two examples built.
14. Type U2 (1912), 115-hp Canton-Unné M 9; a seaplane version built in 1912 with floats was designated HU 2. (See below.)
15. *Marseillaise* (1912), 120-hp Canton-Unné CU. Canard design.
16. Type L3 (1912), 115-hp Renault.
17. Type G2bis Militaire (1912), 80-hp Gnome. Built in February 1912.
18. Type G3 Colonial (September 1912), variant of G2 with 100-hp Gnome; One G 3 was sent to Argentina to serve with the Escuela de Aviacion Militar in 1912.

    Three G3s were used by the RFC:
    a) No.210—Breguet G3 with a 100-hp Gnome. Initially assigned to No.2 Squadron, later to No.4 Squadron. SOC Dec. 1913.
    b) No.211—Breguet G3 with a 100-hp Gnome. Initially assigned to No.2 Squadron, later to No.4 Squadron. SOC Dec. 1913.
    c) There was a third Breguet G 3 (No.214) assigned to the RFC; however, no other details have been found.

    The Breguets were never used by the RFC in France; they spent their entire service lives in England.
19. Type J (1912), 110-hp Salmson engine. One purchased by Sweden and designated B1. Three were purchased by the Siamese Ministry of War in 1913 and assigned to the army air service. A fourth machine was donated to the service by a private citizen.
20. Type HU2 (1912), 110-hp Canton-Unné; floatplane.
21. The RNAS purchased a single "Breguet Tractor Biplane" in 1912. It was powered by an 80-hp Chenu engine and was based at Eastchurch. Assigned serial T6, later No.6.
22. Type HU 4 (1912). The RNAS ordered three HU 4 seaplanes in 1913. Serials 110, 111, and 112 were assigned. The first (No.110) was delivered from France in August 1912. It had a 200-hp Canton-Unné engine and was tested primarily at Calshot. It dose not appear that it was ever used operationally and numbers 111 and 112 were never delivered.
23. Unidentified Breguet—A Breguet tractor bomber was purchased by the RNAS from the French in 1915. It had a 225-hp Sunbeam engine. Assigned serial No.3888, it served with No.1, later No.5 Wings. This may have been a misidentified BM 2.

Breguet U1s at Dijon in 1914. B85.453.

Right: Breguet G2bis 80-hp Gnome trainer with dual controls. B85.313.

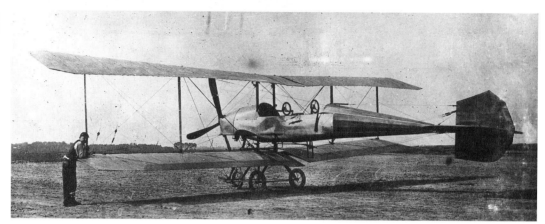

Above: Breguet G2bis No.135 with 100-hp Gnome 9 at Chili. B85.306.

Breguet G3 at Douai. B77.1844.

Above: Breguet U1 of BR 17. MA24624.
Left: Breguet Militaire Type 1912; this airplane was passed to BR 17 from the Breguet military school in late 1913 at Étampes. B86.626.

Breguet U2

The Breguet U2s of 1913 were tractor biplanes whose unusual shape and aluminum covering led to their being labeled (although only by the British Breguet Plane Limited Company) 'tin whistles.' The fuselage, of circular cross-section, had aluminum skin and terminated in a cruciform all-flying tail assembly. Most U2s in service in 1914 had a 110-hp Canton-Unné engine, but some were fitted with a 130-hp engine. The two-bay biplane wing had a single spar and flexed easily. For this reason it was described in contemporary French reports as having l'ailes souples (flexible wings). Twin radiators were stacked vertically on either side of the forward fuselage. The pilot and observer were located well aft, and the occupant of the rear cockpit must have had a particularly limited field of vision because of his location between the wings.

The Breguet U2s were posted to Escadrille BR 17, which was formed on 2 August 1914 and assigned to the 1st Armée in October 1914. The Breguets saw limited service on the Alsace front. A letter from the 1st Armée commander (which had been assigned BR 17) dated 16 August 1914 complained that the type was "incapable of war service" and was flown by its pilots with "great apprehension." He demanded the Breguets be replaced by the more suitable Voisins. The unit re-equipped with Voisin 3s to become VB 2 on 17 November 1914.

Breguet U2 Two-Seat Reconnaissance Plane with 110-hp Canton-Unné
Span 13.50 m; length 8.55 m; wing area 36 sq. m
Empty weight 560 kg (some sources say 798 kg); loaded weight 960 kg (some sources say 1,098 kg)
Maximum speed: 110 km/h

Breguet U2 with Canton-Unné 130-hp at La Brayelle, April 1913. B85.308.

With no further orders for planes, the Breguet factory turned to building Voisins and Morane-Saulniers under license.

Foreign Service
Chile

Four Breguet biplanes were obtained by Chile in 1913. These may have been Type U2s; however, most were powered by 80-hp engines and only one by the customary 110-hp engine of the Type U2. They were based at the Escuela de Aeronautica Militar at IJO Espello and were used for training.

United Kingdom

Two Breguet U2s (one built by British Breguet) were evaluated by the RFC in the military trials of 1912. Neither type was selected for service with the RFC. Later an additional Breguet U2 was purchased by the RFC. Given serial no.310, it had been built by British Breguet. It may have served with No.4 Squadron and was struck off charge in January 1914.

Breguet type 1914 (U2) with 130-hp Gnome piloted by Paul Derome. B85.298.

Breguet 13/X/AG 4

The Breguet 13 first appeared in 1913 (hence the "13" designation which was bestowed by the aviation press). It was powered by a 100-hp Gnome 9J and had a tractor layout. It was tested at Reims in September 1913 but was not adopted for military service. A second version, powered by a 160-hp Gnome engine, was named the Type "X" by Breguet. The press thought the X was a roman numeral so labeled it the Type 10. It was this variant that led to the AG 4.

Louis Breguet had a new design under construction in 1914; it had a 160-hp Gnome and, as mentioned above, appears to have been a development of the preceding Type X. The new plane was a two-bay tractor biplane with wings of equal span, ailerons on both the upper and lower wings, and a four-wheel undercarriage. The end of the fuselage was a steel tube on which the elevator and rudder were carried. There was a small rudder located at the end of the fuselage, and a large fin extended below the fuselage. Because of the lower fin a skid had to be placed in the middle of the fuselage to prevent damage to the fin during landing.

During the German advance on Paris, Breguet moved this machine to Villacoublay and made a voluntary reconnaissance flight during which he detected a change in the advance of the German troops as they turned to the east of Paris. This information was confirmed by aircraft of REP 15 and MF 16, and the 6th Armée attacked the German flank. This resulted in the Battle of the Marne, a major victory for the Allies. Perhaps because of this historic mission, the French officially adopted Breguet's new plane later in 1914. It was later designated the AG 4, which stood for A (tractor), G (Gnome) 4 (160-hp engine). It was given the serial BR 52. Modifications to the aircraft included fitting a much larger fin and rudder. These modifications were carried out by René Moineau and were intended to improve directional control. In addition to these changes, a rack to carry flechettes was mounted on the rear cockpit, and the aircraft was armed with a Hotchkiss machine gun fired by the observer (who sat in the rear cockpit). The aircraft was assigned to BR 17 in September 1914.

A second version of the new tractor was built and given serial BR 53. It, too, was armed with a Hotchkiss machine gun. BR 52 had been struck off charge before the end of 1914; BR 53 made air defense patrols from Le Bourget until it crashed on 26 December 1914.

The AG 4 (serial BR 52) was despised by those who flew it. A letter dated 25 October 1914 from the commander of BR 17 (Capitaine Benoist) stated that the BR 52 was unusable. He cited the fact that the engine had to have five cylinders replaced after only three hours flying time. Barès responded on 20 October 1914 by reassuring the pilots of BR 17 that assigning the type to the unit was an "unfair experiment" and promised the aircraft would be replaced. A letter to Barès from Louis Breguet stated that he believed that the AG 4 project could be salvaged if it were fitted with a new 200-hp engine, and possibly by replacing the wing. It is not known if Breguet carried out these modifications, but no further AG 4s were purchased by the military.

Breguet AG 4 Two-Seat Fighter with 160-hp Gnome
Span 15.35 m; length 8.25 m; height 3.30 m
Empty weight 950 kg; loaded weight 1,350 kg,
Maximum speed: 100 km/h; climb to 500 m in 12 minutes; ceiling 1,500 m
Armament one Hotchkiss machine gun and flechettes
Two built

Breguet AG 4 serial BR.53. It was based at Bourget in 1914. B80.564.

Breguet AG 4; this airplane was fitted with a 160-hp Gnome. B85.281.

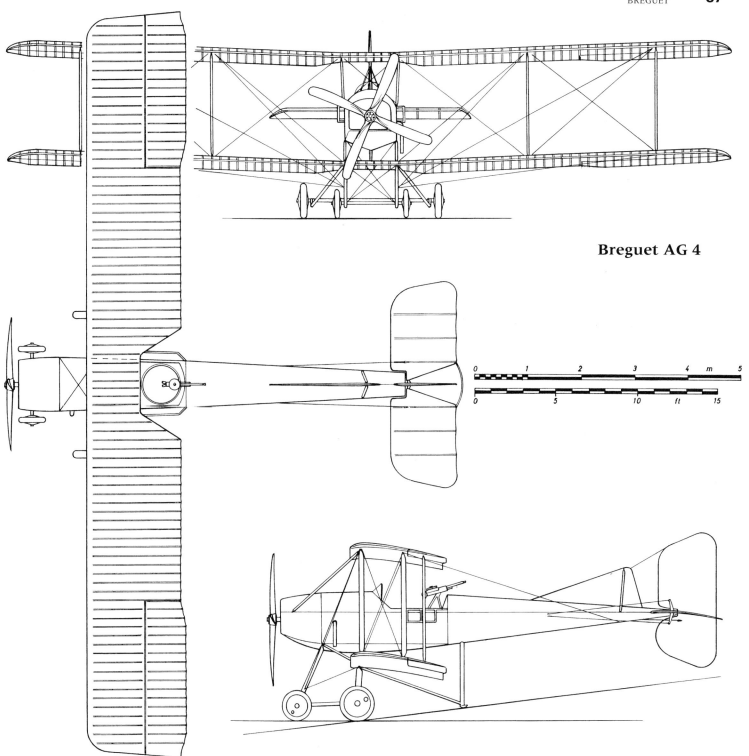

Breguet AG 4

Breguet Pusher Summary

Breguet Type	Designation	Engine	Manufacturer	Role
1	BLC	220-hp Renault	Breguet	Fighter
2	BLM	220-hp Renault	Breguet, Michelin	Bomber
3	BAM	230-hp Salmson A9	Breguet	Bomber
4	BM IV	250-hp Renault	Michelin	Bomber Escort
5	Gamma	250-hp Renault	Breguet, Darracq	Fighter
6	Gamma	230-hp Salmson A9	Breguet	Fighter
7	BUC	220-hp Salmson 2.M7	Breguet	Fighter
8	BC	200-hp Sunbeam	Breguet	Fighter
9	BAC?	230-hp Salmson A9	Breguet	Fighter
10	BUM	220-hp Salmson 2.M7	Breguet, Michelin	Bomber Trainer
11	Delta	3 x 220-hp Renault	Breguet	Bomber
12	BR XII	250-hp Renault	Breguet	Bomber with Cannon

Breguet-Michelin BUM/BLM/BAM

The Breguet-Michelin series of bombers were to become famous because of their participation in the earliest French bombing raids. The series began with the prototype BU 3. It was intended to meet a requirement issued by General Bernard in 1913 for a biplane, with a pusher layout, for reconnaissance and army cooperation duties. However, Breguet decided that his new design could be better utilized as a fighter. The BU 3 was under construction in early 1914 but it did not fly until later in the year because the area where the prototype was to have been tested had to be evacuated because of the rapid German advance through northeast France. The completed aircraft was moved by road to the test center at Villacoublay. The BU 3 later served as the prototype for a new bomber for the Aviation Militaire, completely abandoning the fighter role for which it had been originally designed. (See entry under BU 3/BUC/BLC for further details).

The BU 3 was a three-bay pusher biplane with straight, unstaggered wings. The engine, a 200-hp Canton-Unné 2M7, was mounted as a pusher at the rear of the central nacelle. The crew of two and the main fuel tank were located in the nacelle, which was set on the lower wing. Although Breguet would have preferred to use a tractor layout as on the preceding U2 and AG 4, the Aviation Militaire felt that the pusher layout optimized the gunner's field of fire and improved the view for both crew members. The tailplane and rudder were supported by four booms that extended from the upper and lower wings. Armament was either a Hotchkiss or Lewis machine gun. Bombs were carried on underwing racks capable of holding up to 455 kg. The distinctive undercarriage of the series was first used on the prototype; this consisted of a pair of large main wheels at the rear and two smaller front wheels. At least one aircraft had a 200-hp Canton-Unné A 9 engine; it was designated BAM by Breguet and BM 3 by the STAé. A note dated 20 July 1915 gave details of flight test on BM 2 serial BM 101. The aircraft was tested with a payload of 880 kg. This aircraft could climb to 500 m in 8 min. 35 sec.; 1,000 m in 17 min.; 1,500 m in 27 min. 20 sec.; 2,000 m in 43 min.; and 2,500 m in 60 min. This performance was adequate but changes were suggested for the series. These included: placing a metal screen between the engine and fuel tank, a lever in the cockpit to shut down fuel flow, and changes in the fuselage structure to isolate the fuel pipe from the engine. Problems with engine fires did, in fact, occur in operational service as gas fumes exiting near the engine could ignite. Excessive engine vibration was also suspected as causing the fuel pipes to loosen and leak.

The Breguet BU 3 served as the prototype of the Breguet-Michelin series of bombers. B84.1773.

The Michelin brothers were French patriots who wished to aid their country. They offered to produce 100 bombers for the Aviation Militaire and chose to manufacture the Breguet prototype. As noted above, Breguet had originally intended that it be employed as a fighter, but the Michelin brothers insisted that a bomber would be of more use. Breguet, also in a patriotic gesture, granted the Michelin firm rights to build the type under license free of charge. Production was initiated at the factory in Clermont-Ferrand on February 1915. Fifty planes were built and received the designation BUM (B = Pusher, U = Canton-Unné engine, M = Michelin). Subsequently, the aircraft were designated as Breguet-Michelin 10s by the STAé.

Fifteen were used to form a training school at Camp d'Avord on 20 September 1915. However, there were complaints that they were underpowered. It was noted that the takeoff roll was excessively long and engine failures were frequent. For these reasons, a development of the Breguet BUM employing the same airframe as the BUM but using a more reliable 220-hp Renault 8Gd engine was produced. External changes to the aircraft, designated BLM (L = Renault), included the removal of the fuel tank from the center nacelle, the fuel now being housed

Breguet BU 3 serial BR 54; named *La Negresse*. B85.269.

Breguet-Michelin BM 2. The engine was a 220-hp Renault. B85.428.

in two underwing tanks on either side of the fuselage. The performance of the BLMs was clearly superior to the BUMs. However, there were reportedly two serious accidents, possibly due to the relocation of the fuel tanks resulting in a change in the center of gravity. It was also noted that the absence of shock absorbers on the wheels may have contributed to these accidents. Approximately 50 BLMs were built at the Michelin factory at Clermont-Ferrand and these retained the STAé designation BM 2.

Many of the BLMs produced served with units attached to

BLM (BM 2) Two-Seat Bomber with 220-hp Renault 8Gd

Wing area 70.5 sq. m
Empty weight 3,042 kg; loaded weight 2,142 kg; payload 455 kg
Maximum speed: 128 km/h; climb to 2000 m in 27 min; range 760 km
Armament: a Hotchkiss or Lewis machine gun, a Winchester carbine, and 455 kg of bombs
Approximately 50 built

BUM (BM 10) Two-Seat Bomber with 200-hp Canton-Unné

Wing area 70.5 sq. m
Empty weight 1,315 kg; loaded weight 2,115 kg
Maximum speed: 124 km/h; climb to 2000 m in 40 minutes; range 730 km
Armament: a Hotchkiss or Lewis machine gun and 455 kg of bombs
Approximately 50 built

GBM 5; details of operational missions are given under the entry for the Breguet-Michelin 4. (Some accounts tend to confuse the BUM/BLM series with the Breguet-Michelin 4, but they were different aircraft).

Breguet SN 3

The Breguet SN 3, a further development of the BUM/BLM series, was intended to participate in the 1915 concours puissant. The aim of this competition was to produce an aircraft capable of bombing the city of Essen—hence the "SN" designation. The specifications called for a craft capable of carrying 200 kg of bombs over a 600 km radius (the distance from Nancy to Essen). Speed was to be 120 km/h with a ceiling of 2,000 meters.

The main alteration made to the SN 3 was the change to the unequal span wings, which resulted in a decrease in wing area from 70.5 sq. m (for the BUM/BLM series) to 54 sq. m. The power plant was the same as for the BLM, a 250-hp Renault 8Gd. The aircraft was fitted with a Michelin bomb sight, a mechanical intervalometer, and bomb racks for 30 bombs of 8 kg each.

Only the Paul Schmitt and the Breguet designs were able to meet the requirements of the competition, both being able to fly the required course (Villacoublay-Chartres-étampes-Villacoublay) in four minutes 30 seconds at 132 km/h. The Breguet was preferred because the pusher layout permitted the bombardier/gunner a better view and because its performance was clearly superior to that of the Paul Schmitt.

Although the Breguet SN 3 placed first in the competition, it was not ordered into production because it was felt to be too poorly defended for day attacks and lacked sufficient range. Instead, Breguet began work on an improved SN 3 intended for series production; this would become the Breguet-Michelin 4.

Breguet SN 3 Two-Seat Bomber with 250-hp Renault 8Gd

Wing area 54 sq. m
Empty weight 1,350 kg; loaded weight 2,150 kg
Maximum speed: 135 km/h at 2,000 m; climb to 2,000 m in 28 minutes, range 800 km
Armament: a flexible machine gun and 240 kg of bombs
One built

Breguet SN 3. In order to meet the self-defense requirements of the competition, this aircraft is fitted with a machine gun mounted on a swivel mount. The gunner fired outside the propeller arc. B85.263.

Breguet SN 3

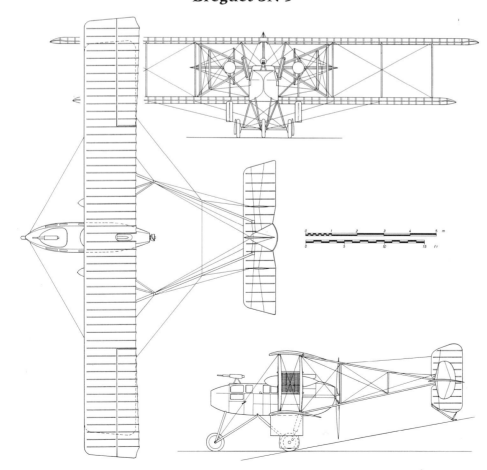

Above: Breguet SN 3. The main alteration made to the SN 3 was the change to the unequal span wings which resulted in a decrease in wing area from 70.5 sq. m (for the BUM/BLM series) to 54 sq. m. B85.247.

Top: Breguet SN 3 which participated in the 1915 concours puissante. B85.268.

Breguet BM 4

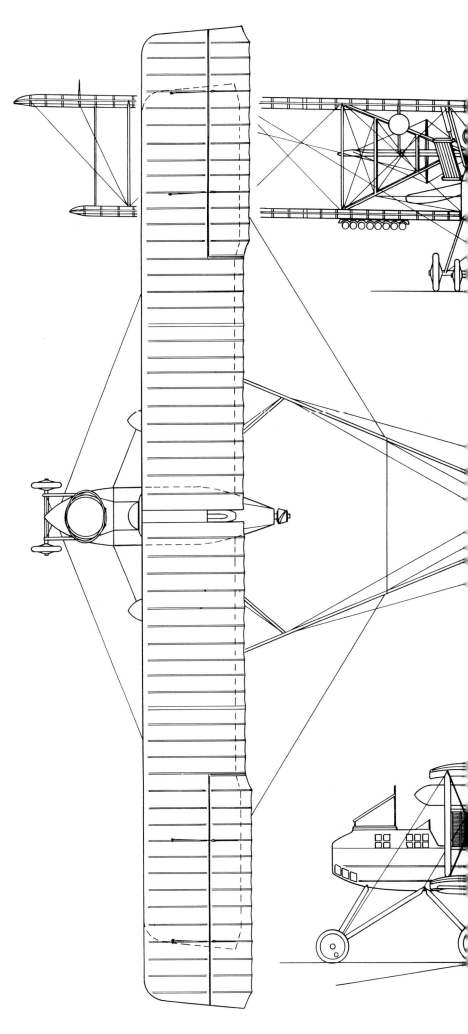

90B

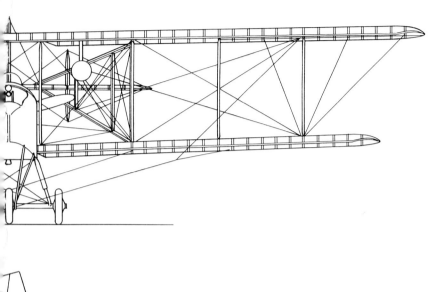
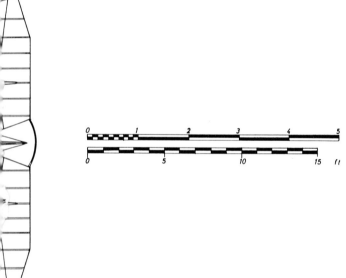
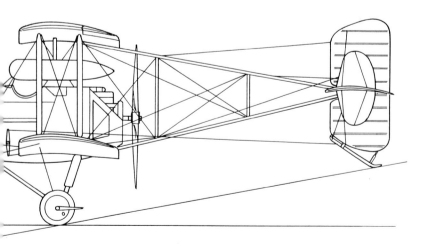

Breguet BLC

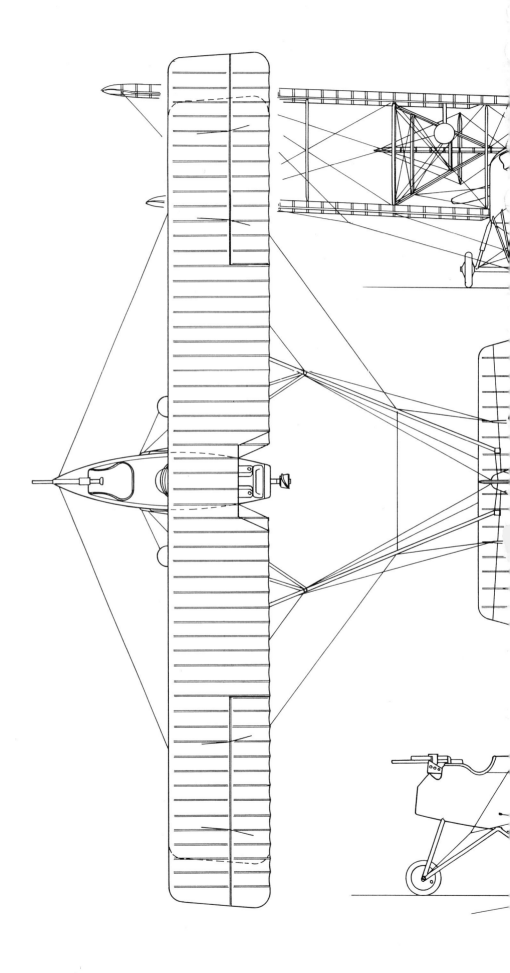

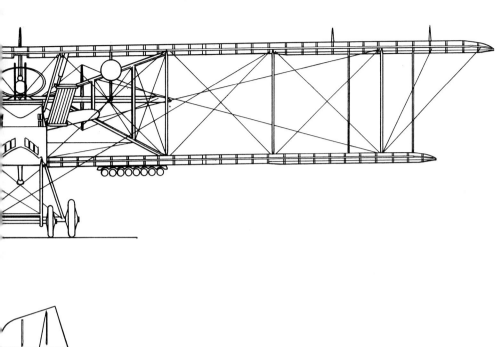

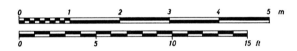
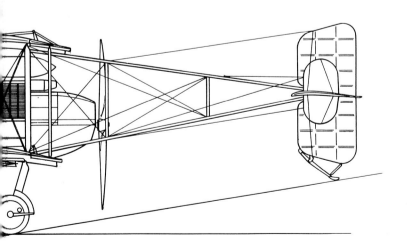

Breguet-Michelin 4 (BM 4)

Developed from the previous SN 3 (which had won the 1915 concours puissant) the BM 4 had a protracted development period that prevented it from entering service until 1916. Even worse, after the aircraft reached the front they were plagued with numerous problems that had to be overcome before they could become operational.

Although intended to be an improvement over the SN 3, the BM 4 was of similar configuration with a Renault 8Gd engine of the same power (250-hp). While the speed was the same as the SN 3, the range and the bomb load were actually reduced. However, the area of the three-bay biplane wing was increased to 66.8 sq. m as compared with 54.0 sq. m for the SN 3. Two hundred of these planes were built at the Michelin factory from April 1916 until May 1917.

After a prolonged development period the aircraft reached the front in April 1916. They were formed into their own bomb group—GBM 5. Some also served in GB 4 along with some of the earlier BUM/BLMs. The philosophy of putting all these aircraft into one unit was, in addition to simplifying logistics, to provide a mobile strike force capable of being sent to areas of critical importance. Such a group was to combine power with mobility. Unfortunately, the BM 2s and BM 4s would provide neither.

The Breguet-Michelins proved unsatisfactory as either day or night bombers, and the type was not well liked by crews. Pilots found the aircraft more difficult to fly than the Voisins. Also, the takeoff and landing runs were still considered too long. Another complaint was that the defensive armament was inadequate for daylight missions.

Operational Service

Originally, 200 Breguet bombers were ordered. These were to be divided into either two or three escadres, each with five escadrilles of ten aircraft each. The usual composition was nine or ten Breguets—eight BM 2s or BM 4s and one or two BUC, BLC, or Breguet 5 fighters (see below). By 21 September 1915 GBM 5 had received 24 Breguet bombers (probably BM 2s) with serials 101–117, 120–121, BR 60–62, and BR 65–66.

GBM 5 was divided into three escadres, each with 30 aircraft, and was commanded by Lieutenant Dutertre. The 1st Escadre (GBM 5) was commanded by Capitaine Yence and was based near Champagne. It consisted initially of three escadrilles: BM 117, BM 118, and BM 119. GBM 5 was supported by aviation park GBM 105. The 2nd Escadre was commanded by Capitaine Gouin and was formed in February 1916 with Escadrilles BM 120, BM 121, and BM 122. It was supported by aviation park BM 106. The 3rd Escadre was never formed because no further aircraft were ordered. Furthermore, BM 122 was never equipped with Breguet bombers; instead, it was formed in 1916 on Caudron G.4s.

In September 1915 the 1st Escadre left Avord for Oiry. However, the escadrilles were all at about half strength. The first raid was made on 30 September, 1915 when two aircraft of BM 115 attacked Guignicourt. The units attempted a number of raids but were soon frustrated by the numerous shortcomings of their aircraft. Usually several aircraft would set out for a target, but only one or two would reach it; most turned back because of engine difficulties. Nevertheless, many railway stations were attacked in October and November. On 1 December 1915 the 1st Escadre moved to Ochey. Engine troubles still plagued the BM 2s. In fact, engine troubles forced the aircraft of Lieutenant Dutertre down close to Metz. After this, the unit returned to control of GBM 5.

The Breguet-Michelin units of GB 4 saw action in the major battles of 1916. On 26 January 1916 a detachment of BM 120 was assigned to GB 4 at Belfort and placed under the control of Capitaine Happe, who also commanded MF 29. Half of the unit's aircraft were fighters (BUC/BLC) and half bombers (BM 2s). The fighters proved to be virtually useless in the escort role and were soon switched to bombing. For example, on 8 March, six M.F.11s of MF 29 bombing the enemy aerodrome at Ensisheim were escorted by BUC/BLC escort fighters. However, the BUC/BLC fighters proved to be slower than the M.F.11s they were assigned to protect, and were therefore consigned to the hangars as useless. The BUC/BLCs were later converted to bombers. Three Breguet-Michelin bombers participated in other raids made by MF 29 and MF 123 but were now escorted by Caudron G.4s of C 34 and C 61. By October the BM 120 unit under Happe's control had 12 aircraft: six bombers (BM 4s) and six fighters (Breguet 5s). GB 4's raid on the Mauser factory at Rothweil on 12 October 1916 consisted of six F.40s of F 29, six F.40s of F 124, and 14 BM 4s and Breguet 5s of BM 120. Despite protection provided by G.4s and Nieuport 11s, BM 120 suffered heavy losses. Three BM 4s and three Breguet 5s were destroyed and an additional BM 4 had to be written off after it limped back to GB 4's base. Thus, almost 60 per cent of BM 120's planes had been destroyed in one raid. After this GB 4, along with its few remaining BM 4s and Breguet 5s, switched to night bombing.

GBM 5 was also active throughout 1916. BM 2s and BM 4s bombed railway stations at Brieulles, Metz, and Thionville in April. However, many were damaged by German anti-aircraft fire. By the end of the month only three to five were reaching

Breguet-Michelin BM 2 No.187 with a 220-hp Renault 12F engine. B85.348.

Breguet BM 4 of BM 20 named *Le Voilà le Foudroyant*. This airplane was forced down at Rustenhart and captured by the Germans during the course of the Oberndorf raid. B88.2537.

the target. The main problem was the Canton-Unné engines of the BM 2s. When GBM 5 at last began to receive the Renault-engined Breguets, serviceability improved dramatically. On 14 May 1916 one plane of GBM 5 dropped 11 75-kg bombs on the airship hangar at Metz-Frescaty. Railway stations were bombed during May; heavy anti-aircraft fire was encountered.

On 16 June 1916 GBM 5 reported it had a total of 22 BM 2s and BM 4s; there were 22 pilots available and 26 in training. By 20 June 18 aircraft could be sent out nightly to attack the Metz and Arnaville stations. The BM 4s were preferred for night missions, being easier to fly than the BM 2 and Breguet 5 fighters; as a result, it was requested that only BM 4s be supplied to the escadrilles of GBM 5.

On 1 July 1916 GBM 6 was disbanded because of the lack of availability (and the poor efficacy) of the Breguet-Michelin series. The groupe's aircraft and pilots were transferred to GBM 5, including aviation park BM 106.

In July BM 117 of GBM 5 bombed train stations. In August GBM 5 continued its campaign against railway stations. For example, on the night of 9/10 August, aircraft of GBM 5 bombed Guny and Apilly; four Breguet 5s with cannon joined the bombers and used their 37-mm guns to attack searchlights.

By early autumn, BM 119 joined BM 117 in attacks on enemy bivouacs and barracks. GBM 5 was now being used primarily as a night-bombing unit, attacking camps, aerodromes, railway stations, and supply depots. Poor weather inhibited sorties in December.

By 1917, the GBM 5 had three escadrilles (BM 117, BM 118, and BM 119) equipped primarily with Breguet-Michelin 4s. BM 120 was still attached to GB 4 but rejoined GBM 5 on 14 April, as did the fifth and final unit, BM 121.

Winter weather continued to inhibit the activity of the bomb groups as 1917 began. GBM 5 made sporadic attacks on German airfields, factories, and railway stations.

By spring, the activity of the escadrilles attached to GBM 5 continued to decline. This may have been because production of the BM 4s and Breguet 5s was being terminated, and the older planes were probably becoming difficult to maintain. In any event, few raids were flown as the Breguet-Michelin escadrilles prepared to receive new aircraft.

In September 1917 the units began to re-equip with the Breguet 14 B2. BM 117 and BM 120 were the first to obtain the new aircraft. Meanwhile, the BM 4s participated in raids against railway stations, the airfields at Chambley, factories in the Moselle valley, and the power plant at Le Couteulx.

By the end of 1917 three of the five Breguet-Michelin escadrilles had taken on the Breguet 14 B2 and one had adopted the Voisin 10. The last unit to give up the Breguet-Michelin 4 was BM 121, which adopted the Voisin 10 in May 1918.

Foreign Service
Romania

Romania acquired a number of Breguet-Michelin 4s and Breguet 5s during the war. On 15 December 1916, there were six available for front-line service. On 9 January 1917 these were organized into Escadrila BM-8, based at Iasi and commanded by Locotenant Armand Delas. It was under the administrative control of the Grand Headquarters of the Romanian Army along with Escadrila F 7. Up until June 1917 it was manned exclusively by French aircrew. On 21 January 1917 there were a total of 12 Breguet-Michelin 4 and Breguet 5s available at the front. As with its French counterpart, Escadrila BM-8 was to be used as a mobile strike force being moved from front to front as needed to supplement the other air units. BM-8 was especially active in the sector of the Russian 6th Army.

BM-8 moved to Galati on 20 March 1917. On 31 March BM-8 attacked warships at Braila Harbor with 16-kg bombs (each aircraft carried five bombs). One warship was claimed as being

---

Breguet BM 4 Two-Seat Bomber with 220-hp Renault 8Gd

Span 18.8 m; length 9.9 m; height 3.9 m; wing area 66.8 sq. m
Empty weight 1,435 kg; loaded weight 2,112 kg; payload 377 kg
Maximum speed: 135 km/h; climb to 2000 m in 28 minutes; ceiling 3,900 m; range 675 km
Armament: 320 kg of bombs, a Hotchkiss 8-mm machine gun, and a Winchester carbine
Approximately 200 built

Breguet BM 4 named *Queen Mary*. The airplane adjacent to the BM 4 is a Nieuport 21. B85.1465.

sunk. Troops and an ammunition depot were attacked on the 31st and 1 April.

Subsequently the aircraft of BM-8 were used for long range reconnaissance missions. On one such flight the crew succeeded in forcing a German airplane to land near Braila. On 1 May BM 8 bombed railway stations, during which a French crew shot down an L.V.G. On 19 May the Breguets were used to attack ships sailing on the Danube and bombed the enemy camp near Ianea.

One of BM-8's most successful raids was on 25 May when Braila harbor was bombed. Hits were scored on an ammunition depot and on an enemy warship.

On 7 July BM-8 moved to Vanatori to escape the attention of German artillery. That same day, the German airfield at Foscani was bombed.

In August 1917 Escadrila BM-8 was assigned to Grupul 3. It continued to support the Russian 6th Army units in the Dobdrudja sector. On 21 August there were night attacks on the railway station and airfield at Foscani; from the 22nd through the 24th BM-8 bombed troops at Marasesti and the airfield at Foscani. Reconnaissance missions were flown along the 1st Army front, accompanied by the aircraft of F 7. On 26 August BM-8 dropped 40 Michelin lance bombs on bivouacs south of Focsani.

In September BM-8 continued its series of raids on Focsani, these missions were flown on the 2nd, 4th, 5th, and 22nd. On 1 October BM-8 bombed enemy barracks at Clipicesti; they were accompanied by Nieuport 11s. On the 15th enemy troops were bombed at Burca, again the Breguets were accompanied by Nieuport 11s. Other missions in October included dropping two "Gros" bombs on a warehouse at Faurei (25th), and attacks on the cantonments at Burca (28th, 29th, and 30th).

November was the last month the B.M.4s and Breguet 5s were active over the Romanian front. On the 5th there was a raid on enemy positions at Clipicesti, and two days later the enemy supply depot at Faurei and barracks at Focsani were bombed. Enemy warehouses and cantonments were attacked on the 12th. The last recorded mission by BM-8 was on 18 November when eight "Gros" bombs were dropped on the German cantonments at Clipicesti.

It seems that none of the 12 Breguet-Michelin 4s or Breguet 5s obtained by Romania survived the war. Known serial numbers include 193, 252, 256, 267, 572, 580, and 587.

Russia

A single example of a Breguet-Michelin 4 was obtained by the Imperial Russian Air Service. The type was felt to be slow and to carry an inadequate bomb load; for this reason only one was purchased. The aircraft was attached to the EVK in the latter half of 1917.

United Kingdom

A Breguet "Tractor" Bomber was obtained from the French in 1915. It had a 225-hp Sunbeam engine. Assigned serial No.3888 it served with No.1, later No.5 Wings. This may have been a mis-identified Breguet 4 bomber.

New Breguet BM 4 serial number 202. This airplane was built at Clermont-Ferrand in April 1916 and was flown by André de Bailliencourt on 24 April 1916. B84.1755.

Breguet BU 3/BUC/BLC/BC

The Breguet-Michelin bombers were not very successful in the strategic bombing role, in part because German single-seat fighters inflicted severe losses on the bomber formations. There is some irony, then, that Breguet's original intention had been to use his plane as a fighter. The prototype of the entire Breguet pusher series was designated BU 3 and was completed and flown in late 1914. The crew of two were seated in tandem with the gunner behind the pilot—he had to stand to fire his gun over the pilot's head. The fuselage nacelle was mounted on the lower of the three bay wings. Twin booms extended from the top and bottom wings, their ends joined by the elevators. On early examples the booms were covered with fabric; later the fabric was removed and two fins added to the elevator. The rudder was mounted in the center of the elevators. The engine was a 200-hp Canton-Unné 2M7 water-cooled radial. The speed of the BU 3 was 125 km/h, impressive considering that the nacelle was armor-plated. The BU 3 was completed in late 1914 and evaluated by Louis Breguet at the C.R.P. It was given serial BR 54. However, as the Michelin brothers specified that they wished to produce a bomber, Breguet converted his design to meet the requirements of the 1915 concours puissant (see BUM/BLM entry).

Aircraft BR 68 was tested at Villacoublay on 22 June 1915. It carried a payload of 380 kg. Flight tests revealed a maximum speed of 138 km/h and climb rates as follows: 500 m in 3 min.; 1,000 m in 6 min. 30sec.; 1,500 m in 10 min. 20 sec.; 2,000 m in 15 min.; 2,500 m in 19 min. 40 sec.; and 3,000 m in 5 min.

The suggestion for re-converting the Breguet bomber into a fighter came from the commander of the Breguet-Michelin bomber unit. He wanted a heavily armed aircraft which could accompany his bombers to and from their objective. Breguet used the preceding BU 3 as the basis for a light-weight fighter design. To save weight he decreased the size of the aircraft and lightened the structure. The plane was to be armed with a 37-mm Hotchkiss cannon.

The new prototype was designated the BUC (Breguet with Canton-Unné 200-hp engine; C= Chasse = fighter). It was smaller than the BUM bomber with a two-bay (instead of three-bay) wing. The lower wing was shorter and had a smaller chord than the upper. The undercarriage was simplified from four wheels (two fore and two aft) to a tricycle arrangement. The nacelle was altered so that the gunner, who now sat in front of the pilot, had a clear field of fire. The STAé designation was Breguet 7.

The availability of a new and more powerful engine resulted in a 220-hp Renault 8Gd engine replacing the Canton-Unné. The aircraft was re-designated BLC (L for Renault). This new engine resulted in the fuel tank being removed from the fuselage nacelle; the fuel on the BUC and most BLCs was in external tanks mounted between the wings. The armament used was either the previously mentioned 37-mm Hotchkiss cannon or a machine gun. These were mounted on movable stands to permit the gunner to fire in all directions. The STAé designation was Breguet 1.

It is unclear how many BUCs and BLCs were produced and it is not known if they had any successes in aerial combat. There were 11 Breguet fighters in service on 1 February 1916, but it is likely that most were Breguet 5s. It is safe to conclude that in 1916 the aircraft were obsolete designs using an outdated form of armament and were quickly withdrawn from service. (For further details see the operational section under BM 4).

United Kingdom

The Royal Naval Air Service obtained some aircraft to escort its bombers. By early 1916 15 Breguets de Chasse were on order for that service. All had been delivered by the end of April 1916. Most were powered by a 225-hp Sunbeam engine. On those the fuel tank was again placed in the fuselage nacelle which had flank radiators on either side. The armament was usually a single Lewis machine gun. The aircraft were called Breguet de Chasse and were designated Breguet 8s by the STAé. Nos. 1390–1394 were assigned to both No.1 Wing (at St. Pol) and to No.3 Wing. Aircraft Nos. 3209–3213 and 3883–3887 were initially assigned to No.5 Wing RNAS at Dunkerque. As with the French, the RNAS found the Breguets too

Two-Seat Cannon or Machine-Gun Armed Fighter			
Specification	BUC (Breguet 7)	BLC (Breguet 1)	BC (Breguet 8)
Engine	200-hp Canton-Unné 2M7	220-hp Renault 8Gd	225-hp Sunbeam
Span, m:	17.95	16.40	—
Wing area, sq. m:	—	54	—
Length, m:	9.2	9.5	—
Height, m:	3.5	3.7	—
Empty weight, kg:	—	1,160	—
Loaded weight, kg:	—	1,600	—
Maximum speed, km/h:	120	138	125
Climb to 1,000 m:	—	6 min. 30 sec.	8 min. 30 sec.
Climb to 2,000 m:	—	16 min. 30 sec.	20 min.
Ceiling, m:	3,000	3,700	3,000
Endurance:	3 hours	3 hours	3 hours
Armament:	one machine gun	37-mm Hotchkiss or one 7.7-mm Lewis	7.7-mm Lewis or two-pounder Davis gun & six 112-lb bombs

Breguet BLC *Liberte* of V 21 at Port-sur-Saône on 13 January 1916. The crewman on the nosewheel gives scale to the size of the airplane. Shortly before this photograph was taken, a BLC took off with the crewman still clinging to the nosewheel. The man was carried 300 meters into the air before the airplane landed safely. MA18235.

Breguet BLC of Detachment de Chasse Breguet at Belfort in February 1916. This unit escorted the bombers of GB 4. This airplane was named *Vérité* and was flown by Cne Gaubert. Note the rectangular air intakes for the lateral radiators. B86.591.

obsolete for use as either fighters or escorts. Some were armed with bombs and at least one was armed with a two-pounder Davis recoilless gun. The aircraft had outlived their usefulness only three months after entering service, and were withdrawn in June 1916.

Breguet 5/6/9/12

The rise of the cannon-armed fighter in France led many manufacturers to turn bombers into fighters by the addition of cannon. Cannon-armed aircraft were seen as being particularly useful in a long-range escort role, and many of these were converted from bombers in service with the Groupes de Bombardement. This meant that the aircraft had comparable performance to those they would be protecting and thus could escort the bombers all the way to the target and back while maintaining station with them. In addition, the cannon-armed aircraft were expected to intercept incoming fighters and airships, attack ground targets, and even destroy enemy searchlights at night.

The cannon-armed version of the Breguet-Michelin bombers that appeared toward the end of 1915 was designated the Breguet 5. Power was supplied by a 220-hp Renault 12Fb engine. Armament was either 20 120-lb bombs or a single 37-mm cannon. There was also provision for a rearward-firing machine gun located on the leading edge of the top wing. Endurance was six hours on the prototype and three hours on production aircraft.

Two hundred Breguet 5s were built in 1916 and 1917. They were built exclusively at the Breguet plant, and not at the Michelin factory; this explains why the BM designation was not assigned to the Type 5 by the STAé. Approximately 11 Breguet 5s were in service on 1 February 1916.

Breguet 5 armed with a 37-mm cannon. The fuel tank mounted underneath the top wing helped reduce the risk of engine fires. This aircraft was assigned to the C.R.P. B88.2761.

Breguet 6. The Salmson A9 was completely enclosed by the engine cowling. B84.1762.

Type 5 Two-Seat Escort Fighter with 220-hp Renault 12Fb

Span 17.5 m; length 9.9 m; height 3.9 m; wing area 57.7 sq. m
Empty weight 1,350 kg; loaded weight 1,890 kg
Climb to 2,000 m in 22 min.; ceiling 3,700 m; endurance 3.5 hrs

Type 5 Cannon Two-Seat Escort Fighter with 220-hp Renault 12Fb

Dimensions: as Type 5
Empty weight 1,394 kg, loaded weight 1,934 kg
Maximum speed:
ground level	131 km/h
2,000 m	128 km/h
3,000 m	124 km/h

Climb to 2,000 m in 15 min. 30 sec., ceiling 4,300 m, endurance 3.5 hrs
Armament: 37-mm Hotchkiss cannon and 7.7-mm Lewis machine gun
200 Breguet 5s were built

Many cannon-armed Breguet 5s were sent to GBM 5, which was composed entirely of Breguet-Michelin escadrilles; however, they were a failure in their intended role. It was quickly discovered that the Breguet 5s were slower than the BM 4s they were escorting, especially after the bombers had dropped their payload. Furthermore, the cannons were found to be of limited usefulness in air-to-air combat. On some occasions the Breguet 5s were flown without the cumbersome cannon, and this necessitated the addition of ballast to the nose for the aircraft to maintain its center of gravity. Fatal crashes occurred when the aircraft was flown without this additional weight. (For further details see the operational section under BM 4).

After it was realized that the aircraft were unable to function as escort fighters, some were used as bombers. Also, the cannon-armed aircraft were employed in the ground attack role.

To prevent shortages of the 220-hp Renault 12 Fb engines, one Breguet 5 was modified to accept a 225-hp Salmson A9. The re-engined aircraft, which was designated the Breguet 6, had an angular cowling completely enclosing the Salmson A9 and which was studded with cooling louvers. Some of the Breguet 6s were also fitted with a 37-mm Hotchkiss cannon. Fifty Breguet 6s were built. A version with a 230-hp Salmson A9 was designated BAC; the STAé designation was Breguet 9.

After the disastrous Oberndorf raid of 1916, the French bomber arm switched to night operations. The remaining Breguet 5s (and possibly 6s) were converted to night attack configuration; this meant widening the undercarriage back to a four-wheeled arrangement to make the aircraft easier to land and taxi at night. This modification resulted in the aircraft receiving a new designation as the Breguet 12. As with the other Breguet-Michelin variants, it could also carry the 37-mm

The uncovered fuselage of a Breguet 5 shows the position of the Renault engine. The main undercarriage wheels were placed directly beneath the engine. B85.333.

Structural detail of Breguet 6 with Salmson (Canton-Unné) A9 and 37-mm cannon mounted in the nose. B85.254.

Hotchkiss cannon. One Breguet 12 was fitted with a searchlight on the right side of the cannon. This aircraft was intended to provide defense against night raids by German airships. The Breguet 12 was assigned to the DCA (air defense units) of the Aviation Militaire defending Paris. These units included Escadrilles 393, 394, 395, 396, and 397. It is not known if aerial victories were obtained, but it is not likely; German airships did not attack Paris in 1916. The last Breguet 12 was in service as late as the summer of 1917.

United Kingdom

The RNAS purchased 59 Breguet 5s. They were assigned serials 1398–1399, 3946, 9175–9200, and 9426–9455 (ordered from the Grahame-White Aviation Co. of Hendon—only ten were built). All but Nos.9175–9200 had 250-hp Rolls Royce engines; the others had 225-hp Renaults. The aircraft were assigned to Nos.3 and 5 Wings.

Above right: Breguet 5 with a 37-mm cannon. The ungainly Breguet 5 proved to be useless in the fighter escort role. B81.685.

Right: Breguet 5 serial number 443 with cannon. B85.276.

Breguet 5 serial number 583 with a Rolls Royce engine. Auxiliary fins are mounted on the stabilizer. B85.245.

Breguet 6 serial BR 592 with 225-hp Salmson A9 engine. Fifty Breguet 6s were built. B85.265.

Above: Breguet 12. This may have been the prototype. B85.250.

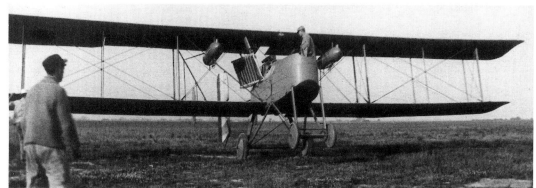

Above: Breguet 12. Aside from the altered undercarriage, the aircraft was virtually identical to the Breguet 5. B83.5705.

Breguet 12. The change to a four-wheel undercarriage was necessitated by the switch to night attack missions. This configuration made it safer to land on poorly-prepared airfields at night. B83.3626.

Breguet Twin-Engine Bomber

Several companies entered aircraft in the 1916 concours puissant for a long-range bomber. Breguet is known to have designed a twin-engined bomber for the competition, but this was apparently destroyed or further development was abandoned shortly before the concours began. Some sources refer to this biplane as the Type E (but they may be confusing the Breguet design with the SPAD E bomber, which was also entered in the 1916 concours). At this point, Breguet's only recourse was to enter his Breguet 11 heavy fighter in the hope that its superior weight-lifting capability would make it a useful bomber.

Breguet 11 Corsaire

The French were desperate to obtain a modern bomber by 1916. The concours of 1915 had failed to produce an acceptable bomber and another concours was to be held in 1916. The specifications called for a twin-engined aircraft; it had to be heavily armed because the bombers were now encountering German fighters that inflicted many losses. For adequate defense, the gunners had to have a clear field of fire both fore and aft. The expected performance included a range of 600 km with a bomb load of 300 kg (in order to attack the German industrial center at Essen). Maximum speed was expected to be 140 km/h. Several aircraft were entered in the concours but few were as ungainly as the Breguet 11.

There exists some confusion as to the role the Breguet 11 Corsaire was designed to fill. It is likely that it was initially intended to function as a destroyer. Presumably, it would attack airships and fighters with its formidable armament. It may also have been intended to have it accompany the bomber formations as a long-range escort. The aircraft had gunners in the nose of the two pusher nacelles with unobstructed fields of fire. A third gunner was seated behind the pilot in the main nacelle and controlled both an upward-firing machine gun and a second gun which could fire below the aircraft. The Breguet 11 was also armed with a 37-mm cannon. However, it was never used in the destroyer or escort role but instead was entered as a bomber in the 1916 concours.

The Breguet 11 was powered by three 220-hp Renault 12Fb engines, two mounted as pushers in nacelles resembling the main fuselage of the Breguet 5. In the center was a conventional fuselage with an engine in the nose. Fuel tanks were mounted beneath the top wing on both sides of the fuselage. The undercarriage had three wheels in a tricycle arrangement under each engine nacelle; again this was nearly identical to that used on the Breguet 5 series. All three nacelles were attached to the bottom wing. The three-bay biplane wings appear to have been of unequal span. The tail had three large rudders.

The Breguet 11 was powered by three 220-hp Renault 12Fb engines. B85.273.

The aircraft performed well, but did not win the competition. It has been stated that the organizers of the 1916 concours were shocked by some of the monsters produced in response to their specifications and it would seem fair to place the Corsaire in that category. The more conventional Morane Saulnier S and SPAD SE were selected as the winners.

Breguet 11 Four-Seat Heavy Bomber/Escort Fighter with Three 220-hp Renault 12Fb Engines
Span 27.65 m; length 11.90 m; height 4.00 m; wing area 105 sq. m
Empty weight 3,100 kg; loaded weight 4,865 kg as a bomber; 4,200 kg as a fighter
Maximum speed: 148 km/h at 2,000 m; endurance 5 hours 35 minutes
Armament: probably never fitted, but planned to include a 37-mm Hotchkiss cannon, three machine guns, or 20 120-mm bombs
One built

Breguet 11 at Villacoublay where it was assembled. B82.920.

Breguet 11 Corsaire

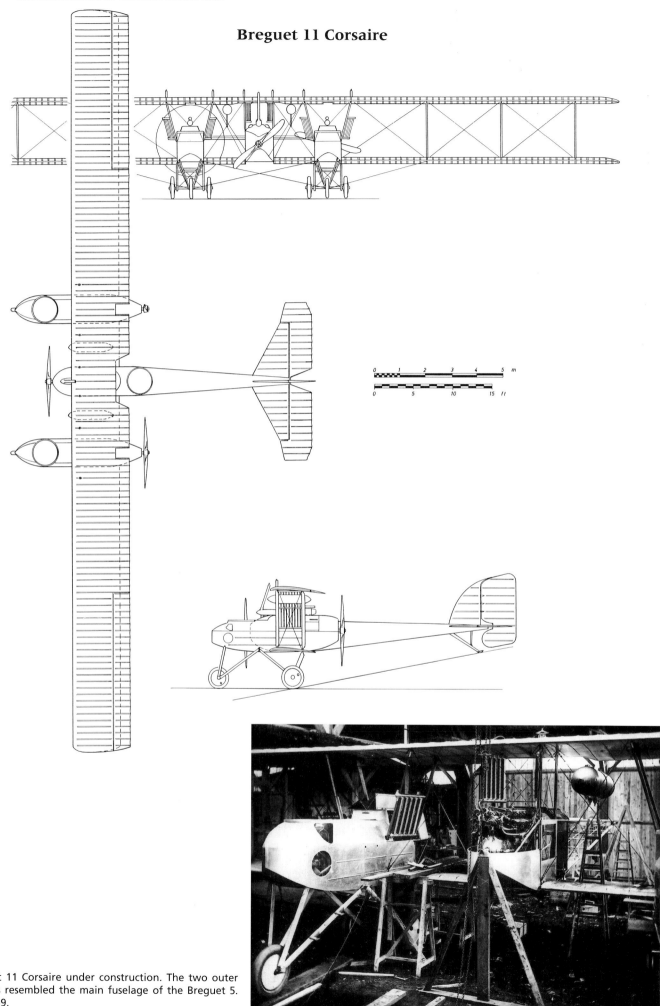

Breguet 11 Corsaire under construction. The two outer nacelles resembled the main fuselage of the Breguet 5. B84.1759.

Breguet 13 and 14

The Breguet 14 became, after the Nieuport and SPAD series of fighters, one of the best known French aircraft of the First World War. It was the most successful bomber developed in France during the war, and when the Breguet 14 entered service it rejuvenated the previously moribund French day bomber force.

After building a small number of BM 4 and Breguet 5 bombers, the Breguet firm had no further orders for aircraft and therefore turned to producing 100 A.R.1s under license. Further development of the Breguet-Michelin bombers was out of the question as this series had proved to be ineffective in their intended roles as day bombers and long-range fighters. Breguet realized that his next design would have to be capable of surviving the hostile environment of the Western Front.

His new design was designated the AV (which may have stood for Avant, or tractor layout). It was equipped with a Renault engine and featured a wing with low wing loading. Breguet returned to the tractor layout he had always preferred (the Breguet-Michelin bombers having been designed as pushers at the request of the Aviation Militaire). Another major innovation was the widespread use of duralumin, which had been employed in the construction of German airships. While this metal was valued for its strength and light weight, it was difficult to work with. Breguet's breakthrough was the discovery of a way to use the alloy in an aircraft intended for mass production. Duralumin was employed in the longerons and spacers inside the fuselage. The longerons and spacers were bolted into welded steel end fittings and braced with piano wire. The engine bearers were made of steel and duralumin tubing. Duralumin tubes also formed the main spars of the wings. The outer struts were reinforced with steel sheaths around the spars. The wing ribs were made of wood and the root ribs were made of poplar. The tail surfaces were of welded steel tubing.

The first prototype was the AV 1, equipped with a 263-hp Renault engine. A second prototype, the AV 2, had a 272-hp Renault, a slightly longer fuselage, and a slightly higher empty weight. Flight tests revealed that the aircraft pulled to the left and this required offsetting the vertical fin to the left. The upper wing was given a slight sweepback; improving stability and giving the rear gunner a better field of fire. After six months of design work and testing, the prototype flew on 21 November 1916. Tests were conducted initially at Villacoublay with Breguet himself at the controls. By January 1917 the initial flight trials had been completed. The tests confirmed that Breguet had produced a bomber that was, for once, more advanced than comparable machines used by the British and Germans. In November 1916 Breguet notified the STAé that the AV prototypes were

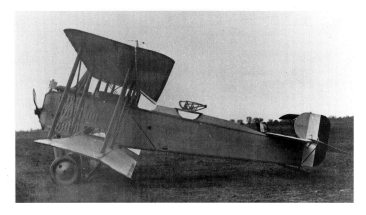

AV 1, also designated the Breguet 13 by the STAé. MA.

ready for testing. Static tests were conducted by the STAé on 26 January 1917 and confirmed the sturdiness of Breguet's design; the wings were found to have a coefficient 5.5 times the total weight of the aircraft. The AV 2 was test-flown by Adjudant Piquet (an STAé pilot) and Lieutenant Lemaitre (a pilot from BM 120). The AV 2 was highly praised, and its usefulness in the day bomber role was readily apparent to the STAé.

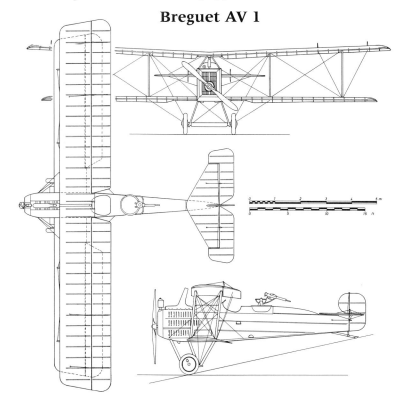

Breguet AV 1

Breguet 14 A2 of BR 220. Pilot is Serrant, observer Dreyfus. The machine gun mounted on the top wing is a field modification. B83.1783.

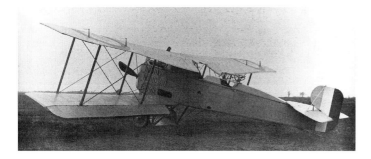

Above and above right: Additional views of the Breguet AV 1 prototype which was developed into the Breguet 14 via the Breguet AV 2. MA.

In November 1916 the STAé had formulated a requirement for four types of aircraft: a two-seat army cooperation plane (A), a three-seat, long-range reconnaissance plane (A1), a two-seat fighter (C), and a three-seat bomber (D). Breguet submitted variants of the AV intended to meet all four categories. The Type A requirement would be met by the Breguet 14 A2 while the two-seat fighter would eventually be developed into the Breguet 17. The three-seat reconnaissance plane (Class A1) and bomber (Class D) version of the AV design were built and tested, but the STAé decided that aircraft in these categories should have two engines and Breguet's design was not developed further. The two-seat fighter variant with a 300-hp Renault engine was tested, but it could not meet the required speed at 3,000 m (180 km/h). Like its competitor, the SPAD 11, Breguet's design was not selected for production (although the SPAD 11 did enter service in the two-seat reconnaissance role).

The STAé had requested that the new bomber be powered by a 200-hp Hispano-Suiza engine. Breguet, instead, decided to use the 220-hp Renault 12Fb that had been the powerplant for the Breguet-Michelin 5. The engine cowling had a plethora of cooling louvers, one of the distinguishing features of the Breguet 14 bombers. The Breguet 14 was an angular biplane. The wings had negative stagger and both were slightly swept back. The upper wing had a greater span than the lower, and ailerons were fitted to the upper wing only. Later machines had horn-balanced ailerons to improve lateral control. The two crew members were seated in tandem. The pilot sat below a cutout in the upper wing and an observer had a separate cockpit just behind him. The observer had a T.O.3 or T.O.4 gun mount with two 0.303 Lewis guns. The pilot had a single, fixed Vickers 0.303 gun mounted on the left side of the fuselage and synchronized to fire through the propeller disc. The undercarriage was strongly braced with the two wheels being separated by a strut with an airfoil cross-section. The early versions of the Breguet 14 B2 had Michelin bomb racks under the lower wings adjacent to the bracing struts of the undercarriage. Thirty-two 11.5 kg bombs could be carried. The A2 reconnaissance versions could carry four bombs, a camera, and a wireless set. There were minor differences between the bomber (designated B2) and reconnaissance (designated A2) versions. The lower wings of the bomber version had a longer span and bungee-sprung flaps were added. These flaps enabled the B2 variants to carry heavier bomb loads; test flights revealed that the aircraft could carry 730 kg of bombs and fuel. Fully loaded, the Breguet 14 B2 could climb to 4,000 m in 26 min and could attain a speed of 165 km/h at that altitude. The observer had an extra set of windows in the fuselage sides; it has been speculated that these were intended to provide enough light to enable him to use the bombsight. There were transparent panels in the underside of the aircraft, which enabled both the pilot and observer to view the ground.

Breguet received an order for 150 Breguet AV 1s

Breguet AV 2. This aircraft differed from the AV 1 by having a 275-hp Renault engine, a slightly longer fuselage, and a fixed fin. Reairche Collection.

(reconnaissance) on 6 March 1917 and on the same date Michelin received an order for 150 AV 2s (bomber version). In fact, initially the STAé designated the AV 1 the Breguet 13 and the AV 2 the Breguet 14. Of course, the Breguet 13 designation was soon dropped and both variants received the Breguet 14 designation. An additional order for 100 aircraft was placed on 4 April, followed by 250 in July and 125 in September. The engine used was primarily the 300-hp Renault 12Fcx. Additional

Breguet AV 2

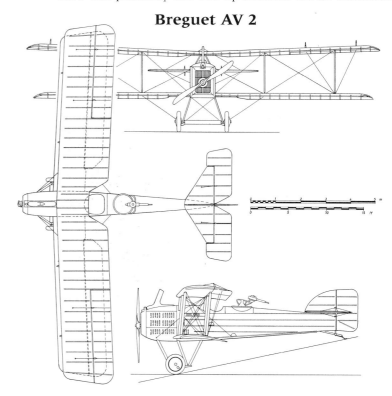

Above: Breguet AV 2 which was the second prototype of the Breguet 14 series. MA70868.

Right: The Breguet AV 2 fitted with balanced ailerons. MA38555.

Below: Breguet 14 A2 at the airfield of No.4 Squadron AFC, 1918. M.T. Cottam via Colin Owers.

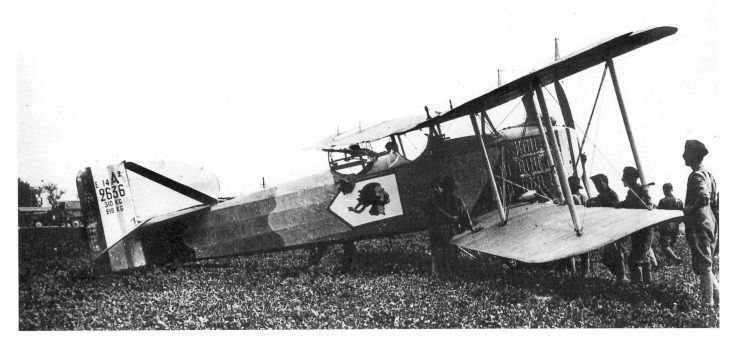

aircraft were built under license. These orders were placed with Darracq (330 aircraft), Farman (220), Paul Schmitt (275), Ballanger (300), and Sidam (300). It is believed that production was evenly divided between the bomber and reconnaissance versions. Serial numbers for the production Breguet 14s began at 1101 (1106 was the first aircraft delivered).

In operational service, the Breguet 14s underwent numerous modifications, including addition of a Lewis gun on the top wing, armored seats, and a gun rigged to fire underneath the fuselage. As mentioned earlier, some Breguet 14s were equipped with horn-balanced ailerons, and on the B2s this change was accompanied by deletion of the lower wing flaps and reduction in the lower wing span and wing tip size.

Variants

Breguet 14 with Fiat Engine—due to a shortage of Renault engines, some versions of the Breguet 14 were fitted with 300-hp Fiat A-12 bis engines. The first example flew in 1917 with a Fiat A-12 engine, and this Breguet 14 could be distinguished by its tapered cowling and underslung radiator. However, these changes reduced the aircraft's performance and, as a result, a modified engine, the Fiat A-12 bis, was developed. This developed the same horsepower but it could be fitted to the aircraft without requiring alterations to the cowling. The Fiat engines were used in the A2, B2, and E2 (trainer) versions of the Breguet 14. The type equipped 24 Breguet 14 escadrilles,

Breguet 14 B2 serial BR 1102 with Michelin bomb racks underneath the wings. This is an early Michelin-built machine. The fuselage windows are lower on the fuselage than on later machines. B84.1085.

Breguet 14 A2 of BR 11. The insignia was a red cocotte. B82.683.

and examples were supplied to Belgium and the United States. Other examples of the Breguet 14 were fitted with the 260-hp Fiat A-12 engine, and at least one with a 600-hp Fiat A-14.

Breguet 14 Ap2—The Ap 2 specification of 1917 called for a high altitude, long-range reconnaissance aircraft. Breguet 14 B2 (serial number 4360) was modified to accept a 400-hp Liberty 12 engine. The aircraft was tested at Villacoubly. Neither it nor its competitor, the Hanriot Dupont 9, was selected by the Aviation Militaire. A similar installation was performed at McCook Field on Breguet 14 B2 AS 94097. As far as can be determined, the installation of the Liberty on the B2 did not affect performance and, had the war continued, it might have proved a useful alternative to the Renault or Fiat engines. On the Breguet 14 A2 the Liberty provided a significant performance edge over the Renault-equipped machines.

Breguet 14 with Lorraine-Dietrich Engine—A Breguet 14 A2, serial 1021, was fitted with a 285-hp Lorraine-Dietrich 8Bd engine. The installation was successful enough to warrant production; some versions of this aircraft had a blunt cowling and an underslung radiator. These were used by French T.O.E. (colonial) escadrilles and may have been designated Breguet 14TOEs, although this designation has not been confirmed. It has been reported that some were supplied to Spain; however, Spanish sources do not confirm this.

Other examples of the Breguet 14 were fitted with 370-hp Lorraine-Dietrich 12Da or 390-hp Lorraine-Dietrich 12E engines. It appears these remained one-off conversions.

Breguet 14 with Panhard Engine—A 340-hp Panhard 12D and a 350-hp Panhard 12C were each fitted experimentally to a Breguet 14 airframe. Series production did not ensue.

Breguet 14 A2 with Renault Engine—A single Breguet 14 A2 was tested with a 400-hp Renault 12k engine in May 1918. This aircraft served as the basis for the Breguet 17 escort fighter.

Other Renault engines fitted to the Breguet 14 included the 310-hp Renault 12Fcy, 320-hp Renault 12Fe, and 350-hp Renault 12Ff.

Breguet 14 AE—This aircraft was designed for use in the French colonies. It first flew in 1920 and carried the registration F-AEEZ.

Breguet 14/400—Version with a 400-hp Lorraine 12Da engine. Seventy were supplied to China and Manchuria during the 1920s.

Breguet 14 C—Version with 450-hp Renault 12Ja engine. It first flew in 1920 and was used as a postal aircraft in the United States.

Breguet 14 H—Floatplane version with 320-hp Renault 12Fe engine. It had a large central float beneath the fuselage and two smaller floats under the wings. At least two were produced and saw service with the Escadrille Indo-Chinoise No.2 based in Bien Hoa in 1925.

Breguet 14 with Supercharger—A number of Breguet 14s were modified to accept Renault 12Fe engines with the Rateau turbo-supercharger. Aircraft with the new engine were found to have a marked improvement in performance. One set a world record in 1923 by climbing to 5,600 m with a payload of 500 kg. Sixteen Breguet 14s with the turbo-supercharged engines were in service with the 34th RAO at Bourget in 1924.

Breguet 14 A2 with 49 sq. m Wing—This aircraft was developed by Breguet in the hope of meeting the STAé's requirement for a maximum speed of 180 km/h at 3,000 m. This version had a reduced wing area of 49 sq. m without the automatic flaps and a more powerful Renault 12Fcx engine.

Breguet 14 B2 serial BR 13005 of BR 134. The observer's windows were on either side of his cockpit. This late Michelin-built machine lacks the trailing-edge flaps on the lower wing. The engine is a Liberty. B89.4797.

Breguet 14 B2 serial BR 1102. B84.1095.

Aircraft 665 was used for official trials. One advantage of the new type was that it could reach an altitude high enough to make it immune to anti-aircraft fire or enemy fighters. In fact, in September 1917 an altitude of 8,000 meters was reached.

Breguet 14 B1—The B1 designation indicates that this was a single-seat version of the Breguet 14 B2 with a wing area of 52 sq. m and automatic flaps. The front cockpit was fitted with fuel tanks and the pilot was relocated to the observer's position. Test pilot De Bailliencourt later tested the Breguet 14 B1 intended for Jules Védrine's raid on Berlin. While flying the aircraft he found that the Solex carburetor froze at altitude. He attempted to land in order not to risk damage to this specially modified machine, but as he descended, the engine stalled. However, he was able to bring the plane down safely. Unfortunately, the prototype of the Blériot 71 apparently crash-landed while attempting to avoid the incoming Breguet 14 B1.

The Breguet 14 B1 could carry a 180-kg bomb load and had a duration of up to seven hours. Two were ordered and were intended to be used on a raid on Berlin. Jules Védrines performed a test flight of 400 km (flying from Paris to Cancale) in July 1917 in preparation for this mission. However, the overall reluctance of the French government to bomb German cities (because of the high risk of German retaliation) combined with the Breguet 14 B1's marginal range and vulnerability to enemy fighters, resulted in cancellation of the plan.

Breguet 14 Floatplane—Another floatplane version of the Breguet 14 was tested in 1924 at the Saint Raphaël Center for naval aviation. This version had twin floats as opposed to the large central float used on the Breguet 14H. The floats used on it were manufactured by Blanchard and could be used interchangeably with the regular undercarriage. At least one Breguet 14 B2 landplane was tested with floatation gear.

Breguet 14 S—The French led the world in the development of ambulance aircraft. An ambulance version of the Breguet 14 was developed to supplement, and later replace, Voisin 10 ambulance planes. These aircraft were designated Breguet 14 S (S = Sanitaire). Initially they were simply modified to carry two stretcher cases in the rear fuselage. Subsequently, a dedicated ambulance version was produced that was a modified version of the Breguet T. In 1918, four Breguet 14 S machines were used over the Aisne front. For postwar use see the section on colonial campaigns.

Foreign variants of the Breguet 14—see the section on foreign service for further details.

Civil Versions of the Breguet 14:

Breguet 14 T—This was a conversion of the basic Breguet 14, capable of carrying two passengers in a cabin in front of the pilot's cockpit. The entrance door was on the starboard side of the fuselage. An ambulance version that had the stretchers located in a cabin ahead of the pilot was also produced.

Breguet 18 T—This was an enlarged 14 T with a 450-hp Renault 11Ja engine. It had an enlarged cabin that could carry four passengers.

Breguet 14 Tbis—This version was a hybrid of the previously mentioned 14T and 18T. It had a cabin that carried three passengers.

Civil operators of the Breguet 14s included Cie des Transports

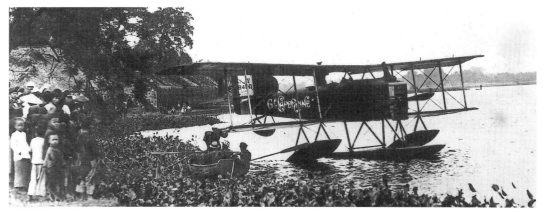

Breguet 14 floatplane of the 1st Escadrille d'Indochine at Hanoi on 20 December 1920. B88.4559.

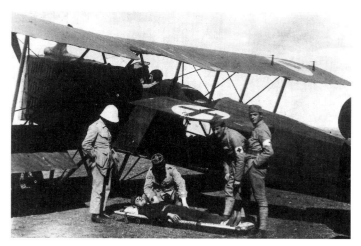

Breguet 14 S ambulance aircraft with a stretcher case being loaded. The Breguet 14 S saw widespread service during the Rif campaign, performing some of the first aeromedical airlifts in history. B83.3084.

Aériens Guyanais—five T bis; Swedish Red Cross—two T bis; Lignes Aériennes Latécoère—106 T and T bis (some of these were converted to carrying mail and were known as the Torpedo or Breguet Latécoère); SNETA—this Belgian airline operated three Breguet 14 A2s with 180-hp engines; Compania Rioplatense de Aviacion—used Breguet 14s on the Buenos Aires-to-Montevideo service in 1921.

Operational Service
Reconnaissance

The equipment of the reconnaissance escadrilles was clearly outdated by late 1917. The aircraft used by these units included Sopwith 1½ Strutters and A.R.1 and 2s. The introduction of the modern Breguet 14 A2s greatly enhanced the effectiveness of these escadrilles. The first unit to receive Breguet 14 A2s was BR 7, which re-equipped with the new aircraft in August 1917.

BR 11 provides an example of the range of missions that Breguet 14 A2 units routinely performed. BR 11 had been formed from C 11 in November 1917; it was among the first escadrilles to receive the new aircraft. It was found that the Breguets were effective in the high-altitude (6,000 m) reconnaissance role. The aircraft also flew liaison and artillery regulation missions. During the Battle of Picardie BR 11 flew photographic missions to aid the planned French counter-offensive in the Ourcq sector. During this offensive BR 11 moved to four different airfields in support of the 2nd Armée. The escadrille was active during the battles near Champagne and photographed the German lines before the Lorraine offensive. Postwar, BR 11 served as part of the Rhine occupation force, assigned to the 33rd Regiment d'Aviation.

Other escadrilles which used the Breguet 14 A2 were:

BR 7, formed from SOP 7 in mid-1917. The Breguet 14 A2s were received at Plessis-Belleville. In June BR 7 was sent to the Vosges front and was assigned to the 6th C.A. Later in 1918, the escadrille was sent to Oise during the Battle of Picardie. The unit next was sent to Bruthécourt and finally to Lorraine. Postwar, BR 7 was based at Dijon; it was redesignated the 7th Escadrille of the 2nd RAO.

BR 9, formed from SOP 9 in May 1918. Initially based at Étampes, the escadrille was assigned to the 19th C.A. and participated in the Battles of Champagne and Second Marne. The unit moved from Noirlieu to Luxeuil in October. BR 9 ended the war attached to the 7th Armée and based at Phaffans. As part of the occupation force, BR 9 was based at Neuf-Brisach. It was disbanded in December 1918.

BR 35, created from AR 35 in November 1917 when it received Breguet 14 A2s. Assigned to the 35th C.A., BR 35 was active in the 1st Armée sector. BR 35's bases included Ferté-sous-Jouarre, Andelys, and Fienvillers during the Battle of Picardie. In May, BR 35 was active in the Battle of Chemin des Dames and moved to Le Tergnier after the battle. It became the 5th Escadrille of the 2nd RAO in January 1920.

BR 43, created from SOP 43 in June 1918. It was assigned to the 68th D.I. in the 7th Armée sector. At the time of its transition to Breguet 14 A2s, the unit was based at Moissy-Cramayel; it ended the war based at Lunéville. Postwar, BR 43 was based at Belfort, Honburg, and Courban. It was disbanded in February 1919.

BR 44, formed from AR 44 in January 1918. Assigned to the 31st C.A. the escadrille participated in the Battle of Picardie and in May the French attacks on the Hindenburg Line. BR 44 was also active in the Battles of Saint-Quentin (September) and Guise (November). Postwar, the escadrille was based at Brassure and Habsheim. It was disbanded in May 1919.

BR 45, formed from AR 45 in March 1918. It was assigned to the 1st Armée and was based at Esquennoy. It was disbanded in March 1919.

BR 104, formed from SOP 104 in May 1918. It was initially assigned to the 31st C.A. and subsequently to the Moroccan division. It became the 1st Escadrille of the 3rd RAO in January 1920.

BR 141, created from SOP 141 in August 1918. The unit was assigned to the 6th C.A. and was based at Lunéville. BR 141 ended the war in the 8th Armée sector based at La Neuville-Dev-Bayon. It was disbanded in April 1919.

Artillery Cooperation

The French also assigned Breguet units to serve exclusively in the artillery cooperation role. These escadrilles were assigned to the heavy artillery (A.L.) units of each Groupe d'Armée (Army Group). The Breguet 14 A2 units routinely performed liaison and artillery regulation. They were:

BR 201, formed from AR 201 in April 1918 at Plessis-Belleville. Initially based at Dugny, it was assigned to the 1st Armée. The unit was particularly active in June during the counter-attack at Matz. At the war's end the BR 201 was based at Ercheu. It subsequently moved to Châlons-sur-Marne, where it became the 2nd Escadrille of the 2nd RAO in January 1920.

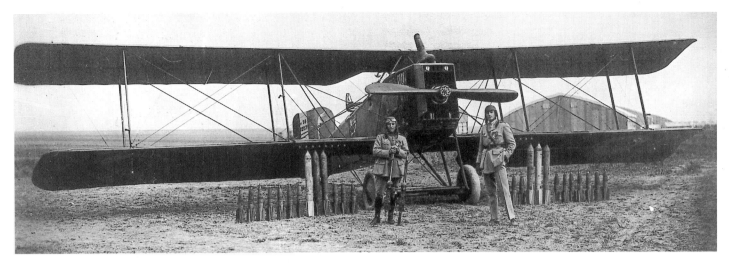

Breguet 14 B2 serial 12079 of BR 123 showing a full complement of 50-kg and 75-kg bombs. The crewmen are identified as the Peyerimhoff brothers. This aircraft was built by Michelin. B86.724.

Breguet 14 lineup of BR 117 July 1918. BR 117 was attached to Escadre 12. B76.1739.

BR 202, created from C 202 in December 1917. Under the command of Lieutenant Trimbach, it was assigned to the 2nd Armée. During the war it received two new commanders (Lieutenant Graff and later Capitaine Saling) and was later assigned to the 1st Armée. BR 202 ended the war at Libermont and was disbanded in April 1919.

BR 205, formed from AR 205 in February 1918. It was assigned to the American 1st Army and was active during the Battle of Saint-Mihiel. BR 205 ended the war based at Courcelles-Saint Aire and was disbanded in March 1919.

BR 206, created from SOP 206 in May 1918 and assigned to the 8th Armée. Based at war's end at Burthecourt, BR 206 was disbanded in December 1918.

BR 207, formed from SOP 207 in April 1918. After a period of training at Étampes, it was assigned to the 4th Armée. BR 207 was based at Cauroy at the war's end. It became the 1st Escadrille of the 5th RAO in January 1920.

BR 208, created from SOP 208 in February 1918. Under the command of Capitaine Proust, and based at Brétigny-sur-Orge, BR 208 performed numerous army cooperation sorties. BR 208 moved to Corbaulieu in April and was assigned to the American 1st Army. It also participated in the Battle of Saint-Mihiel. Based at Vadelaincourt at the war's end, BR 208 was disbanded In March 1919.

BR 209, formed from AR 209 in September 1917. Assigned to the 3rd Armée, BR 209 was based at Croutoy at the war's end. It was disbanded in April 1919.

BR 210, formed from SOP 210 in July 1918. It was active near Verdun and, in October, at Malmaison. Assigned some Caudron R.11s to supplement the Breguet 14 A2s, BR 210 participated in the Battles of Aisne (May) and Montdidier-Noyon (June). The unit was initially assigned to the 10th Armée, later to the 4th Armée, and was based at Herbisse at the war's end. It was disbanded in July 1919.

BR 211, created from AR 211 in March 1918. It was assigned to the 1st American Army and participated in the Battles at Saint Mihiel (September) and Montfaucon (26 September to 15 October). At the war's end, BR 211 was based at Clermont; postwar it had various bases including Tangry and Saint-Dizier. It became the 1st Escadrille of the 4th RAO in January 1920.

BR 213, formed from C 213 in 1918. It was assigned to the 4th Armée and was based at Cauroy at the war's end. It was disbanded in July 1919.

BR 214, created from AR 214 in May 1918. It was assigned to the American 1st Army and was active during the Battle of Saint-Mihiel. Based at Foucaucourt at the war's end, BR 214 was disbanded in December 1918.

BR 216, formed from SOP 216 in April 1918. It was assigned to the 5th Armée and was based at Montbre at the war's end. It was disbanded in March 1919.

BR 217, created from SOP 217 in May 1918. It was assigned to the 3rd Armée and ended the war based at Tigny. BR 217 was disbanded in April 1919.

BR 218, formed from F 218 in late 1917. It was assigned to the 33rd C.A. and ended the war in the 10th Armée sector. It became the 8th Escadrille of the 5th RAO in January 1920.

BR 219, created from SOP 219 in February 1918. It was assigned to the 4th Armée during the Battle of Montagne de Reims and the 2nd Armée during the offensives at Saint-Mihiel and Argonne. It ended the war under the control of the American 1st Army; BR 219 was based at Courcelles-Saint Aire. It became the 4th Escadrille of the 2nd RAO in January 1920.

BR 220, formed from C 220 in October 1917. Commanded by Lieutenant Brédiam and based at Dugny, it became part of Groupement Féquant in April 1918. It joined the Division Aérienne de Duval in June and Groupement Weiller in October (where it was based at Champagne in the G.A.C. sector). It became the 1st Escadrille of the 2nd RAO in January 1920.

BR 221, created from SOP 221 in May 1918. Assigned to the 7th Armée, the escadrille was based at Chaux. BR 221 was disbanded in March 1919.

BR 222, formed from SOP 222 in May 1918. It was assigned to the 5th Armée and based at Montbre. It was disbanded in March 1919.

BR 223, formed from SOP 223 in May 1918. Initially assigned to the 4th Armée and based at Dugny, it was subsequently based at

Right: Breguet 14 A2 of BR 7. B78.917.

Below right: Breguet 14 A2 *Jane IV* of BR 209 flown by Capitaine De La Motte Ango De Flers in 1917. B78.3.

Bottom right: Another view of Breguet 14 A2 *Jane IV* piloted by Mal des Logis Jean Charles Pavy and Lieutenant Hardy of BR 209 in 1917. BR 209 was active over Noyon and Saint Quentin. B77.1864.

Tricon and Bussy-Lettrée. BR 223 participated in the battles at Chemin des Dames, Champagne, and la Serre and was assigned to the 5th Armée. It ended the war at Saint Etienne; postwar, BR 223 moved to Coulonges and was disbanded in early 1919.

BR 224, created from C 224 in October 1917. It was initially assigned to the 35th C.A. Under the command of Capitaine Paul-Louis Weiller, the escadrille was assigned the 10th Armée during 1918. Postwar it served as part of the German occupation force and became the 6th Escadrille of the 4th RAO in January 1919.

BR 226, formed from SOP 226 in April 1918. It was under the command of Lieutenant Ducos de la Haille and assigned to the 3rd Armée. Based at Vaucastille at the war's end, BR 226 became the 5th Escadrille of the 1st RAO in January 1920.

BR 227, created from C 227 in October 1917. It was under the command of Capitaine Poucher and was active in the 4th Armée sector. BR 227 ended the war based at Rumont assigned to the American 2nd Army. In January 1920 the unit became the 7th Escadrille of the 4th RAO.

BR 228, formed from SPA-Bi 228 in July 1918. Initially assigned to the 2nd C.C., it was later assigned to the American 1st Army and based at Prete. Postwar, BR 228 moved to Bicqueley and in 1920 became the 2nd Escadrille of the 4th RAO.

BR 229, created from SOP 229 in May 1918. Assigned to the American 2nd Army, the unit was disbanded in March 1919.

Breguet 14 A2 with 300-hp engine flown by Capitaine Chanaron and Capitaine Garcin of BR 275. Breguet 14s fitted with Fiat engines could be distinguished by their tapered cowling and underslung radiator. B86.2573.

Breguet 14 A2 of BR 209 which was assigned to the 3rd Armée. The escadrille's insignia was a cockatoo. B78.4.

BR 231, formed from SOP 231 in April 1918 and was assigned to the 5th Armée. It was based at Anthienay at war's end and was disbanded in March 1919.

BR 232, created from SOP 232 in May 1918 and assigned to the 4th Armée. Based at Fe Scay at war's end, BR 232 was disbanded in March 1919.

BR 233, formed from AR 233 in April 1918. Assigned to the 4th Armée, and based at Waegemoulin, BR 233 was disbanded in February 1919.

BR 234, created from SOP 234 in May 1918. Assigned to the 1st Armée, the escadrille was based at Libermont at the war's end. It became the 2nd Escadrille of the 1st RAO in January 1920.

BR 235, formed from SOP 235 in May 1918. It was assigned to the 2nd C.A. Italien (an Italian unit serving on the Western Front). BR 235 was disbanded in February 1919.

BR 236, created from SOP 236 in May 1918. It was initially assigned to the 2nd Armée. By the end of the war it was assigned to the American 1st Army and was based at Prete. BR 236 was disbanded in February 1919.

BR 237, formed from SOP 237 in May 1918 and assigned to the 4th Armée. It was based at Fe Scay at war's end and disbanded January 1919.

BR 238, created from SOP 238 in May 1918. Initially assigned to the 6th Armée, it was disbanded in July 1919.

BR 243, formed from elements of Escadrilles 23, 79, and 102 in September 1918. It was based at Brabant-le-Roi and assigned to the 2nd Armée. It became the 5th Escadrille of the 5th RAO in January 1920.

BR 244, created in September 1918 and assigned to the 8th Armée. Under the command of Lieutenant Desaix, it was active over the Lorraine front. The unit was based at Manoncourt at the time of the Armistice. Postwar BR 244 moved to Houdelmont and in January 1920 became the 6th Escadrille of the 2nd RAO.

BR 245, formed in October 1918 from Escadrilles 49, 69, and 76. It was assigned to the 7th Armée and based at Chaux. BR 245 was disbanded in May 1919.

BR 250, formed from SOP 250 in September 1918. It was

A rare photograph showing Breguet 14s in service with the Aviation Maritime. This aircraft was assigned to the Centre d'Instruction at Rochefort-sur-Mer. B78.646.

Below right: Breguet 14 A2 with a 300-hp Fiat engine. The Fiat engine was utilized to help alleviate shortages of the Renault engine. Renaud.

assigned to the 5th C.A. and was based at Morangis. At the end of the war it was under the command of Capitaine Peltier and based at Roucy. Postwar BR 250 was based at La Cheppe and was disbanded in April 1919.

BR 257, formed from AR 257 in July 1918. It was assigned to the 9th C.A. and based at Fe Scay. BR 257 was disbanded in April 1919.

BR 260, formed from SOP 260 in June 1918. It was initially attached to 1st C.A.C. and by the end of the war it was attached to the 1st Armée Artilliere. BR 260 was disbanded in February 1919.

BR 267, formed from AR 267 in July 1918. It was assigned to the 4th C.A. and was based at Menil-Leoinois in the 5th Armée sector. It was disbanded in March 1919.

BR 269, formed from SOP 269 in July 1918. It was assigned to the 2nd C.A. and was based at Burthecourt. It was disbanded in February 1919.

BR 271, created from SOP 271 in July 1918. It was assigned to the 34th C.A. and was based at Wyngene in the 6th Armée sector. It was disbanded in March 1919.

BR 272, formed from AR 272 in July 1918. It was assigned to the 33rd C.A. in the 10th Armée sector and was disbanded in March 1919.

BR 274, created from AR 274 in July 1918. It was assigned to the 16th C.A. and was based at Cerby-les-Bucy in the 3rd Armée sector. It was disbanded in March 1919.

BR 275, formed from AR 275 in April 1918. It was assigned to the 31st C.A. and was based at Libermont in the 1st Armée sector. It was disbanded in August 1919.

BR 279, formed from SOP 279 in June 1918. It was initially attached to the 2nd C.A.; by the end of the war it was assigned to the 22nd C.C. at Bray-Dunes. It was disbanded in December 1918.

BR 281, formed from SOP 281 in August 1918. It was assigned to the 17th C.A. in the American 1st Army sector and was based at Beauzee. It was disbanded in April 1919.

BR 282, created from SOP 282 in September 1918. It was assigned to the 35th C.A. and based at Lavergny in the 3rd Armée sector. It was disbanded in March 1919.

BR 283, formed from SOP 283 in June 1918. It was assigned to the 67th D.I. and based at Feu Du Rendez-Vous in the 1st Armée sector. It was disbanded in May 1919.

BR 287, formed from SOP 287 in June 1918. It was assigned to the 52nd D.I. and as based at Roucy in the 5th Armée sector. It was disbanded in December 1918.

BR 509 NE, formed as an escadrille of liaison for the G.A.N. and, in particular, the Armée of the Nord-Est (North East) in May 1918.

Bombardment

The French day-bomber escadrilles were equipped with Sopwith 1½ Strutters and Paul Schmitt 7s in late 1917. These aircraft were obsolete, and the units using them suffered heavy losses. The Breguet 14 proved to be an excellent basis for a light bomber, the enlarged lower wing span and bungee-sprung flaps enabling the Breguet 14 B2 to carry heavier bomb loads.

The first escadrilles to receive the new bomber were two units still listed as using Breguet-Michelin 4 and Breguet 5s. BM 117 (assigned to GB 5) re-equipped in August 1917, followed by BM 120 (GB 5) in September 1917; they became BR 117 and BR 120.

Two units using Paul Schmitt 6s and 7s received Breguet 14 B2s in November 1917. PS 126 (GB 3) and PS 127 (GB 5) were redesignated BR 126 and BR 127.

Ten of the day bombardment escadrilles were equipped with Sopwith 1½ Strutters. SOP 111 (GB 6) was the first Sopwith-equipped unit to receive Breguet 14 B2s; it became BR 111 in October 1917 followed by BR 66 (GB 6), BR 108 (GB 6), BR 128 (GB 3), BR 107 (GB 3), BR 123 (GB 9), BR 129 (GB 9), BR 131 (GB 4), BR 132 (GB 4), and BR 134 (GB 4).

VB 113 replaced its Voisin 10s with Breguet 14 B2s in March 1918 and was assigned to GB 8 as BR 113.

The Breguet bomber units were organized as follows:

Escadre 12 (day bomber): GB 5: BR 117, 120, 127
GB 6: BR 66, 108, 111
GB 9: BR 29, 123, 129
Escadre 13 (day bomber): GB 3: BR 108, 126, 128
GB 4: BR 131, 132, 134

The introduction of the Breguet 14 B2s greatly enhanced the capabilities of the French day bomber escadrilles. The availability of a modern bomber in large numbers permitted the French to use massed formations against targets in the vicinity of the front. Initially, attacks were concentrated on railway stations and lines of communication. As the war progressed, the Breguet 14s were used to strafe and bomb enemy troops along the front lines. The first attacks began in mid-November 1917 when the railway stations at Juniville and Attigny as well as the iron works at

Breguet 14 A2 of an unidentified escadrille. The A2 variant can be readily distinguished from the B2 by the lack of windows in the observer's compartment. B96.358.

Breguet 14 A2. The A2 variant was used by army cooperation and artillery registration escadrilles. B86.389.
Below: Breguet 14 A2 serial 779 of BR 202 named *Gabrielle*. MA7180.

Neufchatel-sur-Aisne were bombed.

There were only limited numbers of Breguet 14s available in December and this meant that, usually, only a dozen aircraft could be sent to attack. However, it was possible to assemble larger formations of bombers as 1917 came to an end. On 22 December 25 Breguet 14 B2s attacked Montcornet, Asfield, and Bussr-les-Pierpont. The Breguet 14s were also used for night missions; for example, 13 aircraft bombed the railway station at Lumes and the airfield at Maimoisors on the night of 26/27 December.

During 1918 the Breguet 14 B2 equipped all the day bomber groups of the Aviation Militaire. These were Groupes de Bombardement GB 3, 4, 5, 6, and 9. From January 1918 through March 1918 these units were active all along the front. GB 3 was assigned to the 3rd Armée and concentrated on high-altitude (above 5,000 m) bombing attacks, while GB 4 bombed targets in Pechlebon, Rothwreil, Karlsruhe, and Stuttgart. GB 5 flew photographic and bombing missions over Chalons. GB 6 concentrated on railway stations along the front. GB 9 was formed on 11 March, 1918 just before the Battle of Picardie; it was assigned to Escadre 12.

The Battle of Picardie and Flanders began on 21 March and lasted until 19 April 1918. All five of the Breguet-equipped Groupes d'Bombardement were utilized in this battle. GB 3 was based at Maisonneuve. GB 4, as well as all three of Escadre 13s day bomber units (GB 5, 6, and 9), was based at La Villeneuve Le Roi.

The Breguet units began their attacks 23 March, 1918. Special attention was given to attacking troop convoys, train stations, and German troops on the front line. Usually these targets were attacked with bombs, but the Battle of Picardie also saw the first widespread strafing of front-line troops. In addition, the Breguet 14s dropped 75-mm shells modified as anti-personnel weapons. The efforts of all the bomb groups were required in order to sustain the air offensive on this front. The Breguet 14 B2 units attacked troops, bridges, and supply lines in an attempt to slow the German advance. By late March GB 5, 6, and 9 had moved to airfields near Plessis-Belleville, while GB 3 and 4 were at Cramaille. During the first week in April the battle had stabilized enough that the bomb groups were able to resume attacks on railroad and airfield targets.

The raids on German airfields had destroyed significant numbers of enemy aircraft; as a result, the French bombers encountered significantly less aerial opposition. The disruption of the Luftstreitskräfte permitted the bombers to attack command centers and supply dumps far behind the enemy lines. As the Battle of Picardie continued an innovation appeared. Large numbers of SPAD fighters drawn from the Groupes d'Chasse were assigned to escort the bombers. However, the planned rendezvous between fighters and bombers often didn't take place; this resulted in the Breguet 14s proceeding to their targets without fighter coverage. On 3 April GB 3 and 4 moved to Beauvaise, while GB 5, 6, and 9 moved to Montagne. The Breguet escadrilles helped slow the German onslaught on Picardie and Flanders, giving the Allied forces a chance to recover.

The bomb groups were less active during the month prior to the Battle of the Aisne. Two innovations were introduced during this period: fighter groups began to develop effective tactics for escorting bombers to their targets, and diversionary raids were introduced to help draw German fighters away from the major targets. GB 5, 6, and 9 moved to Fouilloy and were assigned to support the 1st and 3rd Armées. GB 3 was based at Fourneuil. GB 4 was at La Villeneuve Le Roi. Between 16 and 27 May large numbers of fighters were assigned to escort the bombers. While these raids were successful, the Commandement de l'Aéronautique complained that the size of these aerial armadas made them unwieldy. Diversionary raids were also introduced during this period. Attacks would be made using a small number of bombers against areas well away from the prime target. This

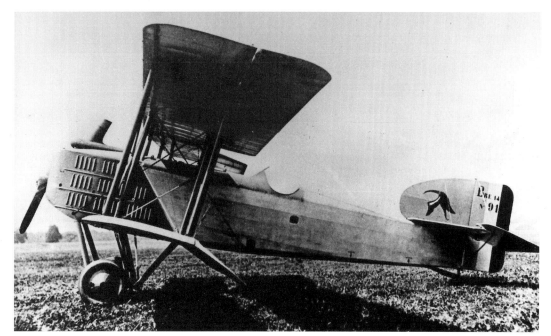

Breguet 14 B1 intended for Jules Védrine's proposed raid on Berlin. However, the aircraft's range proved to be too marginal for the task and the raid was canceled. MA37663.

helped to draw German fighters away from the main bombing force. These diversionary raids proved successful and were frequently repeated during the remainder of the war.

Just before the Battle of the Aisne began on 17 May, GB 3 moved to Champaubert, then Blequecourt. GB 4 was at Le Roi. GB 5 was initially based at May-en-Multien and later moved to St. Dizier. GB 6 was at Villers-en-Lieu. GB 9 was at Founneuil at the start of the battle, later moving to Behoune.

The Battle of the Aisne lasted from 27 May to 4 June. During it, the bomb groups concentrated their attacks on the enemy lines of supply and communication in an attempt to slow the German advance. However, the units had to move (as shown above) because the deteriorating situation at the front endangered their bases. The activity on the eastern portion of the front increased in early June, which necessitated a move of the bomb groups to that area. GB 3 and 4 moved to Fere-en-Tardenois, and GB 5, 6, and 9 concentrated their attacks on troops in the Ourcq Valley.

The Battles of Metz and Soissonnais lasted from 5 June to approximately 15 July. GB 5, 6, and 9 attacked German troops heading for these villages, and, along with the RAF's 9th Brigade, helped to halt the enemy advance by 11 June. GB 3 moved to Blequencourt, while GB 4 was based at Villeneuve-le-Roi. GB 3 and 4 bombed railway stations and German aerodromes. During the latter part of June GB 5, 6, and 9 attacked the frontlines at Ourcy and the Marne. All five bomb groups were allowed periods of rest before the start of the Battle of Champagne.

The first part of July saw a continuation of direct air support for the front-line troops, as well as interdiction of enemy supply lines and troop transport. Just before the beginning of the Battle of Champagne on 25 July, the 2nd Brigade, with GB 3 and 4, was based at Linthelles. The 1st Brigade, with GB 5, 6, and 9, was disposed as follows: GB 5 at Saint-Dizier, GB 6 at Plessis-Belleville, and GB 9 at Behonne.

By the time the Battle of Champagne began on 25 July, yet another innovation had been introduced to the day-bombing campaign. The Breguet 14 B2s were now escorted both to and from the target by Caudron R.11s. These large, twin-engined aircraft had the range to accompany the bombers the full distance and the firepower to defend them. The bomb groups were now fighting on the offensive, concentrating their attacks on escaping troops.

The Battle of the Ile de France took place from 18 July to 4 August. The disposition of the bomber units was:

Below: Breguet 14 B2 serial 4599 showing its 300-hp Renault engine. Via Colin Owers.

1st Brigade
GB 5—Linthelles (to Paris les Romilly 24 July)
GB 6—Linthelles (to Paris les Romilly 24 July)
GB 9—Linthelles (to Gourgançon 24 July)
2nd Brigade
GB 3—Linthelles (to Roissy on 24 July)
GB 4—Linthelles (to Mauregard on 24 July)

From 18 to 22 July these units attacked retreating German troops, especially along the banks of the Marne. At this time the Breguet 14s suffered few losses to enemy aircraft, the Allies having firmly established air superiority. Caudron R.11s and SPAD 13s escorted the bombers. From 22 July to 6 August, the GB units concentrated their attacks on the Vesle and Ardre valleys in support of the 5th Armée.

The next major action was the Battle of Santerre from 8 to 30 August 1918. The bomber escadrilles were to provide support for the 2nd and 3rd Armées. GB 3 was based at Roissy and GB 4 at Mauregard. GB 5 was at the airfield at Hallancourt; GB 6 was at Le Rois; and GB 9 was at Lormaison.

The Battle of Saint Mihiel lasted from 12 to 30 September. The bomb groups were at the following bases:

1st Brigade	2nd Brigade
GB 5 at Martigny	GB 3 at Vombles
GB 6 at Tantonville	GB 4 at St. Dizier
GB 9 at Neufchateau	

All these units actively supported the 4th French Armée and the American 1st Army. With the Luftstreitkräfte all but defeated, the French formations were able to attack any target with relative impunity.

Breguet 14 B2 fitted with a 380-hp Liberty engine. MA1882.

Below: Breguet 14 A2 serial 9060 which was fitted with a Liberty engine. STAé 1223.

During the war's final weeks the bomber units were situated at the following bases:

1st Brigade
GB 5 at Matougues
GB 6 at Bussy-Lettrée
GB 9 at Bury

2nd Brigade
GB 3 at Coupéville
GB 4 at Somme-Vesle

These units supported the French 5th Armée in October 1918.

During the final year of the war the Breguet 14 B2s helped the French and American armies withstand the initial German attacks and then, when the Allied counter-offensives began, the bombers had disrupted the German retreat. The Breguet 14 B2s had made an important contribution to the final Allied victory.

Overseas Units
Serbia
Three Breguet 14 escadrilles were based in Serbia:
BR 522, formed from MF 384 in May 1917. It was disbanded in 1919.
BR 524, formed from MF 398 in December 1917. It was disbanded in 1919.
BR 525, formed from MF 399 in June 1917.

Greece
Three Breguet units served in Greece:
BR 533, created from AR 533 with AR.1s; at the war's end it became the 3rd Squadron of the Royal Hellenic Air Service.
BR 532, created in Greece in November 1917; in 1919 it became the 1st Squadron of the Royal Hellenic Air Service.
BR 534, formed in September 1918. In 1919 it became the 2nd Squadron of the Royal Hellenic Air Service.

Armée d'Orient
Eight Breguet 14 units served with the Armée d'Orient:
BR 501, formed from V 383 with Voisin 8s on 1 July 1917. It was disbanded in 1919.
BR 502, formed from V 385 in July 1917 and was assigned to the 2nd Groupe de Divisions of the A.F.O. It was disbanded in 1919.
BR 503, created from V 386 in June 1917. It was assigned to the 1st Groupe de Divisions Infanterie of the A.F.O. It was disbanded in 1919.
BR 504, formed from V 388 on 14 June, 1917 as a reconnaissance unit for the A.F.O. It became the 5th Escadrille of the 7th RAO in January 1920.
BR 505, formed from V 389 on 14 June, 1917. Assigned as a reconnaissance unit for the A.F.O, it became the 7th Escadrille of the 6th RAO in January 1920.
BR 508, created from Escadrille 508 in August 1917. It used Breguet 14 A2s to regulate artillery fire for the A.F.O. It became the 3rd Escadrille of the 3rd RAO in January 1920.
BR 509, assigned to the S.A.L. of the 2nd Armée d'Orient in August 1917. It became the 1st Escadrille of the 1st RAO in January 1920.
BR 510, formed in September 1917 and used as a bomber unit. Commanded initially by Capitaine de Castex and later by Capitaine Coyne, BR 510 was particularly active during the assault on the German-Bulgarian front on 15 September 1918. It became the 7th Escadrille of the 7th RAO in January 1920.

Postwar Service
The Breguet 14 A2s and B2s remained in French service until the mid-1920s. The aircraft would form part of the German occupation forces, remain a key element of the tactical bomber force, and would be used extensively in the Rif campaigns.

Forty of the Breguet 14 escadrilles were disbanded by 1919. The remaining units were reorganized in 1920 along the following lines:

Left: Breguet 14 A2 serial 16156 with a Renault engine and a Breguet Moreux radiator. MA6424.

Below left: Breguet 14 A2 with 300-hp Renault fitted with a radiator and spinner. Reairche Collection.

Below: Breguet 14 A2 serial 16156 with the Breguet Moreux radiator. MA642.

Breguet 14 B2 Bomber Units

1re RB (Jour) at Metz (1st Day Bombardement Regiment):
1st Group: BR 117 (redesignated Escadrille 201) and BR 120 (202);
2nd Group: BR 129 (205), BR 123 (206);
3rd Group: BR 127 (209), BR 129 (210), and BR 108 (211)

3re RB (Jour) at Avord (3rd Day Bombardement Group):
Initially based at Neustadtand Lachen-Speyerdorf as part of the German occupation forces.
1st Group: BR 107 (201), BR 126 (202);
2nd Group BR 131 (205), BR 132 (206), and BR 134 (207);
3rd Group: BR 128 (209) and BR 66 (210).

Breguet 14 A2 Observation Units:

1st RAO at Tours (1st Observation Regiment):
1st Group: BR 509 (redesignated Escadrille 1), BR 234 (2);
3rd Group: BR 226 (5)

2nd RAO at Dijon (2nd Observation Regiment):
1st Group: BR 220 (l) and BR 111 (2);
2nd Group: BR 201 (3) and BR 219 (4);
3rd Group: BR 35 (5), BR 244 (6), BR 7 (7), and BR 11 (8).
Escadrilles 6 and 8 were attached to the army of the Rhine occupation force.

3rd RAO at Beauvais (3rd Observation Regiment):
1st Group: BR 104 (l);
2nd Group: BR 508 (3)

4th RAO at Bourget (4th Observation Regiment):
1st Group: BR 211 (l) and BR 228 (2);
3rd Group: BR 224 (6);
4th Group: BR 227 (7)

5th RAO at Bron (5th Observation Regiment):
1st Group: BR 207 (l);
3rd Group: BR 243 (5);
4th Group: BR 218 (8)

6th RAO at Toul (6th Observation Regiment):
4th Group: BR 505 (7)

7th RAO at Pau (7th Observation Regiment):
3rd Group: BR 504 (5);
4th Group: BR 510 (7)

In August 1920 there was another major change in unit designations.

The 1re RB became the 11th RAB (Jour). It was based at Metz with the following Breguet escadrilles: 2nd Group: 5 (formerly BR 29), 6 (BR 123), 7 (BR 129); 3rd Group: 9 (BR 127), 10 (BR 44), and 11 (BR 108).

The 3rd RB (Jour) became the 12th RAB (Jour) at Neustadt and was equipped with the following Breguet escadrilles: 1st Group: 1 (BR 107), 2 (BR 126), and 3 (BR 205); 2nd Group: 5 (BR 131), 6 (BR 132), 7 (BR 134), and 8 (BR 9); 3rd Group: 9 (BR 128) and 10 (BR 66).

The Observation Regiments also changed their designations:

The 31st RAO at Tours had Escadrille 13 (BR 226).

The 32nd RAO at Dijon had 11 (BR 220), 12 (BR 111), 13 (BR 201), 14 (BR 219), 15 (BR 35), and 16 (BR 7).

The 33rd RAO at Mayence with the Army of the Rhine had 2 (BR 104), 13 (BR 11), and 14 (BR 244).

The 34th RAO at Bourget had Breguet escadrilles had 15 (BR 227), 16 (BR 224), 17 (BR 228), and 18 (BR 111).

The 35th RAO at Bron had 1 (BR 243), 2 (BR 207), and 12 (BR 218).

The Aéronautique de l'Aviation at Levant had Breguet escadrilles: 52 (BR 117) and 53 (BR 120).

By 1922 approximately 100 Breguet 14s were in service with the Aviation Maritime. The reconnaissance units assigned to the Aviation d'escadre used Breguet 14 A2s, while others were assigned to the training sections at Saint-Cyr, Rochefort, Hourtin, and Sidi Ahmed. The Breguet 14s, however, were not assigned to the *Bearn*.

The Breguet 14 units gradually declined in number after the early 1920s. Some units re-equipped with Potez 15s, but most

Breguet 12

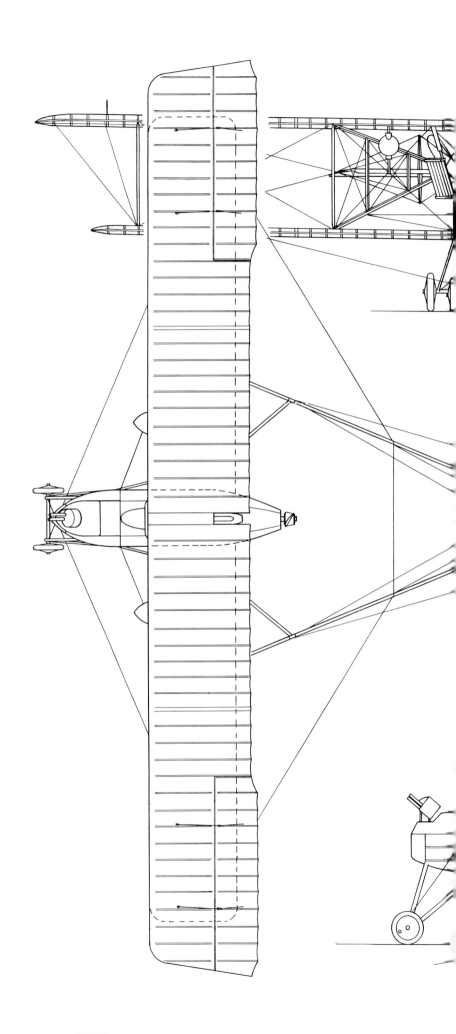

114B

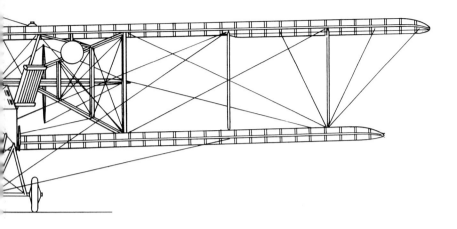

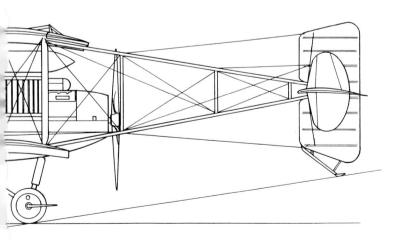

Breguet 5

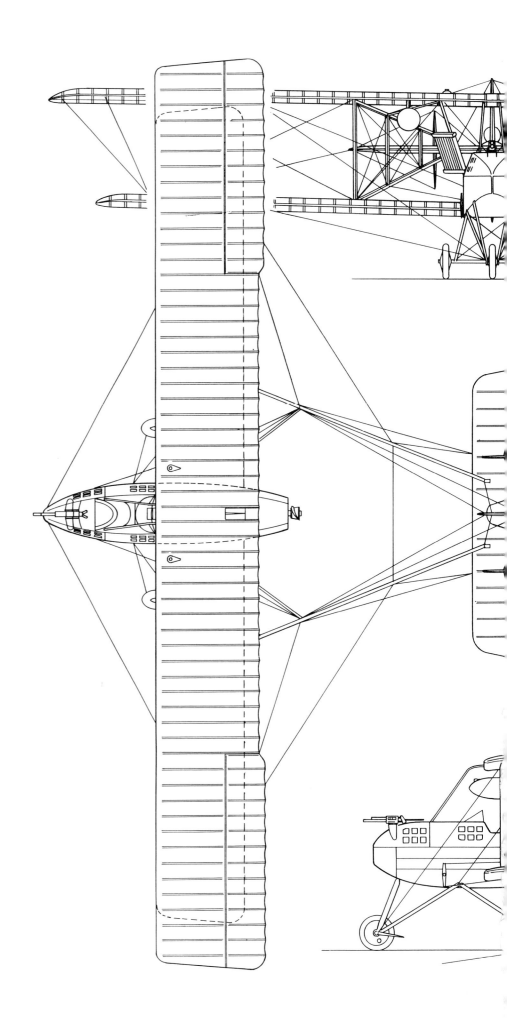

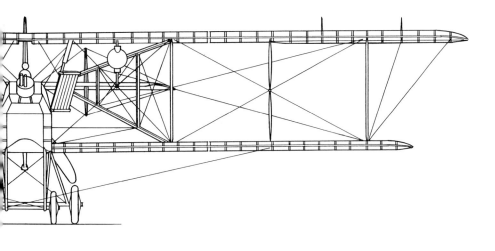

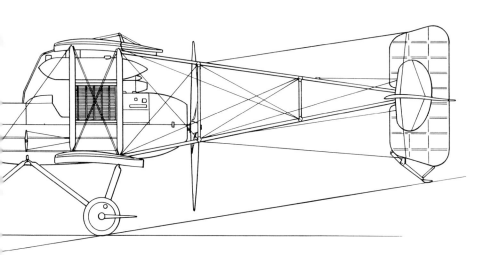

Breguet 14 named *Le Gaulois* of 5eme Escadrille of the 33eme RAO (ex-SPA 96) in 1926. The aircraft are flying over Thionville. B90.0635.

Below right: Breguet 14 Tbis serial 1950 mounted on floats. MA38556.

Below: Breguet 14 A2 serial 6603 based at Istres in 1927. MA16884.

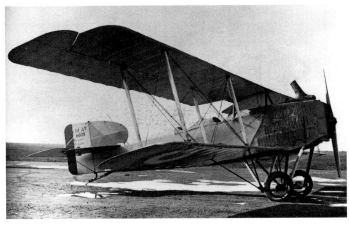

received Breguet 19s. In 1926 there were 376 Breguet 14 A2s, 340 Breguet 14 B2s, and 95 Breguet 14 Tbis (ambulance) still in service. By the late 1920s the Breguet 14s were finally replaced by Breguet 19s, Potez 15s, and Potez 25s.

Colonial Campaigns

The Breguet 14 A2s, B2s, and Ts saw widespread use during the colonial campaigns of the 1920s. After the Armistice, the French forces in Syria were given the new mission of combating rebellious desert tribes. Two escadrilles were available: Escadrille 52 (BR 117) at Rayak and Escadrille 53 (BR 120) with the troops of General Lamothe. During this now-forgotten campaign, the Breguet 14s were used in the reconnaissance, light bombing, and ambulance roles. When used as light bombers they usually carried 12 10-kg bombs.

In 1925 and 1926 all the escadrilles in Syria, including the two Breguet 14-equipped units, were used to attack tribesmen. They flew 6,000 patrols and 4,000 bombing sorties. Aircraft were decisive in the relief of 7,000 encircled French soldiers in the city of Jabal Djebel Druse. From 26 July to 24 September of 1925 the Aviation Militaire maintained contact with units trapped in the city and, in one of the first airlifts in history, flew 200 sorties and dropped 12 tons of food and 54 sacks of mail.

Breguet 14-equipped escadrilles were also used in Morocco. Apparently, all ten escadrilles of the 37th Regiment used Breguet 14 A2, B2, S, and T bis. Four escadrilles of the 37th Regiment participated in the conquest of Ouezzan. From 1922 through 1923 all units of the 37th Regiment were employed in operations around the Moyen-Atlas border. The two ambulance escadrilles, which used Breguet 14S and T bis aircraft, evacuated 870 stretcher cases.

Escadrilles 7, 8, and 10 were employed in 1924 to help deal with uprisings in Northern Morocco, supporting French outposts and forts during the Rif battles.

By 1925 the insurrections in the north had been stopped, but new outbreaks of violence occurred later that year. Again eight bombing/observation and two ambulance escadrilles with Breguet 14s were in action. There were also four Breguet 14 escadrilles on the southern front. To supplement these, six Breguet 14 escadrilles from the Autonomous Groups of Algeria and Tunisia, two escadrilles of Breguet 14 B2s from the 11th Regiment in France, and two of the units based in the south (5th and 6th escadrilles) were transferred to the northern front as temporary reinforcements. In September two Breguet 14 A2 escadrilles from the 32nd RAO in Algeria were also sent to northern Morocco. The units based there flew 5,500 sorties from July 1925 to January 1926.

During the spring of 1926 the 37th Regiment continued to support the force occupying the Rif. The Taza pocket was finally regained after the defenders were subjected to an intense attack during which 12 tons of bombs were dropped.

At the end of the campaign, in June 1927, the 37th Regiment was made up of ten escadrilles: 1st (based at Beni-Malek), 2nd (Taza), 3rd (Bou Denib), 4th (Beni-Malek), 5th (Marrakech), 6th (Kasbah Tadla), 7th (Meknes), 8th (Meknes), 9th (Assaka), and the 10th (Fez). Most of these still had a number of Breguet 14 A2, B2, S, and Tbis on strength. However, by the next year most of these aircraft had been replaced by Breguet 19s.

Foreign Service

Belgium

In 1918 Belgium purchased enough Breguet 14 A2s to equip the 2nd, 3rd, and 5th Escadrilles. These planes had Fiat A-12 engines. The 2nd and 3rd Escadrilles were formed into a Groupe d'Observation and both were based at Moeres airfield. The 5th Escadrille was based at Houtem airfield. The Breguet 14 A2s remained in service until the mid-1920s.

Brazil

Brazil acquired Breguet 14 A2s and B2s in 1919. They were given serial Nos. 1856–1871, 1958–1961, and 1965–1971. Six examples built in Brazil were given serial numbers 1–6. The

Breguet 14 of BR 510 assigned to the A.F.O. in 1917. BR 510 was active during the piercing of the German-Bulgarian front on 15 September 1918. B84.2877.

Right: A Breguet 14 A2 with a Fiat engine in flight. MA37569.

Breguet 14s were assigned to the 1st Esquadrilha de Bombardeio (four aircraft) based at Santa Maria and the 3rd Esquadrilha de Observacao (six aircraft) based at Alegrete. In 1928 these units were disbanded and their aircraft were sent to the Escuela de Aviacion at Campo dos Alfonsos. The Esquadrilha de Aperfeicoamento (Operational Conversion Squadron) was based at Mogi das Cruzes and had six Breguet 14 A2s on strength. This unit was used to combat a revolution at Sao Paulo in July 1924. During this action 11 reconnaissance and bombing sorties were flown and one aircraft was lost in an accident. The aircraft returned to the area from August to September 1924 as part of the Destacamento de Aviacao (Air Detachment). The Breguet 14s were withdrawn in 1927.

China and Manchuria

The Chinese Nationalist Party (KMT) lead by Chiang Kai-shek had its own air force that included approximately 50 Breguet 14/400s. These were used operationally in 1926 when the KMT launched a major offensive to bring local warlords under its control. The Manchurian warlord Chang Hsuch-liang had 16 Breguet 14s as part of his private air force. It is likely that most of these aircraft were destroyed when the Japanese army overpowered the Manchurian forces in 1932.

Czechoslovakia

The Czechoslovakian air corps obtained ten Breguet 14 A2s. They served with the 4. Letecka Setnina at Praha-Kbely and Latecke Dilny at Olomouc in 1919. In 1923 they were assigned to 3. Prozorovaci Rota at Olomouc. The next year they were assigned to 3. Prozorovaci Rota at Olomouc and the Hlavni Letecke Dilny, 81 Bombardovaci Letka at Praha-Kbely.

Denmark

Denmark obtained several Breguet 14 A2s in 1921. They were used by the army flying school from 1920 to 1927.

El Salvador

A single Breguet 14 B2 was obtained from the French in the mid-1920s. After being utilized for good will flights to Honduras and Nicaragua in 1926, it crashed in March 1927 while delivering smallpox vaccine to Nicaragua.

Finland

Finland acquired 22 Breguet 14 A2s between 1919 to 1921. These aircraft were assigned serial numbers 2C 460–471, IC 476–477, IC 481–485, and IA 491–493. These codes were later changed to 3A 1–9 and 3C 11–30. The aircraft were assigned to Flying Division Number 1 at Utti (reconnaissance) and Flying Division Number 2 at Viipuri (bomber). They served until 1927.

Greece

The Breguet 14 B2 entered Greek service in November 1917. These aircraft were used to equip the 532 Mira Vomvarthismou ke Anagnorisseos (532 Bombing and Reconnaissance Squadron). 532 Mira used its aircraft to attack the fortifications of Shra-di-Legen and enemy positions along the Axios River.

A second unit, the 533 Mira Dioxes (533 Fighter Squadron) was formed in June 1918 with a mixture of A.R.1s and Breguet 14 A2s. By the end of 1918 there were 12 Breguet 14s (both A2s and B2s) on strength with the 532 and 533 Mira; these units were designated as the A and C Mira in April 1919. A detachment of four Breguet 14s drawn from these units was sent to Turkey during the Greco-Turkish War. They arrived in June 1919 and joined the D.H.9s of the Greek naval air service. A Mira remained in Orestias, Greece and C Mira was later sent to Turkey. The Breguet 14s were used over Ankara and flew army cooperation and tactical bombing missions. Fears of Turkish reprisal raids led to limits being placed on Greek bombing missions. By the end of August, both Breguet 14 units had returned to Greece after providing air cover for the Greek withdrawal. Both A and C Miras were used to combat the Revolution of Chilos, which began in September 1922.

In 1923 E Mira Dioxes was formed and had a single Breguet 14. It was intended to use this unit in the Greco-Turkish War but the resolution of this conflict in July 1924 resulted in E Mira being disbanded. After this, all the units of the Greek army air force (Stratiotiki Aeroporia) were consolidated into a single unit designated A Mira Aeroplanon and equipped with both Nieuport Nighthawks and Breguet 14s. In 1925 Breguet 19s replaced the Breguet 14s.

Guatemala

A French air mission took three Breguet 14s in shipping crates to Guatemala in 1918. However, when the instructor with the mission died the aircraft were sent back, unassembled, to France.

Japan

At least one Breguet 14 B2 was purchased from the French military mission that visited Japan in 1919. It was used by the Army's Mikatagahara Bombing Team to research bombing techniques. It later became the first aircraft to fly over Mt. Fuji. Nakajima built a Breguet 14 B2 under license as the Nakajima Type B-6. It was powered by a 360-hp Rolls-Royce Eagle VIII. The Nakajima aircraft, however, was never delivered to the army, but was utilized in several long distance flights.

At least one Breguet 14T transport was also acquired by the Japanese.

Lithuania

At least two Breguet 14 A2s were obtained by Lithuania (via Poland) in 1920. They were assigned to the Squadron of the Army of Middle Lithuania.

Persia

The air force of the Shah of Persia obtained two Breguet 14s in 1924.

Poland

The Polish government acquired Breguet 14 A2s and B2s when the French force of occupation redesignated three of its escadrilles as Polish units.

Four views of a Breguet 14 A2 fitted with a 275-hp Lorraine-Dietrich engine. MA38558.

Breguet 14 A2 of the Spanish air force. A spare wheel is attached to the fuselage side. B79.3032.

BR 39 (with 15 Breguet 14 A2s) was initially based at Lublin on the Ukranian front. In September 1919 the French gave this escadrille's aircraft to the Poles, who applied the designation 16th Eskadra (Reconnaissance). By June 1920, however, the unit no longer had any serviceable aircraft and was disbanded.

BR 59 (with 15 Breguet 14 A2s) was also based Lublin and was turned over to the Poles in September 1920. It was redesignated the 17th Polish Eskadra (Reconnaissance). By mid-July the unit was disbanded.

BR 66 was based at Wilno in January 1919. The unit was designated the 4th Eskadra (Reconnaissance). It was disbanded in July 1920 because there were no serviceable aircraft.

The Polish government ordered 70 more Breguet 14s from France and, in November 1919, a new Breguet 14 A2 unit was formed with some of these aircraft; this was the l0th Eskadra (Reconnaissance). The l0th Eskadra was initially based at Lwow, but was sent to Brzesc in August 1919 and then Pozan-Lawica in October 1919. The next year, the unit re-equipped with Bristol Fighters.

More of the Breguet 14s ordered in 1920 arrived in 1921 and these were assigned to the 1st Air Regiment, based in Warsaw. The component units were the 1st Reconnaissance Dyon with numbers 12 and 16 Eskadras and the 4th Reconnaissance Dyon with numbers 3 and 8 Eskadras. By 1924, the Breguet 14s had been replaced by Potez 15s.

Portugal

In 1919 16 Breguet 14 A2s were obtained by the Portuguese Arma de Aeronautica (Air Arm). They were initially sent to the Esquadriha Miste de Deposito (E.M.A.) at Tancos and subsequently assigned to the Gruppo de Esquadrilhas de Aviaco Republica (G.E.A.R. = Republican Group of Aviation Squadrons) based at Amadora. The G.E.A.R was despatched to Angola in 1921 and was redesignated Gruppo de Esquadrihas de Aviaco de Angola (G.E.A.A. = Group of Air Squadrons of Angola). The unit was disbanded in 1923.

Twelve additional Breguet 14 A2s were purchased from France in 1921. These, plus the Breguet 14s from the G.E.A.A., were assigned to the Gruppo Independente de Aviacao de Bombardeamento (Independent Bomber Aviation Group), which was formed at Alverca in 1923.

By 1925 there were still 11 Breguet 14s in service—one with the Escola Militar de Aviacao (Military Aviation School), nine with the G.E.A.R., and one with the Esquadrilha de Treino e Deposito.

Those Breguet 14s still operational were assigned to the Grupo Independente de Aviacoa de Bombardeamento (G.I.A.B. = Independent Bomb Group), which was formed in 1928. The Breguet 14s were at last replaced in 1931 when license-built Potez 15s became available.

A single Breguet 14 T was also purchased by Portugal. Named *Portugal,* it was assigned to the G.E.A.R.

Romania

Romania obtained 20 Breguet 14 B2s after the First World War. They equipped the two bomber squadrons of Grupul 3 based at Bucharest. In the mid-1920s the Breguet 14s were replaced by newer aircraft, including Breguet 19s.

Serbia

Serbia received three French escadrilles equipped with Breguet 14s: BR 522, BR 524, and BR 525. The French used these units to attack enemy camps near Lake Preipa. Mira 532, which was a Greek unit equipped with Breguet 14s, also bombed enemy positions in the Vardar Valley and the railway station at Miletkovo. The French and Greek Breguet 14 units were both active in March and April 1918 attacking railroad stations and enemy position. Enemy airfields at Drama and Hudobva were bombed in May and June, and the Breguet 14s were also used to strafe enemy troops.

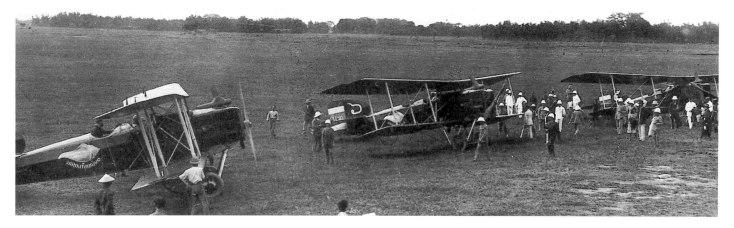
Three Breguet 14s of the Siamese air force at Bach-Mai in the early 1920s. B89.3994.

In July and August there were major raids on the airfields at Hudova and Canatlarsi. September began with heavy attacks on enemy positions. On the 14th there was an Allied breakthrough as the Serbian and French troops penetrated the enemy lines supported by the Breguet 14s. By the end of the month, the Bulgarians surrendered.

The French disbanded BR 522 and BR 524 in 1919. BR 525 became the 1st Squadron of the Serbian air service. In 1920 the surviving Breguet 14s were formed into the 2nd Air Regiment, and in 1923 they were replaced by Breguet 19s.

Soviet Union

Approximately six Breguet 14s were captured by the Soviets in 1919. Almost all of these were damaged, but at least three were used operationally. Breguet 14s were assigned to the Razvedivatel'naya aviaeskadril'ya in Moscow and subsequently with the 14th Otdel'nyi Razvedivatel'nyi Aviaotryad.

Spain

The first Breguet 14 A2s arrived in Spain in July of 1919 as part of a French air mission. Eight more were purchased by Spain before the end of that year. All were powered by Renault 12F engines.

The first two Breguet 14 A2s were assigned to the Tetuan Escuadrilla and both flew combat missions from Tetuan.

In 1921, the first Breguet 14 A2 with a 300-hp Fiat A-12 bis engine was acquired. It appears that the new engine was preferred because it was less expensive than the Renault 14Fs. These new aircraft arrived just in time to take part in the initial stages of the Moroccan campaign, where they were based at Seville. By 1921 there were two grupos using Breguet 14s based in Africa: the 1st Grupo, at Tetuan, and the 2nd Grupo, at LaRache. Each included either one of two escuadrillas equipped with the Fiat-engined Breguet 14s. In 1923 a third escuadrilla of Breguet 14s was assigned to Grupo 1 and based at Larache.

In 1923 the Grupo de Seville was formed with Fiat-engined Breguet 14s; it was redesignated the 22nd Grupo in February 1917. This unit used Breguet 14s until 1931, when they were replaced by Loring R.IIIs. A Grupo Expedicionario composed entirely of Breguet 14s operated in the Melilla and Tetuan area. It was assigned to Grupo 1. By 1926 Grupo 2 had returned to Larach. Later that year a detachment of Breguet 14s was sent from Grupo 2 to southern Morocco. This detachment was then assigned to Grupo 3. A number of Fiat-engined Breguet 14s were sent to the airport at Armilla, in Granada, in October 1926. They were assigned to Grupo 1 in February 1927. These aircraft remained with Grupo 1 until 1930, when they were replaced by Breguet 19s.

Although it may have been cheaper, the Fiat engine proved to be unsatisfactory and plans were made to fit these aircraft with 360-hp Rolls-Royce Eagle VIIIs. These conversions, which required modifications to the airframe, were done at the aircraft park at Sevilla. The two escuadrillas of Grupo 3, based at Larache, were the first to receive the modified aircraft in 1927. These units received their new Breguet 14s in time to utilize them in the last aerial operations of the Moroccan campaign. The Breguet 14s remained in service until finally replaced by Loring R.IIIs in 1931. Some of the Rolls Royce-equipped Breguet 14s were utilized by the Escuadrilla de Sahara from 1928 to 1931.

The final engine to be fitted to Spanish Breguet 14s was the 300-hp Hispano-Suiza 8F. This modification was performed at the aircraft park at Sevilla. On 17 November 1929 a Breguet 14 with the Hispano engine crashed, killing its crew. It is believed that this was the first aircraft to have the new engine.

A total of 140 Breguet 14s were used by the Spanish. It does not appear that any were built in Spain.

Sweden

The Swedish army aviation service (Flygkompaniet) acquired one Breguet 14 in 1919. The aircraft was given serial 9100. In January 1923 it was given civil registration S-AIAA.

Thailand/Siam

Siam served as part of the Allied Occupation Force of the Rhineland. A number of Siamese pilots flew with French escadrilles equipped with Breguet 14s. These pilots returned to Siam in August 1919, taking a number of Breguet 14 A2s and B2s with them. These aircraft were based at Don Muang and served with the 2nd (General Purpose) Group's 1st and 2nd Wings. The 1st Wing was used to carry out reconnaissance and topographic work, while the 2nd Wing was used to fly mail and passengers over Siam's northeast region. Local production of the Breguet 14 began in 1924 in the air service's workshops. The engines were 300-hp Renaults.

It was the rising price of the Renault engines that forced the Siamese air service to consider an alternative engine for its license-manufactured aircraft. It was also hoped that it might be possible to produce a new plane instead of the Breguet 14. It was termed the Boripatra bomber. Its lineage, albeit indirect, from the Breguet 14 is obvious. Several examples were built.

Production of the Breguet 14s began again when Renault engines became available at reasonable prices. These aircraft remained in service until 1933, when license production of Vought V.100s began.

The 2nd Wing provided regular air mail service. Its schedule called for a Breguet 14 to leave the city of Ubol every Tuesday and return every Thursday. Passengers were occasionally carried by this service, and the Breguet 14 Ts were used to transport individuals requiring medical care and medical supplies. The Siamese Royal Family was so impressed with this service that it assisted in raising funds to purchase a single Breguet 14 S ambulance aircraft.

Breguet 14

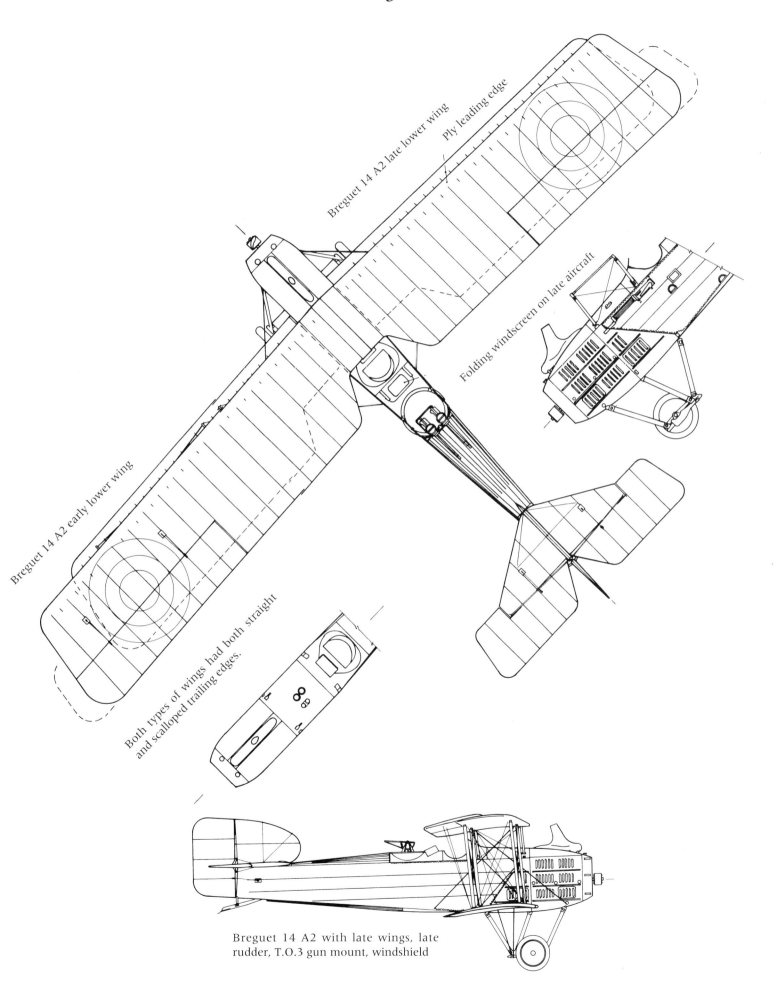

Breguet 14 A2 with late wings, late rudder, T.O.3 gun mount, windshield

Breguet 14

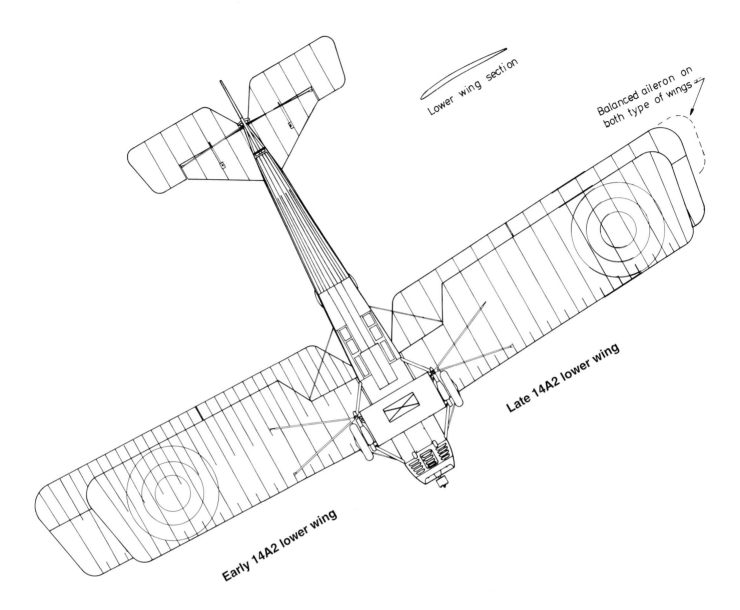

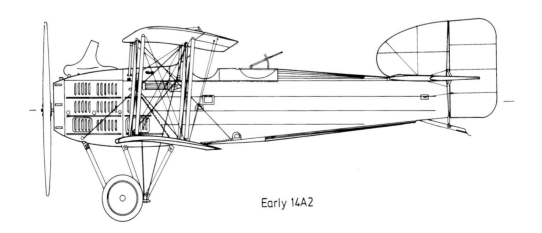

Early 14A2

Breguet 14

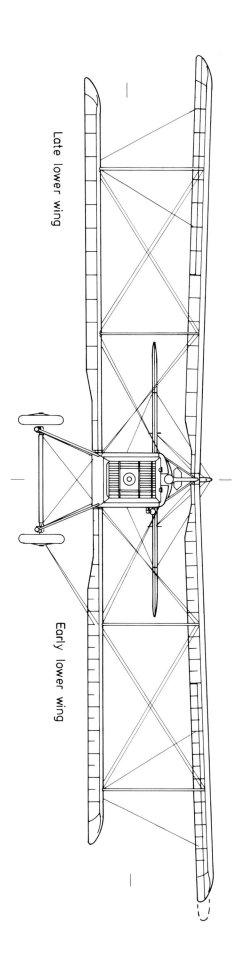

Breguet 14 A2

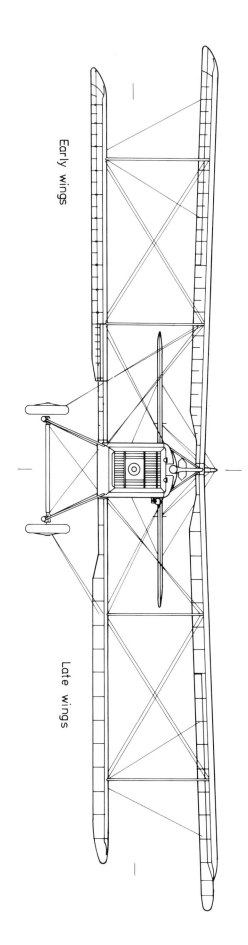

Breguet 14 B2

Breguet 14

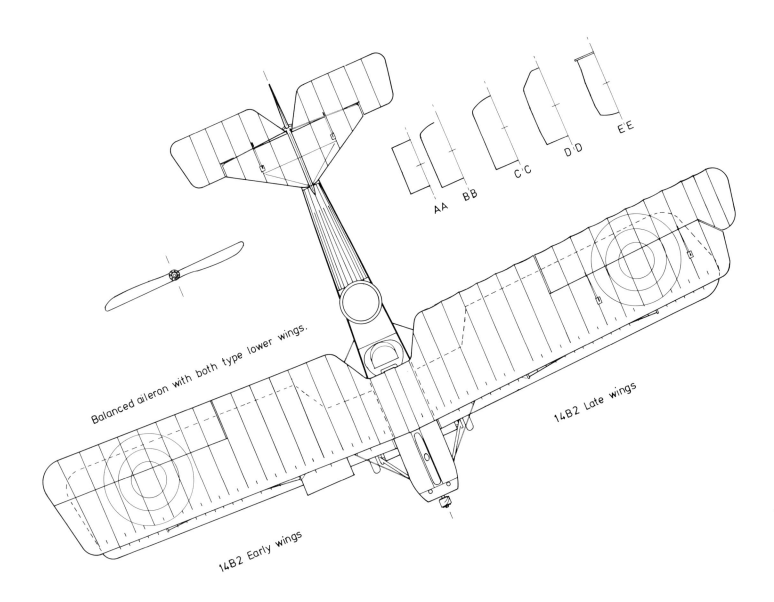

Balanced aileron with both type lower wings.

14B2 Late wings

14B2 Early wings

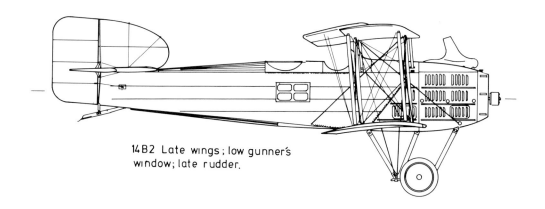

14B2 Late wings; low gunner's window; late rudder.

Breguet 14

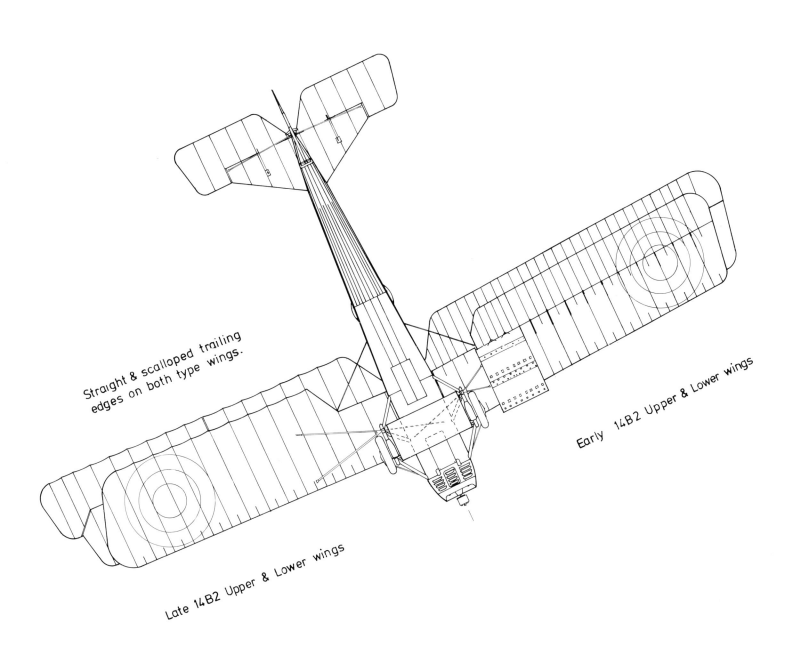

Straight & scalloped trailing edges on both type wings.

Early 14B2 Upper & Lower wings

Late 14B2 Upper & Lower wings

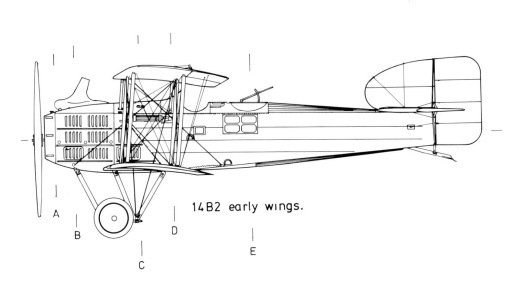

14B2 early wings.

Breguet 14 B2. The aircraft carries the French five-color camouflage scheme. MA2680.

Turkey

Turkey acquired 16 Breguet 14 A2s and 16 Breguet 14 B2s in 1923. The bombers equipped the two bomber companies, while the reconnaissance machines equipped both reconnaissance companies. Breguet 19s replaced the bombers in 1933, while the reconnaissance machines were replaced by Potez 25s and Letov S-16 Ts in 1935. Both versions of the Breguet 14 saw limited action during the Greco-Turkish War.

United States

The United States entered World War I in the belief that a powerful bombing force could shorten, if not end, the conflict. The Air Service had tremendous enthusiasm but no indigenous aircraft with which to conduct bombing raids. The decision, pending the availability of the Liberty-engined D.H.4, was for the Americans to obtain Breguet 14 bombers. The French initially promised to deliver 1,500 of these aircraft, although only 290 had been delivered by the Armistice.

The Americans initially ordered 376 of them: 100 Breguet 14 E2 trainers, 229 Breguet 14 A2s, and 47 Breguet 14 B2s. Approximately half of these had Fiat A-12 or A-12bis engines.

The first American unit to fly the aircraft operationally was the 96th Aero Squadron. Training began at the Michelin brothers testing field on 1 December 1917.

The first operational American day bombardment unit, the 96th Aero Squadron, took ten of its Breguet 14s to Amanty airfield on 18 March 1918. It has been reported that the aircraft were in poor repair and, because French supplies were often unavailable, the unit's mechanics had to make use of modified farm machinery parts. The first mission was flown on 12 June 1918. However, the aircraft were in such poor condition that many raids had to be canceled. Because of these equipment difficulties, no American day bomber units were involved in the Chateau-Thierry campaign of July 1918. On 10 July, however, the 96th Aero Squadron had enough serviceable aircraft on strength to attempt a smaller raid. Six Breguet 14s left to bomb the railroad yards at Conflans. Unfortunately, the inexperience of the American pilots resulted in all six aircraft being captured.

Two days before the Saint Mihiel offensive, the 96th Aero Squadron was assigned to the First Day Bombardment Group. During the offensive, the 96th performed ground attack missions and bombed rail centers. Many Breguet 14s were damaged during crash landings on the muddy airfields, the 96th Aero Squadron losing 16 men and 14 planes in only five days. This was the worst loss rate of any AEF unit. The 96th was reconstituted with new crews and aircraft and was back in action in time for the Meuse-Argonne offensive. Due to bad weather the 96th Aero Squadron was able to fly only two days in November.

The other units of the First Day Bombardment Group used Liberty-engined D.H.4s; however, it was widely believed that the Breguet 14 was the superior bomber. The Gorrel Report notes that the Breguet 14s were faster at altitude, carried a heavier load, had excellent defensive armament, had a protected upper fuel tank and a droppable lower tank, were equipped with the excellent Michelin bomb racks, and that "no stronger ship has probably ever been subjected to hard active service and given such excellent results."

While the 96th Aero Squadron saw the main use of the Breguet 14, the 9th Night Reconnaissance Squadron also used the reconnaissance version. These were employed for day observation as well as night reconnaissance and attack missions, the latter with mixed results. The unit operated over the Toul sector from 30 August to 11 September and took part in the Saint Mihiel and Meuse-Argonne offensives. The 99th Corps Observation Squadron also had Breguet 14s on strength for a brief time.

Uruguay

Uruguay's Escuela Militar de Aeronautica received six Breguet 14 A2s and two Breguet 14 T ambulance aircraft in 1921. An additional Breguet 14 T bis was obtained in 1928. The ambulance aircraft were assigned to the Aviacon Sanitaria in 1930. The Breguet 14 A2s remained with the Escuela Militar until replaced by Potez 25s and Breguet 19s in 1928.

Breguet AV 1 Two-Seat Reconnaissance Plane with 263-hp Renault

Span 14.36 m; length 8.80 m: wing area 50.0 sq. m
Empty weight 1,015 kg, loaded weight 1,525 kg
Maximum speed: 179 km/h at ground level

Breguet AV 2 Two-Seat Reconnaissance Plane with 272-hp Renault

Span 14.36 m; length 8.90 m: wing area 52.0 sq. m
Empty weight 1,020 kg, loaded weight 1,530 kg.
Maximum speed: 181 km/h at ground level; 172 km/h at 2000; 165 km/h at 5,000 m

Breguet 14 A2 Two-Seat Reconnaissance Plane with 300-hp Renault 12Fcx

Span 14.364 m (without horn balances); 14.860 m (with horn balances), length 8.870 m., height 3.330 m; wing area 47.5 sq. m (without horn balances); 49.2 sq. m (with horn balances)
Empty weight 1,030 kg; loaded weight 1,565 kg
Maximum speed: 184 km/h at 4,000 m
Climb:

	2,000 m	6 minutes 50 seconds
	3,000 m	11 minutes 35 seconds
	5,000 m	29 minutes 30 seconds

Ceiling 6,100 m, endurance 2.75 hours
Armament: one synchronized 7.7-mm Vickers machine gun, two 7.7-mm Lewis machine guns on a T.O.3 or T.O.4 ring mount, four 120-mm bombs.

Breguet 14 B2 of the 96th Aero Squadron of the 1st Day Bombardment Group. Drew Eubanks via Tan H. courtesy of Colin Owers.

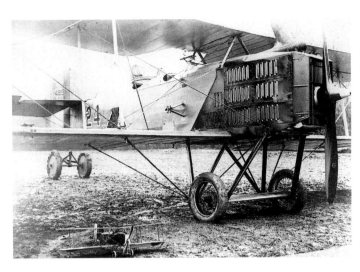

Breguet 14 A2 of the 96th Aero Squadron of the 1st Day Bombardment Group. Drew Eubanks via Tan H. courtesy of Colin Owers.

Breguet 14 in flight. This aircraft carries the French five-color camouflage scheme. Drew Eubanks via Tan H. courtesy of Colin Owers.

Breguet 14 of the 96th Aero Squadron of the 1st Day Bombardment Group in flight. Drew Eubanks via Tan H. courtesy of Colin Owers.

Breguet 14 A2 of the 96th Aero squadron with a model of the Breguet 14 on the wing. Drew Eubanks via Tan H. courtesy of Colin Owers.

Two Breguet 14 A2s of the 96th Aero Squadron of the 1st Day Bombardment Group. Drew Eubanks via Tan H. courtesy of Colin Owers.

Breguet 14 B2 Two-Seat Bomber with 300-hp Renault 12Fcx

Span 14.364 m (without horn balances); 14.860 m (with horn balances); length 8.870 m; height 3.330 m; wing area 50.2 sq. m (without horn balances); 48.5 sq. m (with horn balances)
Empty weight 1,017 kg, loaded weight 1,769 kg, bomb load 355 kg.
Maximum speed: 195 km/h
Climb:

	2,000 m	9 minutes 15 seconds
	3,000 m	16 minutes 30 seconds
	5,000 m	47 minutes

Ceiling 6,200m; endurance 2.75 hours
Armament: one synchronized 7.7-mm Vickers machine gun, two 7.7-mm Lewis machine guns on a T.O.3 or T.O.4 ring mount, one 7.7-mm Lewis machine gun firing through the floor of the aircraft on some B2s, 32 115-mm bombs or equivalent load.

Breguet 14 A2 Two-Seat Reconnaissance Plane with 310-hp Renault 12Fcy

Empty weight 1,040 kg; loaded weight 1,915 kg; bomb load 300 kg
Maximum speed: 195 km/h
Climb:

	2,000 m	7 minutes 40 seconds
	3,000 m	12 minutes 10 seconds
	5,000 m	25 minutes 40 seconds

Ceiling 5,200m; endurance 2.75 hours
Armament: one synchronized 7.7-mm Vickers machine gun, two 7.7-mm Lewis machine guns on a T.O.3 or T.O.4 ring mount, four 120-mm bombs.

Breguet 14 A2 Two-Seat Reconnaissance Plane with Renault 12Fe with Rateau Turbo-Supercharger

Maximum speed: 184 km/h at 3,000 m
Climb:

	2,000 m	10 minutes 1 second
	3,000 m	14 minutes 57 seconds
	5,000 m	28 minutes 16 seconds

Endurance: 3.0 hours

Breguet 14 A2 Two-Seat Reconnaissance Plane with 400-hp Renault 12K

Empty weight 1,202 kg; loaded weight 1,859 kg; bomb load 185 kg
Maximum speed: 203 km/h at 2,000 m
Climb:

	2,000 m	6 minutes 9 seconds
	3,000 m	10 minutes 19 seconds
	5,000 m	22 minutes 28 seconds

Ceiling: 7,600 m; endurance 3 hours 4 minutes

Breguet 14 A2 Two-Seat Reconnaissance Plane with 300-hp Fiat A-12 bis

Empty weight 1,160 kg; loaded weight 1,698 kg
Maximum speed: 167 km/h at 2,000 m; climb to 2,000 m in 11 min. 45 sec.; climb to 3,000 m in 19 min. 28 sec.
Ceiling 5,000; endurance: 3.0 hours

Breguet 14 A2 Two-Seat Reconnaissance Plane with 285-hp Lorraine-Dietrich 8Bd

Empty weight 1,160 kg; loaded weight 1,476 kg
Maximum speed 104 mph at 2,000 m
Climb:

	2,000 m	9 minutes 26 seconds
	3,000 m	15 minutes 11 seconds
	5,000 m	43 minutes 6 seconds

Ceiling 5,600 m; endurance 3.0 hours

Breguet 14 A2 Two-Seat Reconnaissance Plane with 370-hp Lorraine-Dietrich 12Da

Maximum speed: 195 km/h at 2,000 m
Climb:

	2,000 m	5 minutes 20 seconds
	3,000 m	8 minutes 50 seconds
	5,000 m	19 minutes 20 seconds

Ceiling: 7,600 m

Breguet 14 A2 Two-Seat Reconnaissance Plane with 400-hp Liberty 12

Empty weight 1,124 kg; loaded weight 1,713 kg
Maximum speed 126 mph at 2,000 m
Climb:

	2,000 m	5 minutes 49 seconds
	3,000 m	9 minutes 45 seconds
	5,000 m	23 minutes 18 seconds

Ceiling 7,300m; endurance 3 hours

Breguet 14 B2 Two-Seat Bomber with 400-hp Liberty 12

Empty weight 1,124 kg; loaded weight 1,713 kg
Maximum speed: 203 km/h at 2,000 m
Climb:

	2,000 m	8 minutes 57 seconds
	3,000 m	16 minutes 30 seconds
	5,000 m	41 minutes 2 seconds

Endurance: 4.5 hours

Breguet 14 B1 Single-Seat Bomber with 300-hp Renault 12Fcx

Climb to 2,000 m in 14 min. 19 sec.; climb to 3,000 m in 24 min. 44 sec.; endurance 6.0 hours

Nakajima B-6 (Breguet 14 B2) Two-Seat Bomber with 360-hp Rolls-Royce Eagle VIII

Span 14.76 m; length 8.985m; height 3.00 m; wing area 51 sq. m
Empty weight 1,171 kg; loaded weight 1,950 kg
Maximum speed: 191 km/h; climb to 5,000 m in 46 min.; endurance 4 hours
One built

Approximately 8,000 Breguet 14s of all types were produced.

Breguet 15

In the event of the war lasting into 1919, the Aviation Militaire wanted to ensure its technological superiority over the Germans. Toward this end, Breguet redesigned his Breguet 14 to accept a 400-hp Lorraine 12Dd. This new type was designated the Breguet 15 and was intended for both the reconnaissance and bomber role. The aircraft was tested during 1918. However, with the war's end there was no need for the type and further development was abandoned.

Breguet 16 Bn2

The Breguet 16 was a derivative of the Breguet 14 intended to replace the Voisin 8 and 10s used by the night bomber squadrons. The Bn2 specification called for a medium bomber with a crew of two. It should not be confused with the BN2 designation that was for a heavy bomber.

The aircraft was an enlarged version of the Breguet 14 with a considerably increased wing span of 16.96 meters (from 14.36 meters) to permit it to carry enough fuel and weapons for strategic missions. The engine was a 300-hp Renault 12Fe. Armament was three machine guns and 550 kg of bombs.

The prototype was first flown in June 1918 by M. de Bailliencourt, a test pilot for the Breguet firm. His initial comments, before testing, were that changes needed to be made to the position of the rudder bar which was too near the seat. He

The Breguet 16 Bn2 was an enlarged version of the Breguet 14 intended for strategic night bombing missions. B83.1769.

The Breguet 16 Bn2 had a considerably increased wing span (16.96 meters compared to the Breguet 14's 14.36 meters) which permitted the aircraft to carry a heavier payload of fuel and weapons. B76.1677.

also recommended that the throttle be relocated and, because of an inability to see the ground while climbing or diving the aircraft, that indentations be made in the edges of the lower wings. He subsequently made several flights without incident. However, an adjustment to the tailplane was necessary and the plane was returned to the factory so that the change could be made. Bailliencourt reported that Louis Breguet was furious at the delay this caused to the testing program.

Two Breguet 16s were evaluated by GB 1. Series production followed but not in time for the aircraft to see active service.

Breguet 16 Bn2 Two-Seat Medium Bomber with 300-hp Renault 12Fe
Span 17.00 m; length 9.55 m; height 3.42 m; wing area 73.50 sq. m
Empty weight 1,265 kg; loaded weight 2,200 kg
Maximum speed at sea level: 160 km/h; climb to 4,000 m in 51 minutes 10 seconds; ceiling 4,600 meters; range 900 km
Armament: three machine guns and 550 kg of bombs.
Approximately 200 built.

Postwar some were assigned to the 22nd RAB (N) and 21st RAB (N).

A total of about 200 Breguet 16 Bn2s were built by Breguet, Ferbois, Lioré-et-Oliver, and SECM.

A Breguet 16 was tested with a Liberty engine, but as its wing failed during static testing it seems unlikely that it was flown.

Foreign Service

Czechoslovakia

The Czechoslovakian air service purchased a number of Breguet 16 Bn2s; these were assigned to the 1st Air Regiment at Prague (Bohemia).

Portugal

A single example was purchased by the Portuguese Arma de Aeronautica in 1921. The Portuguese took advantage of the type's excellent range to make a long-distance flight from Lisbon to Macao, China. The aircraft, named *Patria,* made part of the flight until it crashed in India and was replaced by a different aircraft type (named *Patria II*). It was struck off charge in 1924.

Above: Breguet 16 Bn2 in the Sahara. B84.2142.

Left: Breguet 16 Bn2 serial 9592. MA20333.

Right: Breguet 16 Bn2 that appears to have a Breguet 14 fin; the pilot is named Walbaum. MA8496.

Below: Breguet 16 with experimental engine installation and four-bladed prop. Reairche Collection.

Above: Another view of the Breguet 16 with an experimental engine. Reairche Collection.

Breguet 17

In November 1916 the STAé formulated a requirement for a two-seat fighter designated the Type C. Breguet submitted a variant of the type AV prototype intended to meet this specification. This two-seater was eventually developed into the Breguet 17. A fighter version of the Breguet 14 bomber would have ensured that the Breguet 14 bomber formations contained escort planes of comparable performance.

In early 1918 Breguet started design work on the new escort fighter. Initial attempts included a Breguet 14 with a 370-hp Lorraine-Dietrich engine. Other designs were greater departures from the classic Breguet 14 layout, including an aircraft with a 400-hp Renault 12J engine and one armed with a cannon in a turret.

The new fighter, designated Breguet 17 C2 (two-seat fighter) appeared in the summer of 1918. Given serial 022, it was powered by a 400-hp Renault 12R. Externally, it was almost identical to the standard Breguet 14 A2. Major changes included a horizontal exhaust stack mounted to starboard in order to improve the pilot's vision, a smaller wing span and area, altered bracing wires, a new fin and rudder, and greatly increased armament. The latter consisted of two fixed 0.303 Vickers machine guns for the pilot and twin Lewis guns on a swivel mount for the observer. In addition, there was a third Lewis gun mounted in the belly of the aircraft that also was used by the observer.

The prototype flew in mid-1918, but its performance seems to have been considered inadequate. Modifications were introduced, including a more powerful 450-hp Renault 12R1 engine and redesigned wings with a deeper under-camber. Although still not of completely satisfactory performance, the Breguet 17 C2 entered series production. A night fighter variant was designated the Breguet 17 Cn 2.

Apparently only about ten of the aircraft were ever built, as the contemporary Hanriot HD.3 was believed to be a superior fighter. The few Breguet 17s built were not available for service until 1919. The aircraft produced in series had horn-balanced ailerons of reduced chord and an enlarged, horn-balanced rudder. As there were not enough aircraft to form a single escadrille, the Breguet 17s were sent to individual escadrilles for record-breaking and long-distance flights. Notable flights include a high-altitude record achieved on 12 February 1923 when Lt. Benoit flew a Breguet 17 to 5,516 m with a 500-kg load, a flight from Paris to Stockholm in 8 hours 15 minutes, and a series of long distance flights in 1923.

Breguet 17 C2 Two-seat Long-Range Escort Fighter with 420-hp or 450-hp Renault 12

Specification	420-hp Renault 12K	450-hp Renault 12K1
Upper/lower span:	13.40 m/12.56 m	14.28 m/12.56 m
Wing area:	41.4 sq. m	45.3 sq. m
Length/height:	—	8.1 m/3.42 m
Empty weight	1,225 kg	—
Loaded weight	1,845 kg	1,840 kg
Maximum speed:	211 km/h at 2,000 m	218 km/h at 2,000m
	206 km/h at 3,000 m	213 km/h at 3,000 m
	198 km/h at 4,000 m	207 km/h at 4,000 m
	185 km/h at 5,000 m	199 km/h at 5,000 m
Ceiling:	6,500 m	7,500 m
Climb to 2,000 m:	7 minutes 6 seconds	5 minutes 45 seconds
Climb to 4,000 m:	19 minutes 49 seconds	14 minutes
Climb to 5,000 m:	29 minutes 43 seconds	20 minutes 41 seconds

Armament: two fixed, synchronized 7.7-mm Vickers machine guns, twin 7.7-mm Lewis guns on a swivel mount, and a third Lewis gun firing through the belly of the aircraft.
Ten Breguet 17 C2s were built.

Breguet 17. Modifications from the Breguet 14 included a more powerful 450-hp Renault 12 K1 engine and redesigned wings with a deeper camber. B83.1801.

Above: Armament of the Breguet 17 C2 two-seat fighter variant of the Breguet 14 included two fixed Vickers guns for the pilot, twin Lewis machine-guns on a swivel mount for the observer, and a third Lewis gun firing downward through the bottom of the fuselage. This aircraft features an enlarged rudder with horn balance. B90.1914.

Left: Breguet 17s at the factory. Cliché Doc.

Below and below left: Four views of a Breguet 17 C2. MA14068.

Breguet 17 C2

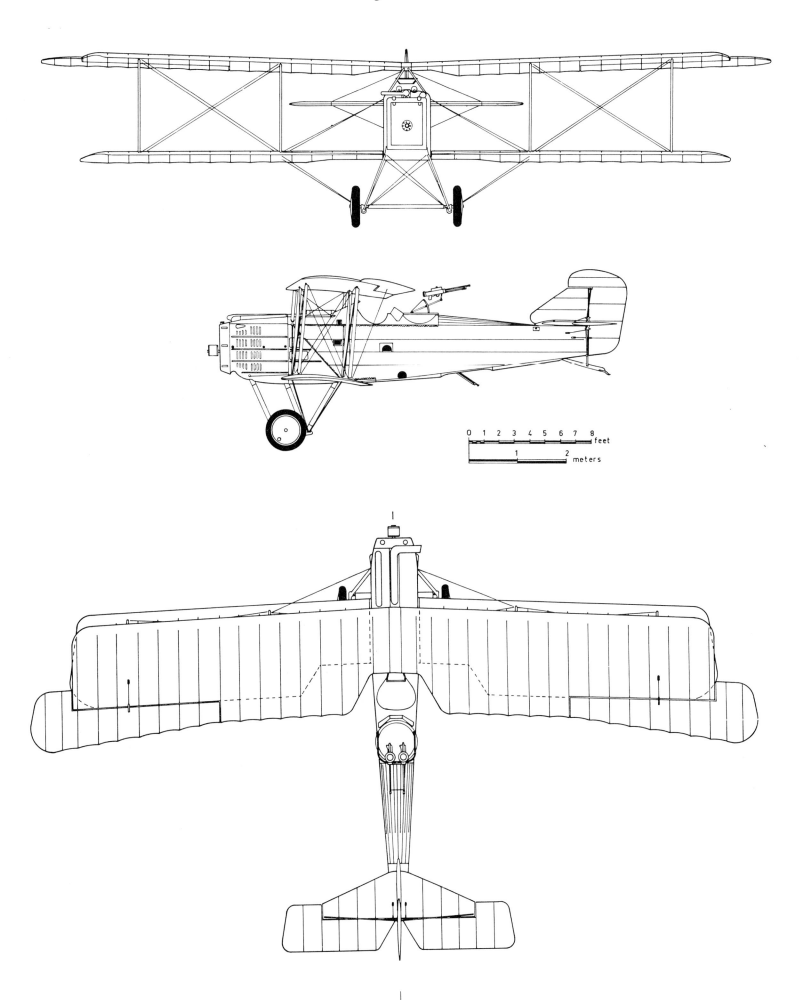

Brun-Cottan

Brun-Cottan Patrol Flying Boat

The Brun-Cottan flying boat was constructed in early 1918, intended for coastal patrols, anti-submarine warfare, and even long-range fighter patrols. It was a three-seat biplane powered by two 235-hp Panhard engines. It underwent testing on 8 April 1918 but never entered service.

The following specifications are the result of testing by the Laborotoire Eiffel.

Brun-Cottan patrol flying boat which was powered by two 235-hp Panhard engines. MA36380.

Brun-Cottan Three-Seat Patrol Flying Boat with Two 235-hp Panhard Engines:
Span 24.80 m; length 14.60 m; height 4.25 m; wing area 107.50 sq. m
Empty weight 2,400 kg; loaded weight 4,400 kg
Maximum speed: 159 km/h at sea level; 152 km/h at 2,000 m; 126 km/h at 4,000 m; climb to 2,000 meters in 15.18 minutes; ceiling 4,300 m; endurance six hours.
Armament: one machine-gun and four 150-kg bombs
One built

Brun-Cottan H.B.2 Flying Boat

A second, larger flying boat was built by Brun-Cottan late in the war. Designated the H.B.2, it had two 350-hp Panhard engines and may have been intended to meet the hydravion bombardement flying boat specification. Built in 1918, the type was undergoing testing in 1919 but, as with the preceding Brun-Cottan flying boat it never entered service.

A contemporary French report gave these specifications.

Brun-Cottan H.B.2 Two-Seat Flying Boat Bomber with Two 350-hp Panhard Engines
Span 24.88 m; length 14.60 m; height 4.25 m; wing area 125 sq. m
Empty weight 2,400 kg; loaded weight 4,400 kg
Maximum speed: 160 km/h at sea level
Endurance: 5 hours 10 minutes
One built

Canton S2

General Duval of the STAé requested in 1918 the development of a heavily-armored aircraft for ground attack. Undoubtedly he had been impressed not only with the Aviation Militaire's own experience using Breguet 14s for strafing and bombing troops but also by the highly effective Junkers J.I sesquiplanes used by the German Luftstreitkräfte. However, while the French had successfully used army cooperation planes and bombers for ground attack, losses to ground fire had convinced the Aviation Militaire of the need for an armored craft. The request for an S class aircraft was submitted on 24 May 1918.

The S2 specification called for a two-seat aircraft with armor capable of carrying 150 kg of bombs and multiple machine guns. Three designs were submitted to meet the classification—the Lioré-et-Oliver S2, the Hochart S2, and the Canton S2.

The Canton 2 was powered by two 230-hp Canton-Unné engines. It featured heavy armor-plating for the pilot, machine gunner, and engines. The wings were unstaggered and of unequal chord, and there were ailerons on the top wing. The rudder and elevators were balanced. The engines were mounted in tractor configuration and were located inside armored nacelles which also contained the fuel and oil tanks as well as the radiators. Armament was planned to be four Lewis machine guns firing ahead and downwards. An additional Lewis machine gun was mounted on a swiveling turrent (possibly for use by a third crew member, although the original specification had only requested a crew of two). Also, the Canton S2 could carry 15 10-kg bombs. Because it was believed that the main threat to the crew would come from ground fire, only the floor of the cockpit was armored. This armor was surprisingly thin, only 4-mm armor plate being used.

Éteve states that the Canton design was rejected because the thickness of the armor was insufficient. However, a more important reason for the rejection was that it was seriously underpowered. Although no photographs of the type seem to have survived, the report of one of the test pilots, Adjudant Leau, still exists. It states that the Canton S2 could barely lift off the ground when fitted with armor. Leau reports near misses with a hangar and power lines close to the field while trying to gain altitude. Even when lightened by 200 kg the plane could barely reach 30 meters. One pilot, unable to climb above a line of trees, stripped the wings off the craft. Further development of

the Canton S2 was abandoned; indeed, all three S2 designs were failures because of their inability to meet the specified performance while carrying the requisite armor.

Canton S2 Two-Seat Or Three-Seat Ground Attack Aircraft (All Performance Specifications Are Estimates)

Loaded weight: 2,775 kg

Maximum speed at ground level: 180 km/h; estimated ceiling 5,000 m (but, in actuality was approximately 30 m)

Armament: five Lewis machine guns and 150 kg of bombs

One built

Caproni Bombers (License-Built)

The 1915 concours to select a heavy bomber had failed to produce a satisfactory plane, and the French now realized that they were behind the other combatants in developing this type. As the Voisin 3s then in service were inadequate, being deficient in both range and bomb load, it now became necessary for the Aviation Militaire to equip bomber escadrilles with a foreign design.

In the fall of 1915 the French accepted an offer from the chief of the Italian air service (Aviazione Militaire) to examine Italian aircraft design and manufacturing techniques. The French were aware of the Caproni bombers being developed and were interested in concluding an agreement for license production of these aircraft in France.

The members of the French team examined the Caproni factory as well as the Fiat (which built the engines) and Pirelli (which produced the tires) plants. They were also taken to the airfield at Pordenone where operational Caproni squadrons were based.

It was decided to arrange for license production of the Caproni 1 and 3 (in the text wartime Italian designations are used for these planes; the postwar designations were Caproni 32 and Caproni 33). They were to be built by the R.E.P. (Robert Esnault-Peletrie) firm. The engines were to be supplied by both the Canton-Unné and Le Rhône factories. The French had intended to purchase Fiat engines but the Italians initially retained these engines for use in their own planes.

The French designations for the Capronis are especially confusing. Unfortunately, some sources have compounded the difficulties by misidentifying the Italian designations. The correct designations are given in the following table:

Wartime Caproni Designation	Postwar Caproni Designation	STAé Designation	Engines
Caproni 1	Caproni 31	CAP.1 B2 (Italian)	Three Fiat 100-hp A-10
Caproni 1	Caproni 31	C.E.P.1 B2 (French)	One 130-hp Canton-Unné and Two 80-hp Le Rhônes
Caproni 3	Caproni 33	CAP.2 B2 (Italian)	Three Isotta Fraschini 150-hp V4A/B
Caproni 3	Caproni 33	C.E.P.2 B2 (French)	Three Isotta Fraschini 150-hp V4A/B
Caproni 5	Caproni 44	CAP.3 BN3 (Italian)	Three Fiat A-12/A-12bis 200-hp
Caproni 5	Caproni 44	C.E.P.3 BN3 (French)	Three Fiat A-12/A-12bis 200-hp

The first produced under license by the R.E.P. was a Caproni 1 (Ca.32) and was designated the C.E.P.1 B2. The initials C.E.P. stood for Caproni Esnault Pelterie.

Twin booms each held an 80-hp Le Rhône in a tractor configuration on a stamped metal plate. The booms had a quadrangular cross-section consisting of four ash longerons braced with wire and covered with fabric. Each boom was attached to the lower wing and held fuel and oil tanks behind the motor. Each boom had an articulated tail skid.

The horizontal stabiliser crossed the top of the booms. The stabiliser was a tubular metal structure braced with sheet metal struts. The outer vertical stabilisers were completely articulated, and only the center rudder had a fixed fin. All three stabilisers were connected to ensure they moved in unison. The rudders were made of metal tubing braced with wood.

The tricycle landing gear consisted of two pairs of wheels suspended beneath each boom by struts attached directly to the lower wing by metal attachments. A pair of nosewheels were attached to the extreme nose. Bungee cords attached the axles to the struts and served as shock absorbers. The nose wheels prevented the aircraft from nosing over when landing. Skids were located at each wingtip and at the end of each fuselage boom. The fuselage skids were flexible, while those at the wing tips were fixed.

Armament consisted of a machine gun mounted in the nose on a transverse mounting. A second machine gun was carried in a mobile mount inside a cupola attached to the top wing. A carbine could be fired beneath the floor of the central nacelle. The bombs were carried inside the central nacelle behind the fuel tanks. A Bowden bomb release system was used.

Due to the unavailability of Fiat engines, various combinations of others were tried, including Lorraine-Dietrich AMs and Canton-Unnés. However, it was eventually decided that production aircraft would be fitted with a single 130-hp Canton-Unné mounted as a pusher and two 80-hp Le Rhônes. Test results included a climb to 1,000 m in 12 min. 50 sec. and to 2,000 m in 33 min. 15 sec. Maximum speed was 110 km/h.

This compared poorly with the Fiats of the Italian-built Capronis which provided horsepower almost double that of the French engines. Tests of the Caproni C.E.P.1 B2s at Amberieu-en-Bugey revealed that it was severely underpowered. Not surprisingly, only 14 examples were built. It is unlikely any of these saw operational service. An order was initially placed with R.E.P. for 50 aircraft but according to correspondence with the SFA dated 12 August 1915, both the Aviation Militaire and the manufacturer agreed that before large-scale production could ensue, a version with more powerful motors would be needed. This aircraft became the C.E.P.2.

R.E.P. next produced a version of the Caproni 3 (Ca.33) powered by three Italian 150-hp Isotta Fraschini V4A engines. Production of the C.E.P.2 B 2 began in 1916 and they soon entered operational service. Later, Caproni-built versions were obtained directly from Italy; these were designated CAP.2 B2.

134 FRENCH AIRCRAFT OF THE FIRST WORLD WAR

Caproni C.E.P.1. The initial production aircraft were fitted with a single 130-hp Canton-Unné mounted as a pusher and two 80-hp Le Rhône engines. B83.5688.

Below: Caproni C.E.P.1. The gunner's cupola on upper wing was unique to the C.E.P.1 and was abandoned on the C.E.P.2. Renaud.

The final version of the bomber used by the French was the Italian-built Caproni 5 (Ca.44). At the insistence of the French government these aircraft had been equipped with three 260-hp Fiat A-12bis engines. Most CAP.3 BN3s, as they were designated, were obtained directly from Italy in mid-1918. However, a small number of Caproni 5s were built at Lyon by R.E.P. The first plane, C.E.P.3 BN3, was sent to Villacoublay for testing but broke down at Chalon and could not get beyond Dijon. The second crashed and was destroyed at Corbeil. The third was tested but was not found to offer a significant improvement over the C.E.P.2 B2s and CAP.2 B2s already in service. General Duval reported there were numerous problems with the C.E.P.3 BN3s including an insufficient bomb load, exhaust flames exiting too close to the fuel tanks, and the fact that the type was difficult to fly because it was nose heavy. It seems likely that problems with the first batch of bombers from R.E.P. convinced the STAé that the Caproni firm in Italy would be a more reliable source. Only a handful of CAP.3 BN 3s were used by the French escadrilles, and all these had been built in Italy.

It was originally anticipated that three Capronis a month would be built, but production quickly fell behind schedule. In 1915 a total of 14 C.E.P. l B2s were built. It is believed that 41 aircraft were built in 1916, although one source suggests as many as 59; all these would have been C.E.P.2 B2s. In 1917 only six C.E.P.2 B2s were built. However, by 1917 it had become apparent that R.E.P. and SAIB (Société Anonyme d'Application Industrielle du Bois) were unable to produce the requisite number of bombers, and arrangements were made to obtain Caproni 3s directly from Italy. As these arrived the Caproni escadrilles changed their designation from CEP to CAP. In 1918 a total of 28 Capronis were obtained. Some of these may have been C.E.P.3 BN3s built by the SAIB. However, official documents show that 20 Caproni 5s (CAP.3 BN 3s) were obtained directly from Italy in early 1918. These were configured to carry two 75-kg and nine 25-kg bombs.

A school was opened at Amberieu-en-Bugey to train French pilots on the Capronis. The instructor was a pilot named Banderieu who would later be responsible for test-flying the Capronis built by SAIB. He had considerable experience with four-engined aircraft, having test-flown the Bleriot 67. The first aircraft used at the center were supplied directly from Italy.

Caproni C.E.P.1; despite the fact that this aircraft carried serial number 36, only 14 examples were built because of the aircraft's poor performance. USAF Museum.

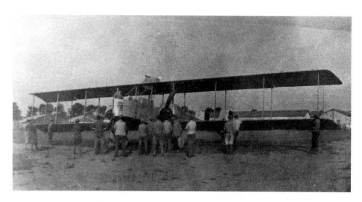

Another view of Caproni C.E.P.1 serial 36. USAF Museum.

Caproni 3 (serial CAP 502) with elevated gunner's mount and Éteve stand. At least three Caproni 3s were built by R.E.P. CC.

C.E.P.2 of CEP 115. This aircraft was flown by Capitaine Lefort. Renaud.

Operational Service

The first unit to receive Caproni C.E.P.2 B2s was CEP 115, formed in February 1916. After a period of training it was assigned to GB 1 in March 1916. GB 1 was based at Malzéville. The unit had initially had 20 C.E.P.2 B2s on strength.

In March CEP 115 participated in night attacks on communication centers and rail lines in the Meuse Valley. During most missions a crew of only two was carried; the gunner's position was eliminated because he was not needed for night operations. The crews of CEP 115 continued training for night missions throughout the summer months. Flights at night were quite hazardous and required highly skilled crews. Accidents were frequent and often had tragic consequences. For example, on the night of 15/16 August a Caproni crashed during takeoff because of engine failure. The aircraft was destroyed and one crewman killed.

A large number of night raids could be carried out beginning in September, for by then most of the unit's crews had been trained for night missions. Furthermore, the longer nights permitted more sorties. Bombers were sent out individually to widely separated targets in order to minimize the chances of mid-air collisions. While this policy was safer, it prevented the planes from concentrating their bombs on a single target.

CEP 115 was also active in October, when it attacked a number of targets including railway stations and the Thyssen ironworks. In retaliation for attacks on its airfield, CEP 115 bombed the airfield at Frescaty on the night of 6/7 November. Bad weather hampered operations in December. Only one major raid was flown the entire month, when on 27 December CEP 115 dropped 150 bombs on various targets.

Inclement weather also prevented CEP 115 from flying any sorties the entire month of January 1917. The unit had mixed results on 9/10 February when four aircraft attacked Mazieres; the raid was successful but one aircraft was lost when a bomb, which had become hung up during the raid, detached while the aircraft was landing. The resulting explosion destroyed the bomber as well as two aircraft inside an adjacent hangar.

CEP 115 was detached from GB 1 on 7 April and assigned to GB 2 at Malzéville. Soon after its re-assignment CEP 115 began to receive the new Caproni 3s (CAP.2 B2s) from Italy and was redesignated CAP 115. The CAP designation indicated that the unit was equipped with the Italian-built Capronis. CAP 115 attacked railway stations, factories, barracks, and enemy airfields.

CAP 115 was considerably less active during the summer months as, once again, the shorter nights meant fewer missions could be flown. Ludwigshafen and Phalsbourg were attacked in July.

The success of CAP 115 resulted in the formation of a second unit in August 1917. It was also equipped with the new CAP.2 B2s and designated CAP 130. It was also assigned to GB 2. At this time both CAP 115 and 130 received new insignia. It seems that while ferrying the CAP.2 B2s across the Alps the pilots had become intrigued by the eagles indigenous to that area. It was decided that both units would be given an eagle emblem: CAP 115's was a green eagle and CAP 130's was blue. The new CAP.2 B2s were very satisfactory and could carry an impressive bomb load. They served alongside the older Caproni C.E.P.2 B2s.

As 1918 began GB 2 moved from Malzéville to Gundrecourt, a move necessitated by persistent German attacks on the former airfield. CAP 115 and 130 attacked Ludwigshafen on 1 February 1918 and again on 25 March. Other targets attacked early in 1918 included Luxemborg and Laon.

Caproni C.E.P.2

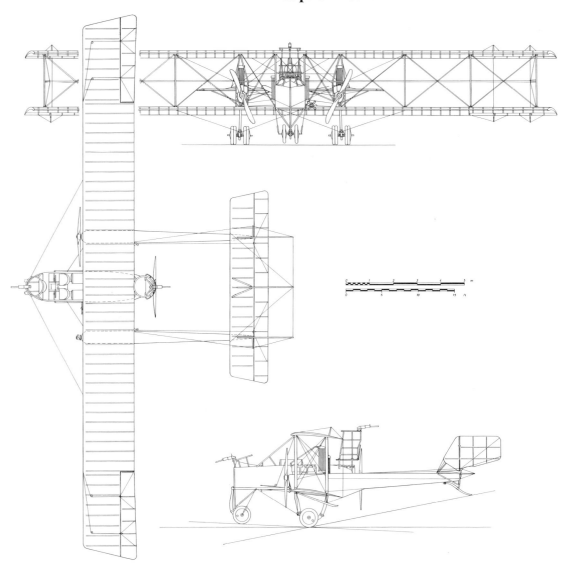

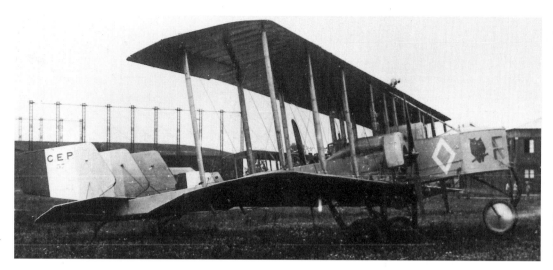

Caproni C.E.P.2 serial CEP 115. The two owls on a branch was the original insignia of CEP 115. B80.1126.

Caproni 33 (C.E.P.2). Power was supplied by three Italian 150-hp Isotta-Fraschini V4A engines. B75.272.

GB 2 received direct support from the Italians themselves when 18 Gruppo was transferred to the Western Front. This unit consisted of three squadriglia: the 5a, 14a, and 15a. Each squadriglia had only four operational Capronis. The combination of GB 2 and 18 Gruppo (which was designated GB 18 by the French) was designated Groupement Villomé, after the unit's commander. It completed 56 missions and dropped 164 tons of bombs on various targets in the Sarre Valley and the city of Ludwigshafen. The Italians had a total of 12 operational aircraft and ten in reserve. The latter were intended to form part of the Inter-Allied Independent Air Force, along with British and American bomber units.

Groupement Villomé attacked a large number of railway stations during the spring of 1918. The German attack at Champagne resulted in the unit moving to les Ferme-des-Greves in April. From this new base GB 2 could attack transportation targets in an attempt to stem the flow of German reinforcements.

In response to the German offensive at Ansfeldhe, units of Groupement Villomé attacked train stations along the Champagne front. In May GB 2 joined British squadrons in attacks on the German airfields at Montcornet, Ville-au-Bois, and Clement-les-Fermes. During May GB 2 was performing quick strikes against targets located by photo reconnaissance aircraft. In many cases these raids were launched as soon as the film had been developed and analyzed.

GB 2 moved to Chateau-Thierry on 28 May. CAP 115 and 130 had a total of 30 Capronis on strength at this time. From 10 to 18 July these Capronis were used to attack German troops advancing in the area of Reims.

During the Battle of Ile-de-France (18 July to 4 August) GB 2 bombed train stations from Guignicourt to Laon and Aisne to Laon. The escadrilles flew up to three sorties per aircraft each night.

Facing page: Caproni C.E.P.2 with revised gunner's mount which was changed from the over-wing position in the C.E.P.1. Renaud.

Caproni C.E.P.1 B2 (Ca.1) Heavy Bomber with Two Crew, a Single 130-Hp Canton-Unné Mounted as a Pusher, and Two 80-Hp Le Rhône Engines

Span 22.20 m; length 10.90 m; height 3.70 m; wing area 100 sq. m
Empty weight 2,000 kg; loaded weight 3,000 kg; bomb load 275 kg
Maximum speed 110 km/h; climb to 1,000 m in 13 minutes; climb to 2,000 m in 26 minutes; endurance seven hours.
Approximately 14 built

Caproni C.E.P.2 B2 (Ca.3) Heavy Bomber with Two Crew and Three 150-Hp Isotta Fraschini V4a Engines

Span 22.20 m; length 10.90 m; height 3.70 m; wing area 100 sq. m
Empty weight 2,000 kg; loaded weight 3,000 kg; bomb load 455 kg
Maximum speed: 120 km/h; climb to 1,000 m in 8 minutes; climb to 4,000 m in 40 minutes; endurance five hours.
Between 50 and 71 built by R.E.P.

Caproni C.E.P.3 BN3 (Ca.5) Heavy Bomber with Three Crew and Three 200-Hp Fiat A-12 Engines

Span 3.4 m; length 12.6 m; height 4.48 m; wing area 150 sq. m
Empty weight 3,000 kg; loaded weight 5,200 kg
Max speed: 138 km/h; climb to 2,000 m in 14 minutes; ceiling 5,000 m; range 560 km
Armament: two 75-kg and nine 25-kg bombs.
At least 20 obtained directly from Italy

Caproni C.E.P.2 of CEP 115. While ferrying the C.E.P.2s across the Alps the pilots had become intrigued by the eagles which were indigenous to that area. It was decided that CAP 115's aircraft would be marked with a green eagle while CAP 130 would use a blue one. B 90.238.

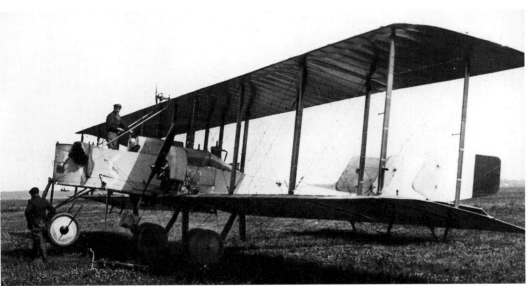

Below left: C.E.P.2. Except for possibly a few Caproni 45s, the Caproni escadrilles were entirely equipped with C.E.P.2s. B85.1163.

The Battle of Santerre (8 to 30 August) saw the Caproni units attacking iron works, troop concentrations, and railroad targets along the Strasburg-Thionville-Hirson line.

Groupement Villomé returned to Epiez for the Battle of Saint Michele (12 September to 30 September). During the Champagne-Argonne offensive (25 September to 11 November) Groupement Villomé remained at Epiez. Since February 1916 CAP 115 had flown 289 sorties and dropped 387 tons of bombs. CAP 130, formed in August 1917, flew 371 sorties and dropped 213 tons of bombs.

The Caproni units rapidly replaced their aircraft after the war ended. CAP 115 re-equipped with Caudron 23s; CAP 130 became Escadrille 211 postwar and soon received Farman F.60 Goliaths.

C.E.P.2. Unlike the two outer rudders which were fully movable, the center vertical surface had a fixed stabilizer portion in addition to the rudder. B85.1162.

Caproni 3 (Caproni 45). SHAA records indicate that 20 were obtained directly from Italy. B88.3414.

Caproni 3. These aircraft were powered by three 200-hp Fiat A12 engines. B84.3750.

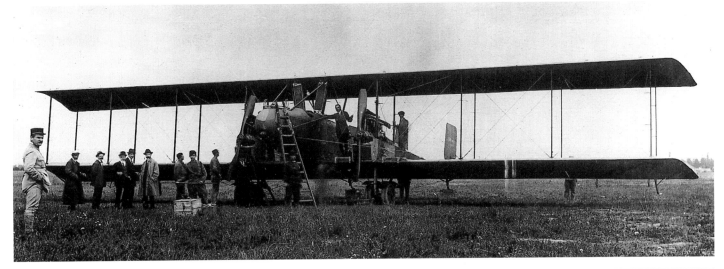

Carroll A2

The A2 specification of 1918 called for a two-seat observation aircraft. It was to have a payload of 450 kg, ceiling of 7,000 m, a normal operational altitude of 3,000 m, maximum speed of 200 km/h, and minimum speed of 90 km/h. The Carroll A2 was produced to meet this specification. Its was a conventional tractor biplane with two-bay wings. The pilot was seated under the top wing and the observer/gunner in a separate cockpit under a cutout in the top wing. The observer was provided with a machine gun mount but photos do not show that armament was fitted to the prototype. The engine was a 240-hp Lorraine-Dietrich 8 Bb. Two prominent fairings were located over the engine compartment. The undercarriage featured large V-shaped struts and a prominent tail skid was fitted. The fin and rudder assembly was elliptical.

The aircraft was in competition with a Breguet 14 A2 with a Renault engine, a Breguet 14 with a Rateau turbo-compressor, and the LeO 2. The Breguet 14 A2 was selected for production; it is likely that the Carroll A2's engine resulted in performance inferior to its competitors. The Breguet 14 A2 was an outstanding success, and no further development of the Carroll A2 was undertaken.

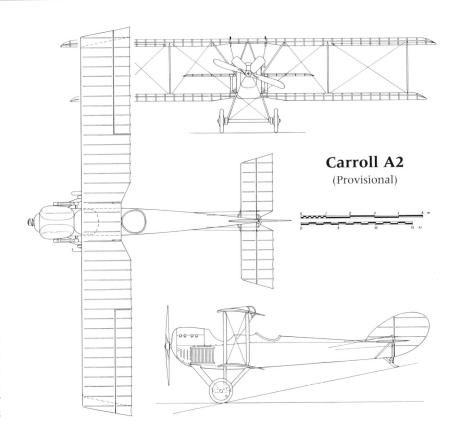

Carroll A2
(Provisional)

Top: Carroll A2. The pilot's position beneath the top wing and behind the bulky engine would have severely restricted his field of view. MA2957.

Above: Carroll A2. MA36383.

Left: Carroll A2. The engine was a 240-hp Lorraine-Dietrich 8 Bb. MA36382.

Caudron

The Caudron brothers (René and Gaston) were inspired by the Wright brothers' first flights in France in mid-1908. Their first design, the Type A, which was to have been powered by two engines, was, instead, flown as a glider because the engines were unavailable. The brothers then decided to produce a smaller aircraft which was intended to use a 25-hp Anzani engine. The Type A2 biplane had a pusher configuration with twin tail booms, twin fins and rudders, and used wing warping. These features would appear on many later Caudron designs even those with tractor-mounted engines. The Caudron A2 took to the air in 1909, its engine driving the propeller by a complicated chain transmission. As a result of troubles with the chain drive, the Caudron brothers repositioned the engine into tractor configuration to eliminate the need for a transmission drive. Thus modified, the aircraft, designated the Type A3, was more reliable and had better performance.

In 1910, the Caudrons established a training school at le Crotoy and an aircraft factory at Rue. The first aircraft to be produced at the factory was a 35-hp Anzani-powered tractor biplane designated the Type A4. It was controlled by a combination of wing warping and auxiliary surfaces mounted on the outer interplane struts. The pilot sat behind the rear spar of the lower mainplanes, where he was completely exposed to the elements.

The next Caudron design, the type B, featured a fuselage nacelle with a 70-hp Gnome or 60-hp Anzani engine in front and the pilot in back. This abbreviated fuselage/nacelle combination would also appear on many of the later Caudron aircraft. The Type B, although a tractor, retained features of the pusher layout; control was entirely by wing warping.

Several other aircraft were produced before the war. These included:

1. Type D trainer (1912) with 45-hp Anzani or 50-hp Gnome. Widely used at the Caudron flight school at Crotoy, also saw extensive service with the Aviation Militaire as a trainer. The New Zealand Canterbury Aviation Company Ltd. also trained its pilots on the Type D during the First World War. The Type D2 was a single-seat biplane with a 35-hp Anzani engine.
2. Type E trainer (1913) with Gnome engine.
3. Type F single-seater (1913) with either 80-hp Gnome or a 60-hp Le Rhône
4. Type G trainer (1912) with 80-hp Gnome or 60-hp Anzani; later developed into G.2.
5. Type H amphibian developed for the Monaco races. Two were later ordered by the navy.
6. Type J floatplane (1913) with 100-hp Gnome.
7. Type K (1913) amphibian with 200-hp Gnome or 100-hp Anzani. Its large wing span (18.15 m) enabled it to serve as a high-seas floatplane. Two pilots were seated side by side in the central nacelle.
8. Type KM (or K2)—A version of the Type K seaplane fitted with armor and a machine gun in the nose of the central nacelle. It had four floats, two beneath the lower wing and two cylindrical floats at the base of the tail booms.
9. Type L (1914) trainer with 100-hp Anzani engine.
10. Type M (1912) monoplane with 50-hp Gnome engine; later version fitted with a 60-hp Anzani. It was built for the Aviation Militaire.
11. Type M2 (1912) monoplane with a 50-hp or 60-hp engine.
12. Type N (1912) racer with 30-hp Anzani engine.
13. A five-seat biplane with a conventional fuselage that retained the wings and tailplane of previous designs. A 70-hp Anzani or 80-hp Gnome was fitted. The design was built in 1912; the type number is not known.
14. A single-seat floatplane flown at Monaco in 1914 with an Anzani engine. The type number is not known.

The Caudron G.2, G.3, and G.4 proved to be excellent reconnaissance aircraft because of their stability, but their pusher configuration made them vulnerable to fighter attacks from the rear. By the middle of the war the Caudron brothers abandoned the pusher layout.

Caudron Type F with a 60-hp Le Rhône engine. B89.3791.

Caudron J Seaplane

The Caudron brothers produced a twin-float seaplane in 1913 and presented it at the Expositions Internationales de l'Aéronautique that same year. The aircraft was a pusher biplane with two-bay wings, with the upper wing being considerably longer than the lower. A pair of canted struts extended from the tip of the lower wings to the edge of the upper wings. The twin floats incorporated a set of small wheels to permit the aircraft to be handled more easily on the ground. The single rudder and elevator were supported by twin booms. The scalloped rudder had an extension on which was mounted a small tail float. Power was supplied by a single 100-hp Gnome. The aircraft was intended for use as a trainer and carried a crew of two.

An example of the Type J was tested in the 1913 naval concours to select an aircraft to serve aboard French ships. A small number of the type Js were obtained by the navy.

Foreign Service
China

A small number of Type Js served with the Chinese air service. For further details see the section on the G.3.

Left: Caudron J seaplane at Saint Raphaël on 7 May 1914. B93.354.

Russia
Two Caudron seaplanes were purchased by Russia and used in the Black Sea.

United Kingdom
The Royal Naval Air Service is also believed to have acquired several Type J amphibians. The aircraft records of the RNAS indicate that the Caudrons were assigned serial Nos. 55–57. Numbers 55 and 56 were powered with 80-hp Gnomes and were assigned to the cruiser HMS *Hermes*. No.57 had a 100-hp Gnome and was based at the Isle of Grain.

> **Caudron J Two-Seat Floatplane with 100-hp Gnome**
> Span 15.10 m; length 9.00 m; wing area 40 sq. m
> Empty weight 500 kg; loaded weight 800 kg
> Maximum speed: 95 km/h, climb rate 33 meters per minute.

Caudron G.2

The Caudron G.2 was a tractor biplane that saw limited service during 1914. It featured the central nacelle, unequal span wings, and twin boom and twin rudders of the early Caudron series. The G.2 had an 80-hp Gnome engine, although in some an 80-hp Le Rhône engine was used. The aircraft was constructed of wood with flexible wings, and lateral control was by wing warping. There were apparently two versions, a single-seater and a two-seater. According to the crews of C 11, the observer's view on the two-seater was very poor. The aircraft were assigned to two escadrilles at the beginning of the war.

At the outbreak of hostilities, the commanding officer of the aviation school at Reims offered his pilots' services to the 5th Armée. On August 16 Escadrille CM 39 (for Caudron Monoplace) was formed. There were only four aircraft on strength. A number of reconnaissance missions were flown over Namur (16 August), Chimay (24 August), Guise (30 August), and during the battle of the Marne (1–10 September). After the front stabilized, the pilots were assigned to various Caudron escadrilles (C 11, C 17, and C 30) with Caudron G.3s. In February 1915 CM 39 re-equipped with Caudron G.3s and was re-designated C 39.

Escadrille C 11, at Crotoy, was the other unit that had the G.2 on strength (alongside some G.3s) when the war began. The pilots were reportedly relieved when their escadrille was completely re-equipped with the more modern G.3 s.

Foreign Service
Australia
Australia acquired several Caudron G.2s for use at flight training schools.

United Kingdom
The RNAS purchased a single G.2 with a 60-hp Anzani engine. It was based at Hendon.

Caudron G.2 serial number C25 at Crotoy. In 1914 examples were assigned to CM 39 and C11. B93.779.

> **Caudron G.2 Single- or Two-Seat Trainer with 80-hp Gnome**
> Wing span 7.25 m, length 12.10 m, wing area 28 sq. m
> Loaded weight 400 kg,
> Maximum speed 106 km/h, climb 166 meters per minute.

Caudron G.3

The Caudron G.3 was a development of the G.2 and retained the twin-boom configuration with a tractor engine. The first G.3 was built at Rue in May 1913 and was flown from le Crotoy. Before the war the Caudron brothers moved their factory to Lyon, where they built sizable numbers of G.3s. Because the G.3 was one of the few French types in large-scale production when hostilities began, the success of the Caudron brothers was assured. Later the brothers opened a second factory at Issy-les-Moulineaux to handle production orders given to them by the Aviation Militaire as well as many other air forces.

The Caudron G.3 was a single-engine, two-seat biplane. The wing spars were made of ash and spruce with reinforcing strips of metal. The spars were ribbed and had no dihedral. The ribs were fitted to the forward spar by slots and were attached to the lower wing by screws. There were 42 ribs in the upper wing and 24 in the lower. Twelve struts held the upper wing in place. The upper wing was longer than the lower, and a pair of struts mounted at an oblique angle connected the tip of the lower wing with the outer portion of the upper wings. This outer portion of the upper wing could be folded back for transport. Roll control was achieved by warping the outer trailing portions of the upper wing. The construction of the stabilizer was similar to the upper wing and initially used warping on the trailing edge; later hinged stabilizers were fitted. The horizontal stabilizer had two spars. The stabilizer was attached to the fuselage by four booms; the two top booms were made of fir and the lower two were made of ash. The end of the lower booms served as landing skids. There were two triangular fins with rudders controlled by foot pedals. The fuselage nacelle was built of fabric-covered ash and attached to the wings by four struts. The landing gear consisted of a pair of wheels attached to the forward part of the lower fuselage booms. Bungees acted as shock absorbers.

The engine was mounted in the front of the nacelle and was separated from the aft fuselage by an aluminum sheet. The

Caudron G.3. The engine used was usually a rotary such as a Gnome or Le Rhône. On training machines a fixed Anzani was used. B87.5030.

engine was usually a rotary such as a Gnome or Le Rhône, but a fixed Anzani was used on trainers. An aluminum engine cowling was sometimes fitted to protect the crew from oil, smoke, and castor oil. The fuel tank was divided into two parts: one section for fuel with a 100-liter capacity, the other for oil with a 5-liter capacity. It was placed on a wooden floor in the fuselage between the observer and the pilot and had a plywood cover. The position of the crew was unsatisfactory; the observer was located behind the engine and beneath the upper wing where his field of vision was extremely limited. The pilot was seated behind the upper wing, where he also had a limited view ahead. Many operational crews switched places, with the pilot in front and the observer behind, but there were complaints that the G.3 was too difficult to land with this arrangement, and in any event the observer's field of vision was unimproved.

The G.3 was adopted for use by the Aviation Militaire as well as by a large number of foreign air forces. A total of 2,450 were built: 1,423 by Caudron and 1,027 by the SFA, Potez, Blériot, and Deperdussin.

Variants

There were several major versions of the Caudron G.3s produced during the war:

G.3 A2—STAé designation for the artillery cooperation version.

G.3 D2—STAé designation for two-seat trainer.

G.3 E1—STAé designation for a G.3 trainer converted to a single-seat configuration.

G.3 E2—STAé designation for rotary-engine trainer.

G.3 L2—STAé designation for a G.3 with an Anzani engine. It was used for liaison.

G.3 R1—STAé designation for a single-seater with a reduced wing span. It was a "penguin" used to train student pilots how to taxi. R signified "rouler."

In service with the Aviation Militaire, the trainer versions of the type G.3 were given the Army designation 12 (XII). Occasionally there is reference to the Caudron G.3 12 (sic).

The observer's position in the G.3, as well as the type's retention of a pusher layout but with a tractor engine, meant that it was virtually impossible to arm the G.3 effectively. Marcel Bloch and Henry Potez were assigned to modify a G.3 so that armament could be carried; Louis Blériot modified the pilot's control system. These changes, carried out on Caudron G.3 No.985, required that a smaller fuel tank be placed near the knees of the observer and that the main tank between the pilot and observer be reduced in size or eliminated. The type was assigned to the C.R.P. and was test-flown by Capitaine Éteve. He was accompanied on the initial test by Lieutenant Frechet, who had supplied the machine gun. The first test flight was made on 15 July 1915. Apparently the modifications were not widely adopted for use in the standard G.3.

A memo from the SFA in 1915 stated that these modifications were being introduced on the G.3 at the Caudron factory:

1. Modification of the painting technique used on the aircraft.
2. Strengthening of the propellers.
3. Reinforced engine mounts.
4. Introduction of tail flaps on some machines.

Caudron G.3. On 8 May 1914. René Caudron flew a G.3 amphibian off the *Foudre*; this was the first shipboard takeoff by the French. R.D. Layman.

Operational Service

The GQG records indicate there were 128 G.3s at the front in August of 1915, along with approximately 31 G.4s. There were a total of 177 G.3s and G.4s by October of 1915, with 161 in service at the front, eight with the R.G.A., seven with training units, and one with the aviation detachment at Vidamee.

By 1 February 1916 there were 141 G.3s in service. They represented 17 percent of the reconnaissance planes still in service and 12 percent of all French types.

C 11 provides an example of the operations of a typical G.3 escadrille. C 11 was formed at Le Brayelle near Douai with six Caudron G.3s. It was stationed here until July 1914, when it was sent to Moltmedy in the 2nd C.A. sector. During August C 11 flew reconnaissance missions in the Battle of the Marne and dropped flechettes on enemy troops. The first wireless experiments were undertaken near Verdun in March 1915. In July C 11 was attached to the R.F.V. and provided reconnaissance to determine the topography of the area and also strafed enemy trenches. In December the escadrille left Verdun and moved to Ancemont near the Meuse. However, poor weather prevented operations in this area, and C 11 was forced to move to a new airfield. On 12 September 1915 the escadrille claimed its first aerial victory, and eight days later C 11 participated in the battles around Champagne, but flights were limited by rain. At this time, the unit's G 3s were used for artillery spotting, long-range reconnaissance, short-range bombing missions against train stations, and aerial combat. In January 1916, despite the aggressive German fighters, three planes of C 11 were able to provide valuable reconnaissance for General Herr of the R.F.V. In April 1916 C 11 was attached to

Caudron G.3 trainer at Juvisy. Note the two men in the moon insignia. B84.1075.

Caudron G.3 trainer at l'Ecole d'Aviation d'Étampes. B86.2483.

the 2nd C.A., with which it remained for the rest of the war. The unit was considered to be one of the most prestigious reconnaissance escadrilles of the Aviation Militaire. During the early part of 1916 numerous aerial victories were claimed, but C 11's main mission was still reconnaissance and artillery spotting. Various targets, including St. Mihiel, St. Maurice, and the steelworks at Joeuf-Homecourt, were bombed in April. By the middle of 1916 the escadrille had completely replaced its G.3s with G.4s.

Other escadrilles that used the G.3 included:

C 6, formed from D 6 in March 1915. Initially assigned to the 5th Armée, C 6 was assigned to the 18th C.A. in July. It was based at Mailly in April 1916 and re-equipped with Caudron G.4s in mid-1916.

C 10, formed from BL 10 in April 1915, assigned to the 35th C.A. and commanded by Capitaine Mercier. In November 1916 it was assigned to the Service du Camp d'Instruction.

C 13, formed from HF 13 in early 1915 and assigned to the 5th C.A. In December 1915 C 13 moved to the 2nd Armée sector and in January moved to the 4th Armée sector at Champagne.

C 17, formed in December 1914 at Bron, commanded by Capitaine Gérard and assigned to the 1st Armée. C 17 moved to Dijon on 10 December. It was especially active over the Lorraine front. C 17's aircraft bombed Metz in February 1915. Under the command of Capitaine Jeannerod it participated in the Battle of Woëvre from April to May 1915.

C 18, formed from BL 18 in February 1915, assigned to the 1st Armée and based at Toul. In January 1916 it was attached to the 30th C.A.

C 27, formed from REP 27 in early 1915. In April C 27 was based at Bethune and was assigned to the 21st C.A. It was also active during the Battle of Artois.

C 28, formed from HF 28 in January 1915, based at Saint-Cyr and assigned to the 11th C.A. In March C 28 was based at Lavillers. The escadrille was commanded by Capitaine Volmerange. C 28 was assigned later to the 56th D.I. It was active along the front at Vauchelles in the area of the Somme in February. After being briefly based at Mondicourt it moved to the 10th Armée sector and was active along the Champagne front in October 1915.

C 30, formed from BL 30 in January 1915. In March 1915 it was assigned to the 6th Armée and was based at Cravancon.

C 34, formed in February 1915 with Caudron G.3s. It was based at Belfort in November 1916.

C 39, formed from CM 39 (with Caudron G.2s) in March 1915. In April it was assigned to the 5th Armée and was based at Villers Allerand. C 39 was active over the Champagne front.

C 43, formed at Lyon-Bron. Commanded by Capitaine Aubry, it was assigned to the 6th Armée. C 42 was initially based at Moreuil and transferred to Cachy on the Somme front in March 1916. In November 1916 it was assigned to the 6th C.A. and based at Bois des Tailles.

C 46, formed in March 1915 at Dijon, commanded by Capitaine Legardeur and assigned to the 6th Armée. The unit was active over Quennevières and supported the 7th C.A. in the Battle of Champagne.

C 51, formed in March 1915, assigned to the 1st C.C. and based at Somme-Bionne. It was active over Champagne and later, Artois. C 51 returned to Champagne in time to participate in the September offensive.

C 53, formed in May 1915 at Lyon-Bron, commanded by Capitaine Sassary and assigned to the 1st C.A. in the 5th Armée sector. C 53 was active in the Godat and LaMiette sectors.

C 56, formed in May 1915 and assigned to the 17th C.A. It performed army cooperation missions over Artois.

The Caudron escadrilles had been largely re-equipped with the superior Caudron G.4s by late 1915/early 1916. A GQG order for 12 July 1916 ordered all escadrilles still possessing G.3s to turn them over the G.D.E. (training establishment). By August 1917 there were only 12 G.3s with front-line escadrilles and in the aviation parks, and seven with the R.G.A. Some G.3s were retained by each unit to serve as pilot trainers. Many G.3s retired to the aircraft parks later saw extensive service in training units, where the docile Caudrons proved quite popular. Indeed, many French pilots obtained their license on Caudron G.3s. The G.3s were to be found at most French schools, including Chartres, Étampes, Le Crotoy, Buc, Amberieu, Châteauroux, Bron, Tours, Istres, and Dijon. It took 50 days to obtain a license for a pilot trained in a G.3. Once the license had been granted the pilot might continue to fly the G.3 in advanced training. In 1917 the order in which a pilot was advanced along the training syllabus was: Maurice Farman M.F.11, Voisin, Caudron G.3, and Nieuport fighter. The G.3s served well into 1917 and some were still operational as trainers as late as 1918. The G.3 was remarkably easy to fly and was very stable in flight. These qualities made it an ideal trainer. Takeoff and climb was normally at a speed of 80 to 85 km/h, cruising speed was between 100 and 105 km/h, descent was at 80 km/h, landing approach was at 70 km/h, and the aircraft landed at 55 km/hr.

A few G.3s were used by the Aviation Maritime. On 8 May 1914 René Caudron made the first takeoff of an airplane from a

Caudron G.3 at the Paris Salon. V662.

French ship, flying a G.3 amphibian off a temporary platform erected on the forecastle of the seaplane carrier *Foudre*. A second attempt on 9 June by a naval aviator, Lieutenant de Vaisseau Jean de Laborde, failed when the Caudron struck a piece of deck gear and crashed into the water. The machine was destroyed but the pilot survived. The accident caused an end to such experiments and the platform was removed from the *Foudre*.

Foreign Service

Argentina

Argentina received its first G.3 from France in 1918. A year later, six more G.3s were sent to Argentina. An additional 15 G.3s were later assembled at El Palomar. The G.3s served with the Escuela de Aviacion Militar (Military Aviation School) which in 1922 became Grupo de Aviacion 1 (1st Aviation Group).

Australia

Australia acquired a number of G.2s and G.3s for service with flight schools. CFS 9 (Central Flying School 9) had one G.3, purchased in 1916 and retired in 1918. G.3s also served with the Australian Half Flight in Mesopotamia. Two were used by IFC 3 and 4 (Indian Flying Corps).

Belgium

About 66 Caudron G.3s with 80-hp Le Rhônes were obtained by Belgium although, surprisingly, not until 1918. The initial batch of 36 arrived in June/July 1918 at the Ecole de L'Aviation Militaire (Military Flight Training School) at Juvisy; the remaining 30 arrived a short time later.

The aircraft were used to equip the 6th Training Group. By May 1920 the 1st, 2nd, and 3rd Escadrilles were equipped with the type. In September 1924 the G.3s were based at Wevelghem and assigned to the 3rd Groupement of the 2nd Groupe d'Ecolage (Training Group). The school at Juvisy continued to use the aircraft until 1928. Some of the 66 aircraft were used at the civil flight training school at Gosselies. It was intended that graduates of this school would provide a potential pool of flyers if war should break out.

China

On 13 July 1913 the Chinese received 12 Caudrons—eight two-seaters powered by 80-hp Gnome engines (which were G.3s) and four smaller single-seaters with 50-hp Anzani engines (which were probably G.2s or Type Js).

The French set up a flight school but found the Chinese to be reluctant students. The Chinese Army Flight School at Nayan was a limited success, for by spring 1914 all but one of the Caudrons were unserviceable. Eventually they were repaired and training resumed in May with a class of 40 Chinese army officers. By early 1916 80 students had been graduated and the school had 20 aircraft (some of them G.3s) on strength.

The Caudrons did see active service in China. Several were sent to take part during an uprising in Mongolia. They were based at Kalgan and armed with bombs. However, according to British sources the aircraft "never left the ground."

Three G.3s with 80-hp Gnomes and a single-seater with a 50-hp Anzani were sent to suppress a bandit uprising on 6 April 1914. Each averaged 150 km of flying over seven weeks and were used primarily for reconnaissance. The aircraft were felt to have turned the tide of battle and the lead airman was promised the head of the rebel leader as a reward.

After this demonstration of the effectiveness of aircraft, several G.3s were assigned to support various Chinese divisions in the field. The aircraft saw little active service throughout the remainder of the war as their condition rapidly deteriorated, at least in part due to poor maintenance. By 1920 those machines that remained were described as "useless."

Colombia

The Escuela Militar de Aviacion (Military Aviation School) became operational in April of 1922. It was located at Flandes and obtained three Caudron G.3s in early 1922. They were withdrawn from service in 1925 when the school was closed.

Denmark

The Danish army air service obtained a single Caudron G.3 in 1914. It was powered by a 60-hp le Rhône engine and remained in service until 1922.

El Salvador

Three Caudron G.3s, which had been built in Italy by the A.E.R. firm (see below), were purchased by the Escuela de Aviacion Nacional in 1924. These had serials C-1, C-2, and C-3. One of them, C-3, participated in maneuvers as part of the First Section of the 1st Squadron in 1924.

Finland

The Finnish air service obtained 12 G.3s in 1920. Six more were built in Finland during 1921–23. Another machine was obtained from a private source. They carried these serials:

1. Twelve Caudron G.3s obtained from France: serials 2A 490–495 (later 1B 1 through 7 and 1D 8 through 12).
2. Six Finnish-built machines: serials 1D 13 and 1E 14–18.
3. A single aircraft obtained from a private source: serial 1B 19.

The Caudron G.3s were not assigned to any of the Flying Divisions but were assigned to the Ilmailukoulu (flying school) at Santahamina. They remained in service until 1924.

Caudron G.3 equipped with skis in service with the Finnish air service. The Finnish G.3s were assigned to the Ilmailukoulu or flying school at Santahamina. B81.2989.

Greece

A number of Caudron G.3s were obtained by the Greek army air service in 1921. They served as trainers at the Sholi Aeroporias Sedes (Sedes Aviation School). One participated in the Campaign in Asia Minor in 1921.

Italy

Italy was in desperate need of adequate combat aircraft when it entered the war in 1915. One of the first types to be obtained was the Caudron G.3. By December 1915 there were five G.3 squadrons on strength. The aircraft were used in the army cooperation role and were organized as artillery Squadriglias.

Caudron G.3s and G.4s were built under license by the A.E.R. plant at Torino. Ninety aircraft were built in 1915, and 80 in early 1916. By the end of the war a total of 250 had been built. Caudron G.3s continued in widespread use as army cooperation aircraft throughout 1916. They were organized as follows:

Gruppo 5 (3rd Armata): 41ª, 42ª, 43ª, and 44ª Squadriglias.
Gruppo 6 (3rd Armata): 45ª Squadriglia.

By early 1917 the G.3s were replaced by other aircraft, including Caudron G.4s. The G.3s were subsequently used as trainers.

A few remaining G.3s were assigned to 41ª Squadriglia in 1917, along with a number of Farmans (probably F.40s). No. 41ª Squadriglia was assigned to 2 Gruppo. In April 2 Gruppo was assigned to the 2nd Armata and in November it was with the 4th Armata.

Japan

A single G.3 trainer with an 80-hp Gnome engine was purchased but the type was not adopted by the Japanese army air service.

Peru

A French mission arriving in Peru in January of 1919 brought a number of aircraft including one Caudron G.3 intended for use as a trainer.

Portugal

The Caudron G.3 saw service with the Portuguese air service. Two G.3s were obtained in 1916 and were assigned to the Escola Aeronautica Militar (Military Flight School) at Vila Nova da Rainha. Four additional G.3s were obtained later. In 1918 the school had been redesignated the Escola Militar de Aviacao (EMA) and a new school had been formed at Alverca, designated the Esquadrilha Mista de Deposito (Joint Training

AER-built Caudron G.3 C507 Captured by Flik 4 in July 1915 and used to perform at least eight reconnaissance missions over the Italian front (in Austro-Hungarian markings). R. Stach via Colin Owers.

and Depot Squadron). It was initially equipped with two G.3s. The G.3s must have been successful trainers, because plans were made to produce them under license by the Parque de Material Aeronautica (Aeronautical Material Park) or PMA. Deliveries began in 1922 and 50 had been produced by 1924. A few PMA-built G.3s were sent to the Grupo de Esquadrilhas de Aviacao de Angola (Group of Air Squadrons at Angola) in 1922. Production of the G.3 was terminated in 1924. Apparently the G.3s were used exclusively for training.

Romania

France supplied a number of combat aircraft to Romania in 1915 to cultivate the Romanian tilt toward the Allies. Among the aircraft were a dozen Caudron G.3s. On 15 September 1915 the Romanian air corps was created and included in its order of battle three Caudron G.3 units. One unit had four G.3s and was assigned to Grupul 1 (1st Group) as an artillery observation squadron; another had four Caudron G.3s, also in the artillery observation role, and was assigned to Grupul 2, and, finally, a third squadron of four G.3s in the artillery cooperation role was attached to Grupul 3.

By 17 August 1916, when Romania entered the war on the side of the Allies, none of the G.3s was serviceable. As 1916 progressed two of the G.3s were repaired; one was assigned to Grupul 2, the other to Grupul 3. By early 1917 the G.3s had been completely replaced in the reconnaissance role by MF.11s.

This view of a Caudron G.3 in flight shows the scalloped trailing edges of the wing and elevator. MA35860.

Russia

Both the Caudron G.3 and G.4 were obtained from the French by the Russian government and served primarily with long-range reconnaissance and army cooperation squadrons. Unlike many other foreign aircraft, neither the G.3 nor G.4 were selected for license production by the Russian government. It appears that about 20 G.3s were obtained. Soviet sources describe them as being antiquated, yet some saw prolonged service with the Russian air service. The main complaints about the type were the same as those listed by French aviators—the G.3 was unarmed and offered the observer a very restrictive field of view. It was, however, easy to fly and had a relatively rapid rate of climb.

Serbia

The Caudron G.3 was introduced into service in Serbia when Escadrille G 89 was assigned by the French to Serbia in 1916. This unit remained under French control and did not become a combined French-Serbian unit. Apparently, Serbia never obtained any G.3s for its own air service.

Spain

The first G.3 was flown to Spain by a French pilot in May/June 1919. The aircraft was intended for use as a primary trainer. About 18 G.3s were obtained by the Spanish government. By December 1919 they were in service with three of the five training centers in Spain, at Getafe, Sevilla, and Los Alcazares. The training syllabus called for students to solo in the G.3 and, after passing their flight exam, advance to the Avro 504. There were 12 G.3s in service in December 1920 (all with 80-hp Le Rhône engines). By 1924 the G.3s had been replaced completely by Avro 504s.

United Kingdom

Caudron G.3s were first obtained from the French in 1915 to help supplement the rapidly-growing RFC. The first aircraft were assigned to No.1 Squadron in March 1915. Although about 140 G.3s were obtained, they never saw widespread use with the British reconnaissance units. No.1 Squadron had four aircraft in May 1915, No.4 Squadron had two, and No.5 Squadron had at least one. The last operational G.3 was removed from service in October 1915.

Several G.3s remained at the 1st Aircraft Park without being assigned to squadrons; the British valued the 80-hp Le Rhône engines more than the airframes to which they were attached. The G.3s were used with greater success in the training role and most of them saw service with the BEF School and 13, 14, 23, and 29 Squadrons. Some were with Nos.1, 3, 4, 6, 9, and 41 Reserve Squadrons. Finally, two G.3s saw service in Mesopotamia in 1915. The aircraft were sent to Basra in July and were subsequently based at Nasani, where they made reconnaissance flights. One was lost on 30 July, the other was lost 16 September to rifle fire.

The Royal Naval Air Service also used the Caudron G.3. A total of 139 G.3s were employed, mostly as trainers in France. At the beginning of the war there was a single G.3 (No.45) at the Eastchurch Station and it moved to Chingford in May 1915. G.3s built by British Caudron were on strength; these included No.40 and No.1372. Aircraft from France were Nos.1592-7; all but No.1595 had Anzani 100-hp engines (No.1595 had a 80-hp Gnome); some or all of these may have been Type Js. Twenty-five more G.3s with 80-hp Gnome engines were obtained from France in 1915. They were allocated serials number 3264-3288. Twenty additional G.3s with 80-hp Gnome engines (Nos.3863-3882) were purchased in France during 1915.

In 1916 more G.3s were purchased for service with the Caudron School. These were built by British Caudron and had serials 8941-8950. These G.3s were used almost exclusively for training. Serials N3050-3099 were based at the RNAS at Vendôme, while N3264-N3288 were based at the Eastchurch and Eastbourne Stations.

Caudron G.3 Two-Seat Reconnaissance Aircraft with 90-hp Anzani

Span 13.40 m; length 6.40 m; height 2.50 m; wing area 27 sq. m
Empty weight 420 kg; loaded weight 710 kg
Maximum speed: 112 km/h, climb to 2,000 m in 18 min.; climb to 3,000 m in 32 min.; endurance 4 hours
Armament: usually flown unarmed although the crew was given a standard issue rifle

A total of 2,450 were built.

Caudron G.3s Built by British Caudron: Engines Included 80-hp Gnome, 80-hp Le Rhône, 70-hp Renault, and 80-hp or 100-hp Anzani

Span 13.26 m; length 6.89 m; height 2.59 m; wing area 28.27 sq. m
Empty weight 435 kg; loaded weight 710 kg (80-hp Gnome)
Maximum speed: 105 km/h; climb to 2,000 m in 27 min.; ceiling 3,050 m; endurance 3.5 hours
Armament: usually flown unarmed although the crew was given a standard issue rifle

Approximately 50 were built.

Caudron G.3s Built by A.E.R.: Engines Included the 80-hp Le Rhône (Although Others Used the 100-hp Anzani)

Span 13.40 m, length 6.40 m, height 2.50 m, wing area 27 sq. m
Empty weight 420 kg, loaded weight 710 kg
Maximum speed: 110 km/h; climb to 1,000 m in 8 min.; climb to 2,000 m in 18 min.; climb to 3,000 m in 30 min.; ceiling 4,000 m; endurance 4.0 hours

A total of 250 were built.

Caudron G.3s Built by PMA with 80-hp Le Rhône

Span 13.40 m; length 6.40 m; height 2.70 m; wing area 27 sq. m
Empty weight 448 kg; loaded weight 663 kg
Maximum speed: 115 km/h; climb 143 m/min.; ceiling 4,000 m; range 400 km; endurance 4.0 hours

A total of 50 were built.

This Caudron G.3 was used as a trainer by the USAS and was based at Issoudun. MA38561.

Caudron G.3

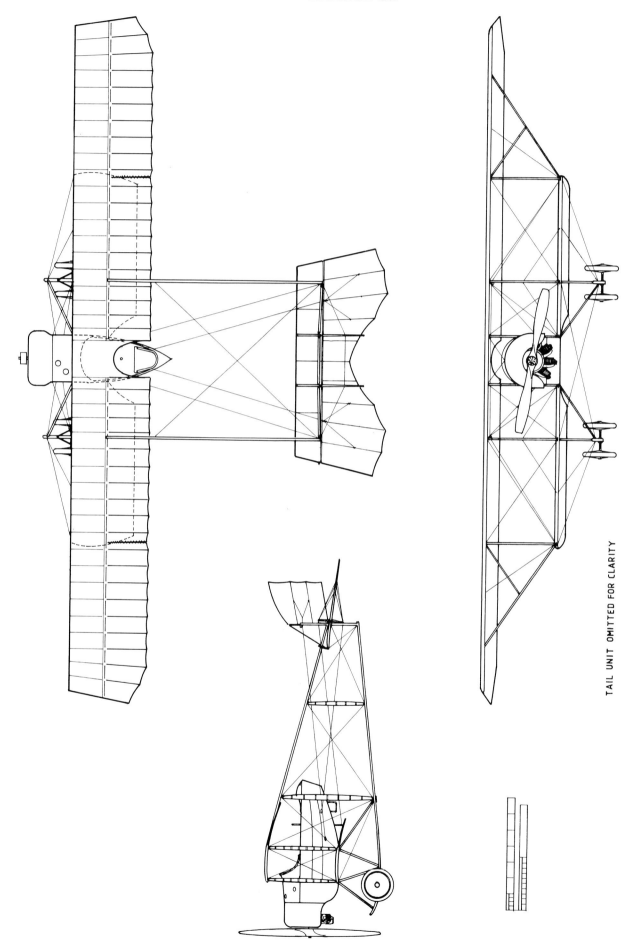

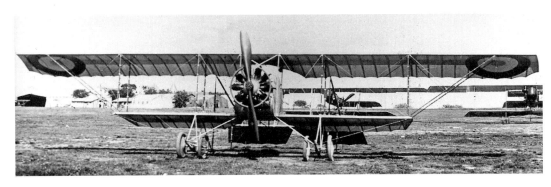

Left: Caudron G.3 with an Anzani engine. Two M.F.11s are in the background. MA34617.

Below: Caudron G.3 in June 1914. MA10993.

United States

Although the Air Service of the A.E.F. never used the Caudron G.3 operationally, many American pilots were trained on them. A total of 192 G.3s were used by the A.E.F. as trainers. They were considered by the American pilots to be "museum pieces," but they were effective in the primary training role. Most of the G.3s were based at Tours. It has been noted that those at Tours were decrepit and it was believed that many of them had been rebuilt from wrecks. There were 50 or 60 G.3s in service at Tours in the fall of 1917. By January 1918 the number had been reduced to around a dozen. By February an additional 30 G.3s/G.4s were available.

Venezuela

When a French air mission arrived at Venezuela in January of 1921 it brought with it 12 Caudron G.3s and G.4s. These aircraft, along with others brought by the French, were used to

establish the Servicio de Aeronautica Militar (Military Air Service). Four additional G.3s were obtained in the mid-1920s.

Caudron G.4

The Caudron G.4 was produced in response to the Aviation Militaire's need for a more powerful army cooperation aircraft which could carry a forward-firing machine gun. By equipping the G.4 with two engines, Caudron increased the aircraft's range and created a position for a nose gunner. The G.3 was redesigned to permit this arrangement; the plane was enlarged and the central crew nacelle was lengthened. The observer fired a machine gun (a Hotchkiss 7-mm or a Lewis gun) on a flexible mounting in the nose. However, this arrangement did not permit the gun to be used to protect against attacks from behind. On some aircraft a crude attempt was made to rectify this by fitting a gun to the top wing fixed to fire to the rear. This arrangement proved ineffective and the gun was soon removed from the aircraft in service. The crew was equipped with a Chauchat gun or a carbine. Some G.4s were fitted with a camera for high-altitude reconnaissance.

To handle the increased weight and also provide the gunner with a better field of fire, two engines (either 80-hp Le Rhônes or 100-hp Anzanis) were placed in streamlined nacelles on either side of the center fuselage. The Le Rhône engines were cowled, but the rotary Anzanis dispensed with the cowlings. The number of rudders was increased from two to four. The prototype G.4's first flight was in March 1915; Caudron built 1,358 G.4s during the war. G.4s were built by Blériot, SPAD, and Caudron.

Variants

The G.4 was built in three major versions: the A2 for reconnaissance, the B2 for bombing, and the E2 for training. The A2 had a wireless set for artillery spotting missions; the B2 could carry up to 220 kg of bombs. The E2 had dual controls; it was intended for training and was powered by either Anzani or Le Rhône engines.

There was an armored version of the G.4 that was sent to some of the best units. It was powered by two 80-hp Le Rhône engines. A GQG memo noted that the armored Caudron G.4s were in great demand, but few were available because of a shortage of iron needed to produce the armor. Another GQG memo to the commander of the 1st Armée, dated 19 July 1916, stated that distinguished escadrilles were being sent the armored version first; these units were C 64, C 66, C 104, C 105, and C 106. These planes received the Army designation IB (B = blindé, or armored). It was eventually planned to supply all Caudron reconnaissance escadrilles with one or two armored G.4s, enabling them to carry out low-level attacks. However, a later memo suggested that the G.4 IBs would be most useful in situations over the front where there was heavy cloud cover (particularly in the winter). Presumably this would have provided the heavy planes with some measure of protection against fighters.

A floatplane version, serial C 2498 and designated *Le Goeland* was also built.

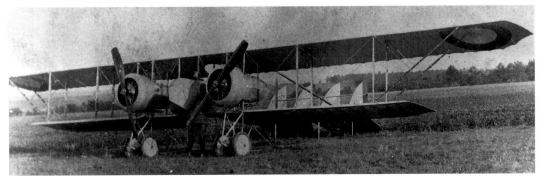

Caudron G.4. The twin engines increased the aircraft's range as well as permitting it to be flown in bad weather. The Blériot 53 was designed to the same specification. B86.724.

G.4 floatplane named *Le Goeland*, one of eight machines evaluated at the Saint Raphaël naval air station. B93.774.

Operational Service

On 15 August 1915 there were 36 G.4s in service with the escadrilles and in the aircraft parks. By 1 February 1916 there were 161 G.4s in service as compared with 141 G.3s. On 1 August 1917 there were 215 Caudron G.4s in service. A total of 139 G.4s were with the front-line escadrilles and aviation parks, with a further 75 available but not yet in the parks. The G.4s at first supplemented the G.3s then in service and by late 1915/early 1916 had replaced the G.3s.

In operational service the G.4 was initially praised for the gunner's improved field of fire, but because of its pusher layout, could not be defended against attack from the rear. Despite this serious limitation, the G.4s were used as bombers and often provided fighter escort for the slower M.F.11s and Voisin 3s and 5s. To overcome the handicaps of their aircraft, G.4 pilots often attacked from high altitude to avoid anti-aircraft fire and fighters. In the attack role the G.4s would dive from high altitude on enemy planes and then try to escape before the Germans who survived could recover. However, the G.4s rapidly became obsolescent and as early as April 1916 the type was being criticized for having mediocre speed, limited maneuverability, poor range, and severe vulnerability to rearward attacks.

Army Cooperation

The Caudron G.4s were used primarily for army reconnaissance and artillery spotting. The typical army cooperation unit also spent a large amount of time performing long-range bombing and ground-attack missions, and even flew combat patrols in addition to routine reconnaissance duties.

C 11 provides an example of a G.4-equipped escadrille. On 29 June C 11 had largely re-equipped with G.4s and had moved to an airfield in the Somme area. C 11 was active in the Somme area as part of the 2nd C.A. Perhaps inspired by the G.4s lack of rearward defense, the escadrille was given the cocette insignia around the middle of 1916. The cocotte's eye was positioned to the rear as if to watch for enemy aircraft. C 11 was extremely active for the next five months, flying reconnaissance, artillery cooperation, ground attack, and night bombing missions. Most of the bombing attacks were against the stations at Ham, Nesle, and Peronne. On 29 December 1916 C 11 moved to Villers-Bretonneiux near Toul. It was assigned to the 88th Division near Delme on 4 February 1917. On 24 April it moved to Marne to take part in the French offensive; it was based at Hourges. However, rain and snow inhibited aerial operations on that front. Later, C 11 moved to Rosnay and performed visual and photo-reconnaissance, bombing, and ground attack operations.

Caudron G.4. On some aircraft a crude attempt was made to provide effective rearward defense by fitting a gun to the top wing or nacelle; the weapon was fixed to fire to the rear. This arrangement proved to be ineffective and the gun was soon removed from the aircraft in service. B77.776.

It was active in this area until June. Despite the fact that German aerial defenses were taking heavy tolls on the unit, C 11 continued low-level attacks on the German trenches to enhance French morale. As 1917 progressed the vulnerable G.4s were supported by the new Letord long-range reconnaissance planes. Capitaine Vuillemin used one of C 11's Letords to great effect in protecting the unit's G.4s from German aircraft when on 29 July he attacked and drove off five German aircraft, claiming one as destroyed. On 11 June C 11 moved to Grigny as part of the 31st C.A., where it flew primarily photo missions. On 10 September it moved to the area of the Meuse and was stationed at Julvecourt. During November 1917 the unit re-equipped with Breguet 14 A2s to become BR 11.

Other units equipped with Caudron G.4s included:

C 4, re-equipped with G.4s in April 1916. It was assigned to the 3rd C.A. It later re-equipped with Caudron G.6s.

C 6, active in the 2nd Armée sector at Verdun in May 1916. It was based at Lemmes and subsequently moved to the Argonne, Saint-Menehould, and Mailly-le-Camp. In November C 6 moved to the 3rd Armée sector at the Somme. It was later assigned to the 10th Armée and based at Fontaine, from where it participated in the Battle of Chemin des Dames. It re-equipped with Salmson 2s in January 1918.

C 9, formed from BL 9 in August 1915 when the unit re-equipped with Caudron G.4s. It was assigned to the 8th Armée and commanded by Lieutenant Escot. C 9 was based at Villers-lés-Nancy in Meurthe-et-Moselle. It re-equipped with Sopwith 1½ Strutters in October 1917.

C 10, created from BL 10 and assigned to the 35th C.A. In April 1917 it was based at Corbeaulieul. Commanded by Capitaine Mercier, C 10 participated in the battles of Aisne, Champagne, Artois, and Picardie. It re-equipped with Salmson 2s in March 1918.

C 13, assigned to the 15th C.A. in the 2nd Armée sector at Verdun. In September it re-equipped with Sopwith 1½ Strutters.

C 17, equipped with Caudron G.4s during 1916 and assigned to the 1st Armée. In April 1916 C 17 was based at Bois des Tailles under the control of the 6th C.A. It later moved to Vitry-le-Francois and then Treux, where it was assigned to the 6th Armée. C 17 was active during the Battle of the Somme. In December it moved to the Flanders sector under the command of Capitaine Taillepied. In April 1917 it was based at Hourges under the control of the 1st C.A. In October 1917 C 17 re-equipped with Sopwith 1½ Strutters.

C 18, assigned to the 30th C.A. in the 2nd Armée sector at

Caudron G.4 C212. This airplane served as a target tow plane for the Ecole Tir de Chasse at Biscarrosse. B78.1163.

Caudron G.4 of C 30. This unit participated in the Somme offensive in July 1916 where it performed bombing missions. B92.1588.

Caudron G.4 preparing for takeoff. The aircraft is being loaded with lance bombs. Renaud.

Verdun. In November 1916 it was based at Grivesnes. The escadrille moved to Montidier in January 1917 and returned to Verdun in June. In March 1918 C 18 re-equipped with Salmson 2s.

C 21, created in December 1916 with several Caudron G.4s, two of them armored. It was assigned to the 6th Armée and was based at Chipilly. Commanded by Lieutenant Bloch, C 21 re-equipped with A.R.1s in March 1918.

C 27, assigned to the 21st C.A. and based at Vadelaincourt in the Verdun sector. It moved briefly to the Champagne area and was active over the Somme in September. In November 1916 C 27 was based at Marcelcave. It returned to the front line in May 1917 and was active on the Aisne front and the Battle of Chemin des Dames. It re-equipped with Salmson 2s in late 1917.

C 28, returned to the control of the 11th C.A. in July 1916, at which time it had a new commander, Lieutenant Pacaud. C 28 moved to Alger and was active over the Somme. In November 1916 it was assigned to 30th C.A. and based at Grivesnes. In February 1917 it moved to the vicinity of Ham. By mid-July C 28 had re-equipped with Sopwith 1½ Strutters.

C 30, re-equipped with Caudron G.4s in July 1916. It was assigned to 6th Armée and participated in the Battle of the Somme. It later moved to Champagne and also received a few Caudron R.4s. In November C 30 was assigned to 37th C.A. and based at Baslieux-les-Fismes. In February 1917 it was based in the 6th Armée sector with the 37th C.A. The aircraft of C 30 were often used as bombers, usually attacking German lines of communication. C 30 re-equipped with Salmson 2s in April 1918.

C 34, based at Romagny and assigned to the 52nd D.I. It re-equipped with SPAD 11s in October 1917.

C 39, assigned to the 5th Armée and in September 1915 re-equipped with Caudron G.4s. It was based at Dugny. In November 1916 C 39 was based at Rosnay in the 5th Armée sector. In the fall of 1917 it was based at Nogent-Sermiers and was assigned to the 38th C.A. In October 1917 it re-equipped with Sopwith 1½ Strutters.

C 42, formed in March 1915 at Lyon-Bron, assigned to the 2nd Division de Cavalerie and based at Lunéville. Later C 42 was attached to the 74th D.I. and performed bombing and reconnaissance missions. It then moved to Nancy to help provide defense for the city. Four G.4s were detached to the Verdun sector during 1916; these later formed the nucleus of the Lafayette Escadrille. In November 1916 C 42 was assigned to the D.A.L. and based at Villers-les-Nancy. In April 1917 the escadrille participated in the Battle of Chemin des Dames.

C 43, assigned to the 6th C.A. It was active during the Battle of the Somme in 1916 commanded by Capitaine Pastier. In August 1917 seven aircraft of C 43 were detached to the 2nd Armée. Based at Lemmes, C 43 was active over the Verdun front. In October 1917 it was placed under the command of Capitaine Simon and later that month re-equipped with Sopwith 1½ Strutters.

C 46, equipped with G.4s in April 1916. It was active in the 6th Armée sector during the Battle of Verdun, based at Vadelaincourt. It re-equipped with Caudron R.4s in June 1916.

C 47, formed in April 1915. In November 1916 it was assigned to the 2nd C.A.C. and was based at Le Hamel. In February 1917 C 47 was assigned to the 2nd C.C. and based at La Cense. Later that month it re-equipped with Salmson 2s.

C 51, assigned to the 1st C.C. and moved to the Somme in February 1916; it suffered heavy losses to German fighters in this area. C 51 was also involved in the Allied attacks at the Somme during July 1916. In November 1916 it was still assigned to the 1st C.C. in the 6th Armée sector and was based at Bougainville. It subsequently transferred to Frétoy, where it re-

Caudron G.4. Two engines (either 80-hp Le Rhônes or 100-hp Anzanis) were placed in streamlined nacelles located on either side of the main fuselage. The Le Rhône engines were cowled, but the rotary Anzani engines dispensed with the cowlings. B86.907.

equipped with Sopwith 1½ Strutters in February 1917.

C 53, re-equipped with Caudron G.4s in June 1916, assigned to the 1st C.A. in the 5th Armée sector and based at Rosnay. C 53 performed long-range reconnaissance and bombing missions. It moved to Oiry near the Marne in February 1917, then Hourges in the 5th Armée sector. Two weeks later it moved to Vadelaincourt. After suffering heavy losses near Verdun, C 53 retired to Baslieux-lés-Fismes. On 21 July the unit moved to Gramaille and the following month to Hamel. It participated in the Battle of the Somme while based at Morlancourt. On 10 October 1917 it moved to Rouvrel and in January re-equipped with SPAD 11s.

C 56, assigned to the 17th C.A. After a period of rest at Lorraine, it moved to the Champagne front in April 1916, where it suffered heavy losses. In February 1917 C 53 was based at Ferme d'Alger in the 4th Armée sector along the Champagne front. In March 1917 the unit re-equipped with Caudron G.6s.

C 61, formed on 8 August 1915 at Lyon-Bron. It was based at Somme-Vesle and was assigned to the 34th C.A. It moved to the Champagne front in November 1916 and subsequently to Belfort, then Fontaine (7th Armée sector) until mid-1917. In July 1917 C 61 moved to Rosnay. It re-equipped with Sopwith 1½ Strutters in October 1917.

C 64, formed in August 1915, assigned to the 4th Armée and based at La Cheppe. In February 1918 C 64 became SPA-Bi 64.

C 74, formed in July 1916 and based at Hondschoote under command of Capitaine Collard. It was assigned to give support to the Belgian army. Capitaine Paillard assumed command in July 1917. C 74 returned to Esquennoy in the vicinity of the Somme in April 1918. At this time it re-equipped with Salmson 2s.

C 104, formed from VB 104 in May 1916, was initially based near Verdun and assigned to the 32nd C.A. In November 1916 C 104 was in the 2nd Armée sector; it was assigned to the 2nd C.A., and was based at Vadelaincourt. In April 1917 the unit participated in the Battle of Chemin des Dames and the next month re-equipped with Sopwith 1½ Strutters.

C 105, formed from VB 105 in early 1916 and assigned to the 2nd Armée from July to December 1916. C 105 was active over the Verdun front and was based at Julvecourt in November 1916. By February 1917 it remained in the 2nd Armée sector but was based at Autrecourt. From March through September 1917 it was assigned to the 5th Armée and participated in the Battle of Chemin des Dames. It re-equipped with Sopwith 1½ Strutters in October 1917.

C 106, formed from VB 106 in June 1916 and assigned to the A.G.L.P. (long-range artillery) of the 10th Armée in November 1917. C 106 was based at Grivesnes. In late 1916 it re-equipped with Sopwith 1½ Strutters.

C 122, formed in 1916 and initially assigned to the 32nd C.A. In November 1917 it was assigned to the 3rd Armée and was based at Grivesnes. It re-equipped with Salmson 2s at the end of 1917.

Artillery Observation

A number of G.4 units were assigned as S.A.L. escadrilles and served in the spotting role for heavy artillery. These units were:

C 202, formed in January 1916 under the command of Capitaine Watteau. It was attached to the 6th (June) and 10th (August) Armées and was active during the Somme offensive. C 202 was assigned to the 5th Armée in February 1917 and was active over the Verdun front. It subsequently moved to the 2nd Armée front. In December 1917 it re-equipped with Breguet 14 A2s.

C 207, formed in 1916 from V 207. In November 1916 it was assigned to 20th C.A. and based at Fricourt on the 6th Armée front. It transitioned to Sopwith 1½ Strutters in October 1917.

C 219, formed from elements of C 21 in January 1917, commanded by Capitaine de Fontenilliat and assigned to the 6th Armée. C 219 participated in the battle of Chemin des Dames from April to May 1917. Dubbed the *Escadrille Espinasse*, it was sent to Italy as S.A.L. unit for the 10th Armée. It re-equipped with Sopwith 1½ Strutters in July 1917.

C 220, formed from V 220 in September 1916 under command of Capitaine de Seguin. In November 1916 C 220 was assigned as a S.A.L. unit of the 2nd Armée and was based at Lemmes. It re-equipped with Caudron G.6s in April 1917.

C 222, formed from V 222 in October 1916. It was assigned to the D.A.L. in November 1916 and was based at Lunéville. It re-equipped with Sopwith 1½ Strutters in 1917.

C 224, formed during 1916 under the command of Lieutenant Goudard. It was assigned to the 2nd Armée and participated in the Battle of Verdun. In October 1916 Lieutenant Roig assumed command. In November 1916 C 224 was based at Lemmes in the 2nd Armée sector. By February 1917 the escadrille was posted to the 15th C.A. In July 1917 it was assigned to the 9th C.A. and based at Hourges. In September 1917 the escadrille was sent to Bonne-Maison where it was stationed with the 35th C.A. in the 6th Armée sector. It re-equipped with Breguet 14 A2s around that time.

C 225, created in January 1917 under command of Capitaine Odic and assigned to the 5th Armée. It re-equipped with Salmson 2s in late 1917.

C 226, formed early in 1917 under command of Capitaine des Isnards and assigned to the 36th C.A. It was based at Bray-Dunes in June 1917. C 226 became SOP 226 in late 1917.

C 227, formed from elements of VB 106 in early 1916 and active in the Battle of Verdun. At the beginning of 1917 the escadrille was sent to Mesnil-Saint-Georges in the 10th Armée sector. Subsequently, C 227 was dispatched to La Cheppe in the 4th Armée sector. In October 1917 it re-equipped with Breguet 14 A2s.

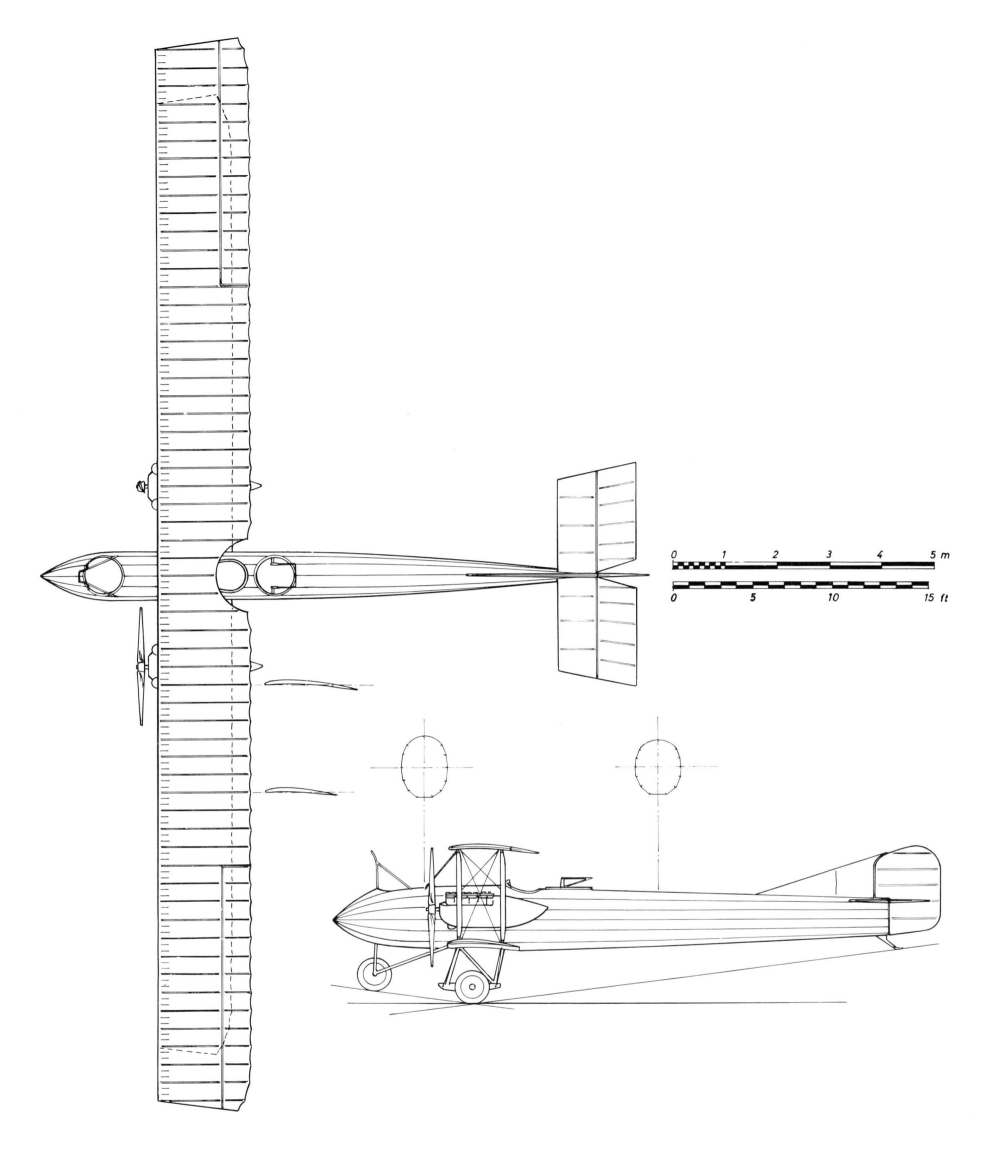
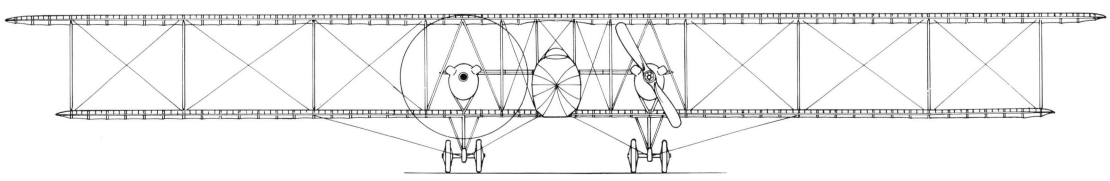

Caudron R.4

154A

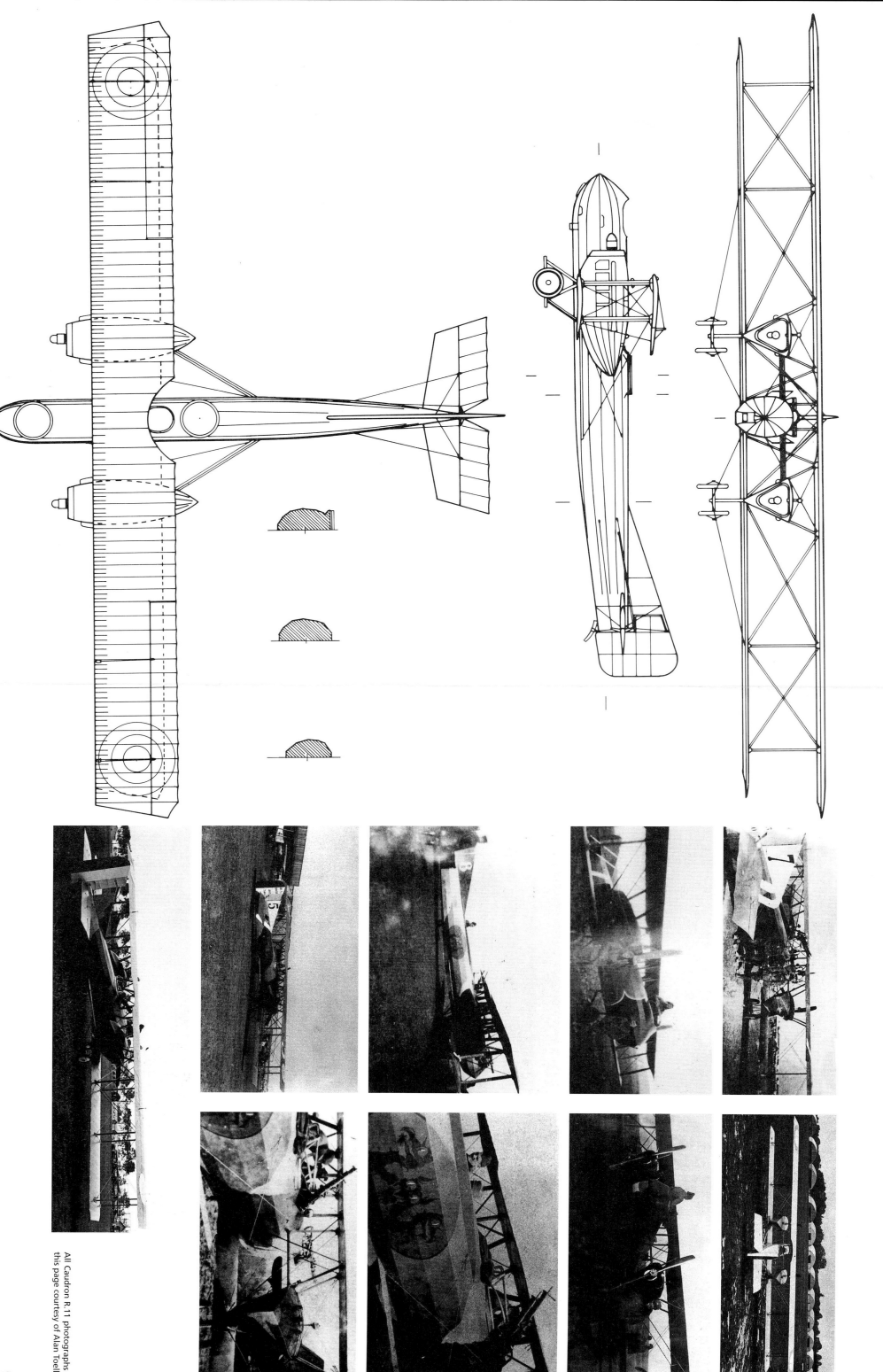

Caudron G.4 of C 66 at Fretoy in 1917. C 66 was formed (along with N 65) from G.C. Malzéville in late 1915 and was assigned to GB 2 at Malzéville. B88.1218.

Bombing

Because of its payload and long range the G.4 was often used as a bomber. Many of the army cooperation units undertook bombing missions and several units became dedicated bomber escadrilles. There were four units designated as bomber escadrilles in the 100 range: C 104, C 105, C 106, and C 122. Despite their designations, two of these units (C 104 and C 122) were used primarily in army cooperation (see above). C 105, C 106, and C 66, however, were assigned to Groupes de Bombardement.

C 66 was formed (along with N 65) from G.C. Malzéville in late 1915 and assigned to GB 2 at Malzéville. Some of its G.4s were used in the escort role and also dropped propaganda leaflets on the Malancourt-Joloncourt area in December. C 66 remained at Malzéville in 1916. In March it provided fighter escort and bombed train stations and airfields. When used as bombers the planes usually did not carry gunners, enabling the G.4s to attack from a height of 4,500 meters, where they were safe from German fighters and anti-aircraft fire. Sorties by one or two G.4s against various targets were flown in April. In May leaflets were dropped on German trenches. High-altitude raids were flown against the Metz-Sablons station and the German airfield at Montange. GB 2 was disbanded on 17 June 1916 and C 66 was assigned to GB 1. By now C 66's commander, Capitaine Henri de Kerllis, had made this an elite unit. The first mission with GB 1 was flown on 22 June when 38 bombs were dropped on the arms works at Karlsruhe. This was a well-defended target and flying the Caudron G.4 in daylight was suicidal. One-third of the nine planes taking part were lost. Capitaine de Kerllis, well aware that his aircraft were no match for German fighters, decided to time his attacks at dawn and dusk, when opposition was less likely. He also used a combination of stealth (such as flying at high altitude) and cunning (avoiding airfields and never flying the same route twice). Utilizing these tactics, C 66 flew numerous missions in July with no losses. On 24 August one pilot of C 66 succeeded in destroying a German aircraft—no small feat in a G.4. On 25 September the unit went to Demuin in the 6th Armée sector, thus ending its association with GB 1.

VB 105 and VB 106 gave up their Voisin 3s and 5s for G.4s and rejoined GB 2 as C 105 and C 106 respectively in early June 1916. C 105 and C 106 flew patrols on 9 and 10 June. When GB 2 was disbanded C 105 was assigned to the 2nd Armée and C 106 to the 6th Armée in the army cooperation role.

Long Range Escort

C 66 often provided fighter escort for the slower Voisins of GB 2; C 66's aircraft often had the dual roles of fighter and bomber on the same mission. On 8 March 1916 three G.4s of C 66 flew escort for a raid on the Brieulles station and even managed to drop 12 bombs on the target. During the latter part of March C 66 flew fighter missions and barrage patrols. The latter mission was necessitated by the fact that the Germans were staging frequent raids on GB 2's airfields, and the Voisins assigned to the bomb group were incapable of intercepting the enemy bombers. On 2 April a G.4 of C 66 attacked a German machine while flying a barrage patrol and fired 150 rounds; the Caudron was hit twice. Another German aircraft was chased off by a G.4's crew on 26 April 1916. Sometimes the crews of the G.4s had other problems to contend with beside Germans; for example, a crew of C 66 was frustrated in an attempt to bring down an enemy plane on 16 May 1916 when their machine gun jammed after only ten rounds had been fired. C 105 and C 106 also flew a few barrage patrols with GB 2 until the Groupe was disbanded on 17 April 1916.

C 34 and C 61 used their G.4s to provide fighter escort for the MF.11s and F.40s of MF 29 beginning in late November 1915. The G.4s were not much faster than the Farmans they were assigned to protect, and the limited efficacy of the G.4 as a fighter was made apparent when a G.4 of C 61 was shot down while escorting a bombing raid. In late January 1916 C 34 and C 61 were withdrawn from GB 4 and assigned to army cooperation duties (see above).

Overseas Escadrilles

Ten Caudron escadrilles served as T.O.E. units. These were G 488 (Greece), G 489 (Côtiére de Mazagan in Morocco), G 490 (Côtiére de Sfax), C 543 (North Africa), C 544 (North Africa), C 545 (North Africa), C 546 (North Africa), C 547 (North Africa), C 548 (North Africa), and C 549 (North Africa).

In a note from the 5th Armée dated 22 September 1916 the G.4 received scathing criticism. While it was acknowledged to have a good climb rate, it was very vulnerable to attacks from the rear. It was recommended that G.4s be limited to artillery spotting and photographic missions conducted over 5,000 m. Another note called attention to the fact that the rudders were suffering structural fatigue secondary to faulty attachment screws. It was recommended that they be reinforced with piano wire.

The vulnerability of the G.4 to rear attack lead the Aviation Militaire to request a new long-range reconnaissance machine. It was to have twin engines in the tractor configuration and a conventional fuselage with gunners fore and aft. This class of aircraft was designated A3 and led to the G.5, G.6, R.4, R.5, and R.11. However, there were significant delays in the development and production of the A3 class.

As late as August 1917 there were 139 G.4s at the front as opposed to 133 G.6s, 53 R.4s, and 89 Letords of all types. Thus it appears that most of the Caudron units had substantial numbers of G.4s on strength until late 1917.

In late 1916 the first G.6s joined those G.4s serving with the Caudron escadrilles. It appears that by 1918 the G.6 had supplanted the G.4s, which were retired to serve beside the G.3 in the training units. Most Caudron units had switched to more modern planes in late 1917. The stop-gap Sopwith 1½ Strutters replaced the G.4s beginning in July 1917.

Caudron G.4

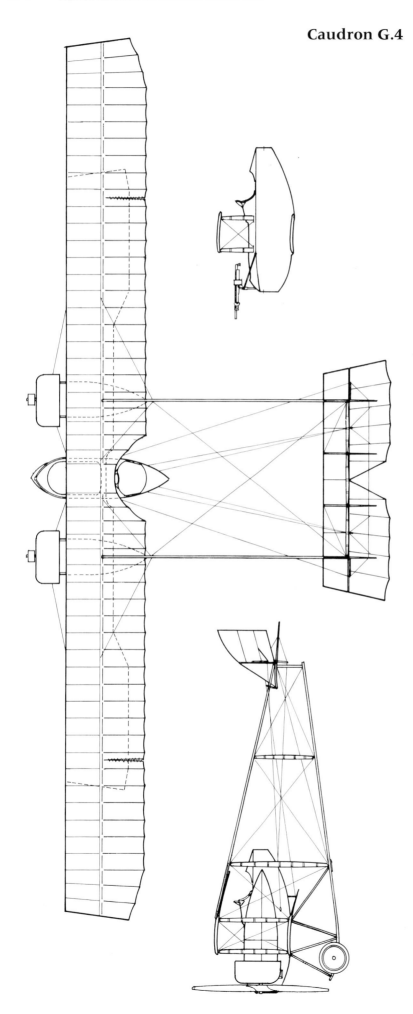
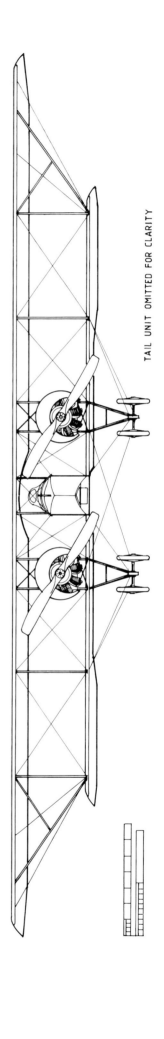

TAIL UNIT OMITTED FOR CLARITY

Caudron G.4 in Belgian service. The Franco-Belgian Squadron No.674 had both French and Belgian pilots. It was equipped with Caudron G.4s. Colin Owers.

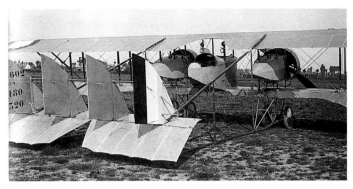

Another view of a Caudron G.4 in Belgian service. Colin Owers.

Apparently a number of escadrilles retained a small number of G.4s for a variety of duties. For example, a GQG memo for 12 January 1918 requested that all units using G.4s for gunnery practice were to exchange them for A.R.1s as soon as possible.

The artillery cooperation units also received new aircraft—Sopwith 1½ A2s, Breguet 14 A2s, and Salmson 2 A2s replaced the G.4s in late 1917.

The Aviation Maritime converted eight Caudron G.4s to floatplanes. Tests were carried out on these at the Saint Raphaël naval base in September 1916.

Foreign Service

Belgium
The Franco-Belgian Squadron No.674 had both French and Belgian pilots. It was equipped with Caudron G.4s.

Brazil
The Brazilian air service obtained two Caudron G.4s from France in 1921. They were intended as photo-survey aircraft and were unarmed. They were assigned to the Servico Geographico de Ejercito (Army Geographical Service) and were withdrawn from service only two years later, in 1923.

Colombia
Colombia obtained four G.4s in early 1922. They were assigned to the Escuela Militar de Aviacion (Military Aviation School), which became operational at Flandes in April of that year. The G.4s were placed in storage at that time and were reactivated in 1925 when a second school at Madrid was opened. They were finally retired in 1929.

El Salvador
A single G.4 was obtained from the Italian aviation mission to El Salvador in 1924.

Finland
A single Caudron G.4 was obtained by Finland in 1922 and was not assigned a serial number. It was apparently donated to the Finnish air service by a private company.

Italy
The G.4s were found by the Italians to have superior qualities and, in particular, a good climb rate and high altitude performance. These two qualities made it especially suitable for service in the Alps. The type was built by the A.E.R. factory in Torino. Forty were built under license in 1916 and 11 in 1917. Italian G.4 units in 1917 included:

3 Gruppo (1st Armata): 46a and 50a Squadriglias.
5 Gruppo (3rd Armata): 42a, 43a, and 44a Squadriglias.
7 Gruppo (6th Armata: reassigned to the 1st Armata in November): 49a Squadriglia.
12 Gruppo (4th Armata): 48a Squadriglia.

By 1918 the Caudron G.4 had been replaced by more effective aircraft such as the Pomilio PE and S.I.A.7 of indigenous origin.

Japan
One Caudron G.4 (sometimes mis-identified in Japan as a G.6) was purchased by the Japanese army air service in 1921. Designated the type Bo.1, it remained in service until 1923.

Netherlands
Five Caudron G.4s were obtained by the Netherlands air service and given serial numbers 427–431. They entered service in June 1918 and remained in service until the mid-1920s.

Portugal
Portugal obtained nine G.4s from France in 1918. It was planned to assign them to the Esquadrilha Expedicionaria a Angola (Angolan Expeditionary Squadron). This unit was intended to see active service in Angola, but the fighting ended before any operational missions could be undertaken. The squadron, now named the Esquadrilha Inicial Colonial (Initial Colonial Squadron), was based at Humpata by October 1918. In 1921 it was redesignated Grupo de Esquadrilhas de Aviacao de Angola (Group of Air Squadrons at Angola) and had one squadron of Caudron G.4s and one with Breguet 14 A2s. The unit was disbanded in 1923.

Romania
The Romanians obtained a number of Caudron G.4s and by July of 1917 eight of these were in service with No.12 Squadron assigned to Grupul 2. In August this squadron was still attached to Grupul 2 and was based at Calmatui in support of the Romanian 1st Army. Apparently none of the G.4s remained in service at the end of the war.

Russia
The G.4 was popular with Russian pilots. With two engines it was much faster, had a better rate of climb, and could be looped more easily than the single-engine G.3s. Between 20 and 40 G.3s were obtained and saw service both during and after the war.

The G.4s were used primarily by what were described as corps air units, each of which had six army cooperation aircraft and two fighters for escort. Soviet sources list 37 G.4s in service in April 1917. This later figure probably also included some G.3s. It would seem that many of the G.4 units were in disrepair, as the reports from air divisions and air fleets in March listed only four Caudrons as serviceable. All four were on the southwest front. By June 1 the situation had improved, with a total of 12 Caudrons listed as serviceable. Eight were on the southwestern and Romanian fronts and four were on the Caucasus front.

One G.4 survived long enough to enter service with the 1st Socialist Air Group—the first Soviet air unit. Six Caudron G.3s and G.4s were still in service as late as 1921.

Caudron G.4 of F 25. The escadrille's marking (a blue star in a white circle) has been painted in the center of the wheel. The Caudron G.4 may have been assigned to the unit to provide fighter escort for the M.F.11s and F.40s. B82.3071.

Saudi Arabia

The kingdom of Saudi Arabia received four Caudron G.4s with 120-hp Le Rhône engines in August 1921. These were apparently built by the A.E.R. firm.

Ukraine

The Ukrainian air service obtained a single Caudron G.4 in 1918.

United Kingdom

A single G.4 (serial 7761) was obtained by the RFC for evaluation. It arrived at No.1 Aircraft Depot at St.-Omer on 14 January 1916. Apparently a second Caudron was also obtained by the same depot on 29 February but was never assigned a serial number. The G.4 was never adopted for service with the RFC.

Twelve G.4s were purchased by the RNAS for use by No.1 Wing in France. The RNAS ordered additional G.4s later in 1915 including:
Serial Nos.3289–3300: 100-hp Anzani—Nos.1, 4, and 5 Wings.
Serial Nos.3333–3344: 80-hp Le Rhône—Eastchurch and No.5 Wing.
Serial Nos. 33894–3899: 80-hp Le Rhône—Eastchurch, Dunkirk, Nos.1, 2, and 5 Wings.

Caudron G.4s purchased in 1916 included an initial batch of 12 (Nos.9101–9131) all with 100-hp Anzani engines. They were assigned to Nos.2, 4, and 5 Wings. A follow-up order for 20 (Nos. 9286–9305) was canceled.

No.5 Wing at Coudekerque had a number of G.4s on strength in 1916, as did No.4 Wing at Petite Snythe. Both units used the G.4s as long-range bombers. Operations from Coudekerque by No.5 Wing began in March 1916. The first target was the airfield at Houttave and Zeebruge. Several G.4s took part in the raid and considerable damage was done to the target. On April 23 No.5 Wing bombed Mariakerke. By May No.4 Wing was able to participate in some raids. On 5 May No.4 and 5 Wings returned to bomb Mariakerke at night. On 19 May the airfield at Ghistelles was bombed, as was Mariakerke again on the 21st.

Attacks were suspended at the end of May but were resumed in August to help divert German aircraft away from the Somme front. On 2 August ten G.4s drawn from Nos.4 and 5 Wings took part in a raid on the St. Denis Westrem airfield. On 2 September Caudron G.4s, again from Nos.4 and 5 Wings, took part in a raid on the Ghistelles airfield and on the 7th St. Denis Westrem was attacked. One G.4 was lost during this raid. Further attacks during September included raids on St. Denis Westrem (17th and 21st), Ghistelles (9th and 23rd), and the Hadzaeme airfield (9th and 24th). Other targets were the Lichtervelde ammunition dump, the Hindenburg batteries, and the airship sheds near Brussels. On November 10, Nos.4 and 5 Wings attacked the base at Ostend. Ten aircraft from No.5 Wing attacked the docks at Ostend again on the 12th. Twenty-two aircraft from Nos.4 and 5 Wings raided Ostend on the 15th.

As 1916 came to a close it was increasingly obvious that the Sopwith 1½ Strutter and Short Bomber were superior to the G.4. The Caudrons were now relegated mostly to night attacks. On 10 February 1917 Nos.4 and 5 Wings attacked Bruges. There were few bombing raids for the remainder of February and March because of bad weather. By April the new Handley Page 0/100 and D.H.4 were on strength and the G.4s ceased to play an active role as a bomber with the RNAS.

G.4s also saw service with the RNAS No.7 Squadron.

United States

The United States purchased ten G.4s for use as trainers. They were never used operationally.

Venezuela

The Venezuelan Centro de Aviacion Naval (Naval Aviation Center) was formed at Palmita in November 1922. It was under the control of the army and it included one Caudron G.4 on strength. The naval center eventually merged with the army air service.

Caudron G.4 Two-Seat Reconnaissance Aircraft with Two 80-hp Le Rhône Engines

Span 16.885 m; length 7.19 m; height 2.55 m; wing area 36.828 sq. m
Empty weight 733 kg; loaded weight 1,232 kg; payload 210 kg
Maximum speed: 130 km/h at sea level, 125 km/h at 2,000 m; 124 km/h at 3000 m; climb to 1,000 m in 6 min. 30 sec.; climb to 2,000 m in 15 min.; ceiling 4,300 m; endurance 5 hours
Armament: one nose-mounted Hotchkiss 7-mm or a Lewis machine gun; the crew was also equipped with a carbine or Chauchat gun.
A total of 1,358 G.4s were built.

A.E.R.-Built G.4 with Two 80-hp Le Rhône Engines

Span 16.885 m; length 7.20 m; height 2.60 m; wing area 36.828 sq. m
Empty weight 845 kg; loaded weight 1,350 kg; payload 505 kg
Maximum speed: 130 km/h at sea level; climb to 3,000 m in 19 min.; climb to 4,000 m in 36 min.; ceiling 4,500 m; endurance 4 hours
A total of 51 G.4s were built by A.E.R.

Caudron G.5

The Caudron G.4 achieved considerable success as a long-range reconnaissance plane, but its deficiencies had been obvious for some time in 1915. In an attempt to improve the type, Gaston Caudron produced a major redesign of the G.4, designated the G.5.

The Aviation Militaire had requested development of a three-seat reconnaissance and light bomber aircraft. This A3 specification was for a large machine with gunners fore and aft for clear fields of fire. Planes designed to meet this requirement were Gaston Caudron's G.5 and G.6, René Caudron's R series, the Morane-Saulnier Type T, and the Salmson S.M.1.

While the G.4 had an abbreviated central nacelle, the G.5 featured a conventional fuselage. This enabled it to carry two gunners as required by the A3 specification. Power was supplied by two 80-hp Le Rhône engines. The plane was built of wood and fabric, with steel fittings. It appears that the G.5 retained a

Caudron G.5. The G.5 was an attempt by Gaston Caudron to meet the requirement for an army cooperation type carrying a crew of three. The aircraft pictured served as a transitional type between the G.4 and G.6; recent research by French scholars suggests that this airplane served as the prototype for the G.6 series. B93.819.

wing similar to the G.4; the surface area of both types was nearly identical.

The aircraft was flown in spring 1915, tested by Gaston Caudron for the SFA. Finding it to be seriously underpowered, he redesigned it into what would be designated the G.6.

Although the G.5 had significant deficiencies, a few were assigned to operational units. In May 1916 G.5 serial number 678 was assigned to S.A.L. 202. Also in May, C 10 and C 27 received two G.5s each. Their crews were armed with Winchester semi-automatic rifles and a Chauchat machine gun. If these reports are accurate, they suggest that at least a small number of G.5s were built before they were replaced by the G.6.

Caudron G.5 Two-Seat Reconnaissance Light Bomber with Two 80-hp Le Rhône Engines

Wing area 38 sq. m
Empty weight 850 kg
Armament: Winchester semi-automatic rifles and a Chauchat machine gun

Below: SHAA files identify this as a Caudron G.5; the aircraft pictured possibly was the prototype for the G.6 series. B93.820.

Caudron G.6

The G.4 had initially proved to be successful in the reconnaissance role, but as 1915 progressed the weaknesses of the design became increasingly apparent. The frequently poor weather severely limited the number of sorties and the G.4's layout prevented adequate defense against attacks from the rear. The Aviation Militaire demanded a new category of aircraft to correct these problems; the A3 class of army cooperation aircraft was to perform long range reconnaissance, bombing, and escort fighter duties. They would need two powerful engines to enable them to fly long-range missions or in high winds. A crew of three was specified: a pilot and a front and rear gunner. Gaston Caudron's response to this requirement was to produce the G.5, while René designed the R.4 series (initially intended for use as a bomber, but actually used in the A3 role).

Limited numbers of G.5s were used at the front, but because they were found to be seriously underpowered, Gaston Caudron had it redesigned to feature a wing of increased area and two 110-hp (later 120-hp) Le Rhône engines. The new aircraft followed the general lines of the G.5, retaining a conventional fuselage with machine gun positions fore and aft. The G.6 was designed by Paul Deville. Although by this time Gaston Caudron had died while testing the R.4, the "G" designation was retained by the firm.

Improved performance was ensured by the two 120-hp Le Rhône engines (although the prototype and early production machines had 110-hp engines). The pilot was seated beneath a cutout in the upper wing. The engines were mounted in nacelles on either side of the fuselage. The landing gear consisted of paired wheels mounted directly beneath each of the nacelles and a rear tail skid. There was a single triangular fin and rudder.

The prototype G.6 had an abbreviated nose barely protruding ahead of the engines. This did not permit an adequate field of fire, and so the G.6 was redesigned with a longer nose of conical profile. The rear gunner was located well aft of the wings in both the prototype and production G.6s. On some G.6s a wind-driven generator was fitted in the nose gunner's position.

The upper and lower wings each had two spars made of ash and spruce reinforced with a strip of steel. The rear longerons had two additional steel strips. The ribs were held to the front spar by a mortis and tenon arrangement and to the rear spar by screws. There were 52 ribs in the upper wing and 32 in the lower, spaced every 0.32 m. The 28 struts were of spruce. The tail had one rudder with a fixed forward fin and a single horizontal stabilizer. The fuselage had four main longerons of spruce. The shape of the fuselage was held by piano wire. The floor was made of plywood sheathed in aluminum and placed on the lower longerons. The G.6 had dual controls. The fuel and oil tanks were behind the engines. The engines had aluminum cowlings. A portion of the engine covering was in fabric and was held in place by laces in order to facilitate removal of the fuel and oil tanks. The undercarriage had two N-struts with an oval cross section. Each strut held a pair of wheels. A runner between each pair of wheels, reinforced with springs, which acted as a brake when the aircraft landed. The rear skid was articulated on a bungee cord to reduce the landing shock.

The aircraft took off and climbed at a speed of 90 to 100 km/h. Cruising speed was around 140 km/h. Descent was at an airspeed of 110 km/h. On approach to landing the speed was to be reduced to 85 km/h and landing was to be at 60 km/h. Pilots were instructed that if the aircraft were to enter a stall they were to avoid violent maneuvers and regain airspeed as quickly as possible. However, pilots found it difficult to recover the G.6 from a stall or spin.

Operational Service

The prototype G.6 first flew in June 1916 and by the end of 1916 50 Caudron G.6s were at the front. A total of 512 were produced. By 1 August 1917 there were a total of 302 G.6s available to the Aviation Militaire—133 with the Caudron escadrilles and in the aviation parks, 162 with the R.G.A., and an additional seven in reserve.

G.6s are known to have served with the following escadrilles:

Army Cooperation

C 4, re-equipped with G.6s in mid-1916. In February 1917 C 4 was assigned to the 3rd C.A. and based at Couroure in the 2nd Armée sector. It took part in the Battle of Verdun under the command of Capitaine Watteau and later Capitaine Plantey. In April the escadrille took part in the Battle of Chemin des Dames. It re-equipped with Salmson 2s in February 1918.

C 6, assigned to the 10th Armée and based in Fontaine. It re-equipped with Salmson 2s in January 1918.

C 9, assigned to the 8th Armée and based at Villers-les-Nancy in February 1917. It re-equipped with Sopwith 1½ Strutters in October 1917.

C 10, assigned to the 35th C.A. In February 1917 it was based at Corbeaulieu. It participated in the battles of Aisne, Champagne, Artois, and Picardie. It re-equipped with Salmson 2s in March 1918.

Caudron G.6 prototype at Issy Les Moulineaux. B93.2110.

The Caudron G.6 prototype had an abbreviated nose. This would not have permitted a nose gunner to have an adequate field of fire, and at some point in its development the G.6 was redesigned to have a longer nose. However, the aircraft pictured was further developed as the C.21 bomber (see pages 170–171). B93.2116.

Caudron G.6 of C 11 at Touljan in 1917 displaying the red cocotte insignia. C 11 was assigned to the 2nd C.A. and was based at Toul in the 8th Armée sector. B83.5656.

C 11, assigned to the 2nd C.A. and based at Toul in the 8th Armée sector in February 1917. It was at Hourgas during the Battle of Chemin des Dames. In November 1917 it re-equipped with Breguet 14 A2s.

C 13, assigned to the 15th C.A. in the 2nd Armée sector and based at Lemmes in February 1917. In September it re-equipped with Sopwith 1½ Strutters.

C 17, based in April 1917 at Hourges under the control of the 1st C.A. In October 1917 it re-equipped with Sopwith 1½ Strutters.

C 18, moved to Montidier in January 1917 and in February was based at Senoncourt, assigned to the 30 C.A. It returned to Verdun in June. In March 1918 it re-equipped with Salmson 2s.

C 21, assigned to the 6th Armée. It re-equipped with A.R.1s in March 1918.

C 27, assigned to the 21st C.A. and based at Luxeuil in February 1917. It was active along the Aisne front and during the Battle of Chemin des Dames. It re-equipped with Salmson 2s in late 1917.

C 28, active in the 3rd Armée sector and assigned to the 14th C.A. In February 1917 it was based at Grivesnes, and then moved to the vicinity of Ham. By mid-July it had re-equipped with Sopwith 1½ Strutters.

C 30, based in the 6th Armée sector with the 37th C.A. The aircraft of C 30 were often used as bombers, usually attacking German lines of communication. It re-equipped with Salmson 2s in April 1918.

C 34, based at Romagny and assigned to the 52nd D.I. in the 7th Armée sector in February 1917. It re-equipped with SPAD 11s in October 1917.

C 39, assigned in February 1917 to the 38th C.A. in the 5th Armée sector and based at Nogent-Sermiers. In October 1917 it re-equipped with Sopwith 1½ Strutters.

C 42, assigned to the 8th Armée and based at Villers-les-Nancy in November 1917. It re-equipped with SPAD 11s in October 1917.

C 43, assigned to the 6th C.A. In August 1917 seven aircraft of C 43 were detached to the 2nd Armée. Based at Lemmes, C 43 was active over the Verdun front. In October 1917, it was placed under the command of Capitaine Simon. In October 1917 it re-equipped with Sopwith 1½ Strutters

C 46, assigned to the 6th Armée during the Battle of Verdun and based at Vadelaincourt. It re-equipped with Caudron R.4s in June 1916.

C 47, active in the 6th Armée sector and based La Cense with the 2nd C.C. It re-equipped with Salmson 2s in 1917.

C 51, assigned to the 1st C.C. and re-equipped with Sopwith 1½ Strutters in early 1917. However, SHAA documents show that it still had ten G.6s on strength in late 1917.

C 53, assigned to the 1st C.A. and moved to Oiry near the Marne in February 1917. Subsequently it moved to Hourges in the 5th Armée sector and two weeks later to Vadelaincourt. After suffering heavy losses near Verdun, C 53 retired to Baslieux-lès-Fismes. On 21 July the unit moved to Gramaille and the next month to Hamel. C 53 participated in the Battles of the Somme while based at Morlancourt. On 10 October 1917 it

Caudron G.6 of C 30; serial 4322. C 30 was assigned to the 6th Armee and participated in the Battle of the Somme. B92.1891.

moved to Rouvel. In January it re-equipped with SPAD 11s.

C 56, re-equipped with Caudron G.6s in March 1917. Commanded by Capitaine Vignon, it was assigned to the 17th C.A. at Champagne. The unit was active during the offensive in this area that began on 17 April. At the end of the battle, C 56 returned to the area of the Meuse and later to Saint-Mihiel. It moved again to Vadelaincourt in November 1917. C 56 re-equipped with Salmson 2s in March 1918.

C 61, assigned to the 34th C.A. In early 1917 it moved to Belfort, then Fontaine (7th Armée sector) until mid-1917. In July 1917 it moved to Rosnay. It re-equipped with Sopwith 1½ Strutters in October 1917.

C 64, assigned to the 4th Armée and based at La Cheppe in February 1917. Exactly one year later C 64 became SPA-Bi 64.

C 74, assigned to the Belgian army and based at Hondschoote under the command of Capitaine Paillard in July 1917. It re-equipped with SPAD 16s in early 1918.

C 104, assigned to the 2nd Armée and based at Behonne in February 1917. In April 1917 the unit participated in the Battle of Chemin des Dames. In May 1917 it re-equipped with Sopwith 1½ Strutters.

C 105, assigned to the 2nd Armée sector in February 1917 and based at Autrecourt. From March through September 1917 it was assigned to the 5th Armée and participated in the Battle of Chemin des Dames. It re-equipped with Sopwith 1½ Strutters in October 1917.

C 106, assigned to the A.G.L.P. of the 10th Armée in November 1916. It was based at Grivesnes. In 1917 it re-equipped with Sopwith 1½ Strutters.

C 122, assigned to the 3rd Armée in November 1917 and based at Grivesnes. It re-equipped with Salmson 2s at the end of 1917.

Artillery Observation

C 202, assigned to the 5th Armée in February 1917 and active over the Verdun front. C 202 later moved to the 2nd Armée front. In December 1917 it re-equipped with Breguet 14 A2s.

C 207, assigned to 20th C.A. and based at Fricourt on the 6th Armée front. It transitioned to Sopwith 1½ Strutters in October 1917.

C 220, re-equipped with Caudron G.6s in April 1917. Under the command of Capitaine de Montaigu, C 220 participated in the Battle of Chemin des Dames. It was successively assigned to the 2nd C.C., 14th C.A., 2nd C.A., and 21st C.A. It re-equipped with Breguet 14 A2s in October 1917.

C 222, assigned to the D.A.L. in November 1916 and based at Lunéville. It re-equipped with Sopwith 1½ Strutters in 1917.

C 224, based at Lemmes in the 2nd Armée sector in February 1917 and assigned to the 15th C.A. In July 1917 C 224 was assigned to the 9th C.A. and based at Hourges. In September 1917 the escadrille was sent to Bonne-Maison where it was assigned to the 35th C.A. in the 6th Armée sector. It re-equipped with Breguet 14 A2s at that time.

C 225, created in January 1917 under command of Capitaine Odic. It re-equipped with Salmson 2s in late 1917.

C 226, formed at the beginning of 1917 under command of Capitaine des Isnards and assigned to the 36th C.A. It was based at Bray-Dunes in June 1917. The unit became SOP 226 in late 1917.

C 227, sent to Mesnil-Saint-Georges in early 1917 and active in the 10th Armée sector. Subsequently, it was dispatched to La Cheppe in the 4th Armée sector. In October 1917 it re-equipped with Breguet 14 A2s.

C 228, formed from F 228 in October 1916 under command of Capitaine Fevre. In October 1916, after a brief rest period at Champagne, C 228 was despatched to Vaux and Douaumont. It was assigned to the 87th Regiment d'Artillerie Lourde in the 2nd Armée sector and based at Sulvacourt. C 227 participated in the battle of Chemin des Dames in April 1917. Three months later it was sent to Flandres. In January 1918 it re-equipped with SPAD 16s.

Caudron G.6. The wing of the G.6 spanned approximately a foot more than the G.4, but was in other respects similar. B83.5710.

Left: Caudron G.6 serial 438. The engines were two 110-hp Le Rhônes; when Anzani engines were used the cowlings were deleted. B80.634.

Caudron G.6 at Port Said on 6 May 1919. The unit was C 575 which formed in August 1918 and was commanded by Capitaine Braquilanges. 93.1476.

Official STAé Performance Data for the Prototype G.6 with Two 120-hp Le Rhône Engines and a Payload Of 500 kg
Span 17.21 m; length 8.660 m; height 2.50 m; wing area 36.90 sq. m; Empty weight 895 kg
Maximum speed: 163 km/h at sea level; 151 km/h at 2,000 m; 146 km/h at 3,000 m; 138 km/h at 4,000 m; climb to 500 meters 1 min. 20 sec.; 1,000 m in 3 min. 20sec.; 3,000 m in 14 min. 55 sec.; and 4,000 m in 20 minutes.

Caudron G.6 Two- Or Three-Seat Reconnaissance Plane and Light Bomber with Two 110-hp Le Rhône Engines
Span 17.22 m; length 8.60 m; wing area 39 sq. m
Empty weight 940 kg; loaded weight 1,440 kg
Max speed: 145 km/h at 2,000 m; climb to 2,000 m in 7.5 min.; ceiling 4,400 m; range 275 km; endurance 3 hours.
Armament: twin 7.7-mm Lewis machine guns fore and aft

C 229, formed in February 1917 and eight months later re-equipped with Sopwith 1½ Strutters.

T.O.E. Escadrilles

G 488, created in August 1917. It was initially assigned as an escadrille côtiére (coastal patrol) intended for service in Tunisia. It was commanded by Lieutenant Bastien and was sent to Patras in Greece in December 1917. From this base it was to protect convoys supplying the A.F.O. in the Mediterranean. It was eventually redesignated Cotiére de Kourtesis. It was disbanded in December 1918.

G 489, created in December 1917 and designated Escadrille Côtiére Mazagan. The unit was based in Morocco and patrolled convoy routes. It was disbanded in December 1918.

G 490, created in March 1918 and designated Escadrille Côtiére de Sfax. Commanded by Capitaine Limasset, it began operations in August and was disbanded in December 1918.

C 543, equipped with Caudron G.6s and based in Algeria where it remained after the war. In January 1920 it became Escadrille 543 of the 3rd Regiment d'Algeria-Tunisie.

C 544, operational in Tunisia beginning in 1918. It became the 5th Escadrille of the Regiment d'Aviation Algérie in January 1920.

C 575, formed in August 1918 and intended for service in Palestine. It was commanded by Capitaine de Braquilanges and based at Port Said in Egypt. It became the 8th Escadrille of the 1st Regiment d'Aviation in January 1920.

C 545, **C 546**, **C 547**, **C 548**, and **C 549** were to be given G.6s and based in North Africa. However, there is no evidence that these escadrilles were ever actually sent to Africa before the war's end.

GQG reports indicate there were serious problems with the G.6. It would go into uncontrolled spirals when performing tight turns. The crews were warned to make careful turns and to coordinate use of the rudder and ailerons because use of the rudder alone was believed to result in the aircraft spiraling out of control. These problems became serious enough that the Aviation Militaire ordered the grounding of all G.6s on 8 June 1917. It was found that there was a problem with the tail fin. Once these were replaced the G.6s were allowed to resume flying in mid-June. It was also discovered it was necessary to replace the turnbuckles for the wing rigging. The numerous difficulties with the A3 series of aircraft in general, and the G.6 in particular, led the STAé to create its own A3 plane, which was produced under license by Letord. The Letord series soon replaced the G.6s in service.

The decision to disperse the A3 series of aircraft to the army cooperation escadrilles also resulted in the G.6 units re-equipping with more modern types beginning in 1917. These were of the A2 (tactical reconnaissance and army cooperation) category, including the Breguet 14 A2, Salmson 2 A2, and SPAD 16 A2.

Above: Caudron G.6 in Egypt being prepared for take-off. Via Colin Owers.

Above right: Caudron G.6 serial 4312. Caudron G.4s are in the background. Via C.C.

Right: Caudron G.6 in Egypt. This view emphasizes the prominent keel which ran along the bottom of the fuselage. Via Colin Owers.

Caudron G.6 of escadrille C 56. C 56 re-equipped with G.6s in March 1917; a year later the G.6s were replaced by Salmson 2s. B75.622.

Caudron R and R.3

The Caudron R was first flown in May 1915. The aircraft, which had been designed by René Caudron, was powered by a single 120-hp Salmson engine. Made of wood and steel, the R had a wing surface area of 40 square meters (which was comparable to the twin-engine Caudron G.5 and G.6) and an empty weight of 700 kg. Only a single example was built.

The identity of the R.2 is not known, but the R.3 was a twin-engine plane powered by two 80-hp Le Rhônes. It is probable that it was the prototype for the entire R series of bombers, reconnaissance planes, and fighters.

Caudron R.4/R.8/R.19

The G.4s in service with army cooperation escadrilles in 1916 were incapable of surviving determined attacks by German fighters. The short fuselage placed the front gunner between the prominent engine nacelles, making use of the forward machine gun difficult. On the other hand, the twin-boom layout of the G.4 meant that the aircraft could not defend against rearward attacks. The Caudron brothers, aware of these deficiencies, designed a twin-engine aircraft with a conventional fuselage that would correct them. Gaston Caudron was responsible for designing the G.5 and G.6, while René devised the R series of multi-purpose planes. Both were probably intended to meet the Aviation Militaire's A3 requirements, although the R.4 was initially intended specifically for use as a bomber (B3).

The prototype Caudron R.4 was completed and flown in June 1915. The upper wing was slightly longer than the lower wing and was supported by three bays of struts. Two 130-hp Renault 12 Db engines were located in nacelles mounted close to the fuselage. There were three cockpits, one for each crew member. The pilot's cockpit was located just aft of the top wing. A gunner was seated in the nose and another was located just behind the pilot; each had twin 7.7-mm Lewis guns on a swivel mount. The aircraft's wings were immense, with a surface area of 70 square meters. The landing gear consisted of two pairs of wheels, one under each of the engines. A single large nosewheel was fitted, as was common practice on French heavy bombers; this prevented the plane from turning over while landing on poorly prepared airfields.

In November 1915 René Caudron presented the R.4 to the Aviation Militaire, proposing it be used as an all-purpose plane which could perform long-range bombing, reconnaissance, and escort missions. The new aircraft was viewed with interest by the SFA, and it was decided to form a commission to evaluate the design. The commission's report was favorable and an order was placed for the R.4 on November 21 1915. However, while the plane looked impressive on paper, testing revealed it to be underpowered, which resulted in a low service ceiling and poor maneuverability. Although it had been originally intended to use the type as a bomber, the R.4's deficiencies resulted in its being used primarily in the reconnaissance role.

Less then one month after the order had been placed Gaston Caudron was killed testing the R.4 prototype. The subsequent investigation discovered that the wing spars had failed near the central portion of the wing. Henri Potez assisted with the redesign of the R.4. Modifications to the wing eliminated the flaw and René Caudron continued testing. The trials were completed in 1916 but the modifications had seriously delayed

Above: Caudron R.4. The nosewheel prevented the aircraft from nosing over on landing. B76.1144.

Caudron R.4. This photograph provides a clear view of the two 130-hp Renault 12 engines. B83.4346.

164 FRENCH AIRCRAFT OF THE FIRST WORLD WAR

Caudron R.4 based at Villers in September 1916. This aircraft is marked with a red devil which was the personal insignia of Von Happe of F 25. B83.5643.

Right: Caudron R.4 at the Caudron factory. B89.3816.

Below: Caudron R.4. Serial number C257?. Serial numbers 2520 through 2567 were assigned to R.4s which were in service in early 1917. B93.280.

the type's entry into service. There was only one R.4 at the front in October of 1916; this was probably assigned to Adjudant d'Aux of C 47, who began training on the R.4 in mid-June. Later in October escadrille A.L.G.P. F.210 (a former S.A.L. unit) had five on strength. Eventually a total of 249 R.4s were built. They were used for long-range reconnaissance, but were found to be surprisingly effective in destroying enemy aircraft. GQG reports indicated that 34 enemy aircraft were damaged by R.4s of R 210 in a two-month period.

R.4s received the STAé designation Caudron 40 A3 but this designation was never used in official documents, in which the planes were always called R.4s. Service planes had 130-hp Renault 12Db or 150-hp Hispano-Suiza 8A engines. Versions of the R.4 with the Renault engines were at one point designated R.19s by the SFA. This appellation is seldom, if ever, used in official documents.

The Caudron R.4 Type 8 was an R.4 airframe fitted with Hispano-Suiza engines. A photograph taken at the Caudron factory in Issy-les-Moulineaux shows that the Type 8 appears to have been three-bay wings. Gunners' positions were located fore and aft, and a large nosewheel was fitted. The Type 8 was probably intended for the A3 or C3 role, but was not selected for use. The R.8 designation mentioned in some documents is not mentioned in the STAé list of approved designations.

Caudron R.4 Type 19. The R.19 designation was unofficial. B93.768.

The Caudron R.4 Type 8 prototype at the Caudron factory at Issy les Moulineaux. The R.8 designation was unofficial. B93.813.

Operational Service

Only a limited number of R.4s were available. In August 1917 there were 63 R.4s available, only 53 of which were either in service with front-line escadrilles or with the aviation parks. One army cooperation escadrille, C 46, re-equipped entirely with R.4s to become R 46. R 46 was assigned to the 6th Armée and was used for long-range reconnaissance over the Somme front. R 46 was commanded by Lieutenant Lecour-Grandmaison. In November 1916 it was assigned to the 6th Armée and based at Chipilly. On the occasion of R 46's 15th victory the unit was cited in the 6th Armée's l'ordre de l'armée. R 46 subsequently passed to GC 12 and re-equipped with Letord reconnaissance aircraft.

Although some sources state that Escadrille 76 was an R.4 unit, the SHAA records show it was formed in July 1916 as an R.4 unit but was soon redesignated as a Nieuport escadrille (although it may still have had a few R.4s on strength).

The R.4s also saw limited service as bombers. For example, 20 R.4s of R 210 attacked the train station at Anisy with incendiary bombs on the night of 19/20 December 1916.

Five S.A.L. units for artillery cooperation used R.4s. These were:

R 207, created from V 207 in May 1917 with a mix of Caudron R.4s and Sopwith 1½ Strutters. It was assigned to the 3rd Armée. R 207 became SOP 207 in October 1917.

R 210, formed from F 210 in October 1916 and assigned to the 5th Armée. On 2 July the unit attacked train stations at Anizy-le-Chateau. In January R 210 was assigned to the 6th Armée on the Chemin des Dames front. Based at Bouleuse, it was active during the Battle, although its main equipment was now Sopwith 1½ Strutters and Letords. The escadrille became BR 210 in July 1917.

R 212, formed from F 212 in March 1917 and assigned to the 5th C.A. in the 4th Armée sector. It was redesignated SPA-Bi 212 in January 1918.

R 214, formed in December 1916 under the command of Lieutenant Trezenem and assigned to the 13th C.A. From 15 July to 28 September 1917 it was based in the 2nd Armée sector. It was involved in the French attacks on the left bank of the Meuse and also performed numerous coastal patrols. In May 1918 it re-equipped with Breguet 14 A2s.

R 217, formed in December 1916 and served under the 3rd and 4th Armées. R 217 became SOP 217 in 1917.

The R.4s had a relatively brief life in escadrille service. Sopwith 1½ Strutters were supplied to S.A.L. units R 207, R 210, R 212, R 214, and R 217 in 1917. The reason for this is given in a GQG memo to the 1st Armée in late 1917 stating that

Above: The Caudron R.4 Type 8. The three-bay wing is emphasized in this front view. B93.772.

Below: The Caudron R.4 Type 8 had serial C 409?. Extensive bracing attached the engine nacelles to the fuselage. B93.773.

Caudron R.4 Three-Seat Reconnaissance Plane and Light Bomber with Two 130-hp Renault 12Db (R.19) or 150-hp Hispano-Suiza 8A Engines
Span 21.10 m; length 11.80 m; height 3.60 m; wing area 63.16 sq. m
Empty weight 1,720 kg; loaded weight 2,337 kg
Maximum speed at 2,000 m: 130 km/h (some sources state 160 km/h); climb to 2,000 m in 18 minutes; range 500 km; ceiling 4,600 m; endurance 3 hours
Armament: twin 7.7-mm Lewis machine guns fore and aft
A total of 249 R.4s were built.

the R.4s were having mechanical problems and the type's poor performance was a source of constant complaints. Many of the aircraft were modified by the R.G.A.s to try to correct these problems. In an attempt to improve the R.4's weaknesses (particularly its poor maneuverability and low ceiling) it was decided to develop a more powerful version. This new aircraft became the successful R.11. Commandant Dorand of the STAé formulated the A3 requirement in response to the deficiencies of the G.6 and R.4 series. It was this requirement that led to the Letord series of reconnaissance, bomber, and fighter planes.

Caudron R.5

The Caudron R.5 was a bomber variant of the R.4 three-seat reconnaissance type. The R.4 had, as mentioned previously, been intended as a B 3 category bomber, but its poor

The Caudron R.5 was a bomber variant of the R.4 three-seat reconnaissance airplane; the engines are 230-hp Renaults. MA12616.

performance with two 130-hp Renault 12Db engines resulted in its being used primarily for reconnaissance. The R.5 was intended to improve performance by being fitted with more powerful engines. The R.5, which had its first flight three months after the R.4, had two 230-hp Renault 12A engines. This increased power was accompanied by a larger wing of 85 square meters area. The empty weight of the R.5 was almost 300 kg more than the R.4. The bomber was entered in the concours puissant of 1916 and SHAA records indicate it was hoped these bombers could replace the Breguet Michelin 4s in service with GBM 5. However, the R.5 was unable to meet the requirements of the competition (the Morane-Saulnier Type S being judged the winner) and was not selected for production.

Caudron R.5 Two-Seat Bomber with Two 230-hp Renault 12A Engines
Wing area 85 sq. m
Empty weight 2,000 kg
One built

Caudron R.9

No details are available concerning the R.6 or R.7, which may have been unbuilt projects. The Caudron R.9 was a two-seater with an 80-hp engine. It was designed by Paul Deville in either 1915 or 1916 after he joined the Caudron firm as chief engineer. It was not selected for service.

Caudron R.10

The Caudron R.10 was a two-seat aircraft designed by Paul Deville. Flown in 1916, it was powered by two 80-hp Le Rhône engines. However, the R.10's performance was felt to be inadequate and it was not selected for production.

Caudron R.11

There is some confusion over the true origin of the R.11. Some French sources suggest the prototype was originally designed to participate in the concours des avions puissants of 1916. This would suggest that the R.11 was a development of the R.5 long-range bomber. Other sources state that the R.11 evolved from a special version of the R.4 that had been equipped with two 150-hp Hispano-Suiza engines; the increased power was said to result in such a dramatic improvement in the R.4's performance that Caudron's chief designer, Paul Deville, created a heavily modified version of the R.4 and equipped it with even more powerful engines.

The Caudron R.11 was probably intended from its inception to meet the Aviation Militaire's requirement for a three-seat long-range escort fighter; the C 3 specification for such a plane was formulated in November 1916. It called for an aircraft to carry light-weight armor as well as two machine guns with 500 rounds each. The fighter was to have a maximum speed of 170 km/h at 2,000 meters and be able to reach an altitude of 3,000 meters in 15 minutes. Some sources credit Capitaine Le Cour Grandmaison with formulating the requirement for a long-range fighter to accompany formations of medium bombers. It is possible that the SPAD SA.3 and Ponnier twin-engine fighters were also developed to meet this requirement.

The new aircraft was initially powered by two Hispano-Suiza 8Ba engines of 200-hp. Its landing gear was similar to the R.4 and G.6, but it dispensed with the nose wheel of the R.4. As with the R.4, the upper wing was longer than the lower. The fuel tanks were located at the rear of the engine nacelles and, in later aircraft, could be jettisoned in flight. To minimize the aircraft's vulnerability, the fuel system was arranged so that either engine could receive fuel from either of the main fuel tanks. The aircraft was well-armed with twin 7.7-mm Lewis guns in the nose and rear fuselage, again as in the R.4. There was an additional machine gun below the nose position to enable that gunner to fire downward. On later aircraft there was provision for dual controls, so that if the pilot was wounded the R.11 could be flown by the observer. Although it could be fitted with a camera for long-range reconnaissance, the R.11's most important role was providing air defense for the Breguet 14 B2s.

The new aircraft was found to have substantially better performance than the preceding R.4s and R.5s. A total of 1,000 were ordered by the Aviation Militaire (of which approximately 370 were built by the end of the war). The R.11s were built by Caudron and under subcontract by Regy and Gremont. The

Caudron R.11 of R 46. R 46 was active over Aisne and the Marne while protecting the units of the 2nd Brigade d'Aviation. B84.592.

Caudron R.11 serial 6078 of R 46 in 1918. R 46 was assigned to protect the bombers of Escadre 13. B76.1940.

STAé designation for the R.11 was Caudron 11 A3 but this was rarely used in wartime reports. Why the three-seat fighter should be designated as a reconnaissance plane is a mystery, but it suggests that the type may originally been planned as a multi-place de combat to be used in the long-range reconnaissance, light bombing, and escort fighter roles. Support for this can be seen in a memo dated 28 November 1917 stating that it was anticipated that Caudron R.11s would replace the Letord series in early 1918.

As with so many other French designs of the period, the type's entry into service was delayed by production difficulties. In this case, the main problem seems to have been secondary to the 215-hp Hispano-Suiza 8Bba or 8Bda engines. In later aircraft a 235-hp Hispano-Suiza 8Beb was fitted. Even though the first R.11 had been completed in 1916, only two were in existence by 1 January 1918. Three months later the total had risen to only 34. At the time of the Armistice there were 54 in service.

Operational Service

Tactics evolved so that these large aircraft could be used in conjunction with both the day bomber forces and the single-seat escort fighters. Originally, the R.11s were to be used to watch for oncoming enemy formations and possibly engage them ahead of the bomber formations. The commander of the First Air Division stated that the R.11s were given to it to serve as a "plane lookout service." The same source stated that because of the increasing threat of the German fighters to the day bomber formations, the R.11s were assigned to provide close-in fighter escort. When SPAD fighter units were available, they were used to provide top cover for the mixed Breguet 14 and Caudron R.11 formations. This meant that the R.11s could travel with the bombers all the way to the target. Because of their fore and aft guns, the R.11s could bring considerable firepower to bear without having to maneuver. The SPADs, on the other hand, could attack the enemy fighters using their superior speed and maneuverability. While this combination proved effective, there were still heavy losses when the formations attempted bombing missions without the SPAD escort. An example of this was the September 1918 raid on the Conflans-Jarny railroad station when Breguet 14s of GB 4 and the R.11s of R 46 encountered heavy opposition over the target. While the aircraft did reach the target and eight German fighters were claimed as destroyed, seven French aircraft were lost.

There was an undeniable élan within the R.11 units, the crews being told that their mission to protect the bombers was "one of sacrifice." It was also noted that specific R.11 crews were associated with specific groupes de bombardement, thus providing a "real moral bond between them." It is interesting to

Caudron R.11 Three-Seat Long-Range Fighter with Two 200-hp Hispano-Suiza 8Ba Engines (Prototype)

Span 17.92 m; length 11.22 m; height 2.80 m; wing area 54.25 sq. m
Empty weight 1,416 kg; loaded weight 2,188 kg

Maximum speed:	ground level	191 km/h
	2,000 m	185 km/h
	3,000 m	180 km/h
	4,000 m	173 km/h
Climb:	1,000 m	3 minutes 50 seconds
	2,000 m	8 minutes 10 seconds
	3,000 m	4 minutes 30 seconds
	4,000 m	22 minutes 25 seconds
	5,000 m	25 minutes

One built

Caudron R.11 Three-Seat Long-Range Fighter with Two 215-hp Hispano-Suiza 8Bda Engines (Production)

Span 17.92 m; length 11.22 m; height 2.80 m; wing area 54.25 sq. m
Empty weight 1,422 kg; loaded weight 2,165 kg

Maximum speed:	2,000 m	183 km/h
	3,000 m	178 km/h
	4,000 m	173 km/h
	5,000 m	164 km/h
Climb:	2,000 m	8 minutes 10 seconds
	3,000 m	14 minutes 30 seconds
	4,000 m	22 minutes 30 seconds
	5,000 m	39 minutes

Ceiling 5,950 m; range 600 km; endurance 3 hours
Armament: five 7.7-mm Lewis guns
Approximately 370 built

Above: Caudron R.11 at Villenveve, spring 1918. Left to right: Lt. A.H. McLanahan, 95th Aero Sqdn., Lt. Eddie Rickenbacker, 94th Aero Sqdn., Lt. Edward Buford, 95th Aero Sqdn., Lt. Russell Hall, 95th Aero Sqdn. University of Texas at Dallas.
Below: Caudron R.11 with cockades under top wing. Alan Toelle.

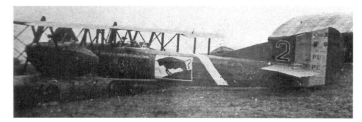

Above: Caudron R.11 with interesting markings. Alan Toelle.
Below: Close-up details of Caudron R.11 above. Alan Toelle.

note that some French air service officers predicted that all fighters of the future would be two-seaters whose "mission spirit" would eliminate the concept of aces.

It is not known how successful the R.11s were in combat against German fighters. Nevertheless, the type was felt to be quite useful in the long-range escort fighter role. At the Armistice R.11s equipped six escadrilles: R 46, R 239, R 240, R 241, R 242, and R 246. Each unit had between 15 and 18 aircraft.

R 46 was the first unit to receive R.11s, being re-equipped in February 1918. This unit was assigned to protect the bombers of Escadre 13 (GB 3 with BR 107, 126, and 128 and GB 4 with BR 131, 132, and 134). R 46 was active over the Aisne and the Marne while protecting units of the 2nd Brigade d'Aviation. It was credited with 37 official victories at the cost of 26 killed and 35 wounded crewmen. The unit was decorated with three citations. R 46 become escadrille 204 of the 3rd RB (Jour) at Avord in 1920.

R 239 was created in May 1918 under command of Lieutenant de Verchéres and assigned to the 4th Armée on the Champagne front. In July it was assigned to protect Escadre 12 (GB 6 with BR 66, 108, and BR 111; GB 5 with BR 117, 120, and I27; and GB 9 with BR 29, 123, and 129). R 239 was credited with four aerial victories. R 239 became the 204th escadrille of the 1st RB (Jour) at Metz in 1920.

R 240 was formed in July 1918 and assigned to protect Escadre 12. The escadrille was under the command of Lieutenant de Durat. R 240 became the 208th Escadrille of the 1st RB (Jour) at Metz in 1920.

R 241 was formed in July 1918. It was active over the 8th Armée's front and was disbanded in December 1918.

R 242 was created in April 1918. It served over the 7th Armée's sector and was disbanded in December 1918.

R 246 was formed from aircraft drawn from R 46, R 239, and R 240 on 2 November 1918. R 246 served along with R 46 in guarding the Breguet bomber units of Escadre 13. R 246, in support of Escadre 13, was active over Rethel and supported the raids on Neufchateau. It became Escadrille 212 of the 3rd RB (Jour) at Avord in 1920.

The last R.11 was withdrawn from service in July 1922.

Foreign Service
United Kingdom
The Royal Flying Corps was interested in using the R.11s as long-range bombers and acquired two examples for evaluation. They were given serials B 8822 and B 8823. The R.11s were evaluated against Sopwith Dolphins. After these tests, one of the R.11s was used as a testbed for various armament installations, the other to test communication systems.

United States
The U.S. Air Service also purchased two R.11s in October 1918. One R.11 was also detached from a French escadrille to serve as an escort fighter for the 88th Aero Squadron in September 1918. The 96th Aero Squadron also had at least one R.11 to escort its Breguet 14 B2s.

Above: Caudron R.11. Via C.C.

Caudron R.11 which suffered a landing accident. MA6466.

Caudron R.12

The Caudron R.12 was a Caudron R.11 re-engined with two 300-hp Hispano Suiza 8Fb engines. This modification was produced at the request of General Duval in 1917 as the new engine was expected to enter production soon, and, assuming it was successful, could endow the R.11 with a much higher performance. The Caudron R.12 first appeared in the summer of 1918. It had a wing area of 52 sq. m, which was identical to the standard R.11 (although some sources give slightly increased wing areas of from 58 to 60 sq. meters). The aircraft had an empty weight of 1,350 kg.

The R.12 first flew in November 1918 and was tested at Issy-les-Moulineaux, where numerous problems were discovered. At least two examples were undergoing testing in July 1919, and although it had been originally intended to equip 12 escadrilles with the R.12, further development was abandoned.

Caudron R.12 Three-Seat Long-Range Fighter with Two 300-hp Hispano-Suiza 8Fb Engines
Wing area 52 sq. m
Empty weight 1,350 kg
Armament: five 7.7-mm Lewis guns
Two built

Caudron R.14

To meet the requirement for a plane capable of carrying a heavy cannon, many companies produced modified versions of heavy bombers. For example, both a Voisin 12 and Farman F.50 were modified to carry cannons. The Caudron R.11 was also fitted with a cannon in August 1918. This new version was designated R.14 Ca3 Canon. Main armament was a 37-mm Hotchkiss cannon, but the five Lewis guns of the standard R.11 were retained for defense. To carry the cannon the new version used the same 300-hp Hispano Suiza 8Fb engines as the R.12. The wings were enlarged with three bays of struts outboard of the engines. Wing area was increased to 63 square meters and the plane had an empty weight of 1,747 kg. An unbalanced rudder was fitted. As with the cannon-armed Voisin 12s and Farman F.50s, development of the R.14 was abandoned soon after the Armistice.

Caudron R.14 Three-Seat Long-Range Cannon Fighter with Two 300-hp Hispano-Suiza 8Fb Engines
Wing area 63 sq. m
Empty weight 1,747 kg
Armament: one 37-mm Hotchkiss cannon and five 7.7-mm Lewis guns
One built

Right: Four views of the Caudron R.14 cannon-armed long-range fighter. The R.14 was an enlarged R.11 with more powerful Hispano-Suiza engines. The R.14 retained the five machine guns of the R.11 in addition to the 37-mm cannon. MA12833.

Caudron C.20

Little is known about the Caudron C.20. At least one source states it was a single-seat army cooperation/light bomber aircraft. Power was supplied by a single 180-hp Gnome Monosoupape engine. The first flight is believed to have been in November 1917. However, this is also when the Caudron O2 fighter first flew; it is possible that the C.20 was an incorrect designation applied to the O2 or that it was a variant of the O2 intended for army cooperation duties.

Single-Seat Army Cooperation/Light Bomber Aircraft with 180-hp Gnome Monosoupape
Wing area 17 sq. m
Empty weight 813 kg

Caudron C.21

The Caudron C.21 was an early attempt to meet the Bn 2 requirement of 1917. The C.21, like the G.6 and unsuccessful R.5, featured a conventional fuselage with two 80-hp Le Rhône engines mounted between the equal-span biplane wings. The abbreviated nose was reminiscent of the prototype G.6. The aircraft was designed by Caudron's chief engineer, Paul Deville. Testing of the C.21 began in November 1917. However, as with the R.10, the engines were not sufficiently powerful; they were subsequently replaced by two 120-hp Le Rhônes. The new plane was designated the C.22.

Postwar, the C.21 was developed into an "aerial limousine." It was labeled the "Monsieur-Madame" and had seats for the chauffeur-pilot and a mechanic-footman in front. Both the crew had open cockpits, but the passengers had an enclosed cabin for two. This new type was labeled the Caudron C.33.

Caudron C.21 which was delivered in 1919 to Bruxelles from Paris. B89.3861.

Caudron C.21 Two-Seat Night Bomber with Two 80-hp Le Rhône Engines
Wing surface area was 39 sq. m
Empty weight was 1,140 kilograms;, loaded weight 1, 480 kg
Max speed 100 miles per hour at 4,000 m; climb to 4,000 m in 22 minutes; endurance 3 hours
One built

Caudron C.22

The Caudron C.22 was a follow-on to the C.21 and, like its predecessor was designed by Paul Deville to fulfill the requirement for a medium bomber capable of tactical attacks with 500 kg of bombs. Other aircraft designed to this specification included the Voisin 11, Farman F.50, Bernard A.B.1, and Breguet 16.

The C.22 was a conventional biplane initially powered by two 120-hp Le Rhône 9Jb engines. Two 130-hp Clerget rotary engines were later fitted, presumably to improve performance. The C.22 Bn2 first flew in November 1917. It could carry six 120-mm and three 155-mm bombs. As with other Bn2 aircraft, defensive armament was light, in this case a single machine gun; the Bn2 aircraft were expected to use the cover of night as the primary means of defense.

The C.22 failed to meet the requirements of the Bn2 category because it was too slow and poorly armed. Further development was abandoned and the Breguet 16 and Farman F.50 were selected for series production.

Caudron C.22 Two-Seat Night Bomber with Two 120-hp Le Rhône 9Jb Engines
Span 18.54 m; length 8.30 m; height 2.52 m; wing area 53 sq. m
Empty weight 950 kg
Maximum speed: 141 km/h at 3,000 m, ceiling 3,352 m
Armament: six 120-mm and three 155-mm bombs, one machine gun.
One built

Caudron C.22. The Caudron C.22 was a follow-on to the C.21 and, like its predecessor, was designed by Paul Deville to fulfill the requirement for a medium bomber. STAé 01144.

Caudron C.22

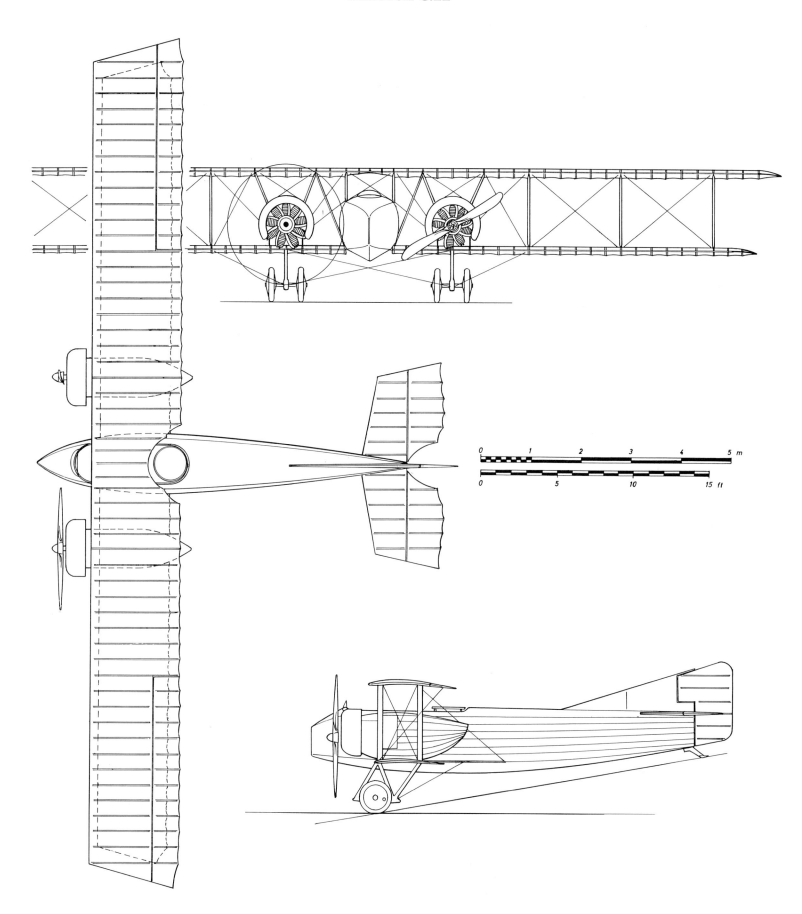

Caudron C.23

A further development of the C.21 and C.22, the C.23 was to prove more successful than its predecessors. The C.23 BN3 was designed to meet the specification for a night bomber with two or three crew members, a bomb load of 1,200 kilograms, and a range to permit attacks on Berlin. Other aircraft competing for the BN3 orders were the Caproni BN3 (Caproni Cap. 5), Letord 9, Delattre BN3, Voisin 12, Renault O2, Delattre BN3, and Sikorsky BN3.

The Caudron C.23 first flew in February 1918. It was of similar configuration to the previous C.21 and C.22. The engines were two 260-hp Salmson CU-9Z water-cooled radials. The aircraft was a biplane with unequal span wings supported by 16 struts. Each wing had two rectangular spars made of three strips of spruce and covered with two strips of steel. The wings were covered in flax and were camouflaged on top and dark brown underneath. the struts were made from spruce or Oregon pine in three segments which were wrapped in strips of ash for added strength. At equal intervals metal strips were used to band the struts for additional strength. Sheet metal was fashioned for the ends of the struts where they were attached to the wing. The ailerons were on the top wing only and had a span of six meters. The tail had a rudder and fixed fin. The rudder had three vertical ribs made of wood, was covered with steel, and attached directly to the fuselage. The horizontal stabilizers were made of two spars supported by ribs and the leading edge was covered in spruce.

The fuselage was constructed from four longerons made of ash and spruce. The nose was flattened, while the cross-section of the mid fuselage was rectangular. The fuselage formers were made of pine strengthened by diagonal bracing wires. The pilot and co-pilot were seated side-by-side ahead of the wing struts. There was a nose turret and a second turret located behind the wing. The lance bombs were carried inside the fuselage. There was a passage between the bomb bay and the front the fuselage. The engines were mounted on sheet metal which was bolted to the upper and lower wing spars. The fuel tank was located behind the engine and was divided into five compartments internally to prevent the fuel from shifting inside the tank during flight. The radiators were supported on an aluminum collar bolted to the base of the motor. The oil tank was located above the fuel tank. Under each engine was an undercarriage of N-configuration with an oval cross section. There were twin wheels, between which was a landing skid used for braking. The skids were made of ash. The axles were mounted on bungee chords. The tail skid was made of ash with a metal tip and was attached to the rudder axle.

Caudron C.23 Three-Seat Heavy Night Bomber with Two 260-hp Salmson Cu-9Z Engines

Span 24.47m; length 12.98 m; height 3.45 m; wing area 106 sq. m
Empty weight 2,341 kg; loaded weight 4,170 kg; payload 500 kg or an 800 kg torpedo
Maximum speed: 144 km/h at 1,000 m; 140 km/h at 2,000 m; climb to 2,000 meters in 20 minutes 37 seconds
Ceiling 4,500 m; endurance 4 hours; range 700 km
Armament: 600 kg of bombs

Projected C.23 with 600-hp Salmson Engines

Wing area 100 sq. m
Loaded weight 3,707 kg; payload 800 kg
Max speed: 170 km/h at 4,000 m; climb to 4,000 m in 25 minutes; endurance 5 hours

The aircraft took off and climbed at a speed of 120km/h. Cruising speed was 140 km/h and the ideal descent speed was 120 km/h. The aircraft was landed at 100 km/h. Pilots were instructed that, should the C.23 enter a spin, they were to apply opposite rudder to stop the spin and then reestablish control after the ailerons became effective. If the aircraft should stall the preferred solution would be to allow the C.23 to build up speed before attempting any correcting maneuvers.

The plane met the BN3 requirements and 1,000 were ordered. Although the design was to prove successful and 54 C.23s had been accepted by the Aviation Militaire before the Armistice, not a single example reached units in time to see action. Postwar, CAP 115 relinquished its Caproni bombers for C.23s and was redesignated C 115. These bombers served with the 22nd RAB (Nuit) at Luxeuil. Crews found the type to be seriously underpowered and it maneuvered poorly. The C.23s were replaced by the superior Farman F.60 in February 1920.

Another version of the C.23 was under development in April 1918. It was to have been powered by two 600-hp Salmson engines and, had it been successful, might have eliminated many of the problems found in the production aircraft. Another C.23 was tested with two 300-hp Hispano-Suiza engines.

Postwar, the C.23 was developed into an airliner and initiated service from Paris to Brussels on 10 February 1919. A version designated the C.23bis had an enclosed cabin and could carry 15 persons.

Caudron C.23 carrying a bomb beneath the fuselage centerline; up to 600 kg of bombs could be carried. B93.281.

Caudron C.23. This appears to be the eighth example produced. The aircraft serial number was probably 08. MA19997.

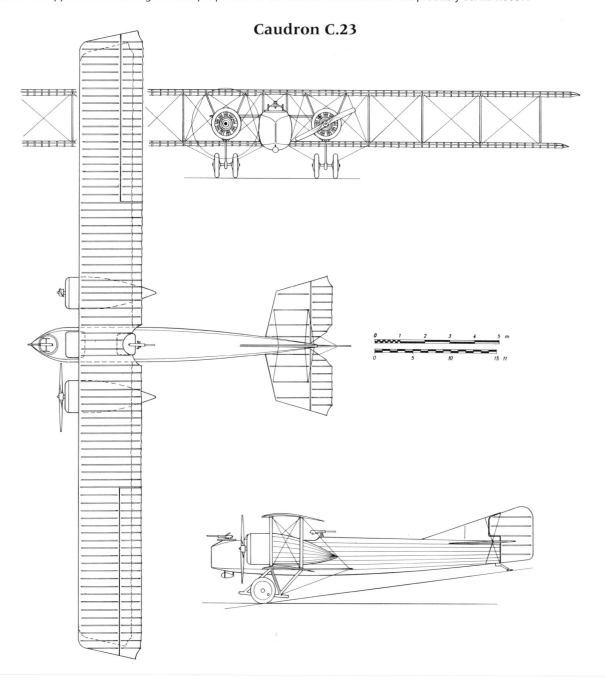

Caudron C.23

Caudron CRB

The CRB (the reason for the unusual designation is unclear) was a twin-engine bomber apparently initiated as a design study for the RFC in 1917. It was to have been equipped with two 200-hp Hispano-Suiza engines. The aircraft was never completed.

Caudron Heavy Bomber Project

A heavy bomber is believed to have been undergoing construction in 1917. It was to have had a bomb load of 600 kg, and the engines were to be 340-hp Salmsons. Apparently the aircraft was never completed.

Caudron O2

The identity of the Caudron O2 is unclear. It has been suggested that the correct designation would have been Caudron 20, which would be more consistent with the designations used by the STAé during the war. However, it also possible that it was originally designated the Type O and that the O2 designated a modified version. It was designed by Paul Deville.

The O2 was a high-altitude fighter with an anticipated ceiling of at least 9,000 meters. It was a biplane with a top wing of greater span than the lower. The airfoil cross section was flat; it was hoped this would improve high-altitude performance. The fully-faired fuselage was constructed of wood and metal. The pilot sat at eye level with the trailing portion of the upper wing.

Pending the availability of the 150-hp Gnome Monosoupape, the prototype was fitted with a 120-hp Le Rhône 9Jb rotary that was almost completely enclosed within a cambered cowling. Armament consisted of a single Vickers machine gun. Another version of the Caudron O2 was fitted with a Le Rhône 9R of either 170 or 180 hp. It is possible that the O2 was also at some point fitted with the Gnome Monosoupape.

The aircraft first flew in November 1917; its performance was inferior to the SPAD 13. The Caudron O2 was not selected for production and further development was abandoned.

Caudron O2 Single-Seat Fighter with 120-hp Le Rhône 9Jb
Wing area 17 sq. m
Loaded weight 625 kg
Maximum speed: 190 km/h at 4,000 m
Armament: one 7.7-mm Vickers machine gun

Caudron O2 Single-Seat Fighter with 150-hp Gnome Monosoupape 9N
Wing area 17 sq. m
Empty weight 400 kg; loaded weight 650 kg
Maximum speed: 210 km/h at 4,000 m
Armament: one 7.7-mm Vickers machine gun

Caudron O2 Single-Seat Fighter with 170-hp Le Rhône 9R
Wing area 17 sq. m
Loaded weight 625 kg
Max speed: 210 km/h at 4,000 m; climb to 4,000 m in 10 minutes; endurance 2 hours
Armament: one 7.7-mm Vickers machine gun

Caudron Type O2. A Caudron C.23 is in the background. B89.3828.

Caudron Type O2. Serial N40. Pending the availability of the 150-hp Gnome Monosoupape, the prototype was fitted with a 120-hp Le Rhône 9Jb rotary engine. B89.3829.

Clément-Bayard

Adolphe Clément-Bayard was a former bicycle and car manufacturer who produced a number of airships for the army. From 1908 through 1914 he produced six airships, two of which were sold to Russia. His last two, the *General Meusnier* and the *Ex-General Meusnier*, were built in 1915 and 1916 respectively. From 1909 through 1910 the firm built Santos-Dumont Demoiselles under license. It was responsible for the development of several aircraft before the war. The firm's aircraft all featured wood and steel construction, wing warping for lateral control, and a common landing gear chassis with two wheels. A single-seat military monoplane was produced in 1914 and had a 50-hp Gnome engine. The Clement-Bayard military monoplane had a high wing and fuselage keel which was similar to contemporary R.E.P. designs. A two-seat monoplane was also produced in the same year; it was powered by an 80-hp Le Rhône engine. Another Clement-Bayard military biplane carried a crew of three. It was a biplane with unequal span wings; the pilot sat in front, with the passengers seated side-by-side behind him.

In 1913 General Bernard formulated requirements for two types of armored aircraft: a single-seat machine for artillery cooperation and high-speed reconnaissance (120 km/h) and a two-seater with a maximum speed of 100 km/h for long-range reconnaissance. Blériot, Breguet, Dorand, Ponnier, Deperdussin, Clément-Bayard, Morane-Saulnier, and Voisin all built machines to meet this demanding specification. The Clément-Bayard firm built an all-steel armored monoplane, with an 80-hp Gnome engine. It appears that only one was built and it was not selected for production by the Aviation Militaire.

Clément-Bayard Bomber

In 1915 the Clément-Bayard firm built a twin-engined bomber probably intended to compete in the 1915/16 concours puissant. It had two 220-hp Renault engines. The only photographs that

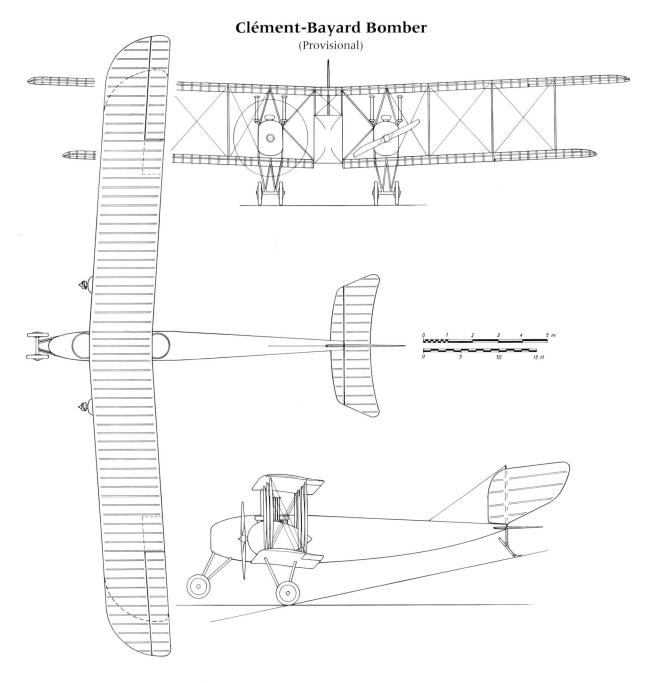

Clément-Bayard Bomber
(Provisional)

have survived show the machine wrecked, but a pair of drawings reveals that it was a biplane with the upper wing larger than the lower. The plane had a tricycle undercarriage with twin nose wheels. The engines were mounted in the tractor position housed in nacelles situated on the upper surface of the lower wing. There was a huge, comma-shaped rudder.

A British report stated that the bomber "rolled 30 yards and then the front wheels collapsed." This observer then wryly noted that "the construction is somewhat reminiscent of early iron bridges, and the machine, as a whole, is not considered of interest." A later report stated that the bomber was considered to be a failure as it had crashed "twice, killing its pilots." Given this background, it is not surprising that further development was abandoned.

For the remainder of the war the Clément-Bayard factory produced aircraft parts and weapons, but no further airplanes or airships. In 1928 Clément-Bayard sold his factory to Citroen.

Clément-Bayard Twin-Engine Bomber with Two 220-hp Renault Engines
Span 24.360 m; length 15.80 m; wing area 109.40 sq. m
Empty weight 2,900 kg; loaded weight 3,600 kg
Maximum speed: 137 km /h; climb to 2,000 m in 21 minutes; ceiling 3,700 meters
One built

Clément-Bayard Single-Seat Monoplane with 70-hp Engine:
Span 9.20 m; length 7.50 m; height 2.0 m; wing area 16 sq. m
Empty weight 320 kg; loaded weight 520 kg
Maximum speed 120 km/h

Clément-Bayard Three-Seat Biplane with 85-hp Engine
Span 16.0 m; length 11.20 m; height 2.0 m; wing area 50 sq. m
Empty weight 650 kg; loaded weight 1,100 kg
Maximum speed 85 km/h

Clément-Bayard Single-Seat Military Monoplane with 50-hp Gnome
Span 9.20 m; length 7.50 m; wing area 16 sq. m
Loaded weight 520 kg
Maximum speed 100 km/h; climb rate 150 meters per minute
One built

Clément-Bayard Two-Seat Monoplane with 80-hp Le Rhône
Span 11.41 m; length 7.45 m; wing area 21 sq. m
Loaded weight 350 kg
Maximum speed 110 km/h; climb rate 125 meters per minute
One built

Clément-Bayard Armored Monoplane with 80-hp Gnome
Wing span 10.20 m; length 6.32 m; wing area 18 sq. m
Loaded weight 375 kg
Maximum speed: 115 km/h; climb rate 100 meters per minute
One built

Coanda-Delaunay-Belleville

Henri Coanda is best remembered for his 1910 design powered by what many feel to have been the first "jet" engine in history. At the outbreak of the First World War Coanda offered his services to France and was subsequently assigned to the aircraft firm of Delaunay-Belleville.

In 1916/17 Coanda designed a unique plane for the Delaunay-Belleville firm. It featured a torpedo-shaped fuselage and two-bay, staggered biplane wings. There was an unconventional tail unit with twin fins and rudders. Power was supplied by a 300-hp liquid-cooled engine located in the middle of the fuselage. Twin shafts extended from the engine to the tail and drove two propellers which were mounted behind each of the rudders. The pilot was seated in the extreme nose just ahead of the wing.

Coanda's machine was tested at the French experimental test center at Étampes in 1917. It is not known how successful this design was, but photographs show that the machine was eventually destroyed in a crash. It appears that no further development was undertaken and Coanda turned his attention to designing a twin-engine bomber for the S.I.A. firm.

The RNAS ordered five machines (as "Delaunay-Bellevilles") and assigned them serial Nos.1395–1399. The order, of course, was later canceled.

Coanda-Delaunay-Belleville Experimental Aircraft. MA2201.

Courtois-Suffit Lescop

Courtois-Suffit Lescop C1

This aircraft is of historical significance as it was one of the first in the world to feature leading edge flaps. It was designed by Roger Courtois-Suffit and Capitaine Lescop. It has been described as being experimental, but French records show that it was intended to meet the C1 specification.

It is strange that an aircraft with such an advanced feature would have an outdated rotary engine, the 160-hp Gnome 9Nc that was also used by the SPAD 15, Nieuport 28, and Morane Saulnier 27 and 28. A more powerful development of the C.S.L.1 with a 250-hp Clerget 11E engine was to follow when that engine became available.

The C.S.L.1 was a small single-bay biplane. Its most impressive feature was the hinged leading edge flaps fitted to the lower wings and the tailplane. On the wings, these flaps were

The CSL C1 is historically significant as one of the first aircraft to feature leading edge flaps. MA353.

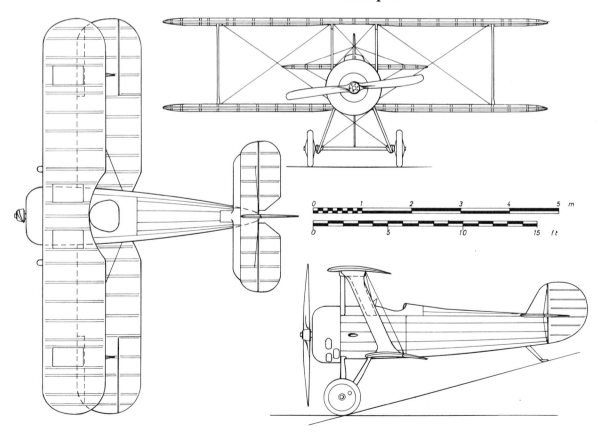

Courtois-Suffit Lescop C1

1.3 m in span and 0.18 m in chord and were hinged along the forward spar. They were constructed of plywood and the hinges of the flaps were covered with fabric. There was a hinged portion which spanned the full length of the tailplane and was 0.15 meters in chord. Their setting was controlled from the cockpit. The upper wing did not feature leading edge flaps, and ailerons were fitted to the lower wings only. The fuselage was fully faired and the engine was housed in a cowling of broad chord. The landing gear had two wheels; each was mounted on a half-axle pivoted at the mid-point of the spreader bars.

The aircraft was built by the S.A.I.B. (Société Anonyme d'Applications Industrielles du Bois) at its factory in Paris. Construction began in October 1917 and the C.S.L.1 was completed and undergoing testing by 1 May 1918. However, since the Gnome Monosoupape 9N was not available, a 140-hp Clerget 9Bf was fitted.

It is not known how successful the C.S.L.1 or its leading edge flaps were. It is known that the aircraft was not selected for production. In fact, none of the planes with the Gnome 9Nc was selected for service.

C.S.L. C1 Single-Seat Fighter with 140-hp Clerget 9Bf
Span 7.80 m; height 2.70 m; wing area 19 sq. m
Empty weight 470 kg; loaded 760 kg
Maximum speed: 220 km/h at sea level (provisional); climb to 4,000 m in 16 minutes; endurance 2.5 hours
Two built

C.S.L. C1 Single-Seat Fighter with 200-hp Clerget 11E (Estimated Performance)
Span 7.80 m; height 2.70 m; wing area 19 sq. m
Maximum speed: 240 km/h at sea level (provisional); climb to 4,000 m in 14 minutes
Project only

Courtois-Suffit Lescop Fighter with Clerget Engine

This aircraft was to have been a refinement of the C.S.L.1 (which was fitted with a 140-hp Clerget to undergo tests, pending the availability of a more powerful engine). While the C.S.L.1 was undergoing tests, development of the more powerful version was continuing. This aircraft was to have had equal-span wings with square tips. This new plane was also to be temporarily fitted with a 150-hp Clerget engine that would be nearly covered by an immense cône de pénétration. The 150-hp Clerget was eventually to be replaced by the 300-hp Clerget 9Bf then under development. However, work on the Clerget 9Bf was never completed, and it appears that the second C.S.L. fighter was never built. This may have been due to the engine problems or because of difficulties with the new C.S.L. itself.

Coutant Flying Boat

A flying boat was designed and built by the Coutant firm that first entered service in late 1917, and was selected for the same patrol/medium bombing specification as the D.D.10 (an aircraft with two 200-hp Hispano-Suiza engines), G.L. HB 2, Tellier T.4, and the Borel-Odier B.O.2. It is interesting to note that the Aviation Maritime chose to purchase a small number of flying boats from each of the five manufacturers rather than standardize on the production of a single type for the patrol/light bomber role. Certainly having such a multiplicity of types in service would only have exacerbated the problems of supply and maintenance at the naval aviation bases.

The Coutant flying boat appears to have been a conventional seaplane but with an exceptionally thin tail. At the end of the upswept empennage was a horizontal stabilizer and a circular rudder. The upper wing was considerably larger than the lower and the 280-hp Renault engine was mounted above the fuselage centerline just below the top wing. The aircraft, which appears to have carried two crew members side by side, was armed with a single machine gun and up to 140 kg of bombs. It is not known how many Coutant flying boats were obtained by the Aviation Maritime, but they saw limited operational service. Only two were delivered from February to September 1917.

Coutant flying boat at the seaplane base at Saint Raphaël. The engine was a 280-hp Renault. According to SHAA records only two examples were delivered to the Aviation Maritime during 1917. B90.0735.

De Bruyère Canard

In April 1917 a canard aircraft of quite unusual configuration was tested at the Belgian training field at Étampes. This field had previously been used by the Farman and Blériot firms to test their aircraft; during the war it was used as a test center at which various experimental aircraft were flown.

The canard was built by French engineer De Bruyère. It had a sleek fuselage with the pilot seated ahead of the wings and a Hispano-Suiza engine (type unknown but possibly a 150-hp Hispano-Suiza 8Aa) was mounted inside the fuselage. The two-bladed propeller was at the extreme end of the tail and was driven by a long shaft from the engine. Canards were mounted in the extreme nose of the aircraft. The aircraft was a sesquiplane with the upper wing extended above the fuselage on struts while the lower wing was suspended below the fuselage. There were prominent fins both above and below the rear fuselage. Twin main wheels were suspended below the lower wing and a large nose wheel was semi-recessed in the extreme nose. J.M. Bruce reports that it may have been armed with a 37-mm Hotchkiss cannon.

On its first test flight the aircraft lifted a few meters and then

The DeBruyère airplane. Pivoting wingtips were used for roll control; if these were over-balanced they may have been the cause for the aircraft's take-off crash. Courtesy J.M. Bruce and G.S. Leslie collection.

rolled over and crashed. Apparently, it was not rebuilt and further development was abandoned.

De Bruyère Canard

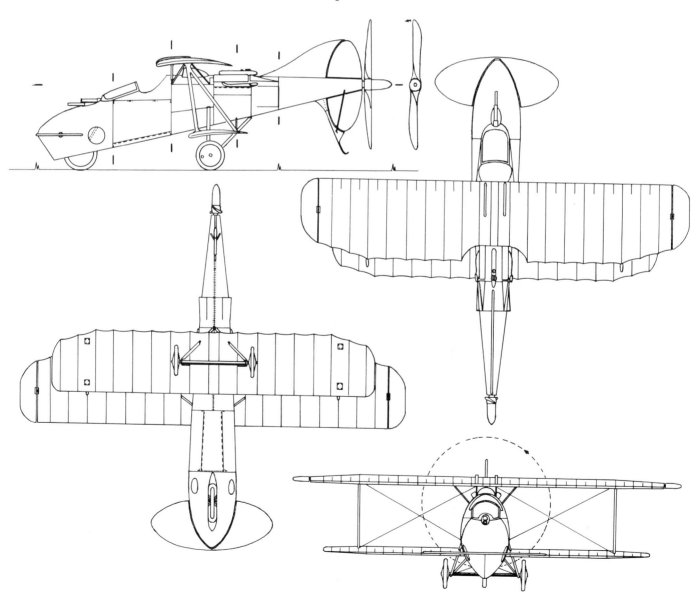

Deconde Flying Boat

This plane was reported in a British Ministry of Munitions document to be under construction in early 1918. It was a two-seat flying boat powered by a single 260-hp Salmson engine. Its design suggests that it was intended for the maritime patrol or "alerte" mission. It was a biplane with a wing area of 245 sq. m. It carried a load of 150 kg, was to have had a maximum speed of 160 km/h at 2,000 m, and climb to 2,000 m in 25 minutes. It is not known if it was ever completed.

Construction Aéronautique Edmond De Marcay

Edmond de Marcay opened two aircraft factories in Paris during 1916 and built SPAD aircraft under license. Perhaps drawing from his experience building the SPAD 7 fighter, de Marcay undertook to design a fighter to meet the C1 specification of 1918.

De Marcay 1 C1 and 2 C1

The C1 category called for a single-seat fighter to be armed with twin machine guns (either 7.7- or 11-mm) or a single 37-mm cannon with a single machine gun. Provisions were to be made for a photo-reconnaissance variant and also a monoplace protégé. The payload was to be between 220 to 270 kg and the maximum ceiling was to be 9,000 m with a service ceiling of 6,500 m. A maximum speed of 240 km/h was requested. A large number of aircraft were submitted to meet the C1 specification. Those using the 300-hp Hispano-Suiza 8F were SPAD 18, 20, 21, and 22; Nieuport 29, Descamps 27, Hanriot-Dupont HD 7, SAB 1, Monineau C1, Semenaud C1, and de Marcay C1.

It had originally been intended that the de Marcay fighter be given a Liberty engine; this was designated the de Marcay 1. However, problems with the Liberty engine resulted in the aircraft being redesigned to take a 300-hp Hispano-Suiza 8Fb. This was designated the de Marcay 2. It was a conventional biplane with the top wing considerably longer than the lower. The extensions of the upper wing were braced by oblique struts fixed at the base of the interplane struts. There was a single bay of bracing struts and the ailerons on the upper wing were horn-balanced. The engine was mounted in an aluminum mount, the front of the fuselage was made of steel and the rear of wood.

De Marcay 2 C1 Single-Seat Fighter with 300-hp Hispano-Suiza 8Fb	
Span 9.25 m; length 6.62 m; wing area 25 sq. m	
Loaded weight 1,704 kg; payload 339 kg	
Maximum speed: sea level	252 km/h
3,000 m	237 km/h
4,000 m	231 km/h
5,000 m	220 km/h
6,000 m	200 km/h
Climb to: 1,000 m	2 minutes
2,000 m	4 minutes 26 seconds
3000 m	7 minutes 26 seconds
4,000 m	12 minutes 3 seconds
5,000 m	18 minutes 16 seconds
Armament two 7.7-mm Vickers machine-guns	
One built	

There were two synchronized 7.7-mm Vickers machine guns.

The aircraft was completed by 1919. The radiator was to have been mounted in the wing. However on production planes the radiator was mounted in the front of the engine and was circular. The engine was closely cowled and the camshaft covers were covered by full-length fairings. A jettisonable fuel tank was

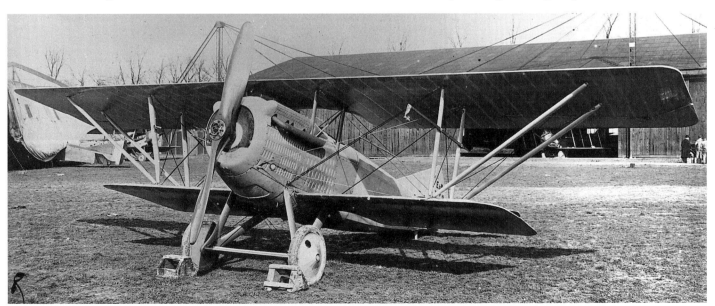

The De Marcay C1 was a single-seat fighter with 300-hp Hispano-Suiza 8Fb. MA26247.

fitted between the landing gear struts. Balanced ailerons were on the upper wing only. Planned armament was two synchronized 7.7-mm Vickers machine guns.

The aircraft had an impressive performance and was the fastest at the 7 March 1919 concours. Bad weather prevented it from being evaluated at high altitude; therefore most of the testing occurred at 500 meters. Most of the comments in the test pilot's reports were positive, but it was noted that the de Marcay 2 C1 had poor turning characteristics. Contemporary notes obtained from the Musée de l'Air show that the 2 C1 was considered adequate as a single-seat fighter. However, its climb rate was inferior to the Nieuport 29, and it was the latter which was selected for series production.

Records in the Musée de l'Air indicated that the de Marcay Type 3 was to have been a single-seat, single-bay biplane. This fighter would also have used the 300-hp Hispano-Suiza engine. The fuselage was to have been of all-metal monocoque construction with the radiator mounted in the front of the cowling. It is possible this design was a refinement of the de Marcay 2 C1 mentioned above. As far as can be ascertained, this type was never built.

Postwar, the de Marcy firm produced another sleek fighter, the de Marcay Type 4, which was rejected primarily because the shoulder-mounted wing inhibited the pilot's view. The Type 5 was a two-seat monoplane with a 370- or 450-hp Lorraine-Dietrich engine and a monocoque fuselage.

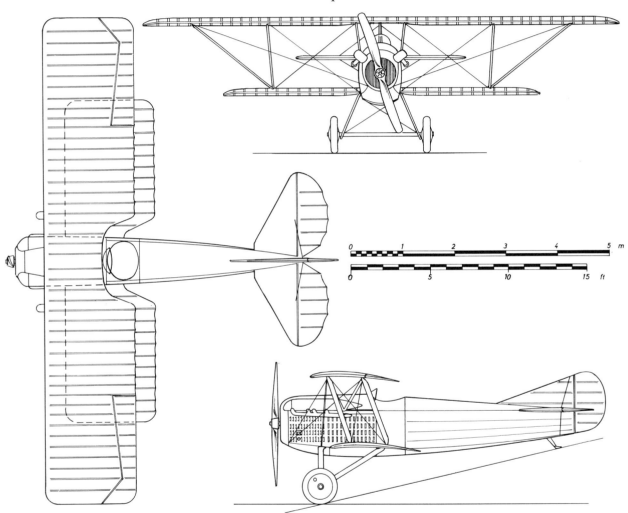

De Marcay 2 C1

De Monge/Buscaylet Experimental Aircraft

Louis de Monge was a propeller designer for the Société Anonyme des Etablissements Lumiere.

In 1914 De Monge built an experimental monoplane that was entered in the concours securité. This was a competition which called for the designers to create an plane that was as safe as possible. De Monge's design utilized a wing with elastic attachments which enabled it to flex under the air flow and to retreat from wind gusts from the side. Springs were used to automatically return the wing to its normal position. It retained the standard Deperdussin fin and rudder. Control was by wing warping. The undercarriage had paired main wheels with a large skid ahead of the main undercarriage (to reduce the risk of nosing over on landing) and a rear skid.

In 1918 de Monge designed an unusual airplane for the Buscaylet firm. It was a single-bay biplane with a "living" wing. Presumably this meant that the forward-swept upper wing was fully articulated. The engine was mounted inside the Deperdussin-designed fuselage and drove a propeller behind the wings. It is not known if the aircraft ever flew.

De Monge founded Etablissements Louis de Monge in the

early 1920s. His later designs included a single seat racer (built in 1921), an experimental type with two 35-hp Anzani 3A2 engines (1923), an experimental plane with a 300-hp Hispano Suiza engine (1924), the Type 101 A2/C2 military plane derived from the Koolhoven F.K.31 with a 420-hp Gnome Rhone Jupiter engine (1924), a twin-engine racer of flying wing configuration with either Bugatti or Vaslin V6B engines (1925), and a reconnaissance plane with a Fiat engine (1926). In mid-1924 de Monge dissolved his firm and joined Buscaylet et Cie.

De Monge parasol entered in the 1914 Concours de Sécurité for a safe airplane. MA.

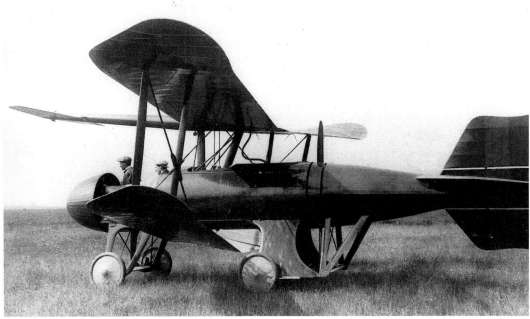

De Monge 1918 aircraft with forward-swept upper wing. Renaud.

This view of the De Monge 1918 aircraft emphasizes the unusual wing layout. 61188 M.

Delattre BN3 Bomber

One of the most interesting and unusual French designs of the 1914–1918 period, the Delattre night bomber was intended to meet the requirements for a heavy bomber comparable to the RAF's Handley Page V/1500.

The aircraft was to have had articulated biplane wings with the fuselage situated between them. The pivot was to be located at the center of gravity; this permitted the fuselage to remain centered while the wings swiveled in a vertical plane to provide lift. The aircraft was to have been powered by two 500-hp Canton-Unné engines buried in the fuselage. The engines were to have had a common shaft so that each supplied power to both the front and rear propellers. The four-bladed propellers were located at the front and rear of the fuselage, giving the aircraft a push-pull configuration. To provide an added measure of safety (the engines of the time not being completely reliable) and to permit economical cruise on one engine, it was possible for the aircraft to fly with one engine off. It was never completed.

Delattre Bomber with Two 500-hp Canton-Unné Engines (project only)
Wing area 260 sq. m
Empty weight 3,870 kg; payload 1,200 kg
Maximum speed: 178 km/h; ceiling 1,300 m; range 1,100 km

Société Provisoire des Aéroplanes Deperdussin

The Société Provisoire des Aéroplanes Deperdussin was created in 1910. Many different types of aircraft were produced by the firm beginning in 1910. These were:
1. Type Canard with 40-hp Clerget engine.
2. Type A monoplace with 30- or 40-hp Clerget engine; two used by Australian CFS as penguins: Serial Nos. CFS 4 and 5.
3. Type B—similar to Type A but with a 50-hp Gnome rotary engine.
4. Military Plane 1911—Three Type Bs each with different engines: 100-hp Clerget, 80-hp Anzani, and 100-hp Gnome. Some examples were acquired by the RFC and RNAS (see entry for Type TT).
5. Type P—with 50-hp Gnome.
6. Type T—1912, with 70-hp Gnome.
7. Deperdussin Monocoque—racer with land and floatplane versions.
8. Type TT—1914 single- and two-seat versions.

In 1913 Deperdussin was jailed on fraud charges and subsequent aircraft were designed by the firm's chief engineer, Bechereau. The firm retained the name SPAD. This time, however, SPAD stood for Société Anonyme pour l'Aviation et ses Derives. The Deperdussin factory also built Caudron G.3s and G.4s under license.

Deperdussin Monocoque

In 1912 the Deperdussin firm produced a sophisticated racer powered by a 50-hp Gnome rotary engine. It was designed by Louis Bechereau, who would become famous for the SPAD series of fighters. Advanced features of this design included its circular, monocoque fuselage, shoulder mounted-wings, and closely cowled engine. The fuselage was faired with plywood and in some military examples there was armor plating beneath the pilot's seat. The wheel spokes were covered by circular

Deperdussin Monocoque flown before the war by Maurice Prevost in excess of 200 km/h. An aircraft based on this type may have been used by DM 36. B83.3483.

fairings to reduce drag. Wing warping was used for lateral control. It was in this type of aircraft that Jules Vedrines won the 1912 Gordon Bennett race. A further development of it was entered in the 1913 Gordon Bennett race and again won. This aircraft had a 50-hp Gnome engine and a maximum speed of 180 km/h.

While Deperdussin's design had been intended for racing, it appears that a development of the type was adopted for military service in 1914. Shortly after the outbreak of war, DM 36 (for Deperdussin Monocoque) was formed. On 15 January 1915 DM 36 had four pilots and four Deperdussin Monocoques with 60-hp Le Rhône engines; at this time DM 36 was under the command of Capitaine Lapeyrouse. The aircraft was named the Deperdussin *Epervier* or *Sparrowhawk*. Despite their fearsome name, the planes could be used only for reconnaissance and artillery spotting; they were incapable of carrying bombs. (For this reason, a single M.F.7 was supplied to DM 36 for use as a bomber.) The escadrille also received a few Deperdussin two-seaters with 80-hp Gnomes (presumably Deperdussin TTs). DM 36 was assigned to the 8th Armée and, in January 1915 was based at Groat Bogaerd. On 19 February 1915, only four months after it had been formed, the unit re-equipped with MF.7s to become MF 36.

Deperdussin Monocoque with 60-hp Le Rhône
Maximum speed: 115 km/h; range 250 km; endurance 3.5 hr.

Deperdussin TT

Armand Deperdussin was responsible for the design of the Deperdussin TT, a monoplane intended for reconnaissance. It succeeded the Type T that had equipped Escadrilles D 4 and D 6.

The TT was a monoplane with a small tail/rudder unit and triangular tail surfaces. It was powered by a 80-hp Gnome rotary engine. The landing gear consisted of a pair of large spoked wheels mounted well below the fuselage on a stalky undercarriage. There were prominent skids to prevent damage to the propeller should the plane nose over during landing. Lateral control was by wing warping and construction was of wood and fabric. Alternate power plants were 100-hp Gnomes or Anzani radials of 45-, 60-, or 80-hp. The crew were seated in tandem, the observer in front of the pilot. Unfortunately, this placed him between the wings and resulted in a very poor downward view. This flaw was part of the reason that the TT was withdrawn from service soon after the war began.

Two Deperdussin TTs were fitted with a single machine gun on top of the cabane struts. The gunner had to stand up to use the gun, and he was provided with an extended windshield to protect him from the slipstream. The raised mounting, while cumbersome and drag-producing, had the singular virtue of enabling the gun to be fired straight ahead without destroying the aircraft's propeller. An armed TT with a 80-hp Le Rhône was tested at Villacoublay by M. Loiseau in 1914. A second aircraft, this one with a 160-hp Gnome engine, was also tested with a gun mount made of steel. The more powerful engine permitted the fitting of armor. This aircraft was sent to the Groupe des Escadrilles de Protection du Camp Retranche de Paris (C.R.P.) but there is no record of its taking part in aerial combats.

Two escadrilles were equipped with the type TT:

D 4, formed in 1912 at Maubeuge and attached to the 5th Armée at the beginning of the war. D 4 was based at Givet on 14 August 1914 and performed a number of reconnaissance sorties over the Sambre Valley. Moving to Romilly-sur Seine, it was again assigned to the 5th Armée. A note from a Lieutenant Adray (who was the commander of D 4) dated 3 September 1914 noted that the machines in his unit had been acquired in 1912 and 1913 and were suffering from fatigue. In October, D 4 moved to Ville-en-Tardenois and was assigned to the 1st C.A. The unit later moved to Bouvancourt where it served with the 3rd C.A. At this time D 4 was commanded by Lieutenant Rochette. It re-equipped with Caudron G.3s to become C 4 on 29 March 1915.

D 6, formed in 1912 at Reims and attached to the 5th Armée in 1914. The escadrille was under the command of Capitaine Aubry. D 6 was active over the Belgian front and during the fighting near the Aisne in October 1914. Although the Deperdussin TTs of D 4 and D 6 were used almost exclusively for reconnaissance and artillery observation, aircraft assigned to D 6 did chase an Aviatik away from the French front on 28 January 1915 and dropped small bombs on an enemy battery on 21 February 1915. Early in 1915 D 6 was based at Baslieux-les-Fismes under the command of Capitaine Desorges. It re-equipped with Caudron G.3s to become C 6 in March 1915.

Foreign Service
Belgium
Two aircraft were acquired by the Belgian air service but lack of spares soon grounded them.

Portugal
In 1912 a single example of what seems to have been a Deperdussin TT trainer was donated to what later became the Portuguese Escola de Aeronautica Militar.

Deperdussin TT at Basra. B80.304.

Deperdussin in Russian service. The Soviet Union still had one Deperdussin on strength with the 1st Socialist Aviaotryad in December 1917. B86.0029E.

Below: Deperdussin TT based at Villacoublay in 1913/14. Two Deperdussin TTs were fitted with a single machine-gun on top of the cabane struts. B84.1697.

Russia/Soviet Union

A Deperdussin TT finished third in the 1913 military aircraft competition and was eventually selected for series production instead of the indigenous S-11 which had placed second. A total of 63 were built (which Shavrov states were erroneously designated "Deperdussin-Monocoques") were built at the Lebedev plant and were used in front-line service as late as 1917. Two floatplane versions were also built at the Lebedev plant. The Soviet Union had one Deperdussin, probably a Type TT, on strength with the 1st Socialist Aviaotryad in December 1917.

Spain

Spain's Escuela Nacional de Aviacion acquired a number of Deperdussins for training. They were a 35-hp Anzani single-seater intended as a penguin to train pilots to taxi, a 50-hp Gnome single-seater, and an 80-hp Gnome two-seater (Type TT). The Deperdussins were later replaced by M.F.7s and Caudron G.3s.

Serbia

The Serbian air service had two Deperdussins TTs during the Balkan wars. They formed part of the escadrille that operated near Skaddar in 1913. No more Deperdussins were purchased and the air service officers stated that Blériot 11s were preferred to the Deperdussins.

Turkey

Turkey acquired two Deperdussins in December 1911; these were the first Turkish military aircraft. One was a single-seater with a 35-hp Anzani engine; the other was a two-seater with an 80-hp Gnome engine. They were assigned to the Yesiköy aviation school in July 1912.

The Deperdussins participated in bombing and reconnaissance missions along the Bulgarian front during the Balkan wars. At the beginning of the First World War a U.S. War Department report of October 29 1914 stated the inventory at Yesilköy included four Deperdussins. However, the Turkish inventory lists only one Deperdussin, named *Osmanli*, used for training. Based at Yesilköy, it observed the Bosphorus and the movements of the Russian fleet. It was one of only two aircraft providing reconnaissance in the area for the Turks. Its end came in early 1915 when it was downed by a Russian Grigorovich M.5, crashing into the Balgrat forest.

United Kingdom

Four Deperdussins were entered in Britain's Military Plane Competition of 1912. Two were built by the French firm, the other two by the British Deperdussin Plane Company.

The French machines were powered by 100-hp Gnomes. They were assigned the competition numbers 26 and 27. However, only No.26 arrived. The French aircraft performed well and was awarded the £2,000 second prize. It was purchased by the RFC and given serial number 258. It was destroyed in September 1912 when it broke up in mid-air, killing its crew. This accident was pivotal in the War Office's decision to place a ban on monoplanes.

The two British machines differed from the French in featuring tandem seating with the pilot in the rear (in the French machine the pilot sat in front). In addition, the British-built Deperdussins deleted the undercarriage skids. One of the British machines was powered by a 100-hp Anzani and was

Deperdussin TT

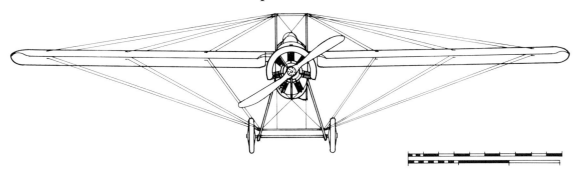

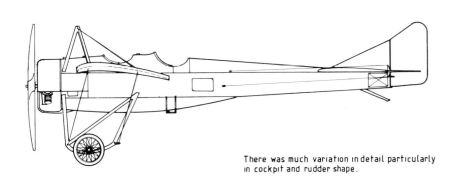

There was much variation in detail particularly in cockpit and rudder shape.

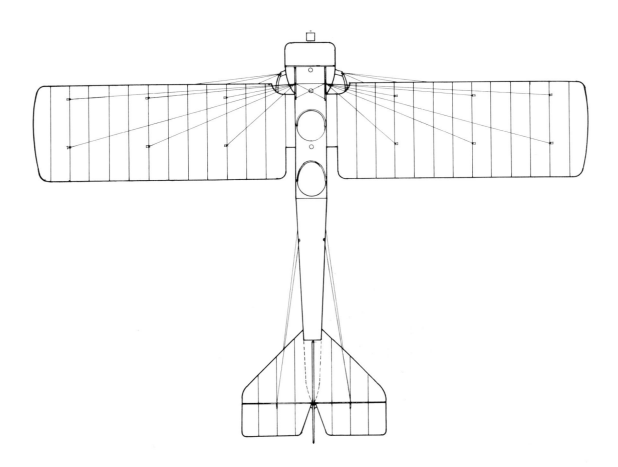

given the competition number 20; it performed poorly and was not purchased by the RFC.

The second British machine had the same engine as the French Deperdussin—a 100-hp Gnome. This was given the competition number 21. It performed well enough to be awarded a £500 prize. It was obtained by the RFC and given the serial number 259. However, it seems that this aircraft was never used operationally, being in poor condition with rusted iron work and wires. It was struck off charge in August 1913.

A Deperdussin with a 60-hp Anzani engine was obtained by the RFC when it was purchased from Captain Patrick Hamilton in June 1912. Originally given the serial number B5, later changed to No.257, it was subsequently assigned to No.3 Squadron. It saw little service and was sent to Farnborough in March and struck off charge (SOC) in November 1913.

Two more Deperdussins were obtained later. No.260 had a 60-hp Gnome engine. It served with No.3 Squadron and was then reassigned to the CFS (Central Flying School), where it was given serial 419. Number 279 had a 70-hp Gnome and stayed at Farnborough; it was never issued to a squadron.

One Deperdussin with a 100-hp Gnome was delivered to the RFC. Assigned serial 280, it was at the Royal Aircraft Factory in the spring of 1913. It had been intended to assign it to No.5 Squadron but, possibly due to the ban on monoplanes, it was not. It was struck off charge in August 1913.

Finally, three more Deperdussins were obtained by the CFS. One of these, No.421, had a 60-hp Anzani engine. Nos. 436 and 437 had engines of unknown types. None was used by the CFS; again, probably due to the ban on monoplanes.

The Royal Naval Air Service acquired several Deperdussin floatplanes and landplanes in 1912 and 1913. These were:
1. Deperdussin 70-hp Gnome—landplane assigned to Eastchurch, Dunkirk, and Dover; SOC December 1914. Serial No.7.
2. Deperdussin 80-hp Anzani—landplane assigned to Eastchurch; SOC 20 February 1915. Serial No.22.
3. Deperdussin 100-hp Anzani—floatplane tested at Grain. SOC 1913 because of defective floats. Serial No.30.
4. Deperdussin 80-hp Anzani—landplane assigned to Eastchurch; crashed May 1914. Serial No.36.
5. Deperdussin 100-hp Anzani—floatplane; twin floats. Serial No.44.
6. Deperdussin 100-hp Anzani—landplane; impressed August 1914 and based at Hendon, later at Chingford and Felixstowe. Serial No.885.
7. Deperdussin 100-hp Gnome—landplane; assigned to Hendon and Chingford in 1915 Serial No.1376.
8. Deperdussin 100-hp Gnome—landplane assigned to Hendon and Chingford in 1915 and later to Felixstowe; crashed July 1915. Serial No.1377.
9. Deperdussin 100-hp Gnome—landplane assigned to Hendon and Chingford in 1915. Serial No.1378.
10. Deperdussin 100-hp Gnome—landplane assigned to Chingford and Hendon in 1915 and later to Felixstowe. Serial No.1379.

Deperdussin TT Two-Seat Reconnaissance Plane with 80-hp Gnome
Span 10.63 m; length 7.30; height 2.70 m; wing span 20 sq. m
Empty weight 420 kg; loaded weight 720 kg
Maximum speed: 115 km/h

Deperdussin Monoplane Built by British Deperdussin with 100-hp Gnome
Span 12 m; length 7.5 m
Empty weight 556 kg; loaded weight 924 kg
Maximum speed: 109.5 km/h; climb to 1,000 m in 3 minutes 45 seconds

Descamps Type 27 C1

In 1918 the STAé issued a new requirement for a fighter to replace the SPAD 13. The C1 requirement called for a single-seat fighter with two machine guns (either 7.7-mm or 11-mm). Alternative armament was to be one machine gun and a 37-mm cannon and, possibly, twin machine guns on a swivel mount. Potential variants were to include photo-reconnaissance versions and an escort fighter (protégé). Choices of engines included the Gnome 9Nc, Hispano-Suiza 8B, ABC Dragonfly, Salmson 9Z, and 300-hp Hispano-Suiza 8Fb. The aircraft designed to use the latter engine included the SPAD 18 Ca, SPAD 20, SPAD 21 and 22, Nieuport-Delage 29, Hanriot HD 7, De Marcay 2, SAB 1, Moineau, Semenaud, and Descamps 27.

Descamps was chief designer with the Anatra factory in Russia. He returned to France later in the war and designed the Descamps fighter, which he designated the Type 27. The plane may have been produced with the cooperation of the Voisin firm. It had a rotund fuselage with a huge spinner. The top wing was conventional but the lower wing had an inverse taper with a negative stagger. The reason for this unusual design is not clear but it may have been intended to provide the pilot with an improved downward view. Horn-balanced ailerons were on the lower wing only. The prototype was armed with two Vickers

Descamps 27 C1. The top wing was conventional but the lower wing had an inverse taper with a negative stagger; this may have been intended to improve the pilot's downward view. MA36381.

machine guns completely enclosed in the fuselage. Two radiators were mounted on either side of the fuselage.

The aircraft was evaluated by the STAé in September 1918. On the basis of STAé testing it was modified. Changes included revisions to the center section struts. Further tests revealed other defects and additional changes were made. The airframe was strengthened, the undercarriage V-struts were repositioned, and Lamblin radiators were fitted.

Flight testing began in 1918 and continued until near the end of 1919. The aircraft performed well, but by the time testing was completed the excellent Nieuport 29 had been selected for production and the Descamps 27 was not developed further.

Descamps Type 27 C1 Single-Seat Fighter with 300-hp Hispano-Suiza 8Fb
Span 9.85 m; length 6.95 m; height 2.57 m; wing area 23.10 sq. m
Empty weight 732 kg; loaded weight 1,071 kg
Maximum speed: 230 km/h at ground level; 172 km/h at 7,000 m; climb to 2,000 m in 4 minutes 46 seconds; climb to 5,000 m in 18 minutes 56 seconds; ceiling 6,800 m; endurance two hours
Armament: two 7.7-mm Vickers machine guns
One built

Descamps 27 C1

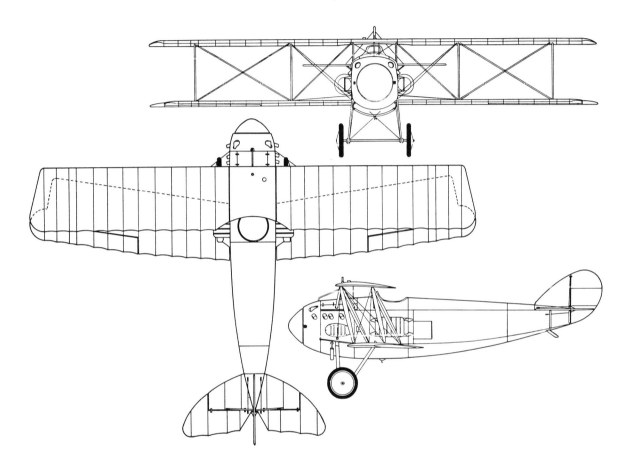

Société des Établissements Donnet-Denhaut

Monsieur Denhaut's gliders featured a sealed fuselage suitable for operation from water. Denhaut had noted that there were a large number of lakes in France from which seaplanes could be operated, and his first plane (built in 1910) was a flying boat racer with a 35-hp Lemasson engine. A pusher biplane, it featured a central step on the fuselage to facilitate takeoff. The aircraft was initially fitted with wheels and tested from the land. When these tests proved successful, the wheels were removed and the aircraft was flown from the water. Three of these planes were built by the Espinosa firm. Denhaut's next two designs were experimental aircraft built by Levasseur in 1912; both were powered by 50-hp Gnome N-1 engines.

In 1912 Denhaut joined forces with Donnet, a Swiss engineer. Donnet evaluated Denhaut's design and, although he could offer no financial assistance, gave Denhaut a 50-hp Gnome. The resulting aircraft, intended initially for civilian use, was designated the Donnet-Lévêque flying boat. The firm of Donnet-Lévêque was established in 1912 to produce Denhaut's designs, and it built a racing version of this plane in 1913 powered by a 70-hp Gnome Y engine. A third flying boat was produced with an 80-hp Gnome 7A. Finally, a fourth version was produced in 1913, designated the Donnet-Lévêque PD; the PD standing for Paris to Deauville (Deauville was a popular resort and it was Donnet-Lévêque's intention to market the plane as an ideal way to reach that city from Paris). Donnet left the company in 1913, and the firm's name was changed to Hydroplane Lévêque. Another racer, produced by the Borel firm and designated the Borel PD, was built in 1913; it was equipped with a 100-hp Anzani 10c engine. Denhaut designed a second aircraft for Borel in the same year, a two-seat sport plane with an 80-hp Gnome and designated the Aéroyacht.

Denhaut joined the army as a pilot in 1914. The authorities

Denhaut's first flying boat which was powered by a 50-hp Gnome engine. Dimensions included a wingspan of 9 m and length of 8 m. B87.1681.

Denhaut flying boat. B76.1075.

eventually realized that he could make a more valuable contribution to the French war effort by designing aircraft rather than simply flying them, and he was quickly released from military service.

After leaving Donnet-Lévêque, Donnet went to work for the Goupy firm for which he designed an all-metal glider and a seaplane with a 100-hp Omega engine. However, at the outbreak of the First World War the Goupy factory was destroyed by a fire and Donnet lost his job. Donnet and Denhaut opened their aviation factory, the Société des Ètablissements Donnet-Denhaut, in May 1915. Their workshop was located on the Isle de la Jatte.

Donnet-Lévêque Flying Boat Types A, B, C

The Donnet-Lévêque flying boat of 1912 was based on Denhaut's earlier flying boat built for Levasseur. It was a biplane with a crew of two. The upper wing was larger than the lower, and there was a single fin and rudder. The engine was a 50-hp Gnome. Designated the Type A, it had a concave hull with a prominent step and two stabilizing floats at the wing tips. The Type B was fitted with a 70-hp Gnome GN Y and ailerons. The Type C had an 80-hp Gnome 7A and an enlarged wing, enabling it to carry three persons. The Type A won the Belgian Coupe du Roi (King's Cup) for seaplanes in 1912. Although it was never used by the Aviation Maritime, the type did see service, albeit in altered form, during the First World War. The Type A was built under license by F.B.A. (See F.B.A. section.)

Foreign Service
Austro-Hungary

The Donnet-Lévêque flying boat was selected for evaluation by Austro-Hungary. The first to be obtained were Type As and Cs of 1912. Four aircraft were purchased in 1912. These were:
1. Serial number 8: Type C purchased 17/12/12 and entered service 1/13.
2. Serial number 10: Type A purchased 11/12/12 and entered service 4/1/13.
3. Serial number 11: Type A purchased 11/12/12 and entered service 1/13.
4. Serial number 12: Type C purchased 17/12/12 and entered service 1/13.

The Austro-Hungarians were impressed by the aircraft and in 1913 three copies were produced by the Pola naval arsenal. These had Gnome engines, an enlarged fuselage, and longer wing span. Since more fuel could be carried, the endurance was increased from 3 to 4.5 hours. These aircraft were:

Serial A 22: ordered 18/8/13 and entered service 18/3/14.
Serial A 23: ordered 9/13.
Serial A 24: ordered 5/13 and entered service 10/14.

Aircraft S 26 was another copy built at Pola with a 50-hp Gnome engine. It was ordered on 20 May 1914 and entered service in August. S 2 and S 4 were also copies of Donnet designs and entered service in 1914. S 17 was a copy of the Type C and had an 80-hp Gnome engine; it had a longer fuselage but a shorter span than the Type A.

Donnet-Lévêque flying boat. Although it was not selected for service with the Aviation Maritime, the Donnet-Lévêque was purchased and developed by the Austro-Hungarian and Danish naval air services. B93.2533.

The fates of these aircraft are as follows: S 2—unknown; S 4—damaged April 1914; No.8—struck off charge 2 September, 1913; No.10—damaged in a crash 6 December 1913; No.11—assigned to Pola and damaged 28 March 1913; No.12—struck off charge after being damaged in an accident 25 January 1913; S 26—damaged 11 April 1915 and struck off charge 11 April 1915; S 22—struck off charge in 1915; A 22—struck off charge May 1915; A 23—struck off charge May 1915; A 24—struck off charge May 1915.

Denmark

The Danish air service obtained two Donnet-Lévêque flying boats in 1913. They were initially stationed at the naval base at Klovermarken, where a hangar and slipway were built specifically for them. A series of flying boats based on the Donnet-Lévêque design was built by the Orlogsvaerftet naval dockyard. These aircraft was designed by H.P. Christensen and built by N.K. Nielsen. They were:

Aircraft Designation	Number Built	Period of Service
F.B.II	8	1914–19
F.B.III	10	1915–20
F.B.IV	2	1917–21
F.B.V	3	1918–21

The Donnet-Lévêque flying boats (which were designated F.B.1s) were used until 1915.

Type	Engine	Span	Length	Wing Area
A	Gnome N1	9.0 m	7.80 m	18.00 sq. m
B	Gnome Y	10.0 m	8.30 m	20.00 sq. m
C	Gnome 7A	10.48 m	8.30 m	22.00 sq. m

Russia

A 1912 Donnet-Lévêque flying boat was purchased by the Russians and delivered to St. Petersburg. It was flight tested in 1913. There are reports that some of the later versions of the craft were delivered to Russia, but these reports cannot be confirmed.

Sweden

One Donnet-Lévêque flying boat was purchased by the Swedish Marinens Flygvasende in 1913. It was assigned serial number S 22. A second aircraft was bought in 1914 and was given serial number S 23. S 22 was struck off charge in 1916; S 23 was given serial number 10 in 1917 and struck off charge in August 1918. Swedish Donnet-Lévêque flying boats had these specifications: wing span of 10.40 m;, length 8.50 m; wing area of 21 sq. m, and maximum speed of 120 km/h.

United Kingdom

England acquired a Donnet-Lévêque seaplane (probably a Type

Donnet-Lévêque Flying Boat

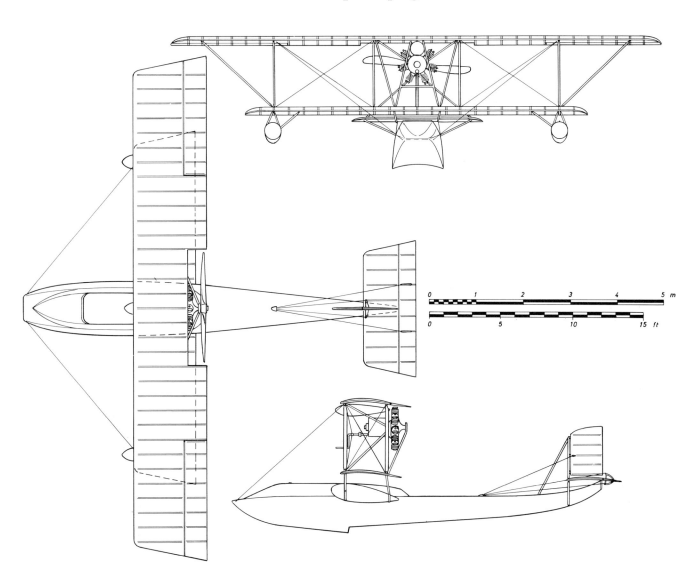

C) in November 1912; it was the first flying boat to be used by the British armed services. Assigned serial H 7 (later No.18) it was powered by an 80-hp Gnome. It was initially based at Eastchurch, then sent to the base at Grain in February 1913. It was struck off charge in June of that year, it having been noted that it had poor controls and was difficult to fly.

D.D.1

The first aircraft to be built by the Donnet-Denhaut seaplane firm was a two-seater powered by an 80-hp RH 9C Le Rhône engine. It flew in late 1915 and was intended to meet the navy's requirement for a short-range coastal patrol aircraft. It was not selected for production.

Donnet-Denhaut D.D.1 Single-Seat Flying Boat with 80-hp Le Rhône RH 9C
Wing span 10.48 m, wing area 22 sq. m
One built

D.D.2

The D.D.2 was the follow-on to the D.D.1 design and reflected the new naval requirement for an anti-submarine aircraft. Flying boats then in service were incapable of dealing with the submarine threat. If an aircraft found a submarine it carried insufficient armament to sink or even damage it. Furthermore, aircraft were unable to summon ships quickly to deal with U-boats because the flying boats did not carry radios, the crews having to rely on carrier pigeons to communicate with their base. Finally, the flying boats were underpowered and easy prey for German fighters; seaplanes often had to be accompanied by landplanes to provide some measure of protection. Realizing that the flying boats of 1914/15 were unable to deal with the growing undersea menace, the navy issued a requirement for a two-seat aircraft capable of carrying a machine gun, two 35-kg bombs, and a radio. It was expected to have an endurance of 4 hours and 30 minutes.

The Donnet-Denhaut firm responded by producing the D.D.2. The first 35 aircraft were powered by 160-hp Canton Unné R9 engines; later production aircraft received the more reliable 150-

Donnet-Denhaut D.D.2 Two-Seat Flying Boat with 160-hp Canton Unné R9
Span 14.20 m; length 10.80 m; height 3.50 m; wing area 36 sq. m
Empty weight 800 kg; loaded weight 1,380 kg
Maximum speed: 150 km/h, climb to 3,000 m in 11 minutes; range 450 km
Armament was a single 7.7-mm machine gun
A total of 36 built

Donnet-Denhaut D.D.2 Two-Seat Flying Boat with 150-hp Hispano-Suiza 8Aa
Span 14.20 m; length 10.80 m; height 3.50 m; wing area 36 sq. m
Empty weight 850 kg; loaded weight 1,450 kg
Maximum speed: 160 km/h (at sea level); range 450 km
Armament: one 7.7-mm machine gun; initially two 35-kg bombs, later two 50-kg bombs
A total of 365 built

Donnet-Denhaut D.D.2 at Toulon in June 1917. The gunner was seated in the nose; the pilot sat ahead of the biplane wings. B90.0722.

Donnet-Denhaut D.D.2

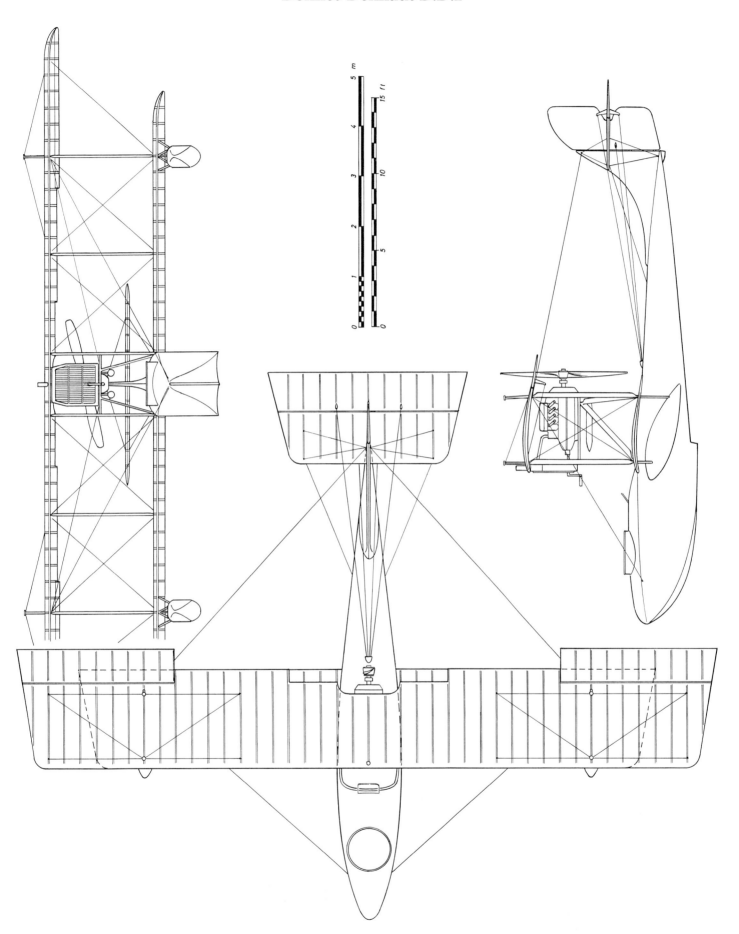

Donnet-Denhaut D.D.2 three-seat flying boat. B80.746.

hp Hispano Suiza 8Aa. One source refers to a version with a 160-hp Lorraine engine, but it is not known how many D.D.2s used it. Aircraft fitted with the Hispano-Suiza engine had four additional struts to support it. The crew of two consisted of a pilot and a machine gunner/ bombardier. The aircraft had an unequal span wing, the upper wing being larger than the lower. There were two bays of struts; earlier aircraft had only a single bay. The engine was mounted at the base of the upper wing. The slim fuselage had a single step and the fin was mounted integrally with the fuselage. The gunner was seated in the nose; the pilot sat ahead of the wings. The tailplane was mounted midway up the fin and the rudder was bisected by the tailplane. Protection was provided by a single 7.7-mm machine gun in the nose. A radio could also be carried.

The D.D.2 was quite successful and about 400 were built. They were assigned to naval aviation bases in the Atlantic, Mediterranean, and North Sea. A total of 211 D.D.2s were delivered from January to October 1917.

D.D.8

The D.D.3, 4, 5, 6, and 7 have not been identified and possibly were unbuilt projects.

When the 240-hp Hispano Suiza 8B engine became available in 1917 the French navy ordered the seaplane manufacturers to submit new designs using this power plant. The F.B.A. firm designed the type S, Levy-Besson submitted a triplane flying boat, and Tellier built a twin-engined flying boat. The Donnet-Denhaut firm redesigned the D.D.2 to take advantage of the more powerful engine. The new design, designated the D.D.8, flew in May 1917 and carried a crew of three (a pilot and two gunners), two 50-kg bombs, and two 7.7-mm machine guns on ring mounts in the nose and behind the wings. A radio was carried, powered by a wind-driven generator mounted on the center section of the upper wing. To support the weight of the crew and engine, the airframe was strengthened and diagonal bracing struts added to the outer wings. The wing was enlarged and now had three bays of struts; the upper wing was still longer than the lower. A version of the D.D.8 with jettisonable wheels to enable it to take off from land and alight at sea was tested at St. Raphaël during the war. Unfortunately, one of the wheels fell off in flight and the plane was destroyed when the pilot attempted to alight on the water.

More than 500 examples of the D.D.8 were ordered in June 1917. They served at naval air stations in the Mediterranean and along the channel coast. A total of 34 D.D.8s were delivered from January to October 1917.

Some served in the patrol and light bombing role along with the contemporary FBA Type S, Levy-Besson 200-hp Triplane, and Tellier T.3.

Foreign Service
Portugal

Portugal purchased 18 D.D.8s in 1918. They initially had 160-hp Lorraine-Dietrich engines; later, some were fitted with 200-hp Hispano Suiza 8 Ac engines. The aircraft remained in service until 1923. The D.D.8s were initially assigned to the Servico de Aviacao de Armada and were used for coastal patrol work.

United States

The U.S. Navy operated D.D.2s, D.D.8s, and D.D.9s. The antiquated D.D.2s were used as trainers because of their limited bomb load. D.D.8s were larger and, since they could carry a useful bomb load, were used in the patrol/bombing role. Some were converted to amphibians by fitting wheels to the hull. Some D.D.9s were also obtained to provide protection for the lightly-armed D.D.8s. The aircraft were based at Ile Tudy and Dunkerque, where they were used for patrols; at Moutchic they were used for training.

The main duty of the aircraft based at Ile Tudy was to escort convoys passing in the vicinity of the base. A pair were assigned to escort each convoy; to escort each the full distance at least eight planes a day were needed. One aircraft would circle over the convoy while the other flew ahead of it.

The first attack on a submarine by lle Tudy's aircraft was made on 23 April 1918 when two of them detected what was later discovered to have been the German *U-108*. Their bombs missed, but they directed the U.S. destroyer *Stewart* to the scene. Her depth charges were believed at the time to have

Donnet-Denhaut D.D.8. Reairche.

Right: Donnet-Denhaut D.D.8 suspended from a hoist. Reairche.

Below: Donnet-Denhaut D.D.8 in its element. Reairche.

destroyed the enemy submarine, but in fact it escaped.

Other aerial attacks on submarines or what were believed to be submarines during 1918 were made on 5 July, 3 August, 10 August, 27 August, 27 September, 22 October, and 25 October. In no case was a U-boat sunk or damaged, but the Ile Tudy aircraft successfully escorted 6,900 ships during the time the station was operational.

Serial numbers for the Donnet-Denhaut flying boats were

> **Donnet-Denhaut D.D.8 Three-Seat Flying Boat with 200-hp Hispano-Suiza 8B**
> Wing span 16.28 m; length 9.50 m; height 3.00 m; wing area 45.77 sq. m
> Empty weight 950 kg; loaded weight 1550 kg; payload 600 kq,
> Max speed at sea level 140 km/h; climb to 2,000 m in 15 minutes; range 500 km
> Armament: two 7.7-mm Lewis guns and two 35-kg (Dunkerque flying boat) or two 50-kg, bombs (patrol class flying boat).
> Approximately 500 built

1219, 1440, 1451,1150, 1151, 1193, 1217, 1218, 1220, 1240, 1241, 1253, 1441, 1442, 1443, 1450, 1454, 1455, 1452, 1453, 1473, 1474, 1475, 1208, 1209, 1110, 1211, 507, 447, 1179, 1180, 1181, 1182, 809, 810, 811, 813, 814, 815, 816, 817, 823, 815, 829, 831, 833, 835, 840, 841, 842, 843, 844, 845, 846, 847, 848, and 849.

D.D.9

The French naval aviation squadrons had suffered heavily at the hands of German fighters. These losses were nowhere more apparent than at the Dunkerque station. In May 1917 alone four flying boats were brought down by enemy aircraft. The navy responded to the increased threat by issuing a specification for a flying boat capable of defending itself and other seaplanes. Donnet-Denhaut responded by producing the D.D.9, a dedicated hydravion de combat.

The D.D.9 was based on the earlier D.D.8 but featured an increased armament and crew complement. It had four machine guns on ring mounts, a pair in the nose and another pair behind the wing. The crew consisted of a pilot, a mechanic, and two gunners. To give the aircraft an adequate performance despite the additional crew and guns, the airframe was enlarged and eventually fitted with a more powerful engine. The span of the lower wing was increased and four bays of struts were used. The rudder was larger and more rounded than that of the D.D.8. Initially the D.D.9s had the same 200-hp Hispano-Suiza 8Ba engines as the D.D.8. However, in the spring of 1918 the 260-hp Hispano-Suiza 8 Fb had become available and was used in the D.D.9. Some were fitted with a 300-hp Canton-Unné Z9 in late 1918. These were designated "Donnet-Denhaut 300-hp." About 100 D.D.9s were built in 1918. The degree of their success is not

Closeup of a Donnet-Denhaut D.D.9 showing the rear gunner's proximity to the propeller. Reairche.

Donnet-Denhaut D.D.9 postwar. Reairche.

Donnet-Denhaut Three-Seat or Four-Seat Flying Boat Fighter with 200-hp Hispano-Suiza 8Ba

Span 16.28 m (some sources state 16.00 m); length 10.75 m (11.00 m); height 3.50 m; wing area 61 sq. m
Empty weight 1,075 kg; loaded weight 1,975 kg (1,800 kg); payload 900 kg
Maximum speed 140 km/h (130 km/h) at sea level; climb to 1,000 m in 6 minutes 30 seconds; climb to 2,000 m in 18 minutes (15 minutes); climb to 3,000 m in 27 minutes; range 200 km
Armament: two to four 7.7-mm guns and two 35-kg "F" bombs
Approximately 100 built (including versions with the 260-hp Hispano-Suiza 8 Fb and 300-hp Canton-Unné Z9)

known but, given that their performance was no better than the D.D.8s they were designed to protect it seems unlikely they were able to provide much defense against German fighters.

They served at four air stations in late 1918, including the aeronautical school and bases in North Africa. Postwar, they were used to make long-distance flights. some D.D.9s were used by the United States Navy (see entry under D.D.8).

D.D.10

In January 1918, the French navy initiated a specification calling for a multi-engine flying boat to carry a 75-mm cannon with 30 shells and two 60-kg bombs. The aircraft should have at least two engines as it was felt that the extra engine provided a higher margin of safety, particularly during extended periods of time over water. It was to have a maximum speed of 140 km/h, an eight-hour endurance, and was to be capable of climbing to 1,000 meters in 15 minutes. The firms of Lévy, Tellier, Farman, Latham, Blanchard, Borel-Odier, and Donnet-Denhaut all submitted aircraft to meet these requirements. The Donnet-Denhaut design, designated D.D.10, was an elegant biplane initially powered by two 200-hp Hispano-Suiza engines. Production D.D.10s were powered by 300-hp Hispano-Suiza 8Fb engines. The engines were mounted back-to-back so that one served as a pusher and the other as a tractor, and each had a

Donnet-Denhaut Three-Seat or Four-Seat High Seas Flying Boat with Two 300-hp Hispano-Suiza 8Fb Engines

Span 22.92 m; length 16.20 m; height 4.20 m; wing area 95 sq. m
Empty weight 2,100 kg; loaded weight 3,700 kg; payload 1,600 kg
Maximum speed at sea level: 155 km/h; climb to 2,000 m in 19 minutes; range 800 km
Armament: four 7.7-mm guns; other armament options included a 75-mm cannon with 30 shells and, in addition, up to 300 kg of bombs
Approximately 30 were built

four-bladed propeller. The engines were suspended between the three-bay wings by a complex arrangement of struts. The wings were of equal span and the ailerons on both the upper and lower wings were connected by struts. As with the D.D.9,

The D.D.10 was based on the earlier D.D.8 but featured heavier armament, twin engines, and a larger crew. B86.2912.

armament consisted of a pair of ring-mounted machine guns in the nose and another pair behind the wings. Thirty aircraft were ordered, but the Armistice prevented any from seeing operational service. Postwar, some aircraft were converted to serve as airliners and were used on the Antibes-Corsica route by the Aéronavale company.

Donnet-Denhaut D.D.10. Production D.D.10s were powered by 300-hp Hispano-Suiza 8Fb engines. B87.5012.

Donnet-Denhaut D.D.10

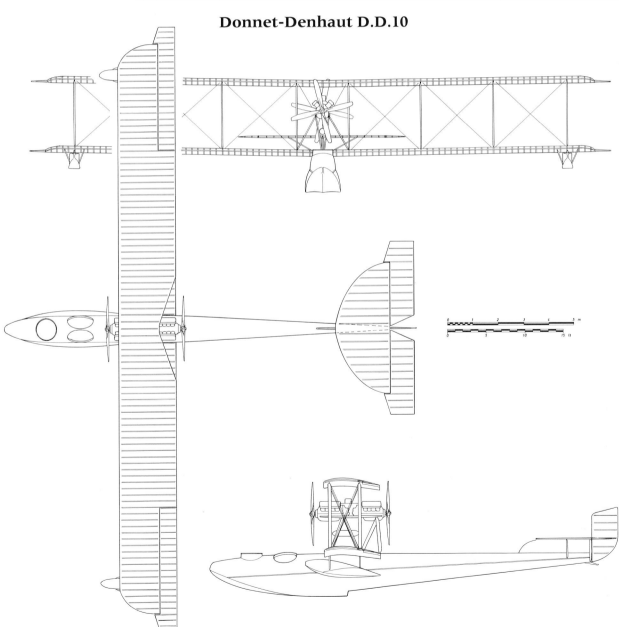

Donnet-Denhaut P.10 and P.15

The French naval authorities decided that they needed a flying boat in the class of the British Felixstowe seaplanes. They issued a specification calling for one having an endurance of eight hours, a large wireless radio with a range of 300 km, a payload of four 120-kg bombs, and two machine guns. The aircraft was to have a maximum speed of 140 km/h and be able to climb to 2,000 m in 25 minutes. Aircraft designed to meet this requirement were submitted by Farman, Nieuport, Tellier, Levy (Besson), Latham, FBA, and Donnet-Denhaut.

A three-engined version of the D.D.10 was developed in 1918 to meet the high seas rquirement. The total power generated was 1,100 hp and was probably supplied by three Hispano-Suiza 8 Fb engines. The plane, designated the P.10, was successful, but it was not adopted by the Aviation Maritime, probably due to the Armistice.

A follow-up Donnet-Denhaut design, the P.15, was to have represented an enlarged and more powerful version of the Donnet P.10. The P.15 had four 450-hp Renault engines. The three-bay wing was to have had duralumin struts, a 75-mm cannon mounted along with a TSF unit in the nose and an enclosed cabin that could carry 12 passengers. Admidships there was to be a cabin for a pilot and a co-pilot seated side by side. In the rear fuselage was room for a rear gunner or up to 18 passengers. The engines were to have been mounted back to back in two streamlined nacelles, each having a compartment for an engineer to permit maintainence in flight. It seems that construction of the aircraft was under way near the end of 1918, but after the Armistice the Aviation Maritime lost interest in the P.15 and only the single example was constructed.

P.15 with Four 450-hp Renaults
Span 33.5 meters; length 22.80 m; height 9.6 m; wing area 355 sq. m
Empty weight 6,700 kg; loaded weight 7,000 kg
Maximum speed: 165 km/h; duration 25 hours
Armament: one 75-mm cannon; probably also four 7.7-mm machine guns
One built

Dorand

Colonel Emile Dorand would play a critical role in the development of French civil and military aviation before and during the First World War. In 1908 he designed a biplane kite later modified to a triplane. Powered by a 43-hp Anzani engine, it became known as the Militaire Avion. The next year he designed the Laboratorie, a biplane powered with a 60-hp Renault and fitted with instruments. A similar but smaller plane was planned with the assistance of Gustave Eiffel. A biplane powered by a 70-hp Renault was designed in 1911. It had a loaded weight of 1,200 kg and a maximum speed of 82 km/h.

Dorand DO.1

In 1914 General Bernard issued a specification for armored planes intended for short- and long-range reconnaissance, bomber, and scout missions. Blériot, Breguet, Clément-Bayard, Deperdussin, Dorand, Ponnier, and Voisin all produced planes to meet this requirement.

Dorand's entry was the DO.1, powered by a single 85-hp Anzani engine. A small number were built but further development was abandoned. As with other armored planes of this period, it was discovered that the engines were not powerful enough to give the DO.1 an adequate performance while

Dorand DO.1 Armored Biplane Two-Seat Reconnaissance Aircraft with 85-hp Anzani
Wing span 19 m; wing area 50 sq. m
Maximum speed 110 km/h at 1,000 m; ceiling 2,800 m; endurance 4 hours

DO.1 flown by Gastinger in November 1914. B84.1626.

Several DO.1s were used to form DO 14 (not to be confused with V 14, which was an entirely different escadrille) at Belfort on 14 December 1914. DO 14 was intended to use the Dorands only on a temporary basis until other planes could be supplied. It was assigned to the 7th Armée at Epinal. DO 14 supported the 7th Armée in its attacks on Hartmannswillerkopf and Richackerhoff. It re-equipped with M.F.11s in early 1915.

Six DO.1s formed Escadrille DO 22 in August 1914. The unit was based at Villacoublay, commanded by Capitaine Leclere. DO 22 was assigned to the 4th Armée and participated in the Battle of the Marne. It flew reconnaissance missions over Châlons-sur-Marne and was based at Suippes. The performance of the DO.1 suggests that the formation of DO 22 was also an interim step pending the availability of a more suitable plane. Indeed, only three months after its formation, DO 22 was re-equipped with M.F.11s and was re-designated MF 22 on 14 November, 1914.

A few of the Dorands were assigned to V 14 to escort that unit's Voisin 3s, but they were also quickly withdrawn from front-line service.

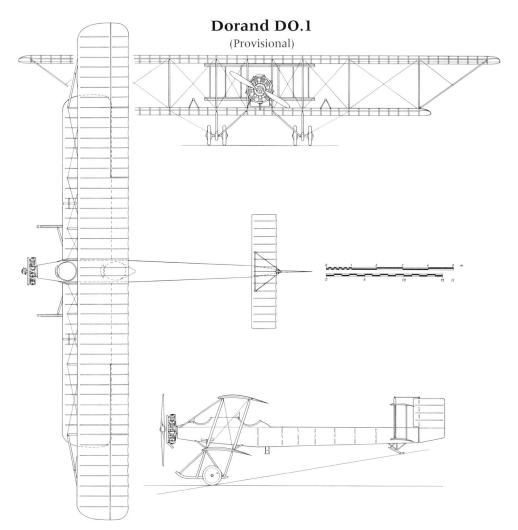

Dorand DO.1
(Provisional)

DO.1 of DO 22. The pilot was named Robillot and the mechanic was named Michou. B86.596.

Dorand Armored Interceptor

In 1913 General Bernard issued a requirement for four types of armored aircraft, one to be an interceptor to destroy enemy planes. Under the original specification produced in 1913, these aircraft were to be armored, carry a crew of three, and have a maximum speed of 100 km/h. Dorand's design was a biplane with huge wings of unequal span; the top wings were slightly shorter than the lower. Two 80-hp Le Rhône pusher engines were situated in nacelles very close to the centerline. The conventional fuselage seated a crew of three in tandem; there was gunner in the nose who had a cannon and a second gunner amidships, behind the pilot and the wings. The aircraft had a biplane tail. The landing gear consisted of a pair of tiny wheels separated by a large landing skid located under each of the engine nacelles. Dorand had shown considerable interest before the war in producing safe, reliable aircraft and in the Dorand interceptor he introduced another safety feature, linking the two

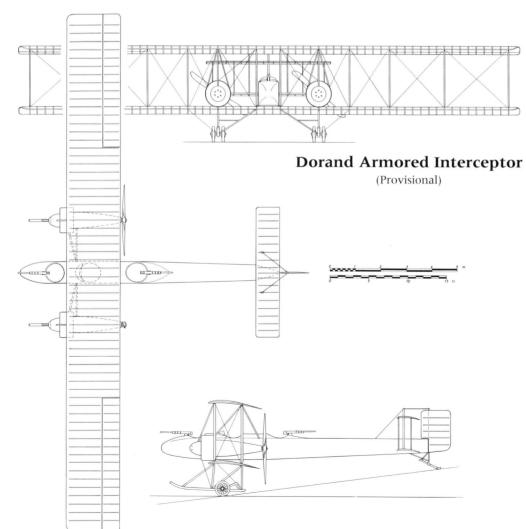

Dorand Armored Interceptor
(Provisional)

Top: Dorand Armored Interceptor. There were two 80-hp Le Rhône engines; the propellers were fitted as a pushers. MA13108.

Above: Dorand Armored Interceptor. The conventional fuselage seated a crew of three in tandem. MA13129.

engines so that should one fail the other would continue to power both propellers. This would eliminate the difficulties associated with asymmetric flight. However, further development of all the armored aircraft was discontinued because of their poor performance while carrying armor. Many of the armored prototypes were sent to the C.R.P., and it is possible that the Dorand interceptor was sent there also.

Dorand Armored Interceptor. Dorand linked the two engines so that should one fail the other would continue to power both propellers. MA13133.

Dorand BU Bomber

Dorand designed a huge triplane bomber to have been powered by two 200-hp Bugatti engines. The twin-boom design had a central nacelle housing a forward-facing gunner and containing the bomb load. A pilot/gunner was seated in each nacelle. The bomber was probably intended to participate in the 1915 or 1916 concours puissant, but was never built.

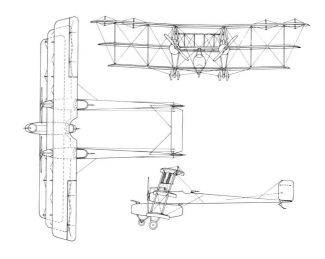

Dorand Flying Boat

Dorand designed a biplane flying boat with an engine buried within the fuselage driving two pusher propellers. It had side-by-side seating and a biplane stabilizer. The type remained an unbuilt project.

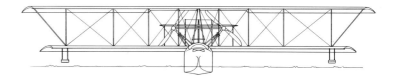

Dormay Hydroplane

This seaplane, designed in late 1916, was powered by a Hispano-Suiza engine. No further details are available.

Doutre Military Biplane

In 1914 the Société Anonyme des Appareils d'Aviation Doutre produced a two-seat biplane intended for service with the Aviation Militare. Similar in layout to the Maurice Farman 7, it featured a 70-hp Renault pusher engine at the rear of a central crew nacelle. The wings were of unequal span, the top wing larger than the lower. Rectangular tail surfaces were located above a crescent-shaped rudder and both were suspended well aft of the wings by prominent booms. A substantial skid assembly extended from the dual wheel landing gear on either side of the fuselage and converged to a point well in front of and above the crew nacelle. An unusual feature was the patented Doutre stabilizer, which helped to compensate for wind gusts or sudden drops in engine power. Lateral control was via ailerons. The type was not selected for use.

Doutre Two-Seat Reconnaissance Aircraft with 70-hp Renault
Span 16 m; length 11.50 m; wing area 52 sq. m
Loaded weight 620 kg
Maximum speed: 100 km/h; climb rate 50 meters per minute
One built

The Doutre military aircraft. A unique feature of the aircraft was the patented Doutre stabilizer which helped to compensate for wind gusts or sudden drops in engine power. 12292.

Dufaux

Dufaux C1

French aircraft manufacturers developed a number of imaginative solutions to the problem of allowing an aircraft to carry a forward-firing machine gun. By far the most imaginative and unusual approach to this problem was employed by an aircraft designed by Armand Dufaux.

Dufaux's solution was to place the engine near the aircraft's center of gravity, driving a propeller mounted in the center of the fuselage. Originally, it seems that he had intended to support the entire rear fuselage from struts extending from the wings and landing gear. However, the Dufaux C1 fighter which appeared in the spring of 1916 was a conventional two-seat biplane, except that the engine was almost completely enclosed within the fuselage and drove the centrally-mounted propeller.

The engine was the 110-hp Le Rhône rotary, mounted in the forward fuselage near the center of the wings, just ahead of the rear spar of the lower wing. The cylinders rotated through a slot in the lower fuselage. The engine was connected to the propeller via a hollow shaft through which the control cables passed. This tube was attached to two star-shaped steel structures on either half of the fuselage. This was the only attachment keeping the fuselage together aside from a long tie rod extending from landing gear struts to the large tail skid. The cockpit was in the extreme nose and the pilot and gunner were

The Dufaux C1 used the 110-hp Le Rhône rotary engine. MA1981.

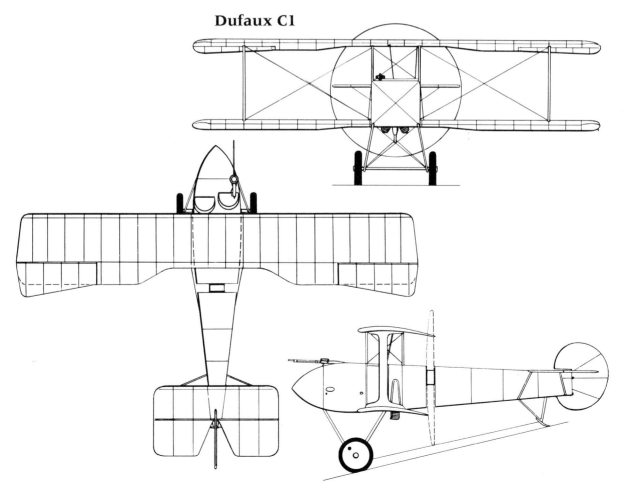

Dufaux C1

seated side by side. A single Lewis machine gun was mounted in the cockpit. The single-bay wings were of equal span. Ailerons were on the top wing only. The struts were of an I configuration and the center struts were made of steel.

The Dufaux fighter was tested at Châteaufort in 1916. By this time the Nieuport 11 with a machine gun mounted on the upper wing and the newly-developed Alkan synchronization gear had made the Dufaux fighter unnecessary.

Dufaux Two-Seat Fighter with 110-hp Le Rhône
Span 7.96 m; length 6.1 m; height 2.8 m
Empty weight 530 kg; loaded weight 740 kg
Maximum speed: 140 km/h; climb to 2,000 m in 13 minutes 7 seconds; ceiling 4,700 m; endurance 2 hours
Armament: One 7.7-mm Lewis gun
One built

The Dufaux C1 undergoing operational trials with Escadrille N 95. N 95 was assigned to the C.R.P. in December 1915. B79.3187.

Dufaux Twin-Engine Fighter

The Dufaux twin-engine fighter was the result of a collaboration between the French ace Nungesser and Armand Dufaux. In 1916 Nungesser had asked Dufaux to design a cannon-armed fighter. Dufaux set out to create a high-performance, very manouverable plane. The 37-mm Hotchkiss cannon was mounted in the center of the fuselage and fired through the hollow propeller shaft. The cannon's breech extended to the pilot's cockpit so it could be reloaded manually. The twin engines would ensure there would be enough power to carry the weight of the cannon while still allowing high performance. The rotary engines were mounted laterally (side by side) in the nose. The propeller was driven by a bevel gearing. The fuselage was fully faired and a large spinner provided a streamlined shape. It is believed that the aircraft was built and flown and may have had a top speed of 200 km/h. However, the SPAD 12 cannon-armed single-seater may have been available at that time and it certainly represented a simpler, if less ingenious, solution to Nungesser's request for a cannon-armed fighter.

Dufaux Single-Seat Canon Fighter with Two Rotary Engines (Unknown Type)
Maximum speed: 200 km/h
Armament: one 37-mm Hotchkiss cannon
One built

Dupperon-Niepce-Fetterer (D.N.F.) Bomber

The initials D.N.F. stood for Dupperon (the designer), Niepce (the manufacturer), and Fetterer (the financial backer). The firm of Niepce and Fetterer was created initially to manufacture aircraft parts under subcontract. The firm also repaired damaged airframes; a total of 1,500 aircraft were repaired during the war.

The D.N.F. aircraft was designed to meet the criteria of the 1916 competition for a heavy, well-armed bomber. It was a heavy bomber powered by three 220-hp Renault engines. As with the Breguet 11, two engines were mounted on the wings in a pusher configuration, with a third in the nose. The horizontal stabilizer was mounted at the tail with a large central rudder. On either side of the rudder two auxiliary fins were mounted on the stabilizer. The landing gear consisted of two wheels mounted in tandem and covered by huge spats. These spats, in turn, were faired directly into the engine nacelle, and a gunner was carried in the forward end of each spat. A large tail skid was mounted directly beneath the rudder.

Test pilots at the concours puissant felt that the D.N.F. bomber was too heavy and therefore difficult to handle. It was rejected for production and further development was abandoned. Some critics of the time stated that D.N.F. stood for "Do Not Fly."

The D.N.F. firm subsequently began construction of what it confidently predicted was to be the world's first "aérobus" because of the new plane's size and load-carrying capacity. The only photographs of it, which was presumably intended to be used as an airliner, show that it was under construction and was probably to have been a triplane. It is not known if construction of the "aérobus" was ever completed.

D.N.F. Bomber with Three 220-hp Renault engines
Span 25.0 m; wing area 123.68 sq. m
Empty weight 3,000 kg; loaded weight 4,700 kg; payload 1,700 kg
Maximum speed: 143 km/h; ceiling 4,800 m
One built

Left: D.N.F. bomber side view. Reairche.

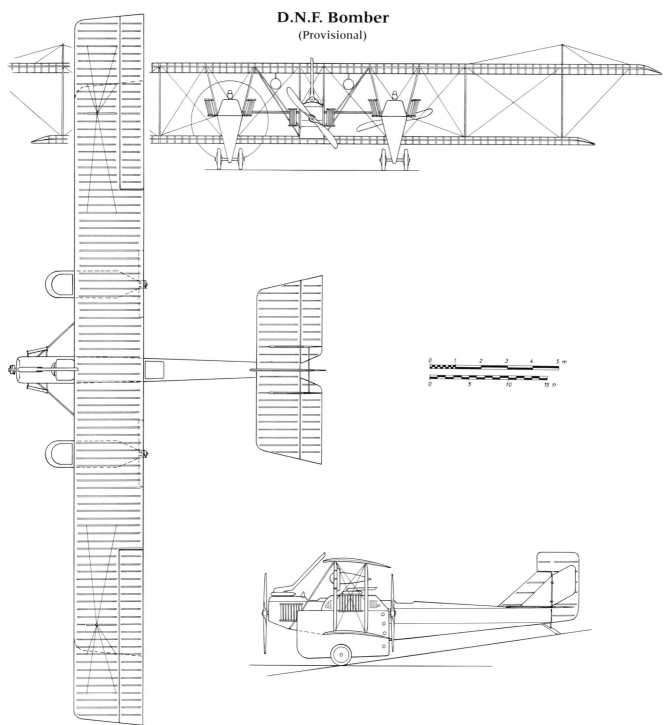

D.N.F. Bomber
(Provisional)

D.N.F. bomber showing the gunner's positions in the wheel spats. STAé 0151.

Rear view of the D.N.F. bomber. STAé 0146.

E.G.A.

Intended to meet the Bn2 requirement, this aircraft was under development in 1918. It was to have been powered by two 220-hp Hispano-Suiza 8Be engines. Work on it was abandoned after the Armistice.

Société Henri et Maurice Farman

The three Farman brothers produced some of the most important French planes of the First World War. The brothers (Henri,[1] Maurice, and Richard) were British citizens who had chosen to live in France. Henri and Maurice established a car factory in 1902. Five years later Henri purchased a Voisin biplane which he modified and entered in the 1907 Archdeacon Cup race. The next year Henri became the first European to fly a one-kilometer circuit.

The association between Farman and Voisin came to an abrupt end when Voisin began to incorporate Henri's modifications in the planes he was selling. The brothers subsequently decided to expand into aircraft production, and in 1909 Henri opened a factory at Mourmelon. His designs used ailerons for lateral control and were considered superior to Voisin's aircraft. Henri's design soon became the best-selling biplane in the world.

Maurice Farman opened his own factory at Mallet. His first plane, the M.F.1, was completed in February 1909. Maurice and Henri continued their separate endeavors until 1909 when they pooled their efforts and founded Avions Henri et Maurice Farman at Billancourt.

Both Maurice's and Henri's planes saw widespread service during the first three years of the war. However, the brothers' reluctance to abandon the pusher configuration when requested to in 1916 resulted in their losing a chance to produce a second generation reconnaissance aircraft for the Aviation Militaire. In fact, the only other Farman design to see operational service was the F.50.

Henri Farman H.F.6 armored military airplane built in 1912; the type was powered by a Gnome engine. MA16870.

Henri Farman Pre-War Planes

1. H.F.1 (1909)—50-hp Vivinus engine.
2. H.F.1bis (1909)—50-hp Gnome Omega (modified H.F.1).
3. H.F.C (1910)—50-hp Gnome Omega with four ailerons.
4. H.F.C-1 (1911)—same as H.F.C but with two ailerons.
5. H.F.2/2 (1911)—50-hp Gnome Omega; monoplane with a tractor motor.
6. H.F.6 (1911)—50-hp Gnome Omega; intended for army cooperation duties.
7. H.F.7 (1911)—50-hp Gnome Omega; streamlined fuselage.
8. H.F.10 (1911)—50-hp Gnome Omega; stabilizer in front of the fuselage.
9. H.F.10-1bis—70-hp Gnome Gamma engine; staggered wings; entered in 1911 concours militaire.
10. H.F.11 (1912)—70-hp Gnome Gamma; army cooperation; stabilizer in front.
11. H.F.11 hydroplane (1912)—70-hp Gnome Gamma. It was probably an aircraft of this type which was purchased by the Royal Naval Air Service in 1912. Designated H4 (later No.11) it had a 70-hp Gnome and at various times was based at Eastchurch, Grain, and Felixstowe.
12. H.F.12 (1912)—experimental plane; could be dismantled in 20 minutes.
13. H.F.14 Hydroplane (1912)—80-hp Gnome Lambda.
14. H.F.15 (1912)—100-hp Gnome Delta; military plane with a single machine gun.
15. H.F.16 (1912)—100-hp Gnome Delta; military plane with sesquiplane layout.
16. H.F.17 (1912)—70-hp Gnome Gamma; dual control trainer
17. H.F.18 hydroplane (1912).
18. H.F.19 hydroplane (1913)—160-hp Gnome double Lambda; Schneider Cup racer.

Maurice Farman Pre-War Planes

1. M.F.1 (1909)—60-hp REP; experimental.
2. M.F.2 (1911)—70-hp Renault 8B; reconnaissance plane entered in 1911 concours militaire.
3. M.F.5 (1911)—reconnaissance plane.
4. M.F.6 (1911)—70-hp Renault 8B.
5. M.F.6bis (1914)—trainer.

H.F.16 MA12544.

(1) At some point Henry Farman changed the spelling of his name to Henri. As the French spelling was used on his aircraft, the authors have elected to use it throughout the text.

H.F.16 serving with the Imperial Russian Air Service. B89.1719.

Henri Farman H.F.20

The H.F.20 was a development of Henri Farman's H.F.16 design of 1912. The H.F.16 was a two-seat reconnaissance plane powered by either a 70-hp or 100-hp Gnome Delta engine. The wings were of sesquiplane layout and some machines were equipped with a machine gun. The H.F.20 retained this basic layout but usually had a 70-hp Gnome Lambda or 80-hp Gnome 7A engine, a slightly reduced wing span, and a longer fuselage nacelle.

The H.F.20 had a two-bay wing with the lower wing being considerably shorter than the upper. The wings were constructed of wood spars covered by cloth. Control was by ailerons on the extended sections of the top wing. Two sets of wooden tail booms extended from the top and bottom wings and a semi-circular rudder was mounted at the point of their convergence. A single elevator was mounted on the top of the boom just ahead of the rudder. This layout gave the machine great stability, which was a key requirement for a military reconnaissance plane. The nacelle was constructed of wood and covered with aluminum. The engine was located at the rear of the nacelle. A large fuel tank was placed in the center of the nacelle, between the crew and the engine. Two pairs of wheels were mounted on short skids. Bungee cords served as shock absorbers. A variety of engines were fitted to H.F.20s in other countries, including Renaults and Le Rhônes.

It was the location of the crew and engine nacelle that ensured that the H.F.20s, as well as subsequent Farman designs, would be the primary French reconnaissance aircraft from 1914 through 1916. The crew were seated in tandem in the central nacelle, which was mounted on the lower wing. This layout permitted the pilot, who was usually seated in the nose, to have a superb view on all sides. A small windscreen helped protect him from the slipstream. The observer sat behind the pilot, where he was in a good position for observing the ground. In those planes fitted with machine guns the pilot's and observer's positions were reversed to provide an excellent field of fire for the gun.

Six escadrilles were equipped with H.F.20s (and its derivatives) in 1914. They were:

HF 1, formed in 1912 at Camp Châlons. At the beginning of the war HF 1 was based at Toul and was assigned to the 33rd C.A. In February 1915 it re-equipped with M.F.11s.

HF 7, formed in 1912 and assigned to Verdun. At the beginning of the war HF 7 flew reconnaissance missions from Verdun under the orders of Lieutenant Gauthier. In August HF 7 was assigned to the 3rd Armée and participated in the battles around the Ardennes forest. Early missions included dropping 75-mm obus (modified artillery shells fitted with fins) and destroying an enemy Zeppelin. In September HF 7 moved to Maulan and supported the 6th C.A. HF 7 also participated in the Battle of Verdun. It re-equipped with M.F.7s in February 1915.

HF 13, formed before the war and in August 1914 was based at Camp Châlons and assigned to the 3rd Armée. In 1915 HF 7 was assigned to the 15th C.A. It re-equipped with Caudron G.3s later in 1915.

HF 19, formed in 1913 at Dijon and participated in the 1913 maneuvers. In August 1914 it was based at Nancy and commanded by Lieutenant Jolain. In January 1915 it was assigned to the 13th C.A. Early in 1915 HF 19 converted to M.F.7s and later to M.F.11s.

HF 28, formed at the beginning of the war and assigned to the C.R.P. It was commanded by Capitaine Mailfert and was based at Issy-les-Moulineaux. In September HF 28 moved to Amiens and in October moved again to Doullens, where it participated in the Battle of Picardie. Two weeks later it was assigned to the 10th C.A. and was based at Léalvillers. Later the escadrille moved to Saint-Cyr and was assigned to the 10th C.A. HF 28 re-equipped with Caudron G.3s in January 1915.

HF 32, formed in September 1914 under command of Capitaine Couret. It was based at Sainte-Menehould and was active over the Argonne and Woëvre. It re-equipped with M.F.11s in September 1915.

The GQG reports for 1914 and 1915 reveal that the H.F.20s were used primarily for reconnaissance and artillery cooperation duties. Almost from the start of the war some H.F.20s were equipped with cameras and T.S.F. units. An example of the types of missions flown by these units can be seen from the 30 January 1915 reports. During that single day HF 7 flew numerous reconnaissance missions over enemy batteries at Harville, took photographs over Éparges and Saint Rely, and used a T.S.F.-equipped machine to direct artillery fire over the 1st Armée sector. The escadrilles were forced to make frequent moves along the front as dictated by the tactical situation.

Although it has been noted that the H.F.20s were too fragile for any missions aside from army cooperation duties, a number of units found more aggressive uses for their planes. Flying over the Ardennes a Lieutenant Roeckel of HF 7 damaged a German

H.F.20. Six escadrilles were equipped with the H.F.20 and its derivatives in 1914. MA37188.

Above: H.F.20 assigned to the C.R.P. in September/October 1914 preparing to leave on a reconnaissance mission. MA26005.

Right: H.F.20s based at Avord. Six escadrilles were equipped with H.F.20s (and its derivatives) in 1914. B84.1125.

Below: H.F.20. B89.1698.

airship on the ground by dropping steel flechettes. Adolf Peqoud, also from HF 7, dropped makeshift 75-mm obus (bombs) on various targets.

The Aviation Militaire was quite satisfied with using the H.F.20s for reconnaissance, as they offered their crews a substantially better field of view than the contemporary R.E.P. Ns, Blériot 11s, and Nieuport 6Ms. However, by 1915 it was clear that the H.F.20s were too fragile for sustained use in the field. They were gradually replaced by superior aircraft beginning in early 1915. HF 1, 7, 19, and 32 re-equipped with M.F.7s and 11s; while HF 13 and 28 received Caudron G.3s.

There were substantial numbers of surplus H.F.20s available and these were assigned to training units. Some of these were fitted with dual controls. The H.F.20s remained in service with some schools as late as October 1917.

Foreign Service
Argentina

Argentina acquired a single Henri Farman with a 50-hp Omega engine in 1912 ; this was probably a H.F.10. It was donated to the Escuela de Aviacion Militaire by the Companes Argentina de Tabacos (Argentine Tobacco Company). Three more Henri Farmans were later obtained, one by direct purchase, the other two manufactured by the Military Aviation Workshop at El Palomar in 1917. All three were probably H.F.20s.

The Argentine navy also utilized a Henri Farman with a 50-hp Gnome engine; it was probably an H.F.10 which had been built by the Workshop of Edmundo Marichal in 1915. This

H.F.20 in Belgian service. The Jero firm produced 24 H.F.20s beginning in 1912.
R. Verhegghen via Colin Owers.

aircraft was destroyed in a flying accident later that year. The Arsenal del Rio de la Plata factory built an refinement of the H.F.20. It had a larger wing and nacelle than the standard H.F.20. Initially powered by a 70-hp Renault engine, it was subsequently fitted with an 80-hp Gnome. The plane was assigned to the Escuela de Vuelo de Fuertr Barragan in 1915 but it, too, was destroyed in an accident that year, It was later rebuilt and continued to be used by the school at Barragan until late 1920.

Belgium

The Belgian army bought a Henri Farman aircraft (described as a Henri Farman 1910; possibly an H.F.10) on 5 May 1911. Assigned to the Ecole d'Aviation Militaire de Brasschatt, it was promptly destroyed the same day. A second Farman (possibly also an H.F.10) was purchased in May and destroyed by the end of July. Two other Farmans were ordered later that year. Four H.F.16s were built under license by the Bollekens firm (later known as Jero) at Antwerp in 1912. These were assigned to the Premier Escadrille. Three H.F.11 reconnaissance aircraft were also built by the firm in 1912. These were designated as Jero-Farmans. The Jero firm produced a total of four H.F.16s intended for the Premier Escadrille and 24 H.F.20s beginning in 1912. One of the H.F.16s became the first armed Belgian plane when a single example was fitted with a Lewis machine gun and tested as a ground attack aircraft in September 1912 at Brasschaat. In 1914 Escadrilles 1 (based at Liege), 2 (Namur), 3 (Brasschatt), and 4 (Brasschatt) all had H.F.20s. In October 1914 two escadrilles left for Ostend and subsequently moved to Saint-Pol. Escadrille 1 moved to the Ans airfield to support the 3rd DA (Third Division of the Belgian Army). Escadrille 2 operated from Belgrade in support of the 4th DA. Escadrilles 3 and 4 were still in the process of being assembled and remained at the Brasschatt airfield. As a result of the rapid German advance, Escadrille 2 left Namur; by this time it already lost two H.F.20s in crashes, and to make matters worse a third crashed during the evacuation from Namur. However, the 1st was soon joined by the 3rd and 4th Escadrilles and all three units flew reconnaissance missions for General Headquarters. Some H.F.20s were fitted with rudimentary radio sets. Escadrille 1 was reinforced with two H.F.20s from the Escadrille 3, and three of the unit's planes were fitted with Lewis machine guns. They proved to be too slow to intercept any German aircraft. Eight of the 3rd Escadrille's H.F.20s were destroyed by a storm in December. By March 1915 there were 12 H.F.20s still in service; six with Escadrille 1 (Saint-Idesbald) and six with Escadrille 4 (Houthem). On 17 April 1915 an H.F.20 piloted by Capitaine Jacquet with Lieutenant Vindevogel as his observer scored the first Belgian air-to-air victory when the crew shot down an Aviatik. The Jero firm built at least six more H.F.20s. By the end of 1915 most of the H.F.20s had been withdrawn from service.

Denmark

Denmark purchased a single H.F.1 and a single H.F.7 for the navy in 1913. In 1913 one H.F.20 was purchased by the Haerens Flyveskole (Army Flying School). It remained in service until 1917.

Greece

Greece purchased its first Henri Farman plane in 1912 when a Greek officer, who had completed flight training at the Farman's school, returned with a single example. The type had a 50-hp engine and may have been an H.F.6. It was named the *Daedaleus* and was used in army maneuvers. It was subsequently changed to a floatplane and used to set a world distance record for floatplanes in its class when it flew from Athens to Hydra and back. Other Greek officers later returned to Greece with an H.F.20 and three "Henri" Farmans (also probably H.F.6s) named *Eagle*, *Vulture*, and *Hawk*. In October 1912 they formed the Proti Mira Aeroplani (1st Aircraft Squadron) at Larissa. In November 1912 they were transferred to Nikopolis and formed into the Apospasma Aeroploais Ipirou (Epirus Aviation Detachment).

During the Balkan Wars the Farmans of the Epirus Aviation Detachment flew reconnaissance sorties over the Turkish positions at Melouna. The Epirus Aviation Detachment was also active over Ioania. Some of the Farmans dropped hand grenades on Turkish troops.

The single H.F.20 is believed to have been destroyed in December. After the Balkan Wars, the remaining Farmans were sent to the Thessalonica airfield for use as trainers. They were retired in 1917.

Italy

The Italian firm of Savoia built the H.F.20 under license in 1913. The first built by the firm was tested by pilot Henri Bille of the Farman firm and Michele Signorini of the Savoia factory in December 1913. Four H.F.20s were built by Savoia and formed Squadriglia Savoia in early 1914. One pilot of this unit made a number of publicity flights to various Italian cities in March 1914. It appears that there were no H.F.20s still in front-line service when Italy entered the war in 1915.

Japan

Japan obtained a Henri Farman (probably an H.F. Type C) in 1910 when a Japanese pilot returned from France after completing flight training. This aircraft was used as a pattern for the Rinji Gunyo Kiku Kenku Kai (Provisional Committee for Military Balloon Research) Kaishiki Number 1 Plane. A single H.F.20 was imported in 1914 and converted to the Kaishiki Number 7 Plane in 1915. Using the experience learned from this conversion, the firm then produced the Army Henri Farman Type Model 4 in late 1915. It was powered by a Japanese version of the 70-hp Renault engine. The aircraft was found to have excellent stability and performance; it entered production in 1916 as the Type MO 1914 (later Type MO 4). The Type MO 4s were built in 1916, 1919, and 1920. A total of 84 were produced.

Netherlands

Three H.F.20s were purchased by the Dutch air service in 1913. They were assigned serial numbers LA 2, LA 3, and LA 4.

Henri Farman H.F.20

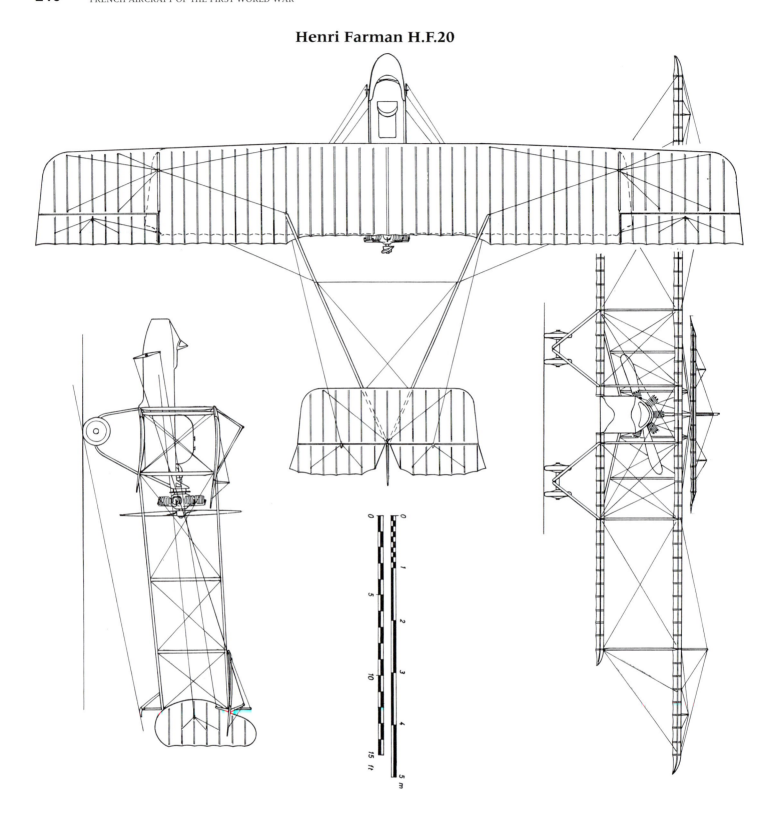

Romania

Romania had 11 H.F.20s on strength at the start of the war. They equipped Escadrillas F-1 and F-2. F-1 was assigned to the 1st Army and was based at Talmaci. F-2 was at Brasov and was attached to the 2nd Army. Most of the H.F.20s had been replaced by M.F.11s when the Corpul Aerian Romana was reorganized on 19 June 1916. The H.F.20s were assigned to two of the escadrillas attached to Grupul 3 and were used for reconnaissance and bombing missions. By the end of the year the H.F.20s were no longer in service.

Russia

Russia used the largest number of Henri Farman planes during the First World War; more than 1,500 Farmans of all types were built in Russia. The Russian-built examples were designated:
1. **Farman 4**: under almost continuous production from 1910 through 1916 and used as trainers; they were powered by a 50-hp Gnome engine. Variants included the Farman 3 (modified Farman 4 with changes to the tailplane and elevator); Farman 4 "Dux": (Farman 4 modified by Dux plant); Christian Farman: (Farman 4 modified by a French aviator named Christian) purchased for use at the Gatchina flying school; Velikii Novgorod (modified Farman 4 with an extended lower wing built by Captain L.E. Vamelkin); Farman 4 Odessa (built at the Odessa naval battalion in 1911); Maslennikov Farman 4 (modified nose fairing);

H.F.20 in Russian service. Approximately 200 H.F.20s were built under license by the Shchetinin and Anatra plants from 1912 until 1922. B85.2415.

Farman-Aviata (60-hp Gnome engine and built at the Aviata plant at Warsaw); Farman 4 Shiukov; Farman Albatros (German-built plane used at Gatchina); Farman 4 Bristol (British-built copies used at the Sevastopol and Gatchina flying schools); the Farman 4 Military type, and the Farman 4 floatplane (produced at the Shchetinin plant in 1912 and fitted with two main floats and a single tail float).

2. **Farman 7**: similar to the basic type 4 but with a single horizontal tail surface. Engines were 50-, 60-, or 70-hp Gnomes or 50-hp Kaleps. The Farman 7s were used in the Balkan Wars. Variants included the Farman 7 Agafonov (built at the Shchetinin plant in 1913) and the Farman 7 Krasilnikov (used as a flying billboard).
3. **Farman 9**: a Farman 7 with a forward elevator and built at the Dux plant.
4. **Farman 11**: used in small numbers by the army.
5. **Farman 15**: similar to the standard H.F.20 but with a larger wing span and powered by an 82-hp Gnome engine. The Dux plant built 18 in 1912. The Farman 15s were used for reconnaissance. One was fitted with a Maxim machine gun in 1913.
6. **Farman 16**: similar to the Farman 15 but with a sesquiplane layout featuring slightly swept outer wing panels. In 1914 these planes were used for reconnaissance. Some were fitted with armor but none carried armament. By 1915 the type had been relegated to training duties. The Farman 16s were used as an interim type before students transitioned to the more demanding H.F.20. Approximately 300 were built in various Russian factories. A floatplane version of the Farman 13 was built near St. Petersburg in 1913.
7. **Henri Farman 20**: these aircraft were built under license by the Shchetinin and Anatra plants from 1912 to 1922. About 200 were built and most served as trainers. At the Gatchina Flying school the H.F.20s were used for ground instruction and initial flight training. A variety of engines were fitted, including 80-hp Gnomes and Le Rhônes and 100-hp Gnomes and Gnome Monosoupapes.

From 1921 to 1922 approximately 30 H.F.20s were in service with Aviation School No.1 at Kacha. The schools at Sevastopol and Tashkent also had H.F.20s on strength. By 1923 there were still 40 H.F.20s in service with the aviation schools. They were probably withdrawn by 1925. The Siberian air force of Admiral Kolchak had one H.F.20.

Serbia

Serbia obtained three H.F.20s in 1913 to be used by the three

H.F.20 Two-Seat Reconnaissance Plane with 80-hp Gnome 7A

Span 13.65 m; length 8.06 m; height 3.15 m; wing area 35 sq. m
Empty weight 360 kg; loaded weight 660 kg
Maximum speed: 165 km/h; climb to 2,000 m in 22 minutes; range 315 km; endurance 3 hours
Armament: one machine gun and/or several 75-mm bombs
It is estimated that 3,310 H.F.20s and all its variants (H.F.21 through H.F.27) were built

AIRCO H.F.20 Two-Seat Reconnaissance Plane with 80-hp Gnome or Le Rhône

Span 13.25 m; length 8.06 m; height 3.15 m; wing area 35 sq. m
Empty weight 360 kg; loaded weight 660 kg
Maximum speed: 105 km/h; climb to 500 m in 8 min.; endurance 3 hours
Armament: one machine gun

Savoia H.F.20 Two-Seat Reconnaissance Plane with 80-hp Gnome

Span 13.60 m; length 8.06 m; height 3.20 m; wing area 35 sq. m
Empty weight 366 kg; loaded weight 621 kg
Maximum speed: 95 km/h; climb to 1,000 m in 13 min.; climb to 2,000 m in 27 min.; endurance 3 hr 30 min.
Four built

Shchetinin and Anatra H.F.20 Two-Seat Reconnaissance Plane with 80-hp Gnome or Le Rhône

Span 13.76 m; length 8.06 m; wing area 35 sq. m
Empty weight 416 kg; loaded weight 675 kg
Maximum speed: 95 km/h; climb to 2,000 m in 55 minutes; ceiling 2,500 m; endurance 3.5 hours
Approximately 200 built

Rinji Gunyo Kiku Kenku Kai Type MO 1914 (MO 4) Two-Seat Reconnaissance Plane with 70-hp Renault

Span 15.50 m; length 9.14 m; height 3.18 m; wing area 58 sq. m
Empty weight 563 kg; loaded weight 778 kg
Maximum speed: 49 knots; climb to 2,000 m in 25 minutes; ceiling 3,000 m; endurance 4.0 hours
Armament: one machine gun
Eighty-four built

Serbian officers who had received flight training at the Farman school in France. However, two were wrecked shortly after their arrival in Serbia. During the Balkan Wars the surviving H.F.20 was assigned to the escadrille at Barbalusu; on 13 March 1913 the first combat mission was flown. By the time the war ended in May 1913 the H.F.20 was no longer in service.

Switzerland

Switzerland had two H.F.20s which were used from 1914 until 1920. One was assigned to the newly-formed Swiss Fliegertruppen in August 1914.

United Kingdom

England bought a single Type III (Type Militaire) in 1910 and used it for army cooperation trials. In 1912 two Henri Farmans (probably Type 11s with 70-hp Gnome Gamma engines) were purchased for use by No.2 squadron. A single H.F.20 was purchased from Claude Graham-White in 1913 and assigned to the C.F.S. In 1913 several H.F.20s were built under license by the Aircraft Manufacturing Company (AIRCO) at Hendon. At least one H.F.20 was modified with the observer's seat moved to the front so that he could fire a machine gun; Hotchkiss, Rexer, and Maxim guns were all test-fired from this aircraft. The

H.F.20s saw limited service with Nos.3, 5, and 6 Squadrons, as well as the CFS. By the beginning of 1915 all had been assigned to training units.

The RNAS acquired 27 H.F.20s (possibly some of these were also H.F.22s). These were: No.31—80-hp Gnome (based at Eastchurch); No.189—80-hp Gnome (Eastchurch); No.940—80-hp Gnome (Hendon); No.1368—80-hp Gnome (Dunkirk, later Eastchurch); No.1374—50 or 60-hp Gnome (Chingford); No.1454—80-hp Gnome—(Gosforth); No.1518 to 1533—80-hp Gnome (3 Squadron, Chingford, 3 Wing); No.1599—80-hp Gnome (Eastbourne); No.3150—125-hp Anzani (Eastchurch); No.3683—140-hp Canton Unné (Dunkirk), and No.3998—140-hp Canton Unné (St. Po).

H.F.20 of the Swiss Fliegertruppe. Switzerland had two H.F.20s which were used from 1914 until 1920. Renaud.

Henri Farman H.F.21

The H.F.21 was a modified version of the H.F.20. Changes included an increased wing span and wing area. These changes did not, however, result in any significant improvement in performance. While some may have been supplied to the Aviation Militaire, the type was not produced in large numbers. The Swiss Fliegertruppe obtained a single example in 1914 when a machine owned by a civilian was donated to that government at the outbreak of war. Given serial number 21, it was used as a trainer until 1918.

H.F.21 Two-Seat Reconnaissance or Touring Plane with 80-hp Gnome Lambda
Span 15.50 m; length 8.20 m; height 3.20 m; wing area 39 sq. m
Empty weight 315 kg; loaded weight 695 kg
Maximum speed: 100 km/h

Swiss Fliegertruppe in 1914; the center aircraft is an H.F.21. Reairche.

H.F. 21 with crew. Reairche.

Henri Farman H.F.22

The H.F.22 was another variant of the H.F.20. Its wing span was larger than that of the H.F.20 but smaller than on the H.F.21. An 80-hp Gnome was usually fitted and the performance was slightly inferior to the H.F.20.

HF 28 used H.F.22s. The Type was noted to have an endurance of four hours and could climb to 2,000 m in 30 to 40 minutes. The H.F.22 could carry one machine gun or two obus. A single armored H.F.22 was also used by the escadrille. The pilots considered the H.F.22 to be a significant improvement over the H.F.20s they had flown previously.

Foreign Service

Argentina
The Argentine naval air service purchased three H.F.22s beginning in 1916; They were allocated to the Naval Balloon and Aircraft Depot and School and were withdrawn from service in 1918.

Belgium
The Belgian air service acquired a single H.F.22 built by the Belgian firm of Jero.

Denmark
Denmark acquired four H.F.22s in 1915. They were given serials HF 1 through HF 4. These machines had been built in Sweden (see entry below).

Greece
Two H.F.22s were obtained by Greek naval air service near the end of 1914. They were based at Paleo Faliro Bay and used as trainers.

Netherlands
The Dutch government purchased H.F.22s that served from

Swedish H.F.22 of the SW 10 series. Courtesy Leonard Andersson.

H.F.22 armed with a machine gun. The H.F.22 could carry one machine gun or two obus. The pilots of HF 28 considered the H.F.22 to be a significant improvement over the H.F.20s they had flown previously. MA38345.

1913 until 1919. They were assigned serials LA 6, 7 (in 1918 given serial HF-10), 8, 9 (HF-11), 10 (HF-12), 11 (HF-13), 15 (HF-14), 16 (HF-15), 17, 18 (HF-16), 19, 20 (HF-17), 21 (HF-18), 26 (HF-19), 27 (HF-20), 30, and 32 (HF-25).

Russia

Russia acquired a small number of H.F.22s in 1915. They were assigned to training units and flown by students who had mastered the H.F.20. At least two were in service with the 1st Military School of Pilots in 1920 and one served as late as June 1923.

Sweden

Sodertelge Verkstaders Aviatikavdelning (SW) produced 14 H.F.22s and H.F.23s (see below) under license in 1914 and 1915. They were designated SW 10s and three were supplied to the Flygkompaniet (Army Aviation Service). The individual aircraft histories are as follows:

H.F.22 Two-Seat Reconnaissance or Touring Plane with 80-hp Gnome Lambda
Span 15.58 m; length 8.20 m; height 3.20 m; wing area 46 sq. m
Empty weight 385 kg; loaded weight 710 kg
Maximum speed: 105 km/h; climb to 2,000 m in 32 minutes

1. SW 10, serial no.8—delivered in 1914; SOC 1916.
2. SW 10, serial no.12—delivered in 1914; destroyed in a crash in February 1916.
3. SW 10—serial no.14—delivered 1914; crashed on 13 May 1916 and subsequently SOC.

United Kingdom

The Royal Flying Corps acquired one H.F.22 in March 1913 when the RFC purchased a number of planes owned by the Graham-White firm. It was assigned serial number 434. The aircraft was sent to the CFS and was later lost in a crash.

Eight H.F.22s accompanied RNAS No.1 Squadron (formerly the Eastchurch Squadron, later No.1 Wing) when it was sent to the Aegean in early 1915 to support the Gallipoli operation. However, they could not carry an adequate payload, were too slow, and suffered from the severe weather; consequently they were confined to single-seat reconnaissance. When six more H.F.22s arrived in May 1915 they were immediately returned as unsatisfactory. In July 1915 the improved H.F.27s replaced the H.F.22s still in service.

H. Farman H.F.22 Floatplane

The floatplane version of the H.F.22 was widely used during the first year of the war. The H.F.22s were equipped with twin floats in place of the standard undercarriage and a third float beneath the rudder. Examples were obtained by the Aviation Maritime.

Foreign Service

Denmark

Denmark received four H.F.22s in 1915.

Greece

The Greek naval air service acquired four H.F.22s in 1914 and used them to form its first naval flight. This unit was transferred to the island of Thasos, where it was redesignated the Thasos Flight. It served alongside No.2 Squadron RNAS and was under British naval command. The H.F.22s raided a number of Turkish targets. In early 1917 the unit returned to Thasos and was redesignated 2nd Flight. By mid-1917 the last H.F.22 had been withdrawn from service.

Italy

H.F.22s were built under license by the Savoia firm and in Italian service were designated H.F.22-H. Examples were assigned to the Regia Marina and were based at Taranto. Others were supplied to the seaplane training base at Trasimento.

Netherlands

The Netherlands naval air service purchased seven H.F.22 floatplanes. The earliest entered service in July 1914 and the last was withdrawn in 1921. The H.F.22s were assigned serials MA-1, M-1, M-2, A-3, A-4, A-5, and A-6.

Russia

Russia acquired a number of H.F.22 seaplanes; these were designated H.F.22bis. They were used by naval units in the Black Sea. During the winter months some of them had their floats replaced by skis.

Switzerland

Switzerland received four H.F.22s in 1915.

H.F.22 (Sometimes Referred to as H.F.22bis) Two-Seat Reconnaissance Floatplane with 80-hp or 100-hp Gnome Lambda
Same dimensions as H.F.22
Maximum speed: 90 km/h; climb to 2,000 m in 40 minutes

Savoia-built H.F.22-H Two-Seat Reconnaissance Floatplane with 80-hp Gnome Lambda
Span 15.50 m; length 8.80 m; height 3.80 m; wing area 44.50 sq. m
Empty weight 525 kg; loaded weight 770 kg
Maximum speed: 90 km/h; climb to 1,000 m in 25 minutes; endurance 3 hours 30 minutes

United Kingdom

The Royal Naval Air Service purchased H.F.22H seaplanes built by the Aircraft Manufacturing Company at Hendon. They were: Nos.96–100: 80-hp Gnome (based at the naval air stations at Southampton and Yarmouth); No.102: 70-hp Gnome; No.110: 100-hp Gnome (Grain); Nos.139–144: 140-hp Gnome (Grain, Felixstowe, and Yarmouth); No.156: 80-hp Gnome (Yarmouth); Nos.886–887: 80-hp Gnome (Eastbourne and Calshot); No.915: 80-hp Gnome (Calshot).

Henri Farman H.F.23

This version of the H.F.20 was powered by a 80-hp Gnome Lambda. It had a considerably larger wing than the standard H.F.20 and was produced in both landplane and floatplane versions. Later versions were fitted with a 160-hp Gnome engine.

Nine H.F.23s were used by the Marinens Flygvasennde (Swedish naval air service). The first was obtained in 1913 and probably was used as a pattern aircraft for the Sodertelge Verkstaders Aviatikavdelning (SW), which built at least seven under license. These were designated SW 11s. An additional two H.F.23s were obtained directly from Farman. Individual aircraft histories are as follows:

1. H.F.23—obtained 1913 and designated F 1. It is believed to have been given serial S 21. Crashed February 1917.
2. SW 11—obtained 1914 and designated F 2. Serial number S 25. Crashed 18 September 1916 and SOC (struck off charge) in October 1916.
3. SW 11—obtained 1914 and designated F 3. It was given serial S 26 and redesignated serial no.3 in 1917. SOC October 1920.
4. H.F.23—obtained 1914 and designated F 4. It was given serial S 27. Crashed 18 July 1915.
5. SW 11—obtained 1914 and designated F 5. It was probably given serial S 244 and was redesignated serial no.5 in 1917. Crashed 12 February 1920 and SOC in October 1920.
6. SW 11—obtained in 1915 and designated F 6. It crashed on 19 October 1916 and was SOC two months later.
7. SW 11—obtained in 1915 and designated F 7. It crashed on 15 July 1915.
8. SW 11—obtained in 1915 and designated F 8. It was given serial no.8 in 1917 and was SOC in October 1920.
9. SW 11—obtained in 1915 and designated F 9. In 1917 it was given serial no.9. It crashed on 20 May 1919 and was SOC in August 1920.

A single H.F.23, given designation F 10, was obtained by the army in 1914. It was withdrawn from service in 1916.

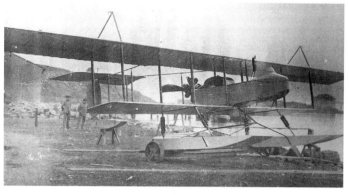

Above and top: H.F.23 floatplanes in Swedish service. Leonard Andersson.

H.F.23 Two-Seat Reconnaissance Plane with 80-hp Gnome Lambda
Span 18.08 m ; length 8.75 m; height 3.20 m; wing area 45 sq. m
Empty weight 390 kg; loaded weight 840 kg
Maximum speed: 95 km/h; climb to 2,000 m in 30 minutes

Henri Farman H.F.24

The H.F.24 was similar to the H.F.20, but the central nacelle was mounted on the top wing instead of the lower. The nacelle held the pilot, engine, and fuel and oil tanks and was also shorter and smaller than those used on previous versions of the H.F.20. It is unclear what advantage this layout would have had over the previous configurations, although the rearward view of the observer and pilot would certainly have been improved. The engine was an 80-hp Gnome Lambda. The plane was built in 1913.

The Farman H.F.24 represented a significant departure from Henri Farman's preceding designs. It was a small sesquiplane intended as a trainer or aerobatic machine (to perform loops). The lower wing span was only 3.75 m. The undercarriage was unique for a Farman design, featuring two independent wheels in order to ensure stability on the ground. As far as can be determined no H.F.24s were purchased by the Aviation Militaire and none is known to have been purchased by foreign air services.

Right: H.F.24. This was an aerobatic trainer intended to teach pilots how to perform a loop. MA38570.
Top of facing page: H.F.24 at the 1913 Paris Salon. MA12070.

H.F.24 Aerobatic Sesquiplane with 80-hp Gnome Lambda
Span 11.50 m; length 8.75 m; height 3.750 m; wing area 23 sq. m
Empty weight 250 kg; loaded weight 501 kg
Maximum speed: 115 km/h; climb to 2,000 m in 16 minutes
Probably only one built

Henri Farman H.F.26

The H.F.25 was designed in 1913 as a two-seat touring plane not intended for military use. The next Henri Farman type was the H.F.26, designed as a two-seat reconnaissance machine. The H.F.26 was built in January 1914 and equipped with a 155-hp Canton Unné R9 engine. It is not known if any H.F.26s were used by the Aviation Militaire or other air services.

H.F.26 Two-Seat Reconnaissance Plane with 155-hp Canton Unné R9
Span 16.08 m; length 8.88 m; height 3.52 m; wing area 39 sq. m

Henri Farman H.F.27

The H.F.27 represented the final development of the basic H.F.20 design. In many ways it represented an attempt to correct some of the flaws discovered in the series during the first few months of the war. The H.F.20s had been found incapable of carrying a satisfactory bomb load and their performance had been mediocre. The wood and canvas structure of the H.F.20s had made them vulnerable to weather and general conditions front-line units faced. In fact, H.F.22s sent to units operating in the Mediterranean area were found to be unable to function in such a hostile climate. The H.F.27 featured an all-metal structure that could resist the damage inflicted by hot climes. To handle the increased weight, a more powerful 140-hp Gnome or Canton-Unné R9 was fitted and the wings were enlarged. Finally, the undercarriage was converted to the quadracycle layout used so successfully by the Voisin bombers. This undercarriage not only gave better support for the aircraft's increased weight, but also reduced the risk of crash landings on soft ground. Armament consisted of a single machine gun and a variable bomb load.

The plane first flew in February 1915 and they were offered

The H.F.27 first flew in February 1915 and was offered to the Aviation Militaire in a letter from the Farman firm dated 5 May 1916. MA38544.

H.F.27 also of the 21st Escadrille of the Imperial Russian Air Service. B85.2444.

to the Aviation Militaire in a letter from the Farman firm dated 5 May 1916. This letter, written by Henri Farman, was addressed to General Hirschauer. Farman pointed out that the H.F.27 was being operated in Russia. In testing it had attained a maximum speed of 145 km/h; carried a payload of 400 kg; and could reach 2,000 m in 12 minutes. In another test an H.F.27 carried seven persons (a total of 600 kg) to an altitude of 2,730 meters. Static testing performed in the presence of French officers, including Chalais Meudon and Capitaine Provost, established that the aircraft had a load coefficient of five. Flight tests revealed the aircraft was easy to land, had good maneuverability, and the crew's field of vision was good. However, the H.F.27s were probably not assigned to French escadrilles. The units assigned to the A.F.O. in 1915 were, instead, equipped primarily with Voisin 5s and M.F.11s.

At least one example of the H.F.27 was tested with a 240-hp Renault engine.

Foreign Service
Greece

Four H.F.27s were used by the Hellenic "Z" Squadron based in the Aegean Sea area. The first Greek pilots to be lost in combat during the First World War (I. Chalkias, B. Lazaris, and D. Argyropoulos) were killed while flying H.F.27s.

Russia

The Imperial Russian Air Service acquired a large number of H.F.27s, many built by the Dux plant. A total of 50 were built in Russia. The Dux H.F.27s were described as having poor maneuverability and the airframe was unable to withstand severe stress. By March 1917 there were 11 H.F.27s still in service distributed as follows: northern front (1), western front (5), southwestern front (3), and Romanian front (2). Three months later the number had increased to 14 with nine on the western front and five on Caucuses front. By the end of the war most H.F.27s were relegated to training duties. At least one pilot, A.K. Tumansky, used his H.F.27 to destroy a German fighter.

United Kingdom

The RFC and RNAS used a number of H.F.27s in the Aegean area. The RFC was the first to obtain H.F.27s and they were assigned to Nos.3 and 5 Squadrons. Six more were obtained from the Admiralty and these were given serials 1801 through 1806. Five went to No.5 Squadron and one to No.3 Squadron. Eight more were ordered in November 1914 and appear to have been sent to No.26 Squadron in mid-1915. Only three could be assembled from the parts that arrived and these were used for reconnaissance from Mbagui and, later, Morogoro. The RFC was satisfied with the aircraft's performance and another 20 were obtained from the RNAS in July 1915. These served with the Air detachment of the Union Expeditionary Force, No.26 (South African) Squadron, Aden Flight of No.31 Squadron, No.31 Squadron in India, and one was used by No.30 Squadron in Mesopotamia. The H.F.27s were used for reconnaissance and bombing and their durability can be assessed from the fact that the type remained in service until 1918.

The RNAS acquired its first H.F.27s in July 1915. The type was also based in the Aegean and used as a bomber. The H.F.27s were assigned to Nos.1, 2, and 3 Squadrons (later Nos.1, 2, and 3 Wings RNAS). One dropped a 500-lb bomb on Turkish barracks in December 1915. Bombing raids were also staged by the RNAS units in Belgium. At least two submarines were attacked and airship LZ-38 was destroyed in its shed by RNAS H.F.27s.

The RNAS aircraft were: Nos.3617–3636: 140-hp Canton Unné (1 Wing); Nos.3900–3939: 150-hp Canton Unné (2 Wing at Imbros and RFC Force D); Nos.8238–8249: 140-hp Canton-Unné (Nos.1 and 3 Wings); No.9099: 160-hp Canton Unné (Calshot); Nos.9134–9153: 160-hp Canton Unné (2 Wing); Nos.9251–9275: 150-hp Canton Unné; only 9251 delivered (two engines ?); N3000–N3049: 150-hp Canton Unné (2 Wing, RFC).

Henri Farman H.F.27

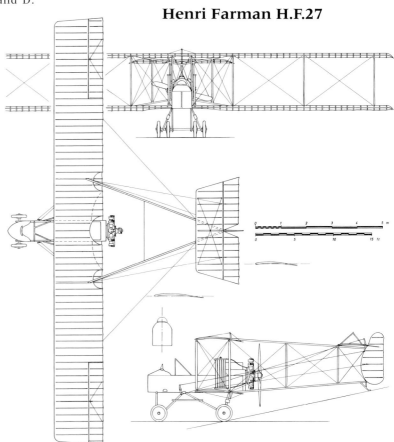

H.F.27 Three-Seat Reconnaissance Plane with 155-hp Canton-Unné R9
Empty weight 770 kg; loaded weight 1,170 kg
Maximum speed: 147 km/h at sea level; 145 km/h at 2,000 m; 142 km/h at 3,000 m; climb to 2,000 m in 12 minutes; climb to 4,000 m in 30 minutes (with a crew of three); ceiling 4,800 m; endurance 2 hours 40 minutes
Armament: one 0.303 Lewis gun and up to 250 kg of bombs

H.F.27 Three-Seat Reconnaissance Plane with 240-hp Renault
Span 16.147 m; length 9.21 m; height 3.45 m; wing area 62 sq. m
Maximum speed: 160 km/h; endurance 7 hours (with 500 kg payload), 8 hours (440 kg), 11 hours (320 kg), 15 hours (200 kg)

Although of poor quality, this rare photograph shows an H.F.27 of the 21st Escadrille of the Imperial Russian Air Service (IRAS) taking off near Baranovitch. B85.2445.

Henri Farman H.F.30

The H.F.30 was a derivative of the H.F.27 and first appeared in December 1915. As with the preceding H.F.27, the H.F.30 was usually powered by a 150-, 155-, or 160-hp Canton-Unné (Salmson) CU X-9 engine. Unlike the H.F.27, the H.F.30 dispensed with the quadracycle gear layout and instead was fitted with a pair of mainwheels underneath the lower wings. The nacelle of the H.F.30 was enlarged and suspended between the two wings, nearer to the top wing. There were two types of fuselage gondolas; one had a blunt nose and seated the observer in front, but the most widely used variant had a more streamlined nacelle and had the pilot in the front seat.

The fuselage nacelle was constructed of four ashwood longerons and covered with plywood and aluminum. The two main wing spars were of steel tube while the wing ribs were of plywood and pine. The wing struts, tail assembly, and undercarriage were made of steel tubing. The radiators were placed on either side of the fuselage. In some H.F.30s a machine gun was mounted on a tubular beam set above the pilot's head; the gun was controlled by the observer. In the variants in which the observer was seated in the nose the machine gun was set on a rotating bracket attached to the observer's seat.

The H.F.30bis was a major variant of the H.F.30. This machine had a 160-hp Salmson engine but instead of having exhausts pipes passing from each of the cylinders there were only two outlet pipes that directed the exhaust fumes upward.

It does not appear that the H.F.30 was adopted for use with the Aviation Militaire, but it did see widespread use with the Imperial Russian Air Service. The type was well-liked by the Russians primarily because of the aforementioned reliability of the Salmson engines; indeed, it appears that approximately 400 were built by Dux during the war. The H.F.30 was one of the most important reconnaissance planes in Russian and Soviet service and remained in front-line units from 1916 to 1921. It subsequently served in training units from 1921 to 1925. There were 70 H.F.30s on strength in 1917 as compared to a total of 40 H.F.27s, M.F.7s, and M.F.11s. On 1 March 1917, 54 H.F.30s were in service with squadrons located on the following fronts: northern front (14), western front (9), southwestern front (23), and Romanian front (8). On 1 June 1917 there were 94 H.F.30s on the following fronts: northern front (25), western front (19), southwestern and Romanian front (35), Caucuses front (15).

In 1921 the remaining H.F.30s (approximately 147) were reported to be on the Caucuses, Turkestan, and Ukrainian fronts. By 1922 only about half this number were still in service and were assigned to the 2nd, 7th, 11th, 16th, and 18th

H.F.30 Two-Seat Reconnaissance Plane with 160-hp Canton-Unné CU X-9
Span 15.80 m; length 10.66 m; height 3.20 m; wing area 49 sq. m
Empty weight 700 kg; loaded weight 1,050 kg
Maximum speed: 155 km/h; climb to 2,000 m in 9 minutes
Armament: one machine gun

H.F.30 Two-Seat Reconnaissance Plane with 160-hp Canton-Unné (Salmson) CU X-9 Built by the Dux Factory in Russia
Span 15.81 m; height 3.20 m; wing area 49 sq. m
Empty weight 830 kg; loaded weight 1,180 kg
Maximum speed: 136 km/h; climb to 1,000 m took 5 minutes; ceiling 4,500 m; endurance 4 hours; range 450 km
Armament: one machine gun
Approximately 400 built by Dux

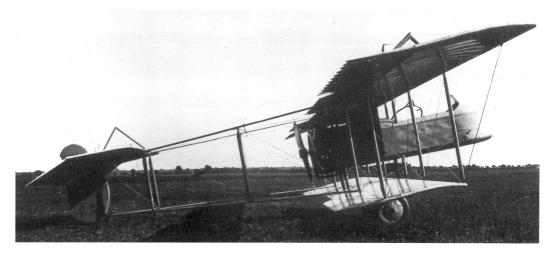

H.F.30. B91.1291.

Henri Farman H.F.30

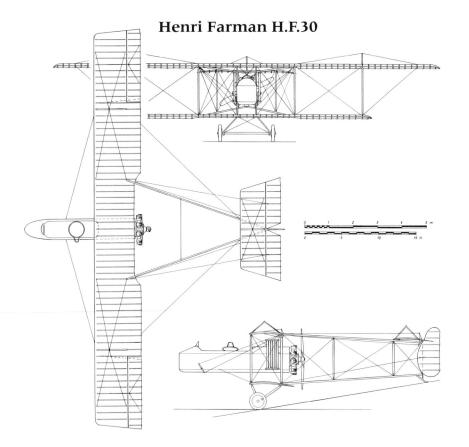

H.F.30 of the Imperial Russian Air Service at Minsk. The H.F.30 was one of the most important reconnaissance airplanes in Russian and Soviet service and remained in front-line units from 1916 until 1921. B85.2563.

Aviaotryady, the Teoreticheskaya Shkola (Leningrad), and No.1 Flying School at Kacha.

There were only 17 H.F.30s on strength in 1924 and these were assigned to the Flying Otryad of the Military School of the KVF (Red Air Fleet), the 1st Military School of Pilots, and the 16th Otdel'nyi razvedivatel'nyi aviaotryad at Irkutsk. A few remained on the civil register as late as 1929.

Examples of the H.F.30 were captured by Czechoslovakia, Poland, Estonia, and the Ukraine. A single H.F.30 was captured by the Estonians in January 1919. It was the first aircraft to be obtained by the Estonian aviation company and was, consequently, given the serial number 1. It was used to bomb Russian positions near Narva and also dropped propaganda leaflets. It was subsequently turned over to the White Russians. The Ukrainian air service had six H.F.30s on strength in 1918. Known serials are 1253, 1424, 1624, 1674, and 1686. The ZUNR (West Ukrainian Peoples Republic) had a single Farman, probably an H.F.30, on strength in 1918. The Siberian Air Fleet of the Far East Republic had two F.30s (serials 8 and 9) on strength in 1920.

H.F.30 of the 21st Escadrille Imperial Russian Air Service in August 1916. B86.0012.

Maurice Farman M.F.7

The M.F.7 was the culmination of Maurice Farman's pre-1914 aeronautical design work. While its wings were similar to the H.F.20, the M.F.7 had a unique forward elevator positioned on curved extensions of the landing skids. The skids were reinforced by angled braces extending from the top and bottom wings. A biplane elevator located at the rear was supported by twin booms attached to the trailing portions of the top and bottom wings. Twin rudders were positioned between the top and bottom elevators. As with the H.F.20, a central nacelle which carried crew, fuel, and engine was attached to the bottom wing. However, in the M.F.7 this nacelle was more rounded and shorter than Henri's design. Like the H.F.20, the crew's location in the nose gave them an unexcelled forward, side, and downward view. Power was supplied by a 70-hp Renault air-cooled engine; sometimes a 100-hp Sunbeam was fitted. The engine was mounted in a pusher configuration in the rear of the nacelle. Dual mainwheels were mounted on the skids and there were rubber shock absorbers. Steel braking skids were located at the rear of the main skids as well as on the bottom portion of the rear elevator. The arrangement of the landing gear made the M.F.7 very easy to land; this would prove of great value when the plane was operating from poorly prepared airfields or was being set down by student pilots. Lateral control of the M.F.7 was achieved through ailerons on both the top and bottom wings. While they drooped when the plane was on the ground, they were held in a neutral position by the airflow during flight. The control stick controlled the elevators and ailerons. Some M.F.7s were fitted with a Éteve machine gun mount in front of the crew nacelle.

The M.F.7 first flew in 1913. Because of its pusher configuration, stable flight characteristics, ease of landing, and structural strength, the M.F.7 was accepted for service with the Aviation Militaire as well as a large number of foreign air services. In French service the standard M.F.7 was designated the Army Type 1. MF 2 noted that the M.F.7s were troublesome in service because excessive engine vibration resulted in damage to the engines.

M.F.7 of Lieutenant Happe in 1913. B88.2988.

Variants

M.F.7bis: This plane was designated Army Type 3. It was a dual-control trainer, later given the SFA designation F 7 D2. The dimensions were identical to those of the standard M.F.7.

M.F.7bis: A version of the M.F.7 powered with an 80-hp De Dion engine and designated Type 4. The Type 5 was identical to the Type 4 but was a dual-control trainer. The Type 5 also carried the SFA designation F 7 D2.

M.F.7 seaplane: This was an M.F.7 with an increased wing span and area. The fuselage was shorter and the overall height was decreased. Twin floats were fitted at the landing gear attachment points. The engine remained the standard 70-hp Renault.

M.F.7ter: This version dispensed with the forward stabilizer and may have been the prototype for the M.F.11. The engine was a standard 70-hp Renault 8b. Dimensions were identical to the standard M.F.7 except the fuselage length was decreased. Deletion of the drag-inducing forward stabilizer and skids resulted in a 5 km/h increase in maximum speed.

Operational Service
Army Cooperation

There were six escadrilles equipped with M.F.7s at the outbreak of the First World War. While these units were equipped with M.F.7s or M.F.11s, there were often a few H.F.20s and Voisin 3s assigned to them. The Farman-equipped escadrilles were:

MF 2, formed in 1912 and based at Verdun at the outbreak of war. It was commanded by Lieutenant Bretey and attached to the 3rd Armée. MF 2 carried out the first French wartime reconnaissance mission. By early 1915 MF 2 was assigned to the 5th C.A. at Clermont near the Argonne.

MF 5, created from VB 5 in August 1914; it had six M.F.7s. The escadrille was operational near Épinal until it moved to Belfort in September. Later, MF 5 moved to Saint-Mihiel. In March MF 5 was assigned to the 1st C.C. and based at Commercy and Toul.

MF 7, formed in February 1914 from escadrille HF 7. It was active over the region of Épinal and was transferred to Belfort in September. Later it was based near Saint-Mihiel. MF 7 was assigned to the 6th C.A. and based at Verdun in March 1915. The next month it was in the 1st Armée sector under command of Capitaine Bordes.

MF 8, based at Nancy at the beginning of the war under command of Capitaine Fassin. At the outbreak of hostilities MF 8 was assigned to the 2nd Armée and moved to the region of Château-Salins. It moved to the Somme front in October. In March 1915 it was based at Vaux.

MF 16, formed in 1914 and mobilized at Camp Châlons under Capitaine Mauger-Devarennes. It was initially assigned to the 3rd Armée but moved to the 6th Armée sector and was based at Picardie at the end of August. In March 1915 MF 16 was assigned to the 6th Armée and based at Vauciennes.

MF 19, formed from HF 19 in early 1915. It was assigned

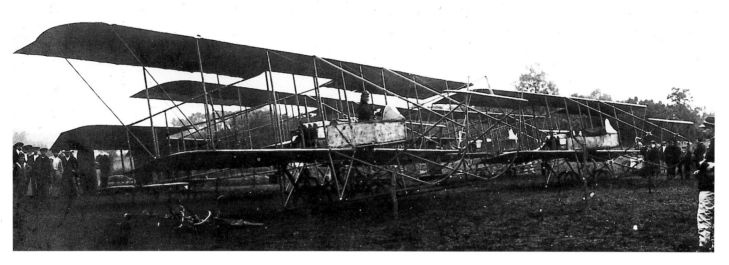

An M.F.7 escadrille commanded by Capitaine Barès in 1912. B87.786.

to the 13th C.A. and based at Amiens. The unit carried out several bombing missions on Montdidier, Saint-Quentin, and Chaulnes. MF 19 was also active in the Verdun sector in March 1916. In November it was assigned to the 10th Armée sector and was based at Pierrefonds.

MF 20, formed in 1913 at Dijon and initially assigned to the 2nd Armée. In March 1915 it was based at Villers les Bretonneux.

MF 25, formed in September 1914 at fort du Buc under command of Capitaine Rossner. It was assigned to the 3rd Armée on the Argonne front, where it performed reconnaissance and artillery spotting missions over the Verdun sector. Gradually, however, MF 25 began to concentrate on bombing missions; in December it dropped a ton of bombs on the German lines. In March 1915 MF 25 was assigned to the 3rd Armée but was functioning as an independent bomber unit. Based at St. Menehould, the escadrille re-equipped with M.F.11s early in 1915.

MF 33, formed in October 1914 at Tours under command of Capitaine Bordage and based at Amiens. In November it was placed under the command of the Belgian army. In January 1915 it was reassigned to the 9th C.A. MF 33 was active during the Battle of Ypres. In April 1915 it was assigned to the 9th C.A. and based at Poperinghe.

MF 35, formed in October 1914 at Saint-Cyr with a mix of Voisin 1914s, H.F.20s, and M.F.7s and commanded by Capitaine Reimbert. It moved to Saint-Pol-sur-Mer, where it participated in the Battle of Yser. On 20 November it was assigned to the 32nd C.A. at Poperinghe. In December MF 35 moved to Hondschoote.

MF 40, created in March 1915. It was assigned to the 4th C.A. and was under command of Capitaine La Morlange.

Some M.F.7s were assigned to escadrilles flying M.F.11s; for details of these units see the entry under M.F.11.

The stability and robustness of the M.F.7 made it an ideal plane for reconnaissance and artillery spotting. The M.F.7s often carried cameras to record enemy troop movements and to assist in the production of accurate maps of the front lines. For artillery spotting the crews often used either flares or colored flags to signal the French artillery units on how to correct their aim. However, T.S.F. units were occasionally carried; these permitted the crews to rapidly signal any necessary alterations of fire. Unfortunately, the transmissions were often unintelligible and in some cases the T.S.F. aerials broke off in flight. An example of the often mundane daily operations of a Farman escadrille can be seen from MF 21's activity report for 3 February 1915. On that day MF 21 flew two reconnaissance missions, three artillery spotting missions, and took several photographs of enemy camps. It should be noted that these missions were often undertaken by front-line units that were seriously under-equipped; for example, MF 20 had only five planes on strength on 10 February 1915.

Bombing

It would be incorrect to conclude that the M.F.7 escadrilles limited their operations to army cooperation duties. Some of the earliest bombing missions were undertaken by these units.

A specialized bombing unit under the command of Capitaine Maurice Happe was formed at Belfort under the designation Escadrille de Bombardement de Belfort. The unit was equipped with M.F.7s. On 19 January 1915 Happe used his machine gun-equipped M.F.7 to destroy a German observation balloon at Remingen and dropped 4,000 flechettes on enemy troops in the area. During an attack on the station at Bollarville on 11 February, Happe's M.F.7 was attacked from behind by a German plane described as an Aviatik. Happe tried desperately to give his observer a clear shot at the German. Finally, the Aviatik was driven off without apparent damage to either plane. This battle pointed out the vulnerability of the M.F.7 because of its pusher configuration and lack of maneuverability. The gunpowder factory at Rothweil was bombed with devastating effectiveness on 3 March 1915. Three days later Happe returned to the powder factory, which was now heavily defended by anti-aircraft gun batteries. Despite the formidable defense, Happe was able to bomb the plant and spend ten minutes over the site to ensure an accurate attack. His plane was hit by 12 projectiles but neither crewman was injured, and the sturdy M.F.7 was able to return safely to the base at Belfort. The Zeppelin works near Friedrichshafen were bombed on 28 April 1915. Two of the six bombs hit a hangar and may have damaged two airships inside. On 29 May 1915 the unit at Belfort was redesignated MF 29. It had six M.F.7s on strength at that time, but by 3 July these had been replaced by the superior M.F.11s.

Fighter

M.F.7s, while never intended for use as fighters, participated in a number of aerial combats during the early part of 1915. The Escadrille de Bombardement de Belfort used M.F.7s to attack enemy planes and provided fighter protection for other M.F.7 bombers. MF 52 flew fighter patrols during April and on the 22nd of that month succeeded in forcing two German planes to return behind their lines.

The M.F.7 remained in front line service until early 1915, but Maurice Farman had developed an upgraded plane, designated the M.F.11. It was an improvement over the M.F.7 and was quickly accepted for service by the Aviation Militaire. By mid-1915 the M.F.11 had superseded the M.F.7 in front-line escadrilles. The M.F.7s were then reassigned to training units. The docile Farmans were used to train student pilots as late as 1917. However, despite the fact that the superior M.F.11s became the main equipment for operational escadrilles, a GQG document dated 1 February 1916 showed there were still 271 M.F.7s listed on strength with the Aviation Militaire.

Foreign Service
Australia
Australia acquired two M.F.7s for the Central Flying School. One, given serial CFS 7, had an interchangeable wheel or float undercarriage. As a floatplane it was shipped to New Guinea in 1914 and remained there through 1915. It was struck off charge in 1917. The second M.F.7 was a landplane designated CFS 15 and used at the Central Flying school beginning in 1916. One M.F.7 was assigned to the half flight in Mesopotamia. Designated 1FC 2, it was destroyed at Baghdad on 13 November 1915.

Belgium
Belgium had several M.F.7s on strength at the beginning of the war. They were assigned to Escadrille 2 at Namur and Escadrilles 3 and 4 at Brasschaat. In October 1914 these units moved to Ostend and 11 days later to Saint-Pol. In March 1915 Escadrille 1 (at Saint-Idesbald), Escadrille 2 (at Saint-Idesbald) and Escadrilles 5 and 6 (at Houthem) had M.F.7s on strength. By 1916 they had been replaced by M.F.11s.

Denmark
The Danish Haerens Flyveskole (Army Flying School) had four M.F.7s on strength during the war. They were given serials MF 1, MF 2, MF 3, and MF 4, and the last machine was struck off charge in 1922.

Greece
The Greek Aposasma Aeroploias Ipirou (Epirus Aviation Attachment) was assigned two M.F.7s in November 1912. The M.F.7s flew reconnaissance missions over Turkish positions during the Balkan Wars and were particularly active during the siege of the Bizani fort. On one such mission a Farman was hit by more than 20 bullets. The M.F.7s flew more than 240 km on some reconnaissance missions. In addition, they dropped hand grenades on Turkish troops. One Farman was destroyed on 11 November 1912. The remaining M.F.7 survived the war and was subsequently assigned to the Aeroporiki Sholi (Aviation School) at Thessalonica. On source suggests that up to six more M.F.7s were purchased in 1916 by the Greek Air Service and served as trainers until 1920.

One Greek M.F.7 was converted to a floatplane and gained a measure of fame in early 1913 (probably 6 February). Piloted by an army officer, Michael Moutoussis, with a navy officer, Ariastides C. Moraitinis, as observer, it flew a reconnaissance mission over the Dardanelles and unsuccessfully tried to bomb Turkish warships. This was almost certainly the first maritime combat mission ever flown by a serving naval officer of any nationality.

Italy
Italy acquired 12 M.F.7s in 1913. Subsequently, the type was built under license by the Societa Italiana Transaerea (S.I.T.). These were equipped with 70-hp Renault engines imported from France. The S.I.T.-built M.F.7s entered service in mid-1913. By the beginning of the war there were four squadriglias formed on M.F.7s: Nos.9ª, 10ª, 11ª, and 12ª. When Italy entered the war in 1915 the M.F.7s (along with M.F.11s) were assigned to the 1ª, 2ª, 4ª, 6ª, 10ª, 11ª, 12ª, and 13ª reconnaissance and bombing squadriglias. Two army cooperation squadriglias, Nos.6ª and 7ª, were also equipped with Farmans. By 1916 most of the M.F.7s had been assigned to training units and were replaced in front-line squadriglia by M.F.11s.

Japan
Japan imported five M.F.7s from mid-1913 through 1914. They were found to have excellent stability and their sturdy construction made them very dependable in service. The Tokyo Army Arsenal built M.F.7s and their 70-hp Renault engines under license, nine being completed by the end of 1914 as two-seat trainers.

Four M.F.7s along with a Nieuport 6M took part in the Tsingato campaign as part of the army's Provisional Air Corps. They flew 86 sorties from 4 September through 6 November, dropping a total of 44 bombs and using T.S.F. to direct artillery fire. On 27 September, in history's first true aeronaval battle, some of these aircraft attempted unsuccessfully to bomb the Austro-Hungarian cruiser *Kaiserin Elisabeth* and two German warships. Other aerial attacks on enemy ships were made later, but the often-repeated claim that a German minelayer was sunk is incorrect.

One of the Farmans may have taken part in one of the war's first aerial combats, a reported exchange of pistol shots with the German garrison's sole airplane, a Rumpler Taube. Reports about this supposed incident, however, are unclear and contradictory, and some sources doubt it actually happened.

The Japanese Navy purchased a three-seat M.F.7 in 1913 and it, together with three two-seat H.F.7s, operated at Tsingtao, initially from the seaplane carrier *Wakamiya Maru*, a crudely converted merchant ship. After that vessel was disabled by striking a mine, the Farmans flew from an improvised beach base. Altogether, the naval Farmans flew 49 sorties and dropped 199 bombs.

After the brief Tsingtao campaign the M.F.7s were used for more pacific exploits. One made a non-stop flight from the Tokorosaw army airfield to Osaka on 1 April 1915 in six hours 30 minutes.

A single M.F.7 was modified to accept a 100-hp Daimler engine by the Provisional Balloon Research Society (PMBRA). One M.F.7 was built by the PMBRA in 1913 and given a 70-hp Gnome rotary; a similar machine was built by the artillery arsenal. These two machines were faster and had a longer range than the other license-built M.F.7s; they were designated Kaishiki No.5 and No.6. A modified M.F.7 was created from one that crashed in July 1914. Designated the 7th Type Mo. 1913, the plane eliminated the front elevator and skids; a change permitting a machine gun to be placed in the nose. This became the first Japanese plane to be equipped with such a weapon. It crashed in May 1915.

Norway
Norway acquired ten M.F.7s, designated FF.1 in Norwegian service. The first three machines were christened, in order of acquisition, *Ganger Rolf*, *Najal*, and *Olav Tryggvasson*. A fourth M.F.7 floatplane was donated to the navy by polar explorer Roald Amudsen. The M.F.7s served until 1921. License-built versions of the M.F.7 were designated FF.2 Shorthorn, FF.3 Hydro, and FF.4 Swan.

Russia
Russia acquired a number of M.F.7s from France and the type may have been built under license. While the M.F.7s were not as widely used as the H.F.20s or H.F.27s, they did see limited service on the western and southwestern Fronts (see section on the M.F.11).

Spain
Spain acquired three Henri Farman Military Types in 1910 and used them as trainers as late as 1914. Pleased with this earlier Farman design, the Servicie de Aeronautica Militar purchased approximately 12 M.F.7s from France and built an equal number under license in Spain. Six M.F.7s arrived in Madrid in

FRENCH AIRCRAFT OF THE FIRST WORLD WAR

M.F.7 (Army Type 1) Two-Seat Reconnaissance Plane with 70-hp Renault 8B

Span 15.52 m; length 11.52 m; height 3.35 m; wing area 60 sq. m
Empty weight 580 kg; loaded weight 855 kg
Maximum speed: 95 km/h; ceiling 4,000 m; climb 15 m/min.; endurance 3 hours 15 min.
Armament: usually a single machine gun and several bombs
A total of 358 M.F.7, 7bis and 7ter aircraft were built

M.F.7bis Two-Seat Reconnaissance Plane (Army Type 2) and Dual-Control Trainer (Army Type 3) with 70-hp Renault 8B

Span 15.52 m; length 11.52 m; height 3.35 m; wing area 60 sq. m
Maximum speed: 95 km/h

M.F.7bis Two-Seat Reconnaissance Plane (Army Type 4) and Dual-Control Trainer (Army Type 5) with 80-hp De Dion

Span 15.52 m; length 11.52 m; height 3.35 m; wing area 60 sq. m
Maximum speed: 95 km/h

M.F.7bis Two-Seat Seaplane with 70-hp Renault 8B

Span 17.92 m; length 9.50 m; height 4.62 m; wing area 65 sq. m

M.F.7ter Two-Seat Civil Airplane with 70-hp Renault 8B

Span 15.52 m; length 9.48 m; height 3.35 m; wing area 60 sq. m
Empty weight 560 kg; loaded weight 860 kg
Maximum speed: 100 km/h

M.F.7 Two-Seat Reconnaissance Plane with 70-hp Renault 8B license-built in Italy by S.I.T.

Span 15.54 m; length 11.50 m; height 3.25 m
Empty weight 580kg; loaded weight 857 kg
Maximum speed: 90 km/h; climb to 1,000 m in 20 min.; endurance 3 hours

M.F.7 Two-Seat Reconnaissance Plane 70-hp Renault 8B Built in Japan by the Tokyo Army Arsenal

Span 15.54 m; length 11.28 m; height 3.45 m
Empty weight 580 kg; loaded weight 855 kg
Maximum speed: 51 kts; ceiling 3000 m; endurance 4 hours
Armament: usually a single machine gun and several bombs
A total of 22 M.F.7s were built by the Army Arsenal; four were built by other manufacturers including PMBRA

M.F.7 Two-Seat Reconnaissance Plane with 70-hp Renault 8B license-built in Spain by Carde y Escoriaza

Span 15.54 m; length 11.35 m; wing area 54 sq. m
Empty weight 580 kg; loaded weight 855 kg
Maximum speed: 95 km/h; climb to 1,000 m in 30 min.; ceiling 4,000 m; endurance 3.30 hours
Twelve built

M.F.7 Two-Seat Reconnaissance Plane with 70-hp Renault 8B License-Built in England by AIRCO and by Brush Electrical Engineering Co.

Span 15.52 m; length 11.52 m; height 3.5 m; wing area 49 sq. m
Empty weight 599 kg; loaded weight 871 kg
Maximum speed: 105 km/h; climb 1,000 m in 15 min.; climb to 2,000 m in 35 min.; endurance 5 hours

May 1913, the first planes to be assigned to the newly created air service. They were given serial numbers 1 through 6. Five of them took part in maneuvers during June 1913 and flew missions from Cuatro Vientos to Villaluenga, Aranjuez, and Guadalajara. In September another airfield was established at Alcalá de Henares to supplement the base at Cuatro Vientos. The M.F.7s were used for flight training, a role for which their flying characteristics made them particularly well suited. The next month another base was formed at Marruecos and four of the M.F.7s were sent there. Operational reconnaissance missions against rebels in Spanish Morocco began on 4 November; on occasion the M.F.7s were used for bombing. One M.F.7 was sent to support the forces under General Silvestre, and a second to the base at Tetuan. Four more M.F.7s arrived in Spain in 1914 and were given serial numbers 7 through 10; they were assigned to Cuatro Vientos. However, by 1915 the M.F.7s had been superseded by the M.F.11 and were relegated once again to training.

The Spanish realized that the beginning of the First World War would severely restrict the availability of planes from France and therefore arranged to build M.F.7s under license by the firm of Carde y Escoriaza at Zaragoza. As previously noted, 12 were built; they were given serial numbers 17 through 28. Toward the end of 1915 some of the M.F.7s were re-engined with 80-hp De Dions. These planes were assigned to the school at Alcala de Henares; the M.F.7s at Cuatro Vientos retained their 70-hp Renaults. By mid-1915 M.F.11s had begun to replace the M.F.7 trainers at Alcala and Cuatro Vientos. Some of the surviving M.F.7s were sent to the school at Los Alcazares. Six

were still in use at the Guadalajara air base as late as 1917. The type remained in service with training units until 1919, the period from 1913 through 1919 becoming known as the "era of the Farman." The M.F.7s were finally replaced by surplus Avro 504s in 1919.

United Kingdom

Great Britain purchased M.F.7s from France and in addition a number were built by the Aircraft Manufacturing Company (AIRCO). They were used primarily as trainers by: 1, 2, 3, 4, 5, 6, 7, 8, 10, 11, 12, 14, 15, 19, 22, 24, 25, 26, 36, 39, 41, 49, 57, and 58 CFS Reserve/Training squadrons. No.8 Depot also had a few M.F.7s on strength. The RFC Plane Flight at Barre was formed in April 1915 and had two M.F.7s on strength; in May two more joined the unit. The planes were used for reconnaissance during May and June.

The Royal Naval Air Service had a number of M.F.7 seaplanes on strength in 1914. These were used as trainers, assigned to the Naval Air Stations at Eastchurch, Felixstowe, Yarmouth, Dundee, and the Isle of Grain.

The M.F.7s in service with the RNAS were Nos.23, 29, 67, 69–73, 95, 113, 114, 115, 116, 117, 146, 188, 888, 909, 910–914, 949, 3001–3012, 8474, 8604, 8605, 8921–8940, N5000–N5059, N5330–N 5439, N5720–N 5749, and N5750–N5759.

Maurice Farman M.F.11

Between the M.F.7 and M.F.11 there were three interim types. The M.F.8 was a seaplane that retained the forward elevator of the M.F.7. The M.F.9 seaplane had a layout similar to the M.F.8 but was smaller. Both the M.F.8 and 9 were intended for civil use. the M.F.10 was a landplane presented at the 1913 Salon. The M.F.8, 9, and 10 were not used by the military services.

The M.F.11 was a refinement of the M.F.7. Built in 1914 (and frequently called the 1914 Farman in contemporary literature), it deleted the forward elevator and elongated landing skids of the M.F.7. It was found that deletion of the drag-inducing forward elevator resulted in an increase in speed and improved maneuverability. The biplane horizontal stabilizer was replaced with a single stabilizer. The wing span was increased by 0.65 m with a concomitant increase in wing area by 3.0 sq. m.

M.F.11. B93.1051.

The length was reduced by 0.50 m as a result of the elimination of the forward elevator and the resulting decrease in the size of the landing skids. The central nacelle was removed from the lower wing and was mounted between the upper and lower wings. Early versions of the M.F.11 were powered by a single 80-hp Renault 8B or De Dion Bouton engine.

The M.F.11s were manufactured as three major subassemblies. The biplane wings were assembled as individual units. The upper wing was 4.00 m longer than the lower wing, and both were fitted with ailerons. The supporting struts were 1.95 m long and the wings had a chord of 2.30 m. The struts were made of ash. The wing spars were constructed of pine and plywood and covered with cotton fabric. The extended portions of the top wing were supported by metal struts mounted at an oblique angle. The shape of the wings was maintained by taut piano wire rigging.

The fuselage nacelle was then attached to the completed wings. The nacelle was constructed of wood over a steel tubular support structure and covered by cotton fabric. The front of the nose was made of aluminum and the forward windscreen of mica. The nacelle was 3.35 m long and 86 cm wide. An 80-hp Renault or De Dion engine was fitted at the rear of the nacelle mounted on two longerons that protruded from the base of the fuselage. The nacelle was fitted 45 cm above the lower wing. A fuel tank with a capacity of 140 liters was placed in front of the engine. The pilot was seated in the rear with the observer/gunner in front. Their leather seats were mounted on a raised platform that was used to store the crew's equipment. A single control stick controlled the ailerons and elevator. Instrumentation consisted of a clock, manometer, and altimeter. There was also a spool on which maps could be stored and then rotated forward to advance the map sheet. The landing gear struts were made of ash. A pair of wheels were attached to each skid and were supported by bungee cords acting as shock absorbers. The wheels were mainly used to maneuver the plane while it was on the ground, while the skids acted to help cushion the shock of landing and also acted as brakes. Each skid was 3.50 m long.

The tail booms were the third major assembly. The booms were made of pine and tightly rigged with piano wire. The twin rudders had a 55 cm chord. The horizontal stabilizer consisted of a fixed portion 5.50 m in length and 1.00 m in chord. A tail skid was attached to the end of each of the booms. Standard armament was a Colt machine gun and a Winchester carbine.

The ability to construct each M.F.11 as three separate subassemblies facilitated production and permitted the Farman factory to build the type in large numbers.

The SFA closely regulated the production of M.F.11s. Modifications in production during 1915 included a new enamel paint and a change in the fabric. The ailerons were displaced lower by altering the pulleys that controlled their movement. The observer was relocated to the front of the plane while the pilot's floorboard was placed over the engine mount. All motors were eventually equipped with mufflers during production.

1914 Variants

1. M.F.11s with 80-hp Renault 8B engines were designated as army types 9, 10, 15. 17, 18, 19, 20, 21, 22, 23, 24, and 25.
2. M.F.11s with 80-hp De Dion-Bouton engines were given army type numbers 6, 9, 12, 14, and 16.
3. M.F.11s with 110-hp Lorraine 6 AM engines were given army type designations 50 and 51.
4. M.F.11bis's with 130-hp Renault 8C engines were given army type designations 28, 29, 30, 31, 32, 34, 35, 36, 37, and 38.
5. M.F.11bis's with 130-hp Renault 8C engines were given army type designation 33. Other M.F.11s with 130-hp Renaults were given Army type designations 28, 29, 30, 31, 32, 33, 34, 35, 36, 37, and 38
6. M.F.11bis's with 100-hp Renault 8Cs were given army designation Type 27.

M.F.11 serial number 470 in August 1915. B85.1189.

1915 Variants

1. M.F.11bis's with 130-hp Salmson A9s were given army designation Type 39.
2. M.F.11bis's with 130-hp de Dion-Bouton 12B engines were given army designation Type 45.

M.F.11 1915 Bomber Versions:

3. M.F.11bis BO with 80-hp Renault R8B had army designation Type 52.
4. MF.11bis BO with 80-hp Renault R8B and a smaller wing was given army designation Type 53.
5. M.F.11bis BO with 130-hp Renault 1R8C was given army designation Type 54.

Operational Service
Army Cooperation

The M.F.11 began to enter escadrille service in early 1915. By 22 August 1915 there were 193 M.F.7s and 11s in service at the front and 12 with training units. A large number of other escadrilles were formed on M.F.11s as 1915 progressed.

MF 1 serves as an example of a standard M.F.11-equipped army cooperation escadrille. It was initially equipped with M.F.11s in February 1915. Assigned to the 33rd C.A. and based at Bruay, it later moved to Villers-Chatel. From this base it participated in the Battle of Artois from 9 through 13 May. The M.F.11s maintained contact with French troops and conducted low-altitude missions. MF 1 was able to locate enemy artillery and machine gun positions by drawing their fire. For these actions it was awarded l'Ordre de l'Armée on 31 May 1915.

During the Third Battle of Artois, the M.F.11s directed artillery fire and reconnoitered behind enemy lines. MF 1 supported the 33rd C.A. during its conquest of Souchez. During an advance near Farbus, MF 1 dropped messages to troops to keep them abreast of the rapidly changing situation.

The slow speed of the M.F.11s resulted in repeated requests by the pilots of MF 1 for more modern machines. When none was forthcoming the escadrille adopted the winged snail insignia as a means of satirizing the performance of their aircraft.

As a result of MF 1's reconnaissance flights, it was discovered that the Germans were concentrating artillery and troops in front of the 33rd C.A.'s lines. In response, a ground attack was launched on 8 February which enabled the 33rd C.A. to retain its position.

MF 1 was active during the Battle of Verdun in 1916 and concentrated its reconnaissance missions primarily on the right bank of the Meuse. For its superlative work MF 1 received a second citation from the Armée.

MF 1 remained active during the Battle of the Somme in July 1916 and gave effective support to the 33rd C.A. In the beginning of 1917 the 33rd C.A. and MF 1 moved to the 10th Armée front. By April the M.F.11s had at last been largely supplanted by A.R.1s and the escadrille's designation changed to AR 1.

MF 2 was assigned to the 5th C.A. It was based at Clermont near Argonne in March 1915, where it remained until March 1916. A German aerial bombardment forced MF 2 to evacuate its airfield in March and move to Autricourt, where it remained until July. Before transferring to the Somme front in September 1916, MF 2 was based at Camp Mailly to practice liaison with the infantry. It remained on the Somme front under the command of Commandant de Vergnettes until November 1916, when it moved to Châlons-sur-Marne. MF 2 was assigned to the 5th Armée and in February 1917 received a new commander, Lieutenant Collard. MF 2 was active in the Battle of Chemin des Dames and re-equipped with A.R.1s in October 1917.

MF 5 was under the command of Capitaine Bordes in April 1915 and in June passed to the 2nd Armée sector near Verdun.

M.F.11 at the Ecole d'Aviation at Étampes in September 1915. B86.2474.

In November 1916 It was assigned to the 2nd C.A. and was based at Lemmes. MF 5 moved to the 4th Armée sector in February 1917 and then to the 33rd C.A. in June 1917. MF 5 re-equipped with Sopwith 1½ Strutters in September 1917.

MF 7 was assigned to the 6th C.A. and based at Verdun in March 1915. Initially based on the Meuse front, the escadrille moved to Champagne where it flew reconnaissance missions beginning in September 1915. In June 1916 MF 7 was based at Courtisols and moved to Verdun where it guided artillery fire against the forts at Vaux and Douaumont. After a brief period of rest at Fère-en-Tardenois, it moved to the Somme in September 1916. In November 1916 it was based at Bois des Tailles. In January MF 7 moved to the 5th Armée sector in preparation for the Chemin-des-Dames offensive; it was under the command of Capitaine Saqui-Sannes. MF 7 re-equipped with Sopwith 1½ Strutters in mid-1917.

MF 8 was assigned to the 2nd Armée and in July 1915 was based at Tilloy. In January 1916 it was assigned to the 11th C.A. and participated in the Battle of Verdun. In November 1916 MF 8 was assigned to the 2nd C.A. and based at Vadelaincourt in the 2nd Armée sector.

MF 14 was given M.F.11s in November 1915. It was assigned to the 7th Armée and was active over Hartmannswillerkopf and Richackerhoff. In November 1916 it was in the 7th Armée sector, where it was assigned to the 76th D.I. and based at Griecourt.

MF 16 was assigned to the 6th Armée and was based at Vauciennes in March 1915. In August it moved to the 10th Armée sector and was active in the Battle of Artois. In July 1916 the unit re-equipped with Farman F.40s.

MF 19 was formed from HF 19 in early 1915; in April it was assigned to the 13th C.A. and based at Amiens. The unit carried out several bombing missions on Montdidier, Saint-Quentin, and Chaulnes. MF 19 was also active in the Verdun sector in March 1916. In November 1916 it was assigned to the 10th Armée sector and was based at Pierrefonds.

MF 20 was assigned to the 2nd Armée and based at Villers les Bretonneux in March 1915. It was assigned to the 14th C.A. in August 1915. In March 1916 the escadrille was sent to Verdun. MF 20 was based at Senoncourt in November.

MF 22 was formed from DO 22 in November 1914. It was active on the Argonne front and was assigned to the 4th Armée from November to December 1914. In March 1915 it was assigned to the 4th Armée and based at Châlons. It was active over the Champagne front. In July 1915 MF 22 was assigned to the 12th C.A. and in November 1916 was based at Le Pilier. It re-equipped with F.40s later that month.

MF 25 was an independent bombing unit stationed near

M.F.11bis of Capitaine Happe on 14 May 1916. By this time it had become clear that the M.F.11 was obsolete and replacement by F.40s had begun. B88.3129.

Argonne, then Vadelaincourt. For further details see below.

MF 29 was an independent bombing unit stationed near Belfort. For further details see below.

MF 32 was formed from HF 32 in March 1915 and was assigned to the 3rd Armée and was based at Hermaville. In September it participated in attacks on the Champagne front during which time it was assigned to the 10th C.A. MF 32 was despatched to the Somme in July 1916; it was subsequently based at Moreuil in the Chaulnes-Chilly sector. In November 1916 it was based in the 10th Armée sector and was assigned to 10th C.A. where it operated from an airbase at Moreuil. MF 32 was active during the battle of Picardie. In April 1917 it participated in the battles at Champagne. MF 32, along with the 10th C.A. moved to Verdun in July 1917. It re-equipped with Salmson 2s in December 1917.

MF 33 was formed in April 1915, assigned to the 9th C.A., and based at Poperinghe. In May the escadrille was based at Verquin, where it participated in the Battle of Artois. In September it was active over the Champagne front. MF 33 moved from Bruay-en-Artois to Brocourt in April 1916. In November 1916 it was assigned to the 9th C.A. and was based at Croix-Comtesse. MF 33 re-equipped with F.40s in late 1916.

MF 35 was assigned to 20th C.A. and based at Poperinghe. It remained attached to the 20th C.A. for the duration of the war. In May 1915 MF 35 moved to Hermaville, where it participated in the Battle of Artois. In August the escadrille, commanded by Capitaine Villa, prepared for the attack at Auve. MF 35 participated in the Battle of Verdun. Later, it moved to the Somme where it operated from fields at Rouvrel, Hamel, Morlancourt, and Sailly-Laurette. In November 1916 MF 35 was based at Fricourt in the 6th Armée sector. After suffering heavy casualties, the unit rested at Crotoy where Lieutenant Fageol assumed command. In February 1917 it was based at Matougues in the 4th Armée sector. It was based at the Centre d'Instruction and became AR 35 in early 1917.

MF 36 was formed from DM 36 in February 1915. It was based at Porthem under the command of Capitaine Picherol until July 1915, when Lieutenant Val assumed command. In November 1916 it was assigned to the 36th C.A. and based at Furnes. In February 1917 MF 36 was assigned to support the Belgian Army. It re-equipped with Sopwith 1½ Strutters in July 1917.

MF 40 was created in March 1915. It was assigned to the 4th C.A. and was under the command of Capitaine La Morlange. In November 1916 MF 40 was based at Rosnay in the 5th Armée sector. In February 1917 it was in the 2nd Armée sector and was based at Couroure.

MF 41 was formed in March 1915, commanded by Capitaine Marquer. It was initially based in the 3rd Armée sector and was active over the Argonne forest. In May 1916 it was assigned to the 2nd Armée in the Verdun sector. MF 41 was active over the Somme front from September to December 1916. In February 1917 it was assigned to the 2nd C.A. and was based at St. Menehould in the 4th Armée sector. It re-equipped with A.R.1s in 1917.

MF 44 was formed in April 1915 at Bron. Shortly thereafter, it moved to Toul under the command of Capitaine Van der Vaero and was assigned to the 1st Armée. In May it participated in the Battle of Woëvre and was particularly active over the Pont-à-Mousson sector. The M.F.11s of MF 44 were also active in the night bombing role. In June 1915 the escadrille was assigned to the 31st C.A. In June 1916 the unit was based at Lorraine and later moved to Froidos in the Verdun sector.

MF 45 was formed in early 1915 with the M.F.11bis with 130-hp de Dion-Bouton 12 B engines (these were given army designation Type 45). These aircraft had been specially ordered by Capitaine Van Duick. In November 1916 MF 45 was assigned to the D.A.L. and was based at Art-sur-Meurthe. In February 1917 it was in the 8th Armée sector and was based at Saint Meurthe. Later in 1917 it converted to A.R.1s.

MF 50 was formed in February 1915 from elements of MF 2. It was equipped with the M.F.11bis with 130-hp de Dion-Bouton 12 B engines. In November 1916 it was assigned to the 16th C.A. and based at Autrecourt in the 2nd Armée sector. It remained here until well into 1917. MF 50 re-equipped with F.40s in August 1917.

MF 52 was formed in April 1915 under the command of Capitaine Prost. It was assigned to Belgian army units near Ypres and Langemark. MF 52 participated in a number of day and night bombing missions. In 1916 it moved to the Somme and was under the command of Capitaine Génin. It was at this time attached to the 2nd C.C. (Morocco). In November 1916 MF 52 was assigned to the 2nd C.A. and was based at Villers-Bretonneaux in the 10th Armée sector. Based at Craonne in April 1917, it was active in the Battle of Chemin des Dames. It re-equipped with A.R.1s in late 1917.

MF 54 was formed in May 1915 under the command of Capitaine Prat. It was assigned to the 10th Armée and based at Bryas. It was active along the front between Arras and La Bassée. In November 1916 it was assigned to 10th C.A. and

M.F.11 of MF 58. MF 58 was under the command of Capitaine Noé and was based at Lunéville and Saint Clément. B87.1140.

based at Moreuil in the 10th Armée sector. It was later assigned to the 8th C.A. and re-equipped with SPAD 16s in January 1918.

MF 55 was formed in May 1915 at Lyon-Bron under the command of Capitaine Lalanne. The escadrille was assigned to the 4th Armée on Champagne front. During the Champagne offensive in September 1915 it supported the 2nd C.C. MF 55 became F 55 in April 1916.

MF 58 was formed in May 1915 under the command of Capitaine Noé. It was based at Lunéville with the D.A.L. and later moved to Saint-Clément. It re-equipped with F.40s in June 1916.

MF 59 was formed in July 1915. In November 1916 it was based at Corcieux and assigned to the 161st D.I. It re-equipped with A.R.1s in October 1917.

MF 60 was formed in August 1915 under the command of Lieutenant Marc. It was assigned to the 14th C.A. later in 1915. In November 1916 it was assigned to the 21st C.A. and was based at Marcelcave. MF 60 began to re-equip with F.40s in 1916.

MF 62 was formed in August 1915 at Lyon-Bron under the command of Lieutenant Horment. It was subsequently assigned to the 6th Armée and based at Brueil-le-Sec near the Somme. In late 1915 MF 62 converted to F.40s.

MF 63 was formed in August 1915 under command of Capitaine Colard. It was assigned to the 2nd Armée in the Verdun sector. It remained there until September 1916 when it moved to Toul. By February 1917 it was assigned to the 1st C.C. and was based at Pierrefonds. MF 63 participated in the Battle of Chemin des Dames in April 1917. It re-equipped with SPAD 16s in September 1917.

MF 70 was formed in November 1915. Its M.F.11s reportedly had Mercedes engines. It was assigned to the 2nd Armée and was based in the Verdun sector. It re-equipped with Farman F.40s in 1916.

MF 71 was formed in November 1915. In December it was sent to the Verdun sector and was assigned to the 2nd Armée. It re-equipped with F.40s in 1916.

MF 72 was formed in November 1915. In November 1916 it was assigned to the 7th C.A. and was based at St.-Menehould in the 4th Armée sector. It re-equipped with Farman F.40s in 1916.

MF 123 was a dedicated bombing unit (see below).

T.O.E. Units

There were four M.F.11-equipped escadrilles assigned to T.O.E. army units.

MF 82 was created from Escadrille No.521. MF 82 was assigned to the A.F.O. forces in Albania on 19 October 1915 under the command of Capitaine Prot. It became MF 382 in September 1916 and re-equipped with A.R.1s in May 1917.

MF 98 *Tenedos* was formed 19 October 1915 for service with French Expeditionary Army to the Dardanelles. In May 1916 it was sent to Serbia, where it was redesignated MF 398.

MF 99 was formed in March 1915 to serve with the French Expeditionary Army to the Dardenelles. On 1 September 1915 it was redesignated MF 399.

MF 384 was formed on 1 September 1916. it was assigned to support the Serbian army but was manned entirely by French personnel.

By February 1916 there were approximately 370 Farmans (mostly M.F.11s) in front-line service, and 100 with training and local defense units.

The primary mission of these escadrilles was army cooperation. This usually entailed photo-reconnaissance missions and direction of artillery fire, the latter was accomplished by either signal flags trailed behind the plane or by T.S.F. The early wireless units were difficult to use and it was necessary to construct a universal code to allow the crews to pass along concise instructions. Usually two M.F.11s were used for artillery direction, each plane carrying a T.S.F. unit in case one should fail. Patterns of dashes and dots were used to signal necessary corrections. More complicated messages were sent via regular Morse code.

During battle the M.F.11s were able to remain in constant contact between the GQG and the rapidly moving army units. This required the troops to use signal panels or smoke to indicate their positions; unfortunately, the soldiers were often afraid that these signals might reveal their location to the enemy and so would not use them.

The M.F.11s flew daily reconnaissance missions and took numerous photos. The main purpose was to discover enemy activity that might indicate an imminent attack or to locate potential targets for the French artillery. The M.F.11s would also cross the enemy lines at low altitude to draw the fire of enemy batteries and machine guns; once the German positions had been revealed, the information was passed along to French commanders. Unfortunately, the distrust of some artillery commanders often meant that these messages were ignored.

However, these duties formed only a part of the functions the M.F.11 escadrilles were required to perform. For example, the records of MF 36 show that from 1 September 1915 through 15 March 1916 it flew 487 reconnaissance sorties, conducted 138 artillery-spotting missions, took 522 photographs, engaged in 97 aerial combats, and flew 32 bombing missions (24 of them at night).

By late 1915 it was obvious that the M.F.11s were vulnerable to enemy fighters, and the commanders of many units demanded better planes. Unfortunately, no suitable design was yet available.

M.F.11 serial number 742 of MF 29. MF 29 was an independent bombing unit using Farmans. It was based at Belfort on 15 July 1915 and was equipped with six M.F.11s. B88.3139.

Bombing

M.F.11 escadrilles flew occasional bombing missions using converted artillery shells or flechettes. The only dedicated bombing units to use to M.F.11s were MF 25, 29, and 123.

MF 25 was formed in 1914 and later became an independent bombing unit. The escadrille was based at Argonne in the 3rd Armée sector. During December 1914 it dropped 888 kg of bombs on various targets in the vicinity of the industrial center of Briey. The next month MF 25 concentrated on attacking enemy balloons and train stations. Early in 1915 the unit moved to Argonne in the 2nd Armée sector and then to Sainte-Menehould. During May and June train stations were attacked with converted artillery shells ranging in size from 90 to 120 kg. In July 2,174 kg of these weapons were dropped on train stations and German airfields. MF 25 flew its first night attack on the night of 25/26 August when the train stations at Challerange, Cernay, and Chatel were bombed. Similar missions were flown in September, although all but one of the raids were flown during the day. In October MF 25 joined with GBM 5 in attacking targets in the vicinity of Vouziers.

At the beginning of the Battle of Verdun on 21 February 1916, MF 25 was the only bomber unit available on that front. Based at Vadelaincourt, the escadrille bombed tactical targets in the vicinity of the front. The next month the unit concentrated on army cooperation duties while it trained for night operations. At this time the inadequacy of the M.F.11 as a day bomber had become apparent and hence there was a switch to night operations. By the 29th September 1916 only eight Farmans were serviceable, and the escadrille had received enough F.40s to replace the M.F.11s.

MF 29 was the other major independent bombing unit using Farmans. It was based at Belfort on 15 July 1915, equipped with six M.F.11s. Captain Maurice Happe, who commanded the unit, arranged his bombers in two vics each flying at different altitudes and guarded from above by one or two M.F.11s armed with machine guns.

An attack on the night of 30/31 July, 1915 against Freibourg resulted in one M.F.11 making a force landing. After this incident, the unreliable 80-hp Dion-Bouton engines were removed from the M.F.11s and replaced by superior 80-hp Renault motors. During a night attack on Cernes on 24 August 1915 it was discovered that many of the bombs failed to explode, and this prompted Happe to insist that more effective bombs be developed. Because of his success in attacking enemy targets, the Germans had put a price on Happe's head of 25,000 marks. Gratified by this response, Happe painted red crosses on his plane and literally taunted the Germans to attack him. A major raid took place on 7 September 1915 when five M.F.11s bombed the Aviatik factory; the strike was so successful that the factory was moved to Leipzig. Other raids during September included attacks on the Lauterbach train station, the Lorrach station, and targets at Rothweil. During the latter mission the bombers were assaulted by a German plane, described as an Aviatik. The tenacious German pilot took advantage of the M.F.11's pusher configuration by staging his attacks from the rear and underneath the French planes. The pilots desperately tried to bring the M.F.11s into a position where their gunners could fire at the German, but the slow and unwieldly Farmans were no match for the German plane. The result was two pilots dead, two taken prisoner when their plane was forced to crashland, and 60 bullets in Happe's aircraft. It was now clear that the usefulness of the M.F.11 as a day bomber was rapidly approaching an end.

M.F.11s with 130-hp engines were now entering service. It was decided that the 130-hp machines would be used as "fighter" escorts, while the standard 80-hp Farmans would serve as bombers. Later C 61, with Caudron G.4s, was assigned to MF 29; the G.4s were often used to provide escort for the M.F.11s. Two M.F.11 fighters and six bombers attacked the poison gas factory at Roessler; this time, however, they were escorted by eight Nieuport 11s of N 49. Despite this protection, two of the M.F.11s were attacked and forced down. On 28 November 1915 MF 29 had five different types of Farmans on strength: eight M.F.11s with 80-hp engines used as bombers, three M.F.11s with 80-hp engines used as trainers, three M.F.11s with 80-hp engines and an enhanced fuel capacity of 290 liters, four M.F.11s with 130-hp engines used as fighters, and seven M.F.11s with 80-hp engines and an enlarged wing span of 18.00 m.

New M.F.11s were sent to MF 29 in January 1916. These were 130-hp versions with enlarged fuel tanks which resulted in the planes being labeled "camel backs." They could carry a payload of 510 kg and could climb to 1,000 m in seven minutes. Later in January MF 29 was attached to GB 4.

One of the most important raids of the war for MF 29 took place on 18 March 1916. A total of 17 M.F.11s, three BM 4s, and three G.4s attacked the Mulhouse station and the Habsheim airfield. German fighters again attacked the nearly defenseless Farmans from behind and at least four M.F.11s were lost in this raid. On 1 April MF 29 moved to Luxeuil along with the rest of GB 4. On that same day MF 29 at last received new planes; unfortunately, these were the only marginally improved F.40.

MF 123 was the only other dedicated bombing unit to use Farmans. In February 1916 MF 123 was formed from personnel and equipment serving with MF 29. The escadrille was commanded by Lieutenant Mouraud. It was initially based at Alsace, but moved to Malzéville in October 1916. By this time it had re-equipped with F.40s.

Fighter

The M.F.11s saw limited service as fighters. In addition to flying bomber escort missions, they would fly "barrage patrols," which meant they would fly along the French lines and attack German aircraft attempting to cross into French airspace. For example, on 26 September 1915 planes of MF 16 attacked three balloons and engaged in three aerial combats. However, by 1916 the M.F.11s themselves required protection by either Nieuport fighters or Caudron G.4 long-range escort fighters.

Most MF escadrilles were eventually re-equipped with F.40s, which, while possessing a more powerful engine, retained the pusher configuration that made them vulnerable to enemy fighters.

Foreign Service

Australia

Australia acquired a number of M.F.11s in 1917. They served with these units:

CFS 16 and 17: two M.F.11s were ordered in 1917 and struck off charge in 1919.

CFS 19 and 20: two M.F.11s were ordered in May 1917 and sold in 1919.

Three half flights in Mesopotamia (1F C 1, 1FC 7, and 1FC 10) used M.F.11s. They performed reconnaissance and aerial supply missions.

No.5 Squadron AFC had 23 M.F.11s used for training. They were given serials A.222, 2233, 4074, 4672, 6897, 7084, B.1957, 1958, 2037, 2222, 4663, 4664, 4671, 4672, 4673, 4722, 4733, 4734, 4735, 4736, 4765, 4788, and 4790.

Belgium

Belgium purchased six M.F.11s with 80-hp Renault engines in late 1914. Later more M.F.11s with De Dion engines were obtained from France. Additional M.F.11s with provision for a machine gun to be mounted in their nose were also acquired.

On March 1915 Escadrille 5 had six M.F.11s with 80-hp Renault engines and Escadrille 6 had six M.F.11s with 100-hp Renaults. Several more with 130-hp engines were purchased from France in 1916. Eventually Escadrilles 1 and 2 at Coxyde and Escadrilles 4 and 5 at Houtem all had M.F.7s and 11s on strength. Escadrilles 1 and 2 were based at Saint-Idesbald and later Maires during 1916. They performed reconnaissance for the Belgian army during the Flanders offensive. Escadrilles 5 and 6 remained at Houtem throughout 1916. Escadrille 6 used its M.F.11s as bombers. As 1916 progressed the M.F.11s were replaced by F.40s.

Greece

The Greek air service obtained six M.F.11s in 1916. The last of these was withdrawn from service in 1920.

Italy

A total of 601 M.F.11s (which were known as Farman 1914s in Italy) were built under license by the Societa Construzioni Aeronautiche "Savoia" at Bovisio-Mombello beginning in May 1914. The Fiat firm at Torino also built them under license beginning in mid-1915. The Fiat-built machines were powered by 100-hp Fiat A-10 engines and carried the designation F.5b while the Savoia M.F.11s had 110-hp Colombo D.110 engines. Finally, the Societa Nieuport-Macchi built 50 M.F.11s.

The Savoia F5.bs were subject to a number of modifications closely resembling those of the Farman F.40s, and the Savoia design may have been inspired by that type. The Savoia Pomilio firm created copies of the M.F.11s designated S.P.1, 2, 3, and 4.

M.F.11 of CFS 17 at Point Cook CFS, Victoria, Australia. All four CFS M.F.11s survived the war and were sold to a civilian operator in 1919. RAAF Museum via Colin Owers.

M.F. 11. B88.0318.

The S.P.1 was essentially a license-built M.F.11. The S.P.2 was also similar to the standard M.F.11 but was more streamlined for better aerodynamics and strengthened so that it could carry the more powerful 260-hp A 12 engine. This meant that the empty weight of the standard S.P.2 was twice that of the standard M.F.11 but the increase in engine power more than compensated for this change. The S.P.2 was now able to carry a camera as well as a Fiat machine gun mounted in the nose. Approximately ten S.P.2s were fitted with 25-mm Fiat cannons; these planes were designated S.P.2bis. A total of 402 S.P.2s was built. However, because of the increase in weight the S.P.2s were found to be unstable in flight and poorly maneuverable. A lightened version with a smaller wing was introduced in the hope of correcting these problems. This version was designated S.P.3 and was found to be marginally faster and more maneuverable. Production of the S.P.3 began in 1917 and a total of 300 were built. However, the S.P.3 was still no match for Austro-Hungarian and German fighters and became known by its pilots as a "coffin for two." Finally, a twin-engine version of the S.P. series was built in 1916; it was a M.F.11 airframe fitted with two Isotta Fraschini V.4b engines mounted as tractors. The A.E.R. firm at Orbassano, near Torino, built 152 examples. The S.P.4 entered service with front-line squadriglias in the fall of 1917 and remained in service until the end of the war.

The M.F.11s equipped 12 squadriglias in November 1916. These units were:

Gruppo 1 (3rd Armata): Squadriglias 27a and 28a.
Gruppo 2 (2nd Armata): Squadriglias 29a and 30a.
Gruppo 3 (1st Armata): Squadriglias 31a and 37a.
Gruppo 4 (2nd Armata): Squadriglias 47a.
Gruppo 7 (1st Armata): Squadriglias 46a (Verona and Asiaso) assigned to the 1st Army, 48 (Belluno) assigned to the 1st Army, and 49a assigned to the 1st Army.
Defense of Udine: Squadriglia 33a.
Albania: Squadriglia 36a (Valona) assigned to the 16th Army in Albania.

Initially the M.F.11s and F5bs were used as bombers, but their light bomb loads resulted in only modest success in this role, and they were shifted to reconnaissance missions.

By 1917 there were 15 Squadriglias which still used M.F.11s and F.5bs. These were:

Gruppo 1 (3rd Armata): Squadriglia 36a.
Gruppo 2 (2nd Armata, 4th Armata): Squadriglias 27a, 30a, and 41a.
Gruppo 3 (1st Armata): Squadriglia 31a.
Gruppo 4 (2nd Armata, 4th Armata): Squadriglias 29a and 45a.
Gruppo 7 (6th Armata, 1st Armata): Squadriglia 32a.
Gruppo 8 (Albania): Squadriglia 34a.
Gruppo 9 (1st Armata): Squadriglia 37a.
Independent Squadriglias 101a (Bari), 102a (Ancona), 104a (Bengasi), 12a (Bengasi), and 7a.

The M.F.11s were withdrawn from front-line service in 1918 and assigned to training units. Some of the S.P. series of M.F.11s remained in Italian service until 1922 and provided support for Italian colonial troops during attacks against rebels.

Japan

Japan built M.F.11s under license designated Army Type Mo-4. The Japanese manufacturer introduced a number of changes resulting in improved stability and maneuverability. The type was put into mass production at PMBRA's factory at Tokorozawa, the army arsenal, and the Atsuta Army Weapon Manufacturing Works of Nagoya. The M.F.11s replaced the M.F.7s in the reconnaissance and bombing roles and served with the balloon company and flight company based at Tokorozawa. Six Mo-6s, as well as eight Mo-4s, served with the air units assigned to the 12th Air Division; they were used in combat in Manchuria and Siberia in 1918. The Mo-4s were also used as trainers, and many were purchased by civilians after the war. It is estimated that 84 were built.

Variants of the standard Mo-4 were built by the Akabane Plane Manufacturing Works; these were the Kishi No.3 (which was identical to the Mo-4 except for a redesigned nacelle), the Kishi No.4 (which had a lower wing of reduced span), and the Kishi No.6 (which was a standard Mo-4 built under license).

The Mo-6 was a Mo-4 fitted with a 110-hp Daimler liquid-cooled engine. Production by the PMBRA began in 1916 and a total of 134 were built by various manufacturers from 1917 to 1921. The Mo-6s experienced numerous problems, usually caused by the Daimler motors, but these were overcome and they served with the 2nd Army Air Battalion at Kagamigahara and the Air Battalions at Tokorozawa and Kagamigahara. Four Mo-6s served with the 2nd Army Air Battalion in Siberia and Manchuria, but their engines proved to be unsuitable for such cold climates. The last example was retired in 1923.

Italian-Built Variants of the M.F.11:

Type	Engine	Span	Length	Area	Weight Empty	Loaded	Max. Speed	Climb	Ceiling	Endur.	No. Built
SP 1	FA 10, 100-hp	16.13 m	9.19 m	56 sq. m	640 kg	920 kg	108 km/h	–	–	–	–
SP 2	A-12, 260-hp	16.74 m	10.70 m	67 sq. m	1250 kg	1700 kg	135 km/h	3,000 m in 26 min.	5,000 m	4 hours	402
SP 3	A-12, 260-hp	14.71 m	10.50 m	60 sq. m	1233 kg	1683 kg	145 km/h	3,000 m in 25 min.	5,000 m	4 hours	300
SP 4	IF V4b, 190-hp	19.80 m	10.70 m	78 sq. m	1700 kg	2500 kg	151 km/h	3,000 m in 18 min.	4,500 m	4 hours	52

An armed M.F.11 in Russian service. Most of the M.F.11s built by the Russians were used as trainers, but a small number were used at the front. B89.1726.

The Army Type 5 Plane was an Mo-4 equipped with dual controls and became the first purpose-built Japanese army trainer. Eleven were built by the Tokorozawa Army Arsenal and Department of Supply in 1919 and 1920.

Norway

Norway purchased approximately ten M.F.11s in 1915. These served until 1921. The tail units from these planes were then fitted to several F.40s that had been purchased, the defective tail units of the F.40s having been the cause of several crashes.

Portugal

Portugal purchased two M.F.11s in August 1916. These were used for pilot training at the flight school at Vila Nova de Rainha. They entered service in September and were retired in 1917.

Romania

At least eight M.F.11s were transferred from the RNAS to the Corpul Aerean Romana. These planes served with the following Romanian units on 10 June 1916:

Grupul 1: (1st Group) three escadrillas.

M.F.11 displaying its radiators on the side of the fuselage. B89.2793.

Grupul 2: three escadrillas.
Grupul 3: two escadrillas.
Grupul 4: three escadrillas.

The M.F.11s were used for reconnaissance and bombing. Several important reconnaissance missions were flown by six F.40s and M.F.11s assigned to Locotenent Cholet. The M.F.11s served well into 1917 alongside the newer Farman F.40s.

Russia

The Imperial Russian Air Service does not seem to have employed many M.F.11s in combat. Most of those built by the Russians were used as trainers, but a small number were sent to the front. On 1 March 1917 there were 25 M.F.11s (along with some M.F.7s) in service—eight on the western front and 17 on the southwestern front. Two months later the number of Farmans had declined to 22 divided into one on the northern front, nine on the western front, and 12 on the southwestern and Romanian fronts.

Saudi Arabia

The Saudis obtained two Maurice Farmans from Italy in August 1921; these were S.P.1s built in Italy.

Serbia

On 27 February 1915 a French escadrille was sent to Serbia to assist in the formation of the Serbian air service. Six M.F.11s were assigned to this unit, which was manned by the French but under the control of the Serbian army command. The planes had serials F.193, F.194, F.195, F.196, F.199, and F.452. The first combat mission was flown on 1 April 1915 and a week later ten reconnaissance missions were flown over Kubin, Shabac, and Roma. Later the M.F.11s moved to a base at Banista outside Belgrade, while two Farmans were sent to Prakova to provide reconnaissance for the region.

Additional M.F.11s arrived in June. The French decided to keep all the M.F.11s in a single unit to simplify supply problems. Most of the planes would be based at Belgrade and employed in strategic reconnaissance, while smaller detachments would be sent to other airfields for tactical reconnaissance.

The type's first air-to-air victory took place 27 May 1915 when an M.F.11 shot down an Austro-Hungarian aircraft near Smedervo. A second victory occurred on 10 July when an

M.F.11 in service with the Serbian air service. Most of the M.F.11s were based at Belgrade and were employed for strategic reconnaissance. B83.3367.

Austro-Hungarian bomber was shot down over Belgrade. During a raid by three M.F.11s on an enemy airfield at Baavanistu in August an M.F.11 scored yet another victory over an Austro-Hungarian plane.

Reconnaissance and bombing missions continued throughout 1915, and the M.F.11s were even used to attack gunboats. Six more M.F.11s arrived in December.

1916 would be the last year that the M.F.11s would be used by front-line units. By 21 April the Serbian air service had been reformed into four escadrilles, all of which were equipped with M.F.11s: MF 82, MF 84, MF 98, and MF 99. By September the M.F.11 units were re-equipped with F.40s.

Spain

Spain acquired M.F.11s in 1914 to replace the M.F.7s in service; because of their abbreviated landing skids they were known as "amputados." Another nickname was "olives" because of the color of their nacelles. At least six M.F.11s were purchased from France; these were powered by 70-hp Renault engines. In 1915 one M.F.11 was based at Arcila and three were there by 1916. The M.F.11s were not as docile as the M.F.7s and there were numerous accidents. Furthermore, the Farmans were plagued by engine trouble until the Renaults were replaced by 80-hp de Dions. Some of the M.F.11s were assigned to the Escaudrilla de Tetuan and one was in use at Arcila as late as 1919. Serial numbers were MF 11 through 16.

Switzerland

Switzerland had two M.F.11s in service from 1915 through 1919. They carried serial numbers 30 and 61. Both were obtained when they landed in Switzerland and were interned. M.F.11 number 30 was used for reconnaissance by Fliegerkompagnie 1 until it crashed on 7 July 1916. No.61 was acquired on 27 June 1916 but was returned to France shortly thereafter.

Ukraine

The Ukrainian air service obtained a single M.F.11 in 1918. It had serial 5407.

United Kingdom

The Royal Flying Corps received its first M.F.11 in 1914 and had eight on strength by the middle of that year. Some M.F.11s were purchased directly from France, while others were built under license by AIRCO and Whitehead. Five were assigned to No.6 Squadron and two were with the CFS M.F.11s also served with Nos.2, 5, 9, and 16 Squadrons; the last shorthorn (as the plane

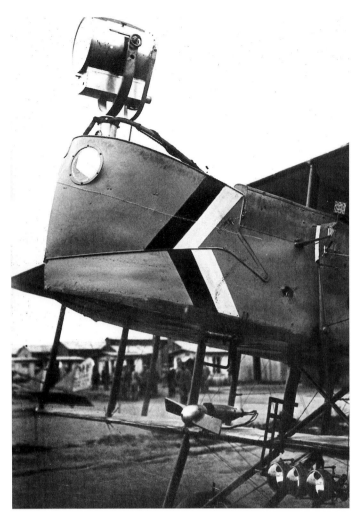

M.F.11bis. The searchlight enabled the aircraft to serve as a nightfighter or attack ground targets at night. B83.4160.

was known in Britain) was retired in November 1915. No.30 Squadron in Mesopotamia had four M.F.11s used for survey duties; all were destroyed by a storm in May 1916. After their retirement these M.F.11s were assigned to Nos.1, 2, 3, 4, 5, 6, 7, 8, 9, 10, 11, 12, 14, 15, 19, 21, 22, 24, 25, 26, 27, 29, 36, 38, 39, 41, 47, 48, 49, 57, and 68 Training/Reserve Squadrons. Others served with Training Depot Squadrons 8 and 204, the Wireless School, the School of Instruction, No.2 Auxiliary School of

M.F.11 flown by Cadet N. Mulroney, AFC, on his first solo at Spittlegate, Lincolnshire, in August 1917. N. Mulroney via Colin Owers.

M.F.11 Two-Seat Reconnaissance Plane with 70-hp Renault 8B	
Span 16.13 m, length 9.48 m, height 3.20 m, wing area 54 sq. m Empty weight 510 kg, loaded weight 810 kg Maximum speed: 100 km/h, climb to 2,000 m in 22 min., endurance three hours Armament: Colt machine gun and a Winchester carbine	

M.F.11 Two-Seat Reconnaissance Plane with 80-hp Renault 8B
Span 16.13 m; length 9.48 m; height 3.20 m; wing area 52 sq. m Empty weight 620 kg; loaded 945 kg Maximum speed: 118 km/h

M.F.11 Two-Seat Reconnaissance Plane with 80-hp De Dion-Bouton
Span 16.13 m; length 9.23 m; height 3.00; wing area 52 sq. m Maximum speed: 118 km/h

M.F.11 Two-Seat Reconnaissance Plane with 110-hp Lorraine 6AM
Span 16.13 m; length 9.23 m; height 3.00 m; wing area 52 sq. m Maximum speed: 118 km/h

M.F.11bis Two-Seat Reconnaissance Plane with 130-hp Renault 8C
Span 18.15 m; length 9.23m; height 3.00 m; wing area 56 sq. m Empty weight 780 kg; loaded weight 1,155 kg Maximum speed: 130 km/h

M.F.11bis Two-Seat Reconnaissance Plane with 130-hp Renault 8C
Span 16.13 m; length 9.23 m; height 3.00 m; wing area 52 sq. m Payload 227 kg Maximum speed: 118 km/h; climb to 2,000 m in 15 minutes; endurance was 3 hours

M.F.11bis Two-Seat Reconnaissance Plane with 100-hp Renault 8C
Span 18.15 m; length 9.23 m; height 3.00 m; wing area 56 sq. m

M.F.11bis Two-Seat Bomber with 130-hp Salmson A9
Span 18.15 m; length 9.23 m; height 3.00 m; wing area 56 sq. m

M.F.11bis Two-Seat Bomber with 130-hp de Dion-Bouton 12B
Span 18.15 m; length 9.23m; height 3.00 m; wing area 56 sq. m

M.F.11bis BO Single-Seat Bomber with 80-hp Renault R8B
Span 18.15 m; length 9.23 m; height 3.00 m; wing area 56 sq. m Empty weight 654 kg; loaded weight 928 kg Climb to 2000 m in 20 minutes.

M.F.11bis BO Single-Seat Bomber with 80-hp Renault R8B
Span 16.13 m; wing area of 52 sq. m

M.F.11bis BO Single-Seat Bomber with 130-hp Renault 1R8C
Dimensions identical to M.F.11bis BO two-seat bomber with 80-hp Renault R 8B

M.F.11 Two-Seat Reconnaissance Plane with 80-hp Renault 8B Built in England by AIRCO and Whitehead:
Span 15.776 m, length 9.30 m, height 3.15 m, wing area 52 sq. m Empty weight 654 kg, loaded weight 928 kg Maximum speed: 116 km/h, climb to 1,000 m in 8 min.; climb to 2,000 m in 20 min.; endurance 3.75 hours Armament: one 0.303 Lewis machine gun and a Winchester carbine

Japanese Army Type Mo-4 Built by PMBRA, Tokyo Army Arsenal and Atsua Army Manufacturing Works with 70-hp or 80-hp Renault 8B
Span 15.50 m; length 9.14 m; height 3.18 m; wing area 58 sq. m Empty weight 563 kg; loaded weight 778 kg Maximum speed: 49 kt; climb to 2,000 m in 25 minutes; ceiling 3,000 m; endurance 4 hours Armament: one machine gun Approximately 84 built

Aerial Gunnery, and No.1 School of Navigation and Bomb Dropping. Eight M.F.11 floatplanes were purchased for use by the Aerial Gunnery School. They were built by AIRCO and were replaced by mid-1917.

The Royal Naval Air Service used approximately 90 M.F.11s; 20 of these were built by the Eastbourne Aviation Company and used as trainers. However, most of the RNAS machines were used for reconnaissance and bombing by No.3 Squadron (later No.3 Wing) in the Aegean. Others served with Nos.1, 2, and 3 Wings. RNAS Serials were: 1127, 1134, 1369–1371, 1380–1387, 1240–1241, 3932–3939, 8106–8117, 8466–8473, 9133, N 1530, N5060–N5079, and N6310–N6329.

Maurice Farman M.F.12

The Maurice Farman 12 was a modified M.F.11bis with an enlarged wing span and a 100-hp Renault 8C engine. First produced in 1914, some M.F.12s were supplied to front-line units and were given army type number 26. They were used in the army cooperation role.

M.F.12 Two-Seat Reconnaissance Aircraft with 100-hp Renault 8C
Wing span 21.00 m, length 9.23 m, height 3.00 m

M.F.12. The Maurice Farman 12 was a modified M.F.11bis with an enlarged wing span and a 100-hp Renault 8C engine. Reairche Collection.

Henri Farman H.F.33

The early fighters of the First World War were developed from reconnaissance types. Examples included the Morane-Saulnier L and Nieuport 10 series. In 1915 Henri Farman attempted to convert his H.F.20 into a two-seat fighter. The engine was an 80-hp Gnome 7A similar to that used on the H.F.20. The wing span was significantly increased from that of the H.F.20, being 6.20 m greater. The H.F.33 was tested in 1915 but apparently was unsuccessful, as no aircraft were selected for production.

Henri Farman H.F.33 Two-Seat Fighter with 80-hp Gnome 7A
Span 19.70 m; length 8.26 m; height 3.05 m
Maximum speed: 110 km/h

Henri Farman H.F.35

The H.F.35 was a huge biplane intended to meet the C3 specification of 1915. This called for a crew of three, armor, and an armament of three machine guns. The C 3 class was intended to escort army cooperation planes, fly barrage patrols, and to attack enemy balloons. Other designs to meet this specification may have included the Ponnier twin-engine fighter and the SPAD SA.3.

Unlike the Ponnier and SPAD designs, which had conventional fuselages with gunners fore and aft, the H.F.35 retained the pusher configuration. For this reason, all three guns were mounted in the front of the central nacelle. Obviously, this greatly limited the usefulness of the H.F.35 as there would be no way for the gunners to fire at aircraft coming from behind.

The H.F.35 resembled a greatly enlarged F.40, retaining the same basic outlines of that classic reconnaissance aircraft. It had a large central nacelle with radiators on either side. The engine was a 220-hp Renault 12Fa. The rudder was supported from the central nacelle by a series of booms. The aircraft was made of metal. The three-bay wings were of unequal span, the upper wing being considerably longer than the lower. The H.F.35 has been estimated as carrying a payload of 1,000 kg. A quadracycle landing gear was fitted, doubtless to manage the aircraft's huge weight and also to prevent nosing over when landing on unprepared fields. One source suggests that the aircraft may have also been intended for use in the attack role. It was flown in December 1915 but apparently was not ordered by the Aviation Militaire.

H.F.35 Three-Seat Escort Fighter with 220-hp Renault 12Fa
Length 10.25 m; height 3.60 m
Payload 1,000 kg
Maximum speed: 165 km/h
Armament: three machine guns
One built

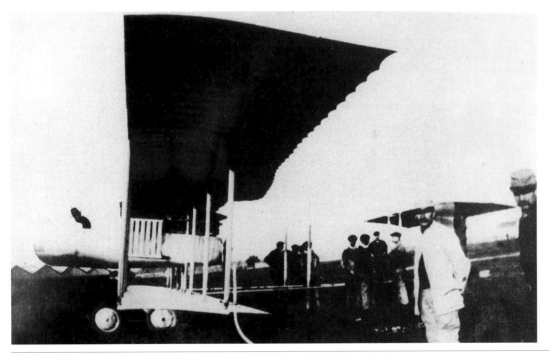

The H.F.35 was a huge biplane intended to meet the C3 category for 1915. All three gunners were seated in the front of the central nacelle. Reairche.

Farman F.40

In 1915 General Hirschauer, the director of military aviation, asked the Farman brothers to design a plane to replace the M.F.11s. The new Farman, designated F.40, was an amalgamation of the design philosophies of both Henri and Maurice Farman. The cockpit was suspended between the wings as in the M.F.7, M.F.11, and H.F.30. Unlike the previous designs, the nacelle was ovoid, giving the F.40 a more streamlined appearance. The sesquiplane wings were also similar to the Maurice Farman designs. However, the tail assembly was that used on Henri Farman's designs—a pair of tail booms converging to a point on which the horizontal stabilizer and rudder were mounted. Because of the combined features of Henri's and Maurice's planes, the F.40s were often referred to as "Horaces." The observer was seated ahead of the pilot in the nacelle and there were prominent radiators on either side of the fuselage. The three-bay wings were of unequal span. The engine most commonly fitted to F.40s was a 135-hp Renault 8C. Armament consisted of a single machine gun, usually a Lewis, mounted on an Éteve stand in the nose, and up to 240 kg of bombs.

F.40s were tested at the front by MF 16 in June 1915. Apparently the pilots complained about the type's poor maneuverability. A decision was made to stop further deliveries until Farman had addressed the problem. However, there was an acute need for a replacement for the M.F.11s then in service, and it was planned to acquire enough F.40s to equip ten escadrilles.

Another memo dated 25 June 1915 praised the improved

performance of the F.40s. It was recommended that the undercarriage be reinforced, the rear landing skid be strengthened, and the rigging of the lower ailerons be tightened. A total of 35 F.40s were produced in August 1915, and 40 more were built in September, enough to equip seven escadrilles.

An evaluation of the F.40s in operational service with MF 1 and MF 22 was written on 17 September 1915. Observations included the fact that the new F.40s had excessive wear on the tail fabric which was attributed to deficiency in the cloth used. Also, the Farman manual for rigging the tailplane was to be disregarded as it was inaccurate.

Large numbers of F.40s were produced and up to ten a day were built. However, it was soon obvious that alterations would be needed to make the F.40 a suitable warplane. A memo from the Ministre de la Guerre dated 19 May 1916 listed necessary changes:
1. The F.40s were nose heavy. To counteract this the nacelle was to be moved back 6 cm.
2. The landing gear needed reinforcement
3. The attachments of the flying wires needed reinforcement.
4. The control stick was to be moved to the left.
5. Dual controls were to be fitted to some F.40s.
6. The tail was reinforced.
7. Strengthening of the tail skid was needed.
8. The aileron pulleys were to be modified.
9. The wing struts were to be reinforced.

The problems with the tail were identified in a GQG memo dated 18 May 1916, reporting that vibrations during flight had resulted in detachment of the hinges holding it in place. This was caused by faulty soldering and was, presumably, to be corrected in the field.

By September 1916 it was obvious to the Aviation Militaire that the F.40s were obsolete. An STAé memo noted that while they could climb well, the F.40s were too slow in a dive and required fighter escort to survive over the front.

Variants

There were a large number of variants of the basic F.40; many of them differed from the standard type only in the type of engine.
1. F.40 with 130-hp Renault 8C. Army Type 42 and 43, Built in 1915. Specifications given below.
2. F.40 with 130-hp Dion-Bouton 12 B. Army Type 46. Built in 1915.
3. F.40H, a seaplane training version of the standard F.40 with 130-hp Renault 8C engine. Built in 1917.
4. F.40bis. Army Type 56 with160-hp Renault 8Gc. Built in 1916.
5. F.40ter. Army Type 57 with 150-hp Lorraine 8A. Built in 1916.
6. F.40 QC with 130-hp Renault 8C. Featured an elongated tail similar to that used on the M.F.11. Built in 1917.
7. F.40P, a variant of the standard F.40 modified to carry Le Prieur rockets. Designed to attack balloons, the F.40P had dimensions identical to the standard F.40.
8. Armored F.40. Some variants of the F.40 were equipped with armor to protect crew and fuel tank. The fuel tank was also lined in rubber. Armored F.40s are known to have been supplied to Escadrilles F 54 and F 60.

Operational Service
Army Cooperation

The F.40s began to replace the obsolescent M.F.11s in late 1915 and served throughout 1916. They were finally replaced in 1917. The F.40 served with the following escadrilles (often alongside the M.F.11 it was intended to supersede):

MF 1, active during the Battle of the Somme in July 1916 and gave effective support to the 33rd C.A. In the beginning of 1917 the 33rd C.A. and MF 1 moved to the 10th Armée front. In February 1917 MF 1 was still assigned to the 33rd C.A. and based at Pierrefonds in the 1st Armée sector. By April the Farmans had at last been largely supplanted by A.R.1s and the escadrille's designation changed to AR 1.

MF 2, assigned to the 5th C.A. and based at Autricourt, where it remained until July. Before transferring to the Somme front in September 1916, MF 2 was based at camp Mailly to practice liaison with the infantry. MF 2 remained on the Somme front under the command of Commandant de Vergnettes until November 1916, when it moved to Châlons-sur-Marne. It was then assigned to the 5th Armée and in February 1917 received a new commander, Lieutenant Collard. MF 2 was active in the Battle of Chemin des Dames and re-equipped with A.R.1s in October 1917.

MF 5, in November 1916 assigned to the 2nd C.A. and based at Lemmes. It moved to the 4th Armée sector in February 1917 and then to the 33rd C.A. in June 1917. MF 5 re-equipped with Sopwith 1½ Strutters in September 1917.

MF 7, in November 1916 based at Bois des Tailles. In January it moved to the 5th Armée sector in preparation for the Chemin des Dames offensive; it was under the command of Capitaine Saqui-Sannes. It re-equipped with Sopwith 1½ Strutters in mid-1917.

MF 8, in November 1916 assigned to the 2nd C.A. and based at Vadelaincourt in the 2nd Armée sector. Assigned to the 6th Armée, it participated in the Battle of Chemin des Dames in April 1917. In October it re-equipped with A.R.1s.

MF 14, November 1916 was in the 7th Armée sector assigned to the 76th D.I. and based at Griecourt. It was re-equipped with A.R.1s in September 1917.

F 16, formed from MF 16 in July 1916. In November 1916 it was assigned to 2nd C.A.C. and was based at Le Hamel in the 10th Armée sector. In December it moved to the 1st Armée sector and participated in the Battle of Chemin des Dames in April 1917. In July 1917 it re-equipped with A.R.1s.

F 19, in November 1916 assigned to the 13th C.A. in the 10th Armée sector and based at Pierrefonds. In February 1917 it was active in the 1st Armée sector, still assigned to the 13 C.A., and was based at Remy. It re-equipped with A.R.1s in early 1917.

Above: Farman F.40H floatplane at Cazaux firing range. C. Cheeseman.

Left: F.40bis, Army Type 51. This variant of the F.40 had a 160-hp Renault 8Gc engine, and was built in 1916. Reairche.

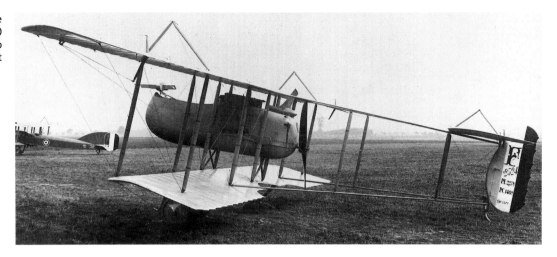

Type 56 was the Army type number for the F.40bis, an F.40 variant fitted with a 160-hp Renault 8Gc engine. This aircraft carries serial F2524. MA38546.

MF 20, in November 1916 was in the 2nd Armée sector and assigned to the 14th C.A. It was based at Senoncourt. In May 1917 it was assigned to the 6th Armée and was active in the Battle of Chemin des Dames. It re-equipped with A.R.1s in August 1917.

F 22, formed when MF 22 re-equipped with Farman F.40s in November 1916. At that time it was assigned to the 12th C.A. and based at Le Pilier in the 10th Armée sector. From March through November 1917 F 22 was active in the 4th Armée sector. In July it re-equipped with A.R.1s.

MF 32, assigned to the 10th C.A. in November 1916. In April 1917 it participated in the attacks at Champagne. MF 32, along with the 10th C.A., moved to Verdun in July 1917. It re-equipped with Salmson 2s in December 1917.

F 33, in November 1916 assigned to the 9th C.A. and based at Croix-Comtesse. It re-equipped with F.40s in late 1916. It was subsequently based at Suippes, where it was active over Champagne. In September it moved, along with the 9th C.A., to the Somme front. In February 1917 F 33 was in the 4th Armée sector assigned to the 9th C.A. and based at Mailly. In April it was based at Fismes and participated in the Battle of Chemin des Dames. It was primarily active over the Vauclerc forest. F 33 re-equipped with A.R.1s in June 1917.

MF 35, in November 1916 based at Fricourt in the 6th Armée sector. After suffering heavy casualties, the unit rested at Crotoy, where Lieutenant Fageol assumed command. In February 1917 it was based at Matougues in the 4th Armée sector. It was based at the Centre d'Instruction and became AR 35 in early 1917.

MF 36, in November 1916 assigned to the 36th C.A. and based at Furnes. In February 1917 it was still based at Furnes but was assigned to support the Belgian army. It re-equipped with Sopwith 1½ Strutters in July 1917.

MF 40, in November 1916 assigned to the 4th C.A. and based at Rosnay in the 5th Armée sector. In February 1917 it was in the 2nd Armée sector assigned to the 4th C.A. and based at Couroure. It re-equipped with A.R.1s during 1917.

F 41, in May 1916 assigned to the 2nd Armée in the Verdun sector. F 41 was active over the Somme front from September to December 1916. In February 1917 it was assigned to the 2nd C.A. and based at Saint-Menehould in the 4th Armée sector. It re-equipped with A.R.1s in 1917.

F 44, converted to F 1,40s in June 1916 and Capitaine Chapelet assumed command. MF 44 moved to the regions of Mort-Homme, then Avocourt. In November it was assigned to the 31st C.A., based at Froidos in the 2nd Armée sector. In June 1917, it re-equipped with A.R.1s.

F 45, in November 1916 assigned to the D.A.L. and based at Art-sur-Meurthe. In February 1917 it was in the 8th Armée sector and was based at Saint-Meurthe. At the end of 1917 MF 45 converted to A.R.1s.

F 50, re-equipped with F.40s in August 1916. It was assigned to the 16th C.A. F 50 re-equipped with A.R.1s in 1917.

F 52, assigned to the 2nd C.A. and based at Villers-Bretonneaux in the 10th Armée sector in November 1916. Based at Craonne in April 1917, F 52 was active in the Battle of Chemin des Dames. It re-equipped with A.R.1s in late 1917.

F 54, assigned to 10th C.A. in November 1916 and based at Moreuil in the 10th Armée sector. It was later assigned to the 8th C.A. and re-equipped with SPAD 16s in January 1918.

F 55, formed from MF 55 in April 1916. It was assigned to the 4th Armée and later to the 5th Armée. Before the Battle of Verdun F 55 operated in support of the 2nd Armée. Capitaine Petit assumed command in October 1916. In November 1916 F 55 was assigned to the 2nd C.A. and based at Lemmes in the 2nd Armée sector. In March 1917 it moved to the 6th Armée sector and participated in the Battle of Chemin des Dames while based at La Cense. Later it moved to Ham in the vicinity of the Somme and was placed under command of Capitaine Hély d'Oissel. At the end of 1917 F 55 re-equipped with Sopwith 1½ Strutters.

F.40, Army Type 42. There were a large number of variants of the basic F.40; many of them differed from the standard F.40 only in the type of engine fitted. Reairche.

F.40H. A small number of these aircraft were used by the Aviation Maritime, and one saw service as an airliner postwar with Compagnie Franco-Bilbaine de Transport Aeronautique in 1919. MA14447.

F 58, formed from MF 58 in June 1916. In February 1917 it was assigned to the 8th Armée and was based at Lunéville. In July 1917 it was sent to Camp Valdahon to assist the American units undergoing training. In September F 58 was sent to Alsace under the command of Capitaine Valleton; later in 1917 it re-equipped with A.R.1s.

MF 59, in November 1916 based at Corcieux and assigned to the 161st D.I. It re-equipped with A.R.1s in October 1917.

MF 60, in November 1916 assigned to the 21st C.A. and based at Marcelcave. In February 1917 it was assigned to the 18th C.A. In late 1917 it re-equipped with Sopwith 1½ Strutters.

F 62, formed from MF 62 in late 1915. It was based at the Somme from 1915 into 1916. In April 1916 it moved to Moreuil and to Cachy in May. It was accompanied by N 37 and N 65. In July 1916 F 62 re-equipped with Nieuport fighters to become N 62.

MF 63, in February 1917 assigned to the 1st C.C. and based at Pierrefonds. It participated in the Battle of Chemin des Dames in April 1917. It re-equipped with SPAD 16s in September 1917.

F 70, formed from MF 70 in 1916. In November 1916 it was based at Toul assigned to the 7th C.A. In December 1916 it was under the command of Capitaine Tranchant. In February 1917 it was assigned to the 8th Armée and based at Toul. Later in 1917 it was placed under the command of Capitaine Lussigny and assigned to the 20th C.A. It re-equipped with A.R.1s in late 1917.

F 71, formed from MF 71 in 1916 when that unit re-equipped with F.40s. In November it was based at Manoncourt. In February 1917 it was assigned to the 8th C.A. and based at St. Menehould in the 4th Armée sector. It re-equipped with A.R.1s at the end of 1917.

F 72, formed from MF 72 in 1916. In November 1916 it was assigned to the 7th C.A. and based at St. Menehould in the 4th Armée sector. The next month F 72 was placed under the command of Capitaine Tranchant. In January 1917 it moved to Bouleuse. In February 1917 F 72 was assigned to the 7th C.A. and based at Bouleuse in the 5th Armée sector. It participated in the Battle of Chemin des Dames. In September F 72 moved to Verdun in the 2nd Armée sector; at that time it re-equipped with A.R.1s.

Artillery Cooperation

F 201, formed as an S.A.L. unit in January 1916 with Farman F.40s. It was under the command of Lieutenant Bouzereau and based at Marcelcave in the vicinity of the Somme. F 201 provided artillery spotting for the 35th C.A. and also flew photographic reconnaissance missions. Lieutenant Fresnay was in command from October 1916 to 23 December 1916, when Capitaine Houdemon took over. Also in October, the unit was redesignated F 201 and reassigned to the 18th C.A. In February 1917 F 201 moved to the 1st C.A. sector in preparation for the forthcoming Battle of Chemin des Dames. During the battle it was commanded by Lieutenant Martin and, like so many other army cooperation units during that battle, suffered heavy losses. In July 1917 it re-equipped with A.R.1s to become AR 201.

F 203, created in January 1916 and initially assigned to the 113th Regiment d'Artillerie Lourde. In November 1916 it was assigned to 2nd C.A.C. and was based at Le Hamel. F 203 was active over the Somme front. In December it was sent to the Aisne and later assigned to the 83rd Regiment d'Artillerie Lourde. It was active during the Battle of Champagne in 1917. F 203 later saw action at Flanders and during the attack on Fort Houthulst. In October 1917 it re-equipped with A.R.1s.

F 204, formed in December 1915. In November 1916 it was assigned to 20th C.A. and based at Fricourt in the 6th Armée sector. It was re-equipped with Sopwith 1½ Strutters in October 1917.

F 205, created in January 1916. In November 1916 it was assigned to the 21st C.A. and based at Marcelcave. It re-equipped with A.R.1s at the end of 1917.

F 206, formed in February 1916. In February 1916 it was assigned to the 12th C.A. and based at Le Pilier. It re-equipped with Sopwith 1½ Strutters in October 1917.

F 208, formed in March 1916, assigned to 9th C.A. and based at Croix-Comtesse in the 6th Armée sector. It re-equipped with Sopwith 1½ Strutters late in 1917.

F 209, formed in October 1916 from V 209. In November 1916 F 209 was based at Ferme Porte in the 1st Armée sector. It re-equipped with Caudron R.4s in January 1917.

F 210, formed in June 1916 from elements of V 210. In July it carried out bombing raids on the train stations at Laon and Saint-Erme-Outre-et-Ramecourt. In November it was based at

A spotlight-equipped F.40 of MF 33. The airplane was captured by the Germans as indicated by the cross on the rudder. B83.4013.

F.40 of F 44. F 44 used Farman F.40s beginning in June 1916 and was based near Verdun. B86.2586.

Arcy-ste-Restitute in the 5th Armée sector. At the end of 1916 it re-equipped with Caudron R.4s to become R 210.

F 211, formed in January 1916. and assigned to the S.A.L. units in the 6th Armée sector on the Somme front. It remained on the Somme front until the end of 1916, under command of Lieutenant Moreau de Bonrepos. In November 1916 it was assigned to the 12th C.A. and based at Le Pilier. In early 1917 F 211 moved to the Champagne front, where it participated in attacks on the Hindenburg line in March and April. In May, under the command of Lieutenant Chamouton, it converted to A.R.1s.

F 212, formed from V 212 in June 1916. It re-equipped with Sopwith 1½ Strutters in March 1917.

F 215, formed in February 1916. In November 1916 it was assigned as an S.A.L. unit for the G.A.N. and was based at Morlancourt in the 6th Armée sector. It became MS 215 in September 1917.

F 216, formed in June 1916 from V 216. In November 1916 it was assigned to 2nd C.A. and was based at Lemmes. In late 1917 it converted to Sopwith 1½ Strutters.

F 218, formed in January 1916. In November 1916 it was with the G.A.R. and training at the Cours d'Artillerie de Beauvais. At the end of 1917 it re-equipped with Breguet 14 A2s.

F 221, formed in January 1916 and in November was assigned to the 2nd C.A. and based at Lemmes. In February 1917 it was assigned to the 16th C.A. and moved to Julvecourt in the 2nd Armée sector. It was later assigned to the 13th C.A. from 15 July to 28 September 1917 and was under the command of Capitaine Bosc. It later served with the 2nd Armée at the Meuse. F 221 re-equipped with Sopwith 1½ Strutters in May 1918.

F 223, formed from V 223 in late 1916. It was based at Saizerais in the 8th Armée sector in November 1916. It remained there until March 1917. In July F 223 was despatched to Lemmes in the 4th Armée sector, where it re-equipped with Sopwith 1½ Strutters.

F 228, formed in February 1916. Although the escadrille initially had M.F.11s on strength it soon equipped with F.40s. F 228 was assigned to the 4th Armée in March 1916 in the Verdun sector, commanded by Lieutenant Verdurand. It was active at Mort-Homme and the Corbeaux Woods. In October it re-equipped with Caudron G.6s.

F 353, formed in April 1917 and commanded by Lieutenant Delbos. In July the escadrille was redesignated F 465. F 465 moved to Bourget, where it remained until September, when it was sent to Bellefontane to support the operations at Verdun. F 465 returned to the C.R.P. in March 1918 and re-equipped with Letords.

T.O.E. Escadrilles

MF 382, assigned to Albania beginning in 1915. It re-equipped with A.R.1s in May 1917.

MF 384, formed from Escadrille 522 in September 1916. It was based in Serbia and converted to Breguet 14s in May 1917.

MF 398, formed from MF 98 *Tenedos* in September 1916. It was based in Serbia and re-equipped with Breguet 14s in December 1917 to become BR 524.

Searchlight-equipped F.40 of F 55. F 55 was active in the 4th, 5th, and 2nd Armée sectors. B89.3273.

F.40 of C 220. The two mushrooms painted on the nose are a personal insignia. The mushrooms had white stems and a red dome. The definitive insignia of Escadrille 220 was a duck walking two ducklings on a leash. B85.506.

MF 399, formed from MF 99 in September 1916. Based in Serbia, it re-equipped with Breguet 14s in December 1917.
F 553 and **F 554** were created for service in Morocco and Algeria.

Bombing

The F.40s saw limited use as day and night bombers. The checkered career of the F.40 bombers can be seen in the records of F 29 assigned to GB 4. A preliminary analysis of the type was issued by the commander of GB 4 on 20 April 1916:

"No F.40 has yet been received by GB 4, although six were to be delivered on the 20th of April. This plane, according to its manufacturer and the Service des Fabrications Aeronautiques (SFA), and its pilots, should be satisfactory as a bomber. Its ability to rapidly climb should enable it to avoid attacks from enemy aircraft. It is able to carry a bombload of 150 kg on short range raids. Its radius of action with 300 liters of fuel and a supplemental oil supply is 300 km."

However, when the pilots received their new planes on 1 May 1916 they were not pleased with their performance or reliability. On 22 June the first major raid by F.40s was undertaken against the barracks and train station at Mulheim. Nine bombers were escorted by three Nieuport 11s and a Sopwith 1½ Strutter. The flight was attacked by eight enemy fighters. During the ensuing battle a German fighter was reported destroyed and a single F.40 was lost. However, it was obvious that without fighter escort the F.40s were almost as defenseless as the M.F.11s they were intended to replace. Furthermore, the engines performed poorly and representatives of the Farman firm were required to go to F 29's airfield and make modifications. No fewer than 14 F.40s were out of commission while being fixed. On 20 September Maurice Happe wrote that "the time of the Farman is passed." He demanded that his unit be relieved of the "burden" of the F.40s and that it be given Sopwith 1½ Strutters. The F.40s were available for a major raid on the Mauser factory on 12 October, 1916. Twelve F.40s from F 29 and F 123 were sent out but only five were able to reach the target. The others either had to turn back due to engine failures or were shot down by enemy fighters or anti-aircraft fire. The German fighters, attacking the F.40s from behind and below, were able to destroy the bombers with relative ease. The diary of GB 4 concluded that "it is not possible for F 29, F 123, and BM 120 to fly long-distance raids without heavy losses and so we advise that our units be re-equipped with Sopwiths and that we combine our operations with the English aviation."

The GQG, shocked by the heavy losses during the Mauser factory raid, concluded that the F.40s were not suitable for front-line use. The Sopwith 1½ Strutters arrived in November and F 29 became SOP 29.

MF 25 also had been a dedicated Farman bomber unit and had received F.40s by the fall of 1916. By that time the F.40s were recognized as being too vulnerable for day missions and most attacks were being made at night. By the end of 1916 MF 25 had 16 planes on strength, mostly F.40s. (The unit retained the MF 25 designation because it had now become famous for its audacious bombing raids.) Targets attacked during 1917 were primarily train stations, but industrial targets and the airship hangars at Coblence, Cologne, and Aux-la-Chapelle were also hit. When the Verdun offensive began in August 1917, MF 25 joined with GB 1 to attack targets along the front. MF 25 was based at Senard with GB 1. A series of German attacks on MF 25's and GB 1's airfield forced both units to move to Bellefontaine, but not before planes from both units attacked German airfields at Caix in retaliation. The escadrille saw little action during the remainder of 1917 as bad weather prohibited operations. On 18 February MF 25 received Voisin 8s and 10s to become V 25.

MF 123 was commanded by Lieutenant Mouraud. It was initially based at Alsace, but moved to Malzéville in October 1916. It re-equipped with Sopwith 1½ Strutters two months later.

Foreign Service
Belgium

Belgium acquired F.40s beginning in 1916. Approximately ten F.40s, F.41s, F.40bis, and F 1,40s were purchased. They served with the 8th Escadrille de Reconnaissance and Bombardement de Nuit based at Coudekerke in 1918. The Belgian air service decided to produce copies of these machines. The basic F.40 design was modified by the Belgian military aviation workshop under the direction of Lieutenant Georges Nelis. The changes introduced included a simplified landing gear and the shape of the nacelle was altered. The Jero firm produced six basic types:
GN 1: modified nacelle and Gnome Rhône engine.
GN 2: streamlined ("tadpole") nacelle; Gnome Rhône engine.
GN 3: 220-hp Hispano-Suiza engine with radiators on either side of the central nacelle. It was used as a night fighter and mounted two machine guns in the nose.
GN 4: Vickers machine gun mounted on the central nacelle; 220-hp Hispano-Suiza engine.
GN 5: 220-hp Hispano-Suiza engine and a circular radiator mounted in the nose of the nacelle.
GN 6: frontal radiator; 220-hp Hispano-Suiza engine.

In mid-1916 Farmans of all types were assigned to the 1st, 2nd, and 3rd Escadrilles based at Saint-Idesbald; in June all three units moved to Maires to support the Flanders offensive. By 1918 two new escadrilles were using Farmans: the 7th at Houthem and the 8th at Coudekerke. The 1st, 2nd, and 3rd Escadrilles had re-equipped with other types by 1918. The remaining F.40s were quickly replaced postwar.

F.40 at Sfax, Tunisia, probably of Escadrille 490. This escadrille later became a G.6 unit designated G 490. Renaud.

F.40 at Sfax, Tunisia, probably also of Escadrille 490. Renaud.

F.40 in Belgian service; approximately ten F.40s, F.41s, F.40bis, and F1.40s were purchased. These types served with the 8th Escadrille de reconnaissance and bombardement de nuit based at Coudekerke in 1918. Via Colin Owers.

F.40 of the 21st Escadrille of the IRAS in February 1917. The Imperial Russian Air Service purchased 20 F.40s in the summer of 1916. B85.2467.

Italy
The Italian Savoia firm built Farman F.40s under license. These served in the Farman squadriglias along with the M.F.11s.

Netherlands
The Dutch air service purchased a single F.40 in 1916. Given serial F 801 (later F 901), the F.40 served until 1920.

Norway
Twelve F.40s were purchased by the Norwegian army air service in 1916, The tail assemblies were later replaced by those from surplus M.F.11s because those on the F.40s had proved to be defective and had caused a number of crashes.

Portugal
Portugal purchased five F.40s in October 1916. They were supplied to the Escola Aeronautica Militar and used for pilot and observer training. One was sent to Mozambique in September 1917 as part of the Esquadrilha Expedicionaria. It was lost in an accident that killed the pilot. The remaining F.40s were retired in 1920.

Romania
Romania had approximately 20 F.40s on strength by 1917. They served with these units:
Grupul 1 (2nd Romanian Army): Escadrillas F-2 and F-6.
Grupul 2 (4th Romanian Army): Escadrilla F-4 (reconnaissance).
Grupul 3 (6th Romanian Army in Russia): Escadrilla F-5.
Escadrilla F-7 (an independent reconnaissance unit under the direct command of General Marele Cartier).

By August 1917 these units had been reorganized as follows:
Grupul 1: Escadrillas F-2 and F-6 assigned to the 2nd Romanian Army and Escadrilla F-4 with the 4th Romanian Army; all on the Russian front.
Grupul 2 (1st Romanian Army): Escadrilla F-7.
Grupul 3 (6th Romanian Army on the Russian front): Escadrilla F-5.

While it had been intended to use the F.40s as reconnaissance aircraft, some of the pilots in the Romanian units converted their planes to bombers by adding makeshift bomb racks. F-2, F-4, F-6, and F-7 all used their F.40s to bomb enemy targets. A crew from F-6 destroyed an enemy fighter on 19 May 1917, becoming the first Romanian crew to shoot down an enemy plane in an F.40. The Armistice was signed four months later and the F.40s were soon retired from front-line units.

Russia
The Imperial Russian Air Service purchased 20 F.40s in the summer of 1916. The type was disliked by the Russians, who noted that versions with the 150-hp Renault had extremely poor stability, while the versions with the 130-hp Renault were considered to be unacceptable for service. The Russians preferred the H.F.30 and so the F.40s were not built in Russia and those received from France were probably relegated to training units.

Serbia
The Serbian air service received F.40s during the latter part of 1916 and they re-equipped units flying M.F.11s—F 82, F 84, and F 99 (all based at Vertekop) and F 98 (based at Gorgop). As 1916 drew to a close, Serbian army units were advancing and the air units followed, soon being based at Verbanc. In November and December the Farman units flew army cooperation and reconnaissance missions. In March 1917 F 84 was redesignated F 384 and became an escadrille of the high command. It was based at Vertekop and flew reconnaissance and bombing missions at the direction of the command. F 99 was redesignated F 389 and assigned to the 1st Army. F 98 became F 398 and was also assigned to the 1st Army. F 82

F.40 Two-Seat Reconnaissance Plane with 130-hp Renault 8C

Span 17.59 m; length 9.25 m; height 3.90 m; wing area 52 sq. m
Empty weight 748kg; loaded weight 1,120 kg
Maximum speed: 135 km/h at 2,000 m; ceiling 4,050 m; climb to 2,000 m in 15 minutes; range 420 km; endurance 2 hours 20 minutes
Armament: a single 0.303 Lewis machine gun and 240 kg of bombs; in the F.40P ten Le Prieur rockets

F.40 Two-Seat Reconnaissance Plane with 130-hp Dion-Bouton 12B

Span 17.59 m, length 9.15 m; height 3.75 m; wing area 52 sq. m
Empty weight 750 kg; loaded weight 1,125 kg
Maximum speed: 135 km/h

F.40 H Two-Seat Seaplane Trainer with 130-hp Renault 8C

Span 17.67 m; length 9.25 m; height 3.95 m; wing area 55 sq. m
Empty weight 770 kg; loaded weight 1,200 kg
Maximum speed: 110 km/h

F .40bis Two-Seat Reconnaissance Plane with 160-hp Renault 8Gc

Span 17.59 m; length 9.15 m; height 3.75 m, wing area 52 sq. m
Loaded weight 1,160 kg
Maximum speed: 140 km/h at 2,000 m; climb to 2,000 m in 12 minutes

F .40ter Two-Seat Reconnaissance Plane with 150-hp Lorraine 8A

Span 17.59 m; length 9.15 m; height 3.75 m; wing area 52 sq. m
Loaded weight 1,160 kg
Maximum speed: 150 km/h

F.40 QC Two-Seat Reconnaissance Plane with 130-hp Renault 8C

Span 17.59 m; length 9.48 m; height 3.75 m; wing area 52 sq. m

became F 382 and was assigned to the 2nd Army.

By mid-1917 the F.40 units had re-equipped with A.R.1s and Breguet 14s.

United Kingdom

The RNAS purchased approximately 50 F.40bis (army Type 56) from France in 1916. They were flown directly from the Farman factory at Paris to the naval base at Great Yarmouth. Twenty of them were the F.40bis with 150-hp Renault engines; these had serials 9155 through 9174 and were based at Hendon, Eastchurch, Grain, and Dunkirk and were assigned to Nos.1 and 2 Wing. A second batch of F.40bis (army Type 56) were purchased in 1916; these were based at Eastchurch.

United States

Thirty F.40s were obtained by the A.E.F. and used as trainers. Apparently, none were sent to the U.S. after the war.

Farman F.40 serial F 4073 in U.S. Air Service markings flown by Lt. Leigh Wade.

Venezuela

A French mission to Venezuela took with it two F.40s. They were used for training.

Farman F.41 and Variants

The F.41 was produced shortly after the F.40. It used an 80-hp Renault 8B engine and was smaller than the standard F.40. The wing span length were reduced, although the height was unchanged. The aircraft also had a more angular nacelle, similar to that used on the M.F.11. The F.41 was 145 kg lighter than the F.40. The F.41 was built in 1915 and was given army type numbers 40 and 41. Some had 80-hp De Dion-Bouton engines. Aside from these changes, the aircraft retained the unequal span wings, tail configuration, and landing gear of the F.40 series.

Variants

1. The F.41H was a seaplane version of the F.41 which was built in 1917.
2. The F.41bis was a modified F.41 with a more powerful 110-hp Lorraine 6AM engine. It was given the army type number 44 and was produced in 1916. The more powerful engine gave it the ability to carry a heavier payload.
3. The F.41bisH was a floatplane version of the F.41bis built in 1917.
4. The Farman F.51 (army Type 51) was a variant of the F.41 and, like that aircraft, was a two-seat army cooperation aircraft. The main difference between the two was that the F.41 was powered by a 80-hp Renault 8B, while the F.51 used a 120-hp Lorraine 6AM. Dimensions were identical to the F.41.

F.41 Two-Seat Reconnaissance Plane with 80-hp Renault 8B

Span 16.32 m: length 9.15 m; height 3.75 m, wing area 49.5 sq. m
Empty weight 605 kg; loaded weight 930 kg
Maximum speed: 128 km/h at 2,000 m; climb to 2,000 m in 18 minutes; range 380 km

The F.41 H Two-Seat Seaplane with 80-hp Renault 8B

Span 16.32 m: length 9.25 m; height 3.90 m, wing area 49.5 sq. m
Empty weight 605 kg; loaded weight 910 kg
Maximum speed: 110 km/h; range 340 km

F.41bis Two-Seat Reconnaissance Plane with 110-hp Lorraine 6AM

Span 16.32 m: length 9.15 m; height 3.75 m, wing area 49.5 sq. m
Empty weight 610 kg; loaded weight 950 kg
Maximum speed: 128 km/h

F.41bis H Two-Seat Seaplane with 110-hp Lorraine 6AM

Span 16.32 m: length 9.25 m; height 3.90 m, wing area 49.5 sq. m
Maximum speed: 110 km/h

Operational Service

The F.41 was used in limited numbers by Farman escadrilles (see entry for F.40) but by 1917 it was considered to be too vulnerable to enemy aircraft and was sent to training units.

The Brazilian naval air service obtained two Farman F.41 floatplanes in 1919. Given serial numbers 20 and 21, they served until 1921.

This aircraft is identified in the SHAA files as an F.44. It is actually an F.41bis, serial F 1758, which was a modified F.41 with a more powerful 110-hp Lorraine 6AM engine. It was given the Army type number 44 (hence the F.44 designation) and was produced in 1916. B89.2812.

Above: F.41, Army Type 51. MA176.
Left: F.41 F-AHMF (c/n 6828) in civilian service after the war. Most aircraft used an 80-hp Renault 8B, although some aircraft had 80-hp De Dion-Bouton engines. B93.2683.

Farman F.1,40 and F.1,41

The F.1,40, intended as an improvement of the basic F.40 design, was a two-seat training variant of the standard F.40 and was produced in 1916. In Farman's designation system the addition of the "1" may have indicated dual control. Most F.1,40s were powered by a 130-hp Renault 8C or 130-hp De Dion engine; some versions were equipped with 150 or 170-hp Renault engines. A diagram (from the Farman brochure) shows that the F1,40 had few changes from the F.40; the wing span was slightly increased over that of the F.40 and the wing area was enlarged by 3.0 sq. m. The length and height were unchanged. The empty weight and loaded weight were higher. The aircraft was designated the army Type 40.

The F.1,41 had the same engine but featured a smaller wing span and reduced wing area. Length and height were unchanged from the F.40. While most F.1,41s were powered by a 80-hp Renault 8B, some had 80hp De Dion engines. The F.1,41 was designated Army Type 70.

Variants

1. F.1,40bis—Floatplane version intended for use as a trainer in 1916. Powered by a 110-hp Lorraine 6AM.
2. F.1,40bis—The F.1,40, but fitted with a more powerful 160-hp Renault 8C engine. This new engine resulted in higher empty and loaded weights and an increased maximum speed of 150 km/h. It was designated army Type 60 and was built in 1916.
3. F.1,40ter—Identical to the standard F.1,40 but powered by a 160-hp Lorraine 8Aby engine. Some had a large camera fitted vertically in the nose. Designated the army Type 61, the F.1,41ter was built in 1916.
4. F.1,41H—Floatplane trainer produced in 1917. It had dimensions similar to the standard F.1,40 but the floats reduced the maximum speed to 110 km/h.
5. F.1,41bis—This was the basic F.1,41 with a 110-hp Lorraine 6 AM engine. It was produced in 1916 and designated army Type 70.
6. F.2,40—Identical to the F.1,40 except for an altered wing profile and a Lorraine 8Bd 275-hp engine. It was built in 1918.
7. Farman 51 E 2—Version of the F.1,41 with wings of altered camber.

F.1,41 Two-Seat Trainer with 80-hp Renault 8B
Span 17.67 m; length 9.15 m; height 3.75 m; wing area 55 sq. m
Empty weight 772 kg; loaded weight 1,190 kg
Maximum speed: 145 km/h

F.1,40bis Two-Seat Floatplane Trainer with 110-hp Lorraine 6AM
Span 17.67 m; length 9.50 m; height 4.31 m; wing area 55 sq. m
Empty weight 740 kg; loaded weight 1,065 kg
Maximum speed: 130 km/h

F.1,40 Two- Or Three-Seat Trainer with 130-hp Renault
Length 9.250 m; height 3.900 m; wing area 55 sq. m
Empty weight 772 kg; loaded weight 1,190 kg
Maximum speed: 110 km/h; climb to 2,000 m in 12 min.

F.1,40bis Two-Seat Trainer with 160-hp Renault 8C
Span 17.67 m; length 9.15m; height 3.75 m; wing area 55 sq. m
Empty weight 780 kg; loaded weight 1,200 kg
Maximum speed: 150 km/h

F.1,40ter Two-Seat Trainer with 160-hp Lorraine 8Aby
Span 17.67 m; length 9.15m; height 3.75 m; wing area 55 sq. m
Empty weight 780 kg; loaded weight 1,200 kg
Maximum speed: 150 km/h

F.1,41 Two-Seat Trainer with 80-hp Renault 8B
Span 16.39 m; length 9.15m; height 3.75 m; wing area 52 sq. m
Maximum speed: 130 km/h

F.1,41H Two-Seat Floatplane Trainer with a 80-hp Renault 8B
Span 16.39 m; wing area 52 sq. m
Empty weight 780 kg; loaded weight 1,200 kg
Maximum speed: 110 km/h

F.1,41bis Two-Seat Trainer with 110-hp Lorraine 6AM
Span 16.39 m; length 9.15 m; height 3.75 m; wing area 52 sq. m

F. 2,40 Two-Seat Trainer with 275-hp Lorraine 8Bd
Span 17.67 m; length 9.15 m; height 3.75 m; wing area 55 sq. m

Army Type 61 serial F 2915. The Farman designation was F.1,40ter, which was identical to the standard F.1,40 except that it was powered by a 160-hp Lorraine 8Aby engine. The F.1,41ter was built in 1916. MA0301.

Farman F.50

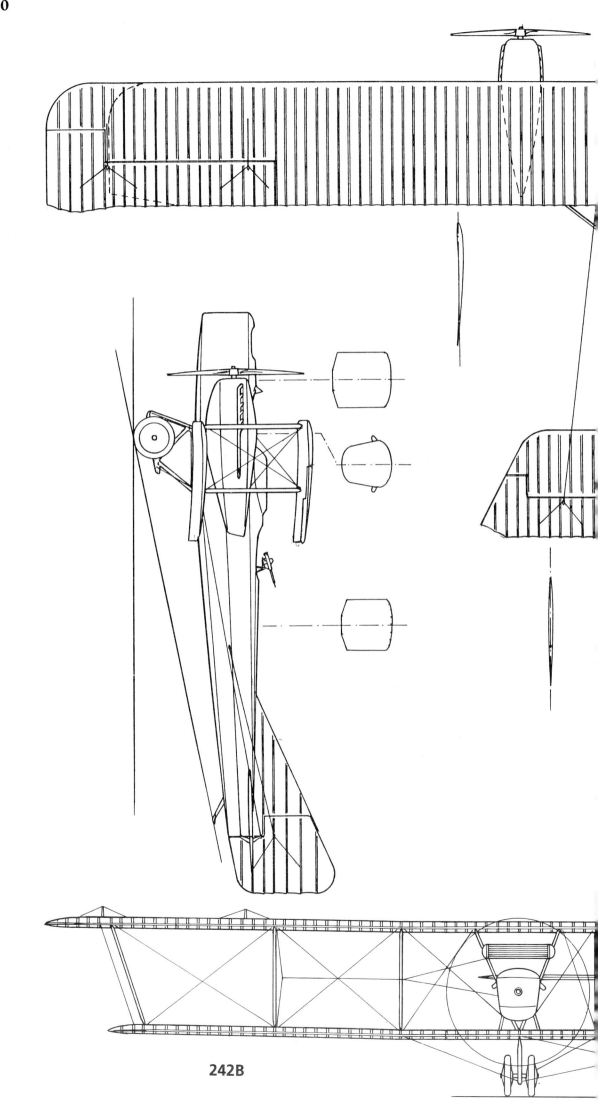

242B

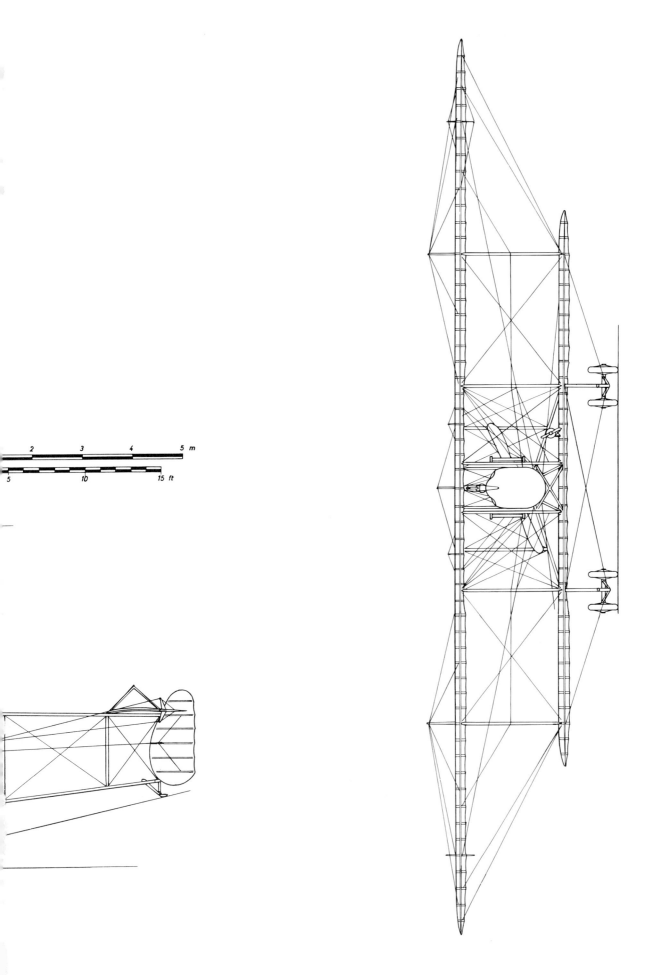

Farman F.40

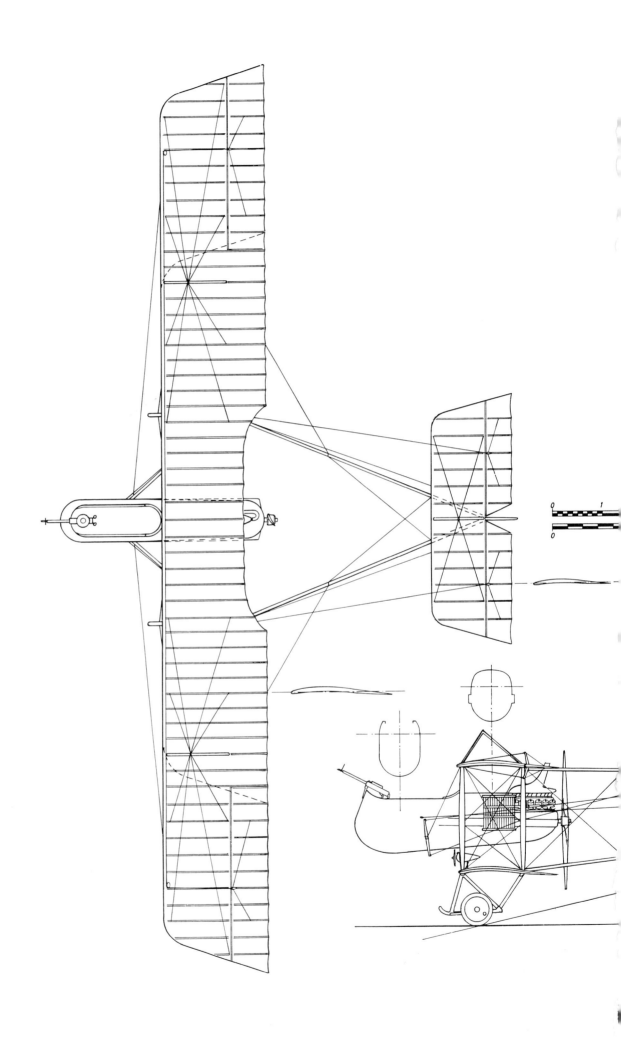

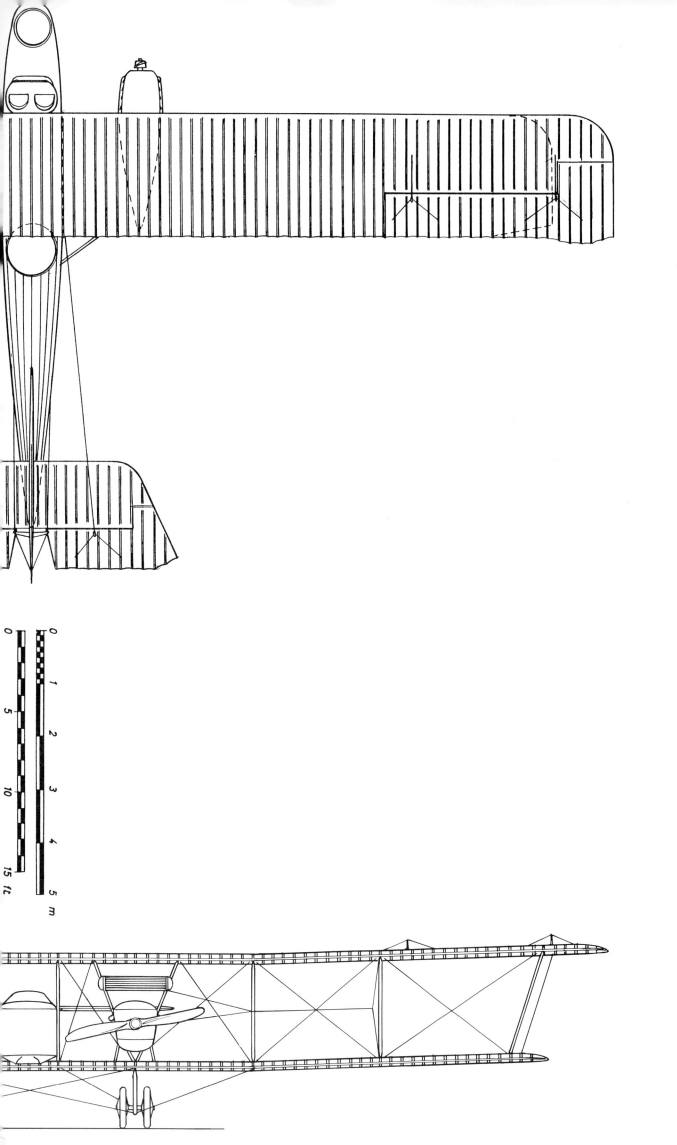

Above: The F.1,40bis was an F.1,40 fitted with a more powerful 160-hp Renault 8C engine. This aircraft was designated as an Army Type 60, and is often referred to as the F.60. MA0299.

Right: Designated the Army Type 60, this was actually a F.1,40bis. The Renault engine gave a maximum speed that was 40 km/h faster than the standard F.1,40. MA38547.

Below: F.1,40ter. The F.1,40ter was identical to the standard F.1,40 except that it was powered by a 160-hp Lorraine 8Aby engine. Some aircraft had a large camera fitted vertically in the nose. Reairche.

Farman F.1,46

F.1,46 was a dual-control trainer (as indicated by the "1" modifier). The F.1,46 differed little from the F.40 on which it was based. The engine was a 80-hp Renault. The price varied between 33,000 to 39,600 FF.

Farman F.1,46 Dual-Control Trainer with 80-hp Renault
Span 17.600 m; length 9.600 m; height 3.750 m; wing area 55 sq. m
Empty weight 675 kg; loaded weight 1,000 kg;
Maximum speed: 100 km/h; climb to 2,000 m in 20 minutes.

Henri Farman H.F.36

The H.F.36 represented a major departure for Henri Farman in many ways. The aircraft was a tractor, unlike the majority of his designs which were pushers. Furthermore, it was made of metal. Farman's only previous all-metal design was the preceding H.F.35. Like the H.F.35, the H.F.36 also featured a quadracycle undercarriage which was required to support the machine's weight. The engine was also a 220-hp Renault 12Fa as used on the H.F.35. The two-bay wings were of sesquiplane configuration, the upper one being considerably longer than the lower. Ailerons were on the top wing only. The fuselage tapered to a point upon which a conventional fin and stabilizer were attached. It appears the crew were seated close together with the pilot's head looking over the top wing. The gunner sat immediately behind him. The upper wing was attached directly to the fuselage, while the lower wing was extended from the fuselage by a series of struts. Intended as a fighter/attack aircraft, it was tested in July 1916 but there were apparently some serious problems. Despite numerous modifications the H.F.36 was not selected for production.

H.F.36 Two-Seat Fighter with 220-hp Renault 12Fa
Span 19.82 m; length 8.60 m; height 3.49 m; wing area 66.5 sq. m
Payload 900 kg
Maximum speed 175 km/h
One built

H.F.36 taxiing. The airplane was intended for use as a fighter/attack aircraft. It was tested in July 1916. MA36409.

Farman F.43

The Farman F.43 was an enlarged and armored version of the F.40. It had the more powerful 220-hp Renault 12Fa engine and was built in 1916. It was given the army type numbers 47 and 48. To accommodate the extra weight of the armor, the wing span was increased by 0.41 m, which increased the wing area 8.0 sq. m. The length was increased by 0.53 m. These changes plus the armor resulted in an empty weight 230 kg higher than the standard F.40, while the loaded weight was 425 kg higher. Despite the increase in size and weight, the F.43 was 15 km/h faster than the standard F.40.

GQG memos mentioned that armored planes were to be used for low-altitude attacks against enemy troops. Obviously, the armor was intended to protect these planes from the intense ground fire they would encounter. The armored Caudron G.4s were preferred because their twin engines provided a better performance with a heavier weapons load. In fact, another GQG memo was sent to Caudron units reminding them that there were F.43s at the R.G.A. that could be used to supplement the G.4s if needed. However, reports clearly show that both the F.43s and Caudron G.4s were of limited use and were to be employed only when the weather (preferably overcast skies) would limit fighter opposition. The Farman F.43s were clearly too slow for use in 1916 and were soon relegated to training units.

F.43 Two-Seat Armored Reconnaissance Plane with 220-hp Renault 12Fa
Span 18.00 m; length 9.68 m; height 3.73 m; wing area 61 sq. m
Empty weight 950 kg; loaded weight 1550 kg
Maximum speed: 160 km/h

F.43 at the Chartres flight school. An armored version of the F.40 intended for ground attack, the F.43's inadequate performance soon caused it to be relegated to training. Reairche.

Farman F.44

The Farman firm continued to experiment with the standard F.40 design despite the fact it must have been apparent that the pusher layout was obsolete. Perhaps in an attempt to improve the type's lackluster performance, the firm fitted a 170-hp engine to an F.40 airframe. The identity of the engine is not known. The resulting plane was designated the F.44, and it retained the same layout as the F.40. By this time the STAé under Dorand had decided to abandon the development of the F.40 series and had, instead, ordered the A.R.1 into production. It is not surprising then that the F.44 was not selected for service use. At last bowing to the inevitable, Farman's next design had a tractor layout.

F.44. Reairche.

Farman F.30 C2

The Farman F.30 (not to be confused with the earlier H.F.30) was developed to meet the C2 specification calling for a two-seat fighter with fixed forward armament and a rear machine gun on a flexible mount. Built and flown in December 1916, it marked a significant departure for the Farman firm as the F.30 used the tractor layout previously spurned.

The F.30A had a number of unusual (and unsatisfactory) features. The pilot was located directly beneath the top wing, where his field of vision was severely restricted. The pilot and gunner were separated by a large, vertical radiator that must have inhibited their communication. The lower wing was suspended beneath the fuselage by a series of struts. The wings were of unequal span, the upper wings being considerably longer. Ailerons were located on the extended portions of the upper wing. The engine was a 160-hp Canton-Unné X-9.

The aircraft was tested at Villacoublay in May 1917. Its flying qualities were described as mediocre and it was rejected by the army. However, the Farmans apparently felt their design had

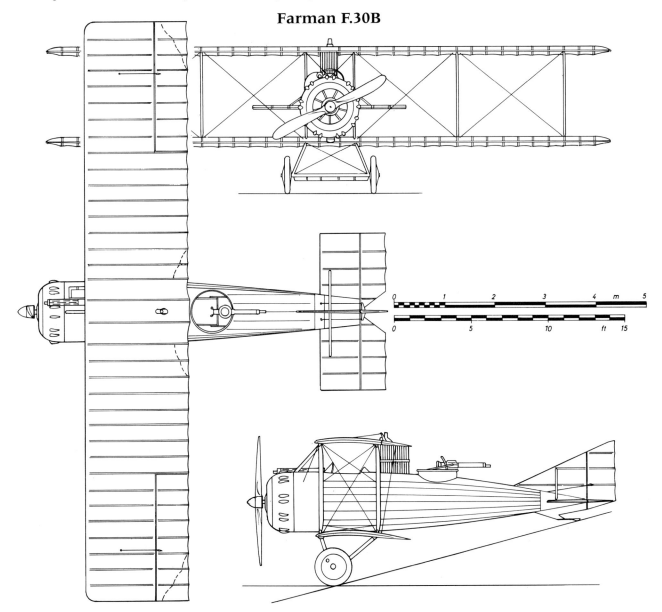

Farman F.30B

merit, as they redesigned the F.30, the new aircraft being designated the F.30 type B. Changes included switching from the sesquiplane layout to equal span wings, although the ailerons still remained on only the top wings. The engine was changed to a 260-hp Canton-Unné 9Za and there were modifications to the engine exhaust system. The observer's gun mount was modified by raising its ring. The tail and rudder area were enlarged and the tailplane was now braced with struts. Armament was a fixed, forward-firing 7.7-mm Vickers and a 7.7-mm Lewis machine gun on a ring mount. The F.30 type B was tested at Villacoublay in July 1917 and as a result, the STAé requested that the surface area of the wings be increased from 34.7 to 54.0 square meters. Apparently these changes were needed because the aircraft's center of gravity was different than had been originally calculated. It is possible that this new version was designated the Farman F.30 type B A.R.2. As the A.R.2 designation stood for Avion Reconnaissance, this would establish that the Farman F.30 was no longer being considered for the C2 fighter role. Further development of the F.30B A.R. series was abandoned in April 1918 because both longitudinal and lateral stability were found to be "irredeemably" deficient, almost certainly as a result of the center of gravity miscalculation. The postwar F.110 was based on certain aspects of the F.30B A.R.2 design.

F.30A Two-Seat Fighter with 160-hp Canton-Unné X-9

Span 11.015 m; length 7.29 m; height 2.96 m; wing area 34.71 sq. m
Empty weight 680 kg, loaded 1,100 kg

Maximum speed:	2,000 m	208 km/h
	3,000 m	204 km/h
	4,000 m	196 km/h
Climb:	2,000 m	6 minutes 35 seconds
	3,000 m	11 minutes
	4,000 m	16 minutes 50 seconds
	5,000 m	24 minutes 50 seconds

Ceiling 6,800 m, endurance 2 hours and 30 minutes
Armament: one synchronized 7.7-mm Vickers machine gun and one 7.7-mm Lewis gun in a swivel mount
One built

F.30B type A.R.2 Two-Seat Fighter Or Reconnaissance Plane with 260-hp Canton-Unné 9Za

Span 14.00 m; length 8.540 m; height 2.96 m; wing area 54.0 sq. m
Empty weight 825 kg; loaded weight 1,375 kg
Maximum speed 210 km/h
One built

F.30A. The Farman F.30 was developed to meet the C2 specification calling for a two-seat fighter. MA36408.

Below: F.30B. Changes from the F.30A included switching from the sesquiplane layout to equal-span wings and the engine was changed to a 260-hp Canton-Unné 9Za. MA01096.

Farman F.45

In 1916 the technical service of the French air service asked the Farman brothers to develop a reconnaissance aircraft and specifically requested a tractor layout. The main flaw of Farman's pusher designs was that it made the aircraft vulnerable to rear attacks—a fact that was exploited by German fighters. Surprisingly, it has been reported that the Farman brothers were reluctant to change to a tractor layout. Commander Dorand of the STAé then requested that Captain Le Pere begin design work on a tractor reconnaissance aircraft which would subsequently become known as the A.R.1 and replace the F.40s.

Apparently, the Farman brothers eventually realized that they would need to abandon the pusher layout that had served them so well and decided to initiate work on a new reconnaissance aircraft. No fewer than five designs with a tractor layout were produced—the F.30, F.45, F.47, F.48, and F.49.

The F.45 was powered by a 170-hp Renault 8Gc engine and was built in 1916. The aircraft had a large exhaust pipe extending from the engine over the top of the wing. The pilot's cockpit was below the top wing with the observer's cockpit immediately behind. The trailing edges of both wings and the stabilizer had scalloped edges. The single-bay wings were of unequal span with the upper being significantly longer than the lower. Ailerons were on the upper wing only. The fuselage, constructed of metal and wood, was suspended midway between the wings.

As with the other tractor aircraft mentioned above, the F.45 had poor flying qualities and inferior performance. It was not accepted by the Aviation Militaire. The Farman company then switched to building A.R.1s under license.

F.45 Two-Seat Reconnaissance Plane with 170-hp Renault 8Gc
Span 15.47 m; length 8.62 m; height 3.79 m; wing area 49 sq. m
Empty weight 760 kg; loaded weight 1,200 kg
Maximum speed: 153 km/h
One built

F.45. The F.45 was powered by a 170-hp Renault 8Gc engine and was built in 1916. B89.2799.

Farman F.46E

The F.46E was intended for use as a trainer in 1916. It featured dual controls with a partition between the student and instructor. It has been estimated that up to 10,000 pilots were trained on it during the war. The aircraft had a layout similar to the F.40 but featured elongated skids with two pairs of wheels mounted on the front of each skid. This was obviously intended to prevent nosing over during landing. The main disadvantage of the aircraft was that it had a pusher layout (unlike most of the other aircraft at the front) and its control system was markedly different from most other aircraft in service. It was powered by an 80-hp Renault 8B.

F.46E Two-Seat Trainer with 80-hp Renault 8B
Span 17.60 m; length 9.60 m; height 3.75 m; wing area 55 sq. m
Empty weight 675 kg; loaded weight 1,000 kg
Maximum speed: 100 km/h; range 400 km

The Norwegian army air corps ordered two in September 1920 after it was learned that there had been numerous accidents with the M.F.7s and F.40s being used at the training schools. The two F.46Es arrived at Kjeller Airdrome in November 1920. Given serial numbers 25 and 27, they remained in service until 1928.

F.46E. The aircraft had elongated skids with two pairs of wheels mounted on the front of each skid. This was obviously intended to prevent nosing-over during landing. Reairche.

The F.46E was an F.40 with 130-hp Dion-Bouton 12 B engine built in 1915. It was intended for training. M2062.

F.46E. The F.46E was intended for use as a dual-control trainer; it has been estimated that as many as 10,000 pilots were trained on it during the war. B89.2778.

Farman F.47

The Farman F.47 was yet another attempt to produce a reconnaissance aircraft with a tractor engine. Powered by a 220-hp Lorraine 8Ba engine, it was produced in 1917. Armament consisted of a Hotchkiss machine gun and bombs on underwing racks. As in the preceding F.45, the fuselage was suspended between the wings. There were separate cockpits, the pilot's under the top wing and the observer behind the wings. No series production was undertaken.

F.47 Two-Seat Reconnaissance Plane with 220-hp Lorraine 8Ba

Span 15.47 m; length 8.18 m; height 3.82 m; wing area 39 m
Empty weight 760 kg; loaded weight 1,260 kg
Maximum speed: 180 km/h
Armament: a Hotchkiss machine gun and bombs
One built

F.47. As on the preceding F.45, the fuselage was suspended between the upper and lower wings. B89.2779.

F.47 Side View

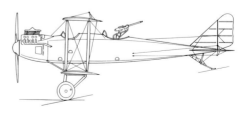

F.47. The aircraft, which was powered by a 220-hp Lorraine 8Ba motor, was produced in 1917. Reairche.

Farman F.48

The F.48 was another design with a tractor configuration intended for reconnaissance missions. It was built in 1917 and, as with the F.47, was powered by a 220-hp Lorraine 8Ba engine. The F.48 appears to have been a modified F.47, the main difference being the much smaller wing on the F.48. The span was 3.40 m less than on the F.47, and the wing area was 2 sq. m smaller. Otherwise, the performance and weights were identical. It is possible that the changes introduced on the F.48 were created in response to problems discovered during flight testing of the F.47. However, because of the type's relatively poor performance, the F.48 was not selected for use by the Aviation Militaire.

F.48 Two-Seat Reconnaissance Plane with 220-hp Lorraine 8Ba
Span 12.07 m; length 8.18 m; height 3.30 m; wing area 37 sq. m
Empty weight 760 kg; loaded weight 1,260 kg
Maximum speed: 180 km/h
One built

Farman F.49

With the decision to replace the F.40 series of reconnaissance planes with the A.R.1, the STAé had delivered a serious blow to the Farman firm. The Farmans attempted to recover by producing a series of two-seat reconnaissance planes with tractor configurations. The Farman F.49 was the last of the series to be built.

The F.45, F.47, and F.48 had all failed to meet the A2 specification because of poor performance. The F.49 may have been intended to correct these problems by having a more powerful engine—a 275-hp Lorraine 8Bd. It was a considerable improvement over the 220-hp Lorraine 8Ba used on the F.47 and F.48. Unfortunately for Farman, the F.49's performance was still inferior, and it was not selected for production. Development was abandoned in 1918.

F.49 Two-Seat Reconnaissance Plane with 275-hp Lorraine 8Bd
Span 13.00 m; length 8.80 m; height 3.10 m; wing area 46 sq. m
Loaded weight 1,288 kg; payload 150 kg
Maximum speed: 130 km/h at 4,000 m; climb to 4,000 m in 16 minutes; endurance 3 hours
One built

Henri Farman Twin-Engine Aircraft

According to a British Ministry of War report, two twin-engine planes were developed by Henri Farman in 1916. It is not known what role they were intended to fill; it is possible that they were to be entered in the 1916 concours puissant for a heavy bomber. One aircraft had twin 130-hp Renault engines, the other two 220-hp Renaults. These machines may have been the same aircraft but with different engines. Neither type was selected for use by the Aviation Militaire and it is possible that their construction was never completed. It is also possible that they served as prototypes for the Farman F.50 bomber.

Farman F.31 C2

The C2 specification of 1918 called for a two-seat fighter armed with a single fixed 7.7-mm or 11-mm machine gun or a cannon, two machine guns or a cannon on a swivel mount, and a machine gun firing to the rear and downward. Aircraft designed to meet this specification were the S.E.A. 4, Borel C2, Breguet 17, Hanriot-Dupont C2, Morane 31 C2, Vickers C2, Morane Bugatti 16 (Type AN), L.D. Ca. 2, and Farman 31.

The Farman F.31 had little in common with the previous Farman F.30, which had proved to have insufficient stability and was rejected by the Aviation Militaire. The F.31 was powered by a 400-hp Liberty 12 engine. Surprisingly, the F.31 had a lower wing suspended from the bottom of the fuselage by a series of drag-producing struts. This particular feature had been present on the F.30 A but had been eliminated from the type B. Probably to help reduce drag, a large fairing was fitted, extending from the fuselage bottom to the lower wing. A significant improvement was made in the crew positions. The radiator that had obstructed communication between the pilot and observer was eliminated. Instead, twin Lamblin radiators were placed under the nose. The pilot was located in a cutout just below the top wing; however, he was seated so far aft that his field of vision over the huge cowling must have been

F.31. The radiator which had obstructed communications between the pilot and observer in the F.30B was eliminated. Instead, twin Lamblin radiators were placed under the nose. B76.121.

minimal. The two-bay wings were of equal span and, as with the F.30, the ailerons were located on only the top wing. A triangular fin/rudder assembly and square tailplanes were fitted. Armament consisted of two 7.7-mm Vickers machine guns and a single 7-mm Lewis gun on a swivel mount.

The Farman F.31 underwent testing at Villacoublay in August 1918. Its speed and rate of climb were considered to be quite satisfactory, but the payload was considered to be too low. Load testing also revealed that the aircraft was unable to meet the requisite load factor; 7 was required but the upper wing spar failed at 5.5. As the Hanriot-Dupont 3 met all specifications, it was selected for production and the F.31 was not developed further.

F.31 Two-Seat Fighter with 400-hp Liberty 12
Span 11.76 m; length 7.35 m; height 2.58 m; wing area 40 sq. m
Empty weight 869 kg; loaded 1,469 kg
Maximum speed: 215 km/h; climb to 2,000 m in 5 minutes 50 seconds; climb to 3,000 m in 9 minutes 40 seconds; climb to 4,000 m in 15 minutes 58 seconds; climb to 5,000 m in 25 minutes 14 seconds; ceiling 6,000 m
Armament: two 7.7-mm fixed Vickers machine guns and a single 7-mm Lewis gun on a swivel mount.
One built

F.31 powered by a 400-hp Liberty engine. The pilot was so far aft that his field-of-vision over the huge cowling must have been minimal. MA1047.

Farman F.31

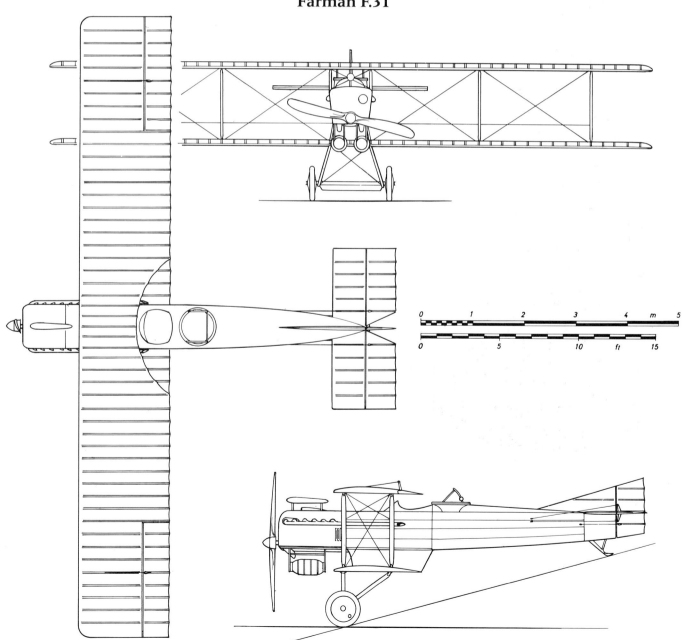

F.B.A. Type C

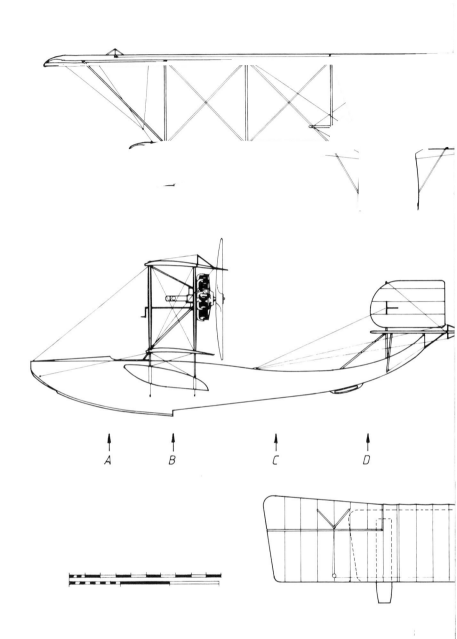

F.B.A. Type C.

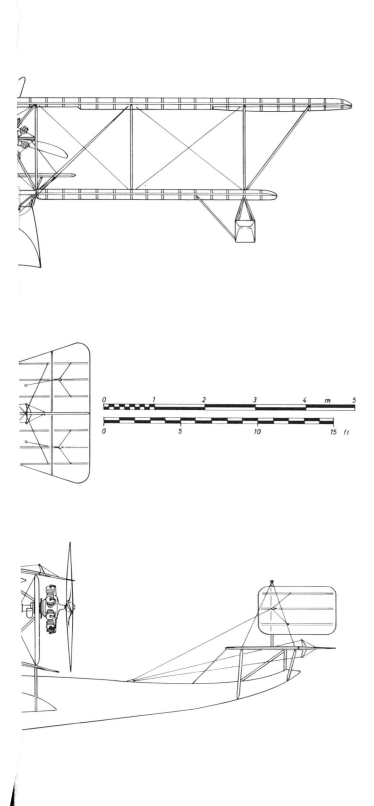

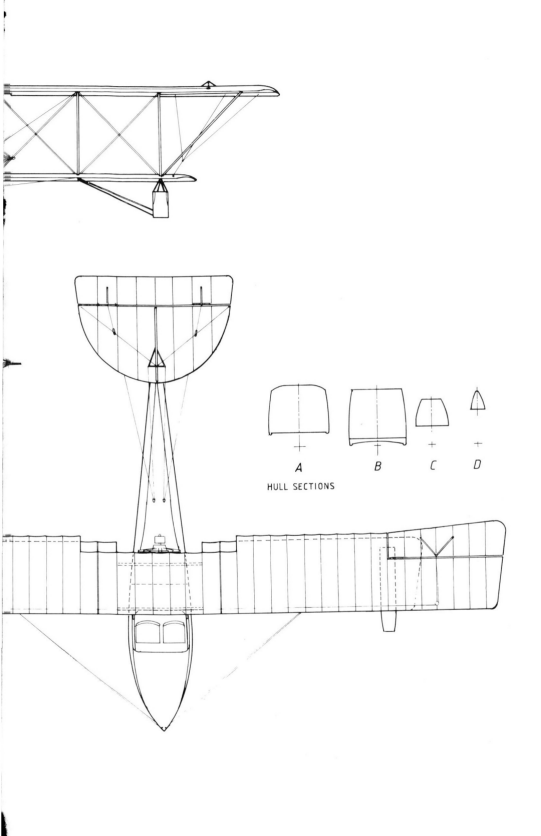
HULL SECTIONS

F.B.A. Type B

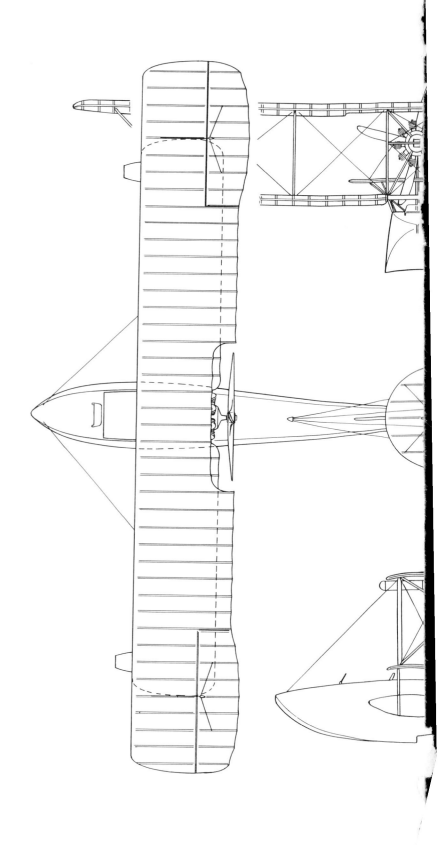

Farman F.50

The French aviation industry had conspicuously failed to design a heavy bomber comparable to those produced by Germany, Britain, and Italy. Despite competitions in 1915 and 1916 there was still no indigenous heavy bomber of comparable quality to the Handley Page V/1500 or Gotha aircraft.

In late 1917 the BN2 specification was formulated calling for a heavy bomber capable of carrying 500 kilograms of bombs over a distance of 1,000 kilometers. It was hoped that enough aircraft would be ordered to equip 36 escadrilles.

Several promising designs were produced to meet the specification. The Farman company, having lost its monopoly on reconnaissance aircraft because of reluctance to abandon the pusher layout, designed a twin-engine tractor biplane designated the F.50.

It was a conventional biplane with a top wing slightly longer than the lower. Two 240-hp Lorraine 8Bb engines powered the prototype, but these were later replaced by 275-hp 8Bds. Radiators were mounted on the tops of the engine nacelles. Behind each engine there was a 330 liter fuel tank. The engine and fuel tank were covered by an aluminum nacelle. The undercarriage consisted of a pair of wheels under each engine nacelle. While both the top and bottom wings had flat center sections, there was a pronounced dihedral on the outer sections of both. Each wing had two spars made of pine and covered in a band of strong, glued fabric, and the triangular leading edge was made of pine. The ribs were made of laminated pieces of white wood. The struts were attached with metal fittings. There were 16 struts made of pine and four V-shaped struts made of steel. Ailerons were on the top wing only. The wings had a load factor of 5.5 when static tests were conducted by the STAé.

The fuselage had a rectangular cross-section made of pine longerons in three pieces. The front fuselage was covered in plywood while the rear was wrapped in fabric. The tail was reinforced with plywood. The structure of the tail surfaces was essentially the same as that employed in the wings.

Pilot and bombardier were located in separate cockpits in front of the wing. There was a machine gun turret in both the front and rear cockpit. A corridor allowed the observer/gunner to move between the two positions and also to enter the cockpit; the observer also had a seat next to the pilot. On operational missions a third crewman was carried to act as a navigator and to help man the defensive armament. This was an important consideration as increasingly frequent attacks by German night fighters made had made an additional gunner a necessity.

The tailskid was made of plywood and there was a metal tip at the end. The F.50 could carry eight 200-mm bombs and nine 120-mm bombs in an internal bomb bay located behind the pilot's seat.

The F.50 underwent testing at Villacoublay in June 1918. On 5 June Lieutenant Boussoutrot flew an F.50 with two 240-hp Lorraine 8Bb engines. The aircraft carried a payload of 1,300 kg. The following results were obtained: climb to 500 m in 2 minutes 35 seconds; 1,000 m in 5 minutes 31 seconds; 2,000 m in 12 minutes 3 seconds; climb to 3,000 m in 22 minutes 38 seconds; and 4,000 m in 44 minutes 51 seconds; maximum speed 151 km/h at 1,000 m; 145 km/h at 2,000 m; 137 km/h at 3,000 m; and 125 km/h at 4,000 m.

On 15 June Lieutenant Canivel flew a similar (possibly the same) aircraft. He described the flight characteristics as follows: response to the controls was rated good; but it was difficult to keep the aircraft in a straight line if one engine was out. A poorly placed rudder bar caused pilot fatigue, particularly when it was flown on one engine. Rudder control was noted to be "a little heavy." The aircraft was also noted to be "heavy" in turns. Overall, the aircraft was described as being mildly fatiguing to fly. The F.50 had good stability in all axes. It was easy to takeoff and land; visibility was described as good.

A third flight occurred on 17 June 1918. The test pilot was Lieutenant Rebourg. He also found that the aircraft was maneuverable but fatiguing to fly.

A fourth test flight carried out on 28 June by Sergeant Lenay confirmed these findings. He concluded that the F.50 was relatively easy to fly for an aircraft of its size.

The F.50 was inferior to its competitor, the Caudron C.23, in terms of bomb load, but had a superior climb rate. The French ordered both types into production. The F.50s were built at the firm's factory at Billancourt and at the factories of Louis Clément, also located at Billancourt.

The cost of the F.50 was 225,000 F. Tests confirmed that the aircraft was underpowered. Test pilot Andre Canivet wrote on 15 June 1918 that it was impossible to maintain level flight on one engine. Some aircraft were subsequently tested with the

Above: The prototype of the F.50. The Farman F.50s were built at the firm's factory at Billancourt. B87.1027.

Farman F.50. (CC).

300-hp Hispano-Suiza. On 1 October 1918 17 F.50s had been built, and by the time of the armistice there were only 45 aircraft in service with GB 1. The F.50 was the only indigenous design of the BN2/3 specification to see combat service.

Variants

1. Farman F.50 T—Designed to fulfill a navy requirement for a torpedo bomber. Two aircraft were modified in 1922 to carry a torpedo. At least one aircraft (7143) was tested at Saint-Raphaël in this configuration, but none was acquired by the Aviation Maritime.
2. Farman F.50 "DCA" (Air Defense)—Produced in 1918, this aircraft was intended to help locate enemy aircraft and signal their location to anti-aircraft batteries. A major modification was the increased armament. The design was not accepted by the DCA.
3. Farman F.50 with modified wing—At least one aircraft was flown with a enlarged wing of 24.05 m span and area of 101.6 sq. m.
4. It appears that a variant of the Farman F.50 to be powered by two 260-hp Salmson engines was under development in early 1918. However, it seems that this remained an unbuilt project.
5. Farman F.50 airliner—The Farman company converted several F.50s to a passenger configuration by adding a cabin behind the two-man cockpit. The cabin seated four or five persons in relative comfort. The gunner's position in the nose was converted to a baggage area and the rear gunner's station was converted into a lavatory. The latter is said to have required the skill of a contortionist to use. The Farman F.50P, as it was designated, was used by four French airlines—Cie des Grandes Express Aeriens (CGEA), Air Union, Lignes Farman, and Compagnie Franco Balbaine.

Operational Service

Small number of the Farman F.50 entered service with VB 110 on 30 July 1918 at Cernon. VB 114 received the new aircraft on 2 August 1918. The units were re-designated F 110 and F 114 respectively. The new F.50s were equipped with the lower-powered 240-hp Lorraine 8 Bb engines.

The first combat mission was flown on the night of 10/11 August in which F 110 contributed one F.50 to the night's raids. By the end of August only a few missions were being flown by the F.50 crews, probably because of difficulties with the Lorraine engines. On 26 August GB 1 moved to join F 114 at Villeneuve. By the end of the month, V 25 had converted to F.50s, becoming F 25.

There were frequent problems involving F.50s. Three force-landed due to engine trouble during the first week of September. By 7 September newer F.50s were sent to GB 1; these had 275-hp Lorraine 8 Bd engines. It was hoped that these engines would prove to be more reliable. On 15 September 1918 32 of the newer 275-hp engines arrived and were divided between F 110 and F 114. Problems concerning the Lorraine engines were addressed in a GQG memo dated 19 October 1918. It was noted that the excessive oil flow could result in damage to the plugs, which would cause a drop in airspeed. Also, engines were being damaged in transit. Crews were warned to check the new engines carefully before placing them in the aircraft. Finally, many problems were the result of inadequate cleaning. Crews were cautioned to make sure the engines were thoroughly inspected and cleaned on a regular basis.

By October small numbers of F.50s were participating in the nightly attacks against train stations. When VB 137 left GB 1 on 8 October the bomb group became an all-F.50 unit. As the month continued the F.50s bombed train stations on the nights of 18/19, 22/23, and 23/24 October. A total of 12,395 kg of bombs were dropped during those attacks. At least four F.50s were lost, most from landing accidents and engine failure. On the night of the 30/31 11 F.50s hit train stations with a total of 4,315 kg of bombs.

During the last month of the war GB 1 continued to attack railroad stations and bivouacs. GB 1's last raid of the war took place on 9/10 November.

Postwar, the Farman F.50 units were:

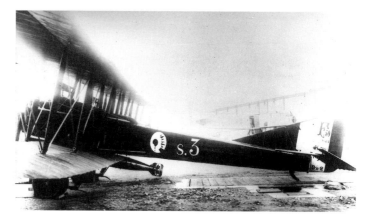

F.50 serial 7143 in service with S.3 of the Aviation Maritime. The aircraft is being tested as a torpedo bomber at Saint Raphaël. B87.4207.

F.50 of F 114. VB 114 received the new aircraft on 2 August 1918. The unit was re-designated F 114. B84.362.

F.50. This aircraft was assigned to GB 1. B77.962.

11th Escadre GB 1: Escadrilles F 25, F 110, and F 114.
11th Escadre GB 7: Escadrilles F 118, F 119, and F 121.
In 1920 the F.50 units were:
Escadrille 201 of the 2nd Group assigned to the 21st RAB (Nuit).
Escadrille 202 of GB 1 assigned to the 2nd RB (Nuit).
Escadrille 204 of GB 2 assigned to the 2nd RB (Nuit).
Escadrilles 205 and 206 of GB 3.

Foreign Service

Argentina

A French military mission to Argentina in 1919 apparently took six F.50s with it. They were based at Palomar and made a number of flights under the direction of Colonel Precardin. There is no evidence that the F.50 was ever acquired by the Argentine air service.

Japan

The Japanese government acquired a single F.50 in April 1920. It was used for research into the techniques of night bombing and was given the designation Type Tei 1. The F.50 was based at the Army Flight School at Tokorosawa in 1922. This aircraft differed from those in French service by being powered by two 230-hp Salmson water-cooled engines. It also had a shorter fuselage and a higher empty weight of 2,336 kg. The more advanced F.60 was selected for service with the night bomber units and no further F.50s were acquired by the Japanese.

Mexico

The Mexican air force acquired six F.50s at the end of 1919; eventually a total of 13 aircraft was purchased. They were assigned to a squadron commanded by Captain Ascension Santana and based at Guadalajara for use against rebel forces. One of them, piloted by Fernando Proal, was in transit when it crashed near Leon. Another arrived at Guadalajara and was used to attack the revolutionaries. The F.50s, along with other Mexican aircraft, required frequent repairs and often were unable to fly. Subsequently, the F.50s returned to their previous base and were divided among three squadrons. Two squadrons were sent to Irapuato to help defend the president. One of the squadrons was active against the rebels, attacking the rebel stronghold in Morelia. In addition, F.50s attacked the railroad at La Piedad. Attacks were also made on a rebel column near Cuitzeo Lake with devastating results. These actions played a significant part in the defeat of the rebels. It is not known precisely when the remaining aircraft were withdrawn from

F.50 serial number 6633. B77.1851.

Farman F.50. (T.A. Dunlap).

> **Farman F.50 Two-Seat Night Bomber with Two 275-hp Lorraine 8Bd Engines**
> Span 22.85 m; length 12.025 m; height 3.30 m; wing area 97 sq. m
> Empty weight 1,815 kg; loaded weight 3,100 kg
> Maximum speed: 150 km/h at 1,000 m; climb to 2,000 m in 12 minutes 30 seconds; ceiling 4,750 m; range 420 km
> Armament: two machine guns and 400 kg of bombs

active service. Serials included 3F 70, 4F 71, 6F 77, 7F 78, 9F 93, 10F 94, 12F 96, and 13 97.

Spain

Two or three F.50s arrived in Spain at Cuatro Vientos on May 9 1919. Those in Spanish service were powered by either 300-hp Hispano-Suiza or Lorraine-Dietrich 8B engines. They were the first aircraft in Spanish service to carry radios. One aircraft was located at the flight school at Valdepenas and another at Seville. One was sent to Africa and based at Tetuan. On January of 1920 a second F.50 arrived at Tetuan. It is not known when the aircraft were withdrawn from service. The Spanish air service felt that the F.50s were inferior to other bombers that were available.

United States

The AEF Air Service acquired two F.50s in March 1918 for evaluation. No further aircraft were ordered.

Henri Farman BN2 Designs

The BN2 specification of 1918 called for a heavy night bomber with a crew of two. It appears that Henri Farman produced two designs to meet this specification. One had two Canton-Unné CU18Z engines. The other had two Canton-Unné CU9Z engines. Both types were abandoned during 1918, probably because of the success of the Lorraine-powered F.50.

> **Henri Farman Twin-Engine Bomber with Two Canton-Unné CU9Z or CU18Z Engines**
> Wing area 135 sq. m
> Empty weight 3,000 kg; loaded wt. 5,000 kg; payload 1,000 kg
> Maximum speed: 140 km/h at 4,000 m; climb to 4,000 m in 30 minutes, endurance 5 hours

Farman BN2 with 400-hp Lorraine Engines

A British Ministry of Munitions report for May 1918 lists a two-seat heavy bomber for night attacks as under development. Power was to have been supplied by two 400-hp Lorraine engines. This large biplane was to have had an empty weight of 2,000 kg, a loaded weight of 5,000 kg, and a military load of 1,000 kg. Wing area was to have been 135 sq. m. Maximum speed was estimated to be 140 km/h at 4,000 m and it was expected to take 30 minutes to reach that altitude. Endurance was estimated at five hours. The type was under construction in April 1918, but no further details are available.

Farman F.51 Flying Boat

This aircraft was designed to meet the same "high seas" flying boat requirement as the Farman three-engine flying boat (see below). It was built in 1918 and featured a hull designed by Blanchard. It was powered by two 275-hp Lorraine-Dietrich 8Bd engines and carried a crew of three. The aircraft was tested at the Saint Raphaël naval air station. However, the Armistice resulted in the French navy's decision not to purchase any additional flying boats in the "high seas" category.

Although the Farman F.51s were not adopted by the French

> **Farman F.51 "High Seas" Flying Boat with Two 275-hp Lorraine 8Bd Engines**
> Span 23.35 m; length 13.880 m; height 4.40 m; wing area 108 sq. m
> Empty weight 2,200 kg; loaded weight 3,650 kg
> Maximum speed: 145 km/h; climb to 2,000 m in 20 minutes

Aviation Maritime, two were purchased by the Brazilian naval air service. These aircraft, given serials 36 and 37, were not uncrated until August 1923; in fact, only aircraft 37 was ever fully assembled.

Farman F.51 flying boat taking off from the water. This photograph is one of the few available of this aircraft. MA38543.

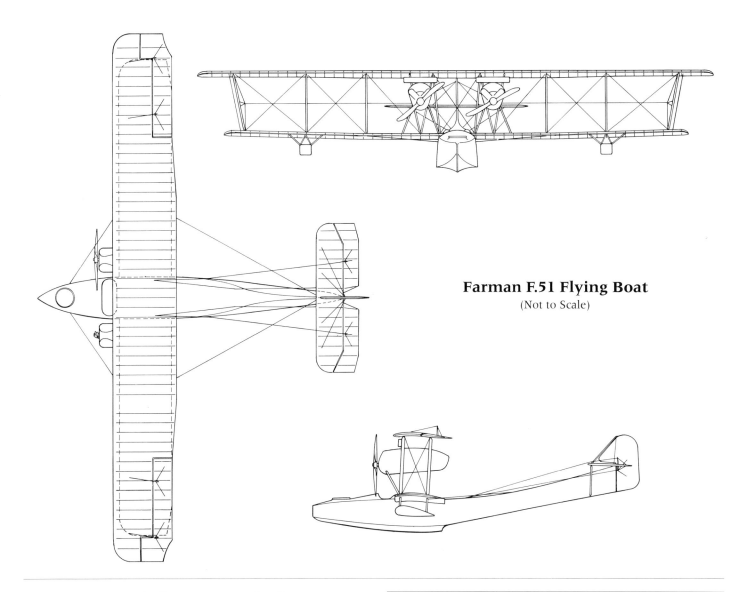

Farman F.51 Flying Boat
(Not to Scale)

Farman High Seas Flying Boat

The French naval requirement for a "high seas" flying boat in the same class as the British Felixstowe flying boats specified many ambitious goals. The aircraft was to have a crew of four, a wireless, a 75-mm cannon with 35 rounds, four I (120-kg) bombs, an endurance of eight hours, a range of 300 kilometers, a maximum speed of 140 km/h, and be able to climb to 2,000 meters in 25 minutes. Apparently the Farman firm submitted two different aircraft to meet this requirement.

The three-engine flying boat appears to have been based on Besson's trimotor flying boat of 1918. In 1918 the Farman firm had built Levy flying boats under license and had considerable experience with flying boat construction in general and the Levy-Besson fuselage design in particular. While the hull of

Farman "High Seas" Flying Boat with Three 330-hp Panhard Engines
Span 33.0 m; length 18.0 m; wing area 200 sq. m
Empty weight 4,500 kg; loaded weight 7,000 kg
Maximum speed: 145 km/h; climb to 2,000 m in 20 minutes;
  Endurance eight hours

Farman's aircraft was similar to Besson's designs, the wings were completely different. Whereas Besson favored a triplane layout, Farman's aircraft was a biplane. Also, the Farman design used three Panhard engines of 330-hp; Besson's aircraft had three 350-hp Lorraine engines. There was a crew of four, with two pilots and an observer seated in the nose and a second observer a separate cockpit behind the wings.

The aircraft was tested at the Saint Raphaël naval air station. It was capable of carrying up to 500 kg of bombs to attack enemy shipping and submarines.

Farman Renault Flying Boat

Around 1918 the Farman firm built a single-engined seaplane with a hull very close in shape to the earlier Besson designs. It was a biplane and had a single 450-hp Renault engine. The hull had an ash framework covered with mahogany planking. Detachable fuel tanks with enough fuel for four hours of flight were carried in the hull. The wing was made to fold backward along the rear spar. Postwar, a development of this aircraft featuring an enclosed cabin was built. It had a single 300-hp Renault engine.

Farman Flying Boat with 450-hp Renault
Span (upper) 18.0 m; span (lower) 14.0 m; length 14.2 m;
  height 3.9 m; wing area 82 sq. m
Empty weight 1,900 kg; loaded weight 2,900m
Maximum speed: 170 km/h; climb to 2,000 m in 15 minutes

Farman Unknowns

A review of the Musée d l'Air archives revealed four previously unknown Farman aircraft.

1. A sleek tractor biplane with a tightly cowled engine. The tail was conventional and the landing gear featured short skids on the inner side of each wheel.
2. A tractor biplane with prominent skids and dual mainwheels under the wing. An additional pair of wheels was located at the end of each skid. It appears to have a communal cockpit. The appearance strongly suggests a primary trainer.
3. Another tractor biplane with the fuselage attached to the upper wing. J.M. Bruce has reviewed the photograph and believes the engine is a Renault. The plane appears to be a two-seat reconnaissance machine or possibly a C2 category fighter.
4. Tractor biplane that may be a variant of unknown #3. The significance of the '43' on the rudder is not known but this type is not an F.43.

Top right: Unknown #2—Farman experimental aircraft—identity unknown.

Right: Unknown #1—Farman experimental aircraft—identity unknown.

Below and drawing at right: Farman unknown #4. Renaud.

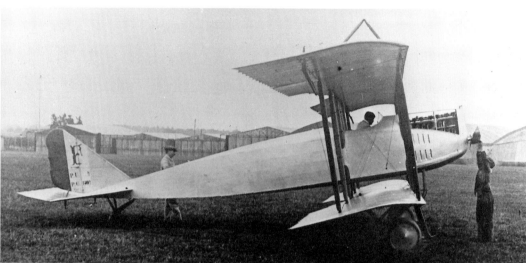

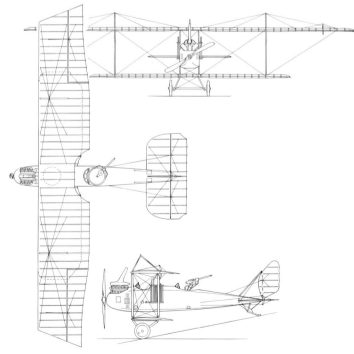

Left and above left: Unknown #3; a Farman tractor biplane with the fuselage attached to the upper wing. The airplane appears to be a two-seat reconnaissance machine or C2 fighter. MA13442.

FLO L and La

The FLO L and La were apparently identical designs featuring the exact same performance parameters and specifications. It is not known if they were designed to meet the requirements for a two-seat fighter (escort with "protégé") or a two-seat army cooperation aircraft. Both aircraft were biplanes, and the specifications for both variants called for a empty weight of 772 kg, a loaded weight of 1,372 kg, and a payload of 200 kg. Wing surface area was 323 sq. m. Endurance was estimated to be three hours. Maximum speed at 4,000 meters was estimated to be 200 km/h and climb to 4,000 m estimated to be in 14 minutes.

The aircraft were reported by the Ministry of Munitions to be under construction in April 1918 but it is not known if they were ever completed; certainly, they were never adopted for service use by the Aviation Militaire.

Hydravions Franco-British Aviation (F.B.A.)

Louis Schreck was a South American representative for the Delaunnay-Belleville automobile firm. In 1911 he joined with Hanriot and an engineer named Gaudard to build the D'Artois flying boat. That same year he took over the Tellier establishment. Using British capital, he later formed Franco-British Aviation (F.B.A.). The firm was based at Argenteuil, but had satellite plants at Juvisy and Vernon. Schreck acquired the patents for the Donnet-Lévêque flying boats before the war. In 1934 F.B.A. was acquired by Bernard.

F.B.A. Type A

F.B.A.'s first flying boat was a single-engine seaplane built in 1913. It was initially known as the F.B.A.-Lévêque because it was heavily influenced by the previous Donnet-Lévêque flying boats, the patents for which, as mentioned above, had been purchased by Schreck. The Type A, as it was later designated, was a petite biplane powered by a 50-hp Gnome Omega engine mounted in pusher configuration on struts between the upper and lower wings. The Type A had side-by-side seating, a wooden hull, and two-bay wings of unequal span.

The aircraft was developed into two variants: one with a wing area of 20 sq. m and a 70-hp Gnome engine, and one with a wing area of 22.2 sq. m and an 80-hp Gnome. The Type A shown at the 1912 Salon had a 50-hp Gnome N1 engine, a span of 9.00 m, a length of 7.79 m, a wing area of 18 sq. m, an empty weight of 309 kg, a maximum speed of 90 km/h, and carried a crew of two. Nine were built in the initial series. At least one example of this type was converted into an amphibian by the addition of retractable wheels to the hull. Another version was used as a racer; powered by a 100-hp Gnome engine, it was entered in the 1914 Schneider Trophy Contest, where it placed second. It appears that the Aviation Maritime never had any Type As on strength, although a development of the aircraft, designated Type B, saw widespread service.

Foreign Service

Austria-Hungary

The Austrians ordered several modifications to the Type A, including an enlarged fuselage to enable more fuel to be carried, a strengthened tail unit, and floats with a circular cross section. These aircraft were fitted with 80-hp Gnome engines. Three were used by the Austro-Hungarian naval air service, given serials A 22, A 23, and A 24. When the First World War began, they were flown as single-seaters to improve their range.

United Kingdom

The RNAS purchased 42 F.B.A. Type As in 1915 and 1916. They were assigned these serials:

1. Serial Nos. 3113, 3114—100-hp Gnome assigned to No.1 Wing.
2. Serial Nos. 3199–3208—100-hp Gnome assigned to No.1 Wing, Dunkerque.
3. Serial Nos. 3637–3656—100-hp Gnome based at Calshot, Dover, Killingholme, Felixstowe, Windermere, and White City.
4. Serial Nos. 9601–9610—100-hp Gnome based at Dover, Windermere, White City, and Calshot.

F.B.A. Type A Two-Seat Flying Boat with 50-hp Gnome N1

Wing span 9.00 m; length 7.70 m; wing area 18.00 sq. m
Empty weight 309 kg
Max speed: 90 km/h
Nine built

F.B.A. Type A Two-Seat Flying Boat with 80-hp Gnome 7A (for Great Britain)

Wing span 11.88 m; length 8.83 m
Loaded weight 570 kg
Max speed: 109 km/h
40 built

F.B.A. Type A. The type was used by the Austro-Hungarian Naval Air Service and the RNAS. MC 12241.

F.B.A. Type B

The F.B.A. Type A was not adopted by the French Aviation Maritime, but an improved version of it saw extensive service with the French and British navies. This version was designated the Type B, and it differed from the Type A in several respects. It had a more powerful 100-hp Gnome engine and larger dimensions. The Type B also had a redesigned tailplane and rudder and some had folding wings. More than 150 Type Bs were produced in France and Britain.

An escadrille of F.B.A. Type Bs was assigned to the naval station at Dunkerque. The first combat mission by Type Bs was flown on 4 February, 1915. They were used primarily for reconnaissance along the Belgian coast but a number of bombing missions were also flown. The F.B.A.s suffered heavily from German air attacks and a number of Type Bs were destroyed or captured. On 26 May 1917 alone, four F.B.A. Type Bs were downed by German aircraft. In June, 1915, six F.B.A. Type Bs were sent to protect the naval base at Venice.

Some aircraft carried two 150-kg bombs, and a 37-mm cannon was fitted to three Type Bs (serial numbers 454, 455, and 457). Most of the Type Bs, however, were unarmed and used as trainers.

Foreign Service

Brazil
Brazil purchased two Type Bs built by the Gosport firm in 1918. They were given serials 7 and 8 and were struck off charge in August 1923.

Portugal
Portugal obtained three Type Bs in 1917, all with 100-hp Gnome engines. They were assigned to the Escola de Aeronautica Militar at Vila Nova da Rainha and remained in service until 1918.

Right: F.B.A. Type B. B87.4041.

F.B.A. Type B. The F.B.A. Type B was an enlarged version of the Type A and had a redesigned tail and a more powerful 100-hp Gnome . MA36410.

Russia
Thirty F.B.A. Type Bs were purchased by the Russians in 1914. An additional 34 were built by the Lebedev plant. They were in service from 1915 through 1916, originally for reconnaissance and later as trainers.

United Kingdom
Fifty-four F.B.A. Type Bs were ordered after the war began. Norman Britain provided 20 machines which were given serial numbers N 1040 to N 1059 and assigned to Flight C at Bognor. They were made in France but assembled in England. The Gosport company also built 60 Type Bs, given serial numbers N 2680 to N 2739. They were used as trainers at the naval aviation training depot at Lee-on-Solent. Three additional aircraft were obtained directly from the French, given serials B 3984 to B 3986. Four Type Bs were supplied to the RNAS via Italy. They were assigned to the RNAS air station at Otranto and later to No.266 Squadron at Malta in April 1918. Many of the Type Bs built by the Gosport firm were used as trainers, while those obtained directly from France were used operationally. Two Type Bs with serial numbers 3648 and 3650 were armed with machine guns and bombs and used for combat patrols over the North Sea. On 28 November 1915 attacks were made by Ensign P.B. Ferrand and mechanic/navigator G.T. Oldfield on a German torpedo boat and the seaplanes accompanying it.

Additional Type Bs were purchased directly from France during 1917. These also had 100-hp Gnome engines and were based at Killingholme and Calshot.

An F.B.A. Type B was used to test the maximum loads that

> **F.B.A. Type B Two-Seat Flying Boat with 100-hp Gnome**
> Wing span 13.71 m; length 9.14 m; wing area 32 sq. m
> Loaded weight 907 kg
> Maximum speed: 96 km/h, range 300 km
> Armament: two 150-kg bombs and a 37-mm cannon
> More than 150 built
>
> **British-built F.B.A. Type B Two-Seat Flying Boat with 100-hp Gnome**
> Wing span 45 feet; length 30 feet (not a metric aircraft)
> Loaded weight 2,000 lb
> Maximum speed: 60 mph, endurance 4 hours
> Armament: two 150-kg bombs or a 0.303 Lewis gun
> 80 built

could be carried by a seaplane. In September 1916 a Type B took off with a gross weight of 907 kg and attained a speed of 98 km/h. In January 1919, 24 Type Bs were still in service with the RAF.

A preserved Portuguese F.B.A. Type B. The Portuguese Type Bs were assigned to the Escola de Aeronautica Militar at Vila Nova da Rainha. MA38562.

F.B.A. Type C

The F.B.A. Type C was a development of the Type B. Type Bs re-engined with 130-hp Clerget 9B rotaries were designated the Type C. The prototype Type C was built in 1915 and series production began in 1916. The dimensions of the Type C were the same as the Type B. The seaplane could carry two 11-kg bombs and a 7.7-mm machine gun. The Type Cs saw extensive service as patrol and anti-submarine aircraft and were operated from many French naval air stations. A total of 78 Type Cs were delivered from January to October 1917. When the F.B.A. Type H was introduced into service, the Type Cs were relegated to training. Several examples were embarked on the seaplane carrier *Campinas*. A single example of the Type C was fitted with two Clerget engines in 1917 but the conversion was not a success. Two additional variants of the Type C were produced postwar: the Type 11 two-seat trainer and the Type 14, another training aircraft. Twenty of the Type 14 were built.

Foreign Service
Italy

The Italian naval air service acquired a number of Type Cs to use against the Austrian Lohner seaplanes (which, ironically, had been heavily influenced by the Donnet-Lévêque flying boat on which the original Type A had been based). In 1916 there were 38 F.B.A. Type Cs in service. A list of the Italian units that flew them is included in the section on the F.B.A. Type H. Some Type Cs were produced under license by the SIAI firm.

> **F.B.A. Type C Two-Seat Flying Boat with 130-hp Clerget 9B**
> Span 13.71 m; length 9.14 m; height 3.05m; wing area 32 sq. m
> Empty weight 640 kg; loaded weight 907 kg;
> Max speed: 110 km/h; climb to 2,000 m in 25 minutes; range 320 km; endurance 2.75 hours
> Armament: two 11-kg bombs and a 7.7-mm machine gun
>
> **SIAI-built Type C Two-Seat Flying Boat with 130-hp Clerget 9B**
> Span 13.70 m; length 8.80 m; height 3.04 m; wing area 30.60 sq. m
> Empty weight 575 kg; loaded weight 890 kg;
> Maximum speed: 110 km/h; climb to 2000 m in 18 min; ceiling 3,500 m; range 450 km
> Armament: two 11-kg bombs and a 7.7-mm machine gun

Russia
The Russian naval air service imported 30 Type Cs and 34 more were built at the Lebedev factory.

F.B.A. Type C taxiing on the water. Type B flying boats re-engined with 130-hp Clerget 9B rotary engines were designated the F.B.A. Type C. B80.749.

F.B.A. Type C. This view shows the 130-hp Clerget 9B rotary engine. MA13214.

Below: A single example of the F.B.A. Type C was fitted with two Clerget engines in 1917, but the conversion was not a success. Reairche.

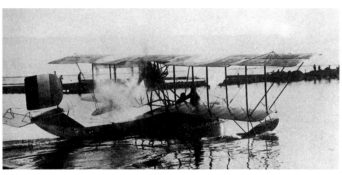

The F.B.A. Type Cs saw extensive service as patrol and anti-submarine aircraft and were operated from many French naval air stations. Reairche.

F.B.A. Type H

In the second half of 1915 an F.B.A. Type B airframe was fitted with a 150-hp Hispano-Suiza 8A engine and flown to Burri. The aircraft had redesigned wings and tailplane and a reinforced hull. It could carry a crew of three. It was successful enough to warrant series production as the Type H and entered service in May 1916. The Type H was intended to fill the Aviation Maritime's requirement for a patrol seaplane capable of dealing with enemy cruisers and submarines. The naval air service required 360 patrol flying boats and most of these were to be F.B.A. Type Hs. The Type H was well armed with a machine gun on a swivel mount in the nose and a 35-kg F series bomb under each of the lower wings. Some of the Type Hs used the 160-hp AM (Lorraine) eight-cylinder engine. The crew of two sat side by side in the cockpit and there was a connecting tunnel to the nose gun compartment. The aircraft were based in the Atlantic, Channel coast, and Mediterranean. A total of 157 Type Hs were delivered from January to October 1917.

At least one Type H was modified to become an amphibian by fitting a pair of wheels on the side of the hull. It was flown on the Colombes race course. Another Type H was modified to accept a 37-mm Hotchkiss cannon and a fixed landing gear and was intended for use as a land-based fighter. (See F.B.A. Type D.)

The Type H was the most numerous F.B.A. flying boat to be produced during the war and was possibly built in greater numbers than any other flying boat in World War One.

Foreign Service
Belgium

The Escadrille Navale acquired five F.B.A. Type Hs during the war. Some were modified to accept 180-hp Hispano 8Ab engines and were redesignated Type N. They arrived on 18 September

F.B.A. Type H based at l'Ecole de Tir at Cazaux. The Type H was fitted with a 150-hp Hispano-Suiza and had redesigned wings and tail and a reinforced hull. B84.610.

F.B.A. Type H in service with the Finnish air force. It is believed that only a single example was purchased by Finland. B81.2996.

1917. From their base at Calais, they were used to patrol the Channel and the North Sea. The escadrille moved to Ostende at the end of the war and was disbanded in 1919. Aircraft No.5 has been preserved and displayed at the Musée de l'Air at Brussels.

Estonia
Estonia acquired a single F.B.A. Type H in the 1920s, given serial number 66.

Finland
Finland acquired at least one Type H.

Italy
The F.B.A. Type H was produced under license in Italy by the Savoia (SIAI) firm. The aircraft differed from the French versions in having a more powerful 170-hp Isotta-Fraschini engine. Also, the dimensions of the Italian seaplanes were larger than the French machines. A total of 982 were built in Italy up until production ceased in late 1918.

In 1917 there were 367 Type Cs and Hs in service. They were assigned as follows on 1 June 1917: Brindisi (12), Valona (18), seaplane carriers (8), and Corfu (6).

At the time of the Armistice, the Italian naval air service had a total of nearly 600 F.B.A.s (Cs and Hs) on strength. These were assigned to the following squadriglias:
Reconnaissance squadriglias: 254, 255 (based at Varano), 256 (Otranto), 257 (Valona), 58 (seaplane carrier *Europa*), 263 (Porto Cosini), and 264 (Ancona).
Patrol squadriglias: 266 (Sanreno), 267 (Porto Maurizio), 268 (Rapallo), 269 (La Spezia), 270 (Palermo), 271 (Civitavecchia), 273 (Livorno), 274 (Piombino), 275 (Ponza), 276 (Napoli), 277 (Sapri), 278 (Terranova Pausiana), 279 (Caqliari), 280 (Milazzo), 281 (Taormina), 282 (Catania), 283 (Siracusa), 284 (Trapani), 285 (Orbetello).

Peru
Three F.B.A. Type Hs were brought to Peru by a French aviation mission in 1919. They were used as trainers.

Serbia
Serbia obtained three Type Hs postwar. They were assigned to Hidroplansk Kommande (Seaplane Command) and assigned codes N 13, 14, and 15. They were used primarily as trainers.

Spain
Two F.B.A. Type Hs were obtained by the Spanish naval air service beginning in 1917. They had been built under license by the Italian SIAI firm. The aircraft were well liked and it is believed that up to six more were purchased by the Spanish. The Type Hs were given serial numbers 1 through 8. There were only three Type Hs in service in 1922, and these were used by the pilots undergoing flight training at the base at Los Alcazares. Aircraft numbers 3, 4, and 6 remained in service until 1927, when, after they were worn out from their extensive use as trainers, they were replaced by Macchi M.18s.

United Kingdom
The Royal Naval Air Service obtained four F.B.A. Type Hs (which had built under license in Italy) in 1917. They were given serial numbers N1075–1078 and were assigned to the base at Otranto; some were also assigned to 6 Wing. Two were sent to Malta in June 1917 and one of them bombed a U-boat on 20 August 1917.

United States
Many of the U.S. naval aviation units based in France used F.B.A. Type Hs but they were rapidly replaced by Donnet-Denhaut D.D.8s and consequently saw only limited action. Three F.B.A.s were also flown by Americans based at Porto Corsini in Italy. The Americans, despite some hesitation, liked the F.B.A.s. An American report describes the hull as being "poorly constructed but has nevertheless given excellent results." The F.B.A.s were all sent directly from the Schreck factory to the U.S. naval base at Moutchic, where they were given serial numbers 295, 296, 297, 801, 802, 806, 807, 808, 813, 814, and 815. All but number 802 were returned to the French at the end of the war; it was destroyed while in U.S. service. These flying boats were used primarily as trainers.

Uruguay
The Uruguayan air service received a single Italian-built F.B.A. with a 220-hp engine, donated by a private citizen named D. Zambra. Consequently the aircraft was named *Zambra*. It was probably a Type H, although the H models usually had engines of lower horsepower.

F.B.A. Type H Three-Seat Flying Boat with 150-hp Hispano-Suiza 8A
Span 14.12 m; length 9.92 m; height 3.10 m; wing area 40.00 sq. m
Empty weight 984 kg; loaded weight 1,420 kg
Maximum speed: 150 km/h; ceiling 4900 m; range 450 km
Armament: a single 7.7-mm machine gun on a swivel mount in the nose and two 35-kg "F" series bombs

SIAI-built Type H Three-Seat Flying Boat with 170-hp Isotta-Fraschini
Span 14.55 m; length 10.20 m; height 3.78 m; wing area 42 sq. m
Empty weight 925 kg; loaded weight 1,400 kg
Maximum speed: 140 km/h; climb to 1,000 m in 8 min.; climb to 2,000 m in 18 min.; climb to 3,000 m in 31 min.; climb to 4,000 m in 47 min; ceiling 5,000 m, range 600 km
982 built

Above: Italian F.B.A. H of the 2ª Squadriglia at Grado in flight. Via Alan Durkota.

Left: Closeup of an F.B.A. H with bombs loaded. Via Alan Durkota.

Below: Italian F.B.A. H serial number 3147 of the 2ª Squadriglia at Grado. Subject of color plate via Alan Durkota.

F.B.A. Type S

The Aviation Maritime's specification for a flying boat for patrol and light bombing was initiated by the availability of the 200-hp Hispano Suiza 8Bb engine in 1917. Levy-Besson submitted a triplane flying boat, Tellier a twin-engine flying boat, and the Donnet-Denhaut firm a redesigned D.D.2. F.B.A. also chose to alter a previous design, the Type H, to accept the new engine. The new plane was designated the Type S.

As well as having a more powerful 200-hp Hispano-Suiza 8Bb, the Type S differed from the Type H in having a larger wing span and a longer fuselage. Other changes included a square radiator and a larger tail unit to compensate for the more powerful engine. The hull was redesigned, with changes to the planing bottom, and the wings were designed to fold for easier storage. While only a single machine gun was carried, the bomb load was increased to two 77-kg bombs or two 110-kg or four 50-kg bombs. A crew of two was standard, but a mechanic could also be carried. Some Type S aircraft had the 210-hp Hispano-Suiza BDd engine.

The aircraft entered service in November 1917 and were based at the naval aviation stations on the Channel, Atlantic, and in the Mediterranean. They remained in service until 1923, when they were finally replaced by Latham and Blanchard flying boats.

Variants

There were developments of the F.B.A. Type S that included one fitted with a 275-hp Hispano engine mounted inside a nacelle. The fuselage was unchanged, but the wings and tail were altered. This experimental aircraft was given serial number 1222. To permit adequate stability, all ailerons were enlarged. The aircraft was destroyed during its second fight and the end of the war halted further development. One Type S was fitted with two rotary engines. Another was designed with a wing chord reduced by 20 cm. It crashed on its first flight.

> **F.B.A. Type S Three-Seat Flying Boat with 200-hp Hispano-Suiza 8Bb or 210-hp Hispano-Suiza 8Dd**
>
> Span 15.60 m; length 10.59 m; height 3.65 m; wing area 46.00 sq. m
>
> Empty weight 1,060 kg; loaded weight 1,600 kg
>
> Maximum speed: 142 km/h; ceiling 4,000 m; range 500 km; climb to 500 m in 2 min. 30 sec.; climb to 2,000 m in 18 min. (some sources say 16 min.)
>
> Armament: a 7.7-mm machine gun in the nose and two 35-kg bombs or, later, two 50-kg bombs (one source states four 50-kg bombs)

The F.B.A. Type S differed from the Type H in having a more powerful 200-hp Hispano-Suiza 8 Bb and larger dimensions. Reairche.

Above: F.B.A. Type S. Changes from the Type H included a square radiator and a larger tail unit to compensate for the more powerful engine. B87.5014.

F.B.A. Type S. The F.B.A. Type S was a development of the Type H with a larger wing span, a longer fuselage, and a more powerful 200-hp Hispano-Suiza 8Bb. B87.5024.

F.B.A. High Seas Flying Boat

The French naval air service issued a requirement in 1918 for what was termed a high seas flying boat. This specification, perhaps inspired by the Felixstowe seaplanes, called for an endurance of eight hours, a large radio with a range of 300 km, four Type I (120-kg) bombs, and two machine guns. The aircraft was to have a maximum speed of 140 km/h and be able to climb to 2,000 m in 25 minutes. Aircraft designed to meet this requirement were submitted by Donnet-Denhaut, Farman, Nieuport, Tellier, Levy (Besson), Latham, and F.B.A.

The F.B.A design had four 250-hp Hispano engines. It was a biplane with a wing surface area of 180 sq. m and was 80 percent complete when the Armistice resulted in cancellation of the project.

F.B.A. Type D Cannon Fighter

The French were strong believers of the value of the cannon-armed fighter. Aircraft of this classification (class D) included the Breguet 5, 6, and 12 as well as the Voisin 4. The D class called for an aircraft with a speed of 125 km/h able to climb to 2,000 m in 14 minutes and carrying a cannon with a wide field of fire. A variation of the D class intended for the destruction of airships called for a speed of 120 km/h and a climb to 2,000 m in 17 minutes.

To meet this specification, Louis Schreck modified one of his Type H flying boats (number 101) into a fighter. This was accomplished by fitting the wings of a Type H flying boat to a fish-shaped monocoque fuselage made of wood. A pair of wheels were attached directly to the aircraft's hull. A conventional fin and rudder were fitted. The aircraft featured a 37-mm Hotchkiss cannon; apparently this did not use the same mount that was patented and designed by Schreck. The Hotchkiss was fitted in the nose and had a limited field of fire. The engine was a 150-hp Hispano-Suiza 8A or a 175-hp 8Aa. Despite its odd appearance, the F.B.A. cannon fighter exceeded the required maximum speed, but its climb to 2,000 meters barely met the 14 minutes that was called for in the D category. The restricted field of fire and the relative suitability of the Voisin cannon series would seem to have made the F.B.A. design superfluous, and further development was abandoned.

F.B.A. Type D Two-Seat Cannon Fighter with 150-hp Hispano-Suiza 8A or a 175-hp 8Aa

Span 14.5 m; length 10.13 m; height 3.35 m; wing area 41 sq. m
Empty weight 761 kg; loaded 1,166 kg
Maximum speed: 135 km/h at 2,000 m; climb to 2,000 m in 14.0 minutes; climb to 3,000 m in 25.0 minutes 50 seconds; ceiling 4,700 m; endurance 3 hours
Armament: one 37-mm Hotchkiss cannon
One built

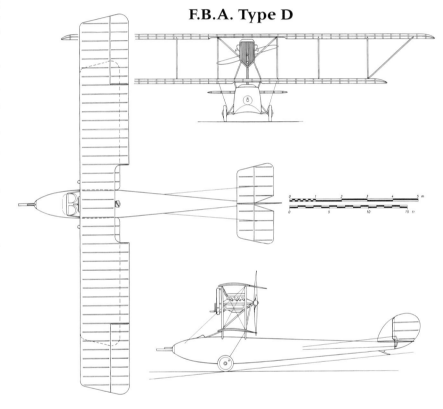

F.B.A. Type D

F.B.A. Type D cannon fighter. The Type D was a Type H flying boat (number 101) converted into a land-based fighter by fitting the wings of a Type H to a fish-shaped monocoque fuselage. Renaud.

F.B.A. Triplane Flying Boat

F.B.A. had a triplane flying boat under development in 1917. This was powered by a single 300-hp Renault (although one source states that a 300-hp Binetti was used). It had been completed by December 1917. It carried a crew of three and was undergoing testing in April 1918, but never entered service. It is unlikely that the type, even if successful, would have been selected for production by the Aviation Maritime. The navy had concluded that triplanes had a number of disadvantages over biplanes, including higher weight and a larger number of supporting struts which made them ungainly, especially when taxiing on the water.

F.B.A. Three-Seat Triplane Flying Boat with 300-hp Renault
Wing area 62 sq. m
Payload 250 kg
Maximum speed: 87 mph at 2,000 m; climb to 2,000 m in 25 minutes
One built

Galvin Floatplane Fighter

Photographs taken during the First World War reveal that a floatplane with an unusual configuration was built in France. The single float appears to have been of conventional size and shape. The tail section was attached directly to the float by what appears to have been a single strut. A crescent-shaped rudder was mounted at the end of the tail section, and the bottom portion of the fin was attached to the end of the float. The forward fuselage was completely separate from the tail and housed the pilot and a 160-hp Gnome rotary engine. There was a huge nose cone which had the same diameter as the forward fuselage and was constructed of a metal alloy. Two N-shaped struts attached the forward fuselage to the central float. There was a shoulder-mounted top wing on the forward fuselage, while the bottom wing was suspended between the forward fuselage and the float. Finally, there were two crescent-shaped floats attached to the ends of the lower wing.

Little information is available concerning this unusual design. It was intended to be used as a fighter, probably to act as an escort for flying boats. It is believed that the Galvin floatplane was flight-tested in 1919.

Galvin Floatplane Fighter with 160-hp Gnome Rotary
Span 8.00 m; length 7.20 m; height 2.30 m; wing area 18.59 sq. m
Empty weight 520 kg; loaded weight 800 kg
Max speed: 200 km/h at sea level; endurance 2 hours
One built

Galvin Floatplane Fighter

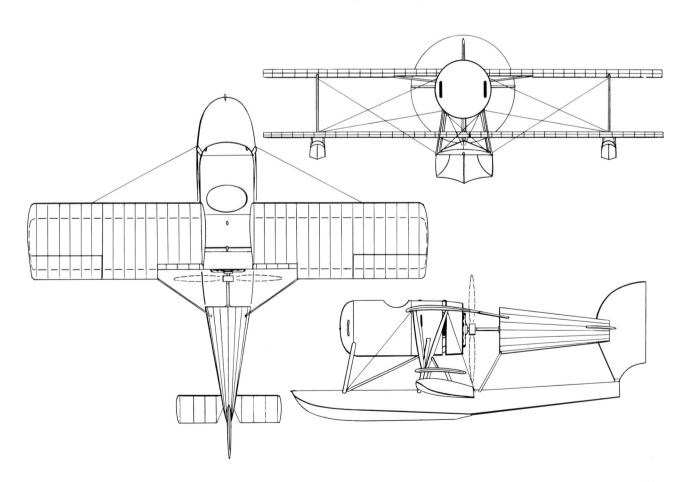

Goupy Aircraft

Ambroise Goupy built his first aircraft, a triplane, in 1908. His second design, a tractor biplane, followed in 1909. Goupy later founded his own company and in 1914 the Etablissements Ambroise Goupy produced two aircraft. Both featured wood construction, ailerons for lateral control, positively-staggered biplane wings, and single mainwheels with skids.

The first was a single-seater powered by an 80-hp Gnome engine. It featured an angular yet slim fuselage suspended by struts above the bottom wing. The aircraft had a large, rectangular tailplane with elevators and a single fin with a large rudder. It had ailerons instead of the usual pivoted wing tips favored by the Goupy firm.

The second aircraft was a three-seat biplane. It had a 100-hp Gnome engine and was equipped with ailerons.

Neither aircraft was selected for service by the military. Subsequently, the firm acquired the services of the Swiss engineer Donnet, who designed a glider and a seaplane. The Goupy factory was destroyed by a fire shortly after the war began and ceased aircraft production.

Goupy Single-Seat Plane with 80-hp Gnome
Span 6.30 m; length 6.80 m; wing area 13 sq. m
Maximum speed: 130 km/h; climb rate 150 meters per minute
One built

Goupy Three-Seat Reconnaissance Plane with 100-hp Gnome
Span 12.75 m; length 9.10 m; wing area 30 sq. m
Loaded weight 390 kg
Maximum speed: 95 km/h; climb rate 35 meters per minute
One built

Two Goupy aircraft at the 1914 Salon. The airplane in the foreground had an 80-hp Gnome; the airplane in the rear was a three-seater with a 100-hp Gnome. MA.

Gourdou-Leseurre C1 and 2 C1

C. Edouard P. Gourdou and Jean A. Leseurre were aeronautical engineers who patented an aerofoil shape intended to reduce aircraft drag in January 1917. An official document shows that in the summer of 1917 a monoplane designed by Gourdou and Leseurre was built at the Wassmer carpentry works in Paris. It featured the new wing that Gourdou and Leseurre had patented (brevet no.504302). The plane appeared in the fall of 1917 and shortly thereafter testing began at Villacoublay. Powered by a 180-hp Hispano-Suiza engine, it attained a maximum speed of 223 km/h. Testing was suspended by December 1917, and the plane underwent a series of apparently minor alterations. The modified design, designated the GLa, retained the 180-hp Hispano-Suiza engine. It was submitted to the STAé to meet the C1 specification for a single-seat fighter with a payload of 110–170 kg, maximum ceiling of 9,000 km, a service ceiling of 6,500 m, and a maximum speed of 140 km/h. Other aircraft designed to meet the C1 category and powered with the 180-hp Hispano-Suiza 8 Ab engine included the Laboratoire Eiffel monoplane.

The Gourdou-Leseurre GLa first appeared in early 1918 and was also built by Société Wassmer. The fuselage was of steel tubing while the engine mounts were constructed of duralumin. The undercarriage struts were made of steel tubing and a conformal fuel tank, which could be jettisoned, was fitted between the landing legs. The Hispano-Suiza 8Ab engine was fully cowled with large covers over the rocker arms. The monoplane wing was suspended by a combination of center section struts and two faired steel struts extending from each side of the fuselage. The wing spars were of steel, the ribs were wooden, and the entire structure was covered in fabric except for the leading edges, which were covered with plywood. This version of the Gourdou-Leseurre fighter had a fixed tailplane.

The GLa received the STAé designation Gourdou-Leseurre C1.

The prototype was heavier than expected and in order to meet the C1 payload specifications the aircraft's structure had to be considerably lightened. As a result of the weight reduction program, the Gourdou-Leseurre C1 was an excellent performer and was faster than any of the other C1 aircraft equipped with the same engine. The STAé pilot's report states the GLa was easy to fly, manoeuvrable, and offered the pilot an excellent field of vision. However, there were problems. The wing support structure was felt to be too weak and the aircraft was substantially revised to incorporate a more extensive array of wing braces. The wing was modified with an increase in span to 9.4 m and elimination of dihedral; in addition, the tail was modified, the fin was reduced in size, and a new rudder was installed. The landing gear struts were revised to handle the increased weight of the aircraft and to permit the belly fuel tank to be jettisoned without difficulty. These changes were extensive enough to warrant the new designation GLb, while the STAé redesignated the aircraft the Gourdou-Leseurre 2 C1. The improved wing bracing enabled the wings to withstand a load factor of ten. Armament was two Vickers 7.7-mm machine guns.

Twenty of these revised aircraft were ordered by the Aviation Militaire. The first was tested at Villacoublay in November 1918. The results revealed a good performance, but the fin and rudder were subsequently enlarged. It is possible that one aircraft was tested with a 230-hp Salmson 9Za; a Le Rhône 9R may also have been fitted. It was also planned to equip one aircraft with the 320-hp A.B.C. Dragonfly engine, but this was probably never done.

The end of the war meant that there was no further need for the Gourdou-Leseurre 2 C1 and plans to produce it for the Aviation Militaire and the United States Air Service were terminated. One example was sold to the Finnish air service in 1923, and 18 more were sent in 1924. One additional aircraft was later assembled from spares. Serial numbers were 8 E 2 and 8 F 5 through 22 (8 F 14 was used twice). Codes were later modified to become GL 5 through 22. These aircraft served with the Havittajaeskaaderi (First Fighter Squadron) commanded by

G-L C1 Single-Seat Fighter with 180-hp Hispano-Suiza 8Ab

Span 9.0 m; length 6.6 m; height 2.3 m; wing area 16.65 sq. m
Empty weight 600 kg; loaded weight 786 kg

Maximum speed:	1,000 m	242km/h
	2,000 m	237 km/h
	3,000 m	230 km/h
	4,000 m	223 km/h
	5,000 m	214 km/h
Climb:	1000 m	2 min 25sec
	2000 m	5 min 16 sec
	3000 m	8 min 43 sec
	4000 m	12 min 44 sec
	5000 m	20 min 10 sec

Ceiling 6000m; endurance 1.5 hours.
One built

G-L 2 C1 Single-Seat Fighter with 180-hp Hispano-Suiza 8Ab

Span 9.4 m; length 6.43 m; wing area 18.8 sq. m
Empty weight 570 kg; loaded weight 850 kg
Maximum speed: 245 km/h at ground level; 220 km/h at 5,000 m; climb to 5000 m in 17 min 30 sec; ceiling 7500 m
Armament: two Vickers 7.7-mm machine guns
Approximately 20 built

Gourdou-Leseurre C1

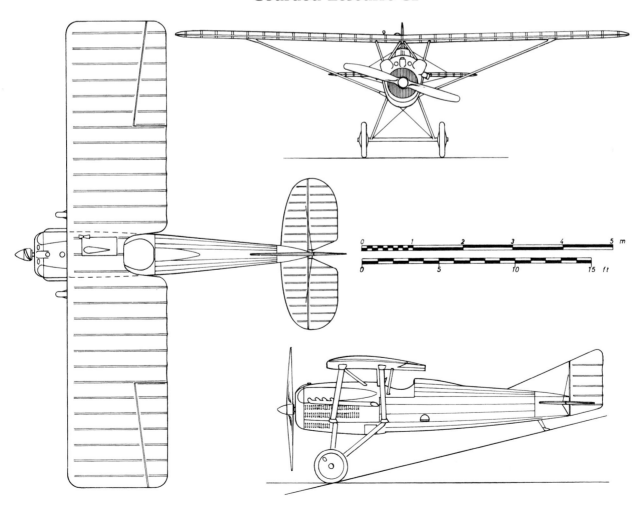

GLa C1 was designed to meet the C1 category and was powered with a 180-hp Hispano-Suiza 8Ab engine. B87.1029.

Captain E. Koni at Utti. The aircraft were withdrawn from service by 1933. Postwar, the Gourdou-Leseurre factory was established at Saint-Maur-des-Fausses.

Gourdou-Leseurre 2 C1

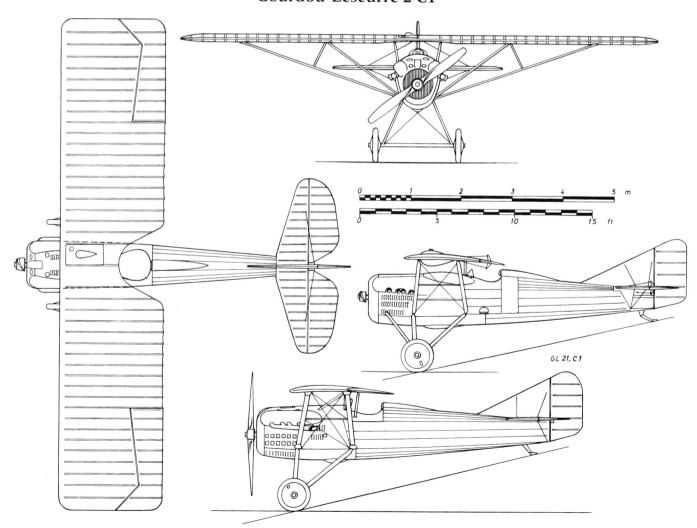

Above: GLa C1. The prototype was heavier than expected and in order to meet the C1 payload specifications the aircraft's structure had to be considerably lightened. B87.1030.

Right: GLb 2C1. The additional wing bracing enabled the wings to withstand a load factor of 10. B92.160.

Below: The Gourdou Lessure GLb C1 employed modifications from the GLa which were extensive enough to warrant the new designation GLb, while the STAé redesignated the aircraft the Gourdou-Leseurre 2 C1.

Aeroplanes Hanriot et Cie

René Hanriot produced several aircraft prior to 1914. These included:
1. D 1—single-seat monoplane; 50-hp Buchet engine.
2. D 2—two- or three-seat monoplane; 100-hp Gnome.
3. D 3—monoplane racer; 100-hp Gnome.
4. D 4—monoplane with all-steel construction; 50-hp Peugeot.
5. D 7—monoplane with 80-hp Gnome.
6. Trainer—35-hp engine.
7. Trainer—45-hp engine.

After the beginning of the First World War Hanriot established Aeroplanes Hanriot et Cie and built Sopwith 1½ Strutters and Salmson 2s under contract.

Hanriot HD.1

The Hanriot firm produced its first aircraft in 1916. The HD.1, as it was designated, was designed by Pierre Dupont. The compact single-seater had an upper wing with dihedral. The lower wing had no dihedral; its span and chord were less than that of the upper wing. The center section struts were similar to those used on the Sopwith 1½ Strutters which Hanriot had built under license. The synchronized Vickers 7.7-mm machine gun was fitted on the port side of the fuselage. The engine was a 110-hp Le Rhône 9Jb driving a Ratmanoff propeller 2.46 meters in diameter.

The fuselage was of wood except for the forward portion extending from the firewall to the cockpit, which was metal. The forward fuselage had three longerons: one behind the firewall, a second supporting the cabane struts and landing gear, and a third at the rear of the cockpit. The lower wing spar passed between the second and third frames. The rear fuselage had four rectangular longerons. The frame for the headrest was situated over the first three frames. At the rear of the fuselage steel tubes supported the tail and landing skid. The upper wing had two parallel spars and a dihedral of four degrees. The lower wing had a single large spar and, as mentioned earlier, no dihedral. The oil tank was in the forward fuselage. There were three fuel tanks: two in front of the cockpit and a third in the headrest fairing. Normally only a single machine gun was carried, but some HD.1s were modified to have two.

The Hanriot HD.1, as the new machine was designated, had an impressive performance which was fully evaluated when it made its first test flights in June 1916. The aircraft had the same engine as the Nieuport 16/17 series, which was probably the reason for its rejection by the Aviation Militaire. At this time the rotary engine's eccentric traits were being widely denounced by fighter pilots, who greatly preferred the SPAD 7 and its water-cooled engine. A 150-hp Gnome Monosoupape engine was later fitted to an HD.1 which also had new fuselage fairings and an altered fin and rudder. Although tests at Villacoublay confirmed its superior performance, no series production followed.

Although rejected by the Aviation Militaire, the Aviation Maritime adopted a modified version of the aircraft designated HD.2. The HD.1 was also widely used by the Italian, Belgian, and Swiss air services (see below).

Foreign Service
Belgium

The Aéronautique Militaire Belge ordered 79 Hanriot HD.1s in 1917. The first example was received in August 1917 by the 1st Escadrille. Others were supplied to most of the fighter units, the 9th Escadrille and 11th Escadrille (at Moeres airfield) being equipped entirely with Hanriot HD.1s during the war.

Despite the fact that the Hanriot HD.1s were found to be lightly armed and relatively slow, they were used with great success by such Belgian aces as Coppens, de Meulemeester, and Olieslagers. Field modifications included fitting two machine guns (which seriously degraded performance), fitting an HD.2 rudder to the HD.1 (which improved maneuverability), and fitting an 11-mm Vickers gun in place of the 7.7-mm weapon.

Postwar, the Hanriot HD.1s were operated by the 7th Escadrille of the 4th Groupe de Chasse based at Schaffen. By 1926 the 7th Escadrille was based at Nivelles. Later that year the unit retired the last of its Hanriot HD.1s.

Italy

The Aviazione Militaire needed a replacement for its Nieuport 17s, which were becoming rapidly obsolete. Examples of the HD.1 had been tested by Italian pilots in Paris who gave favorable reports on the new fighter. The speed, maneuverability, and climb rate were found to be considerably better than that of the Nieuport-Macchi 17s. As a result, an

HD.1 in service with the Belgian air force. The Aéronautique Militaire Belge ordered 79 Hanriot HD.1s in 1917. B87.6226.

Hanriot HD.1

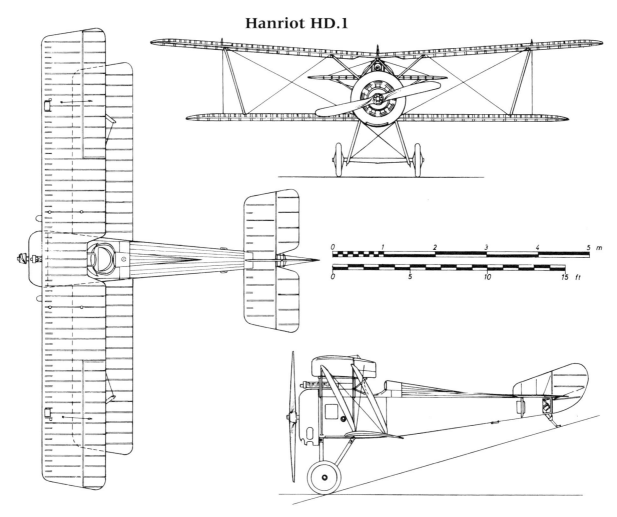

arrangement was made for the Societa Nieuport-Macchi to produce the HD.1 under license.

An initial order for 100 aircraft was placed in late 1916 and production began soon after. Most were powered by a 110-hp Le Rhône engine, although some were fitted with a 120-hp 9Jby. The Vickers machine gun, which originally was offset, was relocated to the centerline of the fuselage to improve the pilot's access to the gun in flight and to make aiming easier.

The Italians were very pleased with the HD.1. They especially appreciated its robustness and agility. It was also marginally faster than the Albatros fighters and Brandenburg D.Is it was fighting.

The initial unit to be equipped with the type was 76 Squadriglia, which received its first in August 1917. The unit was based at Borgnano and assigned to the 6 Gruppo in the 2nd Armata sector. It participated in the 11th Battle of Isonzo on 18 August 1917, where the qualities of the HD.1 became readily apparent.

During 1917 many of the other Italian fighter units received the type. The units equipped with Hanriot HD.1s on 20 November 1917 were:

10th Gruppo (assigned to the Supreme Command): 70a, 82a, and 91a Squadriglias.

13 Gruppo (3rd Armata): 80a and 83a Squadriglias.

6 Gruppo (4th Armata): 76a, 78a, and 81a Squadriglias, two Sezione of HD.1s assigned to defend Padova.

On 26 December 1917 HD.1s of the 6th and the 10th Gruppo Aeroplani participated in the air war over Istrana. By the end of 1917 there had been substantial changes in the dispositions of the HD.1 units. The new organization was:

A colorfully-marked Italian HD.1. The Vickers machine gun, which was originally offset, was relocated to the centerline of the fuselage in order to improve the pilot's access to the gun in flight and make aiming easier. B88.2531.

HD.1 in Belgian service. The Aéronautique Militaire Belge ordered 79 Hanriot HD.1s in 1917. 15631.

Belgian HD.1 serial number 12 of Thistle Escadrille. R. Verhegghen via Colin Owers.

10 Gruppo (Supreme Command): 70a, 82a, and 91a Squadriglias.
3 Gruppo (1st Armata): 72a Squadriglias.
13 Gruppo (3rd Armata): 80a and 83a Squadriglias.
6 Gruppo (4th Armata): 76a, 78a, and 81a Squadriglias.

By the time of the Battle of the Piave in June 1918 the following units had Hanriot HD.1s on strength:
10 Gruppo (Supreme Command): 70a, 80a, and 91a Squadriglias.
16 Gruppo (1st Armata): 71a and 80a Squadriglias.
6 Gruppo (4th Armata): 76a Squadriglia.
9 Gruppo (7th Armata/9th Armatas): 72a and 74a Squadriglias.
15 Gruppo (8th Armata): 78a and 79a Squadriglias.
214 Squadriglia of the Marina Italiana (Italian navy).

The Italian fighter command, perhaps learning from its counterparts in France, concentrated 120 fighters over the battlefield. This "Massa de Caccia," the majority of which were Hanriot HD.1s, proved very successful, and these units claimed 107 enemy aircraft and seven balloons destroyed between 15 and 23 June.

HD.1 units also saw action in Albania (85a Squadriglia Caccia based at Piskupi), Macedonia (73a Squadriglia Caccia based at Negocani), and Venezia Lido (214a Squadriglia of the Italian navy).

The Hanriot HD.1 remained in service throughout the war. By the Battle of Vittorio Veneto on 20 October 1918 the number of squadriglias using the aircraft had increased from 10 to 14 with a total of 144 HD.1s in service. These were as follows:
10 Gruppo (Supreme Command): 70a and 82a Squadriglias.
16 Gruppo (1st Armata): 71a Squadriglia.
3rd Gruppo (1st Armata): 75a Squadriglia.
13 Gruppo (3rd Armata): 80a Squadriglia.
6 Gruppo (4th Armata): 6a and 81a Squadriglias.
24 Gruppo (6th Armata): 83a Squadriglia.
9 Gruppo (7th Armata/9th Armata):72a Squadriglia.
20 Gruppo: 74a Squadriglia.
23 Gruppo (9th Armata): 79a Squadriglia.
8 Gruppo (Albania): 85a Squadriglia.
21 Gruppo (35th Divisione in Macedonia): 73a Squadriglia.
241 Squadriglia (Marina Italiana).

Squadriglias 72a, 73a, 76a, 80a, and 81a were equipped only with HD.1s; the other units operated a mixture of HD.1s, Nieuport 27s, and SPAD 7s.

A total of 1,700 Hanriot HD.1s were ordered from the Nieuport Macchi firm and by the time of the Armistice 831 had been delivered. An additional 70 were delivered after the Armistice.

Postwar, these Italian units still had HD.1s on strength:
3 Gruppo: 75a Squadriglia.
6 Gruppo: 76a and 81a Squadriglias.
8 Gruppo: 85a Squadriglia.
10 Gruppo: 70a and 82a Squadriglias.
13 Gruppo: 80a Squadriglia.
17 Gruppo: 71a and 72a Squadriglias.

20 Gruppo: 74a Squadriglia.
21 Gruppo: 73a Squadriglia.
23 Gruppo: 78a and 79a Squadriglias.
24 Gruppo: 83a Squadriglia.
241 Squadriglia (Marina Italiana) at Venezia.

When the Regia Aeronautica was formed in 1923 there were 48 Hanriot HD.1s on strength, composing 70 percent of the Italian fighter force. There were also 26 HD.1s in service at the Scula Allenamento Caccia (Fighter Training School). By 1924 the structure of the Hanriot HD.1 force was:
6 Gruppo: 70a, 73a, 79a, and 81a Squadriglias.
17 Gruppo: 71a, 72a, and 80a Squadriglias.

By 1926 the Hanriot HD.1s were retired to the training units, having been replaced by the new Fiat C.R.1.

Paraguay

Three HD.1s were obtained by Paraguay postwar for use as advanced trainers.

Switzerland

An Italian Hanriot HD.1 landed in Switzerland in June 1918. The Swiss were impressed enough by it to purchase 16 Macchi-built Hanriot HD.1s in 1921. These were used for pilot training. The aircraft were given serials 651 to 666 and were retired in 1930.

HD.1 Single-Seat Fighter with 120-hp Le Rhône 9Jb	
Span 8.7 m; length 5.85 m, height 2.94 m; wing area 18.2 sq. m	
Empty weight 400 kg; loaded weight 605 kg	
Maximum speed: at ground level 186 km/h; 178 km/h at 2,000 m	
Climb: 1,000 m	2 minutes 58 seconds
2,000 m	6 minutes 3 seconds
3,000 m	11 minutes 3 seconds
4,000 m	19 minutes 30 seconds
5,000 m	32 minutes
Ceiling 6,000 m; endurance 2.5 hours	
Armament: one 7.7-mm Vickers machine gun	
Approximately 100 HD.1s of all variants were built by Hanriot	

HD.1 Single-Seat Fighter with 110-hp Le Rhône 9Jb Built by Nieuport-Macchi	
Span 8.50 m; length 5.85 m, height 2.5 m; wing area 17.50 sq. m	
Empty weight 410 kg; loaded weight 600 kg	
Maximum speed: 183 km/h	
Climb: 1,000 m	2 minutes 40 seconds
2,000 m	6 minutes 40 seconds
3,000 m	11 minutes
4,000 m	16 minutes 30 seconds
Ceiling 5,900 m; endurance 2.5 hours	
Armament: one 7.7-mm Vickers machine gun	
Approximately 900 built under license by Nieuport-Macchi	

HD.1 showing cooling holes in cowling. 01121.

HD.1 serial 301. 01122.

Hanriot HD.2

The HD.2 was a seaplane version of the HD.1 fighter. The aircraft was essentially an HD.1 with a pair of floats suspended beneath the fuselage. A fin beneath the tail of the fuselage was intended to balance the floats. At least one aircraft was tested using a single float, but this arrangement was found to be unsatisfactory and further development was abandoned. The HD.2s usually had the more powerful 130-hp Clerget 9Bs instead of the 120-hp Le Rhône 9Jbs.

To improve stability on the water, the third HD.2 had elongated floats and the fin and rudder were enlarged over the basic HD.1. The HD.2s were armed with two Vickers machine guns and were based at Dunkerque to supply protection for the flying boats based there. A total of 17 HD.2s were delivered from January to October 1917.

The Aviation Maritime also used HD.2s with a conventional wheel undercarriage. These planes may have carried the designation HD.2 C. Examples of these aircraft were flown off a turret of the ship *Paris*. Enseigne de Vaisseau Paul Teste practiced takeoffs and landings from a platform similar to the one fitted to the *Paris*, but located on land at the Saint Raphaël naval base.

HD.2 preparing to takeoff from the *Paris*. MA6723.

Hanriot HD.2

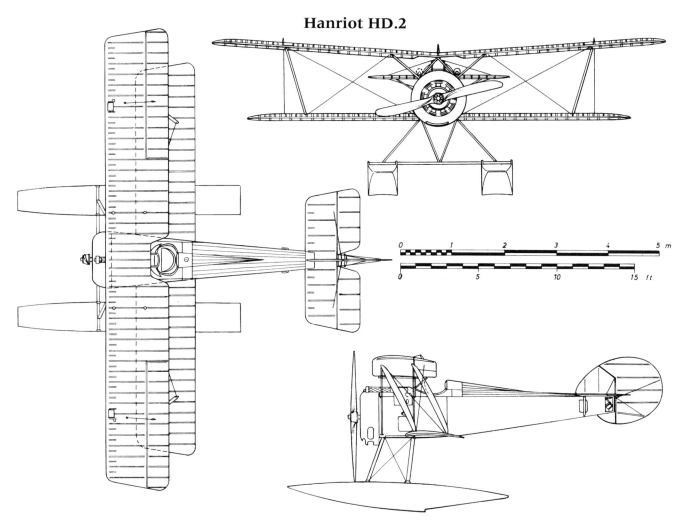

The aircraft used had a 120-hp Le Rhône 9Jb in place of the Clerget engine. After 20 practice flights, the first actual takeoff took place on 26 October 1918 when an HD.2 was successfully launched from the *Paris* and landed at Toulon. A second attempt on 9 November 1918 was unsuccessful because the *Paris* failed to generate adequate wind over the deck, and the HD.2 crashed over the forward deck, fortunately without injury to the pilot. Later, pilots from C 10 were trained to land on a ship's deck by using a steep approach to the ship and had a crude arrestor gear fitted to the HD.2. HD.2s were also flown from the carrier *Bearn* in the 1920s; at that time construction had not been completed on the carrier and it was, in fact, merely a flight deck placed atop the uncompleted battleship.

Approximately 30 HD.2s were sent to the Saint Raphaël naval aviation center at the end of 1918. Here, the HD.2s were converted from floats to wheel undercarriages and were assigned to AC 1 (escadrille de chasse) formed in 1919. The Hanriots served in front-line service until 1925 when they were replaced by Gourdou-Leseurre 22s. They subsequently served as trainers from 1925 until 1928. The temperamental Clerget rotary engines ensured that the HD.2s were not assigned to the *Bearn*; the Clergets were not reliable enough for the demands placed on them by carrier takeoffs and landings.

United States

The U.S. naval aviation service decided to purchase the HD.2 floatplane for use at its Dunkerque naval air station. Aircraft operating from the Dunkerque station had suffered heavy losses due to German fighters, and it was hoped that the Hanriots would provide the Donnet-Denhaut flying boats that were operated from the station with a measure of safety. Twenty-six

Above: HD.2 at Saint Raphaël 19 September 1919. This aircraft was flown by enseigne de Vaisseau Blutel. B87.5028.

Right: HD.2s (and Breguet 14s) of the Aviation Maritime based at Saint Raphaël 1922/23. B88.2717.

F.B.A. Type S

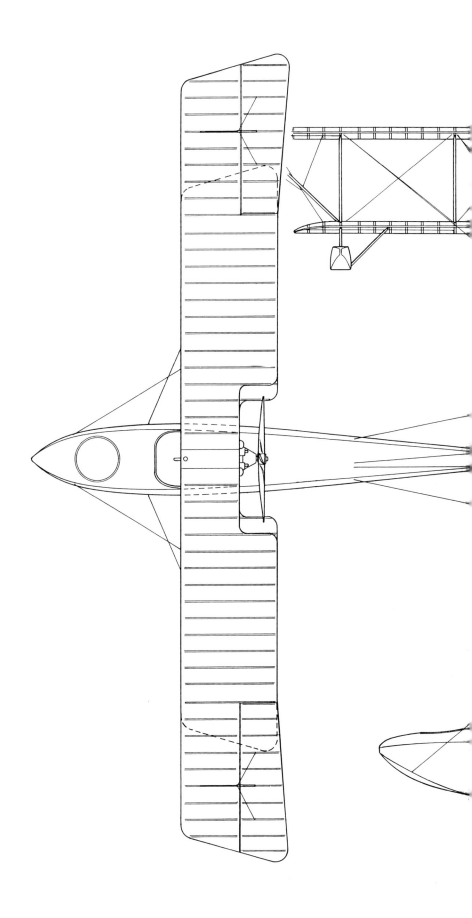

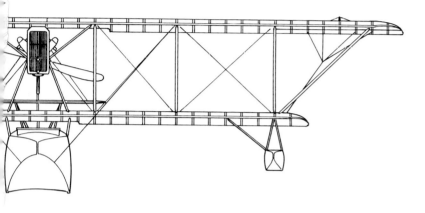
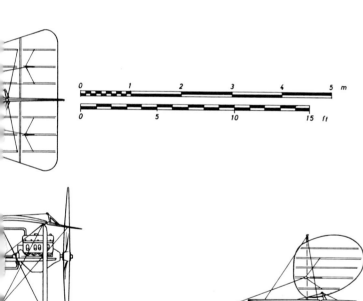

F.B.A. Type H

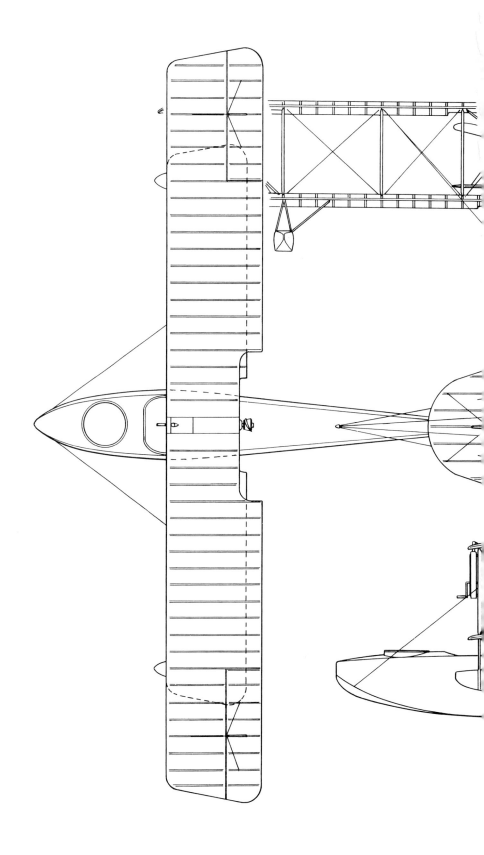

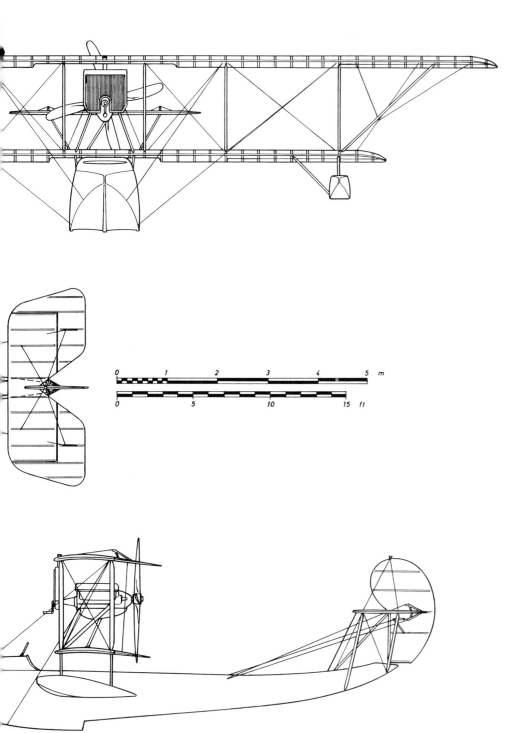

HD.2 of the Aviation Maritime. HD.2s were assigned to AC 1 (escadrille de chasse) formed in 1919. The Hanriots were in front-line service until 1925. B87.5031.

HD.2 Single-Seat Floatplane Fighter with 130-hp Clerget 9B

Span 8.70 m; length 7.00 m, height 3.10 m; wing area 18.20 sq. m
Empty weight 495 kg; loaded weight 723 kg
Maximum speed: 182 km/h; climb to 2,000 m in 6 minutes 30 seconds; ceiling 4,800 m; range 300 km
Armament: two 7.7-mm Vickers machine guns

HD.2 (C) Single-Seat Ship-Based Fighter with 130-hp Clerget 9B

Span 8.70 m; length 5.94 m, height 2.59 m; wing area 18.90 sq. m
Maximum speed: 180 km/h
Armament: two 7.7-mm Vickers machine guns

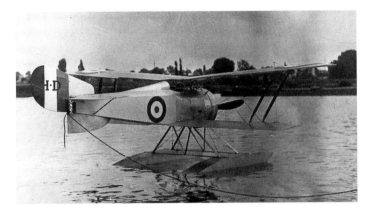

HD.2 based at Dunkerque. B78.1780.

HD.2 A 5621 was one of 26 HD.2s acquired by the U.S. Navy. HD.2s A 5620-A 5629 were converted to landplanes by the Naval Aircraft Factory after the Armistice. B83.2132.

HD.2s were purchased from the French government for the U.S. Navy. Serial numbers were: 211–220, 231–240, 244, 245, 249, 250, 252, and 258.

The HD.2s accompanied the Donnet-Denhaut flying boats on almost 500 patrols. However, no German aircraft were encountered during any of these missions.

After the war, ten of the Hanriots (serial A 5620 to A 5629) were sent back to the United States. The floats were removed and replaced with wheels, flotation bags, and hydrovanes. Modifications were also made to the rudder and a heavier skid was fitted. They were initially used as fighter trainers with two forward-firing 0.303 machine guns. The modified HD.2s were subsequently flown from ramps mounted on turrets of battleships and A-5624 was flown from the cruiser U.S.S. *Mississippi* in August 1919. Four HD.2 landplanes were used by the Second Ship Plane Unit from May to August 1919.

Hanriot-Dupont HD.3 C2 and Cn2

On 24 May 1917 General Duval decided to obtain a two-seat fighter to supplement the single-seat fighters then in service. To meet this specification the C2 category was formulated by the STAé. The specification called for an aircraft with forward-firing machine guns and two flexible machine guns. An option was also requested for the aircraft to be capable for use in the photo-reconnaissance role by fitting a camera in place of the rear machine guns. The aircraft designed for the C2 classification were to have a speed of 220 km/h at 5,000 meters, a minimum speed of 110 km/h, a service ceiling of 8,000 meters, and a payload of 375 kg. The aircraft designed to meet this category were the Borel HS 300, Hanriot-Dupont 5 C2, Vickers Lorraine 370, Farman F.31, Morane Saulnier 16, Morane Saulnier 31, Morane Saulnier 32, Hanriot Dupont C1, Breguet 17, SPAD 20, SEA 4, and the Hanriot-Dupont 3 C2. Only the Breguet 17, SPAD 20, SEA 4, and the Hanriot Dupont 3 were selected for production.

The Hanriot firm had built Salmson 2s under license. Apparently René Hanriot and Dupont were impressed with the 260-hp Salmson 9Za engine used in the Salmson 2, and they

HD.3 of HD 174. Fifteen of these aircraft were in service with HD 174 which was formed in October 1918. B82.1201.

Hanriot HD.3

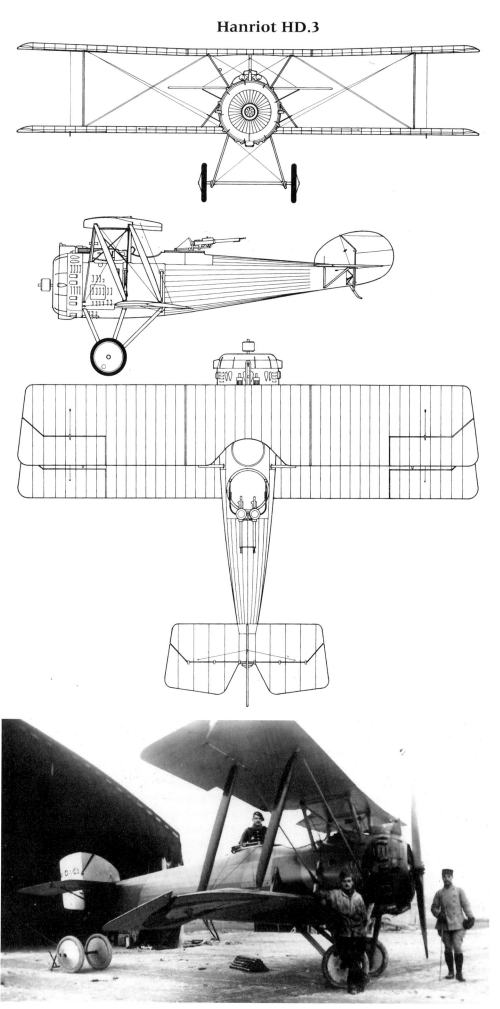

HD.3 of the 2eme Regiment de Chasse at Strasbourg. B84.359.

utilized that engine for the HD.3 C2.

The HD.3 was a single-bay biplane with ailerons on both the upper and lower wings. Each wing had two spars of spruce covered with plywood. The upper wing spar had sections reinforced in duralumin. The ribs were of plywood reinforced by spruce or poplar strips and attached to the spar by wood screws. The upper wing was attached to the fuselage by central cabane struts made of duralumin. The lower wing was attached directly to the fuselage by metal fasteners.

The fixed portion of the rudder had a crescent shape and was constructed of four tubular supports; the rudder also had a metal framework.

The forward portion of the fuselage was made of metal up to the cockpit. The rear portion had four ash longerons with spruce formers reinforced by a network of piano wire. The motor mount consisted of two metal plates connected by a U-shaped former. The fuel tank held 114 liters and the oil tank had a capacity of 24 liters. The fuel tank was within the fuselage, while the oil tank was located on the right side of the fuselage.

The landing gear consisted of three struts on either side which were bolted to the metal portion of the lower fuselage. The metal axle was articulated and encased in a wooden fairing. Bungee chord served as shock absorbers. The tail skid was made from ash and articulated at its attachment point to the lower fuselage.

Armament was two synchronized 7.7-mm Vickers machine guns and two turret-mounted 7.7-mm Lewis machine guns.

The prototype, which carried the military serial 1001, was tested at Villacoublay by a Lieutenant Bourgeois of the STAé on 10 March 1918 and had an impressive performance. The maximum speed was found to be 210 km/h, while the speed at 2,000 m was 207 km/h. The aircraft could climb to 1,000 meters in 3 minutes and 23 seconds.

The satisfactory performance of the HD.3, SEA 4, SPAD 20, and Breguet 17 led to all four aircraft being ordered. By 18 April 120 HD.3s were ordered, but later this number was increased to 300. Changes to the production aircraft included a reduction of the upper

HD.3 serial 1026. This aircraft was assigned to HD 174. MA25493

wing span, which resulted in top and bottom wings of being equal span, and the addition of horn-balanced ailerons.

The first production example left the factory in the fall of 1918. Apparently the performance of the first production examples was inferior to the prototype, but it appears that replacing the Chauviere airscrew by one designed by Ratmanoff resulted in considerable improvement.

The HD.3 entered service very late in the war and, in fact, only 18 were in service at the time of the Armistice. Fifteen were in service with HD 174 which was formed in October 1918. Serials were 1010, 1012, 1013, 1015, 1016, 1020, 1021, 1025, 1026, 1027, 1029, 1035, 1052, 1063, and 1066. The unit was assigned to GC 17. In 1920 HD 174 became the 109th Escadrille of the 3rd Group de Chasse assigned to the 2nd Regiment de Chasse based at Strasbourg. After 1920 the HD.3s were assigned to training units. It has been estimated that about 75 aircraft were built; it is certain that many more would have been obtained had the war continued.

While it appears that only one escadrille was completely equipped with HD.3s, it is likely that the remaining HD.3s were assigned to other units. In a communiqué from the Conferences du Centre d'Études Aéronautiques it was stated that HD.3s were to be used in multiple roles including day bombard-ment and long-range escort for Breguet 14s. Postwar, the type was to be assigned to the aviation units attached to the Corps d'Armée. While serving with these units it was intended that the HD.3s would provide protection for the reconnaissance units and engage in ground attack.

The French were impressed with the German night fighters which had caused not inconsiderable losses to French bombers. A night fighter version of the Hanriot HD.3 was developed and designated HD.3 Cn 2 (two-seat night fighter). A single prototype was constructed by September 1918. Changes from the standard HD.3 included a lower wing of thicker cross-section with heavy under-camber, enlarged ailerons, and an enlarged rudder. The aircraft was not selected for series production.

A single HD.3 was flown by Puget in the 7 September 1920 Michelin Coupe Militaire. It flew from Strasbourg to Angers in 18 hours and 40 minutes, but the race was won by a Breguet (probably a Breguet 17).

Several HD.3s obtained by the Aviation Maritime in 1921 were used to practice arrested carrier landings on the *Bearn*.

Hanriot HD.3 serial 1001. The HD.3 utilized the same 260-hp Salmson 9Za engine employed on the Salmson 2. (CC)

HD.3 Two-Seat Fighter with 260-hp Salmson 9Za
Span 9.170 m; length 6.95 m; height 3.00 m; wing area 25.5 sq. m
Empty weight 723 kg; loaded weight 1,150 kg

Maximum speed:	ground level	210 km/h
	2,000 m	207 km/h
	3,000 m	203 km/h
	4,000 m	196 km/h
	5,000 m	187 km/h
Climb:	1,000 m	3 minutes 23 seconds
	2,000 m	6 minutes 39 seconds
	3,000 m	11 minutes 4 seconds
	4,000 m	17 minutes 9 seconds
	5,000 m	34 minutes 34 seconds
	6,000 m	71 minutes 45 seconds

Service ceiling 5,700 m; range 400 m; endurance 2 hours
Armament: two synchronized 7.7-mm Vickers machine guns and two flexible 7.7-mm Lewis machine guns
Approximately 75 built

HD.3 Cn 2 Night Fighter with 260-hp Salmson 9Za
Span 9.42 m; length 6,92 m; height 3.20 m
Maximum speed: 200 km/h at 2,000 m; ceiling 6,200 m; range 400 km
One built

HD.4 and Navalized HD.3

The development of single-engined landplanes for use in the maritime patrol role was inhibited by the fact that the engines of the period were unreliable. An engine failure while over the sea would mean the definite loss of the plane and the probable death of its crew. On the other hand, fitting the aircraft with floats would seriously degrade performance. A novel solution was tried at the Saint-Raphaël naval air station. Although the aircraft was not flown until after the war, it merits inclusion in this volume as it was a conversion of the HD.3 two-seat reconnaissance/fighter. The wheels were removed from the axle and replaced by twin skis. An inflatable float made of thin rubber was fitted at the base of each wing. Normally, these floats would remain deflated and thus produce minimal drag. In an emergency, however, they could be inflated by compressed air within 30 seconds. The floats were designed by Harry Busteed and weighed 30 kg; the compressed air bottle weighed 7 kg.

The navalized version of the HD.3 had a shorter wing span than the standard HD.3 but otherwise was quite similar. This HD.3 successfully proved the validity of the inflatable float concept. Although it was considered to be a successful design, the war's end as well as the development of naval aircraft with watertight fuselages and more reliable engines led to the abandonment of the plan to use the Busteed floats on naval aircraft. Thus, further development was unnecessary and only the single example was built.

HD.3 No.2003 was sent to the RAF Naval Air Station at Grain in September 1918. Here it was fitted with jettisonable wheels, a hydrovane, and flotation gear to permit operation from naval vessels. A test conducted on 23 October resulted in the aircraft nosing over when ditching in the water. It was subsequently fitted with a revised hydrovane for the main gear and a smaller hydroplane was added to the tail skid. A successful water landing was made on 4 December 1918.

The HD.4 was an HD.3 (serial 2000) fitted with floats, a Chauviére propeller, and an enlarged fin and rudder. While it had been the intention of the French Aéronavale to operate these aircraft, it seems that the HD.2 was preferred over the HD.4 and only the single example was built and tested on the Seine in 1918.

Some sources mention a Hanriot-designed floatplane intended to carry a torpedo. No such aircraft is mentioned in the May 1918 Ministry of Munitions report; it is possible that the floatplane version of the HD.3 was the aircraft referred to. However, it is not possible to confirm that the HD.3 floatplane was ever modified to carry a torpedo.

> **HD.4 Navalized Two-Seat Floatplane Fighter with 260-hp Salmson 9Za**
> Wing span 9.10 m
> Maximum speed: 168 km/h
> One built

The HD.3 serial 2000 was the prototype HD.4. In addition to the twin floats, the aircraft had an enlarged fin and rudder to balance the side area of the floats. MA5923.

HD.5

The HD.5 was a development of the HD.1, intended to fill the C2 requirement for 1918. This category called for a two-seat aircraft with fixed forward machine guns for the pilot and twin machine guns on a ring mount for the observer. The aircraft was also to be capable of high-speed reconnaissance missions carrying a camera in place of the observer's machine guns. It was to have a speed of 220 km/h at 5,000 meters, a minimum speed of 110 km/h, a service ceiling of 8,000 meters, and a payload of 375 kg. A number of aircraft were designed to meet the C2 specification, those using the 300-hp Hispano-Suiza 8F included the Borel-Boccaccio 3000, BAJ fighter, and HD.5.

The main differences from the HD.1 were the fitting of a 300-hp Hispano-Suiza 8F engine, provision for a second crewman, and a larger wing. The two-bay wings were unstaggered and of equal span. The crew were seated in close proximity, the pilot in a prominent cutout in the top wing. The ailerons, rudder, and elevators were horn-balanced. Armament consisted of two fixed 7.7-mm Vickers machine guns and one or two 7.7-mm Lewis guns.

The aircraft was tested in the summer of 1918. Available performance figures include a maximum speed of 213 km/h (slower than the required speed for the C2 class), a ceiling of 6,200 meters (again lower than the 8,000 specified), and a payload of about 450 kilograms (higher than the required 375 kilograms).

Based on these figures, it is not surprising that further development of the HD.5 was abandoned and Hanriot turned his attention to another design to fill the C2 category—the HD.6.

> **HD.5 Two-Seat Fighter with 300-hp Hispano-Suiza 8F**
> Span 10.63 m; length 7.34 m; height 2.54 m; wing area 30.40 sq. m
> Empty weight 800 kg; loaded 1250 kg
> Maximum speed: 213 km/h; climb to 2000 m in 6.2 minutes; ceiling 6,200 m; range 425 to 490 km
> Armament: two fixed 7.7-mm Vickers machine guns and one or two 7.7-mm Lewis guns
> One built

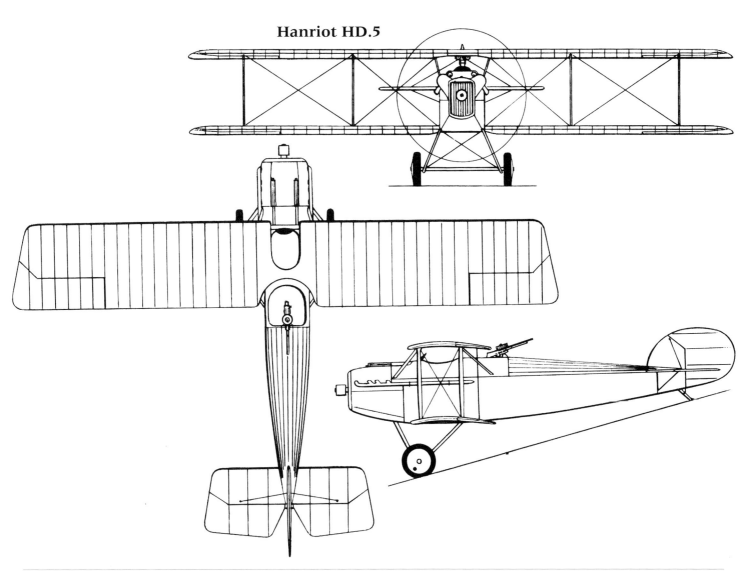

Hanriot HD.5

Hanriot HD.6

The failure of the HD.5 to meet the C2 category requirements spurred Hanriot to try a different design. The HD.6 was given a much more powerful engine—the 500-hp Salmson 18Z air-cooled engine. This engine was created by coupling two Salmson 9 Zs. The HD.5 was the only C2 design to use it. The bulky engine resulted in a rotund fuselage of circular cross-section. The two-bay wings were of equal span and unstaggered. The ailerons were horn-balanced and fitted to both upper and lower wings. The rudder and elevators were also horn-balanced. As with the HD.5, the pilot was seated under a large cutout in the top wing. The pilot and gunner were again located closely together to facilitate communication—this being one of the strong points of the HD.3 and HD.5. Armament was two fixed and synchronized 7.7-mm Vickers machine guns for by the pilot, while the observer had two 7.7-mm Lewis guns on a ring mount and a third Lewis which fired through the floor of the fuselage. Three Lamblin radiators were located under the bottom wing.

HD.6 Two-Seat Fighter with 500-hp Salmson 18Z
Span 13.60 m; length 8.85 m; height 2.90 m; wing area 47.50 sq. m
Empty weight 810 kg; loaded weight 1,250 kg; (another source gives an empty weight of 1,230 kg and a loaded weight of 1,950 kg)
Maximum speed: 225 km/h; climb to 1,000 m in 2 minutes 47 seconds; climb to 3,000 m in 9 minutes 37 seconds; climb to 5,000 m in 19 minutes 20 seconds; ceiling 7,100 m; range 600 km; endurance 3 hours
Armament: two fixed, synchronized 7.7-mm Vickers machine guns, two 7.7-mm Lewis guns on a ring mount, and a third Lewis gun which fired through the floor of the fuselage
One built

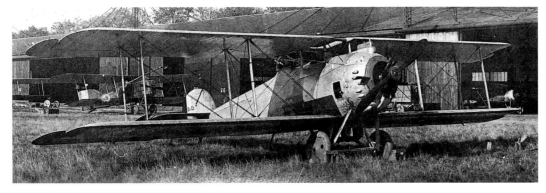

The HD.6. The 500-hp Salmson 18Z water-cooled engine was created by coupling two Salmson 9 Z engines. MA1971.

Hanriot HD.6

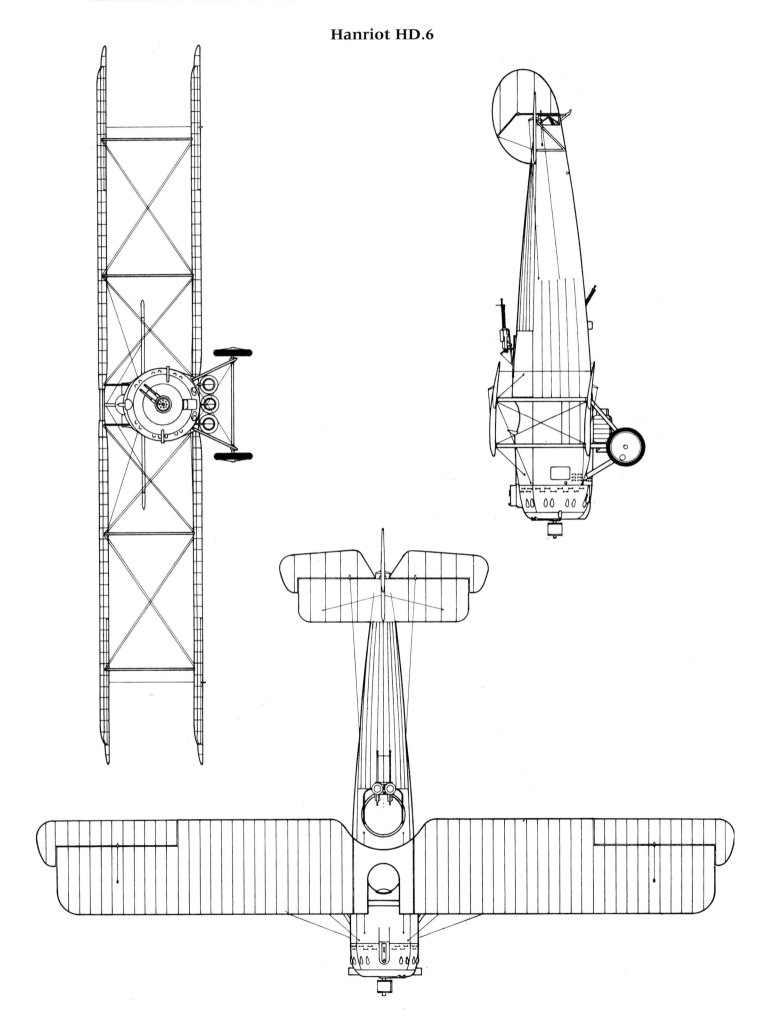

The HD.6. MA1971.

When the fuselage had been completed by November 1918, the new engine was still not available. It appears that the Salmson 18Z (only 25 of which were built) was not fitted until the spring of 1919. Performance figures from testing at Villacoublay showed that both the speed and ceiling were still below that required of an aircraft in the C2 class.

While the HD.6 had an increased maximum speed, improved range, higher ceiling, and a heavier payload than the HD.3, these improvements were modest. It is probable that, with the end of the war, it was felt that the SPAD 20 and HD.3 could perform the C2 missions adequately and further development of the HD.6 was abandoned.

Hanriot HD.7

The Hanriot HD.7 was designed to meet the 1918 C1 classification calling for a single-seat fighter with twin machine guns and the ability to carry a camera for high-speed photo-reconnaissance. Aircraft designed to meet this category and which were also powered by the 300-hp Hispano-Suiza 8F engine included the SPAD 18 Ca.1, SPAD 21, SPAD 22, Nieuport 29, Descamps 27, De Marcay C1, SAB.1, Moineau monoplane, Semenaud fighter, and the Hanriot HD.7.

The Hanriot HD.7 combined features from both the HD.3 and the HD.6 The wings and tail were basically the same as those on the HD.3, and the rudder was similar to that of the HD.6. The fuselage cowling closely surrounded the Hispano-Suiza 8F and twin Lamblin radiators were mounted on pylons which projected the units well below the fuselage. The placement of the radiators so close to the ground virtually assured that they would be damaged or torn off on landing. The fuselage was a redesign of that used on the HD.3, the most obvious changes being the provision of a single cockpit, elimination of the observer's station and machine gun ring, and modifications of the forward fuselage to permit mounting the Hispano-Suiza engine. Armament consisted of twin 7.7-mm Vickers machine guns.

HD.7 Single-Seat Fighter with 300-hp Hispano-Suiza 8F

Span 9.80 m (some sources say 9.00 m); length 7.20 m; height 3.00 m; wing area 28.00 sq. m (25.5 sq. m)
Empty weight 1,230 kg ; loaded 1,900 kg
Maximum speed: 214 km; climb to 5,000 m in 19 minutes 20 seconds; ceiling 7,250 m; range 900 km
Armament: twin synchronized 7.7 mm Vickers machine guns
One built

The aircraft was test flown at Buc in November 1918 and later at Villacoublay. While its performance was respectable, it was inferior to the eventual winner of the competition, the Nieuport 29. No further development of the HD.7 was undertaken.

Above: The Hanriot HD.7 was designed to meet the 1918 C1 requirement. The Nieuport 29 won the competition and was selected for mass production. MA5520.

The Hanriot HD.7 combined features from both the HD.3 and the HD.6. The wings and tail were basically the same as those found on the HD.3, and the rudder was similar to that found on the HD.6. MA5520.

Hanriot HD.7

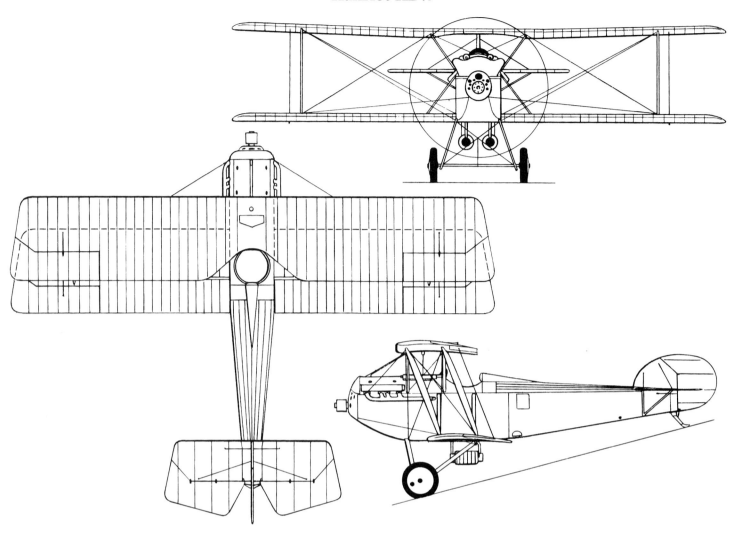

Hanriot HD.8

Considerable confusion exists as to the true identity of the aircraft carrying the HD.8 designation. It was built in 1922, and one source suggests it was based on the HD.7 but powered by a different motor. This version was flown in 1922 and had a maximum speed of 150 km/h. It is listed as being intended to meet the C1 classification for the navy, but further details or specifications and performance are lacking. The same source gives the 80-hp Le Rhône 9C engine as the power plant, but this seems to be a ridiculously outdated engine for use in a fighter.

It has also been suggested that the HD.8 was actually based on the HD.1 but fitted with a 170-hp Le Rhône 9R engine. It seems this engine was quite troublesome and that the HD.8 (if, indeed, that was the identity of the aircraft powered by this engine) was test-flown in early 1918 by the company. As the type was not submitted for official testing, it is probable that the difficulties with the power plant were never actually solved. The specifications given below are for this aircraft.

HD.8 Single-Seat Fighter with 170-hp Le Rhône 9R
Span 9.60 m; length 6.15 m; wing area 25 sq. m
Empty weight 480 kg; loaded weight 690 kg
Maximum speed: 200 km/h at 4,000 m; endurance 2 hours
Armament: two 7.7-mm Vickers machine guns (provisional)
One built

Hanriot HD.9

The HD.9 was designed to meet the Ap.1 category issued by the STAé in 1918. This called for a single-seat photo-reconnaissance plane armed with one machine gun. It was to have an exceptionally high ceiling so that photographs would cover several kilometers of territory. At the same time, it was to have been able to defend itself and thus would need the speed, maneuverability, and firepower of a single-seat fighter. The HD.9 was the only aircraft in this category and was powered by the 260-hp Salmson 9Z. Some sources suggest that the HD.9 was intended to meet the same Ap.2 specification as the Breguet 14. However, as the HD.9 was a single-seater this is obviously not correct.

The aircraft was essentially a single-seat version of the HD.3 fighter with primary changes being deletion of the observer's cockpit, revision of the undercarriage, and the deletion of one of the Vickers machine guns as well as the guns used by the observer. The HD.9 was fitted with a camera placed in the position which had been occupied by the observer in the HD.3.

HD.9 Single-Seat Fighter/High Altitude Photo-Reconnaissance Plane with 260-hp Salmson 9Z
Span 9.00 m; length 6.95 m; height 3.00 m; wing area 25.51 sq. m
Empty weight 710 kg
Maximum speed: 220 km/h; ceiling 7,250 m; range 800 km
Armament: one 7.7-mm Vickers machine gun
At least one but possibly as many as many as ten were built.

Letord 2

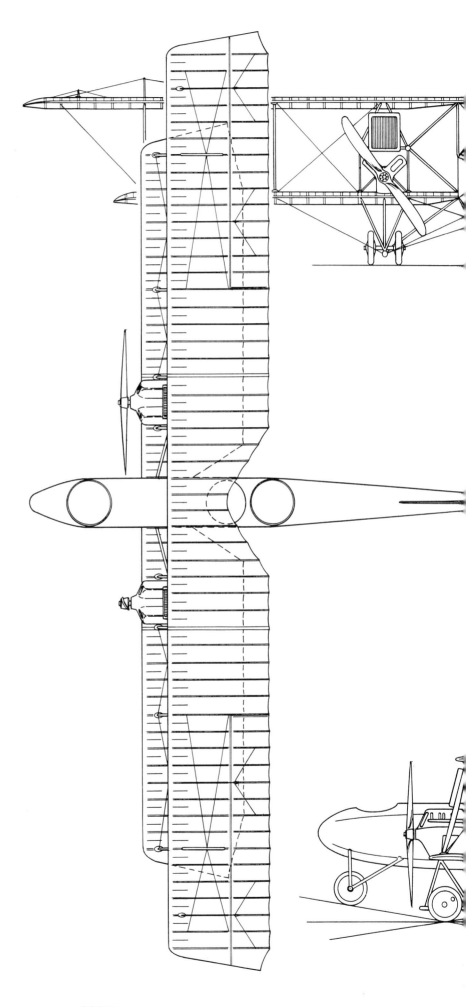

282B

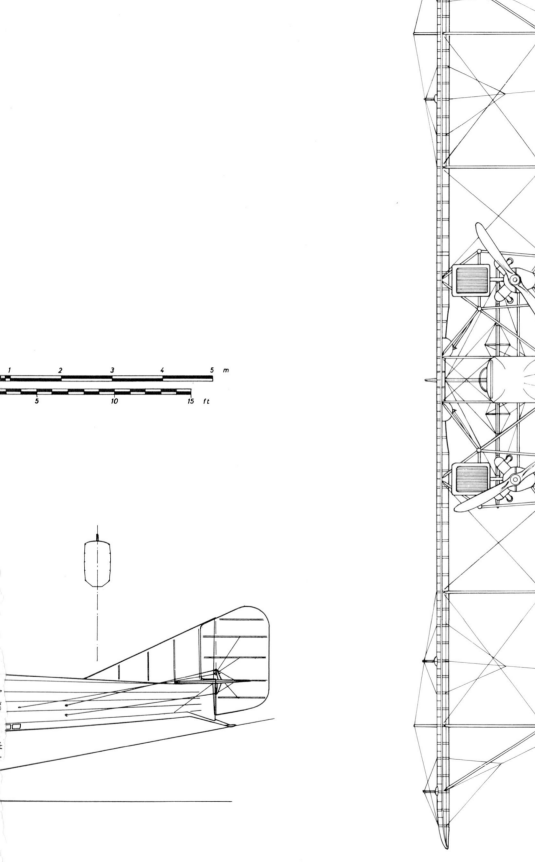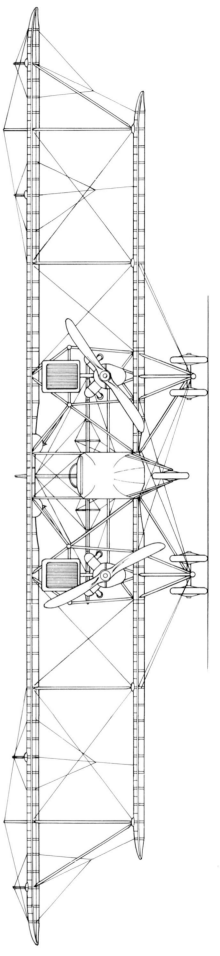

Letord 1

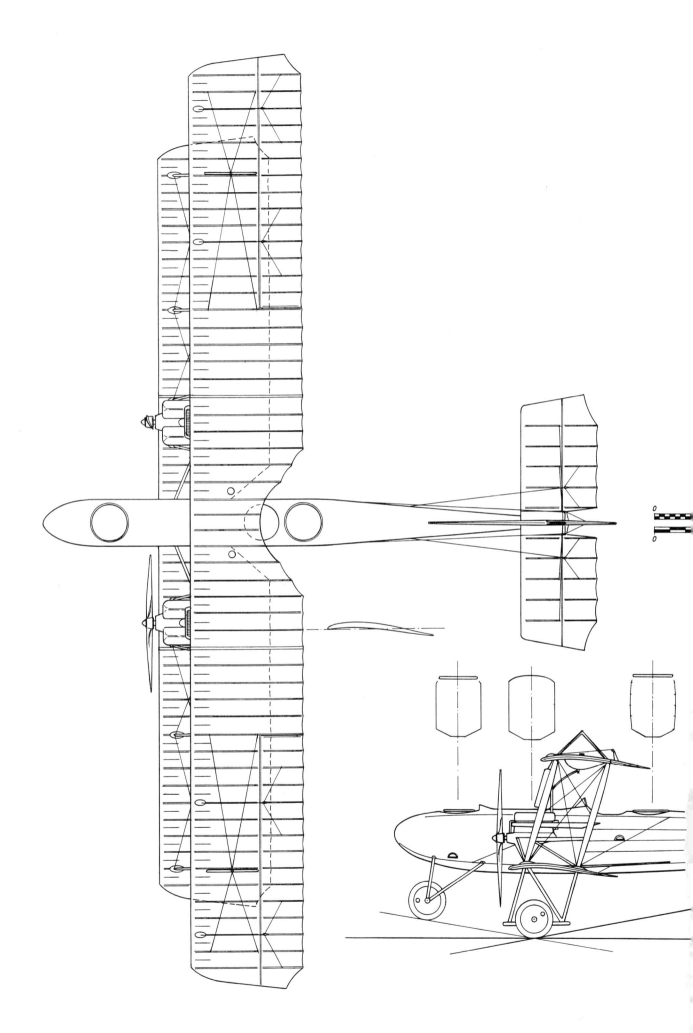

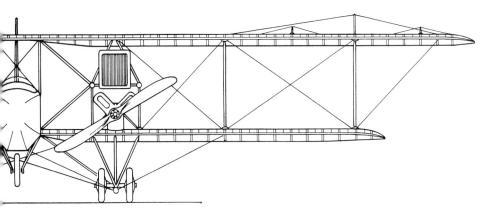

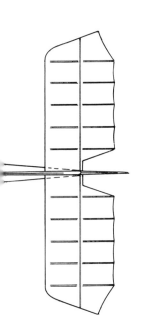
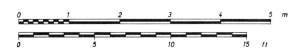

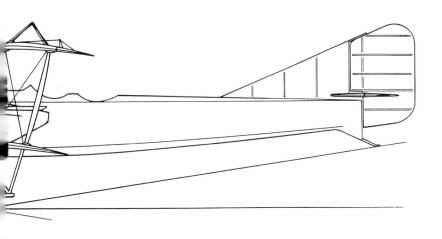

The aircraft's fuel supply was markedly increased to permit long-range reconnaissance

Trials commenced in September 1918 (some sources say November) initially at Buc and later at Villacoublay with the prototype, which had serial number 3001. The aircraft was well-liked and an initial batch of ten was ordered. However, due to the armistice no further orders were placed, and it is not certain how many HD.9s were actually completed.

HD.9. B82 1201.

Hanriot HD.9

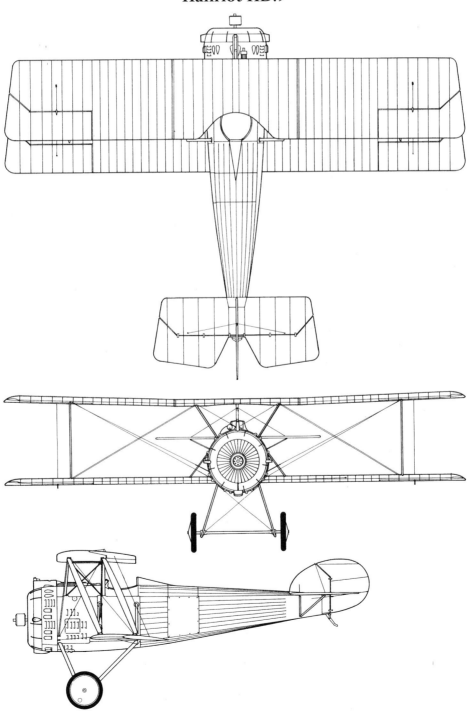

Hanriot HD.12

Although this aircraft did not actually fly until 1921 it falls within the scope of this present work as it represents the last development of that classic First World War design, the Hanriot HD.1. The HD.12 was similar to the HD.1 but with a more powerful engine. It is unclear if the HD.12 was intended for the C1 category (single-seat fighter) or was always planned as a shipboard naval fighter. In any event, the HD.12 was not selected by the Aviation Militaire for production, but it was tested by the Aviation Maritime for possible shipboard use.

The HD.2 and HD.4 had been flown from the turrets of naval vessels. This procedure required that crewmen on the ships literally hold the aircraft down while the pilot brought the engine to maximum power. At a signal from the pilot the personnel would release the aircraft. The HD.12 acquired by the Aviation Maritime was intended to facilitate this arrangement by making takeoff release simpler and more reliable. The primary improvement over the HD.1 was the fact that its more powerful engine meant that the HD.12 required a shorter takeoff run.

The aircraft had a primitive arrestor gear which could be released by the pilot after the engine had been run up. With this new arrangement, the plane could become airborne within only six meters. The aircraft was powered by a 170-hp Le Rhône 9R. Armament was planned to be two synchronized 7.7-mm Vickers machine guns. Other changes from the HD.1 included the deletion of the ailerons from the top wing, being instead fitted to the lower wing. The ailerons were enlarged from those used on the HD.1 and the wings were of equal span.

Initial testing was accomplished at Villacoublay and subsequent testing was performed at the naval base at Saint-Raphaël. Two HD.12s were used to train pilots and were flown from the platform mounted on the *Bapaume*. Although the plane was successful, further development was abandoned in favor of the more modern HD.27.

> **HD.12 Single-Seat Shipboard Fighter with 170-hp Le Rhône 9R**
> Span 8.70 m; length 5.94 m; height 2.59 m; wing area 19.00 sq. m
> Empty weight 480 kg; loaded weight 690 kg
> Maximum speed: 190 km/h at sea level
> Armament: two synchronized 7.7-mm Vickers machine guns
> At least two built

Hanriot Twin-Engine Aircraft

During the war Hanriot is believed to have designed an aircraft with twin 220-hp Renault engines mounted in pusher configuration. Apparently it was never completed; no STAé designation was ever applied to the design.

Hochart S2

The requirement for armored aircraft went back to 1912, when the Artillery Corps decided that all its planes would need to be armored against ground fire. The engines of the time would not permit an aircraft to carry armor and still have an acceptable performance. However, experience during the war with armored Caudron G.4s and, with much less success, F.43s, showed that such aircraft could be of enormous value in attacking ground troops. However, the G.4s and F.43s were too outdated to be successful in this role for long. Indeed, both types were soon relegated to flying when weather conditions would limit fighter opposition. However, the Aviation Militaire still wanted an armored ground-attack plane and in 1918 issued the S2 classification. This category called for a two-seat aircraft with armor and capable of attacking ground troops with machine guns and bombs. The LeO 5, Canton S2, Voisin 12, and Hochart S2 were designed to meet this specification.

Little is known about the Hochart S2 aside from the fact that it had two 200-hp Clerget 11E engines. The wings were of unequal span and were staggered. Armament comprised four machine guns which fired ahead and downward and two flexible guns. Sixteen 10-kg bombs could be carried. The Clerget 11E rotary engines were high-compression, and it was estimated that the aircraft would have poor performance below 2,000 m. Since the Hochart S2 was intended for ground attack, this would appeared to have placed a severe limitation on its usefulness. This, plus the difficulties encountered with the development of the Clerget 11E engine, effectively ended the Aviation Militaire's interest in the type. As far as can be determined, construction of the prototype was never completed.

> **Hochart S2 Two-Seat Ground-Attack Aircraft With Two Clerget 11E Engines (all data provisional)**
> Loaded weight 2,335 kg
> Maximum speed (estimated): 170 km/h at 3,000 m; ceiling between 6,000 and 6,500 m
> Armament: six machine guns and 160 kg of bombs

Ateliers d'Aviation L. Janoir

L. Janoir was a French aviation pioneer who built his first plane in 1912. At the beginning of the First World War Janoir was in Russia working with Vladimir Lebedev, whose firm was building Deperdussins under license for the Imperial Russian Air Service. Janoir helped to organize the first Russian aviation squadrons.

He returned to France in February 1916 and established a factory at Saint-Ouen near the Seine. The Janoir firm concentrated primarily on the repair and manufacture of SPAD fighters.

Janoir Flying Boats

In 1918 Janoir designed at least three different seaplanes. The first was a flying boat with an extremely odd configuration. The slab-sided fuselage, which held the crew of two, was surmounted by two-bay biplane wings. An abbreviated fuselage was supported above the rear hull by several struts. The fuselage had triangular tail surfaces. The engine was located in the nose of the fuselage and a cut-out was provided in the upper wing to enable the propeller to turn. This unusual arrangement had the advantage of keeping the tail surfaces, engine, and propeller clear of water spray. One aircraft was built but details of its flying characteristics and ultimate fate are not known.

Janoir's patent (No. 493.502) showed a second design very similar to the J-3 except a second engine was fitted in front of the propeller; the single propeller was powered by both engines. It does not appear that this second design was ever built.

A third Janoir flying boat was a twin-boom plane with a single in-line engine at the front of each boom. The wings had two bays of struts and appear to have been of equal span. A combination central float and main fuselage was suspended beneath the bottom wing. There was a biplane horizontal stabilizer and four rudders. A wind tunnel model underwent testing at the Laboratory Eiffel, but it appears that this design also remained an unfinished project.

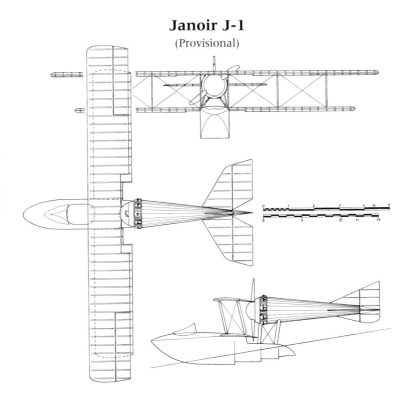

Janoir J-1
(Provisional)

Laboratory Eiffel Fighter

The Laboratory Eiffel fighter is listed as being intended to fulfill the C1 fighter category, but the prototype was actually an experimental aircraft intended to test some of the most advanced aerodynamic features that had been developed by 1918. However, it is interesting to compare the C1 specifications with those actually achieved by the Eiffel fighter. The specification called for a maximum speed of 240 km/h, a payload of 220 to 270 kg, a maximum ceiling of 9,000 m, and a service ceiling of 6,500 m.

Gustav Eiffel was a pioneer of aerodynamics and it was the staff of his aerodynamics laboratory who designed and developed the Eiffel fighter. The aircraft was a low-wing monoplane and the wing was almost completely of cantilever design except for two bracing struts on either side of the fuselage. The two parallel wing spars were of duralumin. The fuselage was remarkably streamlined, with the engine and single machine gun completely enclosed. The ventral radiator was located beneath the nose. The elevators and rudder were balanced. Estimated maximum speed for the aircraft was 265 km/h at 4,000 m, with an estimated ceiling of 8,000 m and climb to 4,000 m estimated in 10 minutes.

While the design was strictly the accomplishment of the Laboratory Eiffel staff, led by W. Margoulis, the aircraft was built by the Breguet firm. Hence some sources give its designation as Breguet LE. The initial prototype was fitted with a 180-hp Hispano-Suiza 8Ab engine and the wing structure had been redesigned because of failures in the spars found during static testing. Auxiliary spars were added to carry the ailerons.

Testing began at Villacoublay in March 9 1918 by Jean Sauclière, a fighter pilot on convalescent leave, who made the first flight but damaged the landing gear on takeoff. After the aircraft was repaired, Sauclière was allowed, with some

The Laboratory Eiffel LE fighter was intended to test some of the most advanced aerodynamic features that had been developed up until 1918. MA.

Laboratory Eiffel LE Fighter

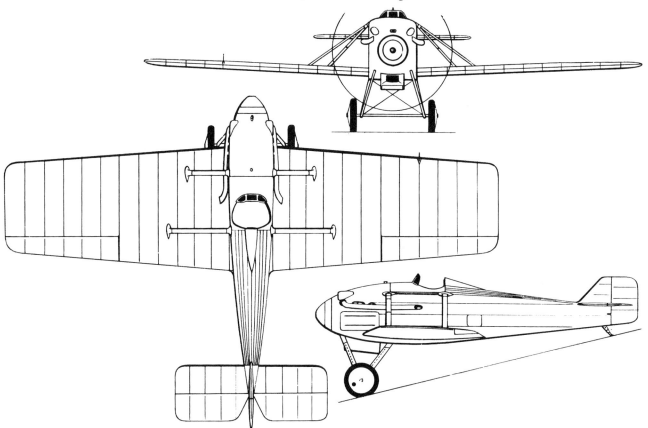

reluctance on the part of Louis Breguet, to attempt the second flight after Breguet's test pilot, Bailliencourt, expressed reservations about flying the LE. The aircraft rose quickly and flew 50 meters before diving into the ground and bursting into flames; Saucliére was killed. The cause for the crash is unknown, but was unofficially attributed to pilot error.

Although tragically brief, this second test flight at least established that the aircraft had an impressive performance, and it was estimated that the maximum speed would be 220 km/h at 4,000 m and that the LE could climb to 4,000 m in ten minutes. These were close to the requirements of the C1 specification.

Based on this estimated performance with only a 180-hp engine, it was planned to re-engine the airframe with a 220-hp Lorraine-Dietrich engine or a 300-hp Hispano-Suiza 8Fb. The airframe for the latter was under construction in January 1918, but apparently the loss of the first prototype resulted in this project being delayed.

At Breguet's insistence, the LE variant with the 220-hp Lorraine-Dietrich was apparently re-designed as a biplane in order to prevent any further structural failures. It was also lightened considerably, possibly also in an attempt to prevent further structural failures.

It seems that the end of the war, coupled with the intense activity at the Breguet plant (which was producing the Breguet 14 and its derivatives), conspired to keep these developments of the Laboratory Eiffel fighter from being completed. The aircraft did leave one lasting impression on French aviation between the wars; because of the catastrophic failure of the Eiffel prototype, further development of the low-wing monoplane was all but abandoned by the French postwar.

LE Experimental Single-Seat Fighter with 180-hp Hispano-Suiza 8Ab

Span 9.78 m; length 6.35 m; height 2.0 m; wing area 20 sq. m
Empty weight 495 kg; loaded weight 700 kg
Maximum speed (estimated): 220 km/h at 4,000 m; climb to 4,000 m in 10 minutes; endurance 2 hours
No armament was fitted to the LE prototype, but two machine guns were planned for production aircraft

Laboratory Eiffel LE fighter. The aircraft was a low wing monoplane and the wing was almost completely of cantilever design except for two bracing struts on either side of the fuselage. MA1630.

L.A.F. *Desmon* Flying Boat

The L.A.F. (Ligue Aéronautique Francaise or French Air League) designed and built a biplane flying boat with a crew of two which began testing in mid-1918. It was powered by two 280-hp Renault engines and was intended as a torpedo plane. The aircraft had twin hulls, each of which had two steps. The first flight was 28 June 1918. On the first two attempts to land a large amount of water sprayed up between the two hulls, making landing difficult. On the thired attempt to land one of the aircraft's hulls failed and filled wtih water. As a result of this the *Desmon* was not flown again and the wreck was condemned in July 1919. Other land-based designs were proposed by L.A.F. but never built. (See Addendum for drawing.)

L.A.F. Flying Boat with Two 280-hp Renault Engines
Wing area 105 sq. m
Payload 500 kg
Maximum speed: 141 km/h at 2,000 meters; climb to 2,000 meters in 25 minutes
One built

Labourdette-Halbronn H.T.1 and H.T.2

The Labourdette-Halbronn H.T.1 and H.T.2 were designed as torpedo bombers. They were the result of a collaboration between Lieutenant de Vaisseau Halbronn of the Section Technique de la Marine, who designed the plane, and Labourdette, who built it. It had twin hulls and triplane wings with a backward stagger. There were two bays of struts splayed outwards. Halbronn's design had wings of unequal length, the lower wing shortest. The H.T.1 was built in 1918 and flew in April 1919. It had two 200-hp Hispano-Suiza engines mounted on the top of the middle wing. A gunner could be accommodated in the nose of each hull, positions providing them excellent fields of fire. The pilot was situated in a raised nacelle which sat astride the two hulls. The torpedo was carried between the twin fuselages. A horizontal stabilizer connected the two hulls at the tail and at the end of each fuselage was a rounded fin and rudder. Apparently the power supplied by the two engines was felt to be insufficient as two additional aircraft, designated H.T.2, were built with two 350-hp Lorraine engines. Tests were conducted at the naval base at Saint Raphaël and continued until 1922. Due to the armistice, no further examples were ordered. An officer of the RNAS who examined the prototype was impressed enough by the twin-hull design that he recommend an example be obtained for study; however, no machines were purchased by the British.

Labourdette-Halbronn H.T.1 Flying Boat with Two 200-hp Hispano-Suiza Engines
Wing area 100 sq. m
Empty weight 2,000 kg; loaded weight 3,200 kg
Maximum speed: 125 km/h
Armament: two machine guns and a torpedo
One built

Labourdette-Halbronn flying boat. The pilot was situated in a raised nacelle which sat astride the two hulls. The torpedo was carried between the twin fuselages. 2476.

Larnaudi

Larnaudi Single-Engine Flying Boat

This Larnaudi flying boat was a single-engine, single-seat biplane under construction in 1918. It was probably intended for the

Larnaudi Flying Boat with 200-hp Hispano-Suiza (all data provisional)
Wing area 18 sq. m
Payload 180 kg
Maximum speed: 180 km/h at 2,000 meters, climb to 1,000 meters in 15 minutes

patrol or "alerte" category of seaplane and was powered by a 200-hp Hispano-Suiza engine. The type was not selected for use by the Aviation Maritime and it is not known if the plane was ever completed.

Larnaudi Twin-Engine Flying Boat

Larnuadi designed a twin-engine flying boat in 1918. It was to have been a triplane with two 500-hp Bugatti engines and a crew of three. As of May 1918 the design had not yet been built. The Bugatti engine, which was to have been built under license by Peugeot, apparently had severe developmental and production difficulties. The Morane-Saulnier Type AN fighter, Dorand flying boat, and the Levy-Besson triplane flying boat had also been designed to accept the new engine. However, the Type AN was eventually fitted with different engines and the Dorand and Levy-Besson designs had to be abandoned at an early stage, possibly due to difficulties with the Bugatti motor. It therefore appears likely that the Larnaudi flying boat also remained an unbuilt project.

Larnaudi Twin-Engine Flying Boat (all data provisional)
Wing area 150 sq. m
Payload 500 kg
Maximum speed: 160 km/h at 2,000 m; climb to 2,000 m in 25 minutes

Latécoère 1 C2

The Latécoère 1 was built in 1918 to meet the C2 specification for a two-seat fighter. Latécoère's design had been influenced by the Salmson 2 which his firm had produced under license. It was designed by Marcel Lemoine and was powered by a 460-hp Salmson Z18 engine. It also had wings of equal span with a forward stagger. It is believed that the Latécoère 1 C2 was constructed of wood and fabric.

However, the Latécoère 1 C2 was not completed until the day of the Armistice, too late for the type to be considered for production. It was evaluated by Didier Daurat of the STAé and was subsequently converted into a postal aircraft. Further development was abandoned.

Latécoère 1 C2 Two-Seat Fighter with 460-hp Salmson Z18
Span 8.60 m; length 5.40 m; height 2.20 m; wing area 22.00 sq. m
Loaded weight 1,150 kg
Maximum speed: 250 km/h
One built

Latham High Seas Flying Boat

Jean Latham founded a company for building seaplanes in 1917. The Société Latham & Cie built 24 Levy HB.2 flying boats before constructing its own design for a large flying boat to meet the new requirement for a "high seas" aircraft.

The French navy decided in 1918 that it wanted a flying boat in the same category as the British Felixstowe series. This category was termed the "high seas" flying boat and the requirements called for a crew of four (two pilots and two observers), a T.S.F. radio with long-range, an endurance of eight hours, a 75-mm cannon with 30 rounds, two machine guns, a bomb load of 120 kg, a speed of 140 km/h, and an ability to climb to 1,000 meters in 15 minutes. Several aircraft submitted designs to meet the "high seas" specification, including Donnet-Denhaut, Farman, Tellier, Levy, Besson, and Latham.

The Latham design featured three Panhard-Levassor P1-12CB engines; two were mounted as tractors and one was a pusher.

Latham CH.1 Flying Boat with Three 350-hp Panhard-Levassor P1-12CB Engines
Span 31.17 m; length 18.05m; height 6.00 m; wing area 176 sq. m
Empty weight 4,700 kg; loaded weight 7,190 kg
Maximum speed: 145 km/h; cruising speed 110 km/h; ceiling 3,000 m; range 1,000 km; endurance eight hours
Four built

The propellers were four-bladed Chauvieres. The tail had a cruciform configuration and there were three rudders between the upper and lower horizontal stabilizers.

Four of these aircraft were ordered and were designated CH.1. However, the "high seas" category of flying boat had been rendered superfluous by the Armistice and no further aircraft were ordered, although consideration was given to using them in the colonial reconnaissance role.

Latham H-5 flying boat. This was a development of the wartime CH-1. The only significant difference between the two was that the CH-1 had three 350-hp Panhard-Levassor P1-12CB engines while the H-5 had four 260-hp Salmson 9Z engines. B87.4469.

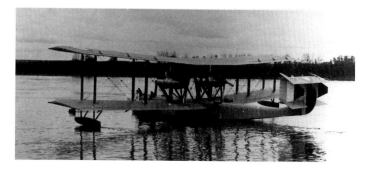

Latham H-5 flying boat. The dimensions were identical to the preceding CH-1, except for the fuselage which was 0.83 meters longer. Both types had the upswept tail with three rudders. B87.4466.

Latham CH.1

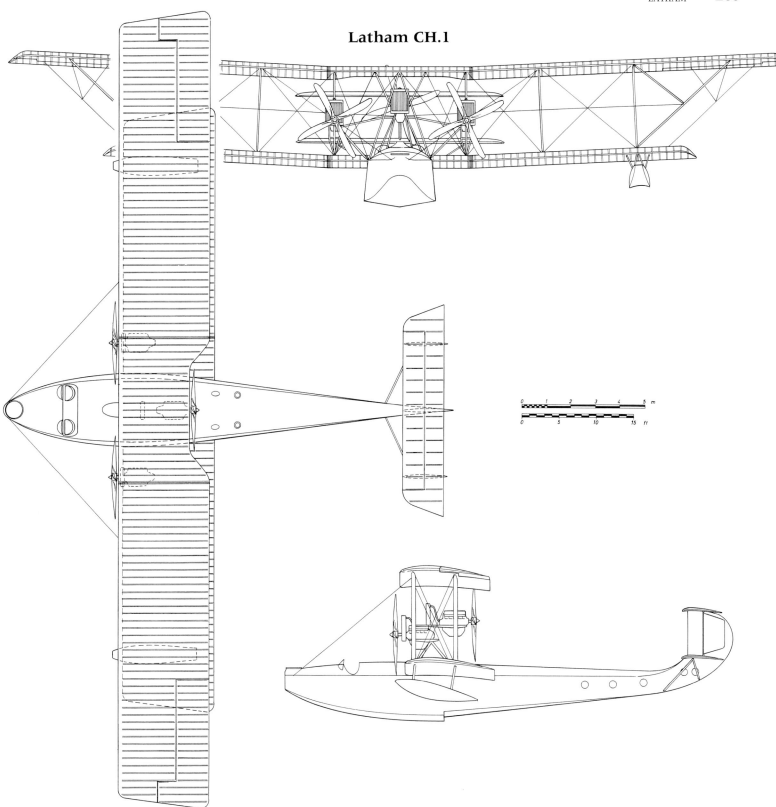

L.D. Ca2

This aircraft was under construction in 1918 to meet the Ca2 specification for a heavy fighter armed with a cannon. Both the Caudron R.14 and Letord 6 were also designed to meet this specification. The L.D. Ca2 (Ca = Cannon; 2 = crew of two) was to have been powered by two Peugeot-built Bugatti engines. The Bugatti's 16 cylinders were arranged in a U-configuration which permitted a cannon to be placed between the two rows of cylinders. Morane-Saulnier and Levy-Besson also submitted designs to use the new Bugatti engine. The presence of two Bugatti engines suggests that the L.D. Ca2 would have been armed with a single cannon in each. However, Morane-Saulnier, for whom the engine was originally intended, had severe reservations about the new motor's ultimate availability. Delays in development of the Bugati led it to re-engine its Type AN prototype. It is likely that development of the L.D. Ca2 encountered similar difficulties, since further work on the design was abandoned in 1918.

Établissements Letord

The Établissements Letord was founded in 1908 to produce balloons and airships. It would become famous, however, for building a series of airplanes based on designs formulated by the STAé. After the war, Letord would build an experimental plane designed by Bechereau. In 1925 the factory was let to Villiers.

Letord 1, 2, 4, and 5

In April 1916 the STAé, at the direction of Colonel Dorand, formulated a requirement for a "triplace corps d'armée" or A3 category plane. This was to replace the other A3 category planes including the Caudron G.6s and later Morane-Saulnier Ts and S.M.1s. The aircraft was to have two 150-hp Hispano-Suiza engines. According to an STAé memo written in late 1917, only the Letord firm expressed an interest in producing a plane to the STAé's specifications. Letord and the STAé worked together in deciding on the appropriate configuration, including the size, profile, and negative stagger of the wings. In fact, in a letter dated 28 October 1917 from Dorand, it was specified that Letord was to follow the STAé's guidelines precisely in producing the new aircraft.

It had been intended to supply front-line units with a three-seat, long-range plane that could perform reconnaissance behind enemy lines as well as undertake light bombing missions. The C.A. and base escadrilles were to have 15 planes, of which four would be three-seaters. The A.L.G.P. and heavy artillery escadrilles were each to have a complement of 15 planes, of which five were to be three-seaters. However, there were severe problems with each of the aircraft types selected to fill this role. The Caudron R.4 and G.6 had proved to have major flaws; in fact, at one point the G.6s had to be grounded because of frequent crashes. The Morane-Saulnier T suffered structural failures of the tail assembly that could result in the loss of the aircraft and its crew. The S.M.1s were hard to maintain in the field and had inadequate performance. Because of these difficulties, Colonel Dorand of the STAé gave high priority to the development of the new Letords intended to meet the requirement for a three-seat, long-range reconnaissance aircraft.

The first of the series to emerge was built under license by the Établissements Letord and hence carried the Letord name. However, it should again be noted that the aircraft was a product of the STAé design bureau, and that Letord was responsible only for producing it to the bureau's specifications.

The Letord 1 was a very large biplane. Its most striking feature was the negatively staggered wings, the top wing being located well aft of the bottom. The wings were of unequal span, the top wing longer. The two 150-hp Hispano-Suiza 8A engines were mounted in nacelles on the bottom wing and were in close proximity to the fuselage. Paired wheels were located below each nacelle. A large nosewheel served to prevent the Letord 1 from nosing over when landing on the poorly prepared airfields of the time. There was also a prominent tail skid. Rectangular tailplanes and a triangular fin were also a feature of the Letord series.

The pilot was seated beneath the trailing edge of the upper wing and a gunner was located just aft of the pilot. Another

Above: Letord 2. Moreau 1716.

Left: Letord 4. The Letord 4 differed from the other recon-naissance variants in having two 160-hp Lorraine-Dietrich 8A engines. B76.1964.

Letord 1 serial number 132 of BR 9. B89.3588.

gunner had a cockpit in the extreme nose. Each had one or two ring-mounted 0.303" Lewis machine guns. Between 130 and 150 kg of bombs could be carried.

By February 1918 there were 125 Letord A3 category planes in service out of 390 ordered.

Variants

1. The Letord 2 was a development of the Type 1, the main difference between the two types being the engines—the Type 1 had Hispano-Suiza engines of 150-hp and the Type 2 had 200-hp Hispano-Suiza 8Ba engines.
2. The Letord 4 retained the main features of the series. It, too, was distinguished by its engines, these being 160-hp Lorraine-Dietrich 8As. The Letord 4 was first tested at the front by LET 46 in December 1917. It was frequently used as a light bomber.
3. The Letord 5 had the most powerful engines of the Letord reconnaissance series—240-hp Lorraine 8Fbs. The Letord 5 may have initially been intended for use in the light bombing role rather than as a reconnaissance type. A distinguishing feature of the Letord 5 was the lack of a nosewheel common to the earlier Letord types. A total of 51 Letord 5s were built.

Operational Service

By 1 August 1917 there were 89 Letords of all types in service at the front, with an additional six at the R.G.A. and three under repair. By November 1917 there were 121 operational on the Western Front. It would appear from available records that most of them were no longer in front-line service by the end of the war.

It had been decided to provide each army cooperation escadrille with three or four Letords rather than establish separate escadrilles made up of only three-seaters. This would ease problems with maintenance and overcrowding that had resulted when these large planes were operated as a single unit. For example, SPA-Bi 20 had seven SPAD 11s, two Sopwith 1½ Strutters, and three Letords (serial numbers 436, 445, and 446). The Letords were designed to provide the escadrilles assigned to the Corps d'Armée and A.L.G.P. with a long-range reconnaissance capability, as well as serving as escort fighters and bombers. It seems that only one escadrille assigned to the armées was ever fully equipped with the type, this being LET 46.

LET 46 was formed from R 46 (with Caudron R.4s) in March 1917. This unit served with the 3rd Armée in December 1917 and the 6th Armée in January 1918. LET 46 was subsequently assigned to the 4th Armée, then the 8th Armée, and, finally, GC 15, by which time it had received Caudron R.11s. Its designation thus changed to R 46.

Many of the other reconnaissance units using other aircraft types were also given Letord aircraft. Units known to have had some Letords on strength include MF 2, C 4, F 8, C 9, C 10, C 16, SPA-Bi 20, C 30, C 39, SAL 40, C 53, AR 58, AR 59, C 61, F 72, SAL 122, MF 211, R 209, SOP 214, SOP 219, SOP 221, SOP 223, SOP 227, SOP 231, and SOP 237. The A.R.1/2s, M.F.11s, F.40s, Salmson 2s, and Sopwith 1½ Strutters used by these units were all short-range reconnaissance types. The Letords were occasionally employed as fighter escorts for the more vulnerable reconnaissance machines, particularly the Farman F.40s, and they were also used as bombers.

The Letord series was evaluated for possible use in providing long-range naval reconnaissance and fighter escort for the vulnerable flying boats and convoys. However, the Letords were too big to be fitted with flotation devices, meaning that they

Letord 1 serial number 155 with modified nose contour, possibly used to evaluate features for the Letord 3 bomber prototype. Renaud.

Letord 1 prototype undergoing testing at Bron. B77.714.

Letord 2 of C 228 which was assigned to the 87th regiment d'artillerie lourde in the 2nd Armée sector. B76.1966.

would immediately sink if forced down at sea. For this reason the Letords were not adopted by the Aviation Maritime. Although some of the coastal escadrilles did receive LET designations, only a small number of Letords were assigned to them; it appears that the majority of aircraft types with the LET escadrilles were Voisins or Caudrons.

It is not known how many aircraft in the Letord series were produced, but a rough estimate would be between 250 and 300 Letord 1s, 2s, 4s, and 5s.

Below: Letord 2 serial number 226. The Letord 2 differed from the Letord 1 mainly in being equipped with 200-hp Hispano-Suiza 8Ba engines. B86.4188.

Above: Letord 4. This is one of the few French airplanes in the First World War to carry a shark-mouth insignia. B88.0926.

Letord 5, serial number 422. The Letord 5 was the most powerful of the Letord reconnaissance series, having two 240-hp Lorraine 8Fb engines. B76.1965.

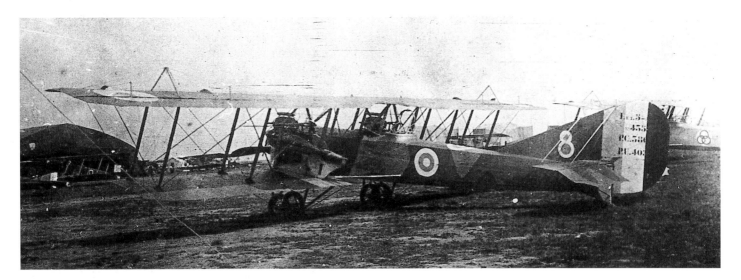

Above: Letord 5. This aircraft has unusual "bullseye" markings which are a personal insignia. Note the interlocking rings painted on the Letord in the background. B84.591.

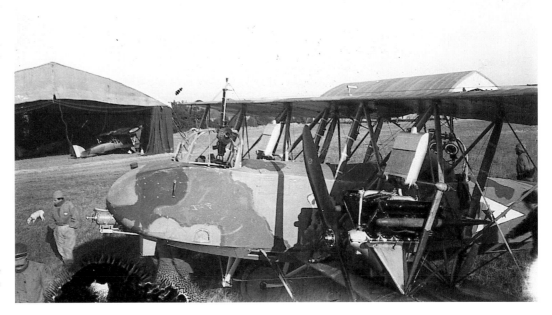

Letord 5 in a factory camouflage scheme. R. Verhegghen via Colin Owers.

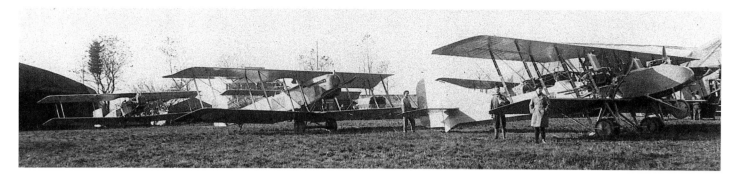

Above: Letord assigned to an AR escadrille. The Letords were used by the army cooperation units for long-range reconnaissance and fighter escort. Renaud.

Right: Letord 5. MA3854.

Below: Letord 5. MA38549.

Below left: Letords in front-line service. The size of these aircraft made it impractical to deploy them in a single escadrille; providing two or three to individual escadrilles proved to be more practical. Brian Flanagan via Colin Owers.

Below: Letord 1 serial number 104. This photograph emphasizes the negative wing stagger that was favored by Dorand.

Letord 1 Three-Seat Reconnaissance Plane with Two 150-hp Hispano-Suiza 8A Engines
Span 17.95 m; length 11.17 m; height 3.660 m; wing area 61.4 sq. m
Empty weight 1250 kg; loaded weight 1900 kg
Maximum speed: 150 km/h at 1,000 m; 148 km/h at 3,000 m; 135.5 km/h at 4,000 m; climb to 1,000 m in 5.20 min; to 3,000 m in 21 min. 30 sec.; to 4,000 m in 36 min.; ceiling 5,200 m; range 350 km; endurance 3 hours
Armament: three or four 0.303 Lewis machine guns and 136 kg of bombs

Letord 2 Three-Seat Reconnaissance Plane with Two 200-hp Hispano-Suiza 8Ba Engines
Span 18.060 m; length 11.170 m; height 3.660 m; wing area 62.0 sq. m
Empty weight 1625 kg; loaded weight 2400 kg
Maximum speed: 157 km/h at 2,000 m; 145 km/h at 4,000 m; climb to 2,000 m in 10 min.; climb to 4,000 m in 29 min.; ceiling 5,400 m; endurance 3 hours; range 370 km
Armament: three or four 0.303 Lewis machine guns

Letord 4 Three-Seat Reconnaissance-Bomber with Two 160-hp Lorraine-Dietrich 8A Engines
Span 18.50 m (some sources say 17.70 m); length 11.83 m; height 3.55m; wing area 61.4 sq. m
Loaded weight 2,186 kg
Maximum speed: 140 km/h at 2,200 m; climb to 2,200 m in 13 min. 40 sec.; ceiling 4,500 m; endurance 3 hours
Armament: three or four 0.303 Lewis machine guns and 150 kg of bombs

Letord 5 Three-Seat Reconnaissance-Bomber with Two 240-hp Lorraine 8Fb Engines
Span 18.06 m; length 11.17 m; height 3.660 m; wing area 63.20 sq. m
Empty weight 1,660 kg; loaded weight 2,445 kg
Maximum speed: 170 km/h at 2,000 m; 152 km/h at 3,000 m; climb to 2,000 m in 10 min.; to 4,000 m in 20 min.; ceiling 4,900 m; range 455 km; endurance 3 hours
Armament: three or four 0.303 Lewis machine guns and 135 kg of bombs
Total of 51 built

Letord 3

The need for a new night bomber led the STAé to request that Letord, Farman, and Caudron undertake the design of a twin-engine bomber designed to use two 240-hp Lorraine engines. The STAé was attracted by the concept of using a modified Letord 1 as the basis for the new bomber as this would speed development time. Modifications to permit the Letord 1 to use the new engine included a new carburetor, rearrangement of engine wiring, and new radiators. Estimated speed was 140 km/h at 2,000 meters and climb to 2,000 meters was expected to take 20 minutes. Estimated weight of the bomber was 3,400 kg and wing surface area was to be 100 square meters (although another memo specified 71 sq. m, which was what was actually built). The bomb load was to be carried internally, which required the fuel tanks to be installed behind the engines. The new plane became the Letord 3, test-flown by Capitaine Villemin. Apparently the prototype was flown as a two-seater with a pilot and rear gunner only. It is likely that the deletion of the front gunner (a dubious asset for night flights) permitted a heavier bomb load, although this cannot be confirmed from available documentation.

The Letord type 3 was the first of the Letord series to be designed from the outset as a long-range bomber, intended to meet the Bn2 specification for a new night bomber. It retained the backward-staggered wing found in most of the Letord series but differed in having equal-span wings with four bays of struts as well as a substantial increase in wing area. Power was supplied by two 200-hp Hispano-Suiza 8Ba engines, the same engines as the Letord 2.

A Letord 3, flown by Maréchal-de-Logis Dupart, was tested by LET 46 in December of 1917. However, it was not selected for production. The Farman F.50 and Breguet 16 were later chosen to meet the Bn2 requirement. An escort version of the Type 3, equipped with cannon, was developed as the Letord 6.

Letord 3 Twin-Engine Night Bomber with Two 200-hp Hispano-Suiza 8Ba Engines
Span 17.95 m; length 11.15 m; height 3.28 m; wing area 71.40 sq. m
Empty weight 1,625 kg; loaded weight 2,400 kg
Maximum speed: 157 km/h at 2,000 meters; climb to 2,000 m in 10 min.; range 370 km
One built

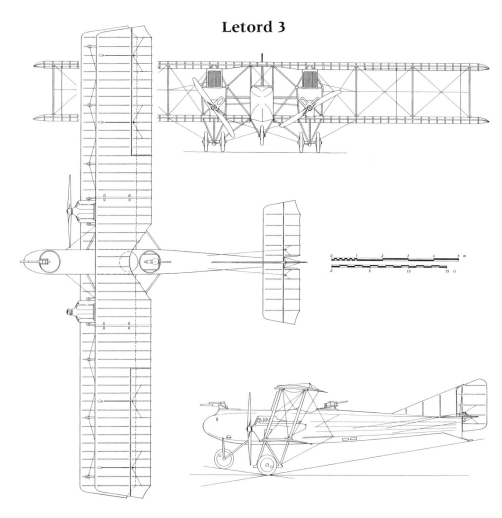

Letord 3. Intended as a medium bomber, the Letord 3 differed from the reconnaissance variants by having equal-span wings with a substantially increased wing area. MA12649.

A Letord 3, flown by Maréchal-de-Logis Dupart, was tested by LET 46 in December 1917. MA38550.

Letord 6

The Letord series of reconnaissance aircraft are well known but there was another development designated the Type 6 Ca.3. The Ca.3 specification called for a three-seat cannon-armed aircraft. It was specified that planes of this class must have sufficient speed and range to allow them to act as escort fighters for the Breguet 14 day bombers. The cannon would allow the escorts to destroy enemy fighters from a distance or be used to devastating effect in close-range combat.

The twin-engine Letord family had been shown to have adequate speed and range for long-range reconnaissance. It would have been logical to consider this aircraft as the basis for a cannon-armed escort fighter. The Letord 3 bomber was modified to serve as the new fighter, presumably because it had a greater range and load-carrying capacity than the reconnaissance variants. The fighter variant was powered by two 220-hp Hispano-Suiza 8Be engines, replacing the 200-hp Hispano-Suiza 8Bas on the Letord 3. The pilot was located just below the upper wing and there was a gunner's station in the nose and a second one just behind the pilot. The nose station contained the Hotchkiss 37-mm cannon.

The Caudron R.11 would vindicate, at least to the French air service's satisfaction, the concept of the long-range escort fighter. In view of this aircraft's success in this role, a cannon-armed variant (the R.14) was under development, thus making the Letord fighter superfluous. Further development of the Letord Ca.3 was abandoned.

Letord 6. The Letord 3 bomber was modified to serve as the basis for this long-range fighter. MA12651.

Letord 6 Cannon-Armed Long-Range Escort with Two 220-hp Hispano-Suiza 8Be Engines
Span 17. 95 m; length 11.05 m; height 3.50 m; wing area 69.00 sq. m
Maximum speed: approximately 150 km/h
Armament: one 37-mm Hotchkiss cannon and one 7.7-mm machine gun
One built

Letord 6

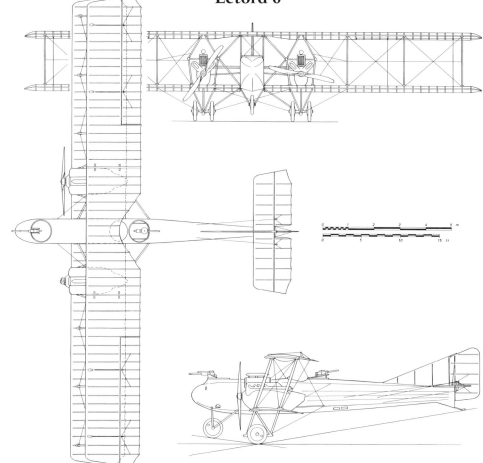

Letord 6. This variant was to be fitted with a 37 mm Hotchkiss cannon in the nose. B84.1660.

Letord 7

Developed from the Letord 3, the Letord 7 was the second aircraft in the Letord family designed as a bomber. As with the preceding Letord 3, it may have been intended to meet the Bn2 classification. Built in 1918, it was a twin-engine biplane with four-bay wings. Unlike preceding planes in the Letord reconnaissance series, the Letord 7 had a wing with increased surface area in order to carry a heavier bomb load. Power was supplied by two 275-hp Lorraine-Dietrich engines, which were slightly more powerful than the 240-hp engines originally planned for the Letord 3. A crew of three was carried. The prototype is believed to have carried serial number 297. The Letord 7 was not selected for production.

Letord 7 Three-Seat Heavy Night Bomber with Two 275-hp Lorraine-Dietrich engines
Span 19.00 m; length 11.35 m; height 3.10 m; wing area 72.0 sq. m
Empty weight 1,760 kg; loaded weight 2,860 kg
Maximum area: 143 km/h at 3,000 meters; climb to 3,000 m in 18 minutes; range 530 km
One built

Above and above right: Letord 7. Unlike other airplanes in the Letord series, the Letord 7 had increased wing area to carry a heavier bomb load. Four views above MA38554.

Right: As with the Letord 3, the Letord 7 had the larger, equal-span wings and was intended to be used as a bomber. MA.

Letord 9

A memo dated 28 November 1917 stated that a development of the Letord series would be the fastest way to replace the Capronis and Voisins in service. The Letord 9 night bomber was designed to meet the BN2/3 requirement for a night bomber capable of carrying 500 kilograms of bombs. The aircraft was markedly different from others in the Letord series. It had an enlarged wing with increased span and surface area. The wings were of equal span, had four bays of struts, and were straight without the negative stagger seen on the earlier Letords. Two 400-hp Liberty 12 engines were mounted in nacelles placed close to the fuselage. Twin main wheels were mounted underneath each of the nacelles. The aircraft had a conventional fin and rudder assembly with a biplane horizontal stabilizer mounted high up on the fin.

Unlike the Letord 7, the Type 9 was designed to carry a crew of only two. The Caudron C.23 was selected to meet the BN2 requirement, and the Letord 9 was not developed further.

Letord 9 Two-Seat Heavy Night Bomber with Two 400-hp Liberty 12 Engines
Span 25.94 m; length 14.95 m; height 4.26 m; wing area 135.0 sq. m
Empty weight 1,243 kg; loaded weight 5,521 kg
Maximum speed: 145 km/h at 2,000 meters; endurance six hours
One built

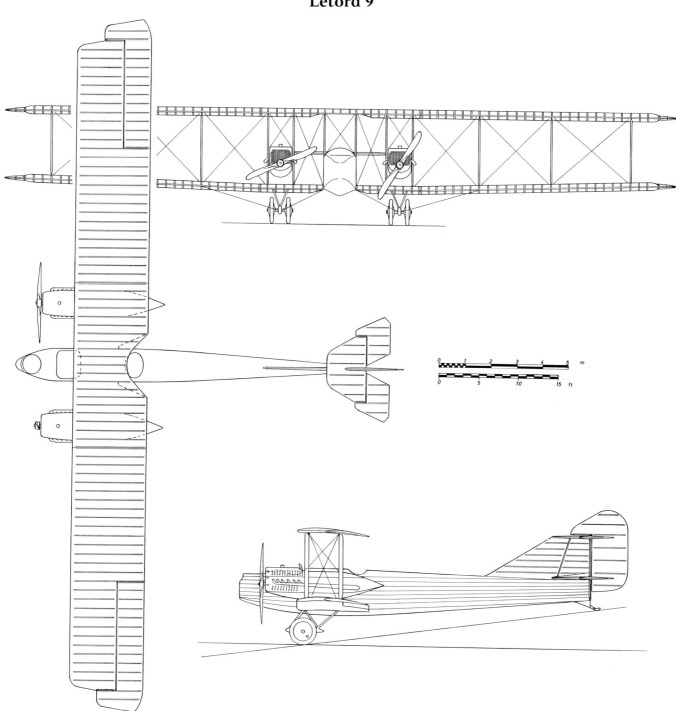

Letord 9

Top: Letord 9 heavy bomber. The Letord 9 was developed to meet the more demanding BN2/3 requirement for a night bomber capable of carrying 500 kg of bombs. As a result, the aircraft was markedly different from the other aircraft in the Letord series. MA13714.

Above: Another view of the Letord 9. MA8564.

Levasseur "Saint Raphaël" Flying Boat

Pierre Levasseur was a French aviation pioneer whose factory built 1,500 propellers and 100 SPAD 7, 12, and 13 fighters during the war. He also built the wings for the Besson H-1 and Levy GL.40, as well as the floats for the SPAD 14, and also played a key role in the design of the Bernard SAB C1 fighter. It has been reported in some publications that Levasseur helped design and build a flying boat constructed from the remains of a captured German LVG. The aircraft in question was built by the CEPA (Technical Section) at the Saint Raphaël naval base. The bottom of the fuselage was modified to form an unsinkable hull and the aircraft's wheels were attached directly to the center section of the hull. Levasseur played a key role in this modification. The aircraft is of some importance, as it was his experience with converting a landplane's fuselage to a unsinkable hull that probably led Levasseur to add this feature to many of his land-based naval aircraft of the 1920s such as the LB.2 and PL.5. This fuselage-hull configuration was to provide an extra measure of safety should the naval aircraft be forced down over the sea. An error that appears in many publications is that the modified LVG was produced in 1918. The conversion was made in 1920; thus, it was actually a postwar design using a World War I airframe.

Levasseur "Saint Raphaël" flying boat. This was a non-flying testbed used to evaluate a watertight hull. Reairche.

Hydravions Georges Levy

Financier George Levy established Constructions Aéronautiques J. Levy in 1914 to build seaplanes. Marcel Besson joined with Levy to help design new seaplanes.

Besson H-1

Marcel Besson obtained his pilot's license in 1910 specifically for the purpose of flight-testing his own designs. His first was a canard monoplane. It was flight-tested at the Champagne aerodrome in 1911. The fuselage was of steel tubing and control was via wing warping with the assistance of ailerons. The 70-hp engine originally fitted proved troublesome and was subsequently replaced by a more reliable Gnome rotary. Besson became interested in maritime aviation and designed an amphibian version of his canard (with three floats and a retractable undercarriage) and a flying boat with folding wings. Neither was built. Shortly before the war, Besson designed a triplane flying boat.

Besson's first wartime design was also a triplane flying boat, probably based on his earlier design. It was designated the H-1 (the H may have stood for *hydravion*). The center of the two-bay wings was substantially longer than the upper and lower ones. The 95-hp engine was in a streamlined nacelle in the center wing. Apparently, Besson felt that this layout would give the aeroplane greater stability by keeping the center of gravity along the line of thrust. The H-1 had a large rudder with a small fin. The H-1 was probably underpowered and this may have been the reason that it was considered for use only as a trainer. However, none was ordered by the Aviation Maritime.

Besson H-1 Flying Boat with 95-hp Engine
One built

Besson 150-hp Flying Boat

Besson's next design was a flying boat with a Renault 150-hp engine. He retained the triplane layout with the large middle wing and the engine mounted in its center. Unlike the H-1, it was a single-bay triplane. A large radiator was mounted in front of the engine. It has been reported that the aircraft was successful in service trials but it was not selected for production. This may have been due to the fact that the F.B.A. and Donnet-Denhaut biplane flying boats were already in service; the Aviation Maritime disliked triplane designs because of their higher weight and structural complexity.

At this time Besson joined forces with Georges Levy, who had the financial resources to develop and produce Besson's designs.

Besson H-1 Flying Boat with 150-hp Renault
One built

Besson-Lesseps Triplane Fighter

A Ministry of Munitions report, dated 1 May 1918, listed the Besson-Lesseps triplane as an unbuilt project. The airplane was intended as a fighter and was to have carried only a single crewman. The engine was to have been a 300-hp Hispano-Suiza. As far as can be determined, construction was never completed.

Besson-Lesseps Triplane Fighter with 300-hp Hispano-Suiza (all data provisional)
Wing area 25 sq. m
Loaded weight 1,100 kg; payload 110 kg
Maximum speed: 220 km/h at 4,000 m; climb to 4,000 m in 10 minutes; endurance 2 hours

Levy-Besson "Alerte" Flying Boat

The French naval command decided in 1917 that a newer class of seaplane was needed, able to make continuous patrols to prevent attacks from German ships or aircraft. Aircraft designed to meet this specification were designated as "Alerte" class seaplanes. They were the progenitor of today's early warning aircraft. As with most other seaplanes at the time, the aircraft

Above: Levy-Besson Triplane. Comparison of this photograph with the Besson Triplane picture at left reveals alterations to Besson's original triplane format; the center wing span was reduced and the interplane struts had been reduced in number and had an "I" shape. It was a single-seat aircraft, possibly intended as a fighter. The MB logo stands for "Marcel Besson." B91.2862.

Left: Besson Triplane. This was an early design with the center wing having a larger span and with a large number of struts. The engine is probably a 300-hp Renault 12Fe. CMA-11.

were to be capable of carrying a T.S.F. wireless radio and a small bomb load. Only F.B.A. and Levy-Besson submitted aircraft to meet this category.

Besson's design was a further refinement of his previous triplane flying boats. The main alteration was the fitting of a more powerful Hispano-Suiza engine. Initial versions of the aircraft had a 200-hp Hispano-Suiza engine; later aircraft had a 250-hp Hispano-Suiza or a Renault. The engine mounting was altered to provide more stability and there was a large radiator with a triangular outline fitted to the front of the engine. The only alteration to the triplane format was that the span of the center wing was reduced and the interplane struts had an I shape. In addition, ailerons were now fitted to both the upper and middle wings. A crew of three was carried. The hull of the flying boat had been designed by Robert Duhamel and built by the Tellier firm. The wings were built at the Victor Depujols shipyards.

More than 100 of these aircraft were built by the Levy firm and they entered service in October 1917. However, they were used as patrol and not "Alerte" class seaplanes (almost all the "Alerte" missions were actually flown by F.B.A. Type Cs). The aircraft saw service primarily in the bombing role (anti-submarine) and could carry two 50-kg bombs as well as a single machine gun.. The Levy-Besson 200-hp triplane easily exceeded the requirements for maximum speed (130 km/h) and useful load including fuel, crew and, armament (600 kg) required of the patrol/light bombing category of flying boat. However, the Aviation Maritime was dissatisfied with all triplanes because their higher weight and structural complexity made them too difficult to maneuver on the water. It was for this reason that relatively few triplanes were purchased. A single-seat fighter version was also produced.

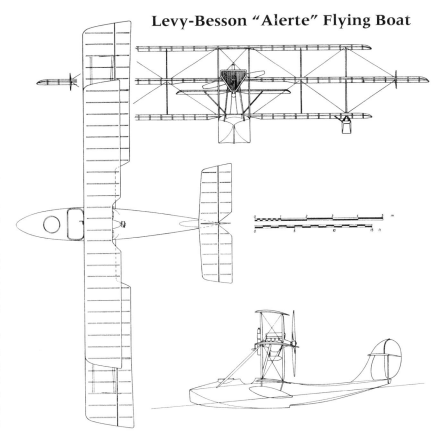

Levy-Besson "Alerte" Flying Boat

Levy-Besson Two-Man "Alerte" Flying Boat with 200-hp Hispano-Suiza
Span (upper) 10.90 m; length 8.40 m; wing area 33.0 sq. m
Empty weight 850 kg; loaded weight 1,330 kg
Maximum speed: 185 km/h; climb to 2,000 m in 11 min. 15 sec.
Armament: two 35-kg bombs
A total of 100 were built.

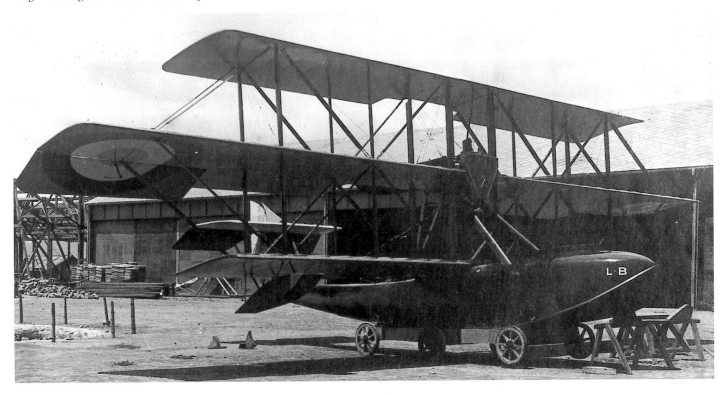

Levy-Besson flying boat at Saint Raphaël. Initial versions of the aircraft had a 200-hp Hispano-Suiza engine; later aircraft had a 250-hp Hispano-Suiza. B88.4151.

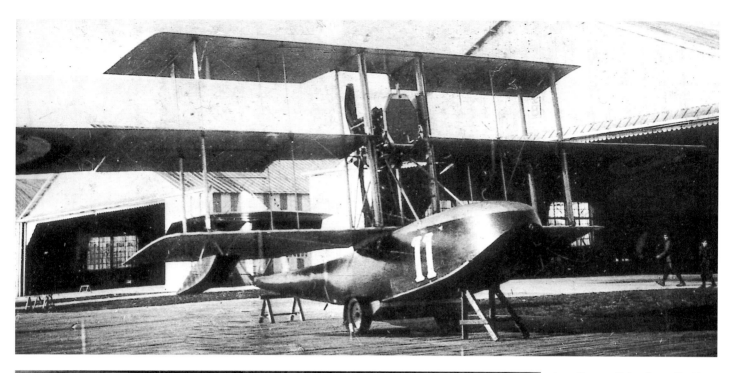

Levy-Besson flying boat No.11 at the Ecole de Tir at Cazaux during the war. B86.339.

Levy-Besson flying boat serial number 68. The hull of the flying boat was designed by Robert Duhamel and built by the Tellier firm. The wings were built at the Victor Depujols shipyards. The engine appears to be a 150-hp Renault. CMA-11.

Levy-Besson 450-hp Flying Boat

The next Levy-Besson design used a single 450-hp Renault engine. The flying boat carried a crew of three and had a triplane layout, in keeping with the Besson formula. Unlike Besson's other designs, the upper and middle wings were of equal length, while the lower wing was shorter. A gunner was carried in the bow cockpit. The Levy-Besson 450-hp flying boat appears to have been intended for the coastal patrol mission, and 12 of them were built. This type may have served as the basis for the LB three-engine "high seas" flying boat.

Levy-Besson Three-Man Flying Boat with 450-hp Renault
Empty weight 2,455 kg; loaded weight 4,000 kg
Maximum speed: 150 km/h
Twelve built

Levy-Besson 300-HP Hispano-Suiza Flying Boat

In April 1918 Levy-Besson had a two-seat triplane under construction, to have been powered by a 300-hp Hispano-Suiza engine. Estimated data include a wing area of 56 sq. m and a military load of 500 kg. The maximum speed was projected to be 160 km/h at 2,000 meters and climb to that altitude would have taken 20 minutes. It is not known if construction on this type was ever completed.

Levy-Besson 500-HP Bugatti Flying Boat

The French government arranged to build the 450-hp Bugatti engine under license in 1917. It was produced by the Peugeot factory and intended for use in the Morane-Saulnier AN (see below). The 16 cylinders were arranged in a U-configuration that permitted a cannon to be placed between the two rows of cylinders. This engine was selected by at least two seaplane manufacturers for use in flying boats—Larnaudi (which see) and Levy-Besson. Levy-Besson's design for a two-seat triplane was probably based on its earlier triplane designs. Morane-Saulnier had concerns about the Bugatti's reliability, and these problems were serious enough to cause it to re-engine the Type AN. It is likely that similar difficulties resulted in delays in the development of the Levy-Besson flying boat, and it is not known if construction of the prototype was ever completed.

Levy-Besson "High Seas" Three-Engine Flying Boat

The "high seas" category of flying boats was intended to be the French equivalent of the British Felixstowe series. The aircraft had to carry a crew of four, a T.S.F. wireless, a 75-mm cannon with 35 rounds, and have a range of 1,120 miles. Utilizing an airframe similar to the previous 450-hp flying boat, the firm fitted three 350-hp Lorraine engines. As with the previous design, the upper and middle wings were of equal span, while the shoulder-mounted lower wing had a reduced span. Because of the huge span, the aircraft had folding wings to permit storage in hangars at French naval air stations. The pilot and observer were seated in the front of the fuselage and both had an excellent field of view. Cutaways in the hull permitted a good view below. The hull was of cedar ply and said by the manufacturer to be a "true boat" that could be operated in heavy seas or bad weather. The aircraft successfully met the specifications for the "high seas" category and 200 were ordered. However, delays in development of the aircraft, coupled with the Armistice, resulted in the order being canceled. One source records that one of the Levy-Besson aircraft was fitted with three 350-hp Panhard engines, although this may have actually been a Farman design that emulated the hull design of the Levy-Besson but was a biplane.

> **Levy-Besson Five-Man "High Seas" Flying Boat with Three 350-hp Lorraines**
> Span 24.90 m; length 17.90 m; height 6.15 m; wing area 185 sq. m
> Empty weight 4,200 kg; loaded weight 7,200 kg; payload 3,000 kg
> Maximum speed: 155 km/h; range 1800 km; endurance 3 hours (up to 12 hours possible in certain configurations)
> A total of 200 were ordered but only a limited number built

Levy-Blanchard "high seas" flying boat. Utilizing an airframe which was similar to the previous 450-hp flying boat, the firm fitted three 350-hp Lorraine engines. B76.108.

Levy-Blanchard "high seas" flying boat. The aircraft was successful in meeting the specifications for the "High Seas" category and 200 were ordered. MA6878.

Georges Levy 40 HB2

The Levy-Besson G.L.40 HB2 was a collaborative effort and may have incorporated some of Marcel Besson's design ideas. However, it was designed primarily by Le Pen and Blanchard. Given the large number of people involved in its design and production, it is not surprising that the aircraft was known by a number of other names. These included Georges Levy 40 HB2, Georges Levy 300-hp Renault, G.L.300, and the Levy-Le-Pen (which was how the seaplane was referred to by the Americans). The HB2 designation probably indicted Hydravion Bombardement with a crew of two.

While the Levy-Besson 200-hp triplane had been successful, the Aviation Maritime's distrust of triplanes probably made it imperative that the new design be a biplane. The Aviation Maritime felt that biplanes offered the dual advantages of providing an adequate surface area without excessive wingspan, and possessed a rigid structure without requiring the large number of drag-inducing struts needed on triplanes.

The aircraft was a two-bay biplane with the top wing

Hydravions Georges Levy. A lineup of G.L.40s have just left the assembly line. MA8023.

longer than the bottom. Diagonal struts connected the base of the upper wing with the outermost interplane struts. The upswept tail had a fixed fin attached to the horizontal stabilizer with aerodynamically balanced elevators and a large unbalanced rudder. The engine was a 280-hp Renault 12 Fe. Consideration was given to fitting the plane with the less powerful 225-hp Lorraine engine, but this does not seem to have been done on operational machines. The engine was mounted as a pusher and was suspended just below the upper wing. A large cutout in the trailing edge of the top wing provided clearance for the propeller arc. There was a pivoted machine gun mounting in the nose. The hull was covered in cedar ply. The fuel tanks were in the hull and Astra pumps were used to supply the engines with fuel. The propeller was built by Levasseur. The crew of two were seated side by side in a small cockpit ahead of the lower wing. There was a stabilizing float under each wing tip.

The aircraft had an impressive weight-carrying ability and was for this reason used primarily as a bomber. Indeed, it was the only French seaplane in 1918 capable of carrying G bombs, which weighed 80 kg and were one of the most effective bombs for anti-submarine warfare. The usual armament load was four 35-kg bombs or two "I" bombs of 120–150 kg and a single Lewis machine gun. Although the aircraft was designed to meet the HB2 category for a two-seat bomber, there was a provision for a third crew member to be accommodated in the extreme nose. From this location the observer had an excellent view and an exceptional field of fire for the machine gun.

Production of the aircraft began shortly after the prototype's first flight, and the type entered service in November 1917. A total of 100 were ordered, built under license by the Farman firm. The G.L.40s were based in metropolitan France, Algeria, Greece, Morocco, Senegal, and Tunisia.

Foreign Service

Finland

Postwar, several aircraft were acquired by the Finnish air service. Twelve Georges-Levy G.L.40 HB2s were obtained although, at least by the Finns, these were given the designation Georges-Levy R. The known serials for these aircraft included B 304 and 400 to 409. They were in service from 1919–23.

> **Georges-Levy 40 HB2 Two/Three-Seat Flying Boat with 280-hp Renault 12Fe**
> Span 18.50 m; length 12.40 m; height 3.85 m; wing area 68.72 sq. m
> Empty weight 1,450 kg; loaded weight 2,350 kg
> Maximum speed: 150 km/h (some sources state 185 km/h); climb to 2000 m in 25 minutes; range 400 km; endurance 6.5 hours
> Armament: a Lewis machine gun and 300 kg of bombs
> Approximately 100 built

Peru

In 1919 a French air mission supplied the Peruvian air service with an estimated three Levy-Le Pen flying boats—possibly G.L.40 HB2s.

Portugal

Portugal's naval air service purchased two G.L.40 HB2s in 1918. They remained in service until 1920.

United States

Twelve examples of the G.L.40 HB2 were used by the U.S. naval air service. The American aircraft had a Lewis gun in the bow. The aircraft used were given serials GL 13, GL 16, GL 20, GL 21, GL 27, GL 28, GL 29, GL 30, GL 60, GL 83, GL 84, and GL 85. They were based at Le Croisic and St. Trojan. The aircraft at St. Trojan escorted convoys beginning on 19 July 1918. However, while a number of missions were flown, it has been recorded that the aircraft's engines had been overhauled two or three times before delivery to the Americans and were therefore difficult to keep repaired. The bomb-carrying gear was also defective and resulted in the loss of an aircraft on 20 August 1918. Although it was also found that reconnaissance flights were of more value than convoy escort missions, the patrols flown from Le Croisic were mainly for convoy escort. GL 27 and GL 84 were sent to the United States after the war and received serials A-5650 and A-5651 respectively.

G.L.40 HB2 serial.435 in 1920. The engine was a 280-hp Renault 12Fe. B93.1745.

G.L.40 HB2 built by Levy. The usual armament load was four 35-kg bombs or two "I" bombs of 120–150 kg and a single Lewis gun. Reairche.

This G.L.40 is based in the Mediterranean; others served at naval air stations in France, Algeria, Greece, Morocco, Senegal, and Tunisia. Reairche.

A total of 100 G.L.40 HB2s were ordered. Some, such as this example, were built under license by the Farman firm. Reairche.

Établissements Lioré et Olivier

The firm created by Fernand Lioré and Henri Olivier was founded in 1906 and began producing airplanes in 1908. In 1908 Lioré collaborated with Witzig and Dutilleul to produce the WLD biplane with a 50-hp Renault engine. The first products of the LeO firm were the Flo-1 monoplane (designed by an engineer named Leflot) an autogyro, a helicopter, and a monoplane designed by Paul Zens. The LeO 2 was a single-engine monoplane tractor with a 40-hp Gregoire engine driving two propellers through a chain drive; it was built in 1911.

Lioré et Olivier 3

During the war LeO built Morane Saulnier aircraft and Sopwith 1½ Strutters under license. Its first original wartime design was the LeO 3, a single-engine aircraft intended for use as a fighter. It was built in 1917 but never flew and further development was abandoned.

Lioré et Olivier 4 and 4/1

The firm of Lioré and Olivier was manufacturing Sopwith 1½ Strutters under license in 1916–17. LeO engineer Leflot designed and constructed an aircraft intended to replace the Sopwith 1½ Strutter in 1917. It was built for the A2 classification, which called for a two-seat observation aircraft capable of carrying a payload of 450 kg at 200 km/h. The service ceiling was to be 3,000 m. Other aircraft designed to this specification included the Breguet 14 with a 300-hp Renault engine and Rateau turbo-compressor and the Carroll A2.

Lioré and Olivier's aircraft resembled the Sopwith 1½ Strutter. It was a two-seat tractor biplane with positively staggered wings of equal length and with two bays of struts, designated the LeO 4.

The most unusual feature of the design was the "umbrella" shape of the upper wing, featuring curved tips and wavy trailing edges. The bottom wing was of conventional design without curved tips and with straight trailing edges. The tail featured an elliptical fin and rudder assembly. Armament consisted of a machine gun synchronized to fire through the propeller arc; the gunner had a separate cockpit fitted with a swivel mount for two machine guns. The engine was a 200-hp Clerget 11E 11-cylinder rotary.

A second version was designated the LeO 4/1. It was broadly similar to the LeO 4 (and may very well have used the fuselage and tail of the LeO 4) but had a more conventional wing shape with four ailerons. A triangular fin and rudder were fitted. It retained the 200-hp Clerget 11E rotary. The plane was built and flown but was considered to be too heavy, demonstrated by the fact that it took 50 minutes to climb to 5,000 m (the Breguet 14 A2, on the other hand, could climb to the same altitude in only 35 minutes). The aircraft was judged to be inadequate by the STAé and the Breguet 14 was selected for production. The sole LeO 4/1 was later used as a civil aircraft (serial F-ABFR) and subsequently as an engine testbed for the company.

Above: Liore et Olivier LeO 4. The umbrella-shaped upper wing with curved tips and wavy trailing edges proved to be unsuccessful and was replaced with a more conventional wing. (CC).

Right: Liore et Olivier LeO 4/1 with more conventional wing and rudder. Reairche.

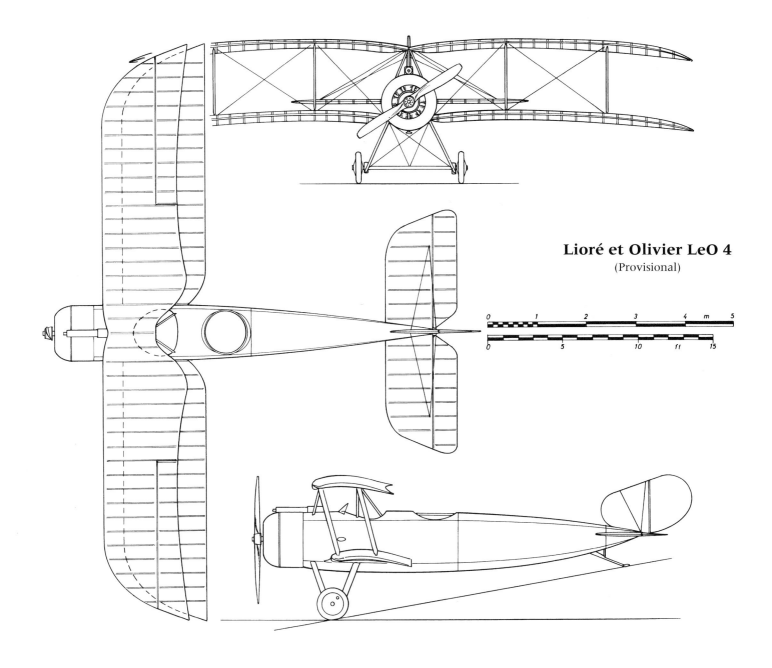

Lioré et Olivier LeO 4
(Provisional)

Lioré et Olivier 5 S2

The need for a heavily armored ground-attack aircraft was formulated as early as May 1918 by the Aviation Militaire. A requirement for a two-seat type was formulated under the S2 classification. Aircraft submitted to meet this specification were the Canton S2, the Lioré-et-Olivier 5 S2, and the Hochart S2. The LeO 5 was a twin-engine biplane powered by two 175-hp Gnome Rhône rotary engines; it was felt that the rotary engines would be less susceptible to small arms fire. The engines were mounted in nacelles suspended between the upper and lower wings by a series of struts. The fuselage featured a short nose that did not extend beyond the propeller spinners. A large fin with rudder and a prominent tail skid were other features. The stabilizer had variable incidence. The two main wheels featured spats and were located under each of the engines. These spats would have helped protect the tires from damage due to ground fire. The crew of two was seated in tandem—the pilot beneath a cut out in the trailing edge of the upper wing and the gunner just behind him. Armament consisted of one machine gun in the nose and two others on a swivel mount. There was a total of 300 kg of armor.

Only one aircraft was built, but because the S2 specifications was not issued until late in the war it did not fly until 1919. When the STAé evaluated the type, it was felt to be more suitable for the Ab2 classification requiring a maneuverable aircraft with light armor for reconnaissance and army cooperation duties. It was developed into the LeO 7, which saw service postwar.

LeO 5 Two-Seat Armored Ground-Attack Plane with Two 175-hp Gnome Rhône Engines
Span was 14.35 m; length 8,30 m; height 47.25 m
Empty weight 1,300 kg; loaded weight 1,900 kg (with 160 kg of bombs)
Maximum speed: 185 km/h; climb to 1,000 m in 4 minutes, range 550 km
Armament: three machine guns
One built

Above left: Although the LeO 5 was not selected to fill the S2 specification, it was further developed into the LeO 7 shown. The LeO 7 was used as a lightly armored army cooperation airplane. B76.125. Above right: LeO 5. MA.

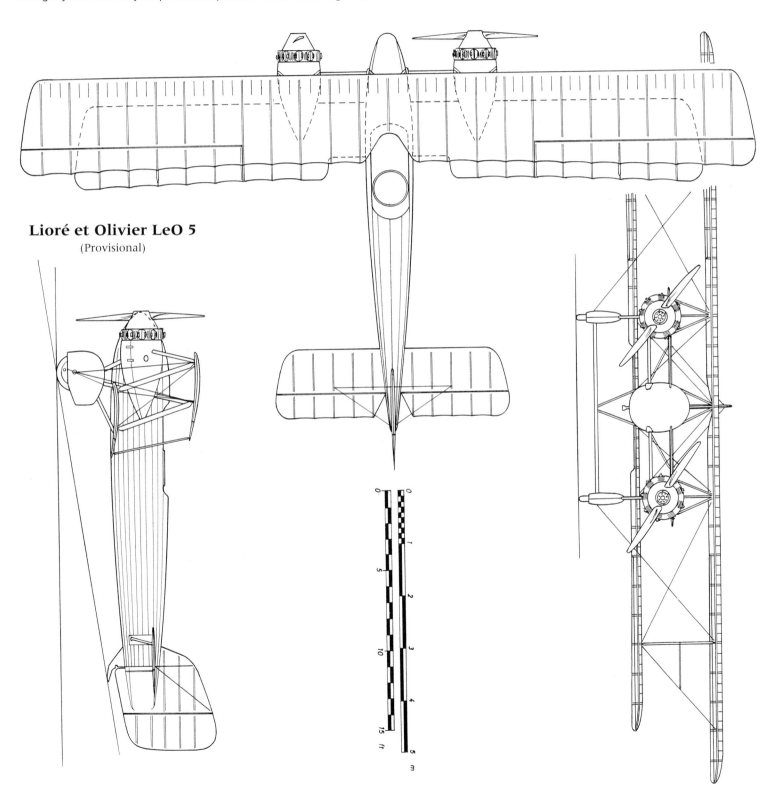

Lioré et Olivier LeO 5
(Provisional)

LeO 5 armored ground attack aircraft. The location of the engine nacelles helped to shield the crew from ground fire. The engines were two 175-hp Gnome-Rhônes, it being felt that the rotary engines would be less susceptible to small arms fire. MA.

Lioré et Olivier S2 Project

The LeO 5 had been intended to meet the S2 category of 1918. However, it was subsequently felt to be more suitable for the Ab 2 category of reconnaissance aircraft carrying light armor. The Lioré et Olivier firm subsequently designed a new plane for the S2 category—a pusher fitted with two 170-hp Renault engines. It appears that it was never built and remained a project only.

Moineau C1 and Pusher

René Moineau was an engineer-test pilot who had tested many of Louis Breguet's aircraft beginning in 1911. In cooperation with the Salmson firm he produced two very unusual designs for reconnaissance aircraft—the S.M.1 and S.M.2. When the STAé formulated the C1 requirement for a fast fighter powered by a Hispano-Suiza engine of 300-hp, Moineau submitted his own design for it.

The specifications called for a single-seat fighter with twin machine guns and which also had to be capable of performing high-speed reconnaissance. Unlike most of the other entries, Moineau's design was a monoplane. It is believed that Moineau's decision to produce a monoplane, believed to have been based upon French analysis of a Fokker D.VIII. Moineau was well aware of the D.VIII and had requested the STAé to supply him with technical reports about it. Moineau's aircraft had an "overhanging wing" (presumably a parasol) and a retractable landing gear housed in the bottom of the wing. J.M.

Bruce believes the wing may have had a cranked (possibly "gull") wing with a horizontal portion attached directly to the fuselage and a bent outer section ("inclinée"). Power was to have been supplied by a 300-hp Hispano-Suiza engine and there was to have been a semi-retractable radiator. Armament was to have been two fixed Vickers machine guns synchronized to fire through the propeller. It appears that construction of the aircraft was never completed.

A second design study by Moineau during the war was a two-seater pusher. No details are available.

Moineau Single-Seat Fighter with 300-hp Hispano-Suiza engine (all data provisional)
Wing area 20.6 sq. m
Loaded weight 1,025 kg
Maximum speed: 250 km/h at 4,000 m; ceiling 8,000 m
Armament: to have been two synchronized Vickers machine guns

Moncassin Flying Boats

The first Moncassin flying boat was a three-seat biplane powered by a 200-hp Hispano engine. Its specifications suggest that it was intended for service in the "alerte" or patrol category of seaplane. The aircraft had a wing area of 193 sq. m. Maximum speed at 2,000 m was estimated at 135 km/h. Climb to 2,000 meters took 45 minutes. Apparently the design proved unsatisfactory, as development was abandoned by April 1918.

The second Moncassin flying boat was a twin-engine design intended for service in the patrol category. It was a two-seat biplane with a single-step hull. Power was supplied by two 300-hp Hispano engines. The wing area was 215 sq. m. As with the Moncassin single-engine flying boat, the twin-engine design was abandoned by April 1918.

Société Anonyme des Aéroplanes Morane-Saulnier

Leon Morane earned his pilot's license (Number 50) in 1910 and set a world speed record in his Blériot monoplane in July 1910. On 3 September 1910 he set a world height record by reaching 2,582 meters. In light of this success, Morane formed an aircraft company with Gabriel Borel and Raymond Saulnier in 1910. Saulnier had worked with Louis Blériot and had a part in designing the Blériot 11. By October 1910 the relationship with Borel was severed and the Société Anonyme des Aéroplanes Morane-Saulnier was formed. The first aircraft produced by the firm was the Type A.

In 1913 the Types G and H appeared and it was these aircraft and their variants that would play a key role in the development of the world's first true fighters—the Types L, H, and N. During the war the firm developed long-range reconnaissance, army cooperation, fighter, and strategic bomber aircraft.

Prewar Morane-Saulnier Aircraft

1. Type A (1911) monoplane with a 50-hp Gnome engine—13 were used at the Morane training school at Villacoublay.

These were later given the designation Army Type 11.
2. Type PP (1911) named for its flight from Pau to Paris.
3. Type C (1911) with an 80-hp Gnome; five re-engined Type A s built for the Imperial Russian Air Service.
4. Type F (1911) with an 80-hp Gnome; two built for Romania.
5. Type G—see text.
6. Type H—see text.
7. Type GA (1912) with a 60-hp Le Rhône; development of Type G.
8. Type GB (1912) with an 80-hp Gnome; development of Type G.
9. Type G seaplane—used as a racer at Monaco.
10. Type BI (1913) with an 80-hp Gnome; two-seat monoplane.
11. Type WR—modification of Type G seaplane built for the Imperial Russian Navy and, also, an armored airplane built for the Aviation Militaire (see text).
12. Hydrobiplane 1912—floatplane of biplane configuration used in the 1912 Monaco seaplane races.
13. Type M (1912) with an 80-hp Le Rhône—a monoplane with light armor to protect the pilot and engine. Three were built for the Air Ministry. (See text.)

Morane-Saulnier WR

There appear to have been two airplanes given the WR designation. The first was a floatplane modification of the Type G which served with the Imperial Russian navy, which designated it the Type WR (see entry for Type G below).

The other Type WR was an armored airplane designed to meet the 1913 specification for a two-seat armored reconnaissance machine. Power was supplied by an 80-hp

Gnome and the machine cost 37,000 FF. It is not known if any of these were in service with the Aviation Militaire at the outbreak of the war.

Morane-Saulnier Single-Seat, Armored Airplane with 80-hp Gnome
Maximum speed: 131 km/h; minimum speed 85 km/h, climb to 1,000 m 5 minutes 55 seconds

---

Morane-Saulnier G

The Morane-Saulnier G was a single-engine monoplane with shoulder-mounted wings. It carried two crewmen seated in an elongated cockpit located in the center of the fuselage between the wings, a position which was later felt to be unsatisfactory because it severely limited downward vision. Those aircraft powered by the 60-hp Le Rhône were designated Type GA; those with the 80-hp Gnomes were designated Type GB.

Ninety-four Type Gs were ordered by the Aviation Militaire. They received the designation MoS.2. The Type G was also manufactured under license by the Grahame-White Company at Hendon; these aircraft were sold to civilian aviators in England, and later to the RFC.

Variants

There were two major developments of the Type G. One was used as a test-bed for the Garros-Hue version of the Saulnier bullet deflection system. A second fighter variant of the Type G was built in the summer of 1915. Changes included a fully faired fuselage, fixed vertical fin, modified landing gear struts, deletion of the observer's position, and fitting of an 8-mm Hotchkiss machine gun and bullet deflectors. The engine was the same as that of the Type H: an 80-hp Le Rhône 9 C. At least one of these aircraft (serial MS497) appears to have been fitted with wings similar to those used on the Type H. These aircraft were not adopted because the Morane-Saulnier Type L and N (which were more effective in the fighter role) were just entering service.

Known training variants include:

1. Type G—80-hp Gnome two-seater, Army Type number 14.
2. Type G—80-hp Gnome two-seater penguin, Army Type number 15.
3. Type G—45-hp Anzani penguin, Army Type number 16.
4. Type G—3 cylinder Anzani single-seater with undercarriage guard (to prevent nose-over), Army Type number 17.
5. Type G—3 cylinder Anzani single-seat penguin, Army Type number 18.
6. Type G—80-hp Le Rhône two-seat penguin, Army Type number 19.

Operational Service

A few of the surviving Type Gs may have been assigned, along with Morane-Saulnier Ls, to MS 23 and MS 26. However, French documents show that Type Ls constituted the bulk of MS 23's and MS 26's equipment. In any event, the location of the Type G's cockpit meant that the crew's downward vision was inadequate, and it was soon replaced in the MS units by the newer Type L parasols. Most of the Type Gs were assigned to training units. Some of those used for training were fitted with either 30-hp or 45-hp Anzani engines. They were designated Morane-Saulnier 16 E1, "E" meaning ecole (or training) aircraft.

Foreign Service
Argentina

The Argentine army obtained two Type Gs in 1912.

Denmark

The Kongelige Danske Flyvevaben (Danish Army Air Service) purchased two Type Gs in 1915. These had been built under

310   FRENCH AIRCRAFT OF THE FIRST WORLD WAR

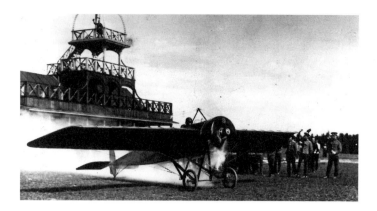

A Morane-Saulnier Type G of the IRAS. In Russian service the Type Gs were initially used in the unarmed reconnaissance role, although on occasion they were used to ram enemy airplanes. B86 0005.

license by the Swedish firm of Thulin and were designated Thulin Bs by the firm. In Danish service they were designated M.S.1 *Hugin* and M.S.2 *Munin*. Originally both had 50-hp Gnomes. *Munin* was struck off charge in 1917. *Hugin* served until 1919 and had its 50-hp Gnome replaced by a 80-hp Gnome in 1917.

Mexico

In 1912 the Mexican government sent five army officers to the Moissant International Aviation School at Long Island, New Jersey. After learning to fly, these officers returned to Mexico, bringing with them two Morane-Saulnier monoplanes (probably Gs). The five, as well as a number of foreign mercenaries, formed the Escuadrilla de Ebano in 1915. The unit was assigned to the Northeast Army and saw action at Tamaulipas, Veracruz, and Yucatan.

Russia

The Imperial Russian Air Service usually employed its Type Gs as single-seaters. The accommodation for the observer (his seat was merely a cushion over the fuel tank) and its location at mid-wing were considered inadequate. There were two versions of the Russian Type Gs. One had a wing with a 9.3 m span; the other had a 10.2 m span. The wing surface areas were 14 sq. m and 16 sq. m respectively. Both wings had identical chords. The smaller wing had two pairs of bracing wires while the larger one had three. Most of the Russian machines used an 80-hp Le Rhône engine. The Russians liked the Type Gs because they could be transported in sections and reassembled in 11 minutes by two mechanics.

The Type Gs were initially used for unarmed reconnaissance although on occasion they were used to ram enemy airplanes. The Type Gs in service with the Imperial Russian Air Service's 11th Air Corps were unarmed except for rifles and pistols. Initially, P.N. Nesterov, who commanded that unit, tried to use his aircraft to maneuver enemy aircraft into the ground. He also attached a blade to the rear fuselage of his aircraft in the hope of getting close enough to cut off his opponent's wings. He even tried to ensnare the propellers of enemy aircraft in a grapple hung from his plane. Finally, Nesterov devised a procedure for destroying enemy aircraft by ramming. These radical tactics were necessitated by Nesterov's inability to find machine guns suitable for mounting on an aircraft. Nesterov tried a ramming attack against an Austrian Albatros; both aircraft and their crews were destroyed in the ensuing crash. However, other Russian aviators were more successful in ramming attacks. The Type Gs remained in front-line service until mid-1915, when they were replaced by the Types L and LA parasols.

After the Gs were withdrawn from operational service they were assigned to training units. These aircraft were often fitted

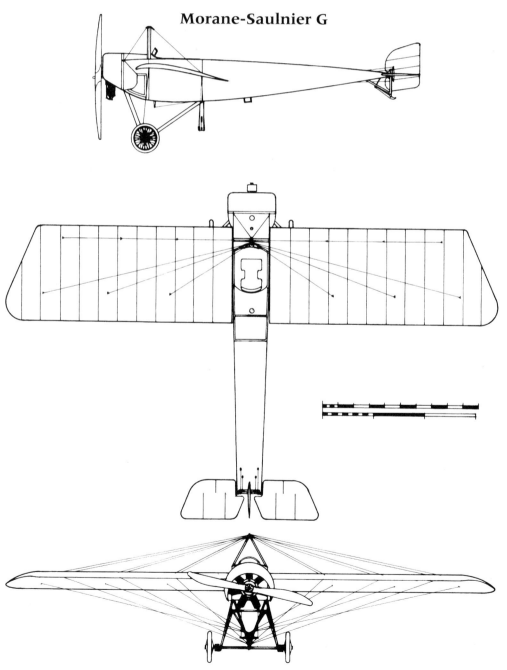

Morane-Saulnier G

with nose wheels to prevent the student pilots from nosing over. Some were only used for taxying, fitted with low-powered engines such as 35-hp Anzanis. There were three basic training versions of the Russian Type G: the Type G produced in France and fitted with the Kachinsky undercarriage, the Slyusarenko trainer, and the Type G modified by a Lieutenant Fride.

The Type WR version of the Type G was built for the Russian navy and had a greenhouse fitted on the fuselage sides ahead of the wing, probably to improve the crew's vision.

In 1916 one Russian Type G was fitted with a smaller wing and a 100-hp Gnome-Monosoupape. It had a performance superior to the standard Type G and set a Russian altitude record of 5,200 m.

Approximately 20 Type Gs survived the war and were in use as late as 1923. They were assigned to the Tashkent Aviation School, the 2nd Higher School of Military Pilots, and the 1st and 2nd Military Schools of Pilots.

Spain

Three Type Gs were obtained by Spain in 1913. They were found to be lighter than the Nieuport 2s also in service and were faster and more maneuverable. Two of the Type Gs were used by the Escuadrilla de Tetuan and later for training at the Escuela Nacional de Aviacion's base at Getafe. The last Type G was withdrawn in September 1919.

Sweden

E. Thulins Aeroplansfabrik produced the Morane-Saulnier Type G under license. As mentioned above, two were built for the Danish Army Air Service and a third was delivered to the Flygkompaniet (Army Aviation service). It was given serial No.5, later redesignated 405.

Swiss

A single Type G was purchased by the Swiss Fliegertruppe in 1914. Given serial No.24, it was used to train pilots in aerial combat. It was withdrawn from service in 1919 and scrapped in the 1930s.

United Kingdom

A single Type G was impressed into British military service at the beginning of the war. It was given serial 482 and served with No.1 Reserve Airplane Squadron and later No.60 Squadron. An order for 12 additional aircraft, a combination of Type Gs and Hs, was placed in 1915. Serial numbers were 587–598.

Morane-Saulnier Type G Two-Seat Reconnaissance Plane with 80-hp Gnome
Span 9.63 m ; length 6.38 m; height 3 m; wing area 16 sq. m
Empty weight 314 kg ; loaded weight 544 kg
Max. speed: 120 m/h; climb to 1,000 m in 7.0 minutes
Armament: usually none
Approximately 94 were built for the Aviation Militaire

Morane-Saulnier Type G Single-Seat Fighter with 80-hp Le Rhône 9C
Span 9.12 m ; length 6.620 m; height 2.54 m
Payload 195 kg
Armament: one unsynchronized 8-mm Hotchkiss machine gun

Morane-Saulnier Type G Two-Seat Reconnaissance Plane with 80-hp Le Rhône and Russian-built 14 sq. m Wing
Span 9.3 m; length 6.7 m; height 3.0 m
Empty weight 340 kg; loaded weight 550 kg
Maximum speed: 122 km/h; climb to 2,000 m in 17.0 minutes; ceiling 3,000 m; endurance 2.2 hours

Morane-Saulnier Type G Two-Seat Reconnaissance Plane with 80-hp Le Rhône and Russian-Built 16 sq. m Wing
Span 10.2 m; length 6.7 m; height 2.3 m
Empty weight 340 kg; loaded weight 625 kg
Maximum speed: 115 km/h; climb to 2,000 m in 25 minutes; ceiling 2,600 m; endurance 2.5 hours

Morane-Saulnier Type G Two-Seat Training Plane with 50-hp Gnome Engine Built by E. Thulins Aeroplansfabrik
Span 10.2 m; length 6.5 m; height 2.55 m; wing area 14.0 sq. m
Empty weight 370 kg; loaded weight 582 kg
Maximum speed: 135 km/h; endurance two hours
Three built

It is believed that two Type Gs were obtained by the RNAS. These had serials 941 and 1242. The first was an impressed aircraft and was based at Eastchurch. Later it was assigned to No.2 Squadron. The second, built by Graham-White Aviation Co., was delivered to Eastchurch in December 1914. It was later sent to Hendon, then Grain.

Morane-Saulnier Type G MS 497 with machine gun. Renaud.

Morane-Saulnier Type H

The Morane-Saulnier Type H was a single-seat version of the two-seat Type G, having a shorter fuselage and wing span. Power was supplied by an 80-hp Le Rhône 9C engine. It was flown for the first time in 1913. About 26 Type Hs were ordered by the Aviation Militaire and given the service designation MoS.1.

A version with a 45-hp Anzani was built, presumably for use as a trainer. This type was later supplied to Portugal. Other versions had 50-hp or 60-hp Gnome engines and an enlarged wing of 18 square meters. Designated army Type 12s, it is likely that they too were used as trainers.

Operational Service

Most of the Type Hs served in Escadrille MS 31, which was formed on 24 September 1914. French documents show that at the time of its formation it had three single-seat Morane-Saulniers (with 60-hp engines) plus two in reserve. The unit was intended as an "escadrille artillerie" and was formed at Dijon-Longvic under the command of Capitaine Yence, who had four pilots. Two days later MS 31 moved to Toul where it was assigned to the 1st Armée. By 6 November 1914 Morane-Saulnier Type Ls had replaced the Type Hs, which were sent to training units.

Two Type Hs were assigned to the C.R.P. (Camp Retranché de Paris) and armed with carbines. At least one of these had a special mount for the gun and may have had deflector plates on the propeller, as well as a modified cowling, spinner, and undercarriage.

Foreign Service
Germany
The Pfalz Flugzeugverke built the Type H under license.

Portugal
The Portuguese air service obtained a single Type H in October 1916. Powered by a 45-hp Anzani engine, it had serial number MS 721 and carried the name *Charge Maxima*. It was assigned to the Escola de Aeronautica Militare at Vila Nova da Rainha and probably used as a trainer.

United Kingdom
The RFC obtained several Type Hs as part of a combined order placed in early 1915 for 12 Gs and Hs. These aircraft arrived in April 1915 and were sent to 1, 2, 4, 10, and 11 Reserve Airplane Squadrons. The Type Hs were used as trainers. A second order for 24 Type Hs was placed with Grahame-White (which produced the type under license in Britain) in 1915. Serials were 5693–5716. Most served with training squadrons but some were assigned to Nos.7, 15, and 60 Squadrons.

Morane-Saulnier Type H Single-Seat Reconnaissance Plane with 80-hp Le Rhône 9C

Span 9.12 m; length 6.28 m; height 2.30 m; wing area 14.0 sq. m
Loaded weight 470 kg
Maximum speed 135 km/h; climb to 1,000 m in 3 minutes; range 280 km
Armament: one 8-mm Hotchkiss machine gun on a modified Type H and a small number of flechettes
Approximately 26 built for Aviation Militaire; between 30 and 35 for the RFC

Morane-Saulnier Type H Single-Seat Training Plane with 45-hp Anzani engine

Span 6.37 m; length 6.37 m; height 2.30 m; wing area 14.0 sq. m
Empty weight 278 kg; loaded weight 438 kg
Maximum speed: 135 km/h; climb to 1,000 m in ten minutes; range 405 km; endurance three hours

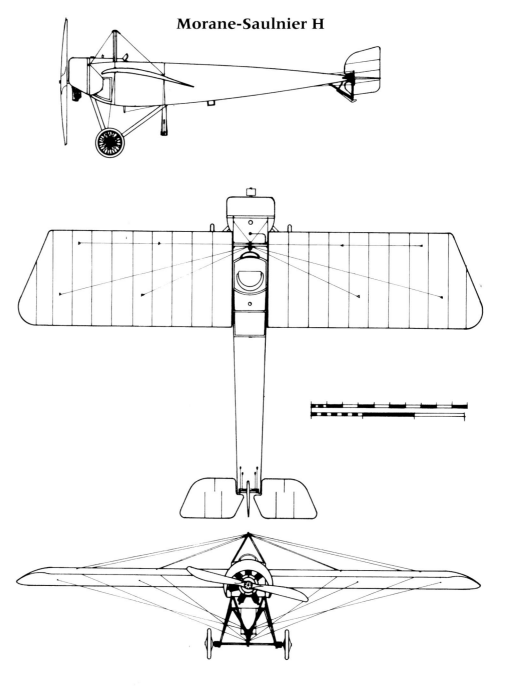

Morane-Saulnier H

A rare photograph showing that at least one Morane-Saulnier Type H was used by the Belgians; no further details are available. R. Verhegghen via Colin Owers.

A Morane-Saulnier Type H at the Ecole d'Aviation at Avord in April 1915. B86.268.

The Morane-Saulnier Type H was a single seat version of the Type G, which had a shorter fuselage and wing span. B92.1315.

Morane-Saulnier Type I

The designation Type I was probably given to this aircraft to avoid confusion with the Nbis (a Type I was developed in 1913 but not sold to the military). The prototype aircraft, completed in March 1916, was probably a modified Type N. The RFC had been pleased with the initial Type Ns it received in late 1915. A decision was made by the RFC headquarters to develop the Type N further by equipping it with a more powerful 110-hp Le Rhône 9J engine. It was also to have an endurance of three hours and, unlike the Type N, was to be capable of being flown by a "moderate pilot." Maximum speed was to be 161 km/h at 1,829 m. The dimensions of the Type I were very close to those of the Type N although the wing span and elevators were slightly larger. Armament was usually a synchronized 0.303 Vickers machine gun.

The RFC's interest in developing the Type N was in marked contrast to the Aviation Militaire's decision to abandon it in favor of the Nieuport 11. However, Type Is did receive the SFA designation MoS.6. It is not known if any Type Is were used by French escadrilles, but it seems unlikely.

Four Type Is were supplied to No.60 Squadron shortly after July 1916. These were assigned serials MS733, MS744, MS735, and MS746 by the French and A198, A199, A202, and A206 by the RFC. Despite the fact that the specification called for an airplane which could be flown by a "moderate pilot," the Type I was apparently a very difficult aircraft to fly. Although fast, the Type Is were disliked by No.60 Squadron's pilots and after October 1916 the two Type Is still serviceable were returned to England for use by training units.

A number of Type I fighters were obtained by the Imperial Russian Air Service; at least 20 were ordered.

Morane-Saulnier Type I was a development of the Type N which was fitted with a more powerful 110-hp Le Rhône 9J engine. B83.1235.

Morane-Saulnier Type I Single-Seat Fighter with 110-hp Le Rhône 9J

Span 8.24 2 m; length 5.815 m; height 2.50 m; wing area 11 sq. m

Empty weight 334 kg; loaded weight 510 kg; payload 187 kg

Maximum speed: 164 km/h at sea level; 156 km/h at 3,000 m; climb to 1000 m in 2 min. 50 sec.; climb to 2000 m in 6 min. 45 sec; climb to 3000 m in 12 min. 40 sec.; ceiling 4700 m; endurance 1.3 hours

Armament: one synchronized 0. 303 Vickers machine gun

Four built for the RFC (not including prototype), no more than 20 built for Russia

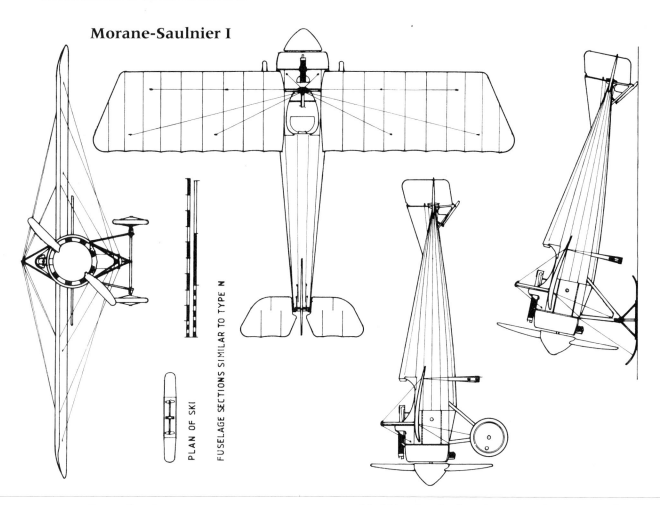

Morane-Saulnier I

Morane-Saulnier Types L and LA

The Type L was to be the first of a large series of reconnaissance aircraft produced for the Aviation Militaire by the Morane-Saulnier firm. It was developed from the prewar Morane-Saulnier Type G, which, after being converted to a parasol configuration, was designated the MoS.19. It retained the slab-sided fuselage of the standard Type G but the undercarriage was modified. The parasol wing also had the same airfoil section of the Type G.

The standard Type L was developed from the MoS.19. One of the most important differences between the two aircraft was the cockpit layout. Unlike the Type G, which had used a bathtub layout, the Type L featured a larger cockpit with separate seats for each crewman. It was designated the MoS.3.

The Type L had a slab-sided fuselage with four spars made of ash and joined by transverse spars of either ash or pine. The wing spars were made of ash. The Type L used wing warping for lateral control, and the cables for supporting landing loads were attached to a prominent central pylon and held in place by iron fasteners. The wing incidence was set at eight degrees. The very small fin and rudder were made of steel tubing and covered with cloth. The rudder was held in place by two metal hinges and the lower portion of the rudder was attached directly to the tail skid support. The horizontal surfaces consisted of two separate balanced elevators that made the Type L sensitive in flight.

There were three fuel tanks; two were located in the nose ahead of the pilot. The tank on the right held 70 liters of fuel while the left held 34 liters. This asymmetry made it necessary to place a counterweight in the right wing. There was also a 34-liter tank in the rear of the fuselage. The undercarriage supports had an M configuration when viewed from the front and there were two V-shaped steel struts for lateral support. The two wheels were supported by a single axis. Each wheel had bungee cords to serve as shock absorbers.

It does not appear that the Aviation Militaire was interested in purchasing the Type L. However, the French government

Morane-Saulnier Type L. The Type L was developed from a Morane-Saulnier Type G which had been converted to a parasol configuration. The Type L retained the same slab-sided fuselage and airfoil section that was used in the standard Type G. B75.140.

Morane-Saulnier T

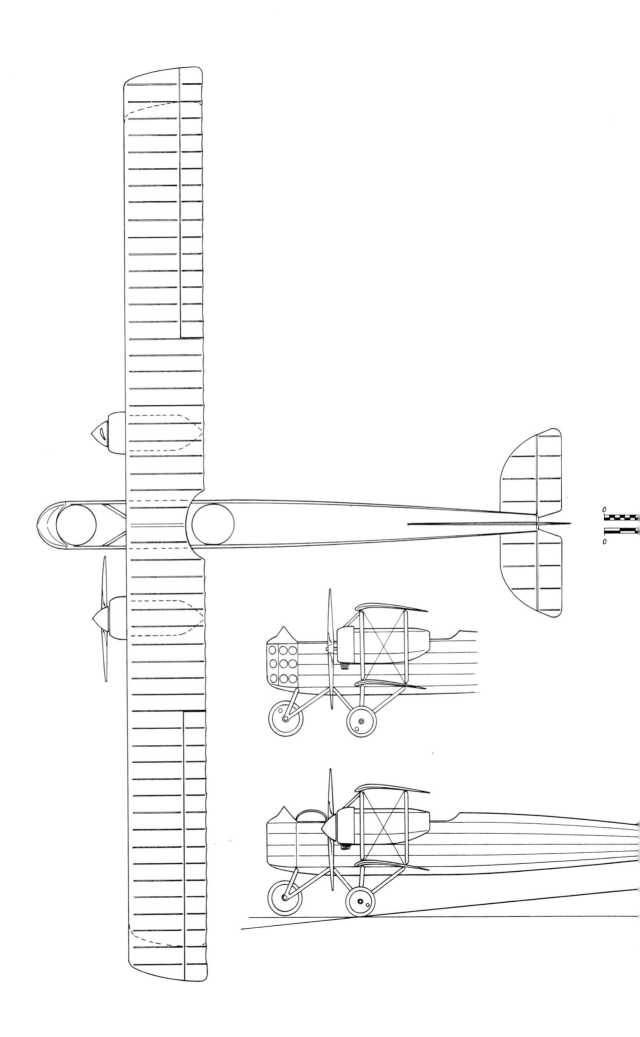

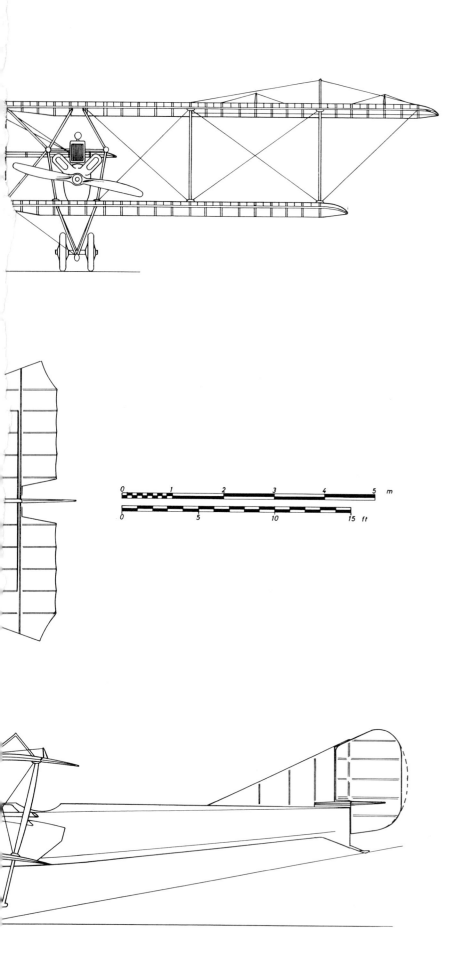

Letord 5

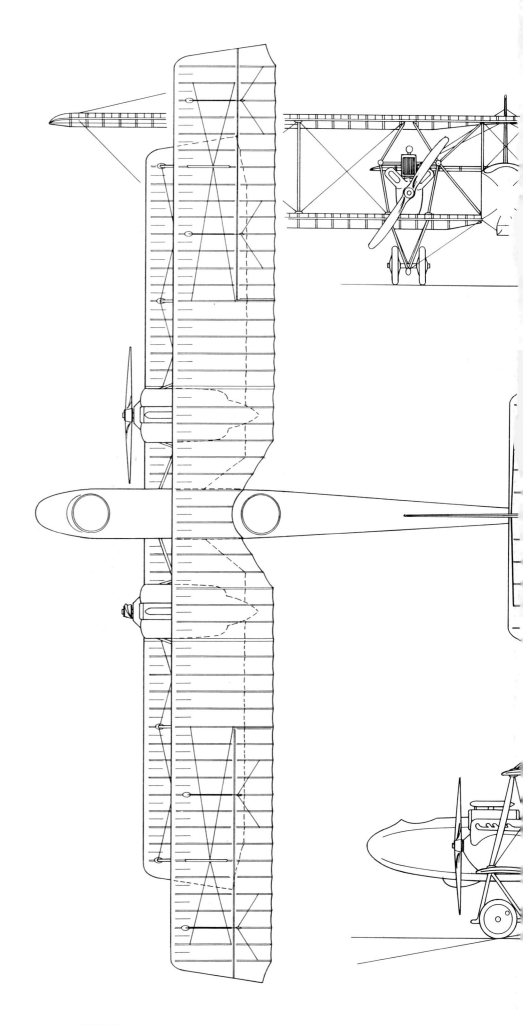

314A

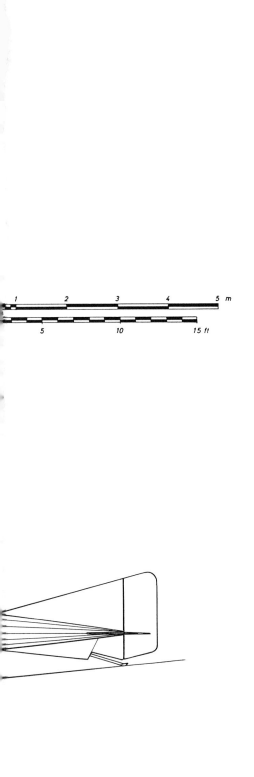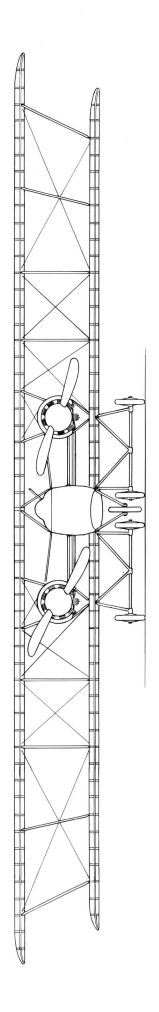

Morane-Saulnier Type L. The crew of de Bernis and Jacotet had forced the German aircraft in the background to land and are posing with the captured plane's crew on 28 April 1915. B81.2671.

gave the Morane-Saulnier firm permission to sell 50 of them to Turkey. They were awaiting shipment when the war began, and they were soon impressed by the French War Ministry. All these aircraft had been powered by 50-hp Gnome engines; in French service many were fitted with 80-hp Le Rhône 9Cs or Gnomes. Approximately 600 Type Ls would be produced by France during the war.

Operational Service

The Type L quickly replaced the Morane-Saulnier G and H. The preference for the Type L was probably due as much to its parasol wing as to its maneuverability. The parasol wing provided the pilot and observer with an outstanding field of view, greatly enhancing the Type L's usefulness in artillery spotting and reconnaissance missions. The escadrilles with Morane-Saulnier Ls were under the direct control of the Groupes d'Armée, while the escadrilles with Caudron G.3s and Voisin 3s were attached to the Corps d'Armée for army cooperation and light bombing missions.

The Morane-Saulnier escadrilles were used for reconnaissance, bombing and, most importantly, fighter patrols. It was the latter mission that would make the Type L famous in 1915.

MS 23 was created on 15 August 1914 and was equipped with some of the ex-Turkish Type Ls. It was initially based at Toul and assigned to the 2nd Armée for the entire time it was equipped with Morane-Saulnier aircraft. MS 23 began flying fighter patrols on 13 March. On 11 May Sergeant Lacrouze forced an enemy aircraft down.

MS 26 was formed on 26 August 1914 with four aircraft and was based at Amiens and then Saint-Souplet, where it participated in the Battle of the Marne. From there MS 26 moved to Anvers, Ostende, Dunkerque, and Saint Pol. It flew a number of fighter patrols over Dunkerque in February 1915. The unit was able to drive off German aircraft on 27 March, 1 April, 13 June, and 16 June. An airplane of MS 26 (at this time serving in the 36th C.A. area) forced an enemy aircraft to land on 15 May. Another German aircraft was forced down on 13 June. A German seaplane was machine gunned by a Morane-Saulnier Parasol on 15 June. On 31 July a German fighter with two machine guns was driven off, apparently without damage to either aircraft.

MS 31 was created on 24 September 1914 at Dijon-Longvic, initially with Type Hs. In November, while based at Toul, MS 31 re-equipped with Type Ls. Three aircraft of MS 31 succeeded in driving off German aircraft on 18 February 1915. On 27 March Sergeant Jensen and Marechal-des-Logis Morel fired at a German airplane and forced it to jettison its bombs. Aircraft of MS 31, which were now armed with Lewis machine guns, drove off two enemy aircraft on 2 April. Other inconclusive attacks took place on 4, 13, 18, and 27 April. Lieutenant Schlumbereger and his observer, Sous-Lieutenant Pardieu, attacked an Aviatik which replied with its machine gun and Pardieu was wounded. The unit's first victory did not occur until 22 September, when Adjudant Bourhis destroyed a German reconnaissance aircraft. He achieved the unit's second, and last, victory while equipped with Morane parasols on 10 October. The unit lost two type Ls to enemy aerial activity.

Most of the time these units flew a combination of reconnaissance, artillery spotting, and light bombing missions. The bombing missions consisted of dropping large numbers of finned darts (flechettes). Although crude, this method of attack was an effective psychological weapon. MS 31 used its aircraft to attack the German airship hangars at Metz-Frescaty.

The Type Ls were also used to attack German balloons, and it was a modified Type L that achieved the distinction of becoming the world's first true fighter. Roland Garros, an aggressive pilot assigned to MS 26, insisted that his observer carry a loaded carbine whenever they flew. After several unsuccessful attempts to shot down an enemy plane, the commanding officer of MS 26 decreed that the unit would henceforth fly unarmed reconnaissance missions.

Despite these early failures, the superb qualities of the Type L were widely appreciated. In October 1914 Colonel Barès, Chef du Service Aéronautique aux Armées, specified that the Type L was the preferred aircraft for aerial combat. The Morane-Saulnier Ls were to be armed with rifles, carbines, or Lewis guns; these weapons were to be fired by the observer.

In 1915 seven more Morane-Saulnier escadrilles were formed. The first of these new units were:

MS 37, also formed in January and based at the airfield at Châteaufort. Assigned to the 3rd Armée, MS 37 participated in the battles at Argonne and Verdun. It flew fighter patrols beginning in mid-February 1915. On 28 March the type Ls drove off a German aircraft and on 4 and 5 June several more German planes were attacked, apparently without significant damage to either side. Capitaine Quilliem and Lieutenant d'Anohold engaged in a ten-minute dogfight with an Aviatik on 6 June; both sides escaped without damage. The next recorded combats occurred on 9, 12 17, 19, and 30 July without result. A Fokker E.III was attacked on 19 August as were two Aviatiks; again, there were no recorded results of either attack. There were more encounters with German aircraft on 13, 14, and 18 September.

MS 38, created on 8 January 1915, based at Châlons-sur-Marne and assigned to the 4th Armée. MS 38's first recorded encounter with a German airplane was on 19 March when an Aviatik was driven off. Other, equally inconclusive, combats took place on 14, 18, and 21 April. While a number of other fighter patrols were flown by MS 38, no victories were recorded.

MS 12, formed in February 1915 when N 12 retired its Nieuport 6s for Type Ls. It was assigned to the 5th Armée. MS 12 began to fly fighter patrols on 5 March. On 1 April Sergeant Navarre and Sous-Lt. Robert attacked an Aviatik and fired several carbine

A Morane-Saulnier Type L serial MS 73 of Escadrille C 30. B92.1602.

A Morane-Saulnier Type L based on the Island of Mavros near Galipoli on 25 October 1915. B84.1514.

rounds at it, forcing it to land. Another combat occurred 14 April when Sergeant Navarre engaged a German aircraft. Enemy aircraft were driven off by MS 12 on 3 June. Other German airplanes were attacked on 15 June and 3 July.

MS 3, formed in March when BL 3 switched from Blériot 11s to the Morane-Saulnier Ls. The commander was Capitaine Brocard who, convinced of the importance of fighter aircraft, would lead his pilots (including Guynemer) to a number of victories. MS 3 was assigned to the 6th Armée. It recorded its first Type L victory on 19 July when an Aviatik C of Feldfliegerabteilung 26 was destroyed. MS 3 did not destroy another German plane until September 1915. However, there were a number of successful fighter patrols that accomplished the unit's mission without having to destroy a German plane. For example, on 27 March and 8 September type Ls of MS 3 attacked German aircraft and forced them to jettison their bombs. There were other encounters with enemy planes which, while inconclusive, resulted in the German aircraft retreating. A number of other combats took place in the fall and winter of 1915 and three victories were scored by Guynemer.

MS 48, created on 29 March 1915 and initially assigned to the Fortress of Verdun. During April the escadrille was assigned to the D.A.L. (Army of Lorraine) with which it remained for the duration of 1915. MS 48's crews engaged in few aerial combats. A type L destroyed a German machine on 15 April; the French gunner was armed with only a carbine. The French, however, were not always the victors; on 22 June a type L was severely damaged by a Fokker E.III. There were a number of inconclusive combats in the spring and summer of 1915; again, these were at least partially successful in that they forced the German aircraft to withdraw. MS 48 also provided fighter cover for Voisin and M.F.11 bombers.

MS 15, created in March when REP 15 retired its R.E.P. Ns in April. It was assigned to the 10th Armée. Along with MF 16 and N 57, MS 15 would perform bomber escort and fighter patrol duties until late 1915. It became a dedicated fighter unit during 1915. Most of the time it flew fighter patrols accompanied by Nieuport 10s and Maurice Farman M.F.11 "fighters" of N 57 and MF 16 respectively. No air-to-air victories were recorded for MS 15 in the GQG daily operational reports.

MS 49, formed in April 1915. Assigned to the D.A.L., it was the first unit to be created specifically as a fighter escadrille. MS 49 flew a number of fighter patrols throughout 1915. In May 1915, a large number of combats occurred, but the first recorded aerial victory did not occur until 4 June when Sergeant Gilbert destroyed an enemy plane. He scored additional victories on 7 and 17 June. Adjudant Peqoud brought down an Aviatik on 11 July.

Although the Type L did not achieve a large number of aerial victories it was used successfully for fighter patrol and bomber escort duties throughout 1915. The Type Ls were at times able to provide fighter coverage over large areas of the front.

The efficacy of the Type L as a fighter was limited by its armament—a single Lewis gun fired by the observer. However, Roland Garros flew a modified Type L which had a forward-firing machine gun and bullet deflectors on the propeller blades. The central wing cutout was deleted. Initial trials had been performed on a Type G, but operational missions were flown with the Type L. As mentioned above, Garros had failed to destroy any enemy aircraft in 1914 but, on 1 April 1915 he brought down a German airplane. There were two more victories, one on 15 April and the third, and last, on 18 April. During the latter Garros' aircraft was brought down by ground fire. In spite of these victories, the French authorities canceled orders that had been placed for Garros-designed deflector-equipped Type Ls. A few Type Ls with deflectors were built but may have been standard machines modified in the field. In any event, by late 1915 the Type L fighters had been replaced by Nieuport 10s.

On 22 August 1915 there were 57 Morane-Saulniers in service at the front; most these would have been Type Ls. There were seven more with training units and six available for service at the front. The total of 70 Morane-Saulniers represented almost 10 percent of all aircraft in service with the Aviation Militaire.

While the Type L had been successful, the Morane-Saulnier firm attempted to improve the design. A modified Type L was designated the Type LA, which stood for Type L with ailerons. The fuselage of the LA was more streamlined with a nearly circular cross-section provided by full-length side fairings and dorsal and ventral deckings. A conical spinner was used and the fixed tail surfaces were enlarged. The engine remained the 80-hp Le Rhône 9C. The wing was entirely new with tapered ailerons. Most aircraft carried a single 7.7-mm machine gun fired by the observer. However, at least one LA was equipped with a Hotchkiss gun that fired through an airscrew fitted with bullet deflectors.

The Type LA was tested in late 1914 and entered service during the summer of 1915. It was given the SFA designation MoS.4. LAs served alongside the Type Ls in MS 3, 12, 15, 23, 26, 31, 37, 38, 48, and 49. Their career was cut short by the appearance of the superior Nieuport 10 and 11 series, which proved to be far more suitable for aerial combat.

In addition to aerial combat, many of the missions flown by the L- and LA-equipped escadrilles were for army cooperation and reconnaissance. The MS units were assigned five main missions during 1915. These were:

1. Fighter patrols to establish aerial superiority and prevent German aircraft from completing bombing and reconnaissance operations. These sorties were initially called vols de barrage (barrage flights).
2. Reconnaissance of the French fortifications.
3. Reconnaissance of German supply and communication lines.
4. Bomber escort.
5. Leaflet dropping.

Another Morane-Saulnier Type L of C 30 which was assigned to 6th Armée. B92.1603.

Below: Morane-Saulnier Type L. The Type L used wing warping for lateral control, and the cables for supporting landing loads were attached to a prominent central pylon and held in place by iron fasteners. B84.1556.

However, the variety of missions undertaken by the MS units is even greater than that shown in this list. For example, MS escadrilles flew Zeppelin patrols and were even used to drop spies behind the enemy lines. Many of the MS units also flew artillery spotting missions during the first few months of 1915. The MS escadrilles were also assigned the task of destroying German observation balloons.

During 1915 the MS units continued to undertake bombing missions, including attacks on the Zeppelin sheds at Friedrichshafen, enemy troop concentrations, and even dropping flechettes on balloons.

The Morane-Saulnier Types L and LA had proved to be remarkably versatile and were effective in virtually any mission they were called upon to perform. However, the pace of aeronautical development was so rapid that by the summer of 1915 both types were obsolescent.

The more modern Nieuport 10 was far superior to the Morane-Saulniers and began to replace the Ls and LAs in the fall of 1915. It was natural that the Nieuport 10 fighters would be supplied first to the MS units as most of these escadrilles had become dedicated fighter units. MS 37 became the first MS unit to re-equip completely with Nieuport 10s when it received the N 37 designation in July 1915. By 20 September 1915 the remaining escadrilles using Type Ls and LAs (MS 3, 12, 15, 23, 26, 31, 38, 48, and 49) had all been designated as Nieuport escadrilles. By February 1916 there were only 59 Type Ls and LAs on strength with the Aviation Militaire. Eighteen were in service at the front and 41 were being used by training units. It is likely that by mid-1916 those aircraft in front-line service had been replaced.

Foreign Service
Belgium
Several examples of the Morane-Saulnier Type L were obtained by the Belgian air service for evaluation. They were used as fighter trainers by the schools at Étampes and Calais.

Czechoslovakia
A single Morane-Saulnier Type L was captured from the Russians and used by the Czech air service postwar.

Finland
Finland acquired two license-built Morane-Saulnier Ls from the Swedish Thulin firm in 1918. These were probably identical to aircraft which had been used by the Swedish army aviation service (see below).

Germany
The Pfalz Flugzeugwerke built the Type L (and Type H) under license. Designated E.Is (and E.IIs when given more powerful engines and longer wings), the Type Hs saw front-line service.

Netherlands
A single Morane-Saulnier Type L was obtained on 22 January 1915 when an RFC machine, serial 1845, landed at Schore op Walcheren. It was assigned serial LA 35, which was later changed to M-23 and then M-4 at the end of 1918.

Peru
Three or four Morane-Saulnier parasols (probably Type LAs) were taken to Peru by the French military mission in November 1919. They were assigned to the Centro de Aviacion Militar at Maranga.

Poland
When Polish personnel formed the Polish Aviation Unit on 23 October 1917 its equipment included a single Morane-Saulnier Parasol (either a Type L or LA which was probably captured from the Russians). The 2nd Polish Combat Aviation Unit also had a single Morane-Saulnier Parasol when it was formed in December 1917. Finally, the 1st Polish Aviation Base had some Type Ls and LAs on strength as late as March 1918.

Romania
In 1915 six Type LAs were purchased from France. Two went to the Escadrilla Allungare Aviatie Inamica (Squadron for Protection Against Enemy Aircraft) assigned to Grupul 1 and two went to Alungare Aviatie Inamica assigned to Grupul 2.

There were only four Type LAs in service when war was declared on 27 August 1916. By late 1916 only one was still serviceable, assigned to Grupul 1.

Russia
The Morane-Saulnier Type L was produced under license by the Dux and Lebedev plants. The Dux plant produced 400 and the

Left: Morane-Saulnier Type LA of the IRAS with skis. B85.2589.

Below: Morane-Saulnier Type LA of the 21st Squadron of the IRAS. B85.2545.

Lebedev plant built 30. An additional 100 were imported from France. They were widely used for armed reconnais-sance. Some were used as fighters and probably served with the 1st Fighter Group (assigned to the 11th Army), the 2nd Group (Southwest Front), and the 3rd Group (Western Front). However, by 1916 the Type Ls were replaced by Nieuport 10s and 11s. The total number of Morane-Saulnier Parasols (Types L and P) still in service in mid-1917 was as follows:

1 March 1917: Northern front (10); western front (17); southwestern front (15); Romanian front (15); Caucasus front (9).

2 June 1917: Northern front (5); western front (8); southwestern/Romanian fronts (7); Caucasus front (35).

Once the Type Ls were retired from the front they were utilized as trainers, and some were in service until 1925. In 1923 the 1st Higher School of Military Pilots in Moscow had 12, and the 2nd Higher School of Military Pilots as well as the 1st and 2nd Military Schools of Pilots also had some on strength.

One Type LA was in use with the Military School of Pilots and Observers and the 2nd Military School of Pilots as late as 1924. The Siberian Air Fleet of Admiral Kolchak had one Type L on strength.

Sweden

A single Morane-Saulnier Type L was obtained in 1914. It was assigned serial number 5 (subsequently changed to 405) and struck off charge in June 1918.

E. Thulins Aeroplansfabrik (later renamed AB Enoch Thulins Aeroplansfabrik) built the Type L under license as the Thulin D. At least five were built in 1917 and given construction numbers D 1 through 5. The initial version of the Thulin D, which was exhibited at the Stockholm aviation exhibit in May 1915, had a 50-hp Gnome engine. Subsequent machines had a 90-hp Thulin

---

Morane-Saulnier Type L Two-Seat Reconnaissance Plane and Fighter with 80-hp Le Rhône 9C

Span 11.20 m; length 6.88 m; height 3.93 m; wing area 18.3 sq. m
Empty weight 385 kg; loaded weight 650 kg
Maximum speed: 125 km/h at sea level; climb to 1,000 m in 8 minutes; climb to 2,000 m in 18 minutes 30 seconds; endurance 4 hours
Armament: a 7.7-mm Lewis gun fired by the observer or rifles and carbines; occasionally flechettes or two 155-mm bombs were carried
A total of 600 were built

Morane-Saulnier Type LA Two-Seat Reconnaissance Plane and Fighter with 80-hp Le Rhône

Span 10.90 m; length 7.078 m; height 3.85 m; wing area 18.3 sq. m
Empty weight 400 kg; loaded weight 650 kg
Maximum speed: 135 km/h at sea level; 135 km/h at 2,000 m; climb to 1,000 m in 6 minutes 10 seconds; climb to 2,000 m in 15 minutes 25 seconds; climb to 3,000 m in 29 minutes 25 seconds; endurance 2.5 hours.
Armament: a 7.7-mm Lewis gun fired by the observer or rifles and carbines; occasionally flechettes or two 155-mm bombs were carried. At least one airplane used a 8-mm Hotchkiss.

Morane-Saulnier Type L Two-Seat Reconnaissance Plane and Fighter with 80-hp Gnome or Le Rhône 9C Engine Produced Under License by Dux and Lebedev

Span 11.20 m; length 6.88 m; height 3.93 m; wing area 18.3 sq. m
Empty weight 375 kg (Gnome); 395 kg (Le Rhône 9C); loaded weight 650 kg; (Gnome); 670 kg (Le Rhône 9C)
Maximum speed: 119 km/h (Gnome) and 127 km/h (Le Rhône 9C); climb to 1,000 m in 8 minutes (Gnome); 1,000 m in 6 minutes (Le Rhône 9C); climb to 2,000 m in 18 minutes 30 seconds (Gnome); 15 minutes (Le Rhône 9C); climb to 3,000 m in 33 minutes; ceiling 3,500 m: endurance 2.6 hours
A total of 400 were built by Dux and approximately 30 by Lebedev

Morane-Saulnier L

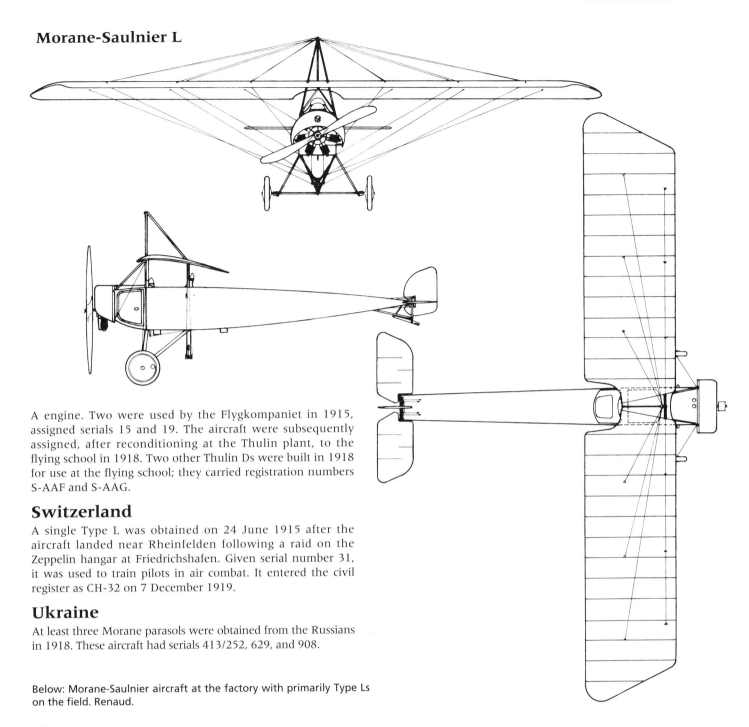

A engine. Two were used by the Flygkompaniet in 1915, assigned serials 15 and 19. The aircraft were subsequently assigned, after reconditioning at the Thulin plant, to the flying school in 1918. Two other Thulin Ds were built in 1918 for use at the flying school; they carried registration numbers S-AAF and S-AAG.

Switzerland

A single Type L was obtained on 24 June 1915 after the aircraft landed near Rheinfelden following a raid on the Zeppelin hangar at Friedrichshafen. Given serial number 31, it was used to train pilots in air combat. It entered the civil register as CH-32 on 7 December 1919.

Ukraine

At least three Morane parasols were obtained from the Russians in 1918. These aircraft had serials 413/252, 629, and 908.

Below: Morane-Saulnier aircraft at the factory with primarily Type Ls on the field. Renaud.

Morane-Saulnier L. The very small fin and rudder were made of steel tubing and covered with cloth. The rudder was held in place by two metal hinges and the lower portion of the rudder was attached directly to the tail skid support. B86 4269. Courtesy Alan Durkota.

Below right: Morane-Saulnier Type L serial MS 274. Unlike the Type G which had used a bathtub layout, the Type L featured a larger cockpit with separate seats for each crewman. Renaud.

United Kingdom

Morane-Saulnier L

More than 50 Type Ls were purchased by the RFC. Many of these were assigned to No.3 Squadron, which had 14 on strength by April 1915, and by the end of 1915 No.1 Squadron had also re-equipped with 13 Type Ls. The Type Ls were replaced by Type LAs in 1916. After being retired from Nos.1 and 3 Squadrons, the Type Ls were allocated to Training Squadrons 15 and 25.

The RNAS acquired 25 Type Ls in 1915. These were assigned to 1 Wing (St. Pol), 2 Wing (Mudros), 3 Wing (Imbros), and 5 Wing (Dover). Serial numbers were 3239–3263 and the airplanes were obtained directly from the Morane-Saulnier firm.

Morane-Saulnier LA

The Type LA began to enter service with the RFC in late 1915. It was used to replace the more antiquated Type Ls in service with 1 and 3 Squadrons and also equipped 7, 12, and 60 Squadrons. The aircraft were found to be difficult to fly and it seems that modifications to the wings had to be made to permit continued service use. As with the Type Ls, the main functions of the Type LAs were reconnaissance, army cooperation, and light bombing. For example, three Type LAs made attacks on German Zeppelin sheds at Brussels on 2 August 1916. The LAs were even used for spy-dropping missions behind enemy lines. The Type LAs were slowly replaced by the Type P in early 1917. The retired Type LAs were then turned over to training units such as the Pilot's School at No.1 Aircraft Depot at St.-Omer and the Reserve Airplane Squadron No.1. By March 1917 it was decided to replace the Type LAs with Bristol Scouts and the aircraft were returned to Britain.

Morane-Saulnier Type M

The Type M was an armored single-seater intended for reconnaissance. The engine was an 80-hp Gnome. Construction was of wood and, as with the preceding Types G, L, and H, control was by wing warping. The dimensions were identical to the Type G, suggesting that the M was a derivative of that design. Three were obtained by the War Ministry in 1913 but it is not known if any were used operationally. The aircraft was given the military designation MoS.13.

Morane-Saulnier Single-Seat Armored Airplane with 80-hp Gnome
Span 10. 20 m; length 6.20 m; wing area 18 sq. m
Loaded weight 490 kg
Maximum speed: 125 km/h; climb rate 143 m per minute
Three built for Aviation Militaire

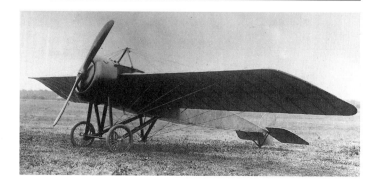

Morane-Saulnier Type M armored aircraft. Three were obtained by the War Ministry in 1913 but it is not known if any were used operationally. B90.2484.

Morane-Saulnier Type N and Morane-Monocoque

The history of the Type N fighter can be traced back to 1912, when Morane-Saulnier and Roland Garros joined forces to design and build a 60-hp monoplane. This aircraft was used by Garros to make a flight from Tunis to Rome. The monoplane was successful enough to warrant further development and a second aircraft with an 80-hp Gnome engine and a smaller wing was built. Fitted with twin floats and designated the Type 0 (see below), it was entered in the 1913 Schneider Trophy race and finished second with an average speed of 92 km/h.

Probably inspired by the Deperdussin monocoque racer, which had finished first in the 1913 Schneider Trophy contest, the Morane-Saulnier firm designed another monoplane with a fully-faired fuselage. The fabric-covered fuselage had a circular cross-section; this streamlining was aided by the huge spinner that almost completely covered the engine cowling. The aircraft was entered in the Aspern meet in June 1914. When war broke out in August the Morane-Saulnier firm developed its sleek racer into a fighter designated the Type N.

The Type N was constructed primarily of wood. The wing was built around two spars with nine ribs on either side of the center section. The wings were covered with fabric except for the roots, which were covered in plywood. Control was, as with the preceding Types H and L, by wing warping. The fuselage was of wood and had a circular cross-section created by wooden stringers covered with fabric. The aircraft was not a true monocoque despite the fact that in official records it was often referred to as the Morane Monocoque.

The cowling and spinner were made of aluminum and closely covered the 80-hp Le Rhône 9C rotary. In fact ,the engine was so tightly enclosed by the cowling that there was insufficient air flow to cool it. The cowling of the prototype Type N was later modified to a more streamlined shape.

The undercarriage used bungee shock absorbers. The fin, rudder, and stabilizer had wooden frameworks covered with fabric. The lower portion of the rudder was supported by the rigid tail skid.

The most important innovation of the Type N was its armament, which usually consisted of a single 8-mm Hotchkiss or 0.303 Lewis machine gun mounted on the fuselage centerline and firing through a wooden propeller fitted with deflector plates. There was always the chance of a catastrophic failure of this system, which could result in the loss of the propeller as well as the aircraft and pilot.

The first Type N to arrive at the front was flown by Roland Garros' friend Eugene Gilbert. Determined to avenge the loss of Garros, who had been captured after crashing behind enemy lines, Gilbert named his aircraft *Le Vengeur*. The Type N was given the company designation Nm. The m (which presumably stood for Militaire as it was an armed variant of the Type N racer) was almost never used in official correspondence concerning the aircraft. The SFA designation was MoS.5 for types with the 80-hp Gnome. This airplane had a fixed 8-mm Hotchkiss gun and an armored airscrew with deflector plates. The 8-mm Hotchkiss had a 25-round clip.

Production machines differed from Gilbert's machine in having a more aerodynamic spinner, a revised head rest fairing, an enlarged rudder, and a fixed fin that had a sharp, rather than curved, leading edge. Part of the rudder extended below the fuselage and was hinged to a small fixed fin.

According to a note dated 5 June 1915, a total of 24 Type Ns were ordered. By June 1915 ten had left the factory and it was anticipated that from 20 June to 5 August an additional 24 were to be built. Maximum speed of the Type N was given as 145 km/h and it could climb to 2000 m in 10 minutes.

After June 1915, the first ten production Type Ns had arrived at the front. These aircraft were usually assigned to MS units to provide escort for the more vulnerable Type L/LAs. MS 12, 23, and 49 are all known to have been equipped with Type Ns; reports suggest that MS 3, 37, and 48 also utilized them. French reports for 1915 rarely mention the Type N specifically, but it is

Morane-Saulnier Type N, probably at the Morane-Saulnier factory. The airscrew is armored and the aircraft is armed with a single Hotchkiss machine-gun. Renaud.

Morane-Saulnier Type N *Le Vengeur* serial MS 388 of pilot Eugene Gilbert. B92.1314.

likely that from July through August 1915 most of the bomber escort missions and barrage flights flown by the MS units included Type Ns.

There were numerous encounters with enemy aircraft during the summer months but most were inconclusive; the German aircraft were usually described as having been forced to withdraw and few victories were achieved.

By September 1915 the Nieuport 10 was becoming widely available, followed a few weeks later by the first Nieuport 11s. In a letter from the Ministry of War dated 7 September 1915, the Type N was preferred over the Type G and L as well as the Nieuport 10. However, when the Type N was evaluated against the Nieuport 11 in September 1915 the results revealed the superiority of Nieuport's design and further development of the Morane-Saulnier's design was abandoned.

The Type N had been found to be a demanding, and at times dangerous, aircraft to fly. The Morane-Saulnier fighter had a tendency to stall above 3000 meters and, perhaps due to its excessive wing loading, was uncomfortable to fly. The Nieuport 10s and 11s, on the other hand, not only had superior performance but were far easier to fly. There were also reports that the Type Ns were difficult to maintain in the field. Finally, the propeller deflection system of the Type N was far less reliable than the machine gun mounting on the top wing utilized on the Nieuports.

The Type Ns continued to see very limited service in the fall of 1915; Jean Navarre (who commanded MS 12) destroyed a German aircraft while flying a Type N on 25 October. However, by the end of 1915 the Type N had been completely withdrawn from front-line units. While the Type N saw only limited service with the Aviation Militaire, it would see more widespread use with the RNAS and RFC.

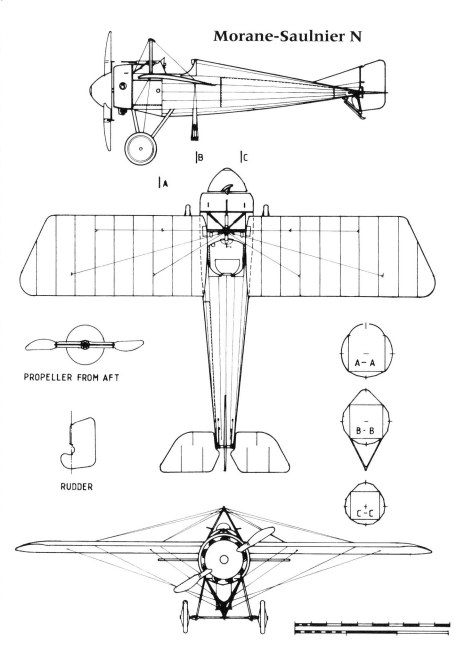

Foreign Service
Russia

The information on the Type Ns used by the Imperial Russian Air Service is distorted by the Russian authors' confusion between the modified Type G fighter, Type I, and the Type N. Shavrov reports on a single-seat "monocoque" monoplane powered by a 110-hp Le Rhône engine. This is almost certainly the Type G of 1915, which had been fitted with a Hotchkiss gun and deflector blades on the propeller. He states that a "small number were used" and that a single experimental aircraft was "built by the Dux plant in 1917." He also reports that Type Ns were obtained directly from France in 1916. It had been planned to produce the Type N at the Dux plant but, for unstated reasons, this plan was never implemented. A number of Morane-Saulnier Type I fighters were also obtained by the IRAS; at least 20 had been ordered.

By mid-1916 the Russians had adopted the French practice of organizing fighter squadrons into larger units. Each of these fighter groups had between four and six squadrons. The groups were stationed at areas of the front where it was deemed essential to establish local air superiority. The 1st Fighter Group (with four squadrons) was assigned to the 11th Army, the 2nd Group served on the southwestern front, and the 3rd Fighter Group was on the western front. While most of these units were equipped with Nieuport 11s, a number of Type Gs, Is, and Ns (collectively and inaccurately known as Morane "Monocoques") were also being used. On 1 March 1917 there were 12 Morane "Moncoques" of all types at the front distributed as follows: northern front (1); western front (3); southwestern front (6); and the Romanian front (2).

By the next month there were 18 "Monocoques" in service, represented 8 percent of all aircraft at the front. As late as July 1917 there were 11 on the Romanian and southwestern fronts. It appears that none was used during the civil war.

Morane-Saulnier Type N serial MS 397 in August 1915. B85.1192.

Morane-Saulnier Type N. B87.6895.

Ukraine

Two "Morane monocoques" were obtained from the Russians in 1918. They had serial numbers 755 and 952.

United Kingdom

The RFC utilized more Type Ns than were employed by the Aviation Militaire; more than half of all the Type Ns produced went to the British. In many ways the RFC's decision to use the Type N was forced upon it by circumstances. The "Fokker scourge" and the limited number of Bristol Scouts available created an urgent need for single-seat fighters. Three Type Ns were ordered by the RFC in mid-September 1915, by which time the Type Ns had been almost completely retired from the front-line escadrilles of the Aviation Militaire. The first three were assigned to Nos.1 and 3 Squadrons. A follow up order for 24 was placed in January 1916 and all had been delivered by mid-June. These aircraft served with Nos.1, 3, and 60 Squadrons. They became popularly known as Morane Bullets or Morane Scouts.

Some of the Type Ns had 11-square-meter wings with a new profile and were similar to the forthcoming Type Is (see above). They were considered to be "excellent practice or transition machines." These were found to be 8 km/h faster than the standard Type Ns but their climb rate was slower. Armament was usually a single 0.303 Lewis machine gun with a 47-round drum. Some Type Ns had a 0.303 Vickers machine gun; in these the gun butt protruded into the cockpit, necessitating a take-up spool to collect the empty belt.

No.60 Squadron was the main user of the Type Ns, receiving its first on 28 May. The aircraft of No.60 Squadron saw action during the Battle of the Somme and while, as with the French, most combats were inconclusive, a number of victories were scored by pilots flying the Type Ns. No.24 Squadron also had some Type Ns, most having been transferred from No.3 Squadron. The British pilots also found the Type Ns difficult to fly and another disadvantage was its close resemblance to the Fokker E.III, which led to recognition errors.

The Type Ns were withdrawn from front-line squadrons in October 1916.

Morane-Saulnier Type N Single-Seat Fighter with 80-hp Le Rhône 9C

Span 8.146 m; length 5.83 m; height 2.25 m; wing area 11 sq. m
Loaded weight 444 kg
Maximum speed: 144 km/h at ground level; climb to 1,000 m in 4 minutes; climb to 2,000 m in 10 minutes; range 185 km; endurance 1.5 hours
Armament: one fixed 8-mm Hotchkiss, 0.303 Lewis, or 0.303 Vickers machine gun
Approximately 44 built; the RFC received approximately 26 directly from the Morane-Saulnier factory

Morane-Saulnier Type N Single-Seat Fighter with 80-hp Le Rhône 9C and Modified Wings

Dimensions: same as standard Type N
Payload: 155 kg
Maximum speed: 152 km/h at ground level; climb to 2000 m in 12 minutes; climb to 4,000 m 45 minutes

Morane-Monocoque Single-Seat Fighter with 120-hp Le Rhône

Span 9.8 m; length 7.0 m; height 2.25 m; wing area 15 sq. m
Empty weight 435 kg: loaded weight 658kg;
Maximum speed: 177 km/h at ground level; climb to 1,000 m in 5 minutes 54 seconds; climb to 2,000 m in 10 minutes 12 seconds; climb to 3,000 m in 17 minutes; ceiling 5,600 m; endurance 2.3 hours
One built by the Dux plant, others obtained directly from Morane-Saulnier. Note: This was probably a modified type G.

Morane-Saulnier N Single-Seat Fighter with 80-hp Le Rhône 9C Supplied to Russia

Span 8.1 m; length 5.8 m; height 2.25 m; wing area 11.0 sq. m
Empty weight 400 kg; loaded weight 575 kg
Production was "started" at the Dux plant but never completed

---

Morane-Saulnier Type O

The Type O was designed by Saulnier for the Monaco aviation rally of 1914. At least two were built. The initial version had a rigid undercarriage and was described as being difficult to fly because of poor lateral control.

In addition to the Monaco rally, the aircraft was entered in the London-Paris-London race in July of 1914; it finished second behind a Morane-Saulnier Type H. The aircraft was modified for the race, being given new, extremely flexible wings and a more practical landing gear with shock absorbers. Unfortunately, these changes did nothing to ease the Type O's handling difficulties.

At least one Type O was in existence at the beginning of the war but it is not known if it was impressed by the Aviation Militaire or Aviation Maritime.

Morane-Saulnier Type P

The success of the Morane-Saulnier L and LA led the firm to produce an improved parasol design in 1916. By September 1915 the Nieuport 10 and 11 series had replaced the Type Ls and LAs in the fighter role. Although the parasols were still proving to be useful for reconnaissance missions, it was felt that with the appearance of the Fokker E.III in 1915 a more powerful version of the L and LA was needed.

The new parasol design was given the Morane-Saulnier firm's designation Type P. It represented a complete redesign of the LA and incorporated a number of changes. The most important was the installation of a 110-hp Le Rhône 9Jb engine. The fuselage of the Type P was fully faired, resulting in a circular cross-section and, presumably, this streamlining contributed to the Type P's higher speed. The tail unit was similar to the preceding LA. Armament was variable and included a synchronized 0.303 Vickers machine gun on top of the wing. There was a mounting for a second machine gun behind the cockpit; it was fired by the observer.

The Type P was tested on 31 March 1916 and was accepted for service with the Aviation Militaire under the designation MoS.21. A total of 565 Type Ps were built.

Operational Service

These aircraft supplemented, but did not replace, the Caudron

Morane-Saulnier Type P of Escadrille C 220. These aircraft supplemented, but did not replace, other types then in squadron use. B85.579.

G.4s, Farman M.F.11s, and Farman F.40s then in squadron use. Many of the French bomber and fighter escadrilles were assigned a few Type Ps, which were used for a variety of missions. For example, Type Ps were assigned to N 12 and N 124 (both of which were fighter units) which used them for short-range reconnaissance, light bombing, and dropping spies behind enemy lines.

Two escadrilles were equipped completely with Type Ps. The first to be formed was MS 140, created in September 1917 and assigned 13 Type Ps (the first three were drawn from F 225). It was assigned to the 4th Armée as an army cooperation unit. The other unit was MS 215, formed from F 215 in September 1917. MS 215 was assigned to the 6th Armée and also had 13 Type Ps on strength. Other escadrilles which had some Type Ps on strength included N 67, N 79, F 203, F 206, C 207, F 215, and F 218. The Type Ps, however, did not stay in service very long as they were clearly obsolescent by mid-1916. Despite the large number of aircraft built, only 118 Type Ps were still in service by August 1917. There were 61 with the escadrilles at the front, 52 with the RGA, three under repair, and two probably serving as a source of spare parts. Both MS 140 and MS 215 had re-equipped with SPAD 16s by early 1918.

An order issued in June of 1917 by the GQG sheds some light on the quick withdrawal of the Type P. It stated that numerous accidents involving the Morane parasols were believed due to structural failure as the result of pilot error or improper handling of the aircraft while being worked on by the ground crew. The order required that a detailed inspection be made of the aircraft before each flight, and particular care was to be taken not to grab the struts of the machine while effecting repairs or entering the cockpit. It was also suggested that the airplanes should not be flown in bad weather and that aerobatics should be avoided. The order went on to state that newer models would be strengthened with the use of metal plates on the struts and tail plane and that the rear spar of the wing was to be reinforced. Despite these precautions, the Type P was withdrawn from service in the fall of 1917. A GQG memo dated 9 August 1917 stated that it was planned to replace the Type Ps with a modified version, the MoS.26.

The MoS.26, an improved version of the Type P, had a synchronized 0.303 Vickers machine gun and another 0.303 Vickers gun fired by the observer. Other changes from the MoS.21 included a strengthened structure, a circular engine cowling, and the addition of an enlarged spinner for improved streamlining. It is not known how many MoS.26s were built, but the figures given for August 1917 (see page 10) almost certainly include some of these aircraft.

There was also a fighter version of the Type P. This had the observer's position faired over and a synchronized 7.7-mm Vickers machine gun mounted in front of the cockpit. Two prototypes were tested and, although the performance was felt to be satisfactory, it was found that the parasol wing obstructed the pilot's vision. A second version of the Type P fighter was built with a lower wing. In addition to this change, the pilot's position was moved aft of the wing, and he was given an

Morane-Saulnier Type P of N 314. The airplane was used as a nightfighter and was based at Malzéville in July 1917. B79.3225.

Morane-Saulnier P

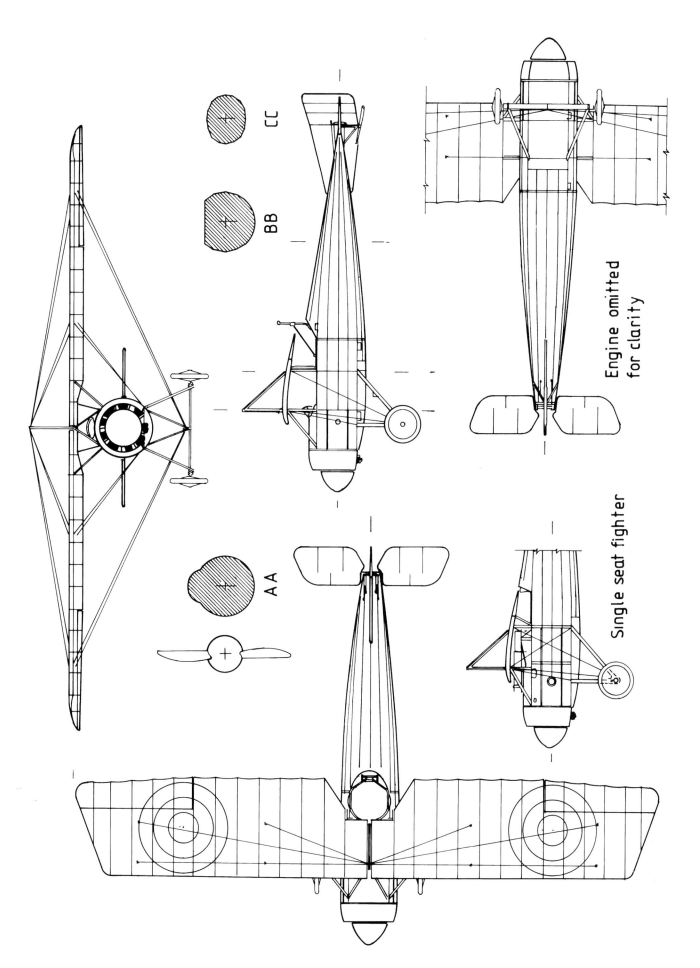

> **Morane-Saulnier Type P Two-Seat Reconnaissance Aircraft with 110-hp Le Rhône 9Jb (MoS.21)**
> Span 11.16 m; length 7.18 m; height 3.47 m; wing area 18 sq. m
> Empty weight 433 kg; loaded weight 730 kg; payload 162 kg
> Maximum speed: 162 km/h at sea level; 155.8 km/h at 2,000 m; climb to 2,000 m in 8 minutes 45 seconds; climb to 3,000 m in 15 minutes 50 seconds; ceiling 4,800 m; range 375 km; endurance 4 hours
> Armament: synchronized 0.303 Vickers machine gun on the top wing and a second machine gun fired by the observer
> A total of 565 were built

> **Morane-Saulnier Type P Two-Seat Reconnaissance Aircraft with 110-hp Le Rhône 9Jb (MoS.26)**
> Dimensions: same as MoS.21
> Empty weight 430 kg; loaded weight 730 kg; payload 205 kg
> Maximum speed: 152 km/h at 2,000 m; climb to 2,000 m in 9 minutes, endurance 2.5 hours; range 375 km
> Armament: synchronized 0.303 Vickers machine gun fired by the pilot and a second Vickers fired by the observer

> **Morane-Saulnier Type P Single-Seat, Single-Gun Fighter with 110-hp Le Rhône 9Jb**
> Span 11.2 m; length 7.2 m; height 3.47 m
> Empty weight 430 kg; loaded weight 610 kg
> Maximum speed: 165 km/h at ground level; climb to 500 m in 1 minute 30 seconds; climb to 2,000 m in 6 minutes 40 seconds; climb to 3,000 m in 11 minutes 20 seconds; climb to 4,000 m in 19 minutes 40 seconds; endurance 2.5 hours
> Armament: one synchronized 7.7-mm Vickers machine gun
> Two built

> **Morane-Saulnier Type P Single-Seat, Twin-Gun Fighter with 110-hp Le Rhône 9Jb**
> Span 11.2 m; length 7.2 m; height 3.27 m
> Empty weight 433 kg; loaded weight 693 kg
> Maximum speed: 156 km/h at ground level; 150 km/h at 2,000 m; climb to 2,000 m in 8 minutes 40 seconds; climb to 3,000 m in 17 minutes 20 seconds; ceiling 4,400 m; endurance 2.5 hours
> Armament: two synchronized 7.7-mm Vickers machine guns
> One built

> **Morane-Saulnier Type P Two-Seat Reconnaissance Aircraft with 110-hp Le Rhône 9Jb (MoS.21) Built in Russia**
> Span 11.16 m; length 7.18 m; height 3.47 m; wing area 18 sq. m
> Empty weight 433 kg; loaded weight 733 kg; payload 162 kg
> Maximum speed: 163 km/h at sea level; climb to 1,000 m in 3 minutes 30 seconds; climb to 2,000 m in 7 minutes 30 seconds; climb to 3,000 m in 12 minutes; ceiling 4,800 m; endurance 3.5 hours

adjustable seat to allow him to see over the wing. Other changes included installation of two 7.7-mm Vickers machine guns and an increase in fuel capacity. The new aircraft's performance was only marginally better than the standard Type P and it was not selected for use.

Foreign Service

Brazil
A single Type P was obtained by Brazil in 1919. It was modified so that it could not fly and was used to train pilots how to taxi. The Type P, given serial 1325, was struck off charge in 1920.

Japan
A single Type P was obtained in 1919 and was used to train pilots how to taxi aircraft.

Russia
The Imperial Russian Air Service obtained only a few Type Ps from France. Most had 110-hp Le Rhône engines, although later aircraft had 120-hp Le Rhônes. The performance of the Type P was clearly superior to the LA, but the aircraft was felt to be considerably more difficult to fly. A few Type Ps were built in Russia but because Russian pilots disliked the type it was never used in large numbers. For details on distribution at the front see the entry under Type L/LA.

United Kingdom
The RFC obtained its first example of the Type P in 1916. The aircraft, serial MS 746, was tested at No.2 Aircraft Depot at Candas, where it was given its new serial A120. Based upon a favorable report on the prototype P, the British placed an order for three more; on some there was to be a synchronized machine gun and a fairing for a camera and wireless. In addition, all aircraft were to have a ring mounting for the observer's machine gun.

Deliveries were delayed because of a shortage of 110-hp Le Rhône engines. However, once this had been resolved an order for 20 more aircraft was placed. By September 1916 this order had been filled and 44 more Type Ps were ordered. While the fairing for the camera and wireless were fitted to some of these aircraft and the observer's gun was provided with a moveable mount, the synchronized gun was not fitted. Instead a Lewis 0.303 gun was mounted on top of the wing for those aircraft assigned to No.3 Squadron. No.1 Squadron also used Type Ps.

There continued to be troubles with the supply of 110-hp Le Rhône engines and on some of the Type Ps an 80-hp Le Rhône was used. These aircraft also had some minor alterations to the wing, struts, undercarriage, fuel tanks, and aileron linkages. Of the total of 36 Type Ps with 80-hp Le Rhônes, at least nine were supplied to No.3 Squadron. After January 1917 all the 80-hp Type Ps had been relegated to training units.

A total of 106 Type Ps (Mos.21 and MoS.26) were ordered. Small numbers of the MoS.26 were used by No.3 Squadron. By October 1917 the MoS.26s had been sent to No.2 Aircraft Depot, and those remaining were relegated to Pilot School No.1 AD at St. Omer, France and 1 Reserve Training Squadron at Gosport.

Morane-Saulnier Type P with fore and aft-mounted machine guns. Renaud.

Morane-Saulnier S

The Morane-Saulnier S was intended as a heavy day bomber. Built in 1915, the aircraft had a crew of three or four; two of which were gunners. There were gunner's stations in the nose and mid-fuselage. The fuselage appears to have been wide enough to accommodate two pilots side by side and was of circular cross-section. Both the fuselage and tail assembly were remarkably similar to those of the later Type P reconnaissance aircraft. There was a rounded fin on each of the elevators. The three-bay wing was of unequal span, the upper wing being slightly longer than the bottom. Two 220-hp Renault engines were mounted in nacelles suspended between the top and bottom wing and were very close to the crew's cockpits. As with many French heavy bombers of the time, the Type S had twin nose wheels to prevent nosing over during landing.

The prototype Type S was given serial MS 625 and designated MoS.10. It was entered in the 1916 concours puissant for a heavy bomber. Only the Type S and the SPAD E were able to meet the requirements of the competition and it was the Type S that was selected for series production. Colonel Barès of the Aviation Militaire requested that 300 be purchased at a cost of 60 million francs. The Parliament felt that this was an excessive sum and reduced the order to 90 aircraft. In response, the order for 90 Type S bombers was canceled by the Aviation Militaire. The decision to produce Caproni bombers under license may have eliminated the need for an indigenous heavy bomber.

It is interesting to note that an STAé document dated 1 November 1916 shows that MoS.10s with two 220-hp Renault engines were in service at the front, but there are no

> **Morane-Saulnier Type S Three-Seat, Twin-Engine Heavy Bomber with Two 220-hp Renault Engines**
> Span 26.00 m; length 12.20 m; area 120 sq. m
> Maximum speed: 140 km/h at 2,000 m; range 1200 km
> Armament: it appears the prototype was unarmed. Planned armament would have included 300 kg of bombs and at least two machine guns.
> One built

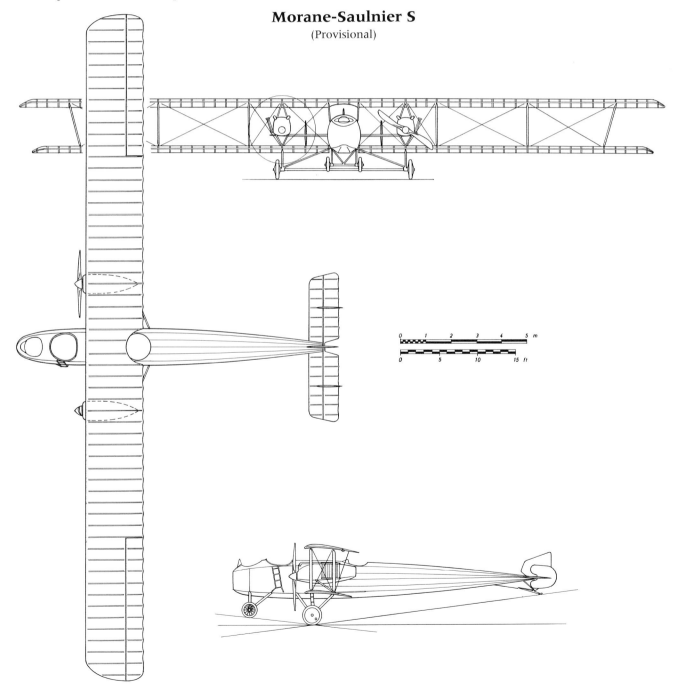

Morane-Saulnier S
(Provisional)

Morane-Saulnier Type S which was entered in the 1916 concour puissant and was selected as the winner. Serial number is 625. MA.

photographs or documents to suggest the Type S ever entered service with the Aviation Militaire. Either the document is referring to a Type S at the front for operational evaluation or to the Type T. The latter would seem to be less likely since the Type T entered service in August 1917 and it used two 110-hp Le Rhône engines.

Morane-Saulnier T

The Type T was designed in late 1914 or early 1915, intended for long-range reconnaissance. It was a large biplane and appeared to be a scaled-down version of the Type S bomber. It has been reported in some sources that the Type T contract was given to the Morane-Saulnier firm as a consolation for the Type S debacle as described above. The original Type T was powered by two 80-hp Le Rhône engines. The angular fuselage originally had a gunner's position in the nose with a large number of portholes. However, the nose was later redesigned and the portholes were eliminated.

The pilot was located beneath the top wing and a second gunner's position was located behind the wings. The rear fuselage tapered sharply and a triangular fin and rudder were mounted on the extreme tail. The undercarriage had an unusual layout with a pair of wheels beneath the center of the fuselage and one wheel beneath each engine nacelle. A single nose wheel, larger than the main wheels, was located in the extreme nose. The Le Rhône engines were tightly cowled and the twin-bladed propellers had huge spinners almost completely covering the engines. The engine nacelles were suspended between the upper and lower wings and were located on either side of the pilot's cockpit.

The Type T was tested by the GDE in late 1916. The first flight was described as "satisfactory" and 90 were ordered on 22 August 1916. However, static testing revealed airframe weakness and it was stated that deliveries would be delayed until this problem was corrected. Considering the problems the aircraft encountered in service it appears that the problem was never entirely eliminated.

The Type Ts were designated MoS.25 A3. The A3 designation denotes a reconnaissance and army cooperation aircraft with a crew of three. Operational Type Ts were equipped with the more powerful 110-hp Le Rhône engines (as compared to the 80-hp engines of the prototype).

The aircraft were delivered on 1 August, 1917, almost exactly one year after the order had been placed. Even then, only 13 had been completed and these were still at the RGA.

It appears that no escadrilles were completely equipped with Type Ts. Rather, the aircraft were supplied to various army cooperation escadrilles. As with the Caudron G.6, R.4, and S.M.1, the Morane-Saulnier Type Ts were intended to give these escadrilles a long-range reconnaissance capability. The Aviation Militaire's policy at this time was to supply the larger A3 airplanes in only small numbers to army cooperation escadrilles. This served to ease the demands on the maintenance crews and prevented overcrowding of airfields. The aircraft are known to have been supplied to C 4, C 11, C 17, C 30, C 39, and C 47.

However, reports indicate that the Type T had numerous problems, resulting in the loss of several. A GQG report from late 1917 stated that there had been many accidents involving the Type Ts. It was reported that the tail skid was located in an awkward position, resulting in debris being thrown against the tail surfaces during landing. This resulted in structural failure of the tail assembly (probably at the elevator hinges, which were found to be weak), often with the loss of the aircraft and crew. The crews were also warned that the nose had to be loaded with ballast if a forward gunner was not carried. They were also warned not to fly the Type T in bad weather. One report

Morane-Saulnier Type T with the Morane Saulnier TRK bomber in the background. MA.

A Morane-Saulnier Type T of C 17 serves as the backdrop for Christmas mass in December 1916. The numerous portholes in the nose were eliminated on later versions of the Type T. B80.36.

suggested that modifications to the tailskids were to be made, but other reports indicate that the Type Ts were being withdrawn from escadrille service in late 1917.

Morane-Saulnier Type T Three-Seat Long-Range Reconnaissance Airplane with Two 110-hp Le Rhône 9Jb Engines
Span 17.65 m; length 10.50 m; wing area 100 sq. m
Loaded weight 3,772 kg
Maximum speed: 136 km/h
Armament: at least two machine guns
90 were built

Morane-Saulnier T. The Type T was supplied to army cooperation escadrilles to provide long-range reconnaissance; the Type T never equipped an entire unit. (CC).

Morane-Saulnier TRK

The Morane-Saulnier TRK was a triplane bomber built in 1915. It may have been intended to participate in the 1915 or 1916 concours for a heavy bomber.

Two 230-hp Canton-Unné A9 engines were mounted inside the fuselage and powered two large propellers by a complex transmission system. One propeller was located on either side of the fuselage between the middle and lower wings. The landing gear consisted of a pair of wheels under each wing and two wheels under the center fuselage; this arrangement was almost identical to that of the Type S and Type T. The pilot and co-pilot were seated side by side in the large nose cockpit and a gunner was located behind the wings. Armament consisted of two 0.303 machine guns. Given the STAé designation MoS.9, only one TRK was built. It was not entered in either the 1915 or 1916 concours, and it is not known if it was ever flown.

Right: Morane-Saulnier Type TRK with an engine exposed. MA32.

Morane-Saulnier Type TRK Three-Seat Bomber with Two 230-hp Canton-Unné A9 Engines
Wing span 20.220 m; length 16.380 m; wing area 140 sq. m
Armament: two 0.303 machine guns
One built

Morane-Saulnier TRK

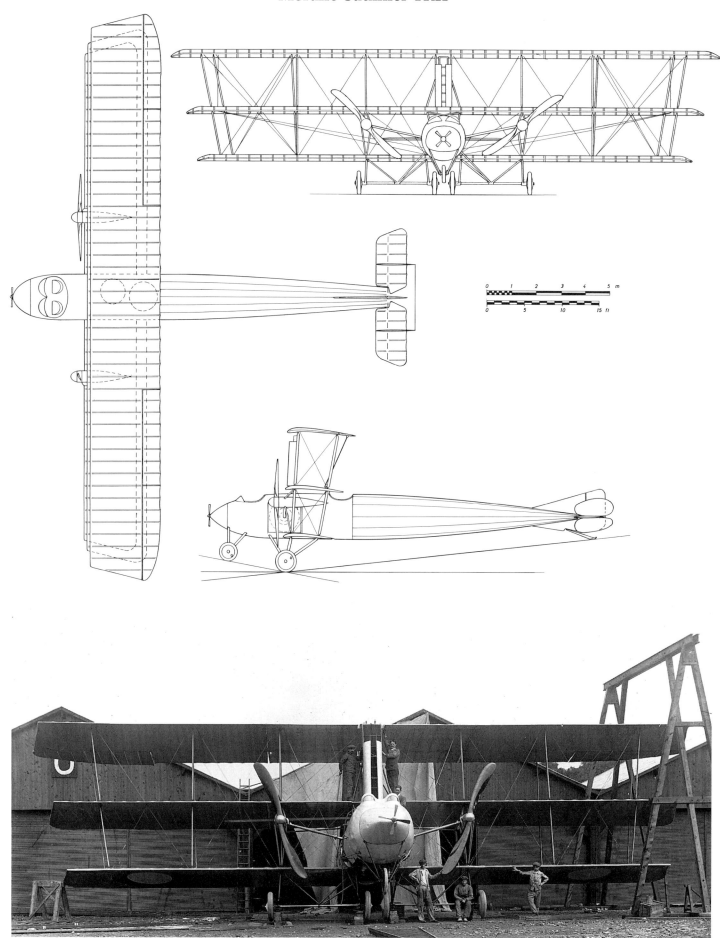

The huge size of the Morane-Saulnier Type TRK is apparent in this photograph. MA33.

Morane-Saulnier Type U

The Type U was a proposed development of the Type V. The Type U was to have a shoulder-mounted wing and control was to have been by ailerons rather than wing warping. The wing was rigidly braced by a complex mesh of struts and wires underneath it. The spinner, undercarriage, and tail were similar to the preceding Type V. One unusual feature was a streamlined windshield, remarkably long and narrow. It is not known if the Type U was built. Given the Aviation Militaire's decision to concentrate on production of the Nieuport 11 and the RFC's plan to withdraw all Morane fighters from squadron service, it would seem unlikely that either service would have been interested in an airplane which was nothing more than an improved Type V.

Morane-Saulnier Type U Single-Seat Fighter with 110-hp Le Rhône 9C (Provisional Specifications)

Span 8.24 m, length 5.813 m; height 2.130 m; wing area 10.962 sq. m
Empty weight 380 kg; loaded weight 530 kg
Maximum speed: 187 km/h; climb to 1,000 m in 2 minutes 55 seconds; climb to 2,000 m in 6 minutes; climb to 3,000 m in 10 minutes; climb to 4,000 m in 16 minutes 10 seconds; climb to 5,000 m in 27 minutes 30 seconds
Armament: one 0.303 Vickers machine gun
Project only

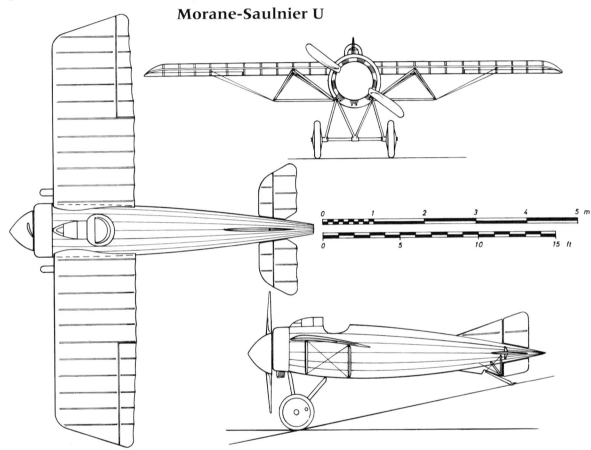

Morane-Saulnier U

Morane-Saulnier Type V

The Morane-Saulnier Type I had been, essentially, a Type N equipped with a more powerful 110-hp Le Rhône 9J. The Type V was to have the same engine as the Type I but incorporate a number of improvements. While the Type I had an endurance of 1.3 hours, the Type V was intended to carry enough fuel for three hours. This requirement came at the insistence of Major General Trenchard, who had been displeased with the Type I's poor endurance. While at least one example is known to have been flown in French markings, further development of the Type V was abandoned by the Aviation Militaire. All the Type Vs produced were used by the RFC.

The most significant change in the Type V was the addition of a 50-liter fuel tank carried within the enlarged fuselage belly fairing. The span and chord of the wings were enlarged, presumably to permit the aircraft to carry this heavier payload. The elevators were also enlarged and the rudder modified. Armament consisted of a synchronized 0.303 Vickers machine gun. As with the Type I, the gun butt protruded into the cockpit, necessitating a take-up spool to collect the empty belt. On the Type V the empty casings ran into a trunk that emptied out through the bottom of the airplane. A redesigned windscreen served to shield both the pilot and the ammunition belt from the slipstream.

Designated Type V by the Morane-Saulnier firm and, possibly MoS.22 by the SFA, 12 were ordered on 1 April 1916. The first was delivered to the RFC on 22 April 1916. It had serial number MS 747 and reached No.1 AD on 16 May and No.3 Squadron on 19 May.

Contemporary reports describe the Type V as being easy to

Morane-Saulnier Type V Single-Seat Fighter with 110-hp Le Rhône 9J

Span 8.75 m; length 5.815 m
Payload 210 kg
Maximum speed: 165 km/h at ground level; climb to 1,000 m in 3 minutes 20 seconds, to 2,000 m in 8 minutes; to 3,000 m in 15 minutes 20 seconds
Armament: one synchronized 0.303 machine Vickers machine gun
Twelve built for the RFC

fly. Assuming these analyses were accurate, the Type V would indeed have represented a significant improvement over the Type N.

The final Type Vs had been sent to the RFC by 26 August 1916. They were initially delivered to No.2 AD and then to No.60 Squadron. These aircraft saw considerable activity with No.60 Squadron. However, while the pilots who had initially evaluated the Type V had praised it as being easy to fly, the pilots at the front found it to be a demanding machine. One pilot stated that it behaved as if it "were doing its best to kill you." Realizing the Type V's deficiencies, Trenchard ordered the remaining aircraft be returned to England.

In October 1916 No.60 Squadron sent its remaining nine Type Vs to England, where they may have seen service with training units.

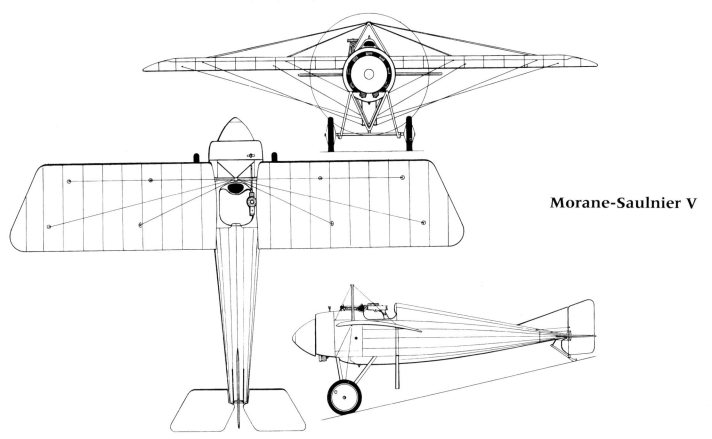

Morane-Saulnier V

Morane-Saulnier Type X

This aircraft is reported to have been designed in 1916 as a heavy bomber. One interesting aspect of the design was that it was intended to use one 300-hp Panhard engine (which may have been buried inside the fuselage) to drive two propellers via a transmission system. It is not known if the Type X was ever flown, and it may have remained an unfinished project.

Morane-Saulnier Type Y

The Type Y was a three-seat reconnaissance aircraft powered by two 220-hp Hispano-Suiza engines. It was built and flight tested in August 1916. There may have been some interest on the part of the Aviation Militaire in purchasing the machine, as the Type Y was given the STAé designation A1. Despite this designation, the Type Y carried a crew of three and, had it been selected for production, would have given the later STAé designation A3 (three-seat reconnaissance/army cooperation machine). However, the Type Y was not ordered and no further examples were built. It may have been a development of the Type S bomber.

Morane-Saulnier Type AC

The Type AC was a monoplane with a shoulder-mounted wing with external bracing. It used ailerons rather than wing warping for lateral control. The fuselage was fully faired with a circular cross-section and the triangular fin and all-moving tailplanes of the Type P. Armament consisted of a single 7.7-mm Vickers machine gun mounted on the fuselage centerline. There was a large headrest for the pilot. The engine was either a 110-hp Le Rhône 9J or a 120-hp Le Rhône 9Jb.

The Aviation Militaire evaluated the aircraft in late 1916 and appears to have been satisfied with it. Thirty were ordered and given the designation MoS.23. This small batch was parceled out to numerous fighter escadrilles and no squadron was ever formed entirely on Type ACs. N 76, for example, is known to have used several Type ACs. The aircraft remained in service until at least the spring of 1917. It would appear that the Aviation Militaire had by this time decided to re-equip its fighter escadrilles with SPAD 7s and no further Type ACs were purchased.

Morane-Saulnier Type AC Single-Seat Fighter with 110-hp Le Rhône 9J
Span 9.8 m; length 7.05 m; height 2.73 m; wing area 15 sq. m
Empty weight 435 kg; loaded weight 658 kg
Maximum speed: 178 km/h at ground level; 174 km/h at 2,000 m; 171 km/h at 3,000 m; ; climb to 2000 m in 5 min. 55 sec; climb to 3,000 m in 10 minutes 15 seconds; ceiling 5,600 m; endurance 2.5 hours
Armament: one synchronized 7.7-mm Vickers machine gun
Approximately 30 built

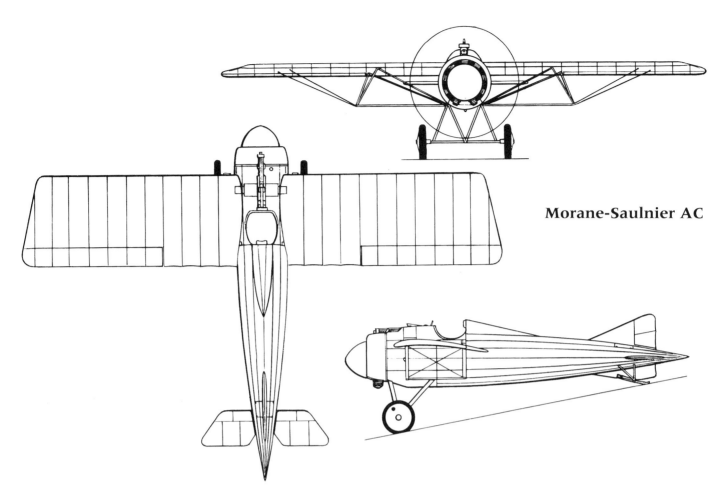

Morane-Saulnier AC

The RFC purchased two Type ACs. In January 1917 both were sent to the British Air Supplies Depot in Paris for evaluation. After testing, they were disassembled and returned to No.1 Reserve Squadron at Gosport at the end of January. They were never used operationally by the British.

Morane-Saulnier Type AC. Thirty were ordered and given the designation MoS. 23; this small batch was parceled out to numerous fighter escadrilles and no squadron was ever formed entirely on the type. B77.88.

Above: A Morane-Saulnier Type AC assigned to N 76. This airplane was named *Viking II* and was flown by Danish volunteer Leith Jenen. A Morane-Saulnier P is in the background. B83.5653.

Left: Morane-Saulnier Type AC at Rosnay assigned to escadrille 76 in March 1917. MA1858.

Below: Morane-Saulnier Type AC. The Type AC remained in service until at least the spring of 1917. MA38571.

Morane-Saulnier Type AC assigned to the Ecole d'Aviation at Pau. B87.5612.

Morane-Saulnier Type AE

The Morane-Saulnier AE was intended as a two-seat reconnaissance aircraft. Little is known about the airplane aside from the fact that it was powered by a 150-hp Le Rhône engine and had a wing area of 22 sq. m.

Morane-Saulnier Type AE two-seat reconnaissance aircraft. Courtesy J.M. Bruce.

Morane-Saulnier Type AF

The limited success of the Type AC did not prevent the Morane-Saulnier firm from continuing to develop fighter aircraft. Its next design was the Type AF, which except for its biplane configuration bore a remarkable resemblance to the Type AC. Although its fuselage was shorter, the Type AF retained the single 7.7 mm Vickers machine gun of the Type AC. The fin and rudder were also similar to those of earlier Morane-Saulnier designs, but the horizontal tailplane was changed to feature fixed stabilizers with conventional elevators. The new wings had negative stagger, the top wing being slightly longer than the bottom. There was a single bay of N-shaped interplane struts with a marked outward rake. There were four horn-balanced ailerons; the gaps between the ailerons and wing were covered by hinge flaps similar to those used on the Nieuport fighters. The engine was a 150-hp Gnome Monosoupape 9Nb.

The Type AF was given the STAé designation MoS.28 C1. Trials were initiated in June of 1917 and from 23 June to 12 July the aircraft accumulated 15 hours of flying time during 34 flights. Pilots had generally favorable comments on the aircraft's performance, and the Type AF's agility and pilot's field of view were singled out for praise. Despite these favorable comments, the Type AC was not selected for production, probably because the SPAD 7s and 13s were already in production.

Morane-Saulnier Type AF Single-Seat Fighter with 150-hp Gnome Monosoupape 9Nb		
Span 7.47 m; length 5.149 m; height 2.35 m; wing area 15.31 sq. m		
Empty weight 421 kg; loaded weight 649 kg		
Maximum speed:	1,000 m	207 km/h
	2,000 m	205 km/h
	3,000 m	200.5 km/h
	4,000 m	194.5 km/h
	5,000 m	185 km/h
climb:	1,000 m	2 minutes
	2000 m	4 minutes 50 seconds
	3,000 m	8 minutes 10 seconds
	4,000 m	12 minutes 40 seconds
	5,000 m	20 minutes 30 seconds
Armament: one synchronized 7.7-mm Vickers machine gun		
One built		

Morane-Saulnier AF

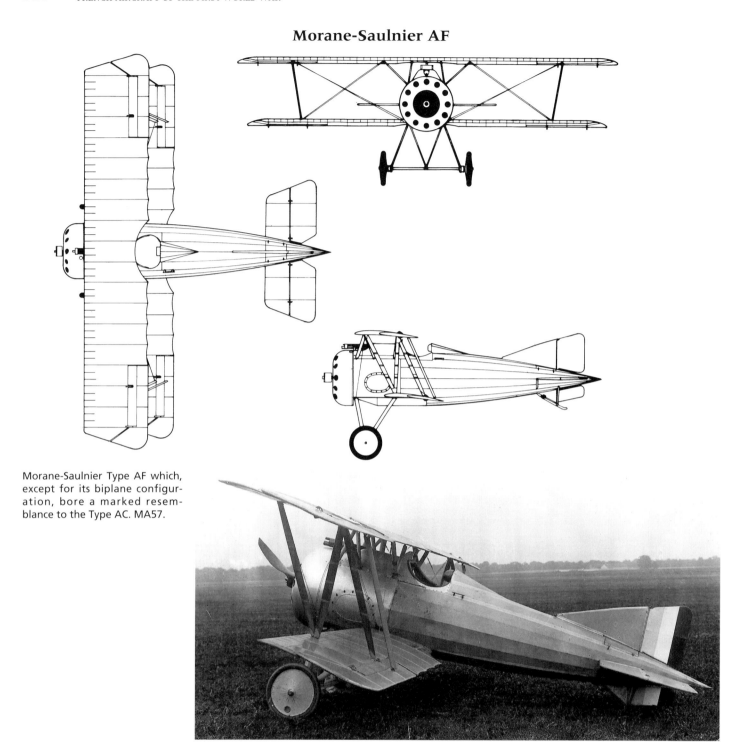

Morane-Saulnier Type AF which, except for its biplane configuration, bore a marked resemblance to the Type AC. MA57.

Morane-Saulnier Type AF. Test pilots had generally favorable comments on the aircraft's performance, and the Type AF's agility and pilot's field of view were singled out for praise. MS-53.

Morane-Saulnier Type AFH

The French did not build an aircraft carrier during the First World War, but they carried out a large number of tests attempting to develop aircraft to be launched from warship turrets. Types developed for this purpose included the Hanriot HD.2, HD.3, and the Morane Saulnier AFH.

The designation AFH indicates that the aircraft was a naval version of the unsuccessful Type AF fighter, the H standing for Hydravion. Design work began in the summer of 1917 and construction was initiated in October of that year. The main change from the basic Type AF was the placement of a thick float almost completely covering the aircraft's wheels. Enough of the tires were left exposed to permit the AFH to make a conventional takeoff from the flight deck or turret ramp of the ship. The tail skid was replaced by a teardrop-shaped float faired into the lower fuselage and fin by a streamlined pylon. The

Morane-Saulnier Type AFH Single-Seat Shipboard Fighter with 150-hp Gnome Monosoupape 9Nb
Span and area identical to the Type AF; length 6.013 m; height 2.840 m
Maximum speed: 196 km/h at sea level; climb to 5000 m in 22 minutes
Armament: one synchronized 7.7-mm Vickers machine gun
One built

lower rudder was enlarged. The floats permitted the aircraft to land in the water and be hoisted back on board the ship.

It is not known if the aircraft was ever flown from a French ship, and no additional AFHs were built. While the Aviation Maritime continued to experiment with launching aircraft from warship turrets, it apparently had been decided that the Hanriot designs were adequate for this purpose.

Morane-Saulnier AFH

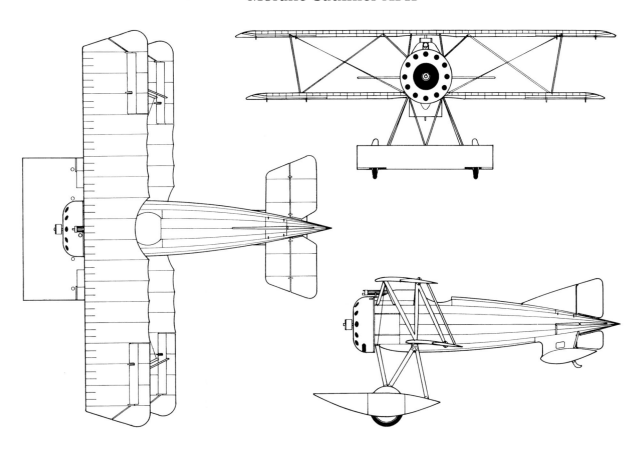

Morane-Saulnier Type AI

The Morane-Saulnier firm designed two fighter aircraft in 1917 that had fuselages based on the Type AC and were powered by the 150-hp Gnome Monosoupape 9Nb rotary engine. One, the Type AF, was a biplane; the Type AI had the monoplane layout preferred by Morane-Saulnier engineers. Although the Type AF did not secure any orders, the Type AI was to prove more successful.

The forward fuselage of the AI was constructed of metal and the rear fuselage was spruce. The Type AI's fuselage was longer than the Type AF and the ventral fin was much smaller. The surface area of the tailplane was slightly reduced and that of the elevator slightly increased. The one-piece wing was of wooden construction and had a slight sweepback. There were horn-balanced ailerons. Two struts on either side of the fuselage braced the wing and were themselves supported by auxiliary struts in mid-span. This arrangement resulted in a very sturdy structure and would be repeated on many of Morane-Saulnier's postwar designs. The landing gear chassis was made of two lateral V-shaped steel tubes and one central V-shaped tube. The tailskid was mounted on a tube which passed through the rudder. There was a small, fixed vertical fin and the horizontal tailplane had fixed stabilizers and mobile elevators. The moving portions of the tail surfaces were made of wood mounted on a metal axle. All control surfaces were actuated by cables.

The prototype was tested at Villacoublay in early August 1917 and the performance was quite good. The aircraft did even better when the Chauviere propeller was replaced by a Levasseur design. Armament was either a single Vickers machine gun with 500 rounds or two machine guns with a total of 800 rounds. The twin-gun Type AI was identical to the single-

Morane-Saulnier AI

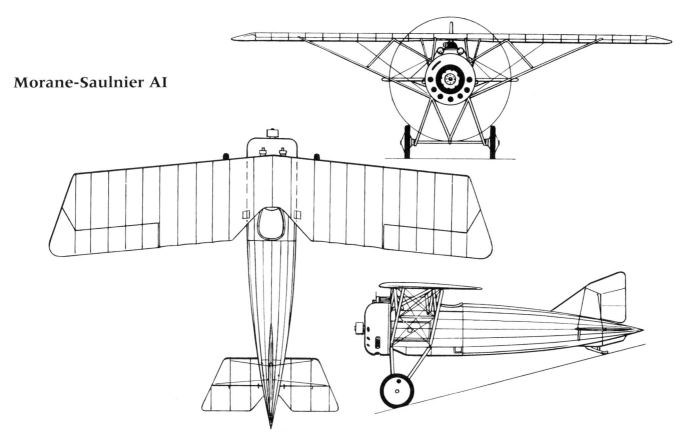

gun version except for having slightly enlarged tail surfaces.

Particularly favorable comments were made about the Type AI's maneuverability, stability, and view from the cockpit. The aircraft was in many ways superior to the SPAD 15 and Nieuport 28 that had the same engine. An order for between 1,100 to 1,300 Type AIs was placed by the Aviation Militaire, which gave the type the designation MoS.27 C1 (for the version with the single gun) and MoS.29 C1 (for the twin-gun version).

Estimated production was between 1,050 and 1,200 aircraft.

Unlike any of the other Morane Saulnier fighters, it was planned to equip entire escadrilles with the new aircraft. N 156 transitioned from Nieuport 24s and 27s and received 15 Type AIs to become MS 156 on 9 February 1918. It was assigned to the 4th Armée. N 158 also changed from Nieuport 23s to become MS 158 on 4 March 1918. Finally, MS 161 was formed from N 161 on 21 February 1918.

The Morane-Saulnier Type AI was powered by a 150-hp Gnome Monosoupape 9Nb rotary engine. Renaud.

MoS.27 Single-Seat Fighter with 150-hp Gnome Monosoupape 9Nb

Span 8.51 m; length 5.65 m; height 2.4 m; wing area 13.39 sq. m
Empty weight 421 kg; loaded weight 649 kg
Maximum speed:
- ground level 225 km/h
- 2,000 m 220 km/h
- 3,000 m 216 km/h
- 4,000 m 210 km/h
- 5,000 m 201 km/h

Climb:
- 1,000 m 2 minutes
- 2,000 m 4 minutes 25 seconds
- 3,000 m 7 minutes 25 seconds
- 4,000 m 11 minutes 15 seconds
- 5,000 m 15 minutes 50 seconds

Endurance 1.75 hours
Armament: one 7.7-mm Vickers machine gun

MoS.29 Single-Seat Fighter with 150-hp Gnome Monosoupape 9Nb

Dimensions identical to MoS.27
Empty weight 413.9 kg; loaded weight 673.9 kg
Maximum speed:
- 220.6 km/h at ground level
- 1,000 m 219 km/h
- 2,000 m 217 km/h
- 3,000 m 214.5 km/h
- 4,000 m 209.5 km/h
- 5,000 m 200 km/h

Climb:
- 1,000 m 2 minutes 20 seconds
- 2,000 m 5 minutes 5 seconds
- 3,000 m 8 minutes 40 seconds
- 4,000 m 12 minutes 45 seconds
- 5,000 m 19 minutes

Armament: two 7.7-mm Vickers machine guns

MoS.27 Single-Seat Fighter with 170-hp Le Rhône 9R

Dimensions identical to MoS.27
Empty weight 421 kg; loaded weight 655 kg
Maximum speed:
- 2,000 m 214 km/h
- 3,000 m 211 km/h
- 4,000 m 205 km/h
- 5,000 m 190 km/h

Climb:
- 2,000 m 4 minutes 30 seconds
- 3,000 m 7 minutes 45 seconds
- 4,000 m 11 minutes 40 seconds
- 5,000 m 16 minutes 30 seconds

Ceiling 7,000 m (approximate); endurance 2 hours 10 minutes
Armament: one 7.7-mm Vickers machine gun

MoS.30 Single-Seat Fighter Trainer with 120-hp Le Rhône 9Jb or 135-hp Le Rhône 9Jby

Dimensions identical to MoS.27
Empty weight 406 kg; loaded weight 526 kg
Maximum speed:
- 2,000 m 197 km/h
- 3,000 m 192 km/h

Climb:
- 2,000 m 5 minutes 12 seconds
- 3,000 m 8 minutes 35 seconds
- 4,000 m 11 minutes

Ceiling 6,000 m; endurance 45 minutes
No armament carried

MoS.30bis Single-Seat Fighter Trainer with 90-hp Le Rhône 9Jby

Dimensions identical to MoS.27
Empty weight 406 kg; loaded weight 526 kg
Maximum speed:
- 2,000 m 150 km/h
- 3,000 m 146 km/h
- 4,000 m 138 km/h

Climb:
- 2,000 m 8 minutes 5 seconds
- 3,000 m 14 minutes 44 seconds
- 4,000 m 26 minutes 27 seconds

Ceiling 4,875 m (approximate); endurance 60 minutes
No armament carried
Approximately 1,100 to 1,300 MoS.27s, MoS.29s, and MoS.30s were built

It seemed that, at last, the Morane-Saulnier firm would be able to sell large numbers of fighters to the Aviation Militaire. However, success would once again elude the company. MS 156 gave up its Type AIs and adopted SPAD 13s to become SPA 156 in March 1918. MS 158 (which had been assigned to the 3rd Armée and been active over the Somme front flying fighter patrols, bomber escort, and trench-strafing missions in the vicinity of Sacy-le-Grand) moved to the airfield at d'Auvillers to re-equip with SPAD 13s on 20 April 1918. MS 161 retained its Type AIs longer than any other escadrille. The unit was based at Boujacourt and later moved to Dugny, where it was assigned to the 5th Armée. By May it, too, had retired its Type AIs for SPAD 13s and was redesignated SPA 161.

The reason for the rapid withdrawal of the Type AIs from front-line service has never been adequately explained. One theory is that troubles with the Gnome engine were the cause of the Type AI's failure. Support for this comes from a GQG memo

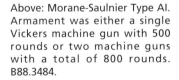

Above: Morane-Saulnier Type AI. Armament was either a single Vickers machine gun with 500 rounds or two machine guns with a total of 800 rounds. B88.3484.

Left: Morane-Saulnier Type AI. Two struts on either side of the fuselage braced the wing and were themselves supported by auxiliary struts in mid-span. Renaud.

Morane-Saulnier Type AI. MA9916.

Right: Morane-Saulnier Type AI of MS 156 which received 15 Type AIs on 9 February 1918. B78.1504.

dated 18 March 1918 stating that units using MoS.27s and MoS.29s were to reduce the number of long-distance flights at high altitude because of engine failures. A solution to the problem was under way by the STAé, but until then the shipment of 150-hp Gnome Monosoupape engines from the factory was suspended. Temple N. Joyce, an American pilot who had the opportunity to fly a Type AI at Issoudun found it to be quite agile and he stated he performed 300 consecutive loops. He also reports, however, that the Gnome Monosoupape engines were prone to catch fire if not handled properly.

There is also evidence that Type AIs might have been subject to structural failures. An RAF major noted in June 1918 that the aircraft had been withdrawn due to "frequent accidents." The fact that a version of the Type AI was produced with lift cables to give additional wing bracing suggests that the wing supports may have been prone to failure.

Morane-Saulnier may have hoped that replacing the Gnome engine with a 170-hp Le Rhône 9R would salvage the Type AI's career, and a single example was fitted with the newer engine. However, the Aviation Militaire did not supply any other front-line escadrilles with Type AIs and relegated them to fighter training units. The Type AIs were subsequently used to teach student pilots aerial combat maneuvering. Type AI trainers were given 120-hp Le Rhône 9Jb or 135-hp Le Rhône 9Jby engines and were designated MoS.30 E1. A dedicated trainer version with augmented wing bracing and a 90-hp Le Rhône 9Jby engine was designated MoS.30bis E1. This aircraft had its armament deleted and also had a reduced fuel capacity.

Toward the end of 1917 a single example of the Type AI was given a wooden monocoque fuselage of circular cross-section. This variant was armed with two 7.7-mm Vickers guns. There were also variants of the AI series fitted with a 150-hp Monosoupape engine or a 170-hp Le Rhône 9R. None was purchased by the Aviation Militaire.

Foreign Service

Belgium
Three MoS.30 E1s are reported to have been supplied to the Aviation Militaire Belge. They were used as trainers and at least one was assigned to the 9th Escadrille.

Czechoslovakia
A single MoS.30 E1 was obtained after the war.

Japan
A single MoS.30 was evaluated by Japan in 1922 for use as an advanced trainer. No further examples were purchased.

Poland
At least one MoS.30 was used at the French Pilot's school at Warsaw in 1920.

Switzerland
At least one MoS.30 was acquired.

United States
Fifty-one MoS.30 trainers were used by the A.E.F. at Issoudun.

Morane-Saulnier Types AN, ANL, ANR, and ANS

The Morane-Saulnier AN series was developed to meet the STAé C2 category of 1917. This specification called for a two-seat fighter to be armed with two synchronized machine guns and two machine guns on a flexible mounting. The latter guns were to be capable of being removed so that a camera could be fitted for high-speed reconnaissance missions. Specifications called for a maximum speed of 220 km/h at 5,000 m, a service ceiling of 8000 m, a minimum speed of 110 km/h, and a payload of 375 kg. Other aircraft designed to meet this specification included the Farman F.30 and 31, HD.5 and 6, Borel-Boccacio Type 3000, BAJ C2, Breguet 17, and SEA 4.

Assembly of the Type AN began in January of 1918 and was completed by summer. The monocoque fuselage was fitted with the bulky 450-hp Bugatti engine. Initially, engine cooling was accomplished by radiators placed in the leading edges of the upper wing; later these were replaced by two Lamblin radiators located under the nose. The first version of the AN had a pointed spinner that was later deleted. The two-bay wings were of equal span and had a backward sweep. The lower wing was faired into the bottom of the fuselage. Both the upper and lower wings had horn-balanced ailerons. Armament consisted of a synchronized 7.7-mm Vickers machine gun and two 7.7-mm Lewis guns on a T0.3 mounting in the observer's cockpit.

The Type AN was tested at Villacoublay and received the STAé designation MoS.31. Evaluation began on 27 October 1918, but the aircraft's performance was to prove inferior to the SEA 4. Despite the Type AN's disappointing performance, particularly its slow rate of climb, it was ordered into production. However, there seems to have been concern that the Bugatti engine might not prove to be reliable and three additional versions of the Type AN were built, each using a different engine. These engines had already been used in other aircraft designed to meet the C2 specification.

The Type ANL (given the STAé designation MoS.32) was powered by a 400-hp Liberty 12 engine. The aircraft had the same basic layout of the Type AN, but the Liberty engine resulted in an even bulkier profile than did the Bugatti. There were minor changes to the wing and tailfin. The two Lamblin nose radiators were retained and there was a long exhaust pipe on either side of the fuselage. Minor changes were made to the upper wing and fin. The armament remained unchanged. The aircraft was tested at Villacoublay in 1919. A version of the ANL that had a thicker wing profile was also tested in 1919 but showed no performance advantage.

The ANR had a 450-hp Renault 12F engine attached to the basic ANL airframe. The aircraft, which was designated MoS.33, had a slightly longer nose and only a single exhaust manifold.

The ANS was the final variant of the AN series. It was

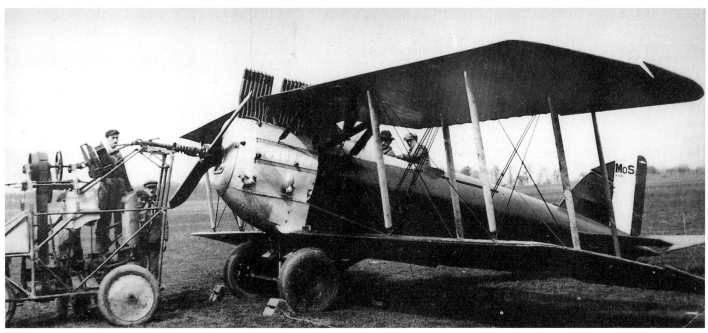

Above: The Morane-Saulnier Type AN was fitted with the bulky 450-hp Bugatti engine. B82.2964.

The Morane-Saulnier ANB which was modified from the prototype; it had two Lamblin radiators beneath the nose. (CC).

The Morane-Saulnier Type ANL was powered by a 400-hp Liberty 12 engine. MS-72.

Morane-Saulnier ANL

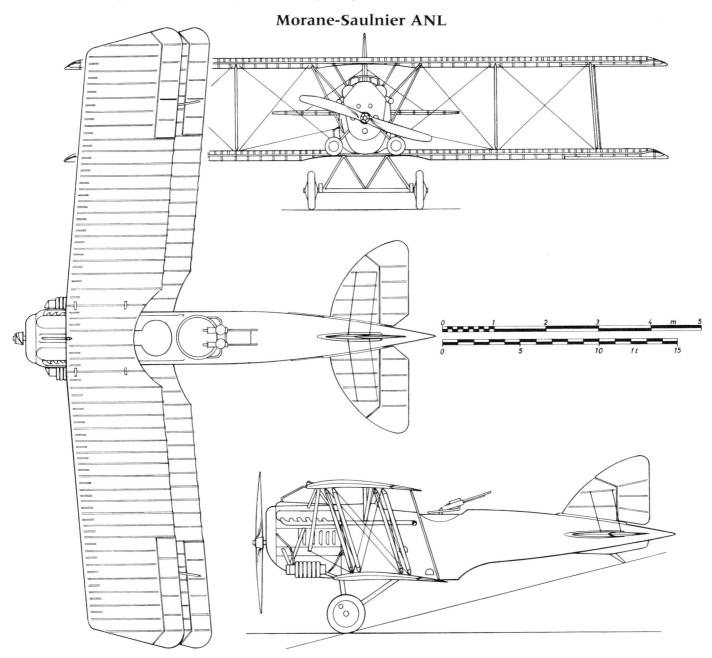

powered by a 530-hp Salmson 18Z and received the STAé designation MoS.34. The ANS was almost identical to the preceding ANL and ANR except for the more streamlined nose profile.

The Types AN, ANL, ANR, and ANS did not enter service, probably because the end of the war reduced the Aviation Militaire's need for new aircraft. Furthermore, the Breguet 17s and SEA 4s were already in production.

Morane-Saulnier ANR

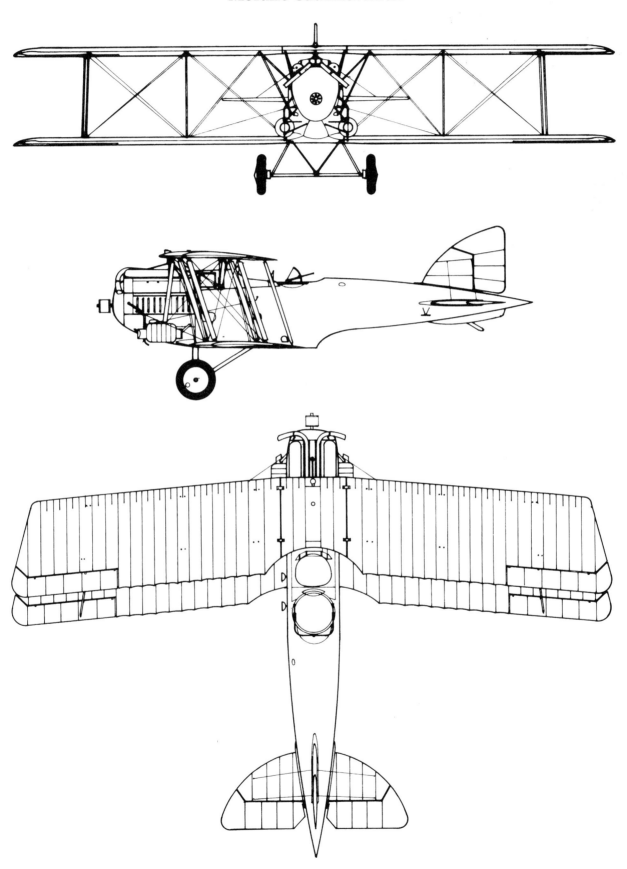

Morane-Saulnier ANS

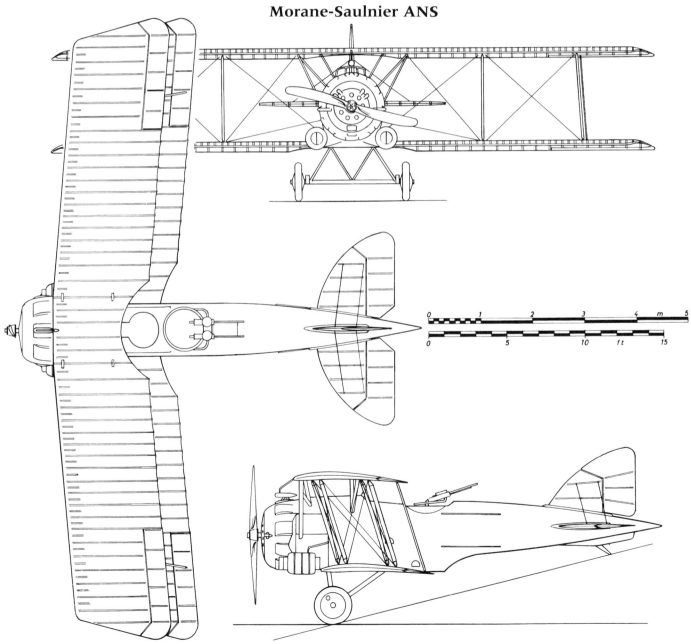

Morane-Saulnier Type AN Two-Seat Fighter with 450-hp Bugatti
Span 11.726 m; length 8.345 m; height 2.77 m; wing area 41.0 sq. m
Loaded weight 1,770 kg
Maximum speed: 225 km/h; climb to 1,000 m in 3 minutes 1 second; climb to 2,000 m in 6 minutes 40 seconds; climb to 3,000 m in 12 minutes 13 seconds; climb to 4,000 m in 23 minutes 9 seconds; ceiling 4,750 m
Armament: one synchronized 7.7-mm Vickers machine gun and two 7.7-mm Lewis guns on a TO.3 mount
One built

Morane-Saulnier Type ANL Two-Seat Fighter with 400-hp Liberty 12
Empty weight 1,190kg; loaded weight 1,766 kg
Maximum speed: 217 km/h; range 650 km; ceiling 6,000 m
Armament: one synchronized 7.7-mm Vickers machine gun and two 7.7-mm Lewis guns on a TO.3 mount
One built

Morane-Saulnier Type ANR Two-Seat Fighter with 450-hp Renault 12F
Maximum speed: 149 km/h at 4,000 m; climb to 4,000 m in 19 minutes 50 seconds; ceiling 7,500 m; endurance 2.5 hours
Armament: one synchronized 7.7-mm Vickers machine gun and two 7.7-mm Lewis guns on a TO.3 mount
One built

Morane-Saulnier Type ANS Two-Seat Fighter with 530-hp Salmson 18Z
Dimensions identical to AN except length was 8.27 m
Armament: one synchronized 7.7-mm Vickers machine gun and two 7.7-mm Lewis guns on a TO.3 mount
One built

Facing page: The Morane-Saulnier Type ANS was powered by a 530-hp Salmson 18Z. MS-81.

Below: The Morane-Saulnier Type ANR had a 450-hp Renault 12F engine MS-76.

Morane-Saulnier Type BB

The Type BB represented a considerable departure for the firm. Previously, the company's line of parasol reconnaissance aircraft had enjoyed considerable success. The new Type BB, however, was a two-seat biplane with a conventional layout built to meet the specification for a two-seat reconnaissance plane. The Type BB was first flown in 1915, powered by a 80-hp Le Rhône rotary engine. The prototype used wing warping for lateral control. The single-bay wings were of unequal span and the fuselage was fully faired with a circular cross-section. There was a balanced elevator and a triangular fin and rudder.

The aircraft was evaluated by the Aviation Militaire and received the STAé designation MoS.7. However, no examples were purchased by the French and the RFC received almost all the aircraft that were produced. Production aircraft dispensed with wing warping and were fitted with ailerons.

Foreign Service
Russia
A small number of Type BBs were supplied to the Imperial Russian Air Service. A single example was built at the Dux plant in 1917. It is described as being armed with two "synchronized" Vickers machine guns.

Spain
In 1915 the firm of CECA (Compania Espanola Construcciones Aeronauticas) acquired the rights to build copies of the Morane-Saulnier Type BB under license. The aircraft they produced were similar to those used by the British, except that a 150-hp Hispano-Suiza 8A engine was used and the radiators were relocated to either side of the fuselage. The first flight of the new aircraft took place at Albericia in January 1916. A total of 12 were produced for the Aeronautica Militaire. By mid-1917 they had been sent to the military airfield at Cuatro Vientos. There apparently were a number of problems with these aircraft and in mid-1918 they were placed in storage at Cuatro Vientos. An attempt to produce an improved version of the CECA Type BB was undertaken by a Luis O'Page. His aircraft, which was completed in 1918, was not purchased by the military.

United Kingdom
One Type BB, serial 3683, was supplied to No.4 Squadron RNAS. The RFC ordered 92 but specified that they were to have 110-hp Le Rhône 9J engines. However, the French refused to supply enough engines so the RFC reduced its order to slightly more than 80. The 110-hp Le Rhônes were often removed from the Type BBs and used on the Type Ps in service with the RFC. Those airframes without engines were then sent to England via No.1 Aircraft Depot.

Those Type BBs that did see service were modified to accommodate a wireless set and camera in the observer's position, although on some these were placed in a fairing on the starboard fuselage. Problems with the ailerons were solved by placing fairing strips over the aileron hinges. The Type BBs were usually armed with two 0.303-inch Lewis machine guns. One was fixed to the top wing to fire upward and was controlled by the pilot; the second gun was on a flexible mount in the observer's cockpit.

The aircraft served with Nos.1 and 3 Squadrons, which were already equipped with Type Ps. It was planned to provide each squadron with a flight of four Type BBs; the aircraft were used in the army cooperation and reconnaissance roles. In addition to Nos.1 and 3 Squadrons, the Type BB was used in limited numbers by Nos.12 and 60 Squadrons, as well as the pilot's school at No.1 Aircraft Depot at St. Omer.

A single Type BB was delivered to the RNAS in October 1915. It was sent to 3 Squadron, 1 Wing and was later sent to Coudekerque as part of 5 Wing.

Morane-Saulnier BB Two-Seat Reconnaissance Plane with 80-hp Le Rhône 9J

Span 8.585 m; length 6.935 m; height 2.615 m; wing area 22.32 sq. m

Empty weight 491 kg; loaded weight 761 kg

Maximum speed: 134 km/h at 3050 m; climb to 1,980 m in 13 minutes, climb to 3,050 m in 26 minutes 48 seconds; ceiling 3,660 m

Armament: two 7.7-mm Lewis machine guns

Approximately 80 built for the RFC

Morane-Saulnier BB Two-Seat Reconnaissance Plane with 80-hp Le Rhône 9J Built at the Dux Factory

Length 7.1 m; height 2.615 m; wing area 23 sq. m

Empty weight 440 kg; loaded weight 625 kg

Maximum speed: 160 km/h at 3,050 m; climb to 1,000 m in 4 minutes 30 seconds, climb to 2,000 m in 9 minutes 30 seconds; climb to 3,000 m in 16 minutes; ceiling 5,700 m; endurance 2 hours

Armament: two "synchronized" Vickers machine guns

One built by Dux

Morane-Saulnier Type BB serial number 581. No examples were purchased by the French; the RFC received almost all the aircraft produced. MS 21.

Morane-Saulnier Type BB (MoS 8). The prototype used wing warping for lateral control. However, production aircraft dispensed with the wing warping mechanism and were fitted with ailerons. B89.1775.

Morane-Saulnier Type BB serial number 3683 of RNAS Capt. Allardyce. Via Ken Molson and Colin Owers.

MORANE-SAULNIER 347

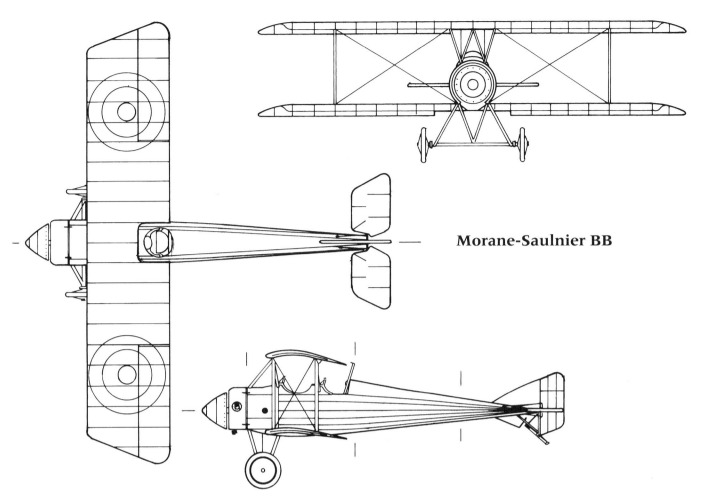

Morane-Saulnier BB

Morane-Saulnier Type BB. The single bay wings were of unequal span. Renaud.

Above: Morane-Saulnier Type BB with over-wing gun. Renaud.

Morane-Saulnier Type BB. The fuselage was fully faired with a circular cross section. Renaud.

Morane-Saulnier Type BH

The Type BH was a modified Type BB built in 1915. It was the first Morane-Saulnier aircraft to use a Hispano-Suiza engine. The 140-hp engine was tightly enclosed by the cowling and cooled by a water jacket around the crankcase. The cylinders, however, were air cooled. Apparently the Type BH was used only as a test bed for the Hispano-Suiza engine and was never placed into production. It was given the STAé designation MoS.8.

Morane-Saulnier Fighter Projects

1. A biplane powered by a 200-hp Gnome Monosoupape engine and of all monocoque construction. Further development was abandoned by 1 May 1918 because the engine proved to be unsatisfactory.
2. A biplane powered by a 200-hp Gnome Monosoupape engine. The fuselage was of all monocoque construction and had fabric-covered wings. The failure of the Gnome engine also resulted in this project being abandoned on 1 May 1918.
3. A biplane powered by a 300-hp Gnome Monosoupape engine and also of all monocoque construction. The engine was a twin-row development of the Gnome 9N but it, too, was a failure. The fighter, which had been built, was abandoned on 1 May 1918.
4. A biplane powered by a 300-hp Gnome Monosoupape engine, with fabric-covered wings and a monocoque fuselage. It was abandoned when the engine proved unsuccessful.
5. An all-monocoque single-seat fighter intended to use the 300-hp Hispano Suiza engine. Development was abandoned by 1 May 1918.
6. A single- or two-seat fighter with a 200-hp Clerget 11E engine. It may have been based on the Gnome Monosoupape designs mentioned above and, like them, was abandoned on 1 May 1918.
7. A two-seat fighter with a 220-hp Le Rhône. This was probably also based on one of the preceding Gnome Monosoupape designs. Development was halted after 1 May 1918.

Société Anonyme des Établissements Nieuport

In 1906 Édouard de Nieport opened a small workshop at Darracq. He initially produced parts for motors, and in 1905 he designed magnetos and starter motors for use in aircraft. In 1909 he decided to construct an aircraft engine and, to test it, built an airplane of his own design. The Société Anonyme des Établissements Nieuport[1] was opened the same year. The new aircraft was a monoplane exemplifying Nieport's belief in simplicity, strength, and streamlining for speed. The fuselage was formed as a half-shell and the pilot's head and shoulders barely protruded above the fuselage. The front of the cockpit was raised slightly to act as a windbreak. The monoplane wings were attached directly to the mid-fuselage. The square rudder, tail fin, and elevators were attached to the fuselage by a series of struts. Power was supplied by a 20-hp Darracq engine, which was to have been replaced by a new engine designed by Nieport and built by Darracq. However, the airframe was destroyed in a flood before Nieport had completed his engine.

Nieport's second design was fitted with his two-cylinder 28-hp engine. This new aircraft was also a sleek monoplane and was tested at the laboratory of Gustav Eiffel. This second design was designated the Nieuport 2N and it flew on 5 January 1910. It was a single-seat monoplane with a conventional fuselage. It was constructed of four ash longerons braced with wooden beams attached by metal fasteners. The exterior of the fuselage was covered in fabric. The pilot was again seated low in the fuselage to minimize wind resistance. The tail surfaces had a semicircular shape and were attached directly to the fuselage. The undercarriage had a long skid extending between the wheels to act as a brake. It was spring loaded, and its narrow track often caused the Nieuport 2N to bounce along the airfield. The wings had a trapezoidal shape with rounded edges and were supported by two ash longerons. They were attached to the fuselage by steel tubes and braced with piano wire. There were multiple versions of the Nieuport 2N. On 10 June 1910 the Nieuport 2N, now fitted with a 30-hp Anzani engine, placed third at the 1911 speed trials at Reims. Édouard Nieport was killed in a plane crash that same year.

The next design, produced under the direction of Charles Nieport, who had assumed control of the company after his brother's death, was the Nieuport 3A, a two-seat version of the 2A. Nieuport's most successful aircraft was the Type 4G. Powered by a 50 hp Gnome rotary engine, a Type 4N placed first in the Gordon Bennett Trophy Race in July 1911; Charles Nieport, flying a similar machine, placed third. The Type 4 was used by ten countries and was built under license in Russia.

Charles Nieport died in a plane crash in January 1913. The company continued to prosper, however, with Gustave Delage serving as chief engineer. The Nieuport 6 seaplane and the Nieuport 10 reconnaissance plane were both produced in 1914 and saw active service during the first year of the war.

Prewar Nieuport Aircraft

1. Nieuport 1—20-hp Darracq engine.
2. Nieuport 2N—28 hp Nieuport designed engine.
3. Nieuport 2A—single-seater with a 40-hp Anzani engine.
4. Nieuport 2B—two-seater with a 20-hp Darracq engine.
5. Nieuport 2G—single-seaters with various Gnome rotary engines.
6. Nieuport 3A—same as Nieuport 2A but with two seats.
7. Nieuport 4G—single-seater with 50-hp Gnome engine.
8. Nieuport 5—unknown.
9. Nieuport 6—military version of type 4; land and floatplane variants.
10. Nieuport 7—unknown.
11. Nieuport 8—unknown.
12. Nieuport 9—unknown. (In Russia the single-seat version of the Nieuport 10 was designated Nieuport 'Niner.')

---

(1) It is not known why the firm founded by Édouard de Nieport was spelled *Nieuport*.

Right: An armored Nieuport at Villacoublay. B84.1698.

Below: Nieuport 4M at the concours militaire 1911 with two Deperdussin TTs in the background. B84.1692.

Nieuport 4M and 6M

The Nieuport 6M was the first Nieuport design to see combat; it was, in fact, one of the first aircraft in history to fly wartime reconnaissance missions. Nieport designed the Nieuport 4M for the military concours of 1911. The aircraft was derived from the Nieuport 2 series but had a 100-hp Gnome engine and a larger wing span. These changes permitted a crew of three to be carried. A version of the type 4 was produced with floats and a crew of one or two. The 1911 concours specified the following criteria:

1. The entire aircraft, including the engine, had to be made in France.
2. The aircraft was to carry a crew of three and a payload of 300 kg over a 300-km distance at a speed of 60 km/h.
3. It had to be capable of taking off from and landing on a variety of airfields.
4. It had to be easily disassembled and reassembled to permit transport by road.
5. Dual controls for pilot and observer were strongly recommended.

The aircraft were required to make three flights from a series of airfields chosen by the organizers of the concours. They were to climb to 500 meters in less than 15 minutes.

Despite these challenging goals, the Nieuport 4M was able to finish first in the competition with a top speed of 116 km/h, well ahead of its nearest competitor. The Nieuport was also far more expensive than its nearest competitor (a Breguet design). The Nieuport cost a steep 780,000 F, more than twice as much as Breguet's aircraft. This amount included the cost of the

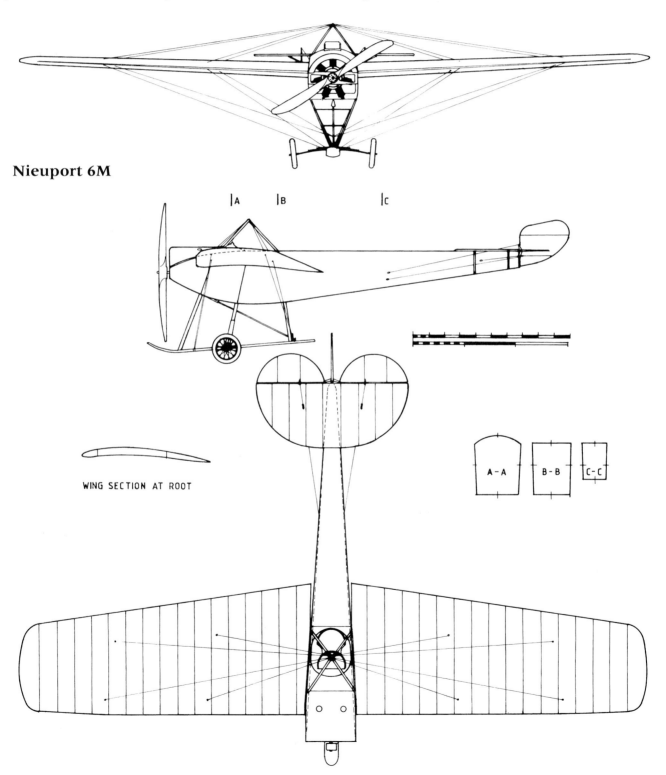

Nieuport 6M

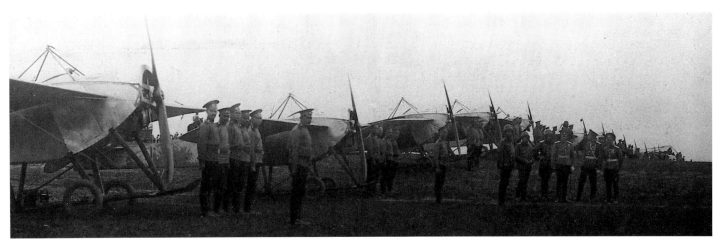

A lineup of Nieuport 6Ms of the IRAS. B89.1741.

prototype (100,000 F), ten examples of the Nieuport 4M (40,000 F each), and a bonus of 280,000 F for having a top speed that was 6 km/h above the 60 km/h specified for the concours. The military agreed to these prices and ten Nieuport 6Ms (the designation for military Nieuport 4Ms) were ordered.

Nieuport 6Ms used as trainers were sometimes fitted with a skid attached to the landing gear to prevent nosing over. Wing warping was controlled by foot pedals, and turning was accomplished by a handle mounted on the side of the cockpit which controlled the elevators and rudder.

The aircraft purchased were formed into a single escadrille, designated N 12. This unit was formed in 1912 at Reims. Initially, the unit's Nieuports had 50-hp Gnome engines, but these were gradually replaced by 80-hp Gnomes. Under the command of Capitaine Aubry, the unit moved to Stenay on 8 August 1914. N 12, with four other escadrilles, was placed under the command of the 5th Armée. For the first few days of the war the aircraft of N 12 carried out routine reconnaissance missions with the observer taking notes on what was seen (no photographic equipment was carried). The aircraft did not usually fly below 1,000 meters because they were vulnerable to ground fire (which was often effective up to 1,500 meters). Reconnaissance missions were usually confined to the enemy's front lines, although on occasion the aircraft of N 12 flew as far as 50 km behind them. N 12 participated in the Battle of Guise on 29 August and the Battle of the Aisne on 15 September, flying missions over Craonne and Brimont. By 28 February 1915 the unit was based at Châlons-sur-Vesles. In combats with German aircraft victories were claimed by pilots Pelletier-Doisy, Navarre, Chambe, and Robere. The GQG had by now concluded that the Nieuport 6Ms were as unsatisfactory as the other monoplanes then in service. It was felt that the shoulder-mounted wings of most monoplanes severely restricted the observer's view of the ground. Furthermore, the climb rate and payload of the monoplanes then in service were unsatisfactory. N 12 re-equipped with Morane Saulnier L parasol monoplanes and N 12 was re-designated MS 12. The Nieuport 6Ms were subsequently sent to training units.

Foreign Service

Argentina
Argentina purchased a Nieuport 2G and a Nieuport 4G (dubbed La Argentina) in 1912. Both aircraft served with the Escuela de Aviacion Militar.

Greece
A Nieuport 4G owned by E. Argyropoulos and christened Alkyon was the first airplane flown in Greece. It first flew on 2 August 1912 and was subsequently assigned to the 1st Flying Squadron at Larissa during the First Balkan War. It is believed to have been destroyed in 1912.

Italy
Italy had purchased several Nieuports based on the Type 4; these aircraft were designated Type 6Ms. Italy had acquired its first examples in 1911 for use with the Expeditionary Corps in Libya. They were assigned to the 1st Flottiglia Aeroplani (Tripoli) and flew some of the first reconnaissance and bombing missions in history. The Italians were impressed enough with the type to arrange for it to be produced under license by the Societa Nieuport-Macchi at Varese. At the beginning of the First World War the Nieuport 6Ms were assigned to the 5th, 6th, 7th, and 8th Squadriglias. However, by mid-1915 they had been replaced.

Japan
Japan acquired a single Nieuport (probably a Type 6M) when a Japanese army officer who had been trained to fly in France purchased a machine in 1913. It was a two-seater with a 100-hp Gnome engine. In 1914 it was sent (along with four M.F.7s) to Tsingtao as part of the Provisional Air Corps Unit and flew reconnaissance missions.

Russia
Russia acquired approximately 50 Nieuport 2s and 3s. They were used as trainers, never operationally. The Nieuport 4 was also built under license in Russia and approximately 300 were constructed. These served in the reconnaissance role as late as 1916 and were used by training units until 1920.

There were five basic versions built in Russia:

1. Nieuport 4G—50-hp Gnome or R.E.P. engines, 60-hp Gnome or Kavalkin engines.
2. Nieuport 4 Espane—80 built by the Shchetinin plant in 1915; powered by a 50-hp Gnome engine. The wings had an S-shaped cross-section and two fewer ribs in each wing half. One aircraft had an 80-hp Gnome.
3. Nieuport 4 "Dux"—Small numbers built in 1912 by the Dux plant; 70-hp Gnome. It was intended to compete in the 1912 military airplane competition.
4. Nieuport 4/6M—Also built at the Shchetinin and Dux plants, this airplane was used by operational units from 1912 through 1915 and served as a trainer from 1915 to 1925. P.N. Nesterov was flying one of these aircraft when he became the first pilot to perform a loop. Engines were either a 70-hp Gnome or a 100-hp Gnome Monosoupape.
5. Nieuport 4 "1914"—Featured a longer wing span and a more powerful 80-hp Gnome or Clerget engine. Built by the Shchetinin plant, this airplane's performance was equivalent to that of the standard Nieuport 4G.

Finally, there were several other Nieuports which were heavily modified by their pilots. These men were N.P. Nesterov

Nieuport 6M Two-Seat Reconnaissance Aircraft with 80-hp Gnome

Span 11.00 m; length 8.4 m; height 2.6 m; wing area 21 sq. m
Armament: none fitted
Approximately ten built for the Aviation Militaire

Macchi-Nieuport 6M Two-Seat Reconnaissance Aircraft with 80-hp Gnome

Span 12.27 m; length 7.70 m; height 2.62 m; wing area 26 sq. m
Empty weight 390 kg; loaded weight 650 kg
Maximum speed: 110 km/h; climb to 1,000 m in 12 minutes; climb to 2,000 m in 30 minutes; range 330 km
Armament: none carried
Approximately 56 built

Japanese Nieuport G Two-Seat Reconnaissance Aircraft with 100-hp Gnome

Span 10.9 m; length 7.8 m; height 2.45 m; wing area 21 sq. m
Empty weight 350 kg; loaded weight 650 kg
Maximum speed: 110 km/h; endurance 4 hours
Armament: none carried
One acquired by Japan

Nieuport 4 Espane Two-Seat Reconnaissance Aircraft with a 50-hp Gnome Built in Russia by Shchetinin

Span 10.7 m; length 7.5 m; height 2.45 m; wing area 22 sq. m
Empty weight 320 kg; loaded weight 520 kg
Maximum speed: 105 km/h; climb to 1,000 m in 16 minutes; ceiling 2,000 m; endurance 4 hours
Armament: none carried
80 built

Nieuport 4 "Dux" Two-Seat Reconnaissance Aircraft with a 70-hp Gnome Built in Russia by Dux

Span 12.3 m; length 8.0 m; height 2.45 m; wing area 22.5 sq. m
Empty weight 422 kg; loaded weight 660 kg
Maximum speed: 104 km/h; climb to 1,000 m in 15 minutes; ceiling 2,000 m; endurance 3 hours
Armament: none carried

Nieuport 4/6M Two-Seat Reconnaissance Aircraft with a 70-hp Gnome Monosoupape Built in Russia by Shchetinin and Dux

Span 12.0 m; length 7.8 m; height 2.45 m; wing area 23.5 sq. m
Empty weight 420 kg; loaded weight 660 kg
Maximum speed: 110 km/h; climb to 1,000 m in 15 minutes; ceiling 2,000 m; endurance 3 hours
Armament: none carried

Nieuport 4 "1914" Two-Seat Reconnaissance Aircraft with a 80-hp Gnome Monosoupape Built in Russia by Shchetinin

Span 12.3 m; length 7.8 m; height 2.45 m; wing area 23 sq. m
Empty weight 445 kg; loaded weight 720 kg
Maximum speed: 110 km/h; climb to 1,000 m in 12 minutes; climb to 2,000 m in 40 minutes; ceiling 2,300 m; endurance 2 hours 10 minutes
Armament: none carried

(1913), V.V. Dybovsky (1913), V.A. Slesarev (1914), V.V. Slyusarenko (1916), and A.A. Krylov/D.D. Feorov (1920). All types used 70-hp Gnome engines. None of these airplanes was, however, selected for series production.

Siam

Siam purchased four Nieuport 4Gs in 1914. They were based at Don Muang airfield and served as trainers.

Spain

Spain had one Nieuport 2N and two Nieuport 2Gs in service with the flying school at Cuatro Vientos in 1912; they were in operation until 1914. In April 1913 five Nieuport 4Gs were purchased and assigned to the Escuela Nieuport de Peu, given serial numbers 4 through 8. The Nieuport 4Gs were used for training flights between Cuatro Vientos and Villalvenga. When the first operational Spanish escaudra was formed at Tetuan on 22 October 1913 it had three Nieuport 4Gs. The escadra moved to Zezulan in May 1914. The Nieuport 4Gs remained in service until 1917.

Sweden

Sweden purchased two Nieuport 4Gs, one in 1912 and the second in 1913. They were the first airplanes supplied to the Flygkompaniet and were assigned serials 1 and 3. Nieuport No.1 was struck off charge in April 1918 and No.3 crashed on 29 May 1916. One Nieuport (possibly a Type 3) was purchased by the Marine Flygvasende in 1913. Assigned serial N 1, it was struck off charge in 1916.

United Kingdom

England purchased five two-seat Nieuport monoplanes beginning in 1911. However, none was operational at the beginning of the First World War.

The RNAS had a single Nieuport 4G on strength in 1912. Assigned serial M3 (later 13), it was later sent to the CFS.

Nieuport 6H

The development of the Nieuport seaplane began in 1912 when Édouard Nieport asked Lieutenant de Vaisseau Gustave Delage to design a set of floats for his Type 4 monoplane. In addition to designing these floats, Delage modified parts of the airframe. The tail float weighed 10 kilograms and had an interior volume of 150 liters. It had an aerodynamic shape and was suspended beneath the tail of the monoplane by a series of struts. The main floats were crescent shaped and had a very thin profile; the tip of each float was made of metal. Each float had a volume of 1.3 cubic meters, a length of 3.2 m (1/3 the length of the fuselage), and a width of 1.10 m. Despite their unorthodox designs, the floats were remarkably effective and enabled the Nieuport to land safely even on rough seas. In fact, there were instances during the war when Nieuport floatplanes based at Port Said were able to land safely on sand. Other modifications included a raised rudder to keep it clear of the sea spray. An elongated fin was placed in front of the rudder and a similar fin underneath the fuselage between the struts supporting the tail float. The pilot and observer sat in tandem in separate cockpits. Some machines carried only a single crewman. Power was supplied by either an 80-hp or 100-hp engine, usually a Gnome or Clerget. There were probably several variants of the floatplanes, with wing spans from 9.50 m to 12.00 m and lengths from 6.50 m to 8.20 m.

Operational Service

In May 1912 the Aviation Maritime purchased a Voisin, Farman, Breguet, and a Nieuport 6H, as the type was now designated, for evaluation. During the maneuvers of 1912 the pilot of the Nieuport successfully spotted an approaching "enemy" fleet near the Lerins Islands. In addition to this important achievement, it was discovered that the Nieuport had an endurance of up to three hours, twice the average of the other machines. On 15 April 1912 France formed the Aviation Maritime and its machines were based at Saint Raphaël.

At the beginning of the First World War it was decided to

Nieuport 6H

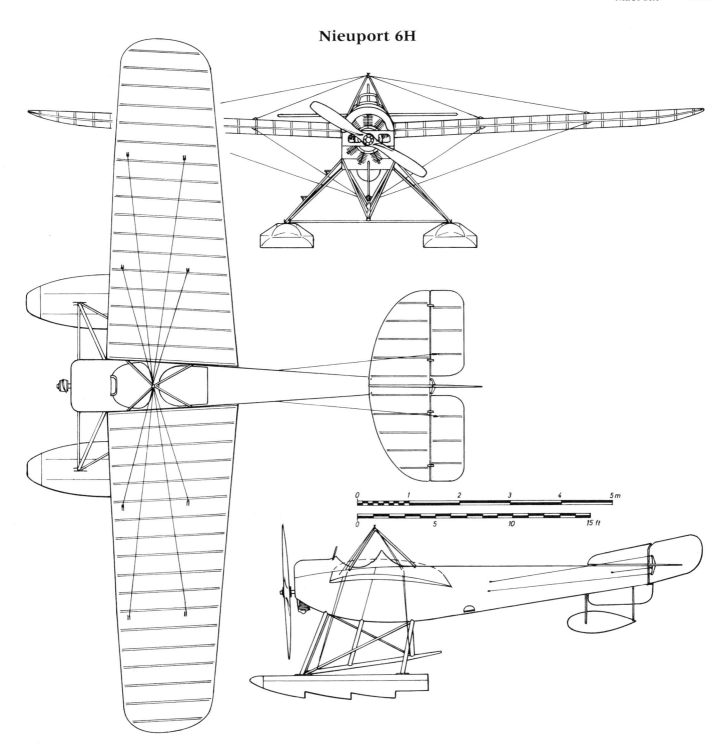

Nieuport 6H of the 1st Escadrille in flight. Via Alan Durkota.

Nieuport 6H. Via Alan Durkota.

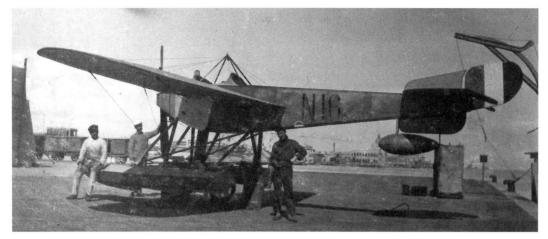

Nieuport 6H serial N16 of the Escadrille Nieuport of the Aviation Maritime at Port Said in April 1915. B 88.3285.

Nieuport 6H being hoisted onboard ship. B93.1995.

transfer six Nieuport 6Hs, which had been part of a Turkish order for 15 aircraft, to Bizerte. These would form the backbone of the Aviation Maritime for the first year of the war, as all the unit's other airplanes were heavily damaged in a storm at Saint Raphaël.

The Nieuport 6Hs were later transferred to Malta, the base used by the Armée Navale, the main French battle fleet, during its early-war operations in the Adriatic. A Nieuport section was sent to Antivarai, Montenegro, transported by the seaplane carrier *Foudre*, but an attack by Austro-Hungarian aircraft on that base forced the French to move the Nieuports to the Lake Scutari. The less dense water, of that fresh-water lake, however, made the Nieuports unable to take off with a full payload. Furthermore, the lake's altitude of 1800 m was too high to enable them to operate effectively. One pilot was forced to return to the *Foudre*.

Disgusted by the poor performance of the aircraft, Admirall [cq] Augusting Boué de Lapeyrère, commander of the Armée Navale, deemed them useless and ordered them and the *Foudre* to Port Said. There, although the escadrille was commanded by Lieutenant de Vaisseau de L'Escaille, it came under the control of Lieutenant General Sir John Maxwell, commander of the British Empire forces defending Egypt and the Suez Canal. At his behest the Nieuports, flown by their French pilots but with British army observers, flew reconnaissance missions along the Levantine coast as far as Smyrna, ranging up to 110 km inland, and were credited with obtaining information crucial to the repulse of the Turkish attack on the canal in early 1915.

The Nieuport operated primarily from the seaplane carriers *Aenne Rickmers/Anne* and *Rabenfels/Raven II*, which functioned originally under their German names and were not renamed until later in 1915 when they formally became Royal Navy vessels, but also from the British cruisers *Minerva* and *Doris* and the Royal Indian Marine auxiliary cruiser *Hardinge*.

Without an adequate source of spare parts, operating in a hostile environment, and subjected to enemy ground fire, the Nieuports had a high attrition rate. Encumbered by their floats, they were unable to climb out of range of enemy fire and consequently sustained considerable damage. To rectify this, they were withdrawn from shipboard use and their floats replaced by wheel undercarriages. Relieved of the weight and drag of the floats, they were able to reach 2,000 m, outside the effective range of small arms fire.

In April 1916 the Nieuports were reconverted into floatplanes, assigned to the French seaplane tender *Campanias* and sent to Agostoli for anti-submarine patrolling. The Nieuport unit was disbanded after 18 months of service. Despite their frail appearance, none of the Nieuport 6Hs experienced a structural or engine failure in flight.

The British were extremely pleased with the unit's performance and noted that "with only a complement of eight seaplanes all our demands were satisfactorily met." However, this remarkable performance was made at a high cost with over 50 percent casualties to the aircrew.

Foreign Service
Turkey

In 1914 the Turks received part of their earlier order for 15 Nieuport 6H floatplanes. One of these became popular because its name, in Latin characters on one side and in Arabic on the other; was *Mahmud Sevkat Pasa*, the name of the Turkish minister of war.

A 1914 Turkish inventory lists two Nieuport 6 floatplanes with 100-hp Gnome engines on strength. The Turks sent a Nieuport 6H to the Dardenelles, and another 6H was used to observe a British squadron near the straits on 25 September. The last reconnaissance mission before war was declared between the Allies and the Ottoman Empire was on 5 November by a Nieuport.

United Kingdom

The Royal Naval Air Service purchased 12 Nieuport 6Hs in 1915. Given serial numbers 3187–3198, they were used as trainers at Calshot, Bembridge, Walney Island, Westgate, and Lake Windermere.

> **Nieuport 6H Two-Seat Reconnaissance Floatplane with 80-hp or 100-hp Gnome or Clerget**
> Span 13.25 m; length 8.70 m; wing area 24.80 sq. m
> Empty weight 545 kg; loaded weight 795 kg
> Maximum speed: 110 km/h
> Armament: none
> Approximately eight built for the Aviation Maritime; 12 for the RNAS, two the Turkish Navy

Nieuport 10

The Aviation Militaire's decision to withdraw the Nieuport 6Ms from service resulted in the Nieuport firm being forced to produce Voisin 3s under license. Fortunately, Gustave Delage had joined the firm as chief engineer and it was he who would play a key part in the development of the Nieuport 10. Many of the design features of the Nieuport 10 would be repeated in virtually all the Nieuport aircraft built during the First World War.

First of all, Delage had to contend with the army's decision that biplanes were ideal in the reconnaissance role as they offered a better view for the crew, had a superior climb rate, and were sturdier than monoplanes. The Nieuport firm, on the other hand, had advocated lightweight monoplanes capable of high speed. Delage's solution was to offer a compromise between a monoplane and a biplane. For the Nieuport 10 he selected a sesquiplane layout; literally, an airplane with one and a half wings. The standard fuselage of the prewar Nieuports was retained and a conventional upper wing was fitted, but the lower wing had a much narrower chord. The lower wing was braced to the upper wing by a V-shaped strut on either side. The "V" layout meant that the lower wings were attached to the solidly braced upper wing at only a single point. This point was actually a split ring that fit around the central wing spar and permitted the lower wing's incidence to be adjusted. While it could not be changed in flight, the incidence of the lower wing could be adjusted on the ground to compensate for the weight of the airplane or its type of engine. Thus the entire lower wing served the same function as the elevator. While this feature does not appear to have seen widespread use in service and was abandoned in later Nieuport designs, the V-shaped struts were retained and became a hallmark of the Nieuport firm. The incidence for the lower wing of a Nieuport 10 with a crew of two was between 2 degrees 30 minutes and 3 degrees, whereas a Nieuport 10 equipped with a 80-hp Le Rhône and carrying only a single crewman would have its lower wing incidence at a setting of 2 degrees 15 minutes The lower wing had dihedral and was slightly swept. Another advantage of this layout was that the reduced chord of the lower wing resulted in a significant improvement in downward view for both the pilot and observer. The sesquiplane combined the maneuverability of a monoplane with the stability of a biplane. Lateral control was via ailerons on the upper wing. The ailerons' front edge was covered with wood and metal to reduce flutter. The fuselage was constructed of pine ash and ply, with metal used at stress points. Tubes were used as the basic structure for the tail section. All parts were covered with linen and doped.

The Nieuport 10 was initially intended to carry a crew of two, as its primary missions were to be reconnaissance and army cooperation. On the Type AV the pilot was seated in front, while in the Type AR the observer was seated in front. Some Nieuport 10 Types AR were equipped with a machine gun fixed to the upper wing and angled to fire outside of the propeller arc. The observer had to stand up, fitting his body through a cut-out in the upper wing, in order to fire the gun. On the other hand, many of the Type ARs were equipped with an Éteve mount and Hotchkiss machine gun, which provided the rear observer a more practical and safer way of firing. Initially the airplanes had 80-hp Le Rhône rotary engines, but these were to prove to be inadequate for a two-seat reconnaissance machine.

A Nieuport 10 was evaluated in the fighter role by MS 26 in June 1915. The aircraft was flown as a two-seater with the gunner sitting ahead of the pilot. In this configuration the Nieuport 10 was described as having several deficiencies. The stabilizer was described as "weak," the tail skid was too heavy, the undercarriage track too narrow, and there was excessive engine vibration. Recommendations included moving the pilot to the front seat and putting the gunner behind him.

The Nieuport 10 was presented to the SFA in early 1915. It was accepted for use as a reconnaissance machine and began to arrive in front-line escadrilles in late April. It was initially supplied to units using the Morane Saulnier Type L/LA parasols that had replaced the Nieuport 6Ms earlier that year. By August there were 81 Nieuport 10s at the front: 42 with front-line escadrilles and the remaining 39 in reserve. There were enough

Nieuport 10 serial N 372. The angle of attack of the lower wing was intended to be adjusted on the ground depending on the weight of the airplane. B78.1231.

Nieuport 10s in service to warrant the redesignation of the following escadrilles as Nieuport units:

N 15, created from MS 15 in the fall of 1915. N 15 was assigned to the 10th Armée. It was based at Verdun from March to April 1916.

N 23, formed from MS 23 in the fall of 1915. Initially assigned to the 2nd Armée, N 23 later moved to the Verdun sector. It was commanded by Lieutenant Louis Robert de Beauchamp.

N 26, formed from MS 26 in the fall of 1915. Initially under the command of Capitaine Thobiet, N 26 was subsequently led by Capitaine Jacques de Sieyes de Veynes, who assumed command in May 1916. N 26 was based on the Belgian front.

N 31, formed from MS 31 in the fall of 1915 and assigned to the 1st Armée. In early 1916 N 31 moved to the Verdun front under command of Lieutenant Lucien Couret de Villeneuve. In October 1916 it was assigned to GC 11.

N 37, formed from MS 37 in July 1915. It was assigned to the 3rd Armée and moved to Lisle-en-Barrois, where it participated in the offensive at Champagne in September 1915. N 37 was based at Sainte-Menehould from late 1915 to early 1916. The unit participated in the Verdun offensive in February 1916.

N 38, formed from MS 38 in September 1915 and assigned to the 4th Armée. N 38 participated in the Battle of Champagne during that month. N 38 was also active during the Battle of the Somme. It moved to Champagne in April 1917 as part of GC 15 (although it was briefly attached to the 4th Armée).

N 48, formed from MS 48 in September 1915. It was based at Froidos and commanded by Capitaine Francois de Thorel d'Orgeix. In August 1916 it moved to the 2nd Armée sector and Lieutenant Georges Matton assumed command in October. The next month N 48 became part of GC 11.

N 49, formed from MS 49 in September 1915. Assigned to the 7th Armée, N 49 was initially commanded by Capitaine Constantin Zarapoff but in May 1916 Capitaine Jules de Boutiny assumed command. N 49 was based at Fontaine in May 1916.

N 57, formed from MS 57 in late 1915. It was initially assigned to the 10th Armée and based at Artois. N 57 was commanded by Capitaine Edouard Duseigneur and served on the Verdun front from March to July 1916. In October 1916 N 57 joined GC 11.

N 65, formed from C 65 in February 1916 and based at Malzéville. However, in addition to Caudron G.4s, C 65 had several Nieuports on strength. It was commanded by Lieutenant Louis Gonnet-Thomas. In October C 65 was assigned to the 7th Armée near Nancy and was redesignated N 65. Movng to Behonnes, the unit was placed under the command of Lieutenant Georges Boillet and then Capitaine Philippe Féquant. In June N 65 joined GC Cachy.

N 67, formed in September 1915 at Lyon-Bron and assigned to the 4th Armée. It was commanded by Lieutenant de La Tour and later by Capitaine de Villepin. In October it moved to the RFV (Region Fortified de Verdun) and was based at La Cheppe. In February 1916 Capitaine de Saint-Sauveur assumed

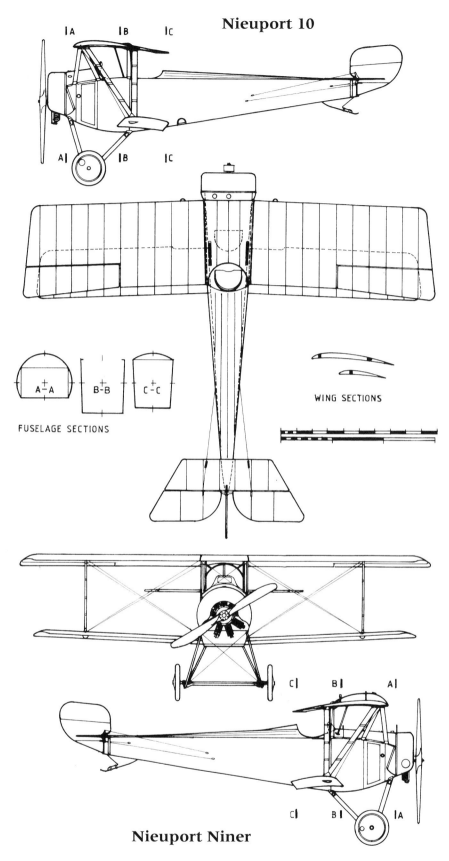

Nieuport 10

Nieuport Niner

command of the unit. N 67 joined GC 13 in November 1916.

N 68, initially an Escadrille de Cavalerie unit September 1915 when it was redesignated N 68 as a fighter unit. It was commanded by Capitaine Garde in November 1916, then Capitaine Lemerle in early 1917. It was assigned to the 8th Armée and in November 1917 became SPA 68.

N 69, created from BLC 5 in September 1915 and assigned to the 10th Armée at the beginning of 1916 as a fighter unit.

The GQG reports for September show that the Nieuport 10s

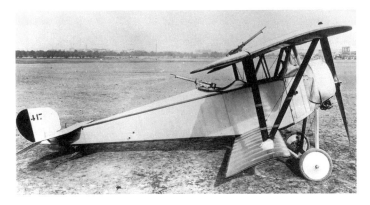

Top, above, and above right: Nieuport 10. University of Texas at Dallas.

Above: Nieuport 10 showing a cowling variation. University of Texas at Dallas.

were initially used in the reconnaissance role. However, their mission soon switched to flying combat patrols along the French front lines. The GQG reports showed that there were increasingly intense combat with German fighters and reconnaissance machines. Usually these were described as inconclusive without loss to either side; however, the enemy airplanes were usually forced to withdraw. In addition to these barrage flights, Nieuport 10s flew bomber escort, reconnaissance sorties, and even dropped spies behind enemy lines. The change in the Nieuport 10's primary mission from army cooperation to fighter resulted from a basic flaw in the original design—with a crew of two the airplane was seriously underpowered. It was found, however, that when the observer was not carried the Nieuport 10 performed well as a single-seat fighter. In this role it had adequate speed and superior maneuverability. On many airplanes, the forward cockpit was deleted and the pilot fired a machine gun (often a Lewis with 47 rounds) fixed on the top wing. Initially, the pilot had to raise his arms above his head in order to fire the gun. Later a Bowden cable system was fitted so that he did not need to change position to use the gun. There were some successes reported even with the two-seat versions of the Nieuport 10. For example, on 9 November 1915 a Nieuport 10 of N 48 carrying a crew of two forced a Fokker to land. The crews were often issued a carbine or Chauchat automatic rifle.

Nieuport 10s were assigned to what may have been the first Groupe d'Combat in the Aviation Militaire. This unit was Groupe d'Combat de Malzéville, which had several Caudron G.4s and two Nieuport 10s and was formed in August 1915. Its function was to guard the bombers of GB 2 during daylight raids and to provide fighter cover for GB 2's base at Malzéville and for the town of Nancy. In December 1915, N 65 with its Nieuport 10's and C 66 with Caudron G.4s were the constituent escadrilles of GC Malzéville. By 25 February 1916 GC Malzéville was disbanded and N 65 moved to Verdun.

The Nieuport 10s continued to serve as fighters until being replaced by Nieuport 11 fighters at the beginning of 1916. After that, the Nieuport 10s continued to serve in small numbers with the escadrilles, usually returning to the task they had previously been intended to perform: reconnaissance. They were gradually replaced by the more powerful Nieuport 12 as 1915 progressed. By February 1916 there were 120 Nieuport 10s and 12s in service but the Nieuport 10s were gradually withdrawn from operational escadrilles during 1916.

Foreign Service
Belgium

Belgium accepted its first Nieuport 10 on 17 June 1915. As the year progressed, the Nieuport 10s were assigned to the following units:

No. 1 Escadrille—1 Nieuport 10 at Coxyde airfield.
No. 2 Escadrille—1 Nieuport 10 at Coxyde airfield.
No. 4 Escadrille—1 Nieuport 10 at Houthem airfield.
　Franco-Belgian Escadrille 674—several Nieuport 10s based at Hondschoote airfield.

It had been intended to use the Nieuport 10s to provide fighter escort for the M.F.11s and Caudron G.4s used by these units. However, it was discovered that the front gunner's position severely restricted his field of fire and that, if he attempted to stand up to fire the gun, the Nieuport 10 would

Nieuport 10. The stand supporting the tailskid has been crudely erased from the original photograph. Renaud.

> **Nieuport 10 Two-Seat Reconnaissance Plane or Single-Seat Fighter with 80-hp Le Rhône**
> Span 7.92 m; length 7.051 m; height 2.67 m; wing area 18 sq. m
> Empty weight 400 kg; loaded weight 650 kg
> Maximum speed: 115 km/h; climb to 2,000 m in 16 minutes; range 250 km
> Armament on the single-seater: one Lewis machine gun and a Mauser pistol
> Armament on the two-seater: one Lewis machine gun and a Winchester and a carbine
> A total of 7,200 Nieuport 10s and 12s were built.
>
> **Macchi-Nieuport 10 Single-Seat Fighter with 80-hp Le Rhône**
> Span 8.03 m; length 6.88 m; height 2.35 m; wing area 18 sq. m
> Empty weight 450 kg; loaded weight 630 kg
> Maximum speed: 140 km/h; climb to 2,000 m in 15 minutes 30 seconds; ceiling 4,000 m; endurance 3 hours
> Armament: one Lewis machine gun
> Approximately 240 built
>
> **Nieuport 10 ("Nieuport Niner") Single-Seat Fighter with 80-hp Le Rhône Built in Russia by Dux and Lebedev**
> Span 9.0 m; length 7.1 m; height 2.35 m; wing area 17.6 sq. m
> Empty weight 430 kg; loaded weight 680 kg
> Maximum speed: 138 km/h; climb to 1,000 m in 6 minutes; climb to 2,000 m in 13 minutes; climb to 3,000 m in 28 minutes; ceiling 3,800 m; endurance 2 hours
> Armament: one machine gun
> Approximately 300 (of all types) built

> **Nieuport 10bis Single-Seat Fighter with 80-hp Le Rhône Built in Russia by Dux and Lebedev**
> Span 8.2 m; length 7.1 m; height 2.35 m; wing area 17.6 sq. m
> Empty weight 430 kg; loaded weight 605 kg;
> Maximum speed 135 km/h; climb to 1,000 m in 7 min.; climb to 2,000 m in 18 min.; climb to 3,000 m in 38 min.; ceiling 3,600 m:
> endurance 2 hrs.
> Armament: one machine gun
>
> **Nieuport 10bis Single-Seat Fighter with 100-hp Gnome Monosoupape Built in Russia by Dux and Lebedev**
> Span 8.2 m; length 7.1 m; height 2.35 m; wing area 17.6 sq. m
> Empty weight 455 kg; loaded weight 630 kg
> Maximum speed: 138 km/h; climb to 1,000 m in 5 minutes 30 seconds; climb to 2,000 m in 12 minutes; climb to 3,000 m in 24 minutes; ceiling 4,200 m; endurance 2 hours
> Armament: one machine gun
> Approximately 20 built
>
> **Nieuport 10 Single-Seat Fighter with 110-hp Le Rhône Built in Russia by Dux in 1917**
> Span 8.2 m; length 7.1 m; height 2.35 m; wing area 17.6 sq. m
> Empty weight 435 kg; loaded weight 610 kg
> Maximum speed: 145 km/h; climb to 1,000 m in 3 minutes 30 seconds; climb to 2,000 m in 8 minutes; climb to 3,000 m in 15 minutes; climb to 4,000 m in 25 minutes; ceiling 4,800 m; endurance 2 hours
> Armament: one machine gun

become unstable. For these reasons, the Nieuport 10s were soon replaced by the Nieuport 11 fighters.

Finland

A single Russian-built Nieuport 10 was captured by Finland in 1918. Later, two Nieuport 10s were acquired by the Finns when two Russian pilots defected with their aircraft. Two of the airplanes were given Finnish codes D 63/18 and D 64/18. The Nieuport 10s remained in Finnish service from 1918 through 1919. They were assigned to Lento-osasto 2 which was based at Antrea. The unit saw action in the Viipuri area.

Italy

In 1915 the Societa Nieuport-Macchi had acquired the rights to build Nieuport 10s under license. The Italians employed the airplane as a dedicated fighter and, for this reason, the observer's cockpit was faired over and a Lewis gun was fitted to the top wing. The first Nieuport 10s arrived from France in July 1915 and were initially assigned to 8a Squadriglia at Aviano. The pilots assigned to this unit were Italians who had been trained in France. A section of two Nieuport 10s was sent to Santa Caterina to provide aerial defense for the Supreme Command. This section saw considerable activity against the Austro-Hungarian Lloyd and Brandenburg reconnaissance aircraft. However, the limited performance of the Nieuport 10s soon became apparent and it has been reported that the Austro-Hungarians and Germans regarded the type as posing a negligible threat. As of January 1916 there were two Italian units using Nieuport 10s: Nos.1 and 2 Squadriglias. During 1916 Nieuport 10s were also used by Nos.70a and 71a Squadriglias.

Teniente Baracca of 70a Squadriglia achieved the first Italian victory with a Nieuport 10 on 7 April 1916 when he destroyed a Austro-Hungarian Brandenburg C.I over Medeuzza. During the second half of 1916 the Nieuport 10s were assigned to provide fighter escort for Caproni bombers during attacks on Fiumee, Gorizia, Trieste, Opcina, and Dornberg.

By the end of 1916, Nieuport 11s and 17s had begun to replace the Nieuport 10s in the fighter role. Nevertheless, as of mid-1917 six units still used Nieuport 10s: 70a, 71a, 75a, 76a, 77a, and 78a Squadriglias. The Nieuport 10 remained in front-line service well into 1917, but in the aerial reconnaissance role. By 1918 most Nieuport 10s had been assigned to training units. The Nieuport-Macchi firm built more than 240 of the type and it is believed that the Italians eventually acquired more than 500 Nieuport 10s before the end of the war.

Russia

Russia received its first Nieuport 10s in late 1915, all two-seaters. Those versions with the pilot in the front cockpit and the observer behind him were designated Nieuport "Nine," while those with the pilot in the rear cockpit were Nieuport 10s. Initially, neither version was armed with machine guns and the observer had to make do with a carbine. The Russians decided that the type would be suitable as a fighter only when the observer's position was deleted and a more powerful engine provided. The Dux and Lebedev plants produced the following versions of the Nieuport 10:

1. Nieuport 10bis: two types of engines—80-hp Le Rhône (only a small number were built) and 100-hp Gnome

Nieuport 10 of the Japanese air service at Gifu, Japan in 1920. B85.1801.

A heavily-armed Nieuport 10 of the IRAS, serial number 1739. The Dux and Lebedev plants produced several versions of the Nieuport 10. B85.2585.

Below: A Nieuport 10 of Escadrille N 561. B85.1234.

Monosoupape (20 built).
2. Nieuport 10 with 110 or 120-hp Le Rhône engines. These were built by the Dux plant and were in production as late as 1917. These airplanes had a larger cowling than Nieuport 10s with 80-hp Le Rhône engines.
3. Nieuport 10bis with 100-hp Monosoupape engines. This version was characterized by five holes in the engine cowling, added to facilitate engine cooling.
4. Nieuport 10bis with 80-hp Le Rhône engine.

Some Nieuport 10s were built as two-seaters but these were used only as trainers. The trainers were in production between 1916 and 1917 and served until late 1920. The Nieuport 10 single-seat fighters remained in production until 1918. The machine guns for these fighters were installed by a sub-contractor in Kiev using gun mounts for the top wing developed by Vasili Vladimirovich, commander of aviation for the 8th Army. In 1916 the Nieuport 10s were supplied to the 2nd, 3rd, 5th, and 12th Fighter units.

On 1 March 1917 Nieuport 10s were in service on the northern front (5), western front (3), southwestern front (16), and the Romanian Front (5). This gives a total of 29 aircraft.

On 1 April 1917 there were still 20 Nieuport 10s in service at the front, representing 10 percent of all aircraft with operational units. As of June 1917 the disposition of the Nieuport 10s at the front was: northern front (3), western front (8), southwestern/Romanian fronts (15), for a total of 29 aircraft. Most were assigned to corps air detachments, having been superseded in the fighter role.

Most of the Nieuport 10s in service in 1917 were being used for reconnaissance. The decision had been made to concentrate on the production of Nieuport fighters and these began to replace the Nieuport 10s on the production line in 1916 and 1917. The remaining Nieuport 10s were used as trainers until the mid-1920s. In 1921 a total of 40 Nieuport 10s were in service. In 1922 some aircraft still served in front-line units; these were with the 4th, 14th, 19th, and 20th Aviaotryady and with the 2nd naval Istrootryad. They remained in service as trainers until 1925.

Commander Salmson, RNAS, in front of his Nieuport 10. The Lewis gun and sight are mounted to enable the gun to fire outside the propeller arc. Via Colin Owers.

Ukraine
A single Nieuport Niner (c/n 1046) and a single Nieuport 10 (c/n 725) were obtained by the Ukrainian People's Republic Air Force in 1918. The Ukrainian air force ceased to exist in 1919 when the Ukraine was occupied by the Red Army.

United Kingdom
The Royal Naval Air Service ordered 24 Nieuport 10s in 1915 (although this number may include some Nieuport 12s). They were allocated serials 3163 through 3186 and were assigned to No.1 Wing. Twelve additional Nieuport 10s were purchased in France and given serials 3962 through 3973. These served with Nos.1, 3, and 4 Wings. Two additional machines, 8516 and 8517, were delivered in 1915. Both were assigned to 4 Wing, and later 1 Wing.

Nieuport 11

The Nieuport 11 was a petite fighter that would become widely known as Bébé by its French pilots. The Nieuport 10 two-seat reconnaissance aircraft had been quite successful in the fighter role when it was flown as a single-seater. The Aviation Militaire needed an aircraft to replace the makeshift Morane-Saulnier L and LA series of fighters and the Nieuport firm readily provided such a machine by the simple expedient of producing a smaller, single-seat derivative of the Nieuport 10.

Both the Nieuport 10 and 11 were sesquiplanes with the top wing larger than the lower wing. Gustav Delage felt this layout permitted the aircraft to have the maneuverability of a biplane and the strength of a monoplane. The lower wing had a wooden spar and had a 6-degree dihedral that could be adjusted on the ground depending on the weight of the machine and the type of engine that was fitted. This machine retained the V-shaped struts, which permitted the lower wing to be pivoted along its axis. The lower wing had a chord of 0.70 m. The top wing had no dihedral and a chord of 1.20 m. It had two main spars and was placed ahead of the lower wing by 0.60 m at an angle of 17 degrees. Ailerons were fitted to the top wing. The tail had horizontal stabilizers 0.45 m deep with a maximum span of 3.25 m attached to the fuselage sides. The rudder was 0.55 m deep; it had no fixed portion. The undercarriage consisted of two simple V-struts with the rear strut canted forward; the wheels were separated by a single axle. The undercarriage track was 1.60 m which proved to be sufficient to ensure stability when taking off or landing. The tail skid was a steel spring that served to brake the aircraft when it landed.

While the Nieuport 11 was in many ways a modern lightweight fighter, its armament was clearly obsolescent. Because the French did not possess an adequate synchronization system, a single Hotchkiss or Lewis machine gun was fixed to the top wing along the centerline and fired above the arc of the propeller. The gun was fitted with a circular ammunition drum containing 97 cartridges and was mounted on a hinged platform so the pilot could tip it backward to replace the cartridges. Some aircraft were fitted with twin Lewis machine guns (one of these machines was used by the Belgian ace Jan Olieslagers). The pilots of the Nieuport 11 were also issued a Mauser pistol.

The engine was either a 50-, 60-, or 80-hp Gnome or Le Rhône 9C rotary. They endowed the aircraft with an impressive performance, but the aircraft was quite dangerous in a lateral skid. Many pilots who had been trained at the Aviation Militaire schools with the stable Caudron G. 3 and Blériot 11 would find the Nieuport 11 an unforgiving aircraft to fly and accidents were frequent. The lower wing tended to fracture under stress, which resulted in many fatalities.

The name Bébé may have arisen from the Nieuport 11's SFA designation as a Class B (light fighter) with a biplane (B) wing. This equaled a Type BB; hence, the Bébé name. Other sources suggest the name originated because of the Nieuport 11's small size.

Variants

A modified Nieuport 11 had standard fuselage and tail but was fitted with a horseshoe-style cowling over the 80-hp Le Rhône 9C engine. It had a widened undercarriage track and the V-interplane struts were canted outward. Aside from a wing area of 13.00 sq. m, no dimensions are available. The type did not enter production. Some sources referred to it as a Type 18, but this designation was applied to another Nieuport design.

On 22 February 1916 a Nieuport fighter was tested with the new 150-hp Hispano-Suiza engine by pilot Albert Etéve. The engine was a success but it was decided to use it on the new SPAD 7, and therefore the Nieuports retained their rotary engines. (It is possible the aircraft was actually a Nieuport 14.)

Operational Service

The first Nieuport 11 arrived at the front on 5 January 1916. By February there were 90 in service with the Aviation Militaire. The aircraft enabled the French escadrilles to regain air superiority from the German units flying the Fokker E.III. While it was a difficult and even potentially dangerous aircraft to fly, the Nieuport 11 was an effective fighter when flown by a skilled pilot. However, aiming was difficult because the gun was situated far above the pilot's head, which required him to angle the aircraft to ensure the bullets struck their target. Still, many of France's greatest aces scored their first victories while flying these aircraft.

However, by early 1916 the records of N 38 show that the Nieuport units were still having to use their Nieuport 10s in the fighter role as there were not enough Nieuport 11s available. For this reason Nieuport 11s were initially supplied only to the unit's most experienced pilots, which, considering how difficult the aircraft was to fly in combat, was probably fortunate. For the Somme offensive N 38 had only five Nieuport 11s available, N 31 had three by October 1916, and N 49 had seven in late March 1916.

The Nieuport 11s were gradually supplied to these escadrilles:

N 3, formed from MS 3 in September 1915. In February N 3 moved to the Verdun sector and was based at Vadelincourt. N 3 joined Groupement de Combat de la Somme in April, and it became part of GC 12 in May 1916.

N 12, formed from Ms 12 in September 1915 and assigned to the 5th Armée. It was under the command of Capitaine Aubry; Capitaine Raymond de Pierre de Bernis assumed command in May 1916.

N 15, based at Verdun from March to April 1916 and assigned to the 10th Armée. It was active during the Battle of the Somme in July 1916. N 15 joined GC 13 in March 1917.

N 23, assigned to the 2nd Armée and active in the Verdun sector. It was commanded by Lieutenant Louis Robert de Beauchamp. It re-equipped with Nieuport 17s at the end of 1916.

N 26, initially under the command of Capitaine Thobie and subsequently commanded by Capitaine Jacques de Sieyes de Veynes in May 1916. N 26 was initially based on the Belgian front; in June 1916 it was assigned to the Groupe de Combat de la Somme.

Italian Nieuport 11 serial N 1615. Via Alan Durkota.

Italian Nieuport 11 serial N 2120. The cowling is painted in the Italian colors of red, white, and green and the undersurfaces of the wings are painted red and green with natural fabric showing between the painted areas. Via Alan Durkota.

N 31, moved to the Verdun front in early 1916, under command of Lieutenant Lucien Courcet de Villeneuve. In October 1916 it was assigned to GC 11.

N 37, moved to Sainte-Menehould from late 1915 to early 1916, initially under the command of Capitaine Quillien; Lieutenant Marcel Feierstein assumed command in April. The unit participated in the Verdun offensive in February 1916 as part of the provisional Groupe de Chasse on that front. In July N 37 was assigned to the 6th Armée as part of GC Cachy.

N 38, assigned to the 4th Armée. It participated in the Battle of Champagne during July 1915. N 38 was also active during the Battle of the Somme. It moved to Champagne in April 1917 as part of GC 15 (although it was briefly attached to the 4th Armée).

N 48, based at Froidos and commanded by Capitaine Francois de Thorel d'Orgeix. In August 1916 it moved to the 2nd Armée sector and Lieutenant Georges Matton assumed command in October. The next month N 48 became part of GC 11.

N 49, assigned to the 7th Armée and initially commanded by Capitaine Constantin Zarapoff. In May 1916 Capitaine Jules de Boutiny assumed command. N 49 was based at Fontaine in May 1916. In December it re-equipped with SPAD 7s.

N 57, assigned to the 10th Armée and based at Artois. N 57 moved to Verdun from March to July 1916 under the command of Capitaine Edouard Duseigneur. In October 1916 N 57 joined GC 11.

N 65, formed from C 65 in February 1916 and based at Malzéville. In addition to Caudron G.4s, C 65 had several Nieuports on strength. It was commanded by Lieutenant Louis Gonnet-Thomas. In October C 65 was assigned to the 7th Armée near Nancy and was redesignated N 65. Moving to Behonnes, the unit was placed under the command of Lieutenant Georges Boillet and then Capitaine Philippe Féquant. In June 1916 N 65 joined GC Cachy.

N 67, assigned to the RFV (Region Fortified de Verdun) and based at La Cheppe. In February 1916 Capitaine de Saint-Sauveur assumed command. N 67 joined GC 13 in November 1916.

N 68, commanded by Capitaine Garde in November 1916, then Capitaine Lemerle in early 1917. It was assigned to the 8th Armée and in November 1917 became SPA 68.

N 69, assigned to the 10th Armée at the beginning of 1916 as a fighter unit. It was commanded by Capitaine Robert Mannenet de Marancour. In February 1916 it moved to the Verdun sector. N 69 was reassigned to the 10th Armée in July 1916.

N 73, formed as the Detachment Nieuport de Corcieux in the 7th Armée sector in April 1916. It became N 73 a short time later and was initially commanded by Lieutenant Pierre Bouny. Lieutenant Honore Lareinty-Tholozan later assumed command. In November 1916 N 73 was assigned to GC Somme, later GC 12.

N 102, formed from VB 102 in June 1916. N 102 was stationed at Malzéville and commanded by Chef de Bataillon Pouderoux. it was assigned to the 2nd Armée.

N 103, formed from VB 103 in February 1916 and assigned to the 6th Armée. It was initially led by Commandant Felix Brocard, then commanded by Capitaine d'Harcourt. N 103 was assigned to GC Somme in July 1916.

N 112, formed from VB 112 in 1916 and assigned to the 4th Armée. It was assigned to GC 15 in March 1916.

Two special Nieuport 11 units were formed during 1916:

Escadrille 92, based in Venice, was formed in July 1916 and was initially designated the Escadrille de Mestre. The unit would see continuous activity during the latter half of 1916. N 92 was active over the Isonzo front and flew escort missions over the Adriatic for Italian torpedo boats, destroyers, and seaplanes. Redesignated N 392 in June 1916, the escadrille succeeded in forcing down three Austro-Hungarian flying boats by the end of 1916.

N 124, formed on 16 April 1916, consisted primarily of American volunteer pilots. The unit later became famous as the Lafayette Escadrille. Initially based at Luxeuil, it moved to Verdun on 19 May 1916.

A.F.O.

N 387 was a combined Franco-Serbian unit (see entry under Serbia for further details).

N 391 was based in Serbia (see entry under Serbia for further details).

The Battle of Verdun began on 21 February, 1916. Nine Nieuport 11 escadrilles were sent to that front in order to achieve local air superiority. Initially, these units (N 3, N 15, N 23, N 31, N 37, N 48, N 57, N 65, and N 69) remained under the direct control of the various Groupes d'Armée. This meant that they spent much of their time escorting reconnaissance and bomber units assigned to the same Groupe d'Armée. Also, the Nieuport units were still required to perform reconnaissance missions, drop leaflets, and drop spies behind the enemy lines. These duties were in addition to fighter patrols and attacking enemy balloons. However, the French would soon realize that it would be more effective to concentrate fighter units into a single group and use them to achieve air superiority over areas of the front. In preparation for the coming Battle of the Somme the Groupement de Combat de la Somme was formed in April 1916: N 3, N 26, N 37, N 62, N 65, N 73, and N 103 all served with this unit during its existence. The mission of this group was to inhibit German reconnaissance aircraft and balloons from detecting the French movement of troops and supplies to the front before the battle began in late June.

By mid-1916 the Nieuport 11s were being replaced by the more powerful Nieuport 16s. It appears that by late 1916 most Nieuport 11s had been withdrawn from front-line service, although many escadrilles retained some on strength. In the field some Nieuport 11s were fitted with 110-hp Le Rhône engines, and some sources indicate that these re-engined aircraft were designated Nieuport 11 Cs. N 38's records show that the escadrille's mechanics were cited in the orders of the escadrille for replacing all the 80-hp Le Rhônes in the unit's Nieuport 11s with the new 110-hp Le Rhônes. As new airplanes arrived with a unit, the older types were usually retained in service. The heterogeneous nature of the fighter escadrille's equipment can be illustrated by the aircraft roster of N 49. In late March 1916 the escadrille had seven Nieuport 11s, five Nieuport 12s, one Nieuport 10, and one SPAD A.2 on strength. Two months later there were five Nieuport 11s, six Nieuport 12s, and one Nieuport 16.

After their retirement from service with the front-line units, the Nieuport 11s were assigned to the training units. In the 1916–17 curriculum a pilot would progress from the M.F.11 to the Voisin 3, to the Caudron G.3, and, finally to the Nieuport 11.

The Nieuport 11 had given France its first modern fighter and enabled the Aviation Militaire to gain ascendancy over the Fokker E.III. At Verdun and the Somme it was the Nieuport 11s that permitted the French air service to control the skies over the battlefield.

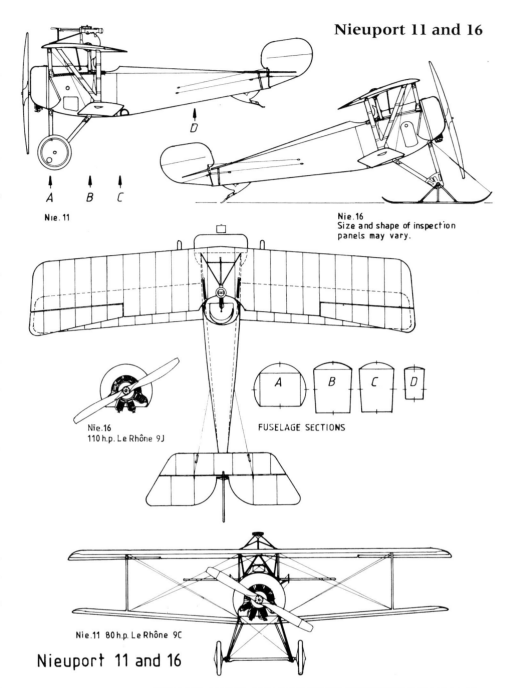

Foreign Service

Belgium

The Belgian air service had discovered that the Nieuport 10s were unsatisfactory in the fighter role and they were replaced in 1916 with Nieuport 11s. It appears that Nieuport 11s were assigned to the reconnaissance units (1st, 5th, and 6th Escadrilles) to supply fighter escort for the slower army cooperation aircraft. In September 1916 the 5th Escadrille became a dedicated fighter unit and was assigned to provide escort for the reconnaissance aircraft of the 1st Escadrille.

Czechoslovakia

When the Czechoslovakian air service was formed in 1919 it incorporated a number of Russian aircraft. A small number of Nieuport 11s were among these Russian machines and these were employed operationally.

Italy

Italy used large numbers of Nieuport 11s to supplement and later replace the Nieuport 10s. The Macchi-Nieuport firm built the type under license and 450 were built by that firm as well as an additional 93 (out of 200 originally ordered) by the Officine Elettro-Ferroviarie in Milano.

The Nieuport 11s began to appear in Italian fighter units during 1916 and 70 Squadriglia, based at Santa Caterina, was

Nieuport 11 of N 561 serving on the Italian front. B85.960.

probably the first squadron to receive them. The unit subsequently moved to Udine, where the supreme command was located. While based in this area 70ª Squadriglia flew

The prototype of the Nieuport 11. The Nieuport firm readily provided the Aviation Militaire with a new fighter by simply producing a smaller, single-seat derivative of the Nieuport 10. B87.0417.

Nieuport 11 serial N 539 of Escadrille N 95 assigned to the CRP in January 1916. B79.3186.

offensive patrols along the front lines. By the end of 1916 most of the 56 enemy airplanes destroyed by Italian fighters had been dispatched by Nieuport 11s. At this time the main complaint about the Nieuport 11 concerned the Lewis gun, which was difficult to reload in flight, and the Nieuport 11s could only carry a small number of additional rounds. By late 1916 some of the Italian Nieuport 11s were fitted with the Le Prieur rocket system for use against enemy balloons.

There were four squadriglias using Nieuport 11s during 1916. These were organized as:

Supreme Command—70ᵃ Squadriglia at Santa Canterina; 9th **Gruppo** (1st Armata)—71ᵃ and 75ᵃ Squadriglias; and 2nd **Gruppo** (2nd Armata)—76ᵃ Squadriglia.

By early 1917 the following units were equipped with Nieuport 11s:

1 Gruppo (3rd Armata)—77ᵃ and 80ᵃ Squadriglias.
2nd Gruppo (2nd Armata, 4th Armata)—76ᵃ and 81ᵃ Squadriglias.
7th Gruppo (6th Armata, 1st Armata)—79ᵃ Squadriglia.
9th Gruppo (1st Armata)—71ᵃ, 75ᵃ, and 78ᵃ Squadriglias.
10th Gruppo (Supreme Command)—70ᵃ and 82ᵃ Squadriglias.

Independent units: Sezione (section) 83ᵃ Squadriglia at Kremain (Macedonia), 85ᵃ Squadriglia at Piskupi (Albania), and the Sezione Nieuport at Belluno.

During 1916 and early 1917 the output of Nieuports from the Macchi-Nieuport firm was between 35 and 45 a month; most of these were Nieuport 11s. However, by early 1917 the Nieuport 11s were being superseded on the production lines by the improved Nieuport 17. By October 1917 only nine squadriglias still had Nieuport 11s on strength:

10th Gruppo (Supreme Command)—70ᵃ, 78ᵃ, and 82ᵃ Squadriglias.
3rd Gruppo (1st Armata)—72ᵃ Squadriglia.
7th Gruppo (1st Armata)—79ᵃ Squadriglia.
9th Gruppo (1st Armata)—71ᵃ and 75ᵃ Squadriglias.
2nd Gruppo (4th Armata)—76ᵃ Squadriglia; and 85ᵃ Squadriglia at Piskupi (Albania).

Only six squadriglias were still using Nieuport 11s at the end of 1917:

3rd Gruppo (1st Armata)—72ᵃ Squadriglia.
9th Gruppo (1st Armata)—71ᵃ and 75ᵃ Squadriglias.
10th Gruppo (1st Armata)—82ᵃ and 91ᵃ Squadriglias.
2nd Gruppo (4th Armata)—76ᵃ Squadriglia; and 85ᵃ Squadriglia at Piskupi (Albania).

The Nieuport 11 had by 1918 been eclipsed by the Nieuport 17. As these newer airplanes reached front-line units, the Nieuport 11s were reassigned to training units. By February 1918 there were only 22 Nieuport 11s still in service. The last unit to have them on strength was 85ᵃ Squadriglia in Albania which flew the Type 11 until 4 November 1918.

Above: Nieuport 11 serial number N571. B83.3506.

Left: Nieuport 11 with pilot de Gavardie in the cockpit. B92.1827.

Netherlands

On 2 February 1917 a Nieuport 11 of A-Squadron Number 1 Wing of the RNAS landed at Cadzand in the Netherlands. The Dutch authorities purchased it from the British and changed its serial from 3981 to LA 40 (the serial was subsequently changed to N 213 and then N 230). Based on their evaluation of the Nieuport 11, the Dutch ordered additional examples of the type be built by the NV Dutch Motor Car and Airplane Factory at Trompenberg. Five aircraft arrived in 1918 and these remained in service until 1925. These five Nieuport 11s were given serials N-215 through N-219. Later in 1918 an additional 20 Nieuport 11s were ordered and given serials N-230 to N-249. Apparently only an additional 12 were completed and these were accepted by October 1918. However, all 12 had been so poorly constructed that they were never flown.

Romania

In 1916 28 Nieuport 11s were transferred from the RNAS to Romania. Eight of these were assigned to Grupul 3 as the Escadrilla Nieuport (Franco/Romana). Eventually four fighter units were equipped with Nieuport 11s:

1st Romanian Army:
Grupul 1—1st and 14th Fighter Squadrons based at Cioara.
Grupul 2—3rd Fighter Squadron (based at Cioara) and the 14th Fighter Squadron.
2nd Romanian Army:
Grupul 1—1st Fighter Squadron based at Borzesti and 6th Russian Army.
Grupul 3—11th Fighter Squadron based at Galati.

The Nieuport 11s supplied air cover for Bucharest and the Russian army at Braila. By 15 December 1916 only 12 Nieuport 11s were still in service, with six undergoing repairs and an additional eight no longer serviceable. The Nieuport 11s were replaced by Nieuport 17s as 1917 progressed.

Russia

The Imperial Russian Air Service had concluded that the Nieuport 10 was only marginally effective as a fighter and found that the Nieuport 11 was a marked improvement. The Nieuport 11s were initially imported from France but a large number were later built in Russia. An order for 200 was placed with the Dux factory. The firm introduced several modifications to the design. For example, pine was used instead of spruce, and flax was used for covering instead of silk. These changes resulted in the Dux-built Nieuport 11s being 30 kg heavier than French-built machines. Unfortunately, the Dux machines were of very poor quality. According to the commander of the Eighth Fighter Unit the Dux Nieuport 11s "...could not be assembled because the parts did not fit and...the bolts fixing the struts with the spars broke into pieces." The Nieuport 11 was later built under license by the Mosca, Anatra, and Shchetinnin factories. Seventy were built during 1916.

As those Nieuport 11s received from France were unarmed, many pilots in front-line squadrons had to supply their own armament in the form of pistols or rifles. However, V.V. Jordan designed a machine gun mount that permitted a weapon to be carried. Although the gun had to be angled so that it fired over the propeller, the mount enabled it to be swiveled vertically.

On 1 March 1917 the disposition of the 37 Nieuport 11s at the front was as follows: northern front (10), western front (5), southwestern front (16), Romanian front (5), the Caucasus front (1).

In April 1917 there were 70 Nieuport 11s in service. By June 1917 48 Nieuport 11s were in service with squadrons on the northern front (12), western front (4), southwestern/Romanian fronts (25), Caucasus front (7).

Although the Nieuport 11 had proved to be a significant improvement over the Nieuport 10, it was quickly replaced in Russian fighter units by the superior Nieuport 17.

A total of 18 Nieuport 11s were in service as trainers as late as December 1921. Three were with the 2nd Military School of Pilots and the 1st Higher School of Military Pilots in June 1923. The last was struck off charge in 1924.

Nieuport 11 Single-Seat Fighter with 80-hp Le Rhône 9C

Span 7.520 m; length 5.50 m; height 2.40 m; wing area 13.3 sq. m
Empty weight 320 kg; loaded weight 480 kg
Maximum speed: 162 km/h at 2,000 m; climb to 2,000 m in 8.50 minutes; climb to 3,000 m in 15 minutes; ceiling 5,000 m; range 250 km; endurance 2.5 hours
Armament: a single Hotchkiss or Lewis machine gun

Macchi-Nieuport 11 Single-Seat Fighter with 80-hp Le Rhône:

Span 7.65 m; length 5.69 m; height 2.40 m; wing area 13 sq. m
Empty weight 352 kg; loaded weight 515 kg
Maximum speed: 155 km/h; climb to 2,000 m in 8.30 minutes; endurance 2 hours and 30 minutes
Armament: a single Hotchkiss or Lewis machine gun
450 were built by Macchi-Nieuport and an additional 93 by the Officine Elettro-Ferroviarie in Milano

Nieuport 11 Single-Seat Fighter with 80-hp Le Rhône 9C Built by the NV Dutch Motor Car and Airplane Factory

Span 7.46 m; length 5.80 m; wing area 13 sq. m
Empty weight 320 kg; loaded weight 480 kg
Maximum speed: 155 km/h; climb to 2,000 m in 8 minutes 30 seconds; endurance 1.5 hours
Armament: a single 7.7-mm Lewis machine gun
17 were built

Nieuport 11 Single-Seat Fighter with 80-hp Le Rhône 9C Built in Russia by Dux, Mosca, Anatra, and Shchetinnin

Span 7.5 m; length 5.6m; wing area 13.5 sq. m
Empty weight 350 kg; loaded weight 525 kg
Maximum speed: 152 km/h; climb to 1,000 m in 4.0 minutes; climb to 2,000 m in 10 minutes; climb to 3,000 m in 19 minutes; climb to 4,000 m in 33 minutes; ceiling 4,500 m; endurance 2 hours.
Armament: one machine gun on a mount designed by V.V. Iordan
Approximately 200 were built by Dux, Mosca, Anatra, and Shchetinnin

Nieuport 11 built by Dux with the prominent headrest. On some of the machines built by Dux pine was used instead of spruce and flax was used for covering instead of silk; this made the aircraft 30 kg heavier than French-built machines. Via G. Haddow and Colin Owers.

Nieuport 11 of Sous Lieutenant de Turenne showing the pilot's personal insignia. B89.1284.

Serbia

The Serbians received extensive support from France's Aviation Militaire and three French escadrilles equipped with Nieuport 11s were active in Serbia in 1916. These units were N 387 (a combined Franco-Serbian unit), N 391 (an all-French unit), and a Nieuport escadrille under the direct control of the Serbian high command. The latter unit was used for reconnaissance and to transport spies behind enemy lines. The Nieuport escadrilles used a combination of Nieuport 10s and 11s.

N 387 was later assigned to the Serbian Second Army and performed reconnaissance over the right bank of the Erna River and the left bank of the Vardar River. It also provided escort for the MF 382.

By March 1917 four Nieuport 12s and Nieuport 11s were assigned to the First Serbian Army. A section of Nieuports was also assigned to the Serbian Second Army. N 387 was now based at Vertekop and was under the direct control of the Serbian high command. By mid-1917 SPAD 7s had supplanted the Nieuports on the Serbian front.

Siam

Siam acquired at least two Nieuport 11s postwar; they were used as trainers until 1933. Serials were 4 and 23.

Ukraine

A single Nieuport 11 (c/n 1057) was obtained by the Ukrainian People's republic Air Force in 1918.

United Kingdom

The Royal Naval Air Service acquired Nieuport 11s. Serials allotted included Nos.3975–3978; and Nos.3980–3994. These airplanes were assigned to No.1 Wing at St.-Pol, and to Nos.1, 2, and 4 Wings. Nos.3975 and 3978 were eventually sent to Romania.

It was planned to provide the RFC with six of the RNAS Nieuport 11s. These were given serial numbers A8738–8743 (corresponding with RNAS serials 3956, 3957, 3958, 3986, 8750, and 8751) but none ever served with the RFC.

Nieuport 12

The Nieuport 10 with its 80-hp Le Rhône engine was too underpowered to be useful as a reconnaissance aircraft. In order to overcome this problem an enlarged version of the Type 10 with a 110-hp Clerget 9Z rotary was introduced in 1915. A pilot was seated under a cutout in the upper wing; the center section of the wing had a transparent covering to provide the pilot with a better upward view. The observer/gunner sat behind the pilot and had a Lewis machine gun on an Éteve mounting. In some aircraft a second gun was fitted to the top wing and was aligned to fire over the propeller. The wings were of unequal span and had the V-interplane struts with an outward rake. Later aircraft had the more powerful 130-hp Clerget 9B engine and a circular cowling faired into the fuselage. This type was designated the Nieuport 12bis and was supplied only to French units. The Nieuport 12bis C2 was a two-seat escort fighter with two Lewis guns.

Operational Service

Contemporary assessments note that the Nieuport 12 had an inferior performance, vibrated excessively, was difficult to fly, and even more difficult to land. Despite the modifications to the top wing, the pilot's view was still described as poor. Nonetheless, the aircraft saw widespread use with French fighter escadrilles and supplanted the inferior Nieuport 10 in the reconnaissance role. Nieuport 12s served primarily with Nieuport 11, 16, and 17-equipped escadrilles being used for short-range reconnaissance and also to transport spies behind enemy lines. The Nieuport escadrilles were:

N 3, moved to the Verdun sector in February 1916. N 3 and was based at Vadelincourt. N 3 joined Groupement de Combat de la Somme in April and it became part of GC 12 in May 1916.

N 12, assigned to the 5th Armée under command of Capitaine Aubry; Capitaine Raymond de Pierre de Bernis assumed command in May 1916.

N 15, based at Verdun from March to April 1916 and assigned to the 10th Armée. It was active during the Battle of the Somme in July 1916. N 15 joined GC 13 in March 1917.

N 23, assigned to the 2nd Armée and moved to the Verdun sector. It was commanded by Lieutenant Louis Robert de Beauchamp. It re-equipped with Nieuport 17s at the end of 1916.

N 26, initially led by Capitaine Thobie, then commanded by Capitaine Jacques de Sieyes de Veynes in May 1916. N 26 was based on the Belgian front. In June 1916 it was assigned to the Groupe de Combat de la Somme. Subsequently it was commanded by Capitaine Victor Ménard in July 1916.

N 31, assigned to the 1st Armée. In early 1916 N 31 moved to the Verdun front under the command of Lieutenant Lucien

Nieuport 12

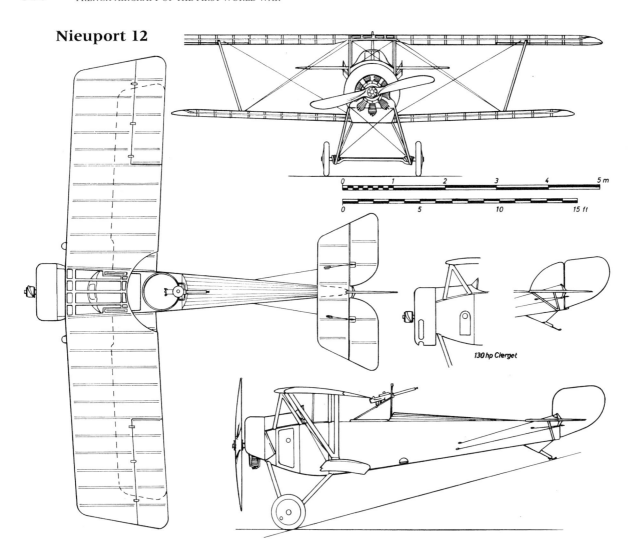

Couret de Villeneuve. In October 1916 it was assigned to GC 11.

N 37, assigned to the 3rd Armée and moved to Lisle-en-Barrois, where it participated in the offensive at Champagne in September 1915. N 37 served at Sainte-Menehould from late 1915 to early 1916. The unit participated in the Verdun offensive in February 1916. N 37 joined the provisional Groupe de Chasse on that front. In July it was assigned to the 6th Armée as part of GC Cachy.

N 38, assigned to the 4th Armée. N 38 participated in the Battle of Champagne and was also active during the Battle of the Somme. It moved to Champagne in April 1917 as part of GC 15 (although it was briefly attached to the 4th Armée).

Nieuport 12 of N 57 which was initially assigned to the 10th Armée and based at Artois. B78.1194.

N 48, based at Froidos and commanded by Capitaine Francois de Thorel d'Orgeix. In August 1916 it moved to the 2nd Armée sector and Lieutenant Georges Matton assumed command in October. The following month N 48 became part of GC 11.

N 49, assigned to the 7th Armée N 49 and initially commanded by Capitaine Constantin Zarapoff; in May 1916 Capitaine Jules de Boutiny assumed command. N 49 was based at Fontaine in May 1916. In December it re-equipped with SPADs.

N 57, initially assigned to the 10th Armée and based at Artois, It was commanded by Capitaine Edouard Duseigneur. N 57 moved to Verdun from March to July 1916. In October 1916 it joined GC 11.

N 62, formed in May 1916 and initially assigned to the 6th Armée and then to GC Somme. N 62 later returned to the 6th Armée where it remained for the rest of the war.

N 65, assigned to the 7th Armée near Nancy in October. Moving to Behonnes, the unit was placed under the command of Lieutenant Georges Boillet and then Capitaine Philippe Féquant. In June 1916 N 65 joined GC Cachy.

N 67, commanded by Lieutenant de La Tour, later Capitaine de Villepin. In October it moved to the RFV (Region Fortified de Verdun) and was based at La Cheppe. In February 1916 Capitaine Henri Constans assumed command. N 67 joined GC 13 in November 1916.

N 68, commanded by Capitaine Garde in November 1916, then Capitaine Lemerle in early 1917. It was assigned to the 8th Armée and in November 1917 became SPA 68.

N 69, assigned to the 10th Armée at the beginning of 1916 as a fighter unit. It was commanded by Capitaine Robert Mannenet de Marancour. In February 1916 it moved to the Verdun sector. N 69 was reassigned to the 10th Armée in July 1916.

This Nieuport 12 is probably a Beardmore-built machine as suggested by the fixed fin at the base of the rudder. University of Texas at Dallas.

Nieuport 12bis. Unlike the Beardmore-built example shown above, the center section of the wing has a clear covering to facilitate the pilot's upward view. University of Texas at Dallas.

Nieuport 12bis. University of Texas at Dallas.

The Nieuport 12bis had a 130-hp Clerget 9B with a circular cowling and lateral fairings. University of Texas at Dallas.

N 73, formed as the Detachment Nieuport de Corcieux in the 7th Armée sector in April 1916. It became N 73 a short time later. It was initially commanded by Lieutenant Pierre Bouny, then Lieutenant Honore Lareinty-Tholozan. In November 1916 it was assigned to GC Somme and later GC 12.

N 75, formed in July 1916 under command of Capitaine Henri de Montfort. It was assigned to the 8th Armée. In April 1917 it became part of GC 14.

N 77, formed in September 1916 at Lyon-Bron. Assigned to the 8th Armée on the Lorraine front, the unit was commanded by Lieutenant Joseph d'Hermite. A notable mission occurred on 23 February 1917 when two aircraft of N 77 bombed the German airbase at Marinois.

N 78, formed at St. Étienne-au-Temple in December 1916. It was commanded by Lieutenant Armand Pinsard and assigned to the 4th Armée. It joined GC 15 in March 1917.

N 79, formed in November 1916. Under the command of Lieutenant Roger Lemercier de Maisoncelle-Vertille de Richemont, N 79 was assigned to the 3rd Armée. In March 1917 Lieutenant Gaston Luc-Pupat assumed command.

N 80, created at Lyon-Bron in December 1916. Commanded by Capitaine Francoise Glaize, N 80 was assigned to the 5th Armée. In January 1917 N 80 moved to Bonne-Maison. It joined GC 14 in March 1917.

N 81, formed at Villacoublay under the command of Capitaine Maurice Mandinaud. N 81 was assigned to the 6th (at Sacy-le-Grand) and then the 7th Armée (Fontaine) in January 1917. Lieutenant Raymond Bailly assumed command in March 1917. It moved to the 4th Armée sector in April 1917 and joined GC 15 a few days later.

N 102, formed from VB 102 in June 1916. N 102 was stationed at Malzéville and commanded by Chef de Bataillon Pouderoux. It was assigned to the 2nd Armée.

N 103, formed from VB 103 in February 1916 and assigned to the 6th Armée. It was commanded by Capitaine d'Harcourt. N 103 was assigned to GC Somme in July 1916.

N 124, formed on 16 April 1916 consisting primarily of American volunteer pilots. The unit later became famous as the Lafayette Escadrille. Initially based at Luxeuil, it moved to Verdun on 19 May 1916.

In addition to the Nieuport escadrilles, Nieuport 12s were also assigned, usually in batches of one to three, to various other French units.

Nieuport 12s often made up a substantial portion of an escadrille's strength. For example, during the Somme offensive N 38 had five Nieuport 12s and five Nieuport 11s in service. N 49 was equipped with 14 airplanes in March 1914; five were Nieuport 12s. By May of that year the six Nieuport 12s on strength constituted one-half of N 49's entire force. Nieuport 12s actually made up the major part of N 77's equipment from its formation in September 1916 until mid-1917.

The records of these escadrilles show that the Nieuport 12s were used for a large number of mundane but important and often dangerous missions. N 77 used its Nieuport 12s not only for reconnaissance, but also for ground-attack, while the Nieuport 12s of N 38 were used for reconnaissance, barrage flights, and even air-to-air combat. In the latter role, however, the Nieuport 12s proved no match for the superior German airplanes and the unit had few victories.

By February 1916 there were 120 Nieuport 10s and 12s in service at the front. However, the Nieuport 12s were replaced as 1916 progressed and by 1917 only four examples with 130-hp Clergets were still in service at the front. A total of 7,200 Nieuport 10s and 12s were built during the war.

The Aviation Maritime also purchased a number of Nieuport 12s; these were based at the various naval aviation centers. The primary missions of the Nieuport 12s were reconnaissance and escort for the more vulnerable flying boats.

Nieuport 12 serial A3291 with French Cockades. J.M. Bruce/G.S. Leslie via Colin Owers.

Foreign Service

Chile
In 1918 the Chilean air service purchased a single Nieuport 12.

Estonia
The Estonaina Aviation Company was formed in 1920 and acquired a single Nieuport 12, probably from Russia. It was given serial number 52.

Poland
The Polish 10th Squadron had several Nieuport 12s on strength in 1920.

Russia
A small number of Nieuport 12s were imported from France and used by the Imperial Russian Air Service. Those with 130-hp Clerget engines were designated Nieuport 12bis. The Nieuport 13 B was a dual-control trainer with a 80-hp Le Rhone engine.

On 1 April 1917 there were only three Nieuport 12s in front-line service. By June that number had increased to ten, four on the western Front and six on the southwestern/Romanian front.

Two Nieuport 12s, one serving with the 1st Military School of Pilots, were in service until 1925.

Siam
Siam obtained at least one Nieuport 12 from France postwar. It was used as a trainer until 1933 and carried serial number 1.

United Kingdom
The Nieuport 12 was also used by the RNAS and RFC. The RNAS had concluded that the Nieuport 10s were too underpowered to be useful as a two-seat reconnaissance airplane. The more powerful Nieuport 12 was purchased in large numbers by the RNAS. In addition to its use as an observation plane, some Nieuport 12s were modified as single-seaters to serve as fighters. A total of 194 Nieuport 12s were used by the RNAS; all were built in France with the exception of 9201–9250. They carried serial numbers 3920–3931; 8510–8515; 8524–8529; 8708–8713; 8726–8744; 8902–8920; N3170–N3173; N3174–N3183; and N 3188. These served with Nos.1, 2, 4, and 5 Wings and in the Aegean.

The Beardmore firm built 50 under license; these featured a fixed fin ahead of the rudder as well as a cowling completely encircling the engine. The Beardmore machines were allotted serials 9201–9250 and saw service with Nos.1, 2, and 5 Wings. Later, many were transferred from the RNAS to the RFC.

A total of 122 Nieuport 12s had 110-hp Clerget engines and the other 72 had 130-hp Clergets.

As noted above, the Nieuport 12s were used by No.1 Wing (Dunkerque), No.2 Wing (Imbros), No.3 Wing (Dardenelles), No.7 Squadron (Petite Snythe) and No.10 Squadron (Dunkerque; later assigned to No.4 Wing). During the Battle of the Somme a number of the RNAS Nieuport 12s were transferred to the RFC. These were assigned to No.46 Squadron

Nieuport 12 serial N1841 named *Le Super Ral Mireli*. B83.4347.

Below: Nieuport 12 serial A5198. The fin shape identifies this as a British-built machine. J.M. Bruce/G.S. Leslie via Colin Owers.

which used the Nieuport 12 from April 1916 until April 1917. The Nieuport 12s were despised by the British crews, who felt

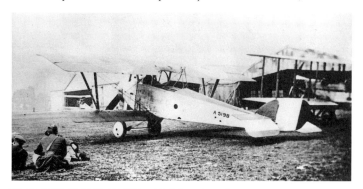

that it was dangerous to fly and had unacceptably poor performance. They were later replaced by Nieuport 20s.

Nieuport 12 Two-Seat Reconnaissance Aircraft with 110-hp Clerget 9Z
Span 9.00 m length 7.10 m height 2.70 m wing area 22.00 sq. m
Empty weight 550 kg; loaded weight 825 kg (120 kg fuel and 180 kg armament)
Maximum speed: 146 km/h at 2,000 m; climb to 1,000 m in 5 minutes 40 seconds; climb to 2,000 m in 14 minutes 15 seconds; ceiling 4,000 m; range 500 km; endurance 3 hours
Armament: one fixed 0.303 Vickers machine gun and one 0.303 Lewis machine gun on an Éteve mount
7,200 Nieuport 10s and 12s were built

Nieuport Reconnaissance Aircraft with 150-hp Le Rhône Engine

This aircraft was built in November 1916 and may have been a development of the Nieuport 12. It was a two-seat biplane powered by a 150-hp Le Rhône engine. The aircraft was not ordered by the Aviation Militaire.

Nieuport Triplanes

The Nieuport triplanes were designed with three wings staggered so that each formed an apex of a triangle. It was believed that this arrangement would eliminate the need for cross-bracing wires, increase stability, and improve lift. All three wings had very narrow chord and each had only a single spar. The staggered arrangement permitted each of the wings to serve as an empennage for the other two, thus producing a lightweight aircraft with improved lift and stability but without heavy, drag-

The first Nieuport triplane, serial N 1118. The unusual wing layout was first tried on a modified Nieuport 10. University of Texas at Dallas.

Top and above: Nieuport triplane two-seater. University of Texas at Dallas.

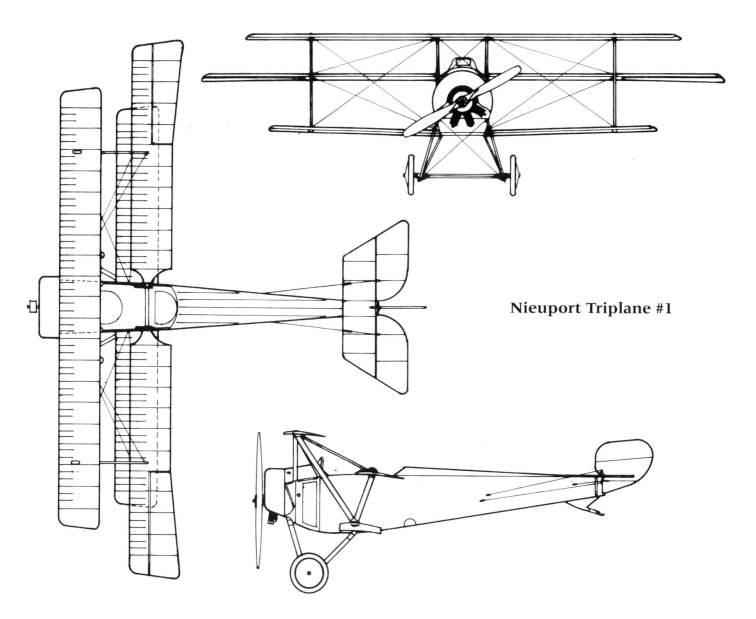

Nieuport Triplane #1

inducing struts. This arrangement was first tried on a Nieuport 10 fuselage with the top wing placed in front of the pilot; the middle wing was at the rear and the bottom wing was staggered slightly ahead of the middle wing. There were two upright center-section struts in the shape of inverted Vs. The engine of this first triplane was a 80-hp Le Rhône. The aircraft was tested in 1916 but the results are not known; however, the type was not ordered into production.

It is quite possible that the tests with the first Nieuport were not unsatisfactory, as another Nieuport triplane was built. This was a single-seater and, unlike the previous design, the upper wing was located to the rear. The middle wing was located ahead of the other wings, being mounted just behind the cowling. The bottom wing was located behind the middle wing. Armament was a single Lewis machine gun. The fuselage seems to have been based on the Nieuport 17. Power was supplied by a 110-hp Le Rhône 9J engine. The propeller had a large "cone de penetration." The type was tested in October 1916 and declared obsolete in November 1916.

The RFC and RNAS obtained an example of a Nieuport

The second Nieuport triplane. The unusual wing layout may have been intended to improve the pilot's view. The wing configuration differs from that of the first triplane. Left B87.0395; right B87.6953.

Nieuport Triplane #2

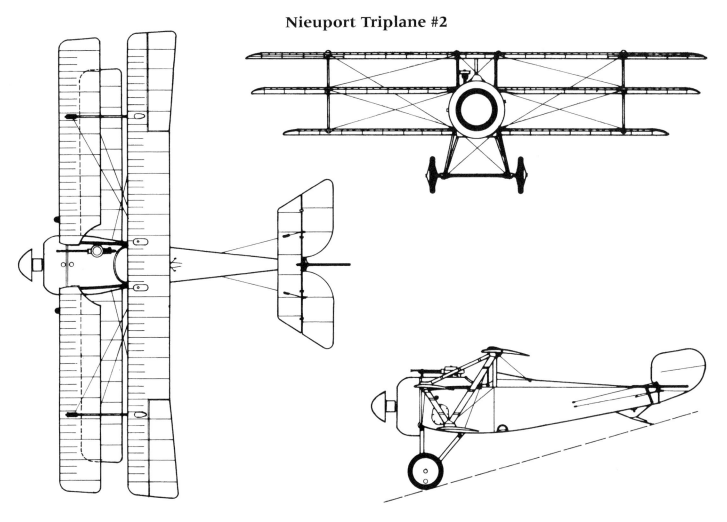

Nieuport Triplanes, Experimental Aircraft; Two with a 110-hp Le Rhône 9J and One with a 130-hp Clerget; Data for Triplane N1388

Span 8.01 m; length 5.85 m; height 2.26 m; wing area 143.16 sq. m
Empty weight 417 kg; loaded weight 629 kg
Maximum speed: 176 km/h at 3,000 m; climb to 3,000 m in 13.6 minutes
Armament: one synchronized 0.303 Lewis machine gun
Approximately three built

triplane. The aircraft was tested on 2 February 1917 and favorable comments were made on the pilot's view from the cockpit and the climb rate. However, a subsequent report dated April 1917 stated that the view directly downward and forward was poor because of the location of the middle wing. Longitudinal stability was described as poor and lateral control

The second Nieuport triplane. University of Texas at Dallas.

Above: Nieuport triplane fighter N 1388 was evaluated by the RFC in 1916. University of Texas at Dallas.

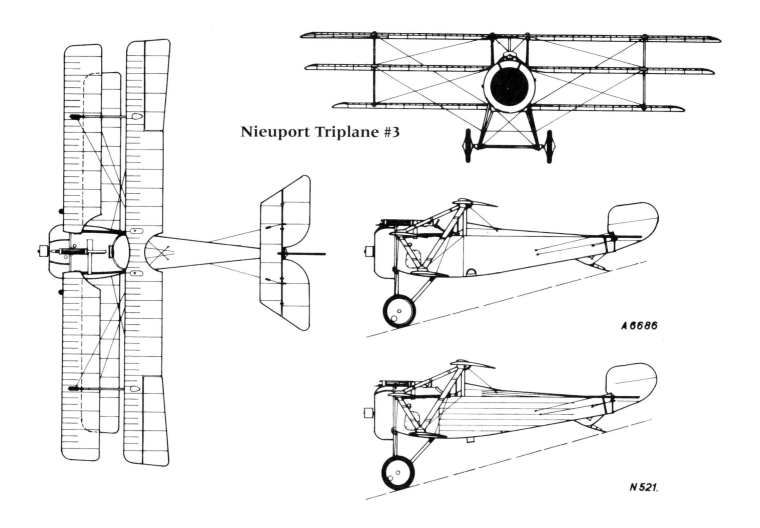

Nieuport Triplane #3

Nieuport triplane fighter N 1388 was evaluated by the RFC in 1916. University of Texas at Dallas.

was only fair. Controllability was good except during taxying; landing was described as difficult because the triplane was prone to "slew round on the ground." After these tests, the RFC, as with the Aviation Militaire, took no further interest in the aircraft.

The RNAS obtained two examples for evaluation. N 521 was powered by a 130-hp Clerget engine, had a faired fuselage, and was assigned to No.11 Squadron. It was deleted on 27 June 1917. A second aircraft, N 532, is believed to have been sent to No.11, and later No.10 Squadrons. It was deleted from service in February 1918.

Nieuport Reconnaissance Aircraft with 150-hp Hispano-Suiza

The Nieuport firm built a two-seat reconnaissance aircraft with a 150-hp Hispano-Suiza engine. The aircraft had an angular cowling with multiple louvers. The tail was upswept and had a comma-shaped rudder. The aircraft was not selected for production.

Nieuport Reconnaissance Aircraft with 180-hp Lorraine-Dietrich

Another version of the 150-hp Hispano-Suiza two-seater was built, powered by a 180-hp Lorraine-Dietrich engine. The lower wings were modified by having sections cut away at their attachment points to the fuselage; this was probably intended to improve the crew's downward view. At least one aircraft, serial N2487, was built. The type was not selected for series production.

An experimental Nieuport two-seater with 180-hp Lorraine-Dietrich engine. University of Texas at Dallas.

Nieuport Reconnaissance Aircraft with 200-hp Hispano-Suiza

Yet another two-seat reconnaissance aircraft was produced, fitted with a 200-hp Hispano-Suiza engine. As with the Lorraine-Dietrich reconnaissance airplane, this Nieuport's lower wings had cutouts at the wing roots to improve downward view. The upper wing had curved tips and horn balanced ailerons. The aircraft had separate cockpits for both crew members, who were seated far apart. Given serial N2354, the type was not selected for production. Some sources indicate that this aircraft may have been designated the Nieuport Type 22.

Another version of this same aircraft had a redesigned lower wing with two spars and a broader chord. It appears that only one was built.

An experimental Nieuport, serial 2354, powered by a 200-hp Hispano-Suiza engine. B87.0396.

Nieuport 13

The Nieuport 13 was a small two-seat biplane with unequal span wings. As with the Nieuport 10 and 12, the Nieuport 13 had a single-bay wing with the V-shaped struts typical of Nieuport aircraft. Power was supplied by an 80-hp Le Rhône engine. The aircraft had a wing area of 23 sq. m. At least one example was fitted with a 150-hp Hispano-Suiza engine and had an Éteve machine gun mount in the observer's cockpit. The Nieuport 13 was probably intended for reconnaissance, but was not selected for production.

Above: Nieuport 13 with an 80-hp Le Rhône engine. MA28515.

A variant of the Nieuport 13 was fitted with a 150-hp Hispano-Suiza engine, and that was probably aircraft N 805 shown here. University of Texas at Dallas.

Nieuport 14

A report of the Commission of Aircraft and Motors dated August 1915 mentioned that two new "fighters" were undergoing evaluation. Both aircraft were two-seaters and were designated "fighters" because they carried machine guns to defend against attacking fighters and to strafe ground troops. The true mission of both aircraft was to perform reconnaissance and artillery spotting missions. One aircraft was the Ponnier M.2, the other was the Nieuport 14. While the M.2 was actually favored over the Nieuport 14 in terms of climb rate, the Nieuport had the advantage of a stronger airframe, better maneuverability, and could be put into production more quickly than Ponnier's design.

However, the Battle of Verdun resulted in the decision to expand the fighter force. It is likely that production of the Nieuport 14 as well as the proposed Nieuport 18/19 bomber suffered as a result of this decision, the Nieuport firm having to allocate its resources to build new fighters. Furthermore, the Hispano-Suiza engines were now to be reserved for the SPAD fighter series. As a result, only a small number of Nieuport 14s were built.

The Nieuport 14 was a two-seat reconnaissance/light bomber powered by a 175-hp Hispano-Suiza 8Aa engine. The motor was enclosed by a bulged cowling with cooling radiators on either side; later examples had modified engine cowlings (see photograph). The two-bay wings were of uneven span with the top wing considerably longer than the lower. The inner pair of struts were placed vertically and the outer pair were slanted

Nieuport 14 Two-Seat Reconnaissance Aircraft/ Bomber with 175-hp Hispano-Suiza 8Aa
Span 11.90 m; length 7.90 m; height 2.65 m; wing area 30 sq. m Loaded weight 1,030 kg
Maximum speed: 138 km/h at 2,000 m; 129 km/h at 3,000 m; climb to 1,950 meters in 15 minutes, endurance 3 hours
Armament: one Lewis machine gun on a ring mount and four 120-mm bombs

outwards. Both were of the standard V configuration. The tailplane had raked tips and there was a balanced rudder. The pilot and gunner sat back-to-back, the latter armed with a single Lewis machine gun.

Two units of GB 1, VB 102 and 103, were dispatched to Plessis in January to re-equip with Nieuports and some sources have recorded that these were to be Nieuport 14s. VB 112 has also been recorded as receiving the Nieuport 14. However, a review of the Aviation Militaire's records suggest that these were in fact Nieuport fighters, and not Nieuport 14s (although it is possible that these units received a few Nieuport 14s in addition to the fighters). The Nieuport 14s had a modest bomb load of up to four 120-mm bombs.

However, a limited num-ber of Nieuport 14s were supplied to army cooperation escadrilles. Little seems to have been recorded of their activities but there were some problems with the type. A note from the Ministry of War dated 22 September 1916 warned that the radiators of the Nieuport 14 were not working adequately and instructed the escadrilles to make alterations to

Nieuport 14, serial N 724, with an operational unit. It appears that the Nieuport 14 was used in limited numbers. MA27871.

them. Furthermore, the radiators were to be drained and the filters cleaned after each flight.

Some Nieuport 14s were subsequently placed in service with training squadrons. Later, the engines were removed from some of the airframes and may have been replaced by 80-hp Le Rhône engines. (See Nieuport 82).

The RNAS ordered 50 but none were delivered.

Below: Nieuport 14. University of Texas at Dallas.

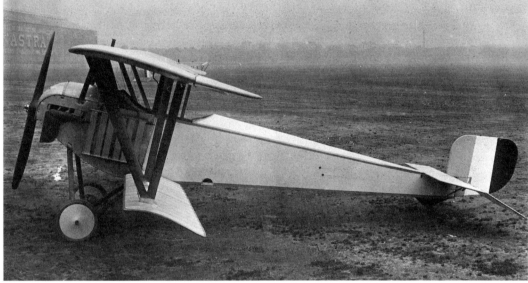

Nieuport 14 prototype. Production machines were powered a 150-hp Hispano-Suiza 8Aa engine. B87.0420.

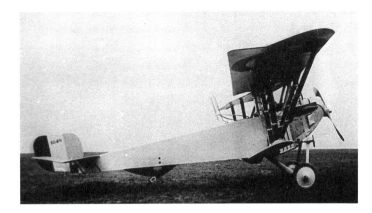

Nieuport 14. The pilot and gunner sat back-to-back, the gunner armed with a single Lewis machine gun. C. Wooley.

Nieuport 14 serial number 2416. J.J. Faye via San Diego Air & Space Museum.

Nieuport 15

Little is known about the Nieuport 15, which was built in 1916. It may have been designed for entry in the 1916 concours puissant intended to select a heavy bomber for the Aviation Militaire. However, it was never entered in that competition. The aircraft was powered by a 220-hp Renault 12F engine. and carried a crew of two in tandem, the pilot seated under the top wing and the gunner behind the top wing, The aircraft was a two-bay biplane with two sets of V-shaped interplane struts. Radiators were located on either side of the fuselage between

Nieuport 15 Two-Seat Bomber with 220-hp Renault 12F

Span 17 m, length 9.50 m, wing area 47.7 sq. m
Loaded weight 1,897 kg
Climb to 2,000 m in 16 minutes 37 seconds; to 3,000 m in 30 minutes 15 seconds
Armament: one or two machine guns on a ring mount; 14 10-kg Anilite bombs
Approximately two built

Right: Nieuport 15 prototype. The aircraft was powered by a 220-hp Renault 12F engine and carried a crew of two in tandem. 87.0397.

Below: Nieuport 15. University of Texas at Dallas.

the pilot's and gunner's positions; at least one version had a frontal radiator. The bomb load, which was apparently carried in underwing racks, has been estimated at 140 kg. At least two aircraft were built, but the type was never accepted for service with the French air service. Ten two-seaters and 60 single-seaters were ordered by the British but these were never delivered. The RNAS assigned serials N5560–N5599 to Nieuport 15s ordered in September 1916, but these were canceled in February 1917.

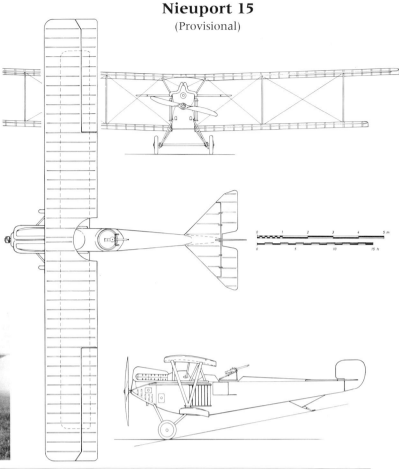

Nieuport 15
(Provisional)

Below: Nieuport 15. The huge wing (47.7 sq. m) was necessary to enable the airplane to attack strategic targets with a bomb load of approximately 140 kg. B83.1190.

Nieuport 16

The Nieuport 16 was virtually identical to the Nieuport 11 aside from the fact that the Type 16 used a 110-hp Le Rhône 9J engine. As 1916 progressed at least some escadrilles fitted their Nieuport 11s with the newer engines in the field. N 38's mechanics, for example, replaced the 80-hp Le Rhônes on all their Nieuport 11s with the newer 110-hp Le Rhônes. It is not clear if these aircraft were ever redesignated as Type 16, or if this designation was only applied to those aircraft shipped from the factory with the 110-hp engines in place. Dimensionally, the Nieuport 16 was nearly identical to the Nieuport 11. Both aircraft used the same wing, which meant that the Nieuport 16 had a higher wing loading. The fuselage remained essentially unchanged.

The aircraft began to appear in early 1916 and production was facilitated by the fact that the Nieuport 16's construction differed so little from that of its predecessor. By mid-1916 the Nieuport 16s were supplementing the Nieuport 11s already in service. N 49, for example reported that in mid-May, it had a single Nieuport 16 on strength.

Nieuport 16s were usually fitted with the same Lewis gun as the Nieuport 11s; these were mounted in the top wing. However, when the Alkan gun synchronization gear became available it permitted some Nieuport 16s to be fitted with a forward-firing Lewis gun mounted on the top decking of the fuselage. The Nieuport 16 could also be fitted with Le Prieur rockets for use against balloons; four of these rockets were fitted to each strut. The aircraft certainly had a better performance than the Nieuport 11; however, it was nose heavy and an engine failure would result in the machine nosing over in a crash landing. It also was sluggish during aerial maneuvers. An obvious solution to these problems

Nieuport 16, serial N 1109, at Girecourt (Vosges) in 1916/17. B79.424.

was to provide increased wing area, which would make the machine less nose heavy and therefore easier to control. These modifications were applied to the next fighter that would come from the Nieuport firm—the Nieuport 17.

Operational Service

It does not appear that the Nieuport 16s were ever used in large numbers and served only as an interim type between the Nieuport 11 and 17. N 49 had only one Nieuport 16 in May 1916 (serving with five type 11s and six type 12s). N 31 did not receive its first Nieuport 16s until October 1916. The Nieuport 16s-equipped escadrilles were active on the Somme front. The Groupement de Combat de la Somme had been formed in April to provide local air superiority over the front. The escadrilles assigned to this Group were N 3 (assigned in May 1916), N 26 (June), N 73 (July), and N 103 (June). Other units that used the type during the Battle of the Somme (24 June to 13 November, 1916) included the seven escadrilles assigned to the Groupement de Chasse Cachy; they were N 15 (10th Armée), N 23 (2nd Armée), N 37 (3rd Armée), N 62 (6th Armée), N 65, N 67, and N 69 (10th Armée). Other escadrilles using Nieuport 16s during the latter half of 1916 probably included N 12 (2nd Armée), N 23 (2nd Armée), N 31 (1st Armée), N 48 (2nd Armée), N 49 (7th Armée), N 57 (10th Armée), N 68, N 73 (7th Armée), N 75 (D.A.L.), N 77 (8th Armée), N 102 (3rd Armée, 4th Armée), N 112 (3rd Armée, 4th Armée), and N 124 (based at Luxeuil). For further details of these escadrilles see the entries under Nieuport 11 and 17.

GC Somme was assigned to support the 6th Armée during the battle. When the Somme offensive began the group acted in the avions d'infanterie role, participating in ground attack as well as attempting to control the skies over the battlefield. When the Germans sent additional aircraft to the Somme front, the Aviation Militaire sent the seven escadrilles of Groupement Cachy to ensure continued air superiority. GC Cachy's primary mission was to provide air cover for the reconnaissance and bombing units assigned to the Somme sector.

However, the Nieuport 16s were soon replaced by the safer and more effective Nieuport 17s. While a few escadrilles, such as N 38, continued to use Nieuport 16s until early 1917, it seems likely that by late 1916 most units had replaced their Type 16s entirely.

Foreign Service

Belgium

Belgium purchased a single Nieuport 16. It was assigned to the 1st Escadrille at Coxyde, where it supplemented the Nieuport 11s already used by that unit.

> **Nieuport 16 Single-Seat Fighter with 110-hp Le Rhône 9J**
>
> Span 7.52 m; length 5.64 m; height 2.40 m; wing area 13.30 sq. m
> Empty weight 375 kg; loaded weight 550 kg
> Maximum speed: 165 km/h at sea level; 156 km/h at 2,000 m; climb to 2,000 m in 5 minutes 50 seconds; climb to 3,000 m in 10 minutes 10 seconds; ceiling 4,800 m; endurance 2 hours
> Armament: one (sometimes synchronized) 0.303 Lewis machine gun; up to eight Le Prieur rockets
>
> **Nieuport 16 Single-Seat Fighter with 110-hp Le Rhône 9J Built by Dux**
>
> Span 8.02 m; length 5.8 m; wing area 14.7 sq. m
> Empty weight 375kg; loaded weight 535 kg
> Maximum speed: 173 km/h; climb to 1,000 m in 2.8 minutes; climb to 2,000 m in 6.4 minutes; climb to 3,000 m in 10 minutes; climb to 4,000 m in 16.5 minutes; climb to 5,000 m in 25 minutes; ceiling 6,300 m; endurance 2 hours
> Armament: one (sometimes synchronized) 0.303 Lewis machine gun

Russia

Russia purchased several examples of the Nieuport 16 and a few were built in Russia by the Dux plant. Examples purchased had 110-hp Le Rhône 9J engines. A single Nieuport 16 with an 80-hp Le Rhône was used as a trainer.

United Kingdom

The RNAS purchased several Nieuport 16s in early 1916 and assigned them to the naval base at Dunkerque. These aircraft were allocated serials 9154 through 9200. However, because of the RFC's desperate need for fighters, these airplanes were transferred directly to the RFC.

Their RFC serials were A 116–A118, A 121, A 125–126, A 130–131, A 133–136, A 164–165, A 184, A 187, A 208, A 210–214, A 216, and A 223–225. The Nieuport 16s were considered to be far superior to the D.H.2 fighters then in service, and Nieuport 16s were assigned to Nos.1, 3, 11, 29, 60, and 64 Squadrons. Armament of the British Nieuport 16s consisted of a Lewis machine gun on a flexible mount; some Nieuport 16s were equipped with the Alkan synchronization system. The Nieuport 16s were soon replaced by Nieuport 17s. The last example to be retired was probably A 131 of No.29 Squadron, withdrawn in April 1917. Many Nieuport 16s were subsequently sent to the Scout School at No.1 Aircraft Depot at St.-Omer.

See Nieuport 11 for Nieuport 16 drawing.

A wrecked Nieuport 16 serial number 976 of Escadrille MF 1 having its Le Prieur rocket launchers removed. B79.1551.

Nieuport 160-hp Renault Bomber

In 1916 the Nieuport firm built a bomber powered by a 160-hp Renault engine. This airplane was of similar size and had a performance comparable to the Nieuport 15. It did not enter service.

Nieuport Twin-Fuselage Aircraft

A twin-engine tractor biplane with two fuselages was built in 1916. It may have been intended for use as a bomber or escort fighter/destroyer in the same F category (Fortress) as the Breguet 11 and the Van Den Born F5. It appears that the RNAS placed an order with Nieuport for this aircraft in 1915. The Nieuports were described as "Twin Biplanes" and were allocated serials 1395–1397. They were to be powered by two 110-hp Clergets (indicating that this order might have actually been for the Nieuport 18/19 described below). The order was canceled in December 1915.

Nieuport Two-Seaters with 120-hp Hispano-Suiza

The Nieuport firm built at least two single-engine tractor airplanes powered by 120-hp Hispano-Suiza engines. The aircraft carried a crew of two, had a square radiator in the nose, a swept back upper wing, and a single bay of struts. Apparently, one had redesigned wings probably to provide improved strength. The aircraft seem to have been intended for ground attack and bombing. Neither of these versions was selected for production.

Nieuport 17

The Nieuport 16 was powered by a more powerful 110-hp Le Rhône 9J engine, but it employed the airframe of the Nieuport 11, which had initially been powered by a 80-hp Le Rhône. This resulted in an aircraft that was nose heavy and sluggish in combat. The obvious answer was to design an aircraft with an enlarged wing; the resulting aircraft became the Nieuport 17.

The Nieuport 17, as with the Nieuport 11, had a single-spar lower wing and an enlarged wing area of 14.75 sq. m (compared with 13.30 sq. m for the Nieuport 16). Transparent panels were fitted to the upper wing sections above the pilot's head. The engine had a circular cowling carefully faired into the fuselage sides. Some early Nieuport 17s had a "cone de penetration" attached to the fixed crankshaft of the 110-hp Le Rhône. The Nieuport 17 engine cowling was usually made of aluminum and built in two halves joined near the centerline. In later aircraft problems with cowling fractures resulted in their replacement by a new cowling of horseshoe configuration. The fuselage had rectangular cross-section frames braced with wire. The front of the fuselage was composed of steel tubes. In front of the cockpit were longerons made of ash while aft of the cockpit they were made of spruce. The top decking was faired with light formers and longitudinal stringers. Behind the cockpit the fuselage was fabric-covered and was reinforced by plywood panels near the stern. The upper wing had the front spars close to the leading edge while the rear spar ran parallel to the lower wing's spar; this placed the interplane struts at a good angle for load carrying. The wing spars were made of spruce. The bottom wing had an incidence that could be adjusted on the ground to facilitate rigging the aircraft for varying loads. The Nieuport 17 retained the sesquiplane layout, resulting in a better field of vision for the pilot. The center section struts were vertical at the front and had an inverted V arrangement. The interplane struts were made of spruce. The leading and trailing edges of the wing were spruce strips. Ailerons were fitted to the top wing only and had increasing chord toward the tip. The tail surfaces were of light steel tubing covered with fabric. The landing gear cross member between the V legs was made of aluminum with a steel tube axle sprung by rubber shock absorbers; the tail skid was a curved steel spring.

A major improvement was made in the Nieuport 17's armament; a synchronized Vickers gun was fixed to the centerline of the forward fuselage. Canvas ammunition belts were carried in drums stored inside the fuselage fairings. The Nieuport 17 could also carry four Le Prieur anti-balloon rockets on each of the main struts. A November 1916 GQG memo reported that some Nieuport 17s were being fitted with Vickers machine guns mounted off-center. This change required relocation of the oil reservoir, the fuel tank, munitions box, machine gun support, engine cooler, exhaust pipe, and windscreen. Some Nieuport 17s were armed with twin Lewis machine guns that were unsynchronized and were mounted on the top wing. The Nieuport 17s were also used in the high-speed reconnaissance role fitted with a 26-cm camera behind the pilot's seat.

Nieuport 17s were fitted with a variety of engines. In some a 120-hp Le Rhône 9Jb was used. One was given a 130-hp Clerget 9B engine and an entirely new tail unit. It would serve as the basis for the Nieuport 17bis, which was basically a Nieuport 17 with either a 130-hp Clerget 9B or 9Z engine. However, the fuselage was fully faired from the engine cowling to the fuselage spacer ahead of the tail skid. The engine cowling was narrower than that of the standard Nieuport 17. At least a few of these

A lineup of Nieuport 17s of Escadrille N 90. The unit insignia was the profile of a rooster. B76.1757.

Above, above right, below, below right: Four views of Nieuport 17 N 1424 with cône de pénétration. University of Texas at Dallas.

aircraft saw active service; one was used by Charles Nungesser. The British experience with the Type 17bis shows that its performance was disappointing and it was difficult to fly. Its weight was high, resulting in a fast landing speed and poor maneuverability. The aircraft probably served in small numbers for a brief period of time, with the French escadrilles; it certainly saw only limited service with the RNAS. Although it had marginally better performance than the standard Nieuport 17, the 17bis was clearly inferior to the SPAD 7 that became available in late 1916. It has been reported that the French aces felt their Nieuports were clearly outclassed by contemporary German fighters and demanded that they be given the new SPAD 7 fighters as quickly as possible. Although attempts were made to improve the Nieuport series of fighters, the SPAD 7 and 13 fighters became the main French fighters for 1917 and 1918.

Nieuport 17s were built under license by CEA (Société pour la Construction et l'Entretien d'Avions), Ateliers d'Aviation R Savary; et H de la Fresnaye, and SAFC (Société Anonyme Francaise de Constructions Aéronautique).

Variants

1. A single Nieuport 17bis derivative had a Clerget rotary engine (type unknown). The fuselage was fully faired. The wings were different from those of the standard Nieuport 17 in that the top wing was not swept back. The fin and rudder were constructed of wood rather than steel. It also had a "cone de penetration" as wide as the fuselage.
2. At least one example of the Nieuport 17bis was fitted with a 150-hp Le Rhône engine. The fuselage was faired and preceded by a huge cone de penetration. The rudder was horn-balanced and the tailplane and rudder were modified from those on the standard 17bis. A single Vickers machine gun was mounted on the upper longeron.

Operational Service
Army Cooperation

Only a small number of Nieuport 17 escadrilles served with the Groupes de Combat. Most units were assigned to support the armées. These units were:

N 12, moved in October 1916 to the Verdun sector and assigned to the 2nd Armée.

N 15, assigned to the 10th Armée; it joined GC 13 in March 1917.

N 23, assigned to the 2nd Armée. Later N 23 moved to the Verdun sector. Although briefly assigned to the 7th

Nieuport 17bis N 2376. This aircraft differed from the Nieuport 17 in having a Clerget 9 engine and a fully-faired fuselage. University of Texas at Dallas.

Armée, N 23 spent virtually the entire war assigned to the 2nd Armée. It re-equipped with Nieuport 17s at the end of 1916. N 23 became SPA 23 in November 1917.

N 26, commanded by Capitaine Jacques de Sieyes de Veynes in May 1916. N 26 was assigned to the Groupe de Combat de la Somme in mid-1916.

N 31, moved to the Verdun front in early 1916 under the command of Lieutenant Lucien Courcet de Villeneuve. In October 1916 it was assigned to GC 11.

N 37, assigned to the 6th Armée in July 1916 as part of GC Cachy. N 37 moved to the 3rd Armée in January 1917 and the 4th Armée in March. It joined GC 15 in March 1917.

N 38, assigned to the 4th Armée and active during the Battle of the Somme. It moved to Champagne in April 1917 as part of GC 15 (although it was briefly attached to the 4th Armée).

N 48, moved to the 2nd Armée sector in August 1916 and Lieutenant Georges Matton assumed command in October. The following month N 48 became part of GC 11.

N 49, assigned to the 7th Armée. It was initially commanded by Capitaine Constantin Zarapoff, but in May 1916 Capitaine Jules de Boutiny assumed command. N 49 was based at Fontaine in May 1916. In December it re-equipped with SPADs.

N 62, formed from MF 62 in 1916. It was active over the Somme in July 1916. In January 1917 the escadrille moved to Fismes and supported the 6th Armée during the Battle of Chemin des Dames. In March 1917 N 62 participated in the battles at La Malmaison.

N 68, commanded by Capitaine Garde in November 1916; Capitaine Lemerle assumed command in early 1917. N 68 was assigned to the 8th Armée and in November 1917 became SPA 68.

N 69, assigned to the 10th Armée in July 1916. Lieutenant Paul Malavaille assumed command in January 1917.

N 75, formed in July 1916 under Capitaine Henri de Montfort. It was assigned to the 8th Armée and initially based at Lunéville but moved to Saiserais in March. In April 1917 N 75 joined GC 14.

N 76, created from R 76 in early 1917. The escadrille was assigned to the 5th Armée and was based at Muizon. Initially commanded by Lieutenant Rene Doumer; Capitaine Jean-Jacques Perrin assumed command in April 1917.

N 77, formed in September 1916 at Lyon-Bron. Assigned to the 8th Armée on the Lorraine front, the unit was commanded by Lieutenant Joseph l'Hermite. A notable mission occurred on 23 February 1917 when two aircraft of N 77 bombed the German airbase at Marinois.

N 78, formed at St. Étienne-au-Temple in December 1916. It was commanded by Lieutenant Armand Pinsard. N 78 joined GC 15 in March 1917.

N 79, formed in November 1916. Under the command of Lieutenant Roger Lemercier de Maisoncelle-Vertille de Richemont, N 79 was assigned to the 3rd Armée. In March 1917 Lieutenant Gaston Luc-Pupat assumed command.

N 80, created at Lyon-Bron in December 1916. Commanded by Capitaine Francoise Glaize, N 80 was assigned to the 5th Armée. In January 1917 it moved to Bonne-Maison. N 80 joined GC 14 in March 1917.

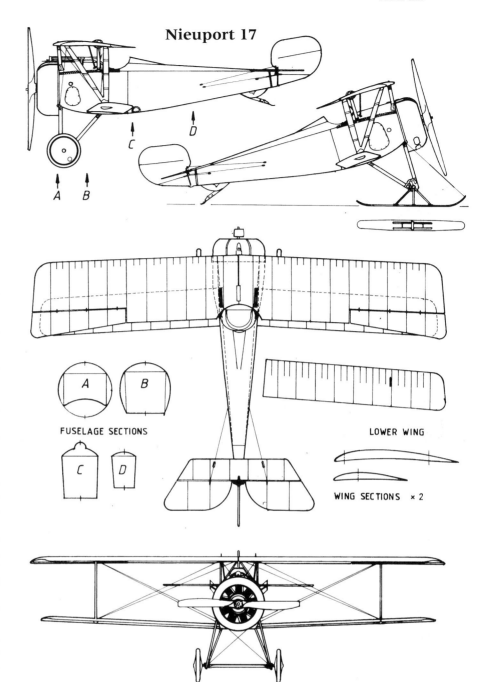

Nieuport 17

Nieuport 17 N 1831 flown by Lt. Santa Maria which was captured by the Germans. The aircraft was brought down by Fl. Abt. 65. J.M. Bruce/G.S. Leslie via Colin Owers.

Guynemer's Nieuport 17 of SPA 3. The cône de pénétration was fixed; it was not a spinner. B88.353.

N 81, formed at Villacoublay under command of Capitaine Maurice Mandinaud. N 81 was assigned to the 6th Armée (at Sacy-le-Grand) and then the 7th Armée (Fontaine) during January 1917. Lieutenant Raymond Bailly assumed command in March 1917. N 81 moved to the 4th Armée sector in April 1917 and joined GC 15 a few days later.

N 82, formed in January 1917 and assigned to the 1st Armée. Based at Sacy-le-grand, the escadrille was commanded by Lieutenant Raoul Echard. It subsequently moved to Fontaine, Chaux, and Bonne-Maison. N 82 briefly joined GC 14 in April 1917. N 82 was later assigned to the 7th and 3rd Armées. N 82 moved to 10th Armée sector in November 1917 and was sent to Italy. It was based at San Pietro di Godego and provided fighter escort during reconnaissance and bombing missions. During the Tomba offensive the units bombed and strafed enemy artillery and machine gun positions. On 1 January 1918 N 82 was based at San Pietro in Gu. It conducted long-range reconnaissance missions over Adige, Brenta, and Piave, and provided fighter escort for the army cooperation units of the 31st C.A. In March 1918 the 10th Armée and its aviation units returned to France. Upon its return N 82 was assigned to the GC 22.

N 83, formed at Lyon-Bron in January 1917 and assigned to the 5th Armée, commanded by Capitaine Charles Béranger. In early 1917 it moved to the 6th Armée sector and was based at Rosnay. N 83 participated in the Battle of Chemin des Dames. In March it moved to Bonne-Maison and joined GC 14 later that month.

N 87, formed in March 1917 at Lyon-Bron and assigned to the 8th Armée.

N 88, formed in March 1917 and assigned to the 7th Armée. Commanded by Capitaine Francois d'Astier de la Vigerie, N 88 moved to the 6th Armée sector at Chemin des Dames in June 1917. It became SPA 88 in September 1917.

N 90, formed from detachments N 504 and N 507 in March 1917 and assigned to the 10th Armée, it was commanded by Lieutenant Pierre Weiss. N 90 became SPA 90 in April 1918.

N 91, formed in April 1917 by combining detachments N 505 and N 507 and assigned to the 8th Armée, based at Lorraine. It was commanded by Lieutenant Jourdain and in January 1918 became SPA 91.

N 92, formed in May 1917 at Chaux from Detachments 502 and 503 and assigned to the 7th Armée, commanded by Lieutenant Edmond George. In May Capitaine Georges de Geyer d'Orth assumed command, and in July N 92 moved to the 2nd Armée sector.

N 93, formed from Detachments N 501 and N 506 in April 1917 and assigned to the 7th Armée in the Vosges sector, commanded by Lieutenant Jean Moreau. N 93 provided support for the bombers of GB 4 and even participated in leaflet-dropping missions. It joined GC 15 in July 1917.

N 102, stationed at Malzéville. Commanded by Chef de Bataillon Pouderoux, it was assigned to the 2nd Armée. In January 1917 it moved to the area of the Marne. Lieutenant Jean Derode assumed command of the escadrille and in July N 102 moved to Flanders.

N 124 began using Nieuport 17s in September 1915 and moved to the 6th Armée sector at the Somme. It was assigned to GC 13 in November 1916. In April 1917 N 124 was assigned to the 3rd Armée at the Somme. N 124 moved to the 6th Armée sector at the Aisne.

N 392 was formed from N 92 in June 1916 and served as an escadrille de protection for the city of Venice.

N 581, with eight Nieuport 17s and seven SPAD 7s, was formed in February 1917. Assigned to the 7th Russian Army, it began operations in May 1917. Based at Boutchatch Galicia, it served alongside Captain Kosakov's Nieuport fighter group. In response to the German offensive of 19 July on the 6th Army front, N 581 was transferred to Buzcacz. In August it moved to the Romanian front where it was assigned to the 3rd Corps. By February 1918 N 581 was re-designated SPA 581 because it had re-equipped with SPAD 7s.

Nieuport 17 N 1490 of Escadrille N 124. As indicated by the fuselage insignia, this airplane was flown by Charles Nungesser after his 15th victory. B83.1298.

Nieuport 17 of Jacques Allez, possibly of N 65. The man-in-the-moon insignia was probably a personal marking. B83.1201.

The escadrilles assigned to the individual armées tended to have less glamorous combat careers. Their records reveal that these Nieuport escadrilles were in almost constant action, flying escort for reconnaissance and bomber aircraft, fighter patrols, and performing reconnaissance sorties (with Nieuport 17s accompanying Nieuport 12s). They also attacked balloons, transported spies behind enemy lines (Nieuport 17s flying air cover with the Nieuport 12s or Sopwith 1½ Strutters acting as transports), and even performing ground attack. Often these units had limited opportunities for aerial combat and scored few victories; therefore, they tend to be ignored by historians. N 90, for example, was assigned to the 8th Armée and was in action from its formation in early 1917 until the end of that year. However, it recorded only a single confirmed victory for 1917 at the cost of one pilot killed and one wounded. Yet these missions were equally as important to the war effort as those performed by the Nieuport escadrilles assigned to the Groupes de Combat.

Two escadrilles equipped with Nieuport 17s were formed in early 1917 and were assigned to the Escadrilles de L'Interieur (D.C.A.). N 311 was formed in February and disbanded in June. N 312 was formed in March and disbanded in July.

The Aviation Maritime also acquired some Nieuport 17s and used them to fly fighter escort for flying boats based at Dunkerque, which had been suffering heavy losses at the hands of German fighters. The short range of the Nieuport 17s meant that they were of only limited use as escort fighters, and they were eventually replaced in the escort role by Letord twin-engine airplanes under control of the Aviation Militaire.

Groupes de Combat

Nieuport 17s were used in large numbers by the French. Every Nieuport escadrille used Nieuport 17s beginning in late 1916 and many were in service until mid-1917. A major revision in the French fighter force was taking place as the Nieuport 17s began to enter service in the fall of 1916. The formation of fighter groups had permitted the French to achieve local air superiority over selected areas of the Somme front. The temporary Groupe de Combat Somme was superseded by the formation of permanent Groupes de Combat. GC 11 was created in November 1916 and was later joined by GC 12, GC 13, GC 14, and GC 15, all of which were equipped with Nieuport 17s.

GC 11 was created at Vadelaincourt and initially had four Nieuport 17 escadrilles—N 12, N 31, N 48, and N 57—assigned to it (N 94 arrived later). GC 11 was assigned to the 2nd Armée and later the G.A.R. (4th, 5th, and 6th Armées). It provided support during the Battle of Chemin des Dames.

GC 12 was also formed in November 1916 with the escadrilles that had been serving with the Groupement de Combat de la Somme, N 3, N 26, N 73, and N 103. GC 12 provided air cover for the G.A.R. (4th, 5th, and 6th Armées) and also the 10th Armée. GC 12's escadrilles participated in the Battle of Chemin des Dames.

GC 13 was formed in November 1916 with N 65, N 67, N 112, and N 124 (N 37 joined later). The group was assigned to the G.A.N. and based at Ravenal and later was assigned to the 3rd Armée. In June GC 13 consisted of N 15, N 65, N 84, and N 124, based at Maisonneuve on the Aisne front in the 6th Armée sector, later moving to the 5th Armée front. GC 13 was active during the Battle of Chemin des Dames. During June its escadrilles received SPAD 7s and only N 124 continued to use Nieuport fighters as primary equipment until it also received the new SPAD 7s in 1917.

GC 14 was formed in March 1917 with component escadrilles N 75, N 80, N 83, and N 86. It was initially assigned to the G.A.R. (4th, 5th, and 6th Armées) and was based at Bonne-Maison. From 16 April to 7 May GC 14's units participated in the Battle of Chemin des Dames and provided support for the reconnaissance units assigned to the 10th Armée. In July GC 14 was assigned to the 2nd Armée.

GC 15 was formed in March 1917 with seven Nieuport escadrilles: N 37, N 78, N 81, N 85, N 93, N 97, and N 112 (later N 102 joined). GC 15 was based at la Noblette in the 4th Armée sector where it provided support during the Battle of Champagne (April 17). By July most of the Nieuport escadrilles had re-equipped with Nieuport 23s and 24s.

The Groupe de Combat Chaux was formed in February 1917, consisting of Nieuport 17 escadrilles N 49, N 81, and N 82. The groupe was assigned to the 7th Armée sector and based at Chaux. In mid-1917 the Groupe was disbanded and the escadrilles dispersed.

Duties for the Groupes de Combat included escorting French day bombers that were attempting to interdict German supply lines. However, because of their relatively poor range, the Nieuport 17s were found to be of little use as long-range fighters. Nieuport 17s also flew escort for army cooperation escadrilles assigned to the Armées. They also intercepted German reconnaissance airplanes and attacked balloons in an attempt to keep the enemy from obtaining an accurate picture of the positions of the French troops. By the spring of 1917 more Nieuport 17s were reaching the escadrilles and this resulted in a significant improvement in the efficacy of the fighter units. However, the Germans were now sending Albatross D.III fighters to their units, and this new type proved to be remarkably successful. It was for this reason that the Nieuport units began to re-equip with the new SPAD 7s as soon as they became available.

Nieuport 17 of Escadrille N 91. The unit's insignia was an eagle carrying a bomb with a lit fuse in its talons. B76.1765.

Foreign Service

Belgium
Belgium purchased 12 Nieuport 17s to equip the 1st and 5th Escadrilles. These units provided fighter escort for reconnaissance squadrons. By late 1917 the Nieuport 17s were replaced by Hanriot HD.1s.

Chile
Chile purchased a single Nieuport 17 in 1919.

Colombia
Colombia acquired four Nieuport 17s in 1921. They served with the Escuela Militar de Aviacion until 1925.

Czechoslovakia
At least one Nieuport 17 was obtained by the Czech air service; it was subsequently captured by the Estonians.

Estonia
The Estonian Aviation Company was formed in 1920 and acquired six Nieuport 17s and 24s from Russia and one from the Czechs. They were assigned serial numbers 42 through 46 and 51.

Hungary
A single Nieuport 17 was captured on 15 May 1919 when a Romanian machine landed behind Hungarian lines. It was pressed into service with the 8th Voros Repuloszazad.

Finland
A single N.17 built by Dux was flown to Finland in 1918.

Netherlands
The Netherlands purchased five Nieuport 17s. They arrived on 24 June 1918 and were given serials N 220 to N 224. N 220 was powered by a 80-hp Thulin motor. The Nieuport 17s were withdrawn in 1925.

Italy
During 1917 the Nieuport 17 replaced the Nieuport 11 in Italian service. Macchi began to build the type under license in December 1916. Approximately 150 were built by the Macchi firm, becoming known as "Super-Bébés." The engine was a 120-hp Le Rhône. The Italians found that the Nieuport 17's lower wing spar broke during certain maneuvers. Because of this, the Italian pilots, along with their counterparts in the RFC, flew the Type 17 with great caution.

During the first half of 1917 the following squadriglias received Nieuport 17s to supplement the Nieuport 11s then in service: 1st Gruppo (3rd Armata): 77a and 80a Squadriglias; 2 Gruppo (2nd Armata): 76a and 81a Squadriglias; 7th Gruppo (1st, 6th Armata): 79a Squadriglia; 9th Gruppo (1st Armata): 71a and 75a Squadriglias; 10th Gruppo (Supreme Command: 70a, 78a, and 82a Squadriglias; and independent units: 1st Sezione 83 Squadriglia based at Kremain (Macedonia) and Sezione Nieuport 11s and 17s at Belluno.

The Nieuport 17 squadriglias flew combat patrols, reconnaissance missions, and escort for Caproni bombers. They were also used in the ground attack role, striking enemy troops with bombs and machine gun fire. The Nieuport 17s proved to be particularly effective in preventing Austro-Hungarian reconnaissance airplanes from penetrating the Italian lines. In addition, many Nieuport 17 seziones (sections) were formed to defend important Italian cities, ports, airship bases, and ammunition dumps.

By October 1917 the following units still had Nieuport 17s: 1 Gruppo (3rd Armata): 77a, 80a, and 84a Squadriglias; 2 Gruppo (4th Armata): 76a Squadriglia; 3 Gruppo (1st Armata): 72a Squadriglia; 7 Gruppo (1st Armata): 79a Squadriglia; 9 Gruppo (1st Armata): 71a and 75a Squadriglias; 10 Gruppo (Supreme Command): 70a, 78a, and 82a Squadriglias; and independent units 1st Sezione 83a Squadriglia at Negocani (Macedonia), 2nd Sezione 83a Squadriglia (Belluno), 3rd Sezione 83a Squadriglia (Cavazzo Carnico), and 85a Squadriglia at Piskupi (Albania).

By the end of 1917 Nieuport 27s, Hanriot HD.1s, and SPAD 7s were replacing the Nieuport 17s still in service. By June 1918 only 2 units had Nieuport 17s: 72a Squadriglia based at Busiago as part of the 9th Gruppo (7th Armata) and 74a Squadriglia based at Castenedolo as part of the 20th Gruppo (7th Armata).

Poland
Poland acquired at least five Nieuport 17s. Two (Nos.6176 and 4233) were captured from the Russians on 7 August 1919. No.6176 served at the Lvov front and, later with the 4th Fighter Group, while No.4233 was sent to the 1st Air Division at Mlodeczno. A Nieuport 17bis was captured in August 1920 and sent to the 2nd Squadron of the 2nd Regiment. A Nieuport 17bis was assigned to the 5th Squadron of the 2nd Regiment in 1920. Finally a Nieuport 17 (or possibly a 21) was assigned to the Moktow area in early 1921.

Romania
Romania purchased 25 Nieuport 17s in 1917, They were assigned to the 1st Fighter Squadron (based at Borzesti) serving with Grupul 1 in the 2nd Romanian Army sector and the 10th Fighter Squadron (based at Bozesti) assigned to Grupul 3 with the 6th Russian Army. Later in 1917 two more squadrons were formed with Nieuport 17s: the 11th Fighter Squadron (Galati) with Grupul 3 assigned to the 6th Russian Army and the 14th Fighter Squadron (Cioara) with Grupul 2 in the 1st Romanian Army sector. Finally, the 3rd Fighter Squadron also had some Nieuport 17s on strength. It was based at Cioara and was assigned to Grupul 2 in the 1st Romanian Army sector.

The Nieuport 17s were used to establish air superiority over sections of the front and to fly bomber escort missions. The aerial activity was not as great as that on the Western Front; there were only 31 air-to-air victories over the Romanian front during the entire war. The 3rd Fighter Squadron accounted for more than one-fifth of these victories, destroying seven enemy aircraft. The Nieuport units also flew reconnaissance patrols. The

Camouflaged Nieuport 17 of Capitaine Auger of SPA 3 in October 1916. B83.5654.

surviving Nieuport 17s saw action in 1919 against the Hungarian Red Airborne Corps.

Russia

The Imperial Russian Air Service purchased a number of Nieuport 17s from the French. Subsequently, the Dux firm, and possibly the Mosca firm, built the type under license. At least 14 were built by the Dux factory. The Nieuport 17 was the first fighter in Russian service to be equipped with a synchronized machine gun. A few Russian Nieuport 17s were fitted with twin machine guns. While most of the Nieuport 17s had a 110-hp Le Rhône, some had 130-hp Clerget engines.

In March 1917 nine Nieuport 17s were assigned: western front (1); southwestern front (2); and Romanian front (6).

The next month there were 14 Nieuport 17s in front-line service. By September there were an additional 23 available—not at the front, but rather in storage at the Central Aviation Warehouse.

Nieuport 17s were used extensively during the civil war by the 1st Red Air Force Fighter Squadron. Nieuport 17s, along with Sopwith 1½ Strutters, also equipped the Slavo-British Squadron commanded by Kozakov.

In December 1920 no fewer than 189 Nieuport 17s were still in front-line service, most based on the Leningrad and Ukrainian fronts. In January 1922 69 remained in service with the 2nd, 3rd, 4th, and 14th Aviaotryady, the 1st and 2nd naval Istrootryady, the Eskadra No.2, and the schools. In 1923 the Nieuport 11 trainers were assigned to the 1st Higher School of Military Pilots in Moscow, the 1st Military School of Pilots at Kacha, the 2nd School of Military Pilots, the Higher School of Aerial Observers, the Military-Technical School, the Training Eskadril'ya, the 16th, 17th, and 47th Otdel'nye razvedivatel'nye aviaotryady, and the Strel'bom school at Serpukhov. The last 29 Nieuport 17s were struck off charge in April 1925. The Siberian Air Fleet of Admiral Kolchak also had a few Nieuport 17s.

Siam

At least one Nieuport 17 was used by the Royal Siamese Aeronautical Service.

Ukraine

The Ukrainians obtained three Nieuport 17s in 1918. One was with the Don Cossack squadron. another had serial No.1437. In 1919 a defecting Czechoslovakian pilot supplied the Ukrainians with a third Nieuport 17.

United Kingdom

The RNAS was the first British service to re-equip with Nieuport 17s. The first three were given serial numbers N 1553, N 1494, and N 1561. Subsequently, the RNAS purchased Nieuport 17bis airplanes which had the 130-hp Clerget 9B and 9Z engines. Most of the Nieuport 17s were built under license by the English Nieuport firm. The Nieuport 17s received serial numbers N 3100–N 3104, N 3184–N3187, N3189–N3197, N 3198–N 3209, and N 5860–N 5909. They were assigned to Nos.6 and 11

Nieuport 17 Single-Seat Fighter with 110-hp Le Rhône 9Ja

Span 8.16 m; length 5.80 m; height 2.40 m; wing area 14.75 sq. m
Empty weight 375 kg; loaded weight 560 kg

Maximum speed:	sea level	165 km/h
	2,000 m	160 km/h
	3,000 m	154 km/h
	4,000 m	137 km/h
Climb:	2,000 m	6 minutes 50 seconds
	3,000 m	11 minutes 30 seconds
	4,000 m	18 minutes 5 seconds

Ceiling 5,300m; range 250 km; endurance 1 hour 43 minutes
Armament: a synchronized 7.7-mm Vickers gun; some were fitted with one or two Lewis guns on an overwing mount; up to 8 Le Prieur rockets could be carried

Nieuport 17bis Single-Seat Fighter with 130-hp Clerget 9B

Span 8.16 m; length 6.00 m; height 2.40 m; wing area 14.75 sq. m
Loaded weight 573 kg
Maximum speed 190 km/h at sea level; 175 km/h at 500 m; 172 km/h at 3,048 m; climb to 3,048 m in 9 minutes 20 seconds; climb to 6,096 m in 32 minutes
Armament: a synchronized Vickers gun; some were fitted with Lewis guns on an overwing mount. Up to 8 Le Prieur rockets could be carried

Macchi-Nieuport 17 Single-Seat Fighter with 120-hp Le Rhône 9Jb

Span 8.20 m; length 5.70 m; height 2.40 m; wing area 15.00 sq. m
Empty weight 410 kg; loaded weight 590 kg
Maximum speed: 170 km/h; climb to 3,000 m in 11 minutes 30 seconds; ceiling 5,500 m; endurance 2 hours
Armament: a synchronized Vickers gun; some were fitted with Lewis guns on an overwing mount
150 built

Nieuport 17 Single-Seat Fighter with 110-hp Le Rhône 9Ja Built by Dux

Span 8.02m; length 5.8 m; wing area 14.7 sq. m
Empty weight 375 kg; loaded weight 560 kg
Maximum speed: 164 km/h; climb to 1,000 m in 3.2 minutes; climb to 2,000 m in 6.8 minutes; climb to 3,000 m in 11.5 minutes; climb to 4,000 m in 19.5 minutes; climb to 5,000 m in 35 minutes; ceiling 5,300 m; endurance 2 hours
Armament: a synchronized Vickers gun; some were fitted with a Lewis gun on an overwing mount
At least 14 built

Squadrons and the Eastchurch Flight School. Additional Nieuport 17s were later ordered and given serials N 6040–N 6079 and N 6530–N 6579. However, both of these additional orders were subsequently canceled.

Two flights of Nieuport 17s were sent to Furnes to provide air cover for reconnaissance airplanes. Nieuport 17s proved to be particularly useful at the Dunkerque naval base and over the Somme. No.3 Wing RNAS operated some Nieuport 17bis planes in the Dardenelles campaign. Machines built by the English Nieuport firm were found to be inferior to those manufactured in France.

On 1 May 1917 there were ten Nieuport 17bis airplanes on strength, assigned serials 3956–3958 and 8745–8751. They were sent to Nos.1 and 4 Wings and some were later transferred to the RFC. The Nieuport 17bis was considered to be difficult to fly and had a higher accident rate than the Sopwith scouts and triplanes. It was primarily for this reason that the Nieuports were withdrawn from RNAS service in mid-1917 and replaced by Sopwith Camels.

Nieuport 17 of Escadrille N 48 which was assigned to GC 11. B82.4620.

The RFC obtained its first Nieuport 17s (N 1494, N 1553, and N 1561) from the RNAS. The RFC Nieuport 17s had a single Lewis machine gun on a Foster mount. A 6678 was tested with a fixed Vickers and an overwing Lewis, but the two guns seriously degraded the aircraft's climb and maneuverability and no other Nieuport 17s were so armed.

The Nieuport 17s were supplied to Nos.1, 11, 29, 40, and 60 Squadrons, and were found to be quite satisfactory and were much sought after. Because of the high attrition in these fighter units, the RFC ordered large numbers of Nieuport 17s from the French. The French were reticent about releasing such large numbers of their best fighters to the British, but they usually relented and supplied the machines. In addition to combat losses, several Nieuport 17s were lost as a result of failure of the lower wings. Modifications to the lower wing spars and their fittings rectified this problem.

The Nieuport 17s served well into 1917 until finally replaced by Nieuport 24s or S.E.5as. At least two surplus Nieuport 17s were subsequently sent to Nos.14 and 113 Squadrons in Palestine. These provided fighter escort for reconnaissance aircraft and also were used as high speed scouts.

United States

The United States purchased 175 Nieuport 17s for use by the A.E.F. The first example was obtained in September 1917. These airplanes were used as fighter trainers. A single sample of a Nieuport 17bis was also obtained for evaluation.

Nieuport 18 and 19

With the failure of the 1915 concours to produce a suitable bomber, it was decided to hold another competition in 1916. Nieuport designed a twin-engine bomber with the engines in a tractor configuration. Two versions were produced. One had two 150-hp Hispano-Suiza engines and was designated the Nieuport 18. The other had two 150-hp Clerget engines and was designated the Nieuport 19. Both carried a crew of three. Known dimensions were a wing span of 18.20 m, a length of 13.30 m, and a wing area of 53 sq. m.

Although no photographs of the aircraft have been found, drawings show a configuration similar to the Gotha G.V. In typical Nieuport fashion, the biplane wings were of unequal span, the top wing substantially longer than the bottom. The bottom wing was also narrower than the top wing. The two-bay

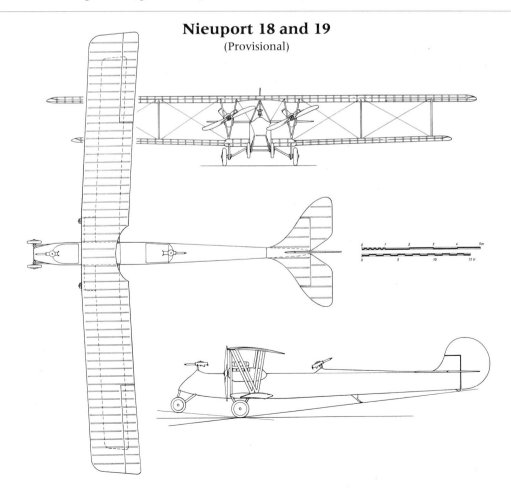

Nieuport 18 and 19
(Provisional)

wing used V-shaped interplane struts. As with most French bombers, the landing gear consisted of a single large wheel under each engine nacelle and a pair of large nosewheels. One gunner was seated in the nose and another was located well aft of the wings. The pilot was seated under the top wing. The very long rear fuselage ended in a oval fin and rudder assembly and the tail surfaces were curved.

The aircraft was not selected for production. It has been suggested that the Aviation Militaire discouraged Nieuport from developing a bomber prototype because it was feared that such a project might interfere with the production of the badly needed Nieuport fighters.

Apparently, the Royal Naval Air Service saw some merit in the aircraft, as 25 were ordered (with crew complement reduced to two) with 110-hp Clerget engines. Serial numbers allocated were 3940–3945, 3953, 3995–3997, and 8475–8486. It is possible that the failure to secure an order from the Aviation Militaire led the Nieuport firm to abandon further development of the Types 18 and 19, and the RNAS machines were never delivered.

Nieuport 20

The Nieuport 20 was a development of the Nieuport 12 produced for the RFC in 1916. While it had the basic Nieuport 12 airframe there were major changes, including a 110-hp Le Rhône engine, a horseshoe cowling, and fuselage fairings. The fuel tank was enlarged and its shape altered. The pilot and observer were seated back-to-back and the observer had a 0.303 Lewis gun. The pilot had a synchronized 0.303 machine gun. Twenty-one (of 30 ordered) were purchased by the RFC.

Some Nieuport 20s were equipped with a 120-hp Le Rhône engine. The airframes were otherwise unchanged. They were designated Nieuport 20bis.

The Nieuport 20 was assigned to Nos.1, 45, and 46 Squadrons. As with the Nieuport 12, the aircraft were not liked by the aircrew and were replaced in early 1917. When these aircraft were withdrawn from service they were assigned to Nos.65 and 84 Training Squadrons, as well as Nos.31, 43, 45, and 55 Reserve/Training Squadrons. Some were also used by No.39 Home Defense Squadron.

> **Nieuport 20 Two-Seat Reconnaissance Airplane/Fighter with 110-hp Le Rhône 9J**
> Span 9.00 m; length 7.00 m; height 2.70 m; wing area 22 sq. m
> Empty weight 453 kg; loaded weight 752 kg
> Maximum speed: 157 km/h at sea level, 152 km/h at 2,000 m; climb to 1,000 m in 5 minutes 12 seconds; climb to 2,000 m in 12 minutes 2 seconds
> Armament: one fixed 0.303 Vickers machine gun and one 0.303 Lewis gun on an Éteve mount
> A total of 21 aircraft were obtained by the RFC; none were used by the Aviation Militaire

Below left: Nieuport 20. Major changes from the Nieuport 12 included the 110-hp Le Rhône engine, a new horseshoe cowling, and the addition of fuselage fairings. Via Colin Owers.

Below: Nieuport 20 used by the RFC. Nieuport 20s were assigned to Nos.1, 45, and 46 Squadrons. Via Colin Owers.

Bottom: Nieuport 20. J.M. Bruce has identified this machine as A259 which was delivered to No.1 Squadron in September 1916 and lost in combat three weeks later. MA.

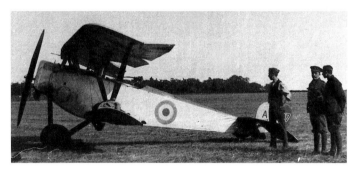

Nieuport 20

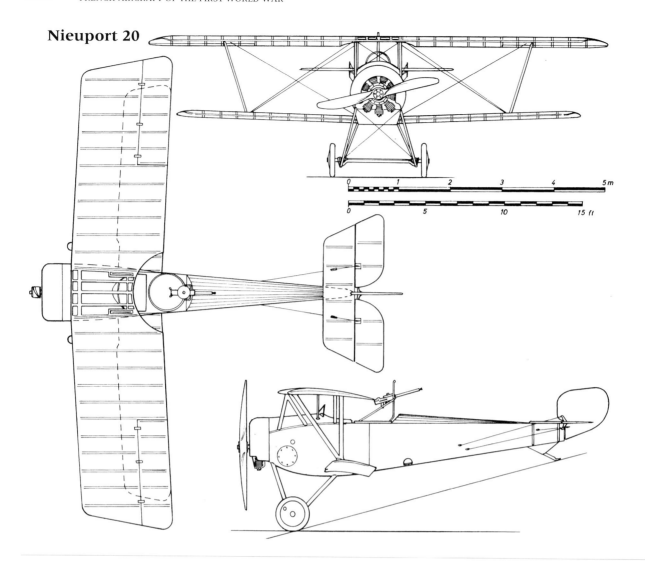

Nieuport 21

The Nieuport 21 combined features of two earlier Nieuport designs. It had the fuselage of the standard Nieuport 11 (but with a lightened airframe) and the wings of a Nieuport 17 (with altered bracing and parallel flying wires). The engine was an 80-hp Le Rhône 9C (compared with the 110-hp Le Rhône used in the standard Nieuport 17).

It has been speculated that this aircraft was to be used as a fighter trainer or a high-altitude bomber escort. The Nieuport firm described it as an avion-ecole, which confirms that its primary mission was that of a trainer. However, some Nieuport 21s were assigned to front-line fighter units and were used operationally alongside the Nieuport 17. A French document showing the disposition of airplanes in the Aviation Militaire on 1 August 1917 stated there were 28 Nieuports with 80-hp Le Rhône engines (this total probably includes Nieuport 10s and 11s). Fifteen were with the front-line escadrilles, one with the RGA, and 12 machines in reserve. The French were able to supply the Imperial Russian Air Service with the lower-powered Nieuport 21s while preserving the 110-hp Nieuport 17s for French escadrilles.

At least one Nieuport 21 was fitted with a 90-hp Le Rhône 9Ga engine. Some may have had a 120-hp Le Rhône 9J, but this has not been confirmed.

Several examples of the Nieuport 21 were acquired by the Aviation Maritime in

Nieuport 21, serial N 1647, of Andre Chainat of SPA 3. The Nieuport 21 had the fuselage of a Nieuport 11 and the wings of a Nieuport 17. The engine was an 80-hp Le Rhône 9C. B.91.4797.

1920. They were used to practice takeoffs from the platform on the *Bapaume* until replaced by Nieuport 32s in 1922.

Foreign Service
Brazil
The Brazilian Escola de Aviacao Militar (Military Aviation School) received 20 Nieuport 21 trainers in 1920. Used as advanced trainers, they were given serial numbers 2101–2104, 2107–2108, 7090, 7096, 7109, 7112, 7114, 7117–7118, 7123, and 7128–7129. Two Nieuport 21s were assigned to the 1st Companhia de Parque de Aviacao. These were used operationally during the San Paulo revolution of June 1924 before returning to Campo dos Afonsos the next month. They were subsequently incorporated into the Destacamento de Aviacao (Air Detachment) and sent back to San Paulo from August to September 1924. The Nieuport 21s were finally retired at the end of 1924.

Russia
Sixty-eight Nieuport 21s were built under license by the Russian Dux firm. They were armed with a single machine gun which on some machines was synchronized. The Nieuport 21 was well-liked by Russian pilots because, despite its weak 80-hp Le Rhône engine, it was extremely maneuverable. In fact, the first controlled aircraft spin in Russia was performed in 1916 by K.K. Artseyulov in a Nieuport 21. The Nieuport 21s were initially used as fighters and equipped the units of the First Fighter Group. In March 1917 there were 43 Nieuport 21s at the front—northern front (4); western front (6); southwestern front (20); Romanian front (12); Caucasus front (1).

By April 1917 there were 68 Nieuport 21s at the front. In June there were 74—northern front (8); western front (15); southwestern/Romanian fronts (37); Caucasus front (14).

In December 1920 there were 52 Nieuport 21s in service, but in 1922 that number had dropped by almost half. In 1922 a few were still in service with the 2nd, 12th and 17th Aviaotryady and the Vozdukheskadra No.2. Subsequently, the Nieuport 21s were employed as fighter trainers. By 1923 the Nieuports were with the 1st Military School of Pilots at Kacha, the 2nd School of Military Pilots at Borisoglebsk, the 1st Higher School of Military Pilots in Moscow, the Higher School of Aerial Observers, the 4th Otdel'naya istrebitel'naya aviaeskadril'ya, the 8th and 47th Otdelnye razvedivatel'nye aviaotryady, and the training eskadril'ya. The remaining Nieuport 21s were retired in April 1925.

Ukraine
The Ukrainian air service acquired two Nieuport 21s in 1918. These had serial numbers 380 and 1317.

United Kingdom
Five Nieuport 21s were acquired by the RNAS.

United States
The United States purchased 197 Nieuport 21s for use as fighter trainers. Nieuport 21s were assigned to the 31st Aero Squadron.

Nieuport 21 Single-Seat Fighter/Fighter Trainer with 80-hp Le Rhone 9C
Span 8.16 m; length 5.80 meters; height 2.40 m; wing area 14.75 sq. m
Empty weight 350 kg; loaded weight 530 kg
Maximum speed: 150 km/h; climb to 2,000 m in 8.45 minutes; climb to 3,000 m in 15.7 minutes; ceiling 5,200 m; endurance 2 hours; range 250 km
Armament: one 0.303 Vickers machine gun

Nieuport 21 Single-Seat Fighter/Fighter Trainer with 80-hp Le Rhone 9C Built by Dux
Span 8.02 m; length 5.8 meters; wing area 14.7 sq. m
Empty weight 370 kg; loaded weight 545 kg
Maximum speed: 150 km/h; climb to 1,000 m in 4.0 minutes; climb to 2,000 m in 8.7 minutes; climb to 3,000 m in 15.7 minutes; climb to 4,000 m in 25.6 minutes; climb to 5,000 m in 46 minutes; ceiling 5,250 m; endurance 2.0 hours
Armament: one 0.303 Vickers machine gun
A total of 68 were built by Dux

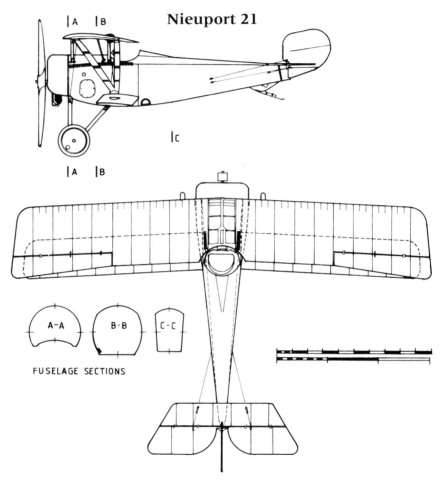

Nieuport Fighter with 150-hp Hispano-Suiza

This 1916 design was an all-new aircraft fitted with a 150-hp Hispano-Suiza engine It had equal span wings with the usual V-shaped interplane struts. The upper wing had a center section with two outer panels (unlike the two-part mainplane of Delage's earlier fighters) and a lower wing of a wider chord. The fuselage was fully faired and much wider than that used on the Nieuport 11/17/24/25 series. There was a neatly-cowled Hispano-Suiza engine with a circular frontal radiator in the nose. The empennage, rudder, and elevators were covered in three-ply wood and fabric and were quite similar to that used on the Nieuport 17bis. It is believed that this type was intended to compete with the SPAD series of fighters; however, it was not developed further and the SPAD 7 was selected for production.

Above right: A Nieuport fighter prototype fitted with a 150-hp Hispano-Suiza engine. University of Texas at Dallas.

Right: Front view of the Nieuport fighter prototype fitted with a 150-hp Hispano-Suiza engine. B87.0395.

Nieuport 23

The next fighter from the Nieuport firm was the Nieuport 23 (the identity of the Nieuport 22 has not been established) which differed from the Nieuport 17 in several minor details. The main difference was the fitting of a new interrupter gear that required the Vickers gun to be fitted a few centimeters to starboard of the centerline and also required modifications to the forward fuselage cross members. The packing pieces of the rear spar of the upper wing were also slightly modified. While some aircraft retained the standard 110-hp Le Rhône 9Ja engine, many Nieuport 23s were fitted with a 120-hp Le Rhône 9Jb. Some Nieuport 23s may have been fitted with the same 80-hp Le Rhône 9C engines used in the Nieuport 21, and it has been reported that these aircraft were intended for use as trainers.

The Nieuport 23s served alongside the Nieuport 17s beginning in early 1917. French escadrilles equipped with Nieuports during the first half of 1917 were assigned to the 1st Armée (N 31, N 82, N 112), 2nd Armée (N 31, N 85), 3rd Armée (N 67, N 79, N 153), 4th Armée (N 38, N 78), 5th Armée (N 23, N 76, N 80), 6th Armée (N 15, N 62, N 65, N 68, N 81, N 83, N 88, N 112), 7th Armée (N 49, N 75, N 81, N 88, N 93), 8th Armée (N 77, N 87, N 91), 10th Armée (N 15, N 69, N 82, N 90), GC 11 (N 12, N 31, N 48, N 57, N 94), GC 12 (N 3, N 26, N 73, N 103), GC 13 (N 15, N 37, N 65, N 67, N 84, N 112, N 124), GC 14 (N 75, N 80, N 83, N 86), GC 15 (N 37, N 78, N 81, N 85, N 93, N 97, N 112), GC Chaux (N 49, N 81, N 82), D.C.A. (N 311, N 312), T.O.E. units in Russia (N 581), T.O.E. units Serbia (N 87/N 387/Escadrille 523, N 37), and the T.O.E. unit assigned to Venice (N 92/ N 392). For further details of these units see the entry for the Nieuport 17.

The Nieuport 23s were replaced by the more reliable Nieuport 24s in mid-1917. This was probably due to numerous accidents involving the Nieuport 23s. A notice from the General Chef de Service Aéronautique to the GQG dated 4 September

Nieuport 23 of N 92. The Nieuport 23 had a new interrupter gear which required the Vickers gun to be fitted a few centimeters to starboard of the centerline. B92.1916.

1917 reported that the Nieuport 23s would be allowed to continue flying only if the following modifications were made:
1. Additional cables to reinforce the wings.
2. Replacement of the lower wings of the Type 23 with the wings from the Type 24.

It was suggested that 150 sets of wings be obtained, which gives a rough estimate of the number of Nieuport 23s that had been purchased by the Aviation Militaire.

It is likely that the lower wings were breaking off when stressful maneuvers were performed. The lower wings of earlier Nieuport airplanes were designed so that their incidence could be adjusted on the ground to accommodate different loads. Although the Nieuport fighters did not use this system, the original fittings for it were retained and this resulted in lower wing failure at the pivoting joints. This would often result in the Type 23s shedding the entire lower wing.

Another GQG memo dated 10 December 1917 called for the replacement of the tailskids because of frequent breakages. The skids were also to be fitted with a rubber shock absorber.

Foreign Service
Belgium
Several Nieuport 23s were sent to Belgium. The 1st Escadrille of the Aviation Militaire Belge had at least one on strength and it is likely that the 5th Escadrille, which was the only other Belgian unit to be equipped with Nieuport fighters, also had a few on strength.

Czechoslovakia
Czechoslovakia acquired several Nieuport 17s and 23s in 1919.

Finland
Two Dux-built Nieuport 23s were given to the fledging Finnish air service when two Russian pilots defected with their airplanes in 1918. These machines, given serials D 61/18 and D 62/18, remained in service until 1920. (One may have been a Nieuport 17.)

Poland
Poland is known to have acquired at least three Nieuport 21s. These were No.3191 (captured in September 1919 from the Russians and assigned to the 1st Squadron), No.3751 (acquired in 1919 and assigned to the 2nd Air Regiment at Luck and later to the training unit at Krakow), and No.4227 (used as a fighter trainer).

Russia
Russia received its first Nieuport 23 in 1917. The type was built under license by the Dux plant. In June 1917 there were 42 Nieuport 23s assigned to the northern front (7), western front (11), southwestern/Romanian front (18), and Caucasus front (6).

On 1 September 1917 there were 70 Nieuport 23s in storage at the Central Aviation Park in Moscow. During the civil war the Nieuport 23s were used extensively by the Russian fighter squadrons and in fact were by that time the most numerous Nieuport type in service.

In December 1920 there were still 72 Nieuport 23s in service; by early 1922 that number had been reduced to 60. These were in service with the 2nd, 3rd, 4th, 8th, 9th, 10th, 12th, and 14th Aviaotryady, the Istrebitel'naya aviaeskadril'ya at Kiev, the Vozdukheskadra No.2, and the 1st naval Istrootryad.

By 1923 the Nieuport 23s had been relegated

Nieuport 23 Single-Seat Fighter Trainer with 80-hp Le Rhône
Span 8.16 m; length 5.80 m; height 2.40 m; wing area 14.75 sq. m
Empty weight 350 kg; loaded weight 535 kg
Maximum speed: 150 km/h; climb to 2,000 m in 8.45 minutes; range 250 km
Armament: one synchronized 7.7-mm Vickers gun; some were fitted with one or two Lewis guns on an overwing mount; up to 8 Le Prieur rockets could be carried.

Nieuport 23 Single-Seat Fighter with 110-hp Le Rhône 9Ja or 120-hp Le Rhône 9Jb
Dimensions identical to Nieuport 23 Trainer
Empty weight 375 kg; loaded weight 560 kg
Maximum speed: 165 km/h at sea level; climb to 3,000 m in 11.5 minutes
Armament: one synchronized 7.7-mm Vickers gun; some were fitted with one or two Lewis guns on an overwing mount; up to 8 Le Prieur rockets could be carried. In RFC service a Lewis gun frequently replaced the Vickers.

Nieuport 23 Single-Seat Fighter with 120-hp Le Rhône 9Jb Built by Dux in Russia
Span 8.03 m; length 6.4 m; wing area 14.7 sq. m
Empty weight 355 kg; loaded weight 547 kg
Maximum speed: 168 km/h; climb to 1,000 m in 2.7 minutes; climb to 2,000 m in 5.8 minutes; climb to 3,000 m in 9.7 minutes; climb to 4,000 m in 15.0 minutes; climb to 5,000 m in 23.0 minutes; ceiling 6,500 m; endurance 1.7 hours
Armament: one synchronized 7.7-mm Vickers gun

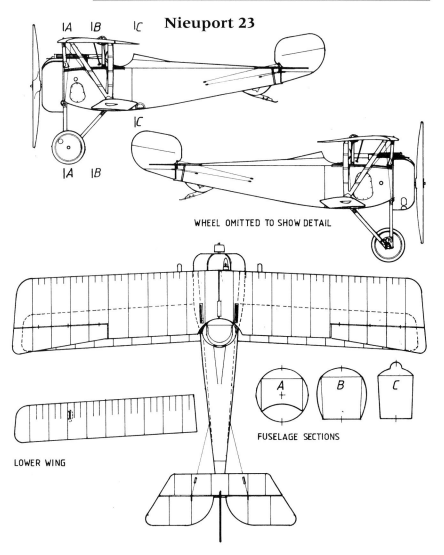

to training duties at the 1st Military School of Pilots, the 1st Higher School of Military Pilots, the Strel'bom school, the 4th Otdel'naya istrebitel'naya aviaeskadril'ya, the 1st Morskoi istrebitel'nyi vozdukhchast' and the 9th, 17th, and 47th Otdel'nye razvedivatel'nye aviaotryady. By 1925 only about 40 remained in service and by the next year only one. The Siberian Air Fleet of Admiral Kolchak had five Nieuport 23s on strength form 1918 to 1920.

Switzerland

Five Nieuport 23s were obtained by the Swiss Fliegertruppe in 1917. Given serials 601 through 605, they became the Fliegertruppe's front-line fighters and remained in service until 1921. In that year No.601 was tested to destruction and it was discovered that the Nieuports could no longer meet the structural requirements of the service; all the type 23s were therefore withdrawn from service.

Ukraine

The Ukrainian air service acquired seven Nieuport 23s in 1918. These had serial numbers 3224, 3226, 3240, 3241, 3246, 3247, and 3731.

United Kingdom

The RFC received 30 Nieuport 23s in August 1917 and, by the end of the war approximately 80 had been received from France. They served alongside the Nieuport 17s with Nos.1, 11, 29, 40, and 60 Squadrons. Later, approximately five Nieuport 23s joined the Nieuport 17s with Nos.14 and 113 Squadrons in Palestine.

United States

Fifty Nieuport 23s were purchased by the United States. These served as fighter trainers. Three had 80-hp Le Rhône engines, the other 47 had the standard 120-hp Le Rhône 9Jb.

Nieuport 24

The Nieuport 24 was an attempt by the Nieuport firm to prolong the success of its line of sesquiplane fighters. Compared with the SPAD 7s coming into service in mid-1917, the Nieuport 17 and 23s had less power, inferior maneuverability, and were considered to be more difficult to fly. It was hoped that the more powerful Nieuport 24 would correct these deficiencies.

The tailplane of the pre-production Nieuport 24 had consisted of spruce spars and interlocking ribs of birch plywood covered with plywood and fabric. However, it is believed that the prototype's tail surfaces had caused problems either during manufacturing or flight testing. While the design of the tail surfaces was being revised a Nieuport 24 was fitted with the tail and balanced rudder of the standard Nieuport 17. The modified airframe was designated the Nieuport 24bis. It was probably hoped that this would serve as an interim type until series production of the definitive Nieuport 24s could begin, and thus prevent the SPAD firm from becoming the sole supplier of fighters to the Aviation Militaire. In a report from the British Aviation Commission dated 11 April 1917 it was stated that the Type 24bis would supersede the Type 17 in production. However, on 10 May another commission report stated the Types 17, 23, and 24bis had been abandoned in favor of the standard Nieuport 24.

The Nieuport 24bis was externally nearly identical to the Nieuport 17bis. Differences included a 130-hp Le Rhône 9Jb engine and a new airfoil section with a pronounced camber to improve lift. The tail of the 24bis was the same as that used on the Nieuport 17 and was constructed of steel tubing and had the "square comma" form of rudder.

A Nieuport with serial number N 3760 may have been the next step in the evolution of the Nieuport 24 series. This aircraft featured two inverted V-struts which supported the upper wing and necessitated the placement of the single Vickers machine gun on the port upper longeron. The cockpit was small, with the scalloped cutout closely faired into a more rectangular opening. The most important feature was its all-wood ply-covered tail surfaces, which served to separate the Nieuport 24 series from the Nieuport 17, 17bis, and 24bis. The landing gear was actually more reminiscent of the Nieuport 27, being a twin half-axle and featuring a completely new tail skid. A fairing was placed over the gap between the aileron and mainplane; this was a length of stiffened canvas attached to the rear spar of the wing (see below). It seems that the N 3760 was a unique design and no further production was undertaken.

The standard Nieuport 24 was to be the only Nieuport fighter in production by May 1917. The main difference between this aircraft and the 24bis was that the former aircraft had tail surfaces similar to those found on the above-mentioned N 3760. The tailplane and elevator had spruce spars and the rudder had a spruce leading edge. Most of the rest of the tail's structure consisted of birch plywood ribs and spars, with plywood covered in fabric. The engine of the Nieuport 24 was a 130-hp Le Rhône 9Jb. The fuselage was fully faired. Both the 24bis and the standard Nieuport 24 had an airfoil section of pronounced camber in the hopes of improving lift. One of the main complaints about the earlier Nieuport fighters had been their poor lateral stability. It was hoped that this could be corrected by the addition of canvas fairing strips over the gap between the ailerons and wing. The Nieuport firm reported that "This alteration has given excellent results as regards ease in handling the machines." However, in service it was found that the strip might spring upward, blanketing the airflow over the aileron and interfering with control of the aircraft. When this band was removed it was found that the sluggish lateral control of the Nieuport 24 improved considerably. Ironically, the relatively quick fix to the Nieuport's stability

A production Nieuport 24, serial N 3961, of N 91. N 91 was assigned to the 8th Armée and based at Lorraine. B86.4235.

Nieuport 24

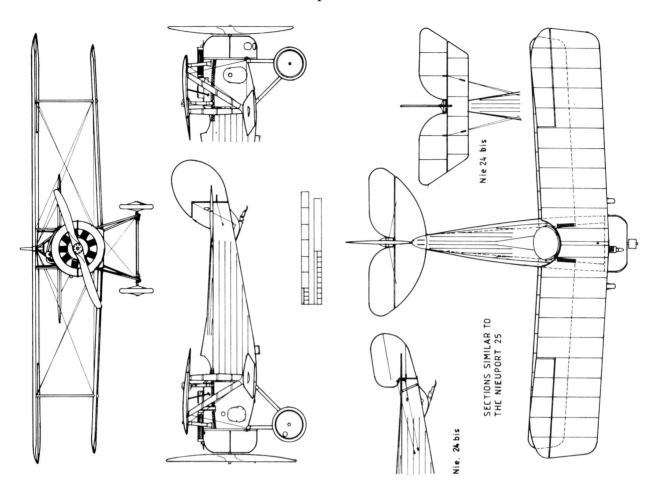

problems that these canvas strips were intended to provide was, in fact, one of the reasons French pilots complained of the Nieuport 24's poor aileron response, unsatisfactory lateral control, and poor landing and turning characteristics. The wing of the Nieuport 24 employed a revised type of aileron with curved tips to correct problems with aileron weakness. The mainplanes had a skin of plywood over the airfoil's upper surface between the leading edge and the main spar. The spars were constructed of lengths of spruce applied to webs of plywood. The fuselage was fully faired with spruce longerons and formers. The landing gear had a one-piece axle as well as the standard Nieuport tail skid. A Vickers machine gun was mounted slightly starboard of the centerline. The ammunition box was also located to starboard, with the spent ammunition belts being fed into a drum inside the port fuselage fairing.

The wing attachments, however, were apparently subject to numerous failures. A GQG memo called for pilots flying the Nieuport 24 or 24bis to avoid aerobatics. Furthermore, the wing attachments were to be "minutely" examined after each flight. Another memo, dated 6 September 1917, called for flying wires of increased size to provide greater strength. These were to be fitted to all Nieuport 24, 24bis, and 27 airplanes.

Operational Service

The Nieuport 24bis entered escadrille service in early 1917. N 84, for example, in March 1917 had predominantly Nieuport 24bis airplanes on strength, along with a few SPAD 7s.

The standard Nieuport 24 reached operational units in June 1917. During the Nieuport 24's period of service from June 1917 to January 1918 approximately one half of the Nieuport escadrilles were still assigned to the armées while the other half were assigned to the Groupes de Chasse.

Army Cooperation

The army cooperation escadrilles provided fighter escort for reconnaissance and bomber units, transported spies behind enemy lines, undertook reconnaissance missions, and flew fighter patrols. These escadrilles were:

N 12, in March 1917 assigned to the 1st Armée sector.

N 23, assigned to the 7th Armée in February 1917, but quickly moved to the 2nd Armée sector. N 23 became SPA 23 in November 1917.

N 38, assigned to the 4th Armée. It joined GC 15 in March 1917.

N 49, assigned to the 7th Armée. It was commanded by Capitaine Jules de Boutiny and was based at Fontaine. It became SPA 49 in December 1917.

N 62, assigned to the 6th Armée. N 62 was based at Fismes in January 1917 and was active during the Battle of Chemin des Dames. In October it participated in the attack at Malmaison. N 62 became SPA 62 in November 1917.

N 68, commanded by Capitaine Garde in November 1916, then Capitaine Lemerle in early 1917. It was assigned to the 8th Armée and in November 1917 became SPA 68.

N 69, moved to Italy with the 10th Armée in October 1917. It was based at Porto Nuova de Verona and in December was redesignated SPA 69.

N 75, based at Saiserais and assigned to the 8th Armée. In April 1917 N 75 joined GC 14 and became SPA 75.

N 76, assigned to the 5th Armée and based at Muizon, commanded by Lieutenant Rene Doumer. Capitaine Jean-Jacques Perrin assumed command in April 1917. He was succeeded by Capitaine Eugene Verdon and then Lieutenant Pierre Vitoux. The unit re-equipped with SPADs in late 1917.

N 77, assigned to the 8th Armée on the Lorraine front and

Nieuport 24bis, serial N 3588. The Nieuport 24bis was externally nearly identical to the Nieuport 17bis, the main differences being the 130-hp Le Rhône 9Jb engine and a wing with a new airfoil section. B.80.35.

commanded by Lieutenant Joseph l'Hermite. A notable mission occurred on 23 February 1917 when two aircraft of N 77 bombed the German airbase at Marinois. N 77 re-equipped with SPADs in June 1917.

N 79, in June N 79 joined the Provisional Groupe de Chasse de Bonneuil based at Catigny. In January 1918 N 79 became SPA 79.

N 82, briefly joined GC 14 in April 1917 and subsequently assigned to the 7th and 3rd Armées. N 82 moved to 10th Armée sector in November 1917 and was sent to Italy later that month. It was based at San Pietro di Godego and provided fighter escort for reconnaissance and bombing missions. During the Tomba offensive the aviation units bombed and strafed enemy artillery and machine gun position. On 1 January 1918 N 82 was based at San Pietro in Gu, where it conducted long-range reconnaissance missions over Adige, Brenta, and Piave, and provided fighter escort for the army cooperation units of the 31st C.A. In March 1918 the 10th Armée and its aviation units returned to France. Upon its return N 82 was assigned to the GC 22.

N 85, formed in March 1917 at Lyon-Bron and assigned to the 2nd Armée under the command of Capitaine Limasset. In April 1917 it joined GC 15.

N 87, formed in March 1917 at Lyon-Bron and assigned to the 8th Armée; it was commanded by Capitaine Pierre Azire beginning in January 1918. In May 1918 N 87 became SPA 87.

N 88, commanded by Capitaine Francois d'Astier de la Vigerie, moved to the 6th Armée sector at Chemin des Dames in June 1917. It became SPA 88 in September 1917.

N 89, formed in March 1917 at Villacoublay. Assigned to the 8th Armée and commanded by Capitaine Guy Tourangin, the escadrille became SPA 89 in February 1918.

N 90, formed from detachments N 504 and N 507 in March 1917, assigned to the 8th Armée and commanded by Lieutenant Pierre Weiss. It became SPA 90 in April 1918.

N 91, assigned to the 8th Armée and based at Lorraine. It was commanded by Lieutenant Jourdain and in January 1918 became SPA 91.

N 92, assigned to the 7th Armée and commanded by Lieutenant Edmond George. In May Capitaine Georges de Geyer d'Orth assumed command and in July N 92 moved to the 2nd Armée sector. It re-equipped with SPADs in May 1918.

N 93, formed from detachments N 501 and N 506 in April 1917. It was assigned to the 7th Armée in the Vosges sector and commanded by Lieutenant Jean Moreau. N 93 provided support for the bombers of GB 4 and participated in leaflet-dropping missions. It joined GC 15 in July 1917.

N 94, formed in June 1917 from detachments N 512, N 513, and N 514 at Melette. It was assigned to the 4th Armée and commanded by Capitaine Edouard Pillett. N 94 joined GC 18 in January 1918.

N 95, created from detachments N 517 and N 519 at Vadelaincourt in April 1917. The escadrille was initially assigned to the 2nd Armée and commanded by Lieutenant Herissant, followed by Lieutenant du Doré. N 92 moved to the 6th Armée sector in July 1917 and the 2nd Armée sector in September.

Nieuport 24bis of N 97. N 97 was formed from detachments N 511 and N 519 in June 1917. The aircraft in the foreground carries serial N4479. Renaud.

Nieuport 24bis serial number 3363 of N 561 escorting an Italian seaplane. N 561 was assigned to the protection of Venice and remained in Italy until the end of the war. B87.3736.

Lieutenant Grandmaison assumed command in October 1917. In February 1918 N 95 joined GC 19.

N 96, formed from N 510 in June 1917. It was attached to the 2nd Armée and was under command of Capitaine Masse. It moved to 4th Armée sector on 9 July and the 5th Armée sector on 22 July. In September Capitaine Eugene Verdon assumed command, followed by Lieutenant Maurice Barthe in January 1918. In February 1918 N 96 joined GC 19.

N 97, formed from detachments N 511 and N 519 in June 1917. It was initially assigned to the 2nd Armée and commanded by Capitaine Francois de Castel. In July N 97 moved to the 4th Armée sector. In September 1918 it moved to the 2nd Armée sector under the command of Lieutenant Herve Conneau. In December N 97 became SPA 97 and joined GC 15.

N 98, formed from elements of N 85 in November 1917. It was assigned to the 2nd Armée sector under the command of Capitaine Leon Bonne. On 1 March 1918 N 98 joined GC 21.

N 99, formed in November 1917 from elements of N 87. It was assigned to the 8th Armée and commanded by Capitaine Roger Lemercier de Maisoncelle-Vertille de Richemont. In February 1918 N 99 joined GC 20.

N 102, commanded by Lieutenant Jean Derode and in July moved to Flanders. In January 1918 N 102 moved to the 1st Armée sector. Capitaine Roger Lemercier de Maisoncelle-Vertille de Richemont assumed command in March and, the next month N 102 became SPA 102.

N 124, in July moved to Flanders in the 1st Armée sector. The next month N 124 moved to the Verdun front and in September to the 6th Armée. In December 1917 N 124 moved to the 4th Armée sector and was disbanded in February 1918.

N 150, formed in July 1917 and was assigned to the 7th Armée. It was under the command of Lieutenant Pierre L'Huillier. In December it was redesignated SPA 150.

N 151, formed at Chaux in July 1917. It was assigned to the 7th Armée and was commanded by Lieutenant Gerard Amanrich. It became SPA 151 in December 1917.

N 152, formed in July 1917 at Lyon-Bron. It was assigned to the 7th Armée and was under the command of Lieutenant Lefèvre. In November Lieutenant Lous Delrieu assumed command and in June 1918 N 152 became part of GC 22.

N 155, formed in July 1917 at Montdésir. It was assigned to the 4th Armée and commanded by Lieutenant Edmond George. In December it became SPA 155.

N 156, formed in January 1918 and assigned to the 4th Armée. It was commanded by Capitaine Emile Paumier. The next month it received Morane-Saulnier Type AIs to become MS 156.

N 157, formed in December 1917 at Chaux. It was assigned to the 7th Armée and commanded by Lieutenant Henri Constant de Saint-Saveur. In March it was assigned to GC 21.

N 158, formed in January 1918 at Bonneuil. It was assigned to the 3rd and then the 4th Armées. N 158 returned to the 3rd Armée in February under the command of Lieutenant Jacques Chaudron. In March N 158 became MS 158.

N 159, formed in January 1918 from elements of N 90. It was assigned to the 8th Armée and based at Lorraine under the command of Lieutenant Albert Robert. In February it became part of GC 20.

N 160, formed in January 1918. It was assigned to the 2nd Armée and based at Brabant-le-Roi under the command of Lieutenant Emile Barès. In May it re-equipped with Morane-Saulnier Type AIs.

N 161, formed in December 1917 and assigned to the 5th Armée. Based at Lhéry, it was commanded by Lieutenant Marcel Parfait. N 161 moved to Boujacourt later in January; followed by a move to Dugny, still in the 5th Armée sector. It re-equipped with Morane-Saulnier Type AIs in February 1918.

N 162, formed in January 1918 at Corcieux from elements of N 152. It was assigned to the 7th Armée and commanded by Lieutenant Daniel Chambariere. It joined GC 20 in March.

N 312, formed in March 1917, but became N 313 in July 1917.

N 313, formed in July 1917 from the personnel of the disbanded N 312. It was assigned to provide fighter protection for naval aircraft operating from Dunkerque.

N 314, assigned to protect the city of Nancy and commanded by Capitaine Perrin. It became SPA 314 in July 1918.

N 315, formed in February 1917 from N 311. It was assigned to the DCA to protect the city of Belfort. It re-equipped with SPADs in July 1918.

N 392, formed from N 92 in June 1916 and served as an escadrille de protection for the city of Venice. In July N 392 became N 561 in July 1917 and remained assigned to the protection of Venice. Commanded by Lieutenant Challeronge, **N 561** remained in Italy throughout the war. Later in 1917 N 561 became SPA 561.

N 523, formed from H 387 in July 1917. It was assigned to the A.F.O. in Serbia and later re-equipped with SPADs.

N 562, created at Corfu, Greece in July 1917. It was assigned to protect the city of Potamos. N 562 disbanded in December 1918.

N 581, with eight Nieuport 17s and seven SPAD 7s, formed in February 1917. It was assigned to the 7th Russian Army and began operations in May 1917. Based at Boutchatch, Galicia, it served alongside Kosakov's Nieuport fighter group. In response to the German offensive of 19 July on the 6th Army front, N 581 was transferred to Buzcacz. In August N 581 moved to the

Left and below: Nieuport 24 N 3778. University of Texas at Dallas.

Romanian front where it was assigned to the 3rd Corps. By February 1918 N 581 was re-designated SPA 581 because it had re-equipped entirely with SPAD 7s.

Groupes de Combat

The Groupes de Combat were used to achieve aerial superiority over selected areas of the front and most of the escadrilles assigned to the GC units in the latter half of 1917 flew Nieuport 23s and 24s.

In June **GC 11** had five Nieuport escadrilles—N 12, N 31, N 48, N 57, and N 94— and was active in the Flanders section; it was later to provide fighter cover for the French army during the Third Battle of Ypres. All five Nieuport escadrilles had some SPAD 7s on strength and often used these for ground attack while the Nieuports flew fighter patrols. By September GC 11 moved to support the 6th Armée during the Malmaison offensive. By early 1918 the Nieuport escadrilles had all re-equipped with SPAD 7s and 13s.

GC 12 had four Nieuport escadrilles: N 3, N 26, N 73, and N 103. It was assigned to support the 6th Armée. However, before the end of the summer three of the escadrilles (N 3, N 26, and N 73) had re-equipped with SPAD fighters. N 103 continued to use its Nieuport 24s until later in 1917.

In June 1917 **GC 13** consisted of N 15, N 65, N 84, and N 124. It flew fighter patrols for the 6th Armée over the Aisne front and participated in the Battle of Flanders in July 1917. Later in 1917, GC 13 was assigned to the GQG and then the 5th Armée. GC 13's mission was to prevent German airplanes or balloons from obtaining information on French troops. N 65 and N 84 re-equipped with SPAD 7s and 13s in June, while N 15 received SPADs the next month. N 124 continued to use Nieuport 24s until November 1917, when it was divided into two components: the American 103rd Aero Squadron and SPA 124.

GC 14 was based at Thiers in June 1917 and had four Nieuport escadrilles: N 75, N 80, N 83, and N 86. In June it was assigned the 10th Armée and was based at Souilly on 5 July. By 7 July GC 14 was assigned to the 2nd Armée on the right bank of the river Meuse at Verdun. From 20 August to 29 September GC 14 provided fighter cover for the French army in the Second Offensive of Verdun. By 14 October GC 14 moved to the 6th Armée sector and was based at Vaubéron; GC 14's escadrilles participated in the fighting at Malmaison from 23 through 27 October. In November GC 14 supported the British army at Cambrai. N 75, N 80, and N 83 re-equipped with SPAD 7s and 13s by the end of 1917. N 86 was the only unit of GC 14 to continue to fly Nieuports in 1918. By December GC 14 had returned to Vaubéron.

GC 15 had six fighter escadrilles assigned to it in June 1917; two units (SPA 37 and SPA 81) used SPAD 7s, the other four (N 78, N 92, N 93, and N 112; later N 85) flew Nieuport 23s and 24s. GC 15 was based at Vadelaincourt and was assigned to the 2nd Armée, supporting operations over the right bank of the Meuse. From 20 August to 8 September GC 15's main task was to provide fighter escort for reconnaissance and bomber escadrilles during the Second Battle of Verdun. By November N 95 joined GC 15 while N 78 and N 93 re-equipped with SPAD 7s and 13s. By the end of 1917 there were only two units assigned to GC 15 which still flew Nieuports: N 85 and N 112. However, by February both units had received SPAD 7s and 13s.

The **Provisional Groupe de Bonneuil** was formed in June 1917 and consisted of N 82, N 153, N 154, and C 46. It was assigned to support the 3rd Armée. On 1 August the Group was disbanded and some of its component units were assigned to the 3rd Armée.

The stability problems of the Nieuport 24s already referred to, as well as the fact that their performance and armament were inferior to that of the SPAD 7 and 13, ensured that the fighter units of the Aviation Militaire converted almost exclusively to SPAD fighters in late 1917 and throughout 1918.

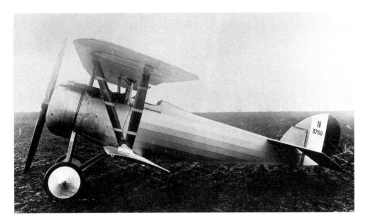
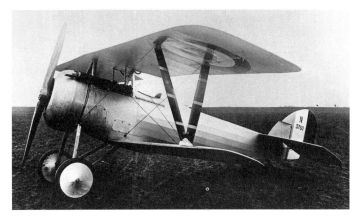

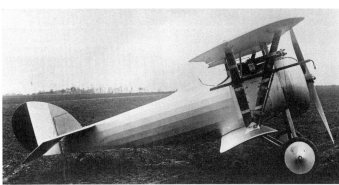

Above, top, and top right: Three views of Nieuport 24 N 3760, which is believed to have been the prototype of the Nieuport 24 series. It had a fully-faired fuselage, an all-wood tail, and a balanced rudder and tailplane. University of Texas at Dallas.

Above right: The prototype of the Nieuport 24. The tailplane of the pre-production Nieuport 24 consisted of spruce spars and interlocking ribs of birch plywood which were covered with plywood and fabric. However, it is believed that the prototype's tail surfaces had caused problems during either manufacturing or flight testing. 87.0394.

Foreign Service

Brazil
Six Nieuport 24s were purchased by Brazil in 1919 and were assigned to the Escola de Aviacao Militar (Military Aviation School) at Campo dos Afonsos. The Nieuports were assigned serials 3042, 3064, 3318, 3889, 4648, and 5149. They were withdrawn from service in 1924.

Bulgaria
A single Nieuport 24 was captured by the Bulgarians in 1917 and pressed into service.

Estonia
Estonia acquired several Nieuport 24s and Nieuport 17s from Russia in 1920. They were assigned serial numbers 42 through 46 and 51.

Greece
Greece purchased approximately 20 Nieuport 24bis fighters and used them, along with some SPAD 7s, to equip 531 Mira (Squadron) in March 1918. The unit was based at Gorgupi. By the end of 1918, 531 Mira had approximately 12 Nieuport 24bis on strength. They were withdrawn from service in 1924.

Japan
Japan obtained several Nieuport 24s from France in 1917. The army was satisfied with the performance of the aircraft and arranged for the Type 24 to be built under license by the Tokorozawa Branch of the Army Supply Depot and later the Nakajima firm. A total of 102 were built from 1921 through 1923 by Nakajima. The aircraft were designated type Ko 3; those with 120-hp Le Rhône engines entered operational service with fighter squadrons in June 1922. The Japanese Nieuport 24s were used to replace the SPAD 13s still in service. They remained in service until 1926, when they were superseded by Nieuport 29s. There were additional Nieuport 24s built which had 80-hp Le Rhône engines; these were used as fighter trainers.

Latvia
Latvia acquired Nieuport 24bis fighters when two were sent to the White Russians by the Allies. Both machines were later seized by the Latvians and used to form an indigenous air arm. After the Russians protested, one of the aircraft was returned. The other was flown on 5 August 1919; this was the first sortie for the Latvian Army Aviation Group. On 27 August 1919 it dropped seven bombs on a railway station. Eventually, 11 Nieuport 24/24bis fighters served with the Latvian air force.

Poland
Poland acquired approximately six Nieuport 24s. No.5424 was captured at Vilnius on 19 April 1919; it had been assigned to the 32nd Bolshevik Artillery. After its capture it was assigned to the Polish 4th Squadron. The remaining five were Nieuport 24bis. No.5086 was delivered by a deserter from the 3rd Bolshevik Squadron. It was subsequently used by the 1st, 6th, and 5th Polish Squadrons. No.4301 was captured on 5 May 1919. It was assigned to the 4th Polish Squadron and later used by the 1st Air Group. Another Nieuport 24bis, serial number unknown, crashed at the airfield of the 14th Polish Squadron. No.9252 was captured in September 1920. It was sent to the 5th Polish Squadron and served until June 1921. Finally, another Nieuport 24bis was obtained by the Polish, but details of its history are not known.

Romania
Romania is believed to have acquired several Nieuport 24s, probably after the end of the war. Some of these were used by the aerobatic team based at Tecuci in 1925.

Russia
The Imperial Russian Air Service received 20 examples of the Nieuport 24 in 1917. Shavrov states that the Nieuport 24s were replaced in operational squadrons by the 24bis, which was built under license by the Aktsionernoe Obshchestvo Dux (Dux) firm

> **Nieuport 24 Single-Seat Fighter with 130-hp Le Rhône 9Jb**
> Span 8.25 m; length 5.870 m; height 2.40 m; wing area 14.75 sq. m
> Empty weight 355 kg; loaded weight 547 kg
> Maximum speed: at ground level 176 km/h
>   2,000 m   171 km/h
>   3,000 m   169 km/h
>   4,000 m   163 km/h
> Climb: 1,000 m   2 minutes 40 seconds
>   2,000 m   5 minutes 40 seconds
>   3,000 m   9 minutes 25 seconds
>   4,000 m   15 minutes
>   5,000 m   21 minutes 30 seconds
> Ceiling 6,900 m; range 250 km; endurance 2.25 hours
> Armament: one synchronized 0.303-in Vickers machine gun; some had a 0.303-Lewis gun on the top wing (the RFC had only the overwing Lewis gun)
>
> **Nieuport 24bis Single-Seat Fighter with 130-hp Le Rhône 9Jb**
> Dimensions same as Nieuport 24
> Maximum speed: at ground level 170 km/h; 170 km/h at 2,000 m; climb to 1,000 m in 2 minutes 40 seconds; climb to 2,000 m in 5 minutes 40 seconds; climb to 3,000 m in 9 minutes 40 seconds; climb to 5,000 m in 21 minutes 40 seconds
> Armament: a synchronized 0.303-in Vickers machine gun; some had a 0.303-Lewis gun on the top wing

> **Nakajima Ko 3 Single-Seat Fighter with 120-hp Le Rhône**
> Span 8.22m; length 5.67 m; height 2.40 m; wing area 15.0 sq. m
> Empty weight 450 kg; loaded weight 630 kg
> Maximum speed: 88 kt
> Armament: one 7.7-mm machine gun
> A total of 101 KO 3s were built
>
> **Nakajima Ko 3 Single-Seat Trainer with 80-hp Le Rhône**
> Span 8.22m; length 5.67 m; height 2.40 m; wing area 15.0 sq. m
> Empty weight 415 kg; loaded weight 595 kg
> Maximum speed: 74 kt
> Armament: one 7.7-mm machine gun
>
> **Nieuport 24bis Single-Seat Fighter with 120-hp Le Rhône 9Ja Built by Dux**
> Span 8.16m; length 6.4 m; height 2.40 m; wing area 15.0 sq. m
> Empty weight 375 kg; loaded weight 567 kg
> Maximum speed: 171 km/h; climb to 1,000 m in 2.7 minutes; climb to 2,000 m in 5.7 minutes; climb to 3,000 m 9.4 minutes; climb to 4,000 m 14.4 minutes; climb to 5,000 m in 21.5 minutes; ceiling 6,800 m; endurance 1.7 hours
> Armament: one 7.7-mm machine gun

from 1917 to 1920. They were widely used during the civil war. A few Nieuport 24s were sent to Afghanistan from Russia in 1921/22. A single example of the Nieuport 24 was built with the wings of a Sopwith 1½ Strutter and an 80-hp Le Rhône engine.

In December 1920 there were 27 Nieuport 24s and 26 of the Nieuport 24bis in service. Production continued at GAZ 1 (the former Dux factory) and a year later there were 79 of both types still in operation. About 140 were built from 1920 to 1923 and most served on the Ukrainian front. By September 1923 there were 125 in operation, dropping to 69 a year later. They were used by the Istrebitel'naya eskadril'naya eskadril'ya in Moscow, the 1st and 2nd otryady of the Istrebitel'naya eskadril'ya at Kiev, the 2nd Otryad of the DVK, the NOA, the 4th, 14th, 18th Aviaotryady, the 4th Otdel'naya istrebitel'naya aviaeskadril'ya at Minsk, the 12th Otdel'nyi razvedivatel'nyi aviatsionnyi otryad at Pervomaisk, the 17th Otdel'nyi razvedivatel'nyi aviatsionnyi otryad at Chita, and the 2nd naval Istrootryad at Odessa.

The Nieuport 24 and 24bis were relegated to training duties in the early 1920s. In this role they served with the 1st Higher School of Military Pilots at Moscow, the Strel'bom school at Serpukhov, the Military-Technical School, the 1st School of Military Pilots, the Academy of the VVS, and the Training Eskadil'ya at Moscow. The last examples served with the 4th and 5th Otdel'nye istrebitel'nye aviaeskadrilii, and the 1st Legkobombardirovochnaya eskadril'ya until retired in 1926.

United Kingdom

The RFC received examples of both the Nieuport 24 and 24bis in July 1917. When they arrived it was noted that the French ferry pilots were critical of the type's lateral control. This problem was confirmed by the British pilots and it was decided that the Type 24s would not be used by front-line squadrons; they were instead assigned to the Scout School at No.2 AD. The control difficulties were corrected by removing the troublesome strips covering the ailerons. After this remedy proved successful, the Nieuport 24s were sent to Nos.1, 29, and 40 Squadrons while the 24bis airplanes were retained by the Scout School. Later, the obsolescent Nieuport 24s were sent to Nos.111 and 113 Squadrons in Palestine.

United States

The United States purchased 121 Nieuport 24s for use as trainers. Twelve had 80-hp Le Rhône engines and 109 had 120-hp engines. In addition to this order, 140 Nieuport 24bis were obtained. Forty of the Nieuport 24s had 80-hp Le Rhône engines while the remainder had 120-hp Le Rhônes. The Nieuport 24bis was used as a fighter trainer, not operationally. Nieuport 24bis airplanes served with the 31st Aero Squadron.

Nieuport 24 of the Japanese air service, based at Gifu, Japan in 1920. A few Nieuport 24s were obtained by Japan. B85.1804.

Nieuport Fighter with 250-hp Clerget

This aircraft was under development in November 1917. It is not known if it was built.

Nieuport Fighter with 230-hp Le Rhône 9L

This single-seat fighter was under study in November 1917. Development had been discontinued by May 1918.

Nieuport 25

The Nieuport 25 represented another attempt to produce a sesquiplane that would be competitive in the combat environment of 1918. The Nieuport 25 had the faired fuselage, wooden tail, rounded ailerons (with the notorious canvas fairing strips over the ailerons), and divided undercarriage of the Nieuport 24. Distinguishing features of the Nieuport 25 included the broad chord fairings of the landing gear struts and an elongated engine cowling. The aircraft, which may have been initially flown with a 150-hp Clerget 9Bd, was later fitted with a 200-hp Clerget 11E.

Data available for the Nieuport 25 suggests that its performance was little changed from the standard Nieuport 24. The type did achieve a small measure of immortality because a single example was used by Charles Nungesser as his personal airplane. Indeed, it is possible that his plane was the only example of the Nieuport 25 built. It was not developed further either because of problems with the intended Clerget 11E powerplant or because it did not offer significantly improved performance over the other Nieuport types to warrant series production.

Shavrov reports that five "Nieuport 25s" with 130-hp Clerget engines and an additional three with different engines (80-, 110-, and 130-hp Le Rhônes) were sent to Russia. To make the matter even more confusing he states that "…they were even called Nieuport 27s." Andersson reports that five Nieuport 25s were in service in December 1920 with units in the Ukraine and that two years later they remained in service with the 12th Aviaotryad and the Istrebitel'nye eskadrilii at Kiev and Moscow. All were struck off charge in 1925, except for one that survived until 1928. Given the casual way in which the Soviets identified the Nieuports in service, it is far from certain that these were actually Nieuport 25s. If they were, it suggests a larger production run than previously estimated for the type.

> **Nieuport 25 Single-Seat Fighter with 200-hp Clerget 11E**
> Climb to 2,000 m in 4 minutes 40 seconds, climb to 3,000 m in 8 minutes; climb to 5,000 m in 20 minutes 10 seconds; ceiling 6,300 m
> Armament: one synchronized machine gun

Nieuport 25 of N 65 with the personal insignia of Charles Nungesser. B83.4015.

Nieuport 25

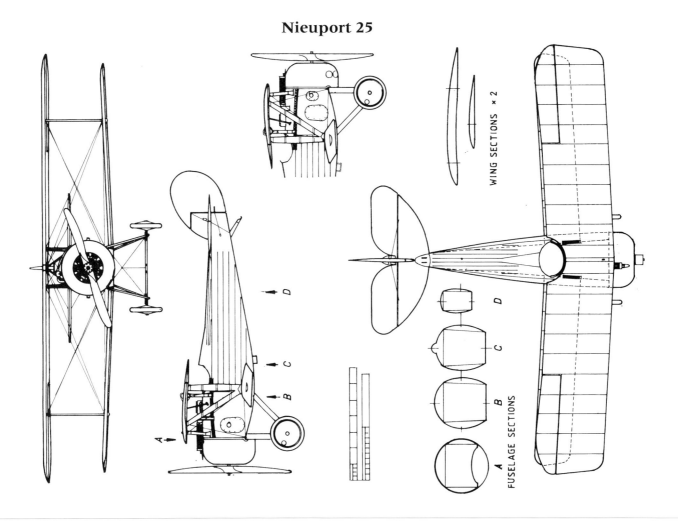

Nieuport 1916 Biplane with 150-hp Clerget

The Nieuport firm developed a biplane in 1916 powered by a 150-hp Clerget engine and believed to have a configuration similar to that of the Nieuport 25. It was not selected for series production.

Nieuport 27

The Nieuport 27 (the Nieuport 26 has not been identified) was almost identical to the standard Nieuport 24 except for a divided-axle landing gear and an internally-mounted tail skid. The wheels each had articulated axles fitted with rubber cords. Initial examples of the Nieuport 27 had the aileron hinge strips and straight-edge ailerons of the earlier Type 24s. Both were eliminated on the later Nieuport 27s, with curved ailerons replacing the straight-edged ones. The aircraft's performance was unchanged from the previous Nieuport 24 series.

Possibly in an attempt to overcome the limitations of the sesquiplane configuration, a Nieuport 27 was fitted with a two-spar wing of broader chord. Thicker V-struts than those used on previous Nieuport fighters were fitted. Except for this, the aircraft had the standard fuselage, tail, landing gear, and engine (130-hp Le Rhône 9JB) of production Nieuport 27s. However, the Nieuport 27, with or without sesquiplane wings, was clearly obsolescent and it is likely that further development was abandoned in favor of the all-new Nieuport 28 and the even more modern Nieuport 29. The Nieuport 27 may have shared the same weaknesses in the wing structure that plagued the later Nieuports as a GQG memo, dated 6 September 1917, called for flying wires of increased size to provide greater strength. These were to be retrofitted to all Nieuport 24, 24bis, and 27 airplanes.

Operational Service

The Nieuport 27 superseded the Type 24 on the production line and was supplied to many of those escadrilles which were still using Nieuport fighters during 1918. The Nieuport sesquiplane was by this time decidedly passé, but the delays in production of the SPAD 13 meant that the Nieuport 24, 24bis, and 27 were still serving alongside SPAD 7s in the fighter escadrilles. N 90, for example, had seven Nieuport 27s, three Nieuport 24s, two Nieuport 24bis, and three SPAD 11s.

The Nieuport 27 was identical to the Nieuport 24 except for alterations to the landing gear. University of Texas at Dallas.

Nieuport 27

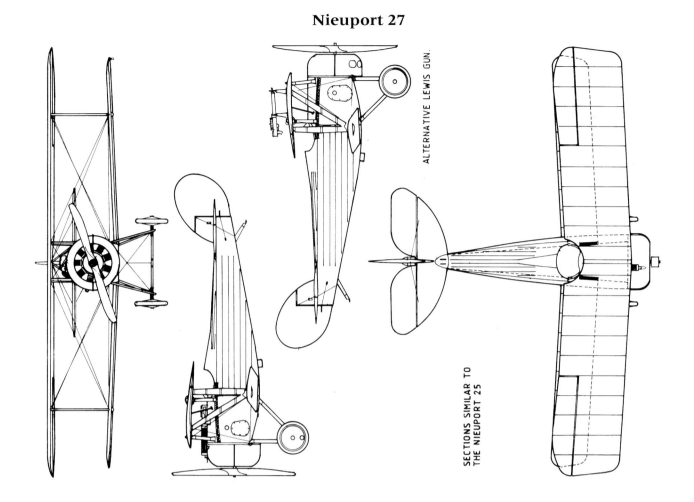

Army Cooperation

The escadrilles assigned to army cooperation duties were:

N 12, moved to Flanders in June 1917 and assigned to the 1st Armée, N 12 was active during the Battle of Malmaison. It re-equipped with SPAD 7s in late 1917.

N 76, assigned to the 5th Armée and based at Muizon. Capitaine Jean-Jacques Perrin assumed command in April 1917. He was succeeded by Capitaine Eugene Verdon and then Lieutenant Pierre Vitou. N 76 re-equipped with SPADs late in 1917.

N 79, joined the Provisional Groupe de Chasse de Bonneuil in June based at Catigny. In January 1918 N 79 became SPA 79.

N 82, based in January 1918 at San Pietro in Gu. It conducted long-range reconnaissance missions over Adige, Brenta, and Piave and provided fighter escort for the army cooperation units of the 31st C.A. In March 1918 the 10th Armée and its aviation units returned to France. Upon its return N 82 was assigned to the GC 22.

N 87, formed in March 1917 at Lyon-Bron, assigned to the 8th Armée and commanded by Capitaine Pierre Azire beginning in January 1918. In May 1918 N 87 became SPA 87.

N 88, commanded by Capitaine Francois d'Astier de La Vigerie, moved to the 6th Armée sector at Chemin des Dames in June 1917. It became SPA 88 in September 1917.

N 90, formed from Detachments N 504 and N 507 in March 1917, assigned to the 8th Armée and commanded by Lieutenant Pierre Weiss. It became SPA 90 in April 1918.

N 91, assigned to the 8th Armée and based at Lorraine. It was commanded by Lieutenant Jourdain and in January 1918 became SPA 91.

N 92, assigned to the 7th Armée and commanded by Lieutenant Edmond George. In May Capitaine Georges de Geyer d'Orth assumed command, and in July N 92 moved to the 2nd Armée sector. Redesignated N 392, the unit re-equipped with SPADs in May 1918.

N 94, formed in June 1917 from detachments N 512, N 513, and N 514 at Melette. It was assigned to the 4th Armée and was under the command of Capitaine Edouard Pillett. N 94 joined GC 18 in January 1918.

N 95, assigned to the 6th, then 2nd, Armée and commanded by Capitaine Francois de Castel. N 95 joined GC 19 in February 1918.

N 96, formed from N 510 in June 1917, attached to the 2nd Armée and under the command of Capitaine Masse. It moved to 4th Armée sector on 9 July and the 5th Armée on 22 July. In September Capitaine Eugene Verdon assumed command, followed by Lieutenant Maurice Barthe, who took control in January 1918. In February 1918 N 96 joined GC 19.

N 97, moved in July to the 4th Armée sector. In September 1918 it moved to the 2nd Armée sector under the command of Lieutenant Herve Conneau. In December N 97 became SPA 97 and joined GC 15.

N 98, formed from elements of N 85 in November 1917 and assigned to the 2nd Armée sector under command of Capitaine Leon Bonne. On 1 March 1918 N 98 joined GC 21.

N 99, formed in November 1917 from elements of N 87 and assigned to the 8th Armée, commanded by Capitaine Roger Lemercier de Maisoncelle-Vertille de Richemont. In February 1918 N 99 joined GC 20.

N 102, moved in January 1918 to the 1st Armée sector. Capitaine Roger Lemercier de Maisoncelle-Vertille de Richemont assumed command in March. The next month N 102 became SPA 102.

N 151, formed at Chaux in July 1917, assigned to the 7th Armée and commanded by Lieutenant Gerard Amanrich. It became SPA 151 in December 1917.

Nieuport 27 serial N 5690 of SPA 87. This aircraft was piloted by Lieutenant Marin, who was killed in combat with a German fighter of Jasta 43 on 5 February 1918. Most of SPA 87's machines had a white cat with a red border; on this camouflaged machine the colors appear to have been reversed. B83.3307.

Nieuport 27. B88.1332.

N 152, assigned to the 7th Armée under the command of Lieutenant Lefèvre. In November Lieutenant Lous Delrieu assumed command and in June 1918 N 152 became part of GC 22.

N 155, formed in July 1917 at Montdésir, assigned to the 4th Armée and based at La Noblette, commanded by Lieutenant Edmond George. In December N 155 became SPA 155.

N 156, formed in January 1918, assigned to the 4th Armée and commanded by Capitaine Emile Paumier. The next month it received Morane-Saulnier Type AIs and was redesignated MS 156.

N 157, formed in December 1917 at Chaux, assigned to the 7th Armée and commanded by Lieutenant Henri Constant de Saint-Saveur. In March it was assigned to GC 21.

N 158, formed in January 1918 at Bonneuil, assigned to the 3rd and then the 4th Armées. It returned to the 3rd Armée in February. N 158 was commanded by Lieutenant Jacques Chaudron. In March it became MS 158.

N 159, formed in January 1918 from elements of N 90, assigned to the 8th Armée and based at Lorraine under command of Lieutenant Albert Robert. In February it became part of GC 20.

N 160, formed in January 1918, assigned to the 2nd Armée and based at Brabant-le-Roi under command of Lieutenant Emile Barès. In May it re-equipped with Morane-Saulnier Type AIs to become MS 160.

N 161, formed in December 1917 and assigned to the 5th Armée, based at Lhéry and commanded by Lieutenant Marcel Parfai. N 161 moved to Boujacourt later in January; followed by a move to Dugny, still in the 5th Armée sector. it re-equipped with Morane-Saulnier Type AIs in February 1918 and became MS 161.

A flight leader of Squadriglia 83ª prepares his Nieuport 27 for a mission from Sovizzo Primavera in 1918. Via Alan Durkota.

Italian Nieuport 27 serial number 5239 in flight. *Ocio!* is written under the cockpit. Via Alan Durkota.

N 162, formed in January 1918 at Corcieux from elements of N 152. It was assigned to the 7th Armée and commanded by Lieutenant Daniel Chambariere. It joined GC 20 in March.

N 313, assigned to provide fighter protection for naval aircraft operating from Dunkerque. It was subsequently designated the Escadrille Dunkerque ou de Saint Pol. In addition to Nieuport 27s it had Sopwith 1½ Strutters on strength. It became SPA 313 in July 1918.

N 314, assigned to protect the city of Nancy and commanded by Capitaine Perrin. It became SPA 314 in July 1918.

N 315, formed in February 1917 from N 311 and assigned to the D.C.A. to protect the city of Belfort. It re-equipped with SPADs in July 1918.

N 392, formed from N 92 in June 1916 and served as an escadrille de protection for the city of Venice. In July 1917 it became N 561 and remained assigned to the protection of Venice. Commanded by Lieutenant Challeronge, N 561 remained in Italy throughout the war. Later in 1917 N 561 became SPA 561.

N 523, formed from H 387 in July 1917. It was assigned to the A.F.O. in Serbia and later re-equipped with SPADs.

N 531, formed March 1918 as a Franco-Hellenic escadrille. It later re-equipped with SPADs.

N 562, created at Corfu, Greece ,in July 1917 and assigned to protect the city of Potamos. It was disbanded in December 1918.

N 581, formed with eight Nieuport 17s and seven SPAD 7s in February 1917. It was assigned to the 7th Russian Army and began operations in May 1917. Based at Boutchatch Galicia, it served alongside Kosakov's Nieuport fighter group. In response to the German offensive of 19 July on the 6th Army front, N 581 was transferred to Buzcacz. In August it moved to the Romanian front, where it was assigned to the 3rd Corps. By February 1918 it was re-designated SPA 581 because it had re-equipped entirely with SPAD 7s.

Groupes de Combat

By January 1918 most of the Groupes de Combat were equipped almost exclusively with SPAD 7s.

GC 11 had one Nieuport unit, N 57, in January 1918, assigned to the 5th Armée. In February GC 11 joined Groupement Féquant at Cramaille; GC 11 was assigned to provide air cover for the Armées of the G.A.N. By March N 57 had received SPAD 13s and became SPA 57.

GC 14 had a single Nieuport escadrille, N 86, which was assigned to the 7th Armée in the G.A.N. sector. By April 1918 N 86 became an all-SPAD unit.

GC 17 was formed in February 1918 and included three Nieuport escadrilles: N 89, N 91, and N 100. All three used a mixture of Nieuport 24s and 27s and were based at Manoncourt-en-Vernois in the 7th Armée sector. The three escadrilles soon received SPAD 13s.

GC 20 was formed in February 1918 with four escadrilles: N 99, N 159, N 162, and SPA 68. The Nieuport escadrilles used a mixture of Nieuport 24bis, Nieuport 24s, and Nieuport 27s. By April GC 20 was based at Manoncourt and was active in the Toul sector. By May N 99, N 159, and N 162 had all been supplied with SPAD 13s to become SPA 99, SPA 159, and SPA 162 respectively.

GC 21 was created in March 1918 and had two Nieuport escadrilles: N 98 and N 157. GC 21 was assigned to the 4th Armée. The next month both N 98 and N 157 gave up their Nieuport 27s for SPAD 13s.

A report from General Pétain dated 22 October 1917 stated that "the Nieuport is inferior to all enemy aircraft. It is essential that it be withdrawn very soon from all escadrilles at the front. It is in my view vital to have by early spring 1918 an Aviation de Chasse composed wholly of SPAD single-seaters or Nieuport Gnome Monosoupape (Nieuport 28) if the trials of the latter type are satisfactory." This spelled the end for the Nieuport line of sesquiplane fighters and by the war's end virtually every fighter escadrille was equipped with SPADs.

Foreign Service
Italy

The Italians purchased Nieuport 27s from the French. In late 1917 and 1918 the Macchi firm was concentrating on license production of the Hanriot HD.1 and therefore, did not contract to build the Nieuport 27 under license. Only about 100 examples of the Nieuport 27 were obtained from the French.

The Nieuport 27s were serving with one unit in November 1917. This was 91ª Squadriglia based at Padova which was assigned to the 10 Gruppo under the control of the supreme command.

By June 1918, nine squadriglias had Nieuport 27s on strength: 3 Gruppo (1st Armata): 75ª Squadriglia; 7 Gruppo (6th Armata): 81ª and 83ª Squadriglias; 9 Gruppo (7th Armata): 72ª Squadriglia; 20 Gruppo (7th Armata): 74ª Squadriglia; 23 Gruppo (8th Armata): 78ª and 79ª Squadriglias; 85ª Squadriglia at Piskupi (Albania); and 73ª Squadriglia at Gazzo.

The Nieuport 27s were used operationally during the Battle of Piave and the offensive at Vittorio Veneto. By October 1918 the number of squadriglias using Nieuport 27s had declined from nine to six. These were: 3 Gruppo (1st Armata): 75ª Squadriglia; 7 Gruppo (6th Armata): 83ª Squadriglia; 10 Gruppo (Supreme Command): 70ª and 82ª Squadriglias; 20 Gruppo (7th Armata): 74ª Squadriglia; and 23 Gruppo (8th Armata): 79ª Squadriglia.

Baccotini and his Nieuport 27 of the 83ª Squadriglia in 1918. Via Alan Durkota.

Below: 83ª Squadriglia lineup at Taliedo in 1918. Via Alan Durkota.

The Nieuport 27s were quickly retired after the war; some were bought by private citizens for recreational flying.

Japan
The Japanese acquired at least one Nieuport 27 after the war. It inspired Tomotari Innagaki to produce a modified version called the Itoh Tsurubane No.2 Aerobatic Plane.

Russia
Two Nieuport 27s were supplied to the Russians. They were serving with the First and Second Military School of Pilots in February 1924.

United Kingdom
The RFC obtained 71 Nieuport 27s in August 1917. In British service the aircraft were armed with a single Lewis machine gun on a Foster mount. They served with Nos.1 and 29 Squadrons, replacing obsolescent Nieuport 17s and 23s. As late as February 1918 No.1 Squadron had a single Nieuport 27, while No.29 Squadron had 19 on strength. The last example of the Type 27 was not withdrawn until April 1918 (although a single non-operational example was serving as late as November 1918). Four Nieuport 27s were sent to Palestine in January 1918. As Nieuport 27s were withdrawn from front-line service they were returned to England and were probably used by the training units.

United States
The A.E.F. purchased 287 Nieuport 27s for use as trainers. Seventy-five had 80-hp Le Rhône engines and 212 had 120-hp Le Rhône motors. At least one of these aircraft was sent to the United States for evaluation. Both the 31st and 37th Aero Squadrons had some Nieuport 27s on strength.

Uruguay
In the late 1920s 24 Nieuport 27s were obtained by Uruguay. They were based at Paso de Mendoza.

Nieuport 27 Single-Seat Fighter with 130-hp Le Rhône 9JB

Span 8.21 m; length 5.87 m; height 2.40 m; wing area 14.75 sq. m
Empty weight 380 kg; loaded weight 535 kg

Maximum speed:	at sea level	172 km/h
	2,000 m	170 km/h
	3,000 m	167 km/h
	4,000 m	165.7 km/h
Climb:	2000 m	5 minutes 40 seconds
	3,000 m	9 minutes 25 seconds
	4,000 m	14 minutes 40 seconds
	5,000 m	21 minutes 30 seconds

Ceiling 6,850 m; range 250 km; endurance 2.25 hours
Armament: one synchronized 0.303-in Vickers machine gun; some had a 0.303-Lewis gun on the top wing (the RFC had only the overwing Lewis gun)

Nieuport 28

The Nieuport series of sesquiplane fighters had been superseded in French service by the SPAD series of fighters. Attempts to enhance the Nieuport designs by redesigning the airframe, fitting more powerful engines, or even changing to a new wing had proved unsuccessful and the SPAD firm's dominance had continued unabated. It was inevitable, then, that the Nieuport firm would at last switch to a conventional biplane with parallel interplane struts. The major deficiency of the sesquiplanes had been low speed and poor climbing ability. An increase in wing area and a more powerful engine was required to correct these defects. Furthermore, shortcomings in the design of the lower wing had resulted in those of the Nieuport 23 ripping off in flight. A new wing had been tried on a Nieuport 27 (possibly in preparation for a similar wing on the Nieuport 28) that had two spars and a wider chord.

The prototype Nieuport 28 had a wire-braced four-longeron fuselage made of wood that was significantly longer than that of the Nieuport 27. The fuselage was also completely rounded; the flat cross section of the fuselage bottom that had been used on the earlier Nieuports was replaced by a completely circular layout. The area from behind the cockpit to the cowling was covered with tulip wood strips rather than metal as on earlier Nieuport designs. The area from the rear of the cowling to the rear of the fuselage was covered by plywood as, apparently, were the tail surfaces. The wings had a wooden framework and two spars, and were covered with fabric. The two pine spars were fitted with wire-braced wooden ribs. The leading edges were covered with plywood veneer. The interplane struts and center section struts were also of wood. The chord of both wings was almost equal; the lower wing was still slightly smaller than the upper. The wing tips were elliptical, and there was no dihedral on the bottom wing. Ailerons were fitted to the lower wing only. The top wing was fitted at the eye level of the pilot. The interplane struts formed a single box-like structure eliminating the need for incidence or stagger wires. The tail was also made of wood and covered with fabric. The undercarriage was of aluminum tube with streamlined fairings and rubber cord shock absorbers. As with the Nieuport 27, the tail skid was internally sprung. Armament consisted of one Vickers 0.303 machine gun mounted to port. The engine was a 160-hp Gnome Monosoupape 9Nc rotary.

The prototype flew in June 1917. The upper wing (with dihedral) was not successful, possibly because the low-set wing inhibited fitting of a second machine gun, and two other configurations were tried. At least two aircraft with full dihedral (N4434 and N6125) were built and flown. Another version was built featuring an upper wing with a slight dihedral of 1.5 degrees. The latter version was to be the only one of the three types to be produced in series.

Production aircraft were fitted with two Vickers 0.303"

Nieuport monocoque, a variant of the Nieuport 28. This aircraft had an upper wing with no dihedral. The engine was a 165-hp Gnome Monosoupape; a single gun was fitted. Photo courtesy of J.M. Bruce.

Nieuport 28 prototype. Reairche.

Nieuport 28 serial N 4434. This early machine features the short center-section struts and has a pronounced upper-wing dihedral. Photo courtesy of J.M. Bruce.

Above, below, below left: Nieuport 28 serial N 4434 which had the upper wing mounted just above the pilot's eye level. University of Texas at Dallas.

Nieuport 28

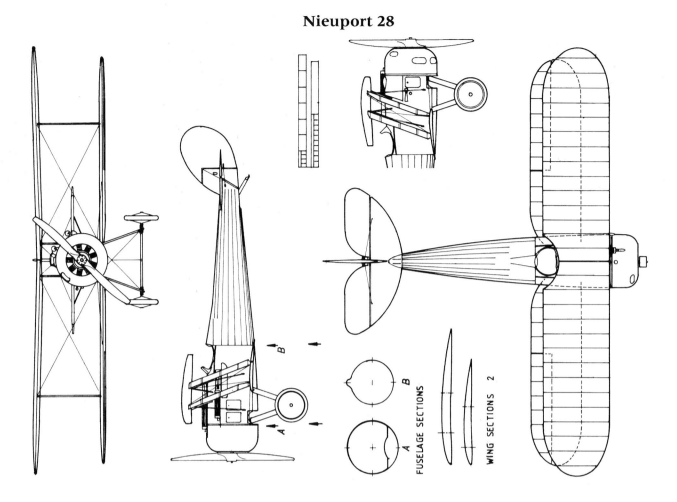

machine guns, one mounted offset to port on the top of the fuselage, the other was attached to a shelf below the port center-section struts. Aside from the reduced dihedral of the upper wing and the twin-gun armament, the production Nieuport 28 seems to have been identical to the prototype. The aircraft was maneuverable and had a rapid climb rate. However, production aircraft were not popular and have been described as tending to shed fabric from the wings when they were steeply dived. Perhaps for these reasons the Nieuport 28 was not selected for production by the STAé. It is interesting to note that all the aircraft designed to fit the C1 specification that used the Gnome 9Nc rotary were also unsuccessful. Of these aircraft (SPAD 15, Morane-Saulnier 27 and 29, and Courtois-Suffit-Lescop), only the Morane-Saulnier designs entered production. These were quickly withdrawn from service, apparently because the Gnome engine was particularly prone to catch fire. In fact, it was reported that earlier versions of the Nieuport 28 were also prone to catch fire because leakage of fuel fumes from the engine would result in fires. Modifications to the exhaust system and engine cowling corrected the problem.

Fortunately for Nieuport, there was still a pressing need for new fighters and the SPAD firm was having difficulties meeting its production goals for the SPAD 13. Also, the entry of the Americans into the conflict meant they would need new aircraft. As the French wished to retain the SPAD 13s for their own escadrilles, it was decided that the Nieuport firm would produce the Nieuport 28 for use by the American air service. A total of 297 were ordered by the A.E.F. Air Service. As mentioned above, the Nieuport 28 was unpopular and was also clearly inferior to the Fokker D.VII. It was retired from front-line service within four months and replaced in A.E.F. service by SPAD 13s.

Nieuport 200-hp Clerget 11E. This aircraft, which was in many respects similar to the Nieuport 28, had a marked positive stagger to the upper wings. 01117.

Nieuport 200-hp Clerget 11E. The upper wings are flat, while the lower wing had dihedral. 01118. See addendum for three-view drawing.

Nieuport experimental fighter with a 200-hp Clerget 11E engine. This airplane is similar to the Nieuport 28, but with a different wing configuration. University of Texas at Dallas.

Nieuport experimental: 200-hp Clerget fighter. Reairche collection.

Variants

1. Nieuport 28 with 175-hp Le Rhône engine: This aircraft was built and tested in November 1917. No further details are available.
2. Version of the Nieuport 28 powered by a 200-hp Clerget 11E engine. The aircraft featured a larger fuselage and increased wing area. The lower wing had dihedral, and there was a pronounced stagger to the wings. It was armed with two Vickers 7.7-mm machine guns. The aircraft had a large cutout into the top wing to facilitate the pilot's view. It underwent trials from November to December 1917. Its performance was comparable to the SPAD 7, which was already in production.
3. The Nieuport firm produced what appears to be a development of the Nieuport 28 but powered by a 165-hp Gnome Monosoupape 9N rotary engine. The aircraft featured a wooden monocoque fuselage with tail surfaces similar to those of the later Nieuport 29. Armament consisted of a single 7.7-mm Vickers machine gun mounted to the port of the centerline struts, which were single-struts mounted in tandem. The wings had no dihedral. The aircraft was tested in December 1917 but development was abandoned by April 1918. Possibly it was rejected because of its rotary engine, which was now falling out of favor. In any event, it offered no significant improvement over the standard Nieuport 28.
4. A second aircraft was tested with a 170-hp Le Rhône 9R engine. It also offered no improvement over the Nieuport 28 and was not selected for production.

Foreign Service

Argentina

Two Nieuport 28s were obtained by Argentina in 1919. One crashed while trying to cross the Andes; the other was reportedly unable to fly because of a "warped" fuselage. Several more Nieuport 28s were obtained in 1919 and remained in service until 1922.

Greece

A unknown number of Nieuport 28s were supplied to the Greek air service. They served alongside SPAD 7s in 531 Mira (Squadron) from 1919 to 1922.

Guatemala

A French military mission took at least one Nieuport 28 to Guatemala in 1920. It was assigned to the Cuerpo de Aviacion

Above: Nieuport 28 serial N 6298. This airplane was tested at McCook Field in the USA, where it was assigned project number P-38. Photo courtesy of J.M. Bruce.

Above, below, below left: Nieuport 28 serial N 6298. University of Texas at Dallas.

Nieuport 28 serial N 6187.

Militar de Guatemala and used as a trainer at the Campo de La Aurora.

Switzerland

In 1918 a single Nieuport 28 was interned when an American pilot landed in Switzerland due to engine trouble. Given the serial number 607, the aircraft remained service with the Swiss Fliegertruppe until 1925, when it was retired due to airframe fatigue. However, the aircraft proved impressive enough to lead the Fliegertruppe to purchase 14 Nieuport 28s in 1923. The unarmed planes were used for pilot training and were given serial numbers 685 to 698. Because of structural fatigue, the aircraft were retired in 1930.

United States

The Air Service of the American Expeditionary Force was in desperate need of fighters and was willing to accept any aircraft, even the Nieuport 28, which had been rejected by the Aviation Militaire. A total of 297 aircraft were delivered to the Americans. Only four units used the Nieuport 28: the 27th, 94th, 95th, and 147th Aero Squadrons.

The 27th Aero Squadron was based at Tours in March 1918. As part of the First Pursuit Group, it participated in the battles in the Toul and Aisne-Marne sectors.

The 94th Aero Squadron received 22 Nieuport 28s between 15 and 21 March 1918. It provided air cover for the 8th C.A. from St. Mihiel to Pont-a-Mousson. By the end of June the 94th had given up its Nieuport 28s for SPAD 7s and 13s. The unit had 18 confirmed victories while flying Nieuport 28s.

The 95th Aero Squadron was based at Issoudun on 16 November 1917. It was particularly active over the Toul sector. By July 1918 the unit had re-equipped with SPAD 13s.

The 147th Aero Squadron was assigned to the 1st Pursuit Group and was based at Tours in March. It was active over the 6th and 8th C.A. sectors. In June the squadron supported the American 1st Army and was based at Toul.

The Nieuport 28s were not well-liked by all the American. pilots, but some men achieved good results with the type. Although the aircraft was quite maneuverable, its engine was prone to catching fire due to leaks in the fuel system. These

> **Nieuport 28 Single-Seat Fighter with 160-hp Gnome Monosoupape 9Nc**
> Span 8.16 m; length 6.40 m; height 2.50 m; wing area 16.00 sq. m
> Empty weight 456 kg; loaded weight 698 kg
> Maximum speed: 198 km/h at 2,000 m; climb to 2,000 m in 5.5 minutes; ceiling 5,180 m; range 400 km; endurance 1 hour 30 minutes
> Armament: two 0.303-inch Vickers machine guns
> Approximately 310 built
>
> **Nieuport 28 Experimental Fighter with 200-hp Clerget 11E**
> Wing area 21.00 square meters; empty weight 530 kg
> Empty weight 530 kg; loaded weight 850 kg
> Maximum speed: 200 km/h at 4,000 m; climb to 4,000 m in 12 minutes; endurance 2.5 hours
>
> **Nieuport 28 Monocoque**
> Loaded weight was 640 kg
> Maximum speed: 198 km/h at 2,000 m; climb to 3,000 m in 7.33 minutes; endurance 2.25 hours

leaks appear to have been caused by vibration. It was also discovered that the Nieuport 28s had a tendency to shed the wing fabric covering the leading edge. There was also a shortage of machine guns which meant that many aircraft were armed with only a single weapon. Although attempts were made to rectify these problems, it was soon decided to replace the Nieuport 28s with other types of fighters. By July most of the Nieuport 28s were relegated to training units.

Postwar, approximately 50 Nieuport 28s were returned to the United States. Most were used by the Army Air Service as trainers. Twelve aircraft went to the United States Navy and were given serial numbers A 5794–A 5805. These aircraft were reportedly flown off platforms fitted to the forward turrets of battleships and were also used to practice formation flying and combat techniques. These Nieuports constituted Combat Squadron Three, which in 1920 was the only fully-armed fighter squadron in the United States Navy. Some of these aircraft were later used as racers.

Nieuport Two-Seat Fighter with 370-hp Lorraine-Dietrich 12Da

In 1917 Nieuport designed and built a two-seat fighter to be powered by a 370-hp Lorraine-Dietrich 12Da. It was reported to be undergoing testing in January 1918. This aircraft would have been a contemporary of the SEA 4 two-seat fighter, but the design has not been listed as one of the aircraft intended to meet the C2 specification of 1918. Apparently, further development was abandoned.

Nieuport Fighter with 275-hp Lorraine-Dietrich 8Bd

This fighter represented another attempt by Nieuport to produce a successful fighter not powered by a rotary engine. The design was far removed from the Nieuport sesquiplane series of fighters. Whereas those aircraft were petite and, indeed, elegant, rotary-engine fighters of this new design were surprisingly awkward. The top wing was mounted far above the fuselage on immense struts. The engine was a Lorraine Dietrich 8Bd of 275-hp (although the 240-hp 8Bb is also a possibility). Huge fairings were placed over the cylinder blocks and a bulky radiator was mounted between them. These fairings probably restricted the pilot's forward view severely. Photographs of the aircraft do not show any armament. It was tested in January 1918.

Apparently the airplane was not completely satisfactory, as a revised version that used the engine appeared in January 1918. This was more aesthetically pleasing and featured swept wings of increased area and equal span. Horn-balanced ailerons were on the upper and lower wings. The fin and rudder assembly was enlarged. The radiator was moved to the top wing center section, which would have improved the pilot's forward view. Flight testing was under way in April 1918. Neither of the Lorraine-Dietrich powered aircraft were selected for production.

Nieuport Single-Seat Fighter with 275-hp Lorraine Dietrich 8Bd
Wing area 21.00 sq. m
Empty weight 535 kg; loaded weight 850 kg
Maximum speed: estimated at 220 km/h at 4,000 m
Armament: two 7.7-mm Vickers machine guns (prototypes appear to have been unarmed)
Two built

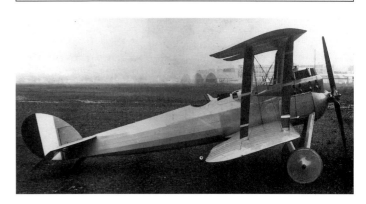

Nieuport biplane with a 275-hp Lorraine-Dietrich 8Bd engine. 87.0393.

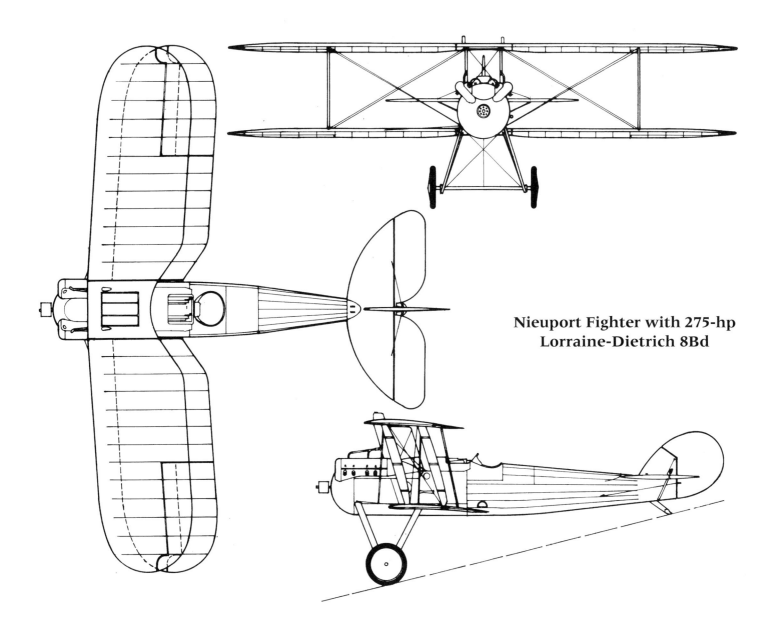

Nieuport Fighter with 275-hp Lorraine-Dietrich 8Bd

Left: Nieuport fighter prototype. This is the second version of this aircraft. B88.1332.

Below and below left: Four views of the experimental Nieuport fighter with a 275-hp Lorraine-Dietrich 8Bd engine. MA6941.

Nieuport Monoplane

A monoplane was constructed by the Nieuport firm in October 1917. The aim of the shoulder-mounted wing was to enhance the pilot's view, reduce drag, and increase structural strength. To provide the pilot with an enhanced downward view, clear panels were placed in the wing roots. The undercarriage was divided by a spanwise fairing of streamlined section which probably provided some lift. It was intended that this surface would serve as an air brake during landing. The landing gear was suspended beneath the center fuselage by N-shaped struts and two sets of bracing struts extended from the middle of each wing to the undercarriage struts. The fuselage was faired and the wings were fabric-covered. The engine was a 150-hp Gnome Monosoupape 9N. The aircraft flew in late 1917/early 1918. Armament was two 7.7-mm Vickers machine guns. The monoplane was not selected for production, but was further developed as the Nieuport 31. A second version was fitted with a 180-hp Le Rhône 9R engine and featured an inverse taper on the inboard trailing edges and an extended fin. Balanced elevators were added later. Development had been abandoned by May 1918.

The Nieuport monoplane sesquiplane. This photograph clearly shows the windows in the wing root which were intended to improve the pilot's downward view. MA24251.

Nieuport Single-Seat Fighter with 150-hp Gnome Monosoupape 9N

Wing area 17.50 square meters
Empty weight 433 kg; loaded weight 703 kg
Maximum speed: 220 km/h at 4,000 m; climb to 4,000 m in 13 minutes; endurance 2.0 hours
Armament: two 7.7-mm Vickers machine guns
Two built

Right: A Nieuport monoplane undergoing operational evaluation at the C.R.P. B82.3562.

Below: A Nieuport monoplane armed with two machine guns. B90.1895.

Nieuport Fighter with 300-hp Hispano-Suiza 8Fb Engine

This aircraft used virtually the same airframe as that employed on the experimental Lorraine-Dietrich engined fighters. The engine in this airframe, however, was the 300-hp Hispano Suiza 8Fb. One change was that the radiators were placed underneath the wings. There were two Vickers 7.7-mm machine guns. The aircraft underwent flight testing in May 1918, but further development was not pursued, preference probably being given to the forthcoming Nieuport 29.

Nieuport Single-Seat Fighter with 300-hp Hispano-Suiza 8Fb

Wing area 21.00 square meters
Empty weight 600 kg; loaded weight 950 kg
Maximum speed: 230 km/h at 4,000 m (est.); climb to 4,000 m in 10 minutes (est.)
Endurance 2.5 hours.
Armament: two 7.7-mm Vickers machine guns
One built

Nieuport biplane fighter with a 300-hp Hispano-Suiza engine. The radiators were located beneath the wings. University of Texas at Dallas.

Nieuport fighter prototype with a 300-hp Hispano-Suiza engine and underwing radiators. B78.1156.

Nieuport 29

With the Nieuport 29, Gustav Delage was at last successful in designing a fighter superior to the SPAD 7 and 13. However, the Nieuport 29 would not enter escadrille service until 1921.

A requirement for 650 fighters to be powered by either a 300-hp Hispano-Suiza 8Fb or a 320-hp ABC Dragonfly engine was formulated on 29 September 1918. The following aircraft were developed to meet this requirement:
1. 300-hp Hispano-Suiza 8Fb : SPAD 18, SPAD 20, SPAD 21, SPAD 22, Nieuport 29, Descamps 27, Hanriot-Dupont HD.7, De Marcay 2, SAB 1, Borel C2, Moineau C1, Semenaud C1, and a re-engined Sopwith Dolphin.
2. 320-hp ABC Dragonfly: Gourdou-Leseurre.

The C1 requirement was quite demanding, calling for a payload of 220 to 270 kg, maximum ceiling of 9,000 m, service ceiling of 6,500 m, maximum speed at 6,500 m of 240 km/h, and a minimum speed at sea level of 120 km/h.

Of the above aircraft only the SPAD 20 and Nieuport 29 would see widespread service with the Aviation Militaire and the Nieuport 29 would become its premier fighter. The prototype Nieuport 29 differed from production aircraft in having single-bay wings and radiators mounted underneath the wings. In addition to this, ailerons were on both the upper and lower wings. The prototype had an impressive performance, but it could climb only to 7,500 m.

The Nieuport 29 was unable to meet the specified maximum altitude, and it was therefore decided to fit it with a redesigned, enlarged wing. The upper wing span was increased on the second prototype, resulting in addition of a second bay of struts instead of the single bay used on the first prototype. A pair of Lamblin radiators were fitted beneath the center fuselage, replacing the wing-mounted radiators. Ailerons were still fitted to the upper and lower wings. The rudder was enlarged and of a high aspect ratio, probably because of control problems. On 14 June 1919 the second prototype was flight-tested by Lieutenant Casale, attaining an altitude of 9,123 m, higher than that called for by the C1 specification.

A number of modifications were made on the basis of tests conducted with the second prototype. These were due to stability problems on the prototype, on which ailerons had been fitted to the upper and lower wings. As a result, the second prototype had the upper-wing ailerons removed and those on the lower wings enlarged to compensate.

It appears that these modifications were first tried on a Nieuport 29 given the number "03," which may have denoted the third prototype.

Production aircraft were fitted with a Hispano-Suiza 8Fb engine in a tight-fitting metal cowling with four exhaust stubs on either side of the nose. The propeller was a Lumiere 144c. The fuselage was streamlined and in front of the cockpit was built up of spiral layers of tulip wood veneer applied over bulkheads made of plywood. The lower wing roots were faired into the bottom of the fuselage. The tail surfaces were covered with tulip wood skin. The wings had two-bays of struts made of wood and covered with fabric. The centerline struts were of an inverted V configuration and there was a 10-degree stagger. The top wing had a slight dihedral and carried two fuel tanks. The undercarriage was made of wood and the axle was hinged. A fuel tank was located behind the engine and there was a wind-driven generator to supply electricity to light the cockpit instruments and heat the pilot's suit. The aircraft was

Nieuport 29 of 7eme Escadrille of the 33Eme Regiment Mixte based at Mayence (ex-SPA 93). B84.2143.

Nieuport 29

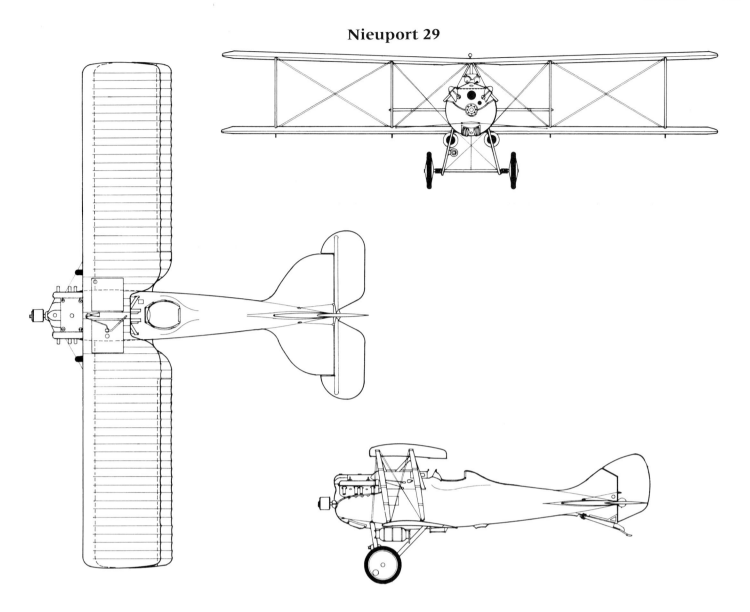

armed with twin Vickers 7.7-mm machine guns mounted in a fairing on the fuselage top decking. Two Lamblin radiators were mounted beneath the fuselage, one on either side of the undercarriage legs. The oil tank was fitted below the engine and was corrugated to facilitate cooling. As a result of all these changes the aircraft did not enter production until 1920, when an order was placed by the Aviation Militaire. The first was delivered on 11 February, 1921. On 9 October 1923 a decision was made to replace the heterogeneous mix of fighter types serving with the escadrilles de chasse with two basic aircraft: the Nieuport 29 and the SPAD 81. These two types remained the primary aircraft of the fighter units until 1929.

Variants

1. Nieuport 29 Fighter Trainer: Five examples of a standard Nieuport 29 were built with 180-hp Hispano-Suiza 8Ab engines to meet the ET 1 requirement for a trainer. The pilot's view was enhanced by the removal of the twin machine guns. The type was entered in the 1923 concours d'avions école and met all the requirements of the competition. However, it was not ordered into series production. Ironically, when the Nieuport 29 C1 s were withdrawn from front-line service in 1928, they were assigned to the training units from 1929 to 1931.
2. Nieuport 29 Bomber: The Rif campaign had showed the utility of airplanes in retaining French control over distant colonies. A group of experimental Nieuport 29s were fitted with six 10-kg bombs for the ground attack role. In a single month the Nieuports dropped a total of five tons of bombs.
3. Spanish Nieuport 29: In 1923 a competition was held in Spain to obtain a fighter aircraft (as well as a bomber and reconnaissance aircraft). The versions for Spain had the wing area reduced to 22 square meters and had the twin radiators placed vertically on the landing gear.
4. Nieuport 29 with 160-hp Gnome 9Nc engine: Two examples were produced by the firm at Issy, designated Type Gs. One was re-equipped with a Hispano-Suiza 300-hp engine and flown in the Grand-Prix at Monaco in 1920 fitted with twin floats.
5. Nieuport 29bis: A version of the Nieuport 29 fitted with a 180-hp Le Rhône 9R rotary engine. It was intended to be operated from aircraft carriers. The single aircraft built never went into production because the Aviation Maritime selected the Dewotine 1 as its standard fighter.
6. Nieuport 29 D: A high-altitude version of the Nieuport 29. This aircraft had a lengthened fuselage with a triangular configuration. Originally designated the Type 40, it was not ordered as there was felt to be no requirement for a high-altitude fighter. The aircraft was instead used in an attempt to set a high-altitude record by adding a Rateau turbo-compressor and reinforcing the wing struts. On 19 March 1920 Lieutenant Jacques Weiss attempted to set an altitude record with it. However, the pilot lost consciousness at 7,000 meters because of a failure in the oxygen system. Fortunately, he regained consciousness at 3,400 meters and

Second version of the Nieuport 29 prototype serial N 12002. University of Texas at Dallas.

Nieuport 29 serial N12104. University of Texas at Dallas.

Nieuport 29. University of Texas at Dallas.

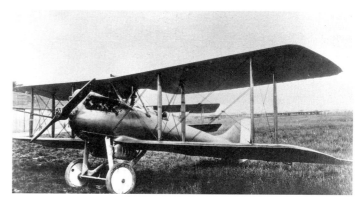
Nieuport 29. University of Texas at Dallas.

was able to pull out of the dive, albeit with the loss of part of the rudder.

7. Nieuport 29 Racer: A modified Nieuport 29 which was entered in the Gordon Bennett Cup races of 1920. The aircraft was modified by having the wing span reduced with single-bay bracing and an increase in span of the lower wing until it was 21 inches longer than the upper. This aircraft won the race and went on to win the 1922 Deutsche Cup competition.

8. Nieuport Type 22 m (29bis): A standard Nieuport 29 was fitted with a wing of reduced surface area. It was displayed at the Salon de l'Aéronautique in 1922 but no production was undertaken.

A total of 250 Nieuport 29s were constructed in France from 1922 to 1924. In addition to those built by the Nieuport firm, others were built by Schreck, Levasseur, Potez, Blériot, Letord, Farman, and Buscaylet. It was also built under license in Spain, SABCA (Belgium), Nakajima (Japan), Caproni (Italy), and Macchi (Italy).

Operational Service

The 1st Regiment d'Chasse with Nieuport 29 Escadrilles 101–109 was based at Thionville. The 2nd Regiment d'Chasse with Nieuport 29 Escadrilles 101–108 (109 was a HD.3 unit) was based at Strasbourg. The 3rd Regiment d'Chasse with Nieuport 29 Escadrilles 101–107 was based at Châteauroux. A total of 24 escadrilles were equipped with Nieuport 29s, although some also had SPAD 81s on strength. The Nieuport 29 remained in front-line service until 1928.

Foreign Service

Argentina
A small number of Nieuport 29s were acquired from France and assigned to Gruppo de Aviacion 1.

Belgium
After the war, 21 Nieuport 29s were purchased by the Belgian air service. In addition to these, 87 were built under license by SABCA from January 1924 to October 1926. They were used primarily by the 9th ("Chardons") and 10th ("Cometes") Escadrilles although some were assigned to other units.

Italy
The Nieuport 10, 11, 17, and 27 had played a key role with the Italian fighter units and it is therefore not surprising that the Italians would select the Nieuport 29 as the first new fighter to be ordered by the Regia Aeronautica. The Nieuport 29 was clearly superior to the other fighters then in Italian service. Initial orders placed were for enough aircraft to equip five squadriglias. They were built under license by the Macchi and Caproni firms. The serial numbers were as follows:

1. M.M. 34, 35, 36, 37, 58, and 59, the experimental serial numbers applied to the initial examples of the Nieuport 29.
2. 1000–1049: Fifty built by Macchi delivered and between July 1923 and March 1924.
3. 1050–1099: Fifty built by Caproni and delivered between August 1923 and June 1924.
4. 1100–1199: Sixty built by Macchi and Caproni (each built 30) in 1924.
5. 1700–1714: Fifteen built by Macchi in 1929.

A total of 175 were built in Italy.

The Nieuport 29 had a relatively brief life in Italian service. The first unit to receive it was 84[a] Squadriglia, which accepted its first aircraft in October 1924. The units using the Nieuport-Macchi 29 in 1925 were 7 Gruppo: 76[a], 84[a], and 91[a] Squadriglias; and 23 Gruppo: 70[a], 74[a], and 75[a] Squadriglias.

The order of battle for 1926 was: 7 Gruppo: 84[a], 86[a], and 91[a] Squadriglias; 8 Gruppo: 92[a] squadriglia with N.29s and A.C.2s; and 23 Gruppo: 70[a], 74[a], and 75[a] Squadriglias.

The Nieuport 29s were withdrawn from front-line squadriglias in 1927, replaced by the indigenous A.C.2s, A.C.3s, and Fiat CR.20s.

Japan
Nieuport 29s were built under license in Japan by Nakajima. The

Nieuport 29 Prototype with 300-hp Hispano-Suiza 8Fb
Span 9.80 m; length 6.65 m; height 2.56 m; wing area 26.75 sq. m
Empty weight 767 kg; loaded weight 1,190 kg
Maximum speed: 226 km/h; range 500 km; ceiling 8,200 m
Armament: two 7.7-mm Vickers machine guns
Two built

Nieuport 29 Production with 300-hp Hispano-Suiza 8Fb
Span 9.70m; Length 6.49 m; height 2.56 m
Empty weight 760 kg; loaded weight 1,150 kg
Maximum speed: 236 km/h; range 580 km; ceiling 8,500 m
Armament: two 7.7-mm Vickers machine guns
1,250 built

Nieuport 29 SHV Racer with 300-hp Hispano-Suiza 8Fb
Span 8.00 m; length 7.30 m; height 3.10 m; wing area 22 sq. m
Empty weight 925 kg; loaded weight 1,220 kg
Maximum speed: 250 km/h
Armament: none
Four built

Nieuport 29 ET 1 Fighter Trainer with 180-hp Hispano-Suiza 8Ab
Span 9.70 m; length 6.49 m; height 2.50 m; wing area 22.00 sq. m
Empty weight 750 kg; loaded weight 1,080 kg
Maximum speed: 200 km/h; range 340 km; ceiling 6,000 m
Armament: two 7.7-mm Vickers machine guns
Three built

Nieuport 29bis Carrier-Based Fighter with 300-hp Hispano-Suiza 8Fb
Span 8.00 m; length 6.50 m; height 2.50 m; wing area 22.00 sq. m
Empty weight 761 kg; loaded weight 1,100 kg
Maximum speed: 250 km/h; range 500 km
Armament: two 7.7-mm Vickers machine guns
One built

Nieuport 29 V Racer with 320-hp Hispano-Suiza 8Fb
Span 6.00 m; length 6.20 m; height 2.50 m; wing area 12.30 sq. m
Empty weight 600 kg; loaded weight 936 kg
Maximum speed: 303 km/h
Armament: none
One built

Nieuport 29 Vbis Racer with 330-hp Hispano-Suiza 8Fb
Span 6.00 m; length 6.20 m; height 2.50 m; wing area 13.20 sq. m
Empty weight 470 kg; loaded weight 805 kg
Maximum speed: 313 km/h
Armament: none
One built

Macchi Nieuport 29 with 300-hp Hispano-Suiza 8Fb
Span 9.70m; Length 6.44 m; height 2.66 m; wing area 26 sq. m
Empty weight 832 kg; loaded weight 1,190 kg;
Maximum speed: 230 km/h; range 580 km; ceiling 7,500 m; climb to 4,000 m in 9 minutes 44 seconds
Armament: two 7.7-mm Vickers machine guns
175 built

Nakajima Ko 4 (Nieuport 29) with 300-hp Mitsubishi/Hispano-Suiza 8Fb
Span 9.70 m; Length 6.44 m; height 2.64 m; wing area 26.80 sq. m
Empty weight 825 kg; loaded weight 1,160 kg;
Maximum speed: 126 kts; ceiling 8,000 m; climb to 4,000 m in 13 minutes 30 seconds; endurance 2 hours
Armament: two 7.7-mm Vickers machine guns
608 built

Japanese army air service had purchased several Nieuport 29s from France in 1923; they were impressive enough that 608 were built as the Ko 4 army fighter. Production continued until 1932. The Nieuport 29s replaced the Nieuport 24s and SPAD 13s then in squadron service and were used operationally in the Manchurian and Shanghai battles. The aircraft were assigned to the following Japanese army squadrons:

1st Air Regiment (3rd Ground Division) with two squadrons; 3rd Air Regiment (16th Ground Division) with three squadrons; 4th Air Regiment (12th Ground Division) with two squadrons; 7th Air Regiment (20th Ground Division) with one squadron; and 8th Air Regiment (Formosa Ground Division) with one squadron. The Nieuport 29s remained the prime equipment for Japanese fighter units until 1933, when they were replaced by the Nakajima Type 91.

Spain
Spain acquired 40 Nieuport 29s. Although some sources state that ten were built under license, Spanish sources indicate that all were purchased directly from France. These aircraft were assigned to the Grupo de Caza of Getafe (under J. Gonzalez Gallarza) based in Madrid. It was redesignated the 11th Gruppo in 1927. Other Nieuport 29s replaced Martinsyde F.4s with the Escuadrilla de Caza at Melilla. Attrition substantially reduced the 11th Gruppo's size by 1930. The Nieuport 29s were struck off charge around 1931.

Sweden
Ten Nieuport 29s were purchased by the Flygvapnet in 1926. They were designated J 2s and all were assigned to F 3 based at Malslatt. During 1928 and 1929 the aircraft were sent to training squadron F 5 at Ljungbyhed.

Thailand
Nieuport 29s were obtained in 1920 by Thailand (Siam) and, along with a number of SPAD 13s, equipped the 1st Pursuit Group into 1930.

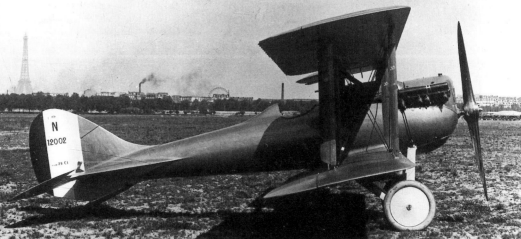

Nieuport 29 prototype, serial number N 12002. The twin Lamblin radiators are barely visible beneath the fuselage. B88.1322.

Nieuport 30

The Nieuport firm produced an aircraft to meet the type B2 specification of 1917. This called for a day bomber with a crew of two and capable of carrying 300 kg of bombs. It was intended as a replacement for the Breguet 14 and was to have been powered by a single 600-hp Renault 12M. The bomber was to have utilized the same duralumin structure as the Breguet 14. The wings had a pronounced backward stagger with two-bays of struts on either side, and the upper wing was shorter than the lower. Three Lamblin radiators were fitted under the engine. The fuselage was constructed of wood and had a rectangular cross section. The aircraft was in competition with the Breguet 14 with a 300-hp Renault engine fitted with a Rateau turbo-compressor. The Breguet design was selected by the Aviation Militaire, but was built only in small numbers due to the Armistice. It is not known how far construction had progressed on the Nieuport 30 before the Armistice, but no photographs or drawings have been found that show it in the B2 configuration.

Nieuport 30T.1 with 450-hp Renault
Span 13.50 m; length 10.80 m; height 4.20 m; wing area 65 sq. m
Empty weight 1,500 kg; loaded weight 2,400 kg; payload 300 kg
Maximum speed: 190 km/h; range 550 km; climb to 2,000 meters in 18 minutes
Armament: none carried on transport version but the B2 specification required 300 kg of bombs
Seven built with Sunbeam engine, one with 420-hp Darracq 12A

With the end of the First World War, the Nieuport firm foresaw the need for aerial transports. The Nieuport 30 design was converted to an airliner by an engineer named Pillon. The interior of the fuselage was fitted with a small cabin for six to eight passengers seated below and directly behind the pilot. The engine was now a 450-hp Renault. A Nieuport 30 with a 420-hp Sunbeam Matabele engine was displayed at the Sixth Salon de Aéronautique between 19 December 1919 and 4 January 1920. The first Nieuport 30 had serial F-CGAT and entered service with Compagne General Transaerienne (CGT) in 1920. Six additional Nieuport 30 T.1s later joined the CGT fleet.

Nieuport 30T.1 of Compagnie General Transaerienne F.CGAT. This airliner was developed from the Nieuport 30 B2 day bomber after the end of the war. B87.4541.

Nieuport 80 Trainer

As noted earlier, the Nieuport 10 was a two-seat reconnaissance airplane. When it was replaced in service by the Nieuport 11, the Nieuport 10s were turned over to the training schools. These aircraft, two-seaters with communal cockpits, were ideal in the training role. A few modifications were made to many of them and they received a designation in the "80" series. The reason for the unusual numbering has never been adequately explained. It is believed that the number denoted the fact that the aircraft was powered by an 80-hp engine. The designation does not appear in the STAé list of approved aircraft appellations.

The Nieuport 80 was the first two-seat trainer in this series and appeared in 1917. Its dimensions were identical to the earlier Nieuport 10, although it was slightly heavier. While some Nieuport 80s had dual controls and were designated E2s, others were single-seaters designated as Nieuport 80 E1s.

In the French training system the Nieuport 80s were used as primary trainers. From these (as well as M.F.11s, Caudron G.3s, and G.4s) the pilot would move on to the advanced flying schools. Training took place at numerous airfields including Buc, Amberieu, Châteauroux, Bron, and Tours.

Nieuport 80 Single-Seat or Two-Seat Trainer with 80-hp Le Rhône
Span 7.90 m; length 7.00 m; wing area 18 sq. m
Empty weight 440 kg; loaded weight 690 kg
Maximum speed: 140 km/h; climb to 2,000 in 17 minutes; range 280 km

Foreign Service

Brazil

A French aerial mission to Brazil in November 1918 brought ten of the single-seat Nieuport 80 E1s. These were used at the Escola de Aviacao Militar (Military Aviation School) in Campo dos Afonsos. They were given serials 8001 through 8010. The Nieuport 80 E1s were struck off charge in 1924.

Portugal

The Escola de Aeronautica Militar (Military Aviation School) of the Portuguese Aeronautica Militar received three Nieuport 80 E2s in 1919.

Japan

A single example of the Nieuport 80 trainer was purchased by the Japanese after the war.

United States

The A.E.F. Air Service purchased 147 Nieuport 80 E1s. The American training syllabus (which was closely modeled on the French system) called for the student pilot to graduate from the Breese Penguin to the Nieuport 80. Primary training centers were located at Foggia, Tours, Avord, Châteauroux, Voves, and Vendôme.

Nieuport 80 of the Japanese air service in 1919. This was a variant of the Nieuport 10 equipped with an 80-hp Le Rhône engine. B85.1872.

Nieuport 81

The Nieuport 81 was the second in a series of two-seat trainers based on earlier Nieuport designs. It appears to be closely based on the Nieuport 12, although its wing area was one square meter larger. The aircraft had two separate cockpits with dual controls. The 110-hp Clerget engine was replaced by an 80-hp Le Rhône, which may explain the "80" series designation.

The Nieuport 80s were used as primary trainers, the pilots then moving on to the more sophisticated Nieuport 83s, Nieuport 11s or 17s, or Sopwith 1½ Strutters. The Camplan School at Bordeaux-Merignac retained its Nieuport 81s until 1923.

Foreign Service

Belgium
Belgium purchased a single Nieuport 81 for evaluation.

Brazil
Brazil received Nieuport 81 E2s in 1919. They were given serial numbers 7513 and 8101–8108 and assigned to the Escola de Aviacao Militar at Campo dos Afonsos. Some of the Type 81s were later converted to taxi trainers ("penguins"). The Nieuport 81s remained in service until 1924.

Japan
Forty Nieuport 81 E2s were brought to Japan in 1919 by a French aviation mission. Mitsubishi built Nieuport 81 E2s under license for the army, the first being completed by May 1922. The Japanese designation was Ko 1. The trainers were used at the Tokorozawa Flying School beginning in 1922. Others served with various air regiments and with the Kagamigahara Airfield.

United States
The A.E.F. Air Service purchased 173 Nieuport 81s. These were probably used as primary trainers in the same manner as the Nieuport 80s (see above).

Nieuport 81 Single-Seat or Two-Seat Trainer with 80-hp Le Rhône
Span 9.05 m; length 7.14 m; wing area 23 sq. m
Empty weight 490 kg; loaded weight 760 kg
Maximum speed: 130 km/h; climb to 2,000 m in 19 minutes; range 260 km

Nakajima Ko 1 Nieuport 81 Single-Seat or Two-Seat Trainer with 80-hp Le Rhône
Span 9.20 m; length 7.20 m; height 2.60 m; wing area 23 sq. m
Empty weight 490 kg; loaded weight 760 kg
Maximum speed: 70 kts; ceiling 4,000 m
57 built

Above: Nieuport 81 serial N 9754 at Villacoublay. MA36443.

Nieuport 81. This aircraft was based on the Nieuport 12 with a wing area increased by one square meter and an 80-hp Le Rhône. Renaud.

Nieuport 82

The Nieuport 82 appeared to be a modified (probably rebuilt) version of the Nieuport 12. It featured a much larger wing span than the Nieuport 12 with the wing area increased by nearly 40 percent. This aircraft also had large wheels added ahead of the main landing gear to prevent it from nosing over on landing. As with all the aircraft in the "80" series the engine was a 80-hp Le Rhône. There was a single cockpit for both the pilot and the student. Some reports suggest these were rebuilt from Nieuport 14 army cooperation aircraft that were rejected as unsuitable for front-line service; other reports suggest that the Nieuport 82 was developed from the unsuccessful Nieuport 15 bomber. The Nieuport 82 was nicknamed *La Grosse Julie*.

The Nieuport 82s, along with the Nieuport 80 and 81, served as primary trainers.

Brazil purchased seven Nieuport 82s. Given serials 2380, 4767, 8049, 8064, 8065, 8069, and 8070, they remained in service at the Escola de Aviacao Militar at Campo dos Afonsos until 1924.

Nieuport 82 Single-Seat or Two-Seat Trainer with 80-hp Le Rhône
Span 12.10 m; length 7.88 m; wing area 30 sq. m
Empty weight 550 kg; loaded weight 820 kg
Maximum speed: 110 km/h; range 220 km

Nieuport 82 serial N 8065. The two front wheels were added to help prevent landing accidents. 3568.

Left: Nieuport 82. University of Texas at Dallas.

Nieuport 83

The Nieuport 83 was the last in the line of two-seat trainers powered by 80-hp Le Rhône engines. It had dimensions nearly identical to the Nieuport 80 and was probably a rebuilt Nieuport 10. The aircraft featured the communal cockpit of the Nieuports 10 and 82. In addition, the struts were mounted vertically and did not have the outward cant that preceding Nieuports used. Also, unlike the other "80" series trainers, the Type 83 had straight ailerons with no washout. Some did not have dual controls.

The French probably used the Nieuport 83s as primary trainers.

Foreign Service
Brazil
The Brazilian Escola de Aviacao Militar at Campo dos Afonsos received 14 Nieuport 83 E2s (dual controls) in 1919. They were given serial numbers 8877, 8889, 8890, 8894, 8901, 8902, 8908, 8909, 8910, 8915, 8917, 8919, 8921, and 8922. They remained in service until 1924.

Italy
A version of the Nieuport 83 was produced by the Nieuport-Macchi firm, designated the Type DC 10.000.

Japan
Japan imported several Type 83s in 1919. The type was license-built by Nakajima. Designated Ko 2, the Nieuport 83 E2s were used by the Tokorozawa Army Flying School, the air regiments, and at the Kagamigahara Airfield until 1926.

Portugal
Portugal purchased seven Nieuport 83 E2s in 1916. They were assigned to the EMA (Escuela Militar de Aviaco).

United States
The A.E.F. Air Service purchased 244 examples of the Nieuport 83 E2. The syllabus called for the student pilot to graduate from the Nieuport 80 or 81 primary trainers to the Nieuport 83 E2. Advanced training for American pilots was undertaken at Avord, Foggia, and Issoudon. From the Nieuport 83, the pilot advanced to the Nieuport 24.

Uruguay
Uruguay obtained six Nieuport 83s in 1921. They were based at Paso de Mendoza.

Nieuport 83 Single-Seat or Two-Seat Trainer with 80-hp Le Rhône
Span 7.90 m; length 7.0 m; height 2.70 m; wing area 18 sq. m
Empty weight 450 kg; loaded weight 700 kg
Maximum speed: 140 km/h; climb to 2,000 m in 26 minutes

Nakajima Ko 2 Nieuport 83 Single-Seat or Two-Seat Trainer with 80-hp Le Rhône
Span 8.10 m; length 7.035 m; height 2.9 m; wing area 18.40 sq. m
Empty weight 440 kg; loaded weight 710 kg
Maximum speed: 76 kts; ceiling 5,000 m; endurance 2 hours
40 built

The Nieuport 83 had struts that were mounted vertically instead of having the outward cant that preceding Nieuport trainers used. 4956.

Nieuport-Tellier Designs

The Nieuport firm acquired a controlling interest in the Tellier firm in late 1918. Tellier's designs were redesignated as follows:

Tellier T.6—Nieuport S
Tellier T.5—Nieuport BM (Bimoteur or twin-engine aircraft—see above).
Tellier T.8—Nieuport TM (Tri-moteur).
Tellier *Vonna*—Nieuport 4R 450 (four 450-hp Renault engines—see Tellier).

Nieuport 4R 450 Long-Range Flying Boat

This aircraft was intended to be used in a postwar attempt to cross the Atlantic Ocean. The aircraft seems to have drawn heavily on the Tellier *Vonna* flying boat designs as it featured the same hull design. The influence of Tellier is not surprising as that firm was gradually absorbed by the Nieuport works from 1918 to 1919. The aircraft was to have had four 450-hp Renault engines mounted in tandem pairs on side-by-side nacelles on the center wing. In 1918 two aircraft were under construction, but it is not known if any were completed. Only a drawing of the design is known to exist. This shows that it was planned to use the aircraft as an airliner. Again, this may have been based on the Tellier design for the four-engine *Vonna* flying boat, although the Tellier design was to have used four 325-hp Panhard and Levassor V-12 engines (see Tellier).

Nieuport 4R 450 with four 450-hp Renault Engines
Span 40 m; length 21.5 m; height 9.4 m; wing area 342.3 sq. m; (Vonna specifications were identical except for a wing area of 285.3 sq. m)
Empty weight 9,200 kg; loaded weight 15,000 kg
Maximum speed: 125 km/h (estimated)
Two under construction; it is not known if any were completed

Nieuport S

The Nieuport S represented an amalgamation of Tellier's and Nieuport's design philosophies. It was an amphibian with the fuselage of the T.3, a wing similar to that used on the T.4, and a tail configuration similar to preceding Nieuport designs. The engine was to have been a 430-hp Darracq-Coatalen 12A or 450-hp Sunbeam, but it does not appear that this aircraft was ever built. (See entry under Tellier for illustration.)

Nieuport S with 430-hp Darracq-Coatalen 12A or 450-hp Sunbeam (all data provisional)
Span 23.8 m; length 14.1 m; height 4.24 m; wing area 95 sq. m
Empty weight 2,540 kg; loaded weight 3,800 kg
Maximum speed: 140 km/h; climb to 1,000 m in 9 min. 30 sec.; climb to 2,000 m in 22 min. 30 sec.; endurance 4 hours
It is not known if any were built

Nieuport 31

The Nieuport 31 was the last of the wartime Nieuport designs. It had a shoulder-mounted wing with a monocoque fuselage. The undercarriage was suspended beneath the fuselage by prominent struts and the wheels were partially covered by a large airfoil that actually served as a second wing. This auxiliary airfoil made the aircraft, for all practical purposes, a sesquiplane. The engine was a 170-hp Le Rhône 9R. The pilot sat behind and above the main wing, where he had a good field of view at least in part due to the large cut-outs in the trailing edge of the wing. Two pairs of large struts connected the main wing with the auxiliary wing/undercarriage. The fin and rudder assembly, as well as the monocoque fuselage, was similar to the Nieuport 29 series. Armament was intended to be two Vickers 7.7-mm machine guns. This airplane was not developed further, possibly due to the fact that the rotary engine had fallen into disfavor with the Aviation Militaire. It is also possible that the success of the more conventional Nieuport 29 obviated the need for the Nieuport 31.

Nieuport 31 Single-Seat Fighter with 170-hp Le Rhône 9R
Span 8.60 m; length 6.60 m; height 2.40 m; wing area 17.62 sq. m
Empty weight 500 kg; loaded weight 780 kg
Maximum speed: 230 km/h (196 km/h at 3,000 m); climb to 3,000 m in 10 minutes; ceiling 5,750 m; range 450 km; endurance 2 hours
Armament: two Vickers 7.7-mm machine guns
One built

Nieuport 31

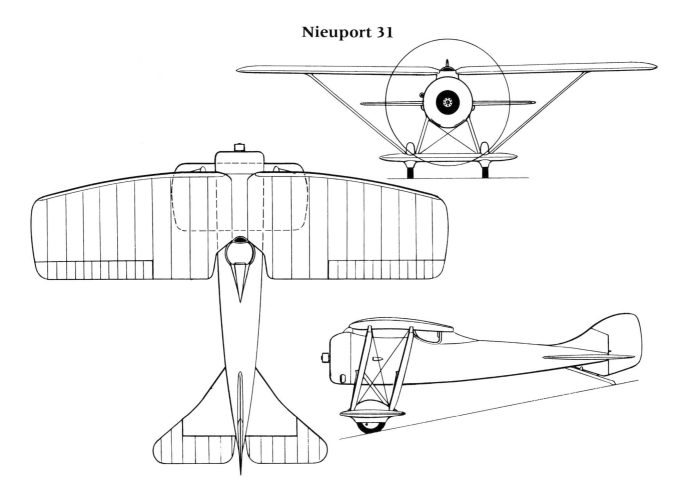

Nieuport 31. The wheels were partially covered by a large airfoil which, for all practical purposes, made the Nieuport 31 a sesquiplane. STAé 224.

Noel Reconnaissance Aircraft

In 1914 the firm of Noel designed a reconnaissance aircraft powered by an 80-hp Gnome rotary engine. It was a two-bay biplane; the top wing was considerably longer than the lower. The layout was similar to the contemporary Caudron G.3, although the horizontal stabilizer was mounted atop of the rudder. It has been reported that the plane was intended for reconnaissance, but other sources indicate it may have also been offered for use in the école (trainer) role. In any event, it was not selected for use by the Aviation Militaire.

The Noel reconnaissance aircraft of 1914. MA97099.

Papin-Rouilly Helicopter

The helicopter designed by Papin and Rouilly featured a single hollow wooden blade and a fan acting as a counterweight. It was powered by an 80-hp Le Rhône rotary engine. The pilot was in a nacelle located between the blade and counterweight. The engine was started by a pulley system. The pilot operated the machine by three foot pedals as follows: one pedal opened the valve to admit air to the hollow blade, where it was ejected into a nozzle placed 90 degrees to the blade. A second pedal engaged or disengaged the engine. The third pedal allowed air into a rod, providing control of pitch and yaw. The escaping air left the blade at a speed of 328 feet per second. The blade was to turn at 60 revolutions per minute but managed only 47 during testing. The helicopter weighed close to 600 kg, considerably more than the 400 kg originally planned.

The construction of a prototype was started in February of 1914 and completed in June. It was stored in a hanger next to Lake Cercey, close to Pouilly-en-Axois where the tests were to take place. The onset of war in August deterred any tests until the winter of 1914–15 when permission to proceed was given by the authorities.

A test was carried out on 31 March 1915 on Lake Cercey. The helicopter seemed to break from the water, but became unstable during flight because the center of gravity had been miscalculated. A military commission observing the test determined that such a machine could not become operational, and halted further evaluation. The machine remained at the lake until sold for scrap wood in 1919.

Above: Papin-Rouilly helicopter undergoing testing at Lake Cercey. MA37165.

Right: Drawing of the Papin-Rouilly helicopter showing key components:

1. Rotary motor.
2. Air vent.
3. Central passage where compressed air is separated between the blade and the rudder.
4. Lifting blade.
5. Air ejection nozzle.
6. Fixed nacelle.
7. Rudder air rod.
8. Directional air nozzle.

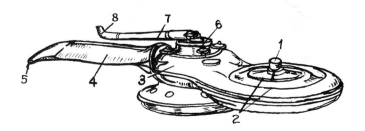

Avions Ponnier

In 1913 Louis Alfred Ponnier built a racing monoplane that placed second in the Coupe Internationale at Reims. Other designs included the L.1 Eclareur with an 80-hp engine; a touring monoplane with an 80-hp Le Rhône; a two-seat monoplane with an 80-hp engine; a biplane with a 50-hp Gnome, and a racer with a 160-hp Gnome. The racing aircraft, designated the D.3, was flown by Jules Vedrines; it finished second in the Gordon-Bennett Cup Race of 1913.

Ponnier monoplane D3 in 1913. MA30595.

Ponnier L.1

In 1914 Avions Ponnier produced the Ponnier Scout, a small, single-engine aircraft with an abbreviated nose, biplane wings of unequal span, and a comma-shaped rudder. The engine was an 80-hp Gnome. No armament was fitted. The aircraft was probably designed as a sporting aircraft, but because of its relatively high speed was presented to the military as a single-seat scout. One feature of this aircraft, which may have carried the designation L.1, was that it could be easily taken apart and reassembled. The aircraft was not selected by the army, but a single example may have been impressed into service with the C.R.P. in November of 1914.

Ponnier L.1 with 80-hp Gnome
Span 8.0 m; length 5.25 m; wing area 12.00 m
Loaded weight 215 kg
Maximum speed: 115 km/h; climb 200 m per minute
Armament: none
One built

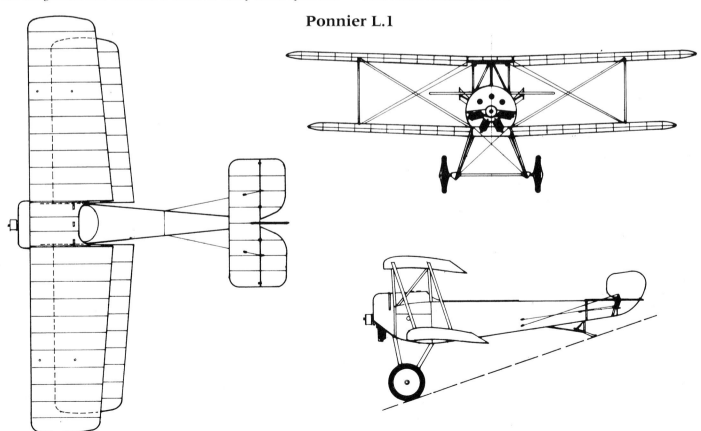

Ponnier L.1

Ponnier L.1 on 4 July 1914. Although it was not selected for service with the Aviation Militaire, a single example may have been assigned to the C.R.P. at the beginning of the war. 13224.

Ponnier Armored Reconnaissance Aircraft

The Ponnier armored biplane was built to meet General Bernard's 1913 specification for an armored two-seat, short-range reconnaissance machine. Blériot, Breguet, Clement-Bayard, Deperdussin, Dorand, Ponnier, and Voisin all produced prototypes to meet this demanding specification.

The Ponnier built in 1914 was a biplane powered by an 80-hp Gnome. It was not selected for use because the engine was not sufficiently powerful to give the heavy aircraft an adequate speed and climb rate. The prototype may have been designated the Ponnier D.8.

By 1914 Turkey had purchased a Ponnier armored reconnaissance plane with an 80-hp engine. It was used by De Goys who was serving as an adviser to the Turkish air service. The Ponnier was sent to Syria in February 1915 and was lost while landing at night, its loss depriving the 4th Army of aerial reconnaissance. It is not known if this was the armored biplane or the Ponnier D.3 shown on the preceding page.

Ponnier Armored Reconnaissance Aircraft
Span was 10.55 m; length 5.30 m; wing area 16.0 sq. m
Loaded weight 340 kg
Maximum speed: 130 km/h; climb 160 m per minute
Armament: none
One built

Ponnier M.1 and M.2

The Ponnier M.1 was designed by Emile Eugene Dupont and built by the Ponnier firm in late 1915. It was a single-bay biplane with wings of unequal span. The type's most noticeable feature was the immense cône de pénétration covering the two-bladed propeller; it was in marked contrast to the petite tail surfaces. Power was supplied by a Gnome 9C rotary engine. The fuselage and wings were of wood with fabric covering. A 7.7-mm Lewis machine gun was mounted over the top wing.

Apparently the prototype was flight-tested by a number of pilots, including Charles Nungesser, who flew the M.1 on 29 January 1916. During that flight the aircraft crashed and Nungesser broke both legs and his jaw. It appears that the M.1 was unstable, perhaps because of its tiny tail surfaces. The Ponnier was not adopted for production by the French. Some

Ponnier M.1 Single-Seat Fighter with Gnome 9C Rotary
Span 6.18 m; length 5.75 m; height 2.30 m; wing area 13.5 sq. m
Empty weight 304 kg; loaded weight 464 kg
Maximum speed: 167 km/h at sea level; Climb to 1,000 m 4 minutes 40 seconds
Armament: one 7.7-mm Lewis machine gun

M.1s were sent to the training schools, but none equipped opera-tional units.

However, the Belgians needed a modern fighter aircraft and were having difficulty in obtaining Nieuports from France. A Major Tournay, who was in charge of purchasing aircraft for the Belgian air service, ordered 30 M.1s. They were built by La Société Anonyme Francaise de Constructions Aéronautique,

Ponnier M.1 fighter; showing the diminutive tail. MA32840.

Ponnier M.1

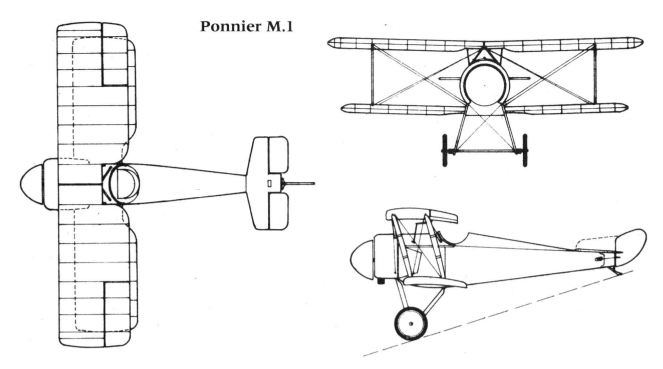

which had superseded Avions Ponnier.

The Belgians soon discovered that the M.1s were difficult to fly. Some were modified by having the cône de pénétration deleted, the tailplane and elevators enlarged, and a fixed fin fitted. It does not appear that these changes resulted in any significant improvement in flying qualities; Belgian ace Willy Coppens noted that the M.1s remained unstable even after these alterations. Consequently, the order for 30 was reduced to ten (some sources say 18). There is no evidence these aircraft ever were used operationally by the Belgians.

Ponnier M.1. The engine was an 80-hp Gnome 9C rotary. B85.405.

Ponnier M.1 B86.4238.

Ponnier M.2 Two-Seat Reconnaissance-Fighter with Gnome 9C Rotary (as recorded by STAé)

Span 8.10 m; length 7.10 m; height 3.10 m; wing area 24 sq. m
Empty weight 510 kg; loaded weight 810 kg

Maximum Speed:	at ground level	167.5 km/h
	1,000 m	163.7 km/h
	2,000 m	162 km/h
	3,000 m	149.5 km/h
Climb:	500 m	1 minutes 50 seconds
	1,000 m	4 minutes 30 seconds
	1,500 m	7 minutes 25 seconds
	2,000 m	10 minutes 45 seconds
	2,500 m	15 minutes 25 seconds
	3,000 m	20 minutes

Ceiling: 4,750m

Ponnier M.2 Two-Seat Fighter (all data provisional)

Wing span 7.52 m; length 6.45 m; wing area 20 sq. m
Empty weight 380 kg; loaded weight 650 kg
Maximum speed: 165 km/h; climb to 2,000 m in 10 minutes
Armament: one 7.7-mm Lewis machine gun mounted on the top wing angled to fire outside the propeller arc and another Lewis on a swivel mount for the gunner

Below: Artists sketch of the Ponnier M.2, which was probably intended for use in the two-seat reconnaissance role and was test flown against a two-seat Nieuport with a Hispano-Suiza engine (probably a Nieuport 14). Although the M.2 performed well, it was felt the Nieuport had the advantage of a stronger airframe, better maneuverability, and could be put into production more quickly.

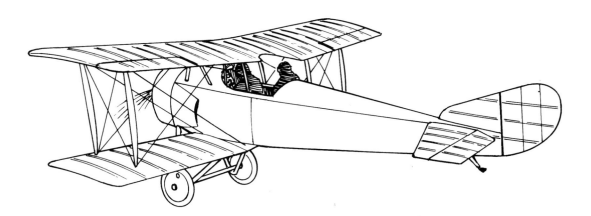

Major Tournay was subsequently removed from his position as aircraft procurement officer, at least in part due to the M.1 debacle.

An enlarged version of the M.1, with a two-man crew and designated the M.2, was proposed. The engine was probably a 110-hp Le Rhône or Gnome Monosoupape. The M.2 was tested in 1915 and, according to the Commission of Aircraft and Motors Report dated August 1915, was actually favored over a two-seat Nieuport (possibly a Nieuport 14) in terms of climb rate. However, the Nieuport design was stronger, had better maneuverability, and could be produced more quickly than the Ponnier. Therefore, the M.2 was not adopted for service. The report gives the aircraft's wing area as 24 sq. m, which was slightly higher than the 20 sq. m reported by other sources. It is possible that the aircraft tested was a modified M.2.

Ponnier M.2
(Provisional)

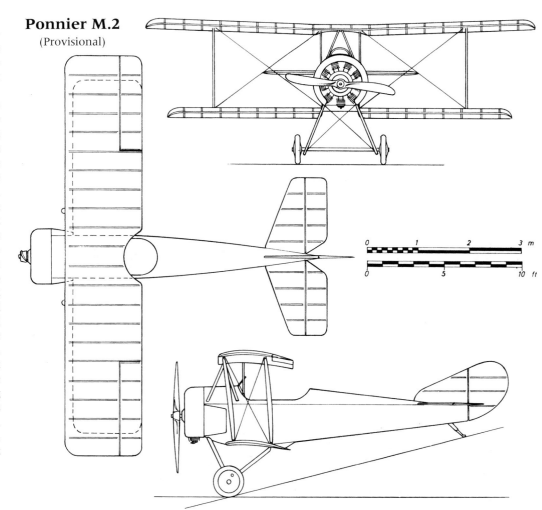

Ponnier M.1. Some M.1s were modified by having the cône de pénétration deleted as in this photo; others had the tailplane and elevators enlarged and a fixed fin fitted. B79.780.

Ponnier P.I

Ponnier designed a twin-engine fighter carrying a crew of two and fitted with dual controls. The intention was that one crewman would fly the plane while the other used his machine gun. This sounded fine in theory, but the crew would be hard pressed if they were to be attacked simultaneously from front and rear. The two-bay wing had diagonally placed struts near the end of the lower wing. The engine nacelles featured the huge spinners favored by the Ponnier firm. However, the location of the engines would have severely restricted the fields of fire of both crew members. The horizontal stabilizers had small fins on the top and bottom surfaces. The aircraft had been completed by January 1916, but little is known of its subsequent fate. It is

Ponnier P.I
Span 12.70 m; length 7.250 m; wing area 30 sq. m
Empty weight 630 kg; loaded weight 1,000 kg
Maximum speed: 160 km/h
Armament: two machine guns

Above: Ponnier P.I. It appears from this view that the field of fire for both the front and rear gunners was hampered by the large nacelles. MA4208.

Ponnier twin-engine fighter P.I. The concept of two pilots, each manning a machine gun, was also tested on the SPAD SA.3. MA4197.

Ponnier P.I
(Provisional)

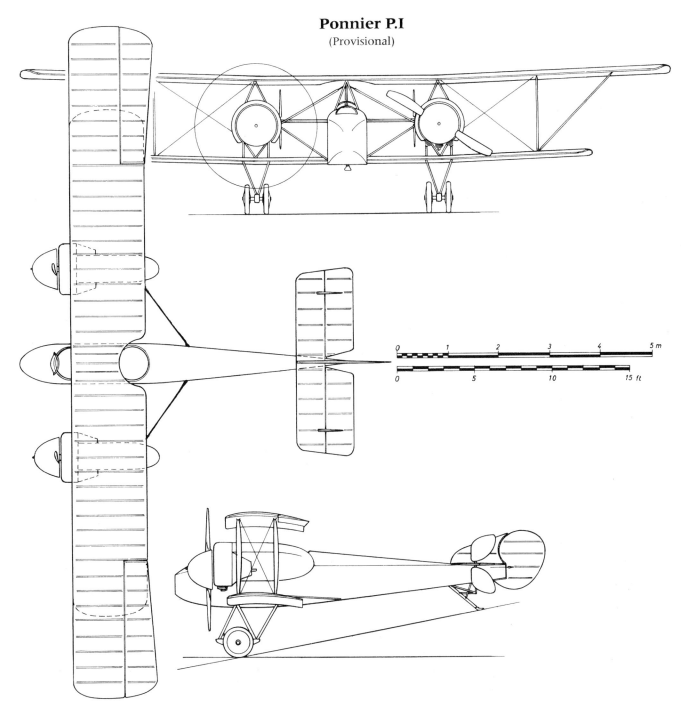

interesting the note that the SPAD SA.3, which had a similar concept of gunners with dual controls, flew in February 1916. It is not known if two such similar designs arose from a similar STAé requirement (although no such specification has been found) or if the appearance of both aircraft at nearly the same time was merely coincidental. The Blériot 65 may also have been based on this requirement.

Ponnier P.I. This view shows the location of the rear gunner's cockpit. MA4289.

Ponnier Pusher

Little is known about this unusual aircraft, and it is not possible to verify that it was a Ponnier design. The upper wing was longer than the lower and had a slight sweep-back. The pilot was seated in the rear with the observer/gunner in front. According to the memoirs of the pilot (nicknamed "Zozo") which were published in 1933, the aircraft had originally been fitted with a 120-hp engine which was replaced by a 45-hp Anzani. An attempt was made to fly the aircraft in either June or July of 1916. Presumably due to the reduced power the aircraft stalled on takeoff and crashed, badly injuring its unfortunate pilot.

Ponnier Pusher. Reairche.

Rausser Flying Boats

Two flying boats were designed by Rausser during the war. Little is known about the designer, although contemporary reports suggest that Constructions Aéronautiques J. Levy actually built the boat, presumably to Rausser's specifications. The first plane had a 180-hp Hispano-Suiza engine and carried a crew of two. Known specifications include a wing area of 50 sq. m and a military load of 100 kg. Maximum speed was estimated at 150 km/h at 2,000 m. It was ready for testing in November 1917, but further details are lacking.

A second flying boat (which may have been a modification of his 180-hp design) was built by Rausser in late 1917. It was a three-seat biplane powered by two 200-hp Hispano-Suiza engines. The hull had three steps. The wing surface area was 80 sq. m, payload was 300 kg, and the maximum speed at 2,000 meters was estimated to be 141 km/h. It carried a crew of three, suggesting that, unlike the 180-hp aircraft, it was designed for long-range patrol/reconnaissance.

Neither aircraft were ordered by the Aviation Maritime, and it is not known if construction of either was completed.

Renault

Renault O1 Bomber

The Renault O1 was designed to fill the BN 2 specification for a night bomber capable of carrying 500 kg of bombs. The aircraft was to be powered by two of the firm's own 420-hp engines. The project was delayed for several months, but construction began later in 1918. By this time the Renault firm had decided to equip the bomber with 450-hp engines.

Construction was undertaken at Renault's O factory at Billancourt; hence, the bomber was designated the O1. It was completed near the time of the Armistice and stored in a hangar at Villacoublay. It is not known if the O1 ever flew. However, the end of the war appears to have brought further development of the plane to an end. It was returned to Billancourt, where it was later destroyed.

Renault O1 Bomber with Two 450-hp Renault Engines
Loaded weight 6,500 kg, payload 1,200 kg
Maximum speed (estimated): 160 km/h at 4,500 m
One built

Renault O1 bomber. This photograph was taken at Renault's factory located at Billancourt. MA4460.

Renault O1
(Provisional)

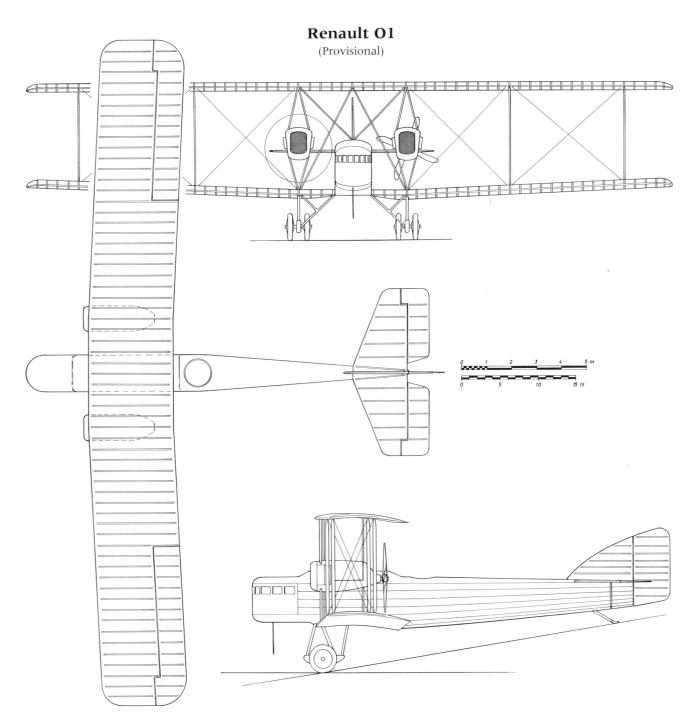

Renault O1 bomber. Power was supplied by two 450-hp Renault engines. MA16610.

Renault Fighter

The British Ministry of Munitions report for May 1918 lists a single-seat biplane fighter as being under construction by the Renault firm. The engine was to have been a 400-hp Renault 12J. Known specifications included an empty weight of 1,700 kg, loaded weight of 2,280 kg and a military load of 150 kg. Estimated performance included a maximum speed of 220 km/h at 4,000 m, and the fighter was to have been able to reach 4,000 m in 12 minutes. It is not known if construction of was completed.

Établissements D'aviation Robert Esnault-Peletrie (R.E.P.)

Robert Esnault-Peletrie built his first powered plane in 1907. Designated the R.E.P.1, it had a 25-hp R.E.P. engine. It was followed by the similar R.E.P.2 and R.E.P.2bis with a 30-hp R.E.P. engine. In 1908 the Association des Industriels de la Locomotion Aerienne was founded. Subsequent designs included the R.E.P. Type D (60-hp R.E.P.); R.E.P. Type K (80-hp Gnome); and a two-seat floatplane. R.E.P.'s patent rights were purchased by Vickers in 1911.

R.E.P. Parasols

In 1914 the R.E.P. firm produced two parasol designs. Both were of steel construction, and had fuselages with triangular cross-sections. Control was by wing warping. One version was a single-seater and had a 60-hp Le Rhône; the other a two-seater with an 80-hp Gnome. Given the Aviation Militaire's insistence that reconnaissance machines provide the crew with the best possible field of vision, it would seem that the R.E.P. parasols would have been favored over the R.E.P. N with its shoulder-

R.E.P. Single-Seater Parasol with 60-hp Le Rhône
Span 10.25 m; length 6.74 m; wing area 19 sq. m
Empty weight 263 kg; payload 150 kg
Maximum speed: 120 km/h; climb 150 m/min.

R.E.P. Two-Seater Parasol with 80-hp Gnome
Span 11.25 m; length 7.67 m; wing area 22 sq. m
Empty weight 295 kg; payload 250 kg
Maximum speed: 125 km/h; climb 107 m/min.

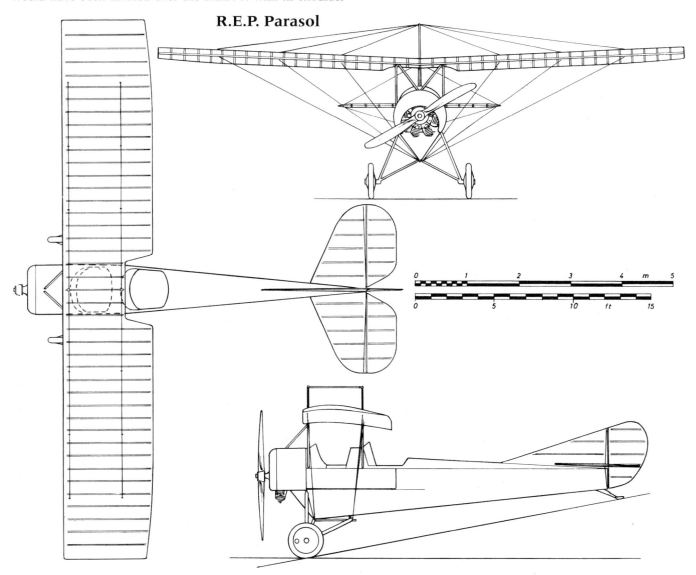

R.E.P. Parasol

mounted wings. However, the R.E.P. parasols were not adopted; instead, the R.E.P. N was selected.

Approximately 12 R.E.P. Parasols were ordered by the RNAS. They were assigned serials 8454 to 8565 and were fitted with 70-hp engines. The first two, 8454 and 8455, arrived in August 1915 and were assigned to No.4 Wing. No.1 Wing also received at least one.

One of No.1 Wing's aircraft became lost and was interned in the Netherlands in 1916. It was purchased from the British and given Dutch serial LA 23. It was at some time armed with a forward-firing Hotchkiss gun mounted laterally so that its field of fire was outside of the propeller's arc. It was later assigned serial REP 3.

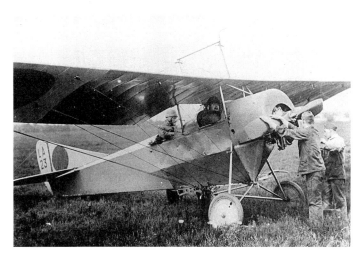

Right: RNAS R.E.P. Parasol 8460 interned in Holland and later given serial LA 23. Via F. Gerdessen and Colin Owers.

R.E.P. parasol reconnaissance airplane. Approximately 12 R.E.P. Parasols were ordered by the RNAS. MA3135.

R.E.P. N

Only one aircraft designed and produced by the R.E.P. firm saw action with the Aviation Militaire during the war. This was the R.E.P. N, a two-seat reconnaissance aircraft with a shoulder-mounted wing and powered by a 80-hp Gnome or Le Rhône engine in a semi-enclosed cowling. The fuselage had a triangular cross section with a "keel" along its base. The plane was constructed of fabric-covered steel tubing. There was a very small rudder and a large tailplane/elevator assembly. Lateral control was by wing warping, the wires for which were supported by two large pylons in front of the observer's and pilot's cockpits.

Two units used the R.E.P. N operationally. The first was **REP 15**, formed in 1912 at Reims. Initially the escadrille was based at Mailly in 1913. At the outbreak of war it was assigned to the General Aviation Reserve at Saint-Cyr in the 5th Armée sector. The R.E.P. Ns flew reconnaissance missions during the Battles of Charleroi and the Marne. In October 1914 the unit was assigned to the 10th Armée sector near Artois where it remained until April 1915. On that date MS 15 was formed with Morane-Saulnier Type Ls.

The other unit was **REP 27**, assigned to the 10th Armée at the outbreak of the War. It, too, participated in the Battle of the Marne. According to an SHAA document, the personnel of REP 15 and REP 27 were joined into a single escadrille to form C 27 with Caudron G.3s on 8 January 1915. However, apparently a few R.E.P. Ns were retained on strength, as one belonging to escadrille 15 is recorded as destroying an Aviatik near Lanevin on 2 March 1915 with a carbine.

No other escadrilles were formed on the R.E.P. Ns and further development was abandoned, probably because the shoulder-mounted wing obstructed the crew's visibility—a fatal flaw in an aircraft intended for reconnaissance duty.

After the failure of the R.E.P. N, the firm undertook the license production of Voisin aircraft, then Caproni bombers.

Foreign Service

Serbia

Two R.E.P. Ns destined for Turkey were seized in transit by the Serbian authorities at the outbreak of the First Balkan War. An third R.E.P. N was later obtained when a Turkish example was captured. The Serbians appear to have made only limited use of these planes; most combat missions were flown by Blériot 11s purchased from France.

Turkey

By 1912 the Ottoman air service had 17 aircraft on strength, most of them R.E.P. Ns, which were considered by the Turks to be a sturdy monoplane. Records indicate there were several 60-hp monoplanes, various two-seat trainers with 80-hp engines, and a two-seat reconnaissance machine with a 70-hp engine. Before the First Balkan War an R.E.P. N was sent to

Thessalonica, where it was later captured by the Serbians, as was a second sent there at the outbreak of the war. Numerous sorties were flown by the two remaining R.E.P.s over Thessalonica, Fesa, Nuri, Salim, Fazil, Midhat, Refik, and Fevzi during both Balkan Wars. A U. S. War Department report of October 29 1914 listed four R.E.P. Ns at the Yesilköy flight school. By the time the First World War began the surviving R.E.P. Ns had been relegated to training.

R.E.P. Two-Seat Reconnaissance Plane with a 80-hp Gnome or Le Rhône
Span 9.2 m, length 6.6 m
Empty weight 270 kg
Maximum speed: 116 km/h
Armament: limited to hand-held weapons improvised by the crew

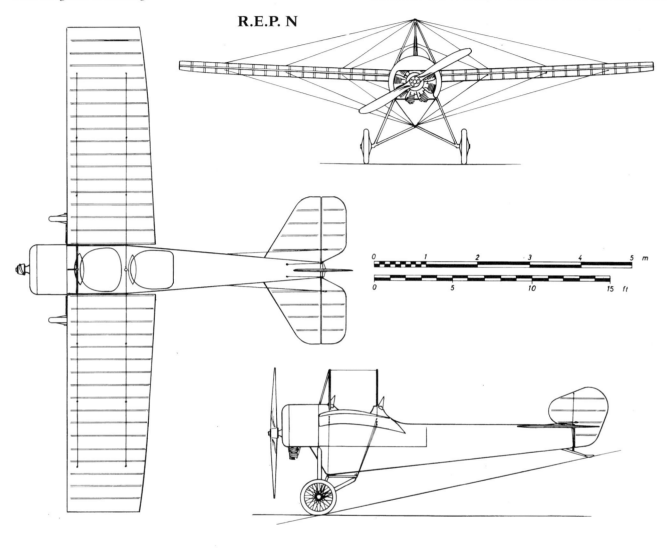

An R.E.P. Type N. MC 9725.

Above: The R.E.P. Type N in service with the Aviation Militaire. B77.659.
Left: R.E.P. Type N at Avord. Lateral control was by wing warping, the wires for which were supported by two large pylons located in front of the cockpit. B92.1558.

R.E.P. C1

The R.E.P. firm constructed a single-seat fighter to meet the Aviation Militaire's C1 specification. Powered by a 230-hp Salmson 9Z engine, it was a single-bay biplane of steel tube construction with fabric covering. It had an abbreviated rudder and large tailplane/elevators similar to those of the R.E.P. N and Parasol. The lower wing was attached to the bottom of the fuselage by a prominent fairing. This arrangement permitted the upper wing to be placed just above the pilot's cockpit. The pilot's lateral view was good because his head projected thorough a large cutout in the wing. His downward vision, however, was very limited. Armament consisted of two Vickers machine guns, one mounted just to the right of the center section strut and the other farther down the left side of the cowling. The R.E.P. C1 was tested at Villacoublay in March and April of 1918. However, its maximum speed and climb rate were virtually identical to the SPAD 13 already in production, and it was not selected by the Aviation Militaire.

R.E.P. C1 Single-Seat Fighter with 230-hp Salmson 9Z		
Span 8.38 m; length 6.35 m; height 2.53 m; wing area 20.466 sq. m		
Empty weight 658 kg; loaded weight 968 kg		
Maximum speed:	1,000 m	217 km/h
	2,000 m	212 km/h
	3,000 m	207 km/h
	4,000 m	196 km/h
Climb:	1,000 m	2 minutes 7 seconds
	3,000 m	8 minutes 31 seconds
	4,000 m	13 minutes 51 seconds
	5,000 m	20 minutes 59 seconds
Endurance 1.5 hours		
Armament: two 7.7-mm Vickers machine guns		
One built		

R.E.P. C1

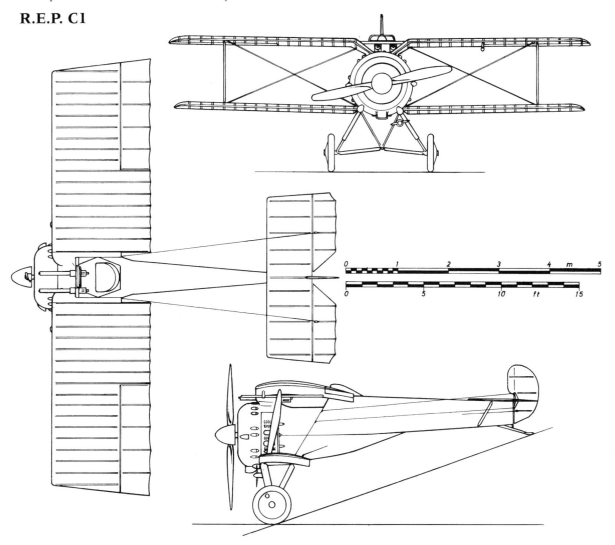

434 FRENCH AIRCRAFT OF THE FIRST WORLD WAR

Above: Three views of the R.E.P. C1 fighter of 1918. The lower wing was attached to the bottom of the fuselage by a prominent fairing. The upper wing was placed just above the pilot's cockpit. MA2149.

Société de Moteurs Salmson

Salmson-Moineau S.M.1

The Société de Moteurs Salmson specialized in producing aircraft engines, many of them water-cooled radials. In 1916 the firm produced an aircraft designed by the famous prewar aviator René Moineau and intended to meet the A3 requirement for a three-seat reconnaissance airplane. Moineau selected the 240-hp Salmson 9A2c engine for his design. However, this engine was bulky and Moineau feared that mounting it in the nose would result in excessive drag. His solution was to mount the engine transversely within the airframe. A complicated series of gearboxes and shafts transmitted the output of the engine to two propellers mounted inboard of each wing on X-shaped struts with conical supports. The gearing also insured that the propellers turned in opposite directions to minimize torque. The fuselage was mounted between the two wings. The radiators were mounted in the nose and provided a boxy pulpit for the gunner. A prominent exhaust stack was placed on the starboard side of the fuselage and extended from the engine bay to the top wing. The massive undercarriage had a tricycle layout and was made of steel tubing. This arrangement is believed to have contributed to a number of accidents. When pilots attempted to land at too steep an angle the aircraft would either flip over when the nose gear collapsed, or it would cause the airplane to bounce into the air.

The pilot was located beneath a cutout in the upper wing where his field of view was severely limited. The rear gunner sat just behind the upper wing and had an excellent field of fire. Both the front and rear gunner had APX 37-mm cannons. There was a triangular fin and rudder on the prototype. On later aircraft the rudder was in the form of a parallelogram. The wings were made of wood and the upper wing was longer than the lower. The wings were wire braced and had square tips. Ailerons were fitted to the top wing only.

> **S.M.1 Three-Seat Long-Range Reconnaissance Aircraft with 240-hp Salmson 9A2c**
> Span 17.475 m; length 10.0 m; height 3.80 m; wing area 70.0 sq. m
> Loaded weight 2,050 kg; payload 370 kg
> Endurance 3 hours
> At least 155 built

Above: Salmson-Moineau S.M.1 serial SM 12 of SOP 43. Pilots were warned never to allow the aircraft to roll on the nose landing gear as this was only to be used to prevent the aircraft from nosing over. B74.31.

Salmson S.M.1 serial number 20. MA26250.

Detail of the S.M.1's laterally-mounted Salmson A9 engine. B77.1086.

Below: Salmson-Moineau S.M.1. This photograph shows the complicated drive train needed to power both propellers. B77.1087.

The new aircraft was given the company designation Salmson-Moineau A92H. The A9 indicated the Salmson A9 engine while the 2H denoted two propellers. The aircraft was tested at Villacoublay in 1916, and it was successful enough to warrant production. It was given the STAé designation S.M.1 A3. When the aircraft designation system was changed on 1 May, 1918, the S.M.1 was redesignated the Sal. 1 A3. A total of 100 S.M.1s had been ordered by 11 November 1916. (Based on the tail code numbering sequence it appears that at least 155 may have been built.) However, its performance was clearly inferior to the Sopwith 1½ Strutter and it was even suggested that the type be flown as a two-seater to improve its performance.

The S.M.1 flight manual instructed the pilot to begin with a takeoff into the wind until there was sufficient airspeed for rudder control. At 30 km/h the S.M.1 would lift off its tailskid and the aircraft would be rolling on its main landing gear. The pilot was warned never to allow the aircraft to roll on the nose landing gear as this was only to be used to prevent the aircraft from tipping over. Once the aircraft was level the flow of fuel would ensure sufficient acceleration of the aircraft to lift off horizontally; a steep angle of attack was not needed for take-off. When flying speed was attained the tail lifted 2 degrees to the horizon at 1,500 m altitude. The S.M.1 was to be landed lightly on the main wheels and then allowed to slide back onto the tail skid.

While there were easily enough S.M.1s built to equip an entire escadrille, it was decided to send small numbers to front-line escadrilles. It was common practice to provide each army cooperation escadrille with a few long-range reconnaissance aircraft of the A3 classification. The S.M.1s were provided to F 2, F 19, F 41, F 45, F 58, F 63, F 71, F 72, C 219, F 223, and AR 289 in mid-1917. By August there were 32 Salmson S.M.1s in escadrille service. SM 229 had ten on strength in early 1917; it combined with

Salmson-Moineau S.M.1 of F 58 on 24 August 1917. B83.3256.

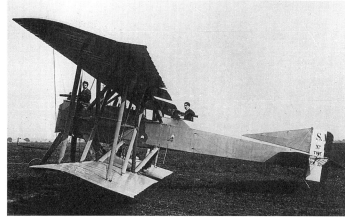

Above: S.M.1s.

S.M.1 of F 58. B86.4578.

C106 in early 1917. Some documents refer to the resulting unit as SM 106.

The S.M.1s caused great difficulties in front-line service. As mentioned above, accidents were frequently caused by the unusual landing gear layout. The complex engine transmission resulted in numerous breakdowns, and the aircraft was probably a mechanic's nightmare. It is also likely that the numerous drag-producing struts and supports seriously restricted the S.M.1's performance.

The S.M.1s were found to be of limited use in operational service, and those that survived were sent back to the aircraft parks. Despite their plane's flaws, some units were able to carry out their missions. A 10th Armée memo noted that the commander of the 37th C.A. had praised the crews of SM 106 for the valuable reconnaissance and photographic missions that were flown. At least one machine, belonging to AR 289, was still in use in 1918. This S.M.1 flew reconnaissance in support of the 66e Division des Chasseurs a Pied until late 1918.

The Imperial Russian Air Service received two S.M.1s in 1917. The Russians also found the S.M.1s to be of little use and they were quickly withdrawn.

There are reports of at least one S.M.1 being fitted with a 160-hp Salmson P 9 engine. However, the type was not developed further.

Salmson S.M.1s were assigned to army cooperation units in groups of three or four aircraft. Only one (or possibly two) escadrilles was ever formed entirely on the type. MA7587.

Left: Three-quarter rear view of S.M.1 serial SM 10. A captured German aircraft and a Farman F.40 are in the background. B86.1165.

Below: S.M.1. The radiators were located in either side of the front gunner's cockpit. B85.1685 via Tom Darcey.

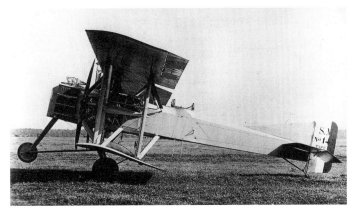

S.M.1 serial SM 14. B75.134.

Salmson-Moineau S.M.2

The S.M.2 was an enlarged and more powerful version of the S.M.1. An additional 240-hp SAL 9A2c engine was placed in the nose and was used to drive a single propeller. The laterally mounted SAL 9A2c was retained and this engine still drove two propellers through a complex transmission system. The wing span was enlarged, there were additional wing struts, and the undercarriage was reinforced. The airplane was intended for the S2 ground attack role. The crew was reduced to two, the nose gunner's position having been replaced by the additional engine. The airplane was tested in 1918 but there were serious problems with the engine cooling system, and only a single S.M.2 was built.

Salmson S.M.2. 0112.

Above: The ungainly Salmson-Moineau S.M.2 which was intended to fulfill the S2 specification for a ground attack aircraft. 0103.

Salmson 2 A2

It soon became clear to the Aviation Militaire that the reconnaissance units' A.R.1s and A.R.2s were underpowered and that the Sopwith 1½ Strutters were too fragile and carried an inadequate payload. A requirement for a two-seat reconnaissance aircraft capable of carrying a crew of two, several machine guns, a camera, and a T.S.F. unit was announced in 1916. Several aircraft were developed to meet this requirement; including the Breguet 14 and the Salmson 2 A2.

The Société des Moteurs Salmson, which had previously concentrated on the production of aero engines, had gained considerable experience in aircraft manufacturing when it had built Sopwith 1½ Strutters under license. To meet the new requirement it designed a completely new biplane, drawing upon the knowledge acquired in producing the Sopwith machines.

The Salmson 2 was a biplane powered by a 230-hp Salmson 9Za. The wings were of equal span with ailerons on both the upper and lower wings. The wings had two hollow spars of spruce. The ribs were plywood and the bottom was covered with poplar. The spars of the lower wing were attached to the fuselage longerons by metal fittings. The upper wing was held in place on the fuselage by the cabane struts and was attached to the lower wing by two bays of struts.

The fuselage had a rectangular structure composed of four longerons with several fuselage formers which were attached by hinges and reinforced by a latticework of piano wire. At the rear were vertical tubes to which the rudder and vertical stabilizer were attached. The longerons were made of ash sheathed in aluminum at their attachment points. There were two 100-liter fuel tanks, both held in cradles in the lower fuselage. The oil tank held 47 liters and was located on the right side of the fuselage beneath a rounded fairing. The engine was mounted on two metal holders which were held together by a U-shaped support. The motor mount was attached to the fuselage longerons by bolts. The armament was two flexible 7.7-mm Lewis guns and a synchronized 7.7-mm Vickers.

The undercarriage had six struts (three per side) attached to the motor mount and the rear spar. The articulated axle was wrapped in bungee chord to act as a shock absorber. The tail skid was a tube of duralumin attached to the lower fuselage.

A Salmson 2 underwent STAé testing on 29 April 1917. With a military load of 510 kg the following results were obtained:

Maximum Speed	Time to Climb
—	1,000 m in 3 minutes 10 seconds
187 km/h at 2,000 m	2,000 m in 6 minutes 45 seconds
185 km/h at 3,000 m	3,000 m in 11 minutes 15 seconds
177.5 km/h at 4,000 m	4,000 m in 17 minutes 40 seconds
168 km/h at 5,000 m	5,000 m in 27 minutes 30 seconds

The performance was good and the test pilot stated that the Salmson 2 responded well to the controls but the nose tended to wander ("hunt") when coming out of a turn. It was recommended that the bungee chords attached to the rudder be tightened. A drawback in combat was the considerable distance between the pilot and gunner.

It had not been planned to equip the aircraft with bomb

Salmson 2 A2 serial number 513. B91.6141.

Experimental Salmson 2 fitted with a 260-hp engine. The aircraft underwent testing in 1917. MA15424.

racks; however, as the war progressed modifications were carried out to some Salmson 2s to enable them to carry 230 kg of bombs for ground attack.

Production totaled 3,200 aircraft. Salmson produced 2,200 and the remaining 1,000 were built by Latécoère, Hanriot, and Desfontaines.

Variants

The aircraft underwent remarkably few changes during its service career. Latécoère, which built the type under license, produced a number of interesting variants. The Latécoère Type 2 was a Salmson 2 modified to conform more closely to the Latécoère production process. Only six were built; the other aircraft produced at that plant were assembled without modification. The Latécoère firm also modified one aircraft to enable it to carry a torpedo; however, no production of this type was undertaken.

The Salmson 2 D2 was a two-seat training version of the Salmson 2. The D designation stood for double command (dual control). In addition to having dual controls fitted, the Salmson 2 D2s were fitted with 130-hp Clerget 9B engines. Several were built in 1917.

Operational Service

It appears that the first escadrille to be equipped with Salmson 2s was SAL 122, formed from C122 (equipped with Caudron G.6s) in October 1917. The A.R.1 and A.R.2 had proved to be mediocre aircraft and many of the escadrilles that used them were re-equipped with Salmson 2s. The Sopwith 1½ Strutter had performed well but was clearly outdated by late 1917; the Salmson 2 was far superior, especially in terms of its ability to carry a much heavier payload over greater distances. The Salmson 2-equipped escadrilles used their new aircraft for short-range reconnaissance, artillery spotting, ground attack, and even leaflet-dropping missions.

The following escadrilles used the Salmson 2:

SAL 1, which had previously used A.R.1s, was re-equipped with Salmson 2s in January 1918. At that time it was assigned to the 3rd C.A. and based at Belfort. The unit saw action in the Second Battle of Picardie and flew reconnaissance, artillery spotting, and leaflet-dropping sorties in the vicinity of Alsace. SAL 1 supported both the 17th C.A. and the 1st American Army during the Saint Mihiel offensive. The escadrille's main task during this time was reconnaissance over the front. In 1920 SAL 1 became the 5th escadrille of the 4th RAO at Bourget.

SAL 4, created in February 1918 and assigned to the 3rd C.A. It was initially based at La Cheppe and subsequently moved to Coucy and Goussancourt. At the time of the Armistice it was based at Manoncourt. After serving as part of the occupation force, SAL 4 was disbanded in July 1919.

SAL 5, formed from SOP 5 in July 1918. Commanded by Capitaine de Peyronnet, SAL 5 was assigned to the 6th C.A. until disbanded in August 1919.

SAL 6, formed in December 1916 when C 6 gave up its G.4s for six Salmson 2s (and three Sopwith 1½ Strutters). Assigned to the 18th C.A. and commanded by Lieutenant Latour, it was disbanded in July 1919.

SAL 8, formed from AR 8 (with A.R.1s and 2s) in February 1918. It was commanded by Lieutenant Wiedeman and assigned to the 11th C.A. The unit ended the war at La Cheppe under the command of 5th Armée. SAL 8 was disbanded in December 1919.

SAL 10, formed from C10 in September 1918 when it re-equipped with seven Salmson 2s. The unit remained attached to the 35th C.A. under command of Capitaine Pène. After the Armistice SAL 10 was assigned to the 3rd Armée. It was redesignated the 2nd Escadrille of the 31st Escadre in January 1920.

SAL 13, formed from SOP 13 in early 1918. It was assigned to 15th C.A. and was active in the 2nd Armée sector. SAL 13 was

Salmson 2 A2 of SAL 32. The escadrille was assigned to the 10th C.A. and was initially based at Montdidier. B 76.1736.

Salmson 2 of SAL 33; serial number 49. The pilot's name was Delavenne. MA23874.

under command of Capitaine Pecquet. In November it was assigned to the 1st Armée. The unit was disbanded in July 1919.

SAL 14 converted from A.R.1s to Salmson 2s in mid-1918 around the time of the Third Battle of Flanders. Moving to the vicinity of Champagne on 15 July, it participated in the Battles of the Marne and Picardie from July to September 1918 under the command of Capitaine Dezerville. The escadrille was assigned to the 46th D.I. At the end of the war, it was based at le Meuse under command of the 1st Armée. It became the 3rd Escadrille of the 4th RAO in January 1920.

SAL 16, created from AR 16 in February 1918 when ten Salmson 2s were given to the escadrille. Assigned to the 10th Armée and under command of Lieutenant Boudreaux, SAL 16 was active over the Italian front. After leaving Italy in March, it moved to Plessis-Belleville to participate in the Battle of Picardie. A month later SAL 16 moved to Bovelles. At the war's end, the escadrille was based at Coucy-lès-Eppes. It was disbanded in July 1919.

SAL 17, created from SOP 17 in mid-1918. Assigned to the 1st C.A., it was based in the 10th Armée sector. SAL 17 was particularly active in the Battle of Chemin des Dames. In July it participated in the Second Battle of the Marne. Postwar, SAL 17 served as part of the occupation force in Germany and was based at les Vosges and Mayence. It became the 6th Escadrille of the 5th RAO in January 1920.

SAL 18, formed from C18 in January 1918. It was supplied with seven Salmson 2s and was assigned to the 30th C.A. SAL 18 was active during the battle of Aisne in May and the Marne in July. It also participated in the battles of Picardie (August) and Vauxaillon (September). After the war it was based at Lys and then Escaut (Belgium). It was redesignated the 4th Escadrille of the 4th RAO in January 1920.

SAL 19, formed from AR 19 in April 1918. The unit had ten Salmson 2s and was assigned to the 13th C.A. It was based in the 5th Armee's sector at the war's end. In January 1920 SAL 19 became the 4th Escadrille of the 33rd RAO.

SAL 22, created from AR 22 in May 1918. The escadrille had ten Salmson 2s (known serials were 619,682, 5 037, 3 0833, and 3 101). Assigned to the 12th C.A., it was active in the Battle of Piave. SAL 22 was frequently used in the light bomber role and served on the Italian front until the Armistice with Austro-Hungary. It was disbanded in April 1919.

SAL 24, formed from SOP 24 in March 1918. Assigned to the 2nd C.A., it was disbanded in January 1920.

SAL 27, formed from C 4 in 1918. Assigned to the 21st C.A., it participated in the aerial battles over the Marne, Chemin des Dames, and Champagne. At the War's end SAL 27 was based at Seraincourt. In January 1920 it became the 6th escadrille of the 3rd RAO.

SAL 28, formed from SOP 28 in early 1918 under the command of Lieutenant Seyer. Assigned to the 2nd C.A., SAL 28 was active during the Battles of Picardie and Saint-Mihiel. It ended the war at Frescaty and was disbanded in July 1919.

SAL 30, created from C 30 in April 1918 when it gave up its G.6s for ten Salmson 2s. Known serial numbers include 572, 573, 583, 718, 854, 3025, 30129, and 4111. Assigned to the 1st C.C., the unit was commanded by Lieutenant Mendigal. It was disbanded in December 1918.

SAL 32, formed from AR 32 in February 1918. It was assigned to the 10th C.A. and was initially based at Montdidier. SAL 32 was commanded by Capitaine Sourdillon. After a brief rest period in September, the unit was preparing to participate in the Lorraine offensive when the war ended. Based at Alsace at the war's end, SAL 32 became the 3rd Escadrille of the 5th RAO in January 1920.

SAL 33, created from AR 33 in December 1917 and assigned to the 9th C.A. It was active in the Battle of Artois in August based at Montdidier. It moved to Faucoucourt in September, and spotted for artillery fire at Argonne. The escadrille moved to Auve at the war's end. It became the 3rd Escadrille of the 3rd RAO in January 1920.

SAL 39, formed from SOP 39 in February 1918 and assigned to the 38th C.A. It became the 16th Escadrille of the 31st RAO in 1921.

SAL 40, formed from AR 40 in March 1918 when ten Salmson 2s were assigned. It was assigned to the 4th C.A. and was disbanded in March 1919.

SAL 41, created from AR 41 in June 1918 when ten Salmson 2s were assigned. It was assigned to the 32nd C.A. Postwar, SAL 41 participated in the occupation of Germany and was based at Saarrbruck. It was disbanded in March 1919.

SAL 47 had only a brief existence. Assigned to the 2nd C.C., C 47 replaced its G.6s with Salmson 2s and Sopwith 1½ Strutters to become Escadrille SAL 47 in December 1917. It

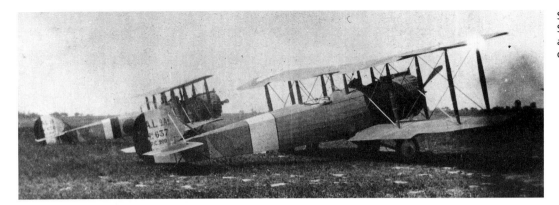

Salmson 2 serial number 637 of SAL 59. This escadrille was assigned to the 10th Corps d'Armée. MA17378.

subsequently received SPAD 11s later that month to become SPA-Bi 47.

SAL 50, created from AR 50 in February 1918. It had seven Salmson 2s and three Letords. Assigned to the 16th C.A., it was disbanded in March 1919.

SAL 51, formed from SOP 51 in December 1917. Assigned to the 1st C.A.C., it was disbanded in July 1919.

SAL 52, created in March 1918 from AR 52. It was given ten Salmson 2s and assigned to the 38th D.I. The escadrille suffered heavy losses: a dozen pilots and observers were injured or killed. However, SAL 52 also recorded two aerial victories during this time. Postwar the escadrille was based at Fougerolles and later Colmar. On 20 January 1920 it became the 4th Escadrille of the 5th RAO.

SAL 56, created from C 56 in March 1918. Its G.6s were replaced by seven Salmson 2s and three Letords. SAL 56 was assigned to the 17th C.A.; based in the Douaumont sector, it participated in the pursuit of the retreating Germans during the French offensives of 1918. The unit became the 8th Escadrille of the 3rd RAO in January of 1920.

SAL 58, created from AR 58 in February 1918 when it was sent ten Salmson 2s. Assigned to the 73rd D.I., it was active during the Battle of the Somme in April. It was based at Meaux in early June and subsequently moved to Le Férté-Gaucher, where it was active over the Marne. It later moved to Verdun and then Baccarat, where it was based at the time of the Armistice. SAL 58 was disbanded in December 1918.

SAL 59, formed from AR 59 in April 1918 with ten Salmson 2s. Assigned to the 10th C.A., the escadrille was sent to Poland after the war and became the 10th Escadrille of the 6th RAO in January 1920.

SAL 61, formed from SOP 61 in early 1918. It assigned to the 34th C.A. and based at Bouleuse in the vicinity of the Marne. Later it moved to Resny (Oise) and participated in the Battles of Picardie (March) and the Somme (August). SAL 61 was sent to Belgium at the war's end. It was disbanded in March 1919.

SAL 70, created from AR 70 in April 1918. Assigned to the 20th C.A., it was initially based at Dugny but soon moved to Étampes. It was disbanded in April 1919.

SAL 71, formed from AR 71 in April 1918 with ten Salmson 2s. It was based at Dugny and was assigned to the 5th C.A. It was disbanded in February 1919.

SAL 72, created in March 1918 when AR 72 received ten Salmson 2s. It was based at Plessis-Belleville and was assigned to the 7th C.A. It was disbanded in April 1919.

SAL 74, created from C 74 in April 1918. It was based at Esquennoy near the Somme and was assigned to the 36th C.A. Seven Salmson 2s (and possibly some SPAD 7s) were assigned to the unit, which subsequently moved to the Dunkerque region. By June 1918 it was based near Nancy. SAL 74 was disbanded in March 1919.

SAL 105, created from SOP 105 in early 1918. It was assigned to the 5th C.A. and was active in the Battles of the Aisne (May/June), Champagne (July), and the Marne, Serre, and Meuse. It became the 4th Escadrille of the 7th RAO in January 1920.

SAL 106, formed from SOP 106 in May 1918.

SAL 122, formed from C122 (equipped with Caudron G.6s) in October 1917. It was assigned to the 32nd Corps d'Armée. It was disbanded in March 1919.

SAL 203, formed from AR 203 in February 1918. Assigned to the S.A.L. 2nd Armée, it was active over Noyon and Saint-Quentin in August. In January 1920, SAL 203 became the 4th Escadrille Levant of the 1st RAO.

SAL 204, formed from SOP 204 in February 1918. It was attached to the S.A.L. of the 2nd Armée. It was disbanded almost exactly one year later.

SAL 225, created from C 225 (with G.6s) in late 1917. It was assigned as a S.A.L. unit for the 5th Armée and helped to spot for long-range artillery fire. It was disbanded in March 1919.

Salmson 2 of Col. Hamonic. MA7779.

Salmson 2 serial 22. MA15423.

SAL 230, formed from AR 230 in March 1918 as an S.A.L. unit for the 5th Armée. It was disbanded in March 1919.

SAL 251, formed from SOP 251 in August 1918. It was assigned to the 11th C.A. under command of Lieutenant Delavigne. It was disbanded in February 1919.

SAL 252, formed from SOP 252 in May 1918. It was assigned to the 21st C.A. and was disbanded in March 1919.

SAL 253, formed from AR 253 in May 1918. Assigned to the 20th C.A., it was disbanded in April 1919.

SAL 254, formed from AR 254 in August 1918. It was assigned to the 12th C.A. and was based at Verona on the Italian front and later moved to Nove. It participated in the French breakthrough at Piave in October. Sal 254 was disbanded in April 1919.

SAL 256, formed from AR 256 in May 1918 and assigned to the 9th C.A. SAL 256 was disbanded in April 1918.

SAL 259, formed from AR 259 in May 1918. Assigned to the 7th C.A., it was based at Ormeaux, where it flew reconnaissance sorties before the Second Battle of the Marne. After participating in the French counter-offensives during the summer, it was active in the offensives near Noyon, based in the vicinity of Blérancourt and Audignicourt. The escadrille was particularly active over the Coucy forest during the French drive to the Hindenburg line. Moving to an airfield at Bray-Dunes, SAL 259 participated in the Flanders offensive and was later active over Lys and l'Escaut. At the end of the war, it was active in Belgium. It was disbanded in April 1919.

SAL 262, formed from AR 262 in July 1918 and assigned to the 8th C.A. The unit quickly saw action during the German offensive at Champagne. It later performed low-altitude reconnaissance during the Battle of Saint-Quentin from 27 September to 10 October. The unit remained assigned to the 8th C.A. until disbanded in February 1919.

SAL 263, formed from SOP 263 in May 1918 and assigned to the 14th C.A. The unit was disbanded in early 1919.

SAL 264, formed from AR 264 in August 1918, and assigned to the 13th C.A. It was active over the Vesle, Aisne, and Meuse, where its Salmson 2s were used for low-altitude reconnaissance and strafing. After serving as part of the German occupation force at Gosenheim-Mayence, SAL 264 was disbanded in March 1919.

SAL 270, formed from SOP 270 in July 1918. Assigned to the 15th C.A., it was disbanded in April 1919.

SAL 273, formed from SOP 273 in June 1918. Assigned to the 38th C.A., it was disbanded in March 1919.

SAL 277, formed from SOP 277 in May 1918. Assigned to the 2nd C.A.C. and commanded by Lieutenant Lellouche, it participated in the Battle of Saint-Mihiel. The unit was cited for its action during the Battle. It was based at Metz-Frescaty at the war's end. Postwar it moved to Many and later Tours. SAL 277 became the 7th Escadrille of the 1st RAO.

SAL 280, formed from SOP 280 in June 1918. Assigned to the 3rd C.A., it was disbanded in March 1919.

SAL 288, formed from AR 288 in May 1918. Assigned to the 45th D.I., it was disbanded in February 1919.

Postwar Service

Postwar, the number of Salmson 2 escadrilles declined rapidly, but a substantial number of these units were included in the 1920 reorganization of the Armee d l'Air. These included:

1st RAO: SAL 10, 203, 277.
3rd RAO: SAL 19, 27, 33, and 56.
4th RAO: SAL 1, 14, 18, and 259.
5th RAO: SAL 17 and 32.
6th RAO: SAL 39 and 59.
7th RAO: SAL 8 and 105.
RA in Algeria and Tunis: SAL 71.
31st RAO: SAL 10 and 277.
33rd RAO: SAL 6, 17, 19, and 33.
34th RAO: SAL 1, 18, 34, and 254.
35th RAO: SAL 32 and 52.

Salmson 2 A2 Two-Seat Reconnaissance Aircraft with 230-hp Salmson 9Za	
Span 11.750 m; length 8.500 m; height 2.90 m; wing area 37.270 sq. m	
Empty weight 780 kg; loaded weight 1,290 kg	
Maximum speed:	sea level 188 km/h
	2,000 m 186 km/h
	3,000 m 181 km/h
	4,000 m 173 km/h
	5,000 m 168 km/h
Climb:	1,000 m 3 minutes 18 seconds
	2,000 m 7 minutes 13 seconds
	4,000 m 17 minutes 20 seconds
	5,000 m 27 minutes 30 seconds
Ceiling 6,250 m; range 500 km	
Armament: one synchronized Vickers 7.7-mm machine gun and two ring-mounted 7.7-mm Lewis machine guns	
A total of 3,200 were built	

Army Type Otsu 1 (Kawasaki-Salmson 2) with 230-hp Salmson 9Za

Span 11.767 m; length 8.624 m; height 2.90 m; wing area 37.27 sq. m
Empty weight 903 kg; loaded weight 1,500 kg
Maximum speed: 101 kts at 2,000 m; climb to 3,000 m in 11 minutes 42 seconds; ceiling 5,800 m; endurance up to 7 hours
Armament: one synchronized 7.7-mm machine gun and one or two 7.7-mm machine guns on a ring mount
300 built by Kawasaki and approximately 300 built by Tokorozawa Branch of Army Supply Depot.

Salmson 2 D2 Dual-Control Trainer with 130-hp Clerget 9B

Span 11.75 m; length 8.76 m; height 2.86 m; wing area 37.27 sq. m
Maximum speed: 145 km/h

Salmson 2 A2 of the 1st Aero Squadron assigned to the 1st Corps Observation Group. B76.967.

37th RA (North Africa): SAL 8 and 105.
36th Groupe Autonome d'Observation in Algeria/Tunisia: SAL 58 and 253.
1st Groupe at Rayak: SAL 16.
2nd Groupe at Mouslimie: SAL 24.
3rd Groupe at Rakka: SAL 40 and 203.

By September 1923 the Salmson 2s had been completely replaced by Potez 15s and Breguet 19s.

Postwar, the Aviation Maritime purchased a few surplus Salmson 2s to equip the reconnaissance unit assigned to the Aviation d'Escadre (Fleet Aviation). Unlike the pilots of the Aviation Militaire, the naval pilots found the Salmsons difficult to fly. Because of their poor maneuverability, Salmson 2s were not operated from the aircraft carrier *Bearn*. Instead, they were relegated to training duties and were based at the CAM St. Raphaël. They were replaced by Breguet 14s in 1921.

Approximately 36 Salmson 2s found their way on to the civil register after the war. Modified versions saw limited service as airliners. Latécoère built 20 Salmson 2s incorporating a two-seat passenger cabin. They were used by the Lignes Aeriennes Latécoère. However, the cabins, which were located in the position previously occupied by the observer, were never satisfactory. Oil frequently entered the cabin through the ventilation ports. In addition, there were numerous failures of the propeller bolts, and the carburetors were unreliable. The latter problem was so severe that Latécoère modified an airframe to accept a new, more reliable carburetor. A slight modification to the old carburetors obviated the need for this modification. Salmson 2 airliners were used by Lignes Aeriennes Latécoère (approximately 40 aircraft), Cie Franco-Roumaine (14), CIDNA (5), CMA (2), Cie Gle Transaerienne, Cie Air Transport (1), Air Union (1), and Aero-Transports Ernoul (14).

Foreign Service

Belgium
The Belgian air service obtained a single "war orphaned" Salmson 2 confiscated by the commander of the 4th Escadrille based at Hondschote.

Czechoslovakia
The Czechoslovakian air service purchased 50 Salmson 2s in 1919. They equipped 3 Letecka Setnina (Olomouc), 6 Letecka Setnina (Vajnory), Letecke Dilny (Olomouc), and Letecke Sklady (Praha). In 1923 they were assigned to 1 Pozorovaci Rota (Olomouc), 2 Pozorovaci Rota (Praha-Kbely), 6 Pozorovaci Rota (Vajnory), 7 Pozorovaci Rota (Kosice), 8 Pozorovaci Rota (Nitra), Letecka Pluk, Dilna (Nitra); Letecka Pluk Sklad (Nitra), and Hlavni Vzduchoplavecky Sklad (Olomouc). They were withdrawn from service by 1924.

Japan
The Japanese used the Salmson 2 in large numbers. After examining one taken to Japan in 1919 by the Mission Francaise d'Aéronautique, arrangements were made to produce the aircraft under license. They were initially built by the Tokorozawa Army Air Base Arsenal in Nagoya beginning in 1920; later the Kawasaki Aircraft Company built about 300. The aircraft equipped one Japanese army air service reconnaissance battalion and served as part of the reconnaissance component of the four mixed battalions. Known units include the 2nd Air Regiment (Kagamigahara), 4th Air Regiment (Tachiarai), 5th Air Regiment (Tachikawa), 6th Air Regiment (Chijo, Korea), and the 8th Air Regiment (Heito, Formosa). Some of the Salmson 2s were assigned to bomber units and saw action in the Siberian, Manchurian, and Shanghai incidents. They were replaced in 1933 by the Type 88 reconnaissance aircraft. Between 600 and 1,000 Salmson 2s were built in Japan. One (J-BAEF) was supplied directly to Japan Air Lines and the Asahi Press Group by the Japanese army. Eventually, a total of 12 Salmson 2s were used by Japan Air Lines.

Peru
Peru acquired four Salmson 2s when a French air mission arrived in November 1919. They were used as advanced trainers.

SALMSON

Salmson 2 A2

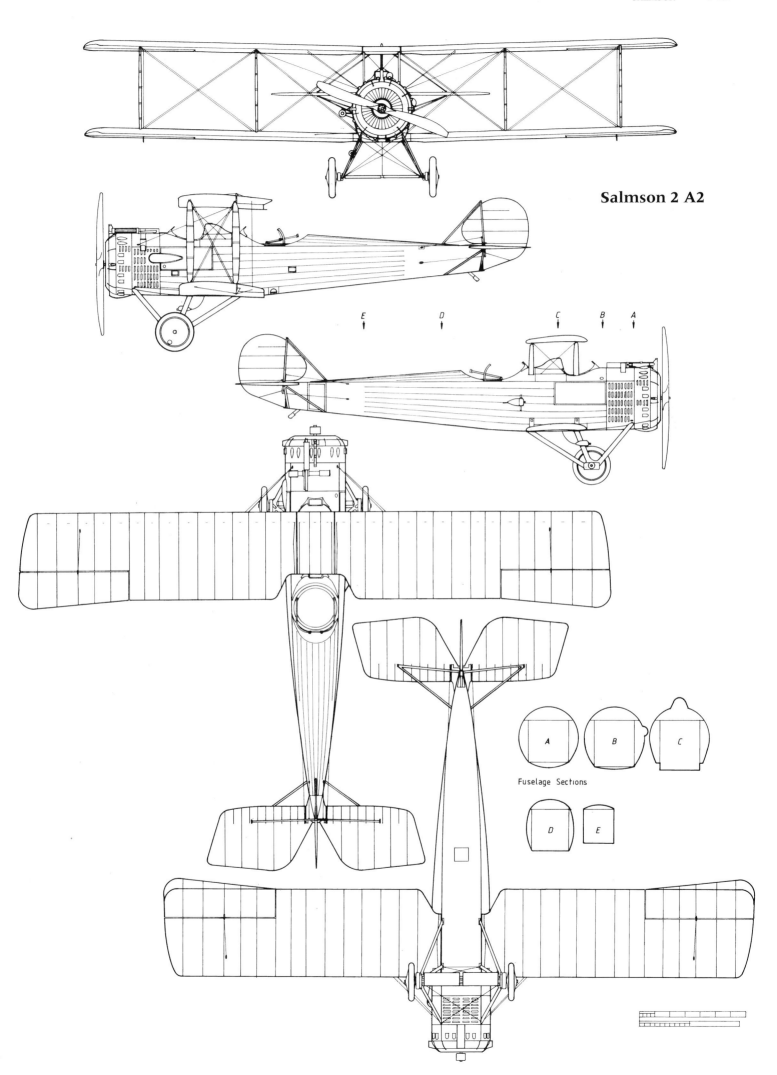

Fuselage Sections

Poland

Poland acquired its Salmson 2s directly from France when the French government turned over the equipment of three Salmson units to the Polish government. Beginning in June 1919 SAL 580, 581, and 582 gave their Salmson 2s to the fledgling Polish air service. SAL 580 arrived first, in June 1919, flying directly from France to Priotrkow. In March 1920 its aircraft formed the 18th Polish Squadron. However, they were probably in poor shape as the unit had to disband in July of that year because of a lack of usable planes. SAL 581's aircraft equipped the Polish observer's school. SAL 582 arrived in Warsaw in December 1918 and by September 1919 its aircraft were being used by the Polish 1st Squadron. By August 1920 that unit had re-equipped with Bristol F 2Bs.

Russia/Soviet Union

In 1919 one Salmson 2 was captured in the south and a second is known to have been captured in Caucasia in 1922. By 1923 nine Salmson 2s were on the strength of the RKKVF (Worker's and Peasant's Red Air Fleet). They served with the 17th And 18th Otdel'nye Razvedivatel'nye Aviaotryady at Chita and the 5th Postoyannaya Aviabaza in Siberia. The last aircraft was probably stuck off charge in 1926. The Siberian Air Fleet of Admiral Kolchak had 23 Salmson 2s on strength in 1920. The Siberian Air Fleet of the Far Eastern Republic had four Salmson 2s in 1920; they had serial numbers 1, 2, 3, and 4.

Spain

Spain acquired a single Salmson 2 in February 1920 for evaluation by the Servicio d'Aeronautica Militar. No others were purchased by the Spanish.

United States

The A.E.F. air service purchased 705 Salmson 2s. The first 18 arrived in April 1918 and were used to replace A.R.1s and Sopwith 1½ Strutters. These units of the 1st Army used Salmson 2s:

1st Corps Observation Group (1st, and 12th Corps Observation Squadrons)

3rd Corps Observation Group (88th and 90th Corps Observation Squadrons).

5th Corps Observation Group (99th and 104th Corps Observation Squadrons).

7th Corps Observation Group (258th Corps Observation Squadron).

1st Army Observation Group (24th and 91st Army Observation squadrons).

The 2nd Army had one unit, the 168th Corps Observation Squadron, that used Salmson 2s.

The American Salmson units saw action in the Aisne-Marne, St. Mihiel, and Meuse-Argonne offensives. The Salmson 2s were popular with the aircrew and were vastly superior to the A.R.1s, Sopwith 1½ Strutters, and SPAD 16s. The first Salmsons supplied to the 1st and 12th Aero Squadrons had a Lewis gun mounted on the upper wing. However, this position degraded the aircraft's performance so severely that the guns were removed. Some Salmsons were re-equipped with modified Lewis guns produced by the Savage Arms Co. The Vickers machine gun (synchronized to fire through the propeller) was replaced in some airplanes by Marlin machine guns. The Salmson 2 was found to be faster than Pfalz and Albatros fighters and also had a superior climb rate. However, its performance was inferior to that of the Fokker D.VII. One complaint noted by the aircrew was that there was room for only 48 photographic plates. This meant that only a relatively limited area could be photographed; to cover a larger area several Salmson 2s would have to be sent. However, overall, the Salmson 2s performed well in U.S. service.

Salmson 3 C1

The Salmson 3 was designed to meet the C1 classification of 1918. As with the Hanriot HD.9 C1, the Salmson 3 was fitted with a 230-hp Salmson 9Z engine and a Ratmanoff CUH propeller. The prototype Salmson 3 was constructed of wood and covered with cloth. It was a two-bay biplane with equal span wings. The triangular tail surfaces featured a large rudder. The fuselage had four longerons made of steel. The tail was constructed from steel tubes. The airframe was noted to have a load coefficient of eight. The fuel and oil systems were the same as those utilized on the Salmson 2.

The Salmson 3 was tested at Villacoublay on 2 May 1918 by a pilot named Barbot. He described its visibility as poor and complained that it was difficult to fly. The test pilot's reports mentioned the following points: the plane turned well but required a large amount of pressure on the controls, while climbing was easy. The plane was tiring to fly. Stability was good. Cockpit visibility was poor because the seat was too low

Salmson 3 Single-Seat Fighter with 260-hp Salmson 9Zm	
Span 9.85 m; length 6.40 m; height 2.48 m; wing area 23.936 sq. m	
Empty weight 696.7 kg; loaded weight 1,026.7 kg	
Maximum speed: 2,000 m	215 km/h
4,000 m	207 km/h
5,000 m	202 km/h
6,000 m	190 km/h
Climb: 1,000 m	2.73 minutes
2,000 m	5.43 minutes
4,000 m	13.28 minutes
5,000 m	21 minutes
6,000 m	34.10 minutes
Ceiling 7,000 m; range 350 km	
Armament: two 7.7-mm Vickers machine guns	
One built	

and the pilot was located between the wings. The large fuselage and location of the struts caused problems, contributing to pilot fatigue. Other criticisms of the Salmson 3 included the inaccessibility of the machine guns; it was recommended they be moved back 15 centimeters.

Later, the Salmson 3 was fitted with a 260-hp Salmson 9Zm engine but that did not result in a significant performance improvement. While the aircraft's speed at altitude was marginally better than the SPAD 13, its climb rate was decidedly inferior to the SPAD's and it was not selected for production. Only the single example was built.

Salmson 3 fighter which was fitted with a 230-hp Salmson 9Z and a Ratmanoff CUH propeller. STAé 02329.

Salmson 3

Salmson 3 three-quarter forward view.

Salmson 4

In 1918 the STAé issued the Ab 2 specification calling for a two-seat aircraft with light armor intended for ground attack. It was to be capable of carrying a T.S.F. and camera. Two aircraft were submitted to meet this specification—the Vendôme biplane and the Salmson 4.

The Salmson 4 was basically an enlarged version of the Salmson 2 reconnaissance aircraft. The wing span was increased and the number of interplane struts was raised to 12. The Salmson 4 was powered by a 260-hp Salmson 9Z engine. The prototype was fitted with a Regy 822 propeller and was flown by test pilot Georges Barbot.

The Salmson 4 was selected for series production over the Vendome design. An order for several machines was placed on 3 May 1918 and by the time of the Armistice, 12 were assigned to front-line escadrilles. No others were built and the type was finally withdrawn from service in 1920.

Salmson 4 Two-Seat Ground Attack Aircraft with 260-hp Salmson 9Z

Span 15.20 m; length 8.80 m; height 2.96 m; wing area 49.28 sq. m
Empty weight 1,410 kg; loaded weight 1,935 kg

Maximum speed:	ground level	168 km/h
	1,000 m	147 km/h
	2,000 m	143 km/h
	4,000 m	137 km/h
Climb:	500 m	3 minutes 18 seconds
	1,000 m	8 minutes 21 seconds
	2,000 m	20 minutes 14 seconds
	4,000 m	43 minutes 27 seconds

12 built

Above: Salmson 4. At the time of the Armistice, 12 of these aircraft were assigned to front-line escadrilles. Colin Owers.

Left: Salmson 4. The Salmson 4 was an enlarged version of the Salmson 2 reconnaissance aircraft and was powered by a 260-hp Salmson 9Z engine. B87.1038.

Below: Salmson 4. Compared to the Salmson 2, the wing span of the Salmson 4 was increased and the number of interplane struts was raised to 12. B87.1041.

Salmson 5

The Salmson firm tested a modified version of the Salmson 2 in 1917. Designated Salmson 5, it had positively staggered wings with a slight sweepback. The shape of the rudder was slightly altered and enlarged elevators were attached to a fixed tailplane. The latter was a marked departure for aircraft produced by the Salmson firm.

The Salmson 5 was intended to meet the requirements of the A2 category for artillery spotting. However, it was 23 km/h slower than the standard Salmson 2 and does not appear to have offered any significant advantages over that aircraft. Only a single Salmson 5 was built.

Salmson 5 Two-Seat Reconnaissance Aircraft with 230-hp Salmson 9Za

Span 12.20 m (some sources state 11.75 m); length 7.80 m (some sources state 7.70 m); height 2.91 m (some sources state 2.82 m)

Maximum speed: 169 km/h at 2,000 m

One built

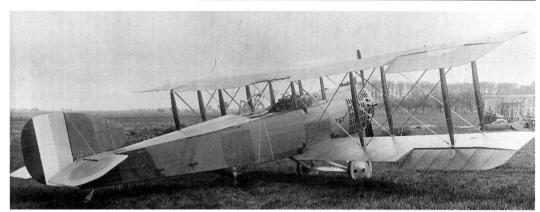

Salmson 5. The rudder shape was unique to the Salmson 5. MA 1272 via Colin Owers.

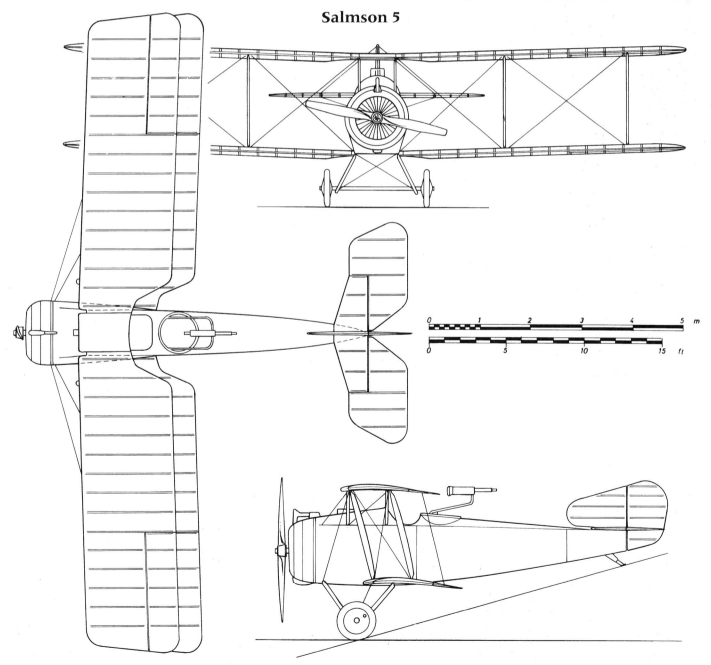

Salmson 5

Salmson 6

The Salmson 6 was designed to the 1918 STAé A2 specification calling for a single-engine army cooperation aircraft. It was based on the standard Salmson 2 and powered by a 230-hp Salmson 9Za engine. Only a single example was built. It may be the aircraft shown on page 440.

Salmson 7

The Salmson 7 was intended to meet the A2 specification for 1918 calling for a payload of 450 kg, a ceiling of 7,000 m, a maximum speed of 200 km/h, and a minimum speed of 90 km/h. The new A2 design was to carry a T.S.F. unit. The Breguet 14 with a Rateau turbo-compressor and the SEA 4 were also designed to meet this requirement.

The Salmson 7, possibly drawing from the experience with the preceding Salmson 5 and 6, was also a development of the standard Salmson 2. New features included an enlarged wing with a slight sweepback. The Salmson 7 had the same 230-hp Salmson 9Za engine as the Salmson 2. One major difference was the location of the crew members' cockpits. The pilot and observer were seated far apart on the Salmson 2, and this made communication between them virtually impossible. To rectify this problem, the crew of the Salmson 7 were seated back-to-back in a single cockpit tub. The Salmson 7 underwent STAé static testing on 19 August 1918; a coefficient of 6.5 was attained.

The Aviation Militaire planned to order enough Salmson 7s to equip 74 escadrilles. However, the termination of hostilities resulted in only 20 being delivered to army cooperation escadrilles.

Salmson 7 Two-Seat Army Reconnaissance Aircraft with 230-hp Salmson 9Za
Span 12.34 m; length 8.83 m; height 2.86 m; wing area 38.70 sq. m
Maximum speed: 189 km/h
Twenty built

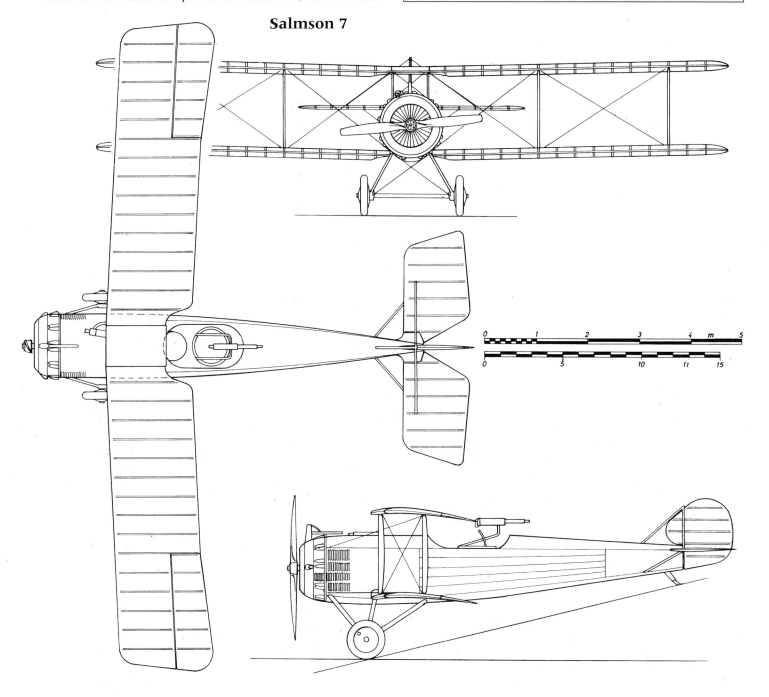

Salmson 7

Salmson 7 army cooperation aircraft. Twenty examples were delivered to army cooperation escadrilles. MA4261.

Salmson 7 front view. 02818.

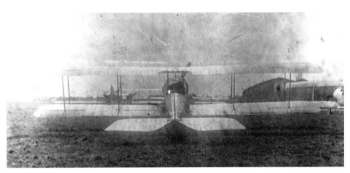

Salmson 7 rear view. 02821.

Ateliers de Constructions Mécaniques & Aéronautiques Paul Schmitt

Paul Schmitt was fascinated with variable incidence wings. In 1904 he began to experiment with kites featuring variable incidence. Schmitt built his first aircraft in 1910; this was a tailless design powered by a 70-hp engine. Three subsequent airplanes had elevators but no rudders. His seventh design (c/n 7) had a Renault engine and a conventional rudder.

Schmitt was one of the most innovative pioneers of French aviation. However, relatively few of the aircraft he designed would see operational service. This was due primarily to production delays resulting in Schmitt's airplanes being obsolescent by the time they entered service.

Despite the limited success with his own designs, Schmitt produced up to five airplanes a day under license, the most important being the Breguet 14.

Paul Schmitt 3

The Paul Schmitt Type 3 was the ninth (c/n 9) airplane built by Schmitt. The landplane version was powered by a 150-hp Canton Unné P9 engine. One source states that 160-hp Gnome engines were used, but this has not been verified. The airplane entered service with the Aviation Militaire in 1915. It is estimated that six were built and these probably carried military serials P.S.3 through P.S.8. The Type 3 was originally intended as a bomber, but was actually used only by training units.

A floatplane variant of the Type 3 may have been purchased by the Aviation Maritime. It had the same basic layout and P9 engine of the landplane. The three-bay wing was retained, but the surface area was reduced. The navalized Type 3 had a single float under the center fuselage and two smaller floats at the wing tips. The radiator was relocated from beneath the fuselage (where water ingestion would have been a problem) and placed along the centerline of the top wing. The floatplane was built for an American M. Belmon and shipped to the United States in 1916. The Type 3 floatplane also carried the designation "9"; this may have been the construction number.

Paul Schmitt 3 Two-Seat Bomber/Trainer with Either a 150-hp Canton Unné P9 or a 160-hp Gnome
Span 17.50 m; length 10 m; height 3.15 m; wing area 49 sq. m
Empty weight 650 kg; loaded weight 1,100 kg
Maximum speed: 116 km/h; range 460 km
Approximately 6 built

Above: Paul Schmitt P.S.3. MA12323.

Left: Paul Schmitt P.S.3. The Type 3 was originally intended as a bomber, but was actually used only by training units. MA32827.

Below: Paul Schmitt floatplane. This was a navalized Type 3 which had a single float under the fuselage and two smaller floats at the wing tips. The radiator was relocated from underneath the fuselage to the centerline of the top wing. MA36448.

Paul Schmitt P.S.3 in flight. MA36444.

Paul Schmitt S.B.R.

Aircraft c/n 6 built by Paul Schmitt in 1913/14 featured a 160-hp Gnome engine and a variable incidence wing. In 1914 this machine established a series of records for altitude, speed, range, and climb rate for an aircraft carrying three or more passengers. The variable incidence wing could be pivoted up or down by 12 degrees. This permitted the aircraft to takeoff, climb, and land at a steeper angle than could be achieved by conventional designs. The variable incidence wing reduced the takeoff run and enabled the airplane to land on shorter fields. The design so impressed the judges at the international concours of 1914 that they awarded Schmitt a prize of 30,000 francs.

At the outbreak of war Schmitt placed his award-winning aircraft at the disposal of the army. Used by the same pilot (named Garaix) who had flown it during the 1914 concours, it was lost on 15 August 1914 during a bombing raid on Metz.

A variant of the pre-war design was built by Schmitt in 1915. This machine was initially termed the S.B.R. (Schmitt Bomber Renault) and was entered in the concours puissant of 1915. The S.B.R. retained the variable incidence wing but had a more powerful 200-hp Renault engine. The variable incidence wings were controlled through a series of wheels and chains which pulled the wing forward, causing it to rotate along the axis of a steel tube within the wing. The normal setting was 5 degrees for flights when the maximum payload of 450 kg was carried. A setting of 6 degrees was used for flight above 2,000 meters. Finally, a setting of 7 degrees was used for landing. An incidence indicator was provided for the pilot so he could set the wings at the recommended angles. The aileron cables were arranged to pass through the wing's axis of rotation; this enabled the wing incidence to be changed without causing the aircraft to change direction. The airplane flew the 600-km circuit at an altitude of 2,000 meters while carrying 350 kg of fuel and oil as well as a payload of 450 kg. The S.B.R. and the Breguet SN 3 were the only two aircraft to meet the requirements of the competition.

However, the French authorities were dissatisfied with Schmitt's design. The main complaints were that the maximum speed of 120 km/h was too slow, the defensive armament too weak, and the variable incidence wing was felt to be of little value in a bomber aircraft. The Breguet SN 3 was judged the winner of the 1915 concours puissant.

However, there were those who felt the S.B.R. provided the basis for a useful bomber, especially as the Breguet design also had numerous deficiencies. Eventually, the Aviation Militaire decided to purchase modified versions of both the SN 3 and S.B.R. They were to serve as interim types until a more suitable bomber, to be selected at the 1916 concours puissant, became available.

Paul Schmitt S.B.R. (Type 4) at the 1915 concours puissant. The aircraft had an all-metal fuselage. MA36447.

Paul Schmitt 6

The S.B.R. was modified by Schmitt to meet the requirements of the Aviation Militaire. The most significant change was the deletion of the variable incidence wing—no operational Schmitt bomber had this feature. The lower wing was suspended beneath the fuselage; for this reason, Schmitt changed the designation from S.B.R. to B.R.A.B., which stood for Bomber Renault Aile Basse (low wing). A more powerful 250-hp Renault engine was fitted. The prototype is believed to have been the 22nd aircraft built by Schmitt (c/n 22). It was given the STAé desig-nation Paul Schmitt Type 6. Approximately ten were built and these were probably given military serials 22 through 31. The Type 6 was soon replaced by the Type 7.

Paul Schmitt 6 Two-Seat Bomber with 250-hp Renault
Span 17,50 m; length 9.50 m; wing area 48.40 m
Empty weight 1, 300 kg; loaded weight 1, 975 kg
Maximum speed: 125 km/h at 2,000 m; climb to 2,000 m in 19 minutes; range 600 km
Approximately 10 built

Paul Schmitt P.S.6 at the Ecoles d'Aviation at Chartes in 1916. 82.4896.

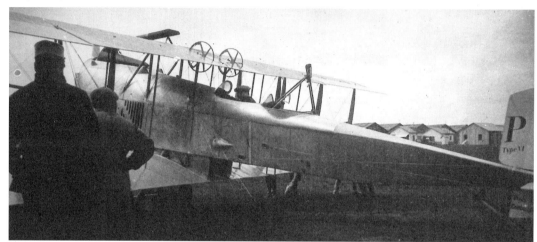

Right: Paul Schmitt P.S.6 at the Ecoles d'Aviation at Chartes in 1916. The rudder shows this airplane carries the designation Type VI. The lower wing was now suspended beneath the fuselage. Approximately ten Type 6s were built and these were probably given military serials 22 through 31. B82.4876.

Paul Schmitt 7

The Paul Schmitt 7 was a development of the Type 6 bomber. In most respects it was identical to the Type 6 and both used the 250-hp Renault engine. However, the lower wing of the Type 7 was attached directly to the fuselage. Thus the Type 7 carried the Schmitt designation B.R.A.H., standing for Bomber Renault Aile Haute (high wing). The wing area was so large that it led to serious problems when the Type 7 was introduced into operational service. As with the Type 6, the Type 7 dispensed with the variable incidence wing. The wings were constructed in two parts mounted to a fixed central section. The center section was attached to the fuselage by four steel tubes. The fuselage was constructed entirely of metal and covered with aluminum. As with the Type 6, the radiator was located in the nose. The landing gear was altered on the Type 7; whereas the Type 6 had featured a quadracycle undercarriage or nose skids, most Type 7s had a single pair of wheels beneath the fuselage. The landing gear was supported by ash runners and bungee cords which served as shock absorbers. Wing spars were made of ash and pine. The wing formers were of plywood. There were 16 struts; four were of metal, the others of pine. Ailerons were on the top

Paul Schmitt 7 Two-Seat Bomber with 250-hp Renault
Span 17.50 m; length 9.50 m; wing area 48.40 m
Empty weight 1,298 kg; loaded weight 2,098 kg
Maximum speed: 135 km/h at 2,000 m; 131 km/h at 3,000 m; 121 km/h at 4,000 m; climb to 2,000 m in 21.45 minutes; climb to 3,000 m in 36.5 minutes; ceiling approximately 4,000 m; endurance 5 hours
Armament: two 7.7-mm machine guns fired by the gunner and 150 kg of bombs
Approximately 150 built

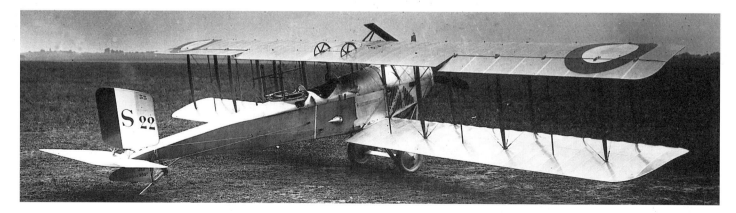

Paul Schmitt P.S.7 serial PS 22. MA13423.

Paul Schmitt P.S.7. In most respects the Type 7 was identical to the Type 6; however, the lower wing of the Type 7 was attached directly to the fuselage. B78.1545.

wing only. There was a fuel tank of of 450 liters capacity.

The B.R.A.H. was given the STAé designation Paul Schmitt Type 7, and approximately 150 were ordered. They are believed to have had Schmitt construction numbers 33 to 180 and carried military serials P S 22 to P S 170.

The Paul Schmitt 7/4 was the only major variant of the P.S.7. It featured dual nosewheels, presumably added to prevent the aircraft from nosing over during landing. This was a common problem with French night bombers of the period; the heavily laden aircraft frequently crashed when they attempted night landings on soft or muddy soil. The addition of the nosewheels would have become necessary when the role of the Paul Schmitt 7 was changed from day to night bomber.

The government contract for the P.S.7 was to prove a mixed blessing for Schmitt. While he now had the opportunity to produce a new bomber for the Aviation Militaire, he had yet to create a factory to produce the planes. Like many other French aviation pioneers, Schmitt had very limited financial and manufacturing resources. He was able to secure economic assistance from diverse sources, including a French minister of parliament and an American banker who anticipated selling large numbers of P.S.7s to the fledgling American air service.

However, Schmitt's problems were far from over—he still needed to construct a factory to assemble the bombers. Huge sums of money were spent building the Schmitt establishment, obtaining the requisite raw materials, and hiring a skilled work force. The result was that the Schmitt bomber, which had been selected as an interim day bomber in 1915, did not reach front-line units until April 1917. By 1 August 1917 only 16 were in service. The delay was to prove fatal both for Schmitt's plans to produce an armada of P.S.7s and for the unfortunate crews who had to fly the obsolete machines in combat.

Operational Service

On 8 April 1917 PS 125 and PS 126 were formed on the Paul Schmitt 6 and 7. Two additional units were formed, PS 127 on 14 April and PS 128 on 15 May 1917. All four were assigned to GB 3 on 7 July 1917.

It was soon apparent that the new bomber had serious deficiencies. The aircraft were found to be too slow and carried inadequate protection for daylight missions (this problem had been noted as far back as the 1915 concours puissant). The flying characteristics of the P.S.7 were so poor that the aircraft became unstable if the maximum bomb load was carried. The bomb load had to be reduced even further so that more defensive armament could be carried, the aircraft being particularly vulnerable to fighters because of its slow speed. Formation flying was virtually impossible because of the P.S.7's poor maneuverability, due to the huge wing. The aircraft, because they were unable to fly in tight formation, could not concentrate their firepower and were therefore easy prey for German fighters.

The PS escadrilles of GB 3 were based at Plessier Saint-Just in April, operating a mix of P.S.6s, P.S.7s, and P.S.7/4s. After the PS escadrilles had been attached to GB 3, the group moved to Pierrefonds. From this base the PS units were to bomb targets in the Somme and Aisne sectors. Later in the month the Paul Schmitt bombers were active over the Serre Valley. On 2 May the PS units successfully bombed Essigny-le-Petit. On 27 May a major raid was flown on the Nouvion-Catillon aerodrome. The next month the units spent most of their time practicing formation flying. They returned to action on 13 July when stations and bivouacs

Paul Schmitt 7/4 in May 1917. The Paul Schmitt 7/4 was the only major variant of the P.S.7. It featured a dual nosewheel which was presumably added to prevent the aircraft from nosing over during landing. B87.899.

P.S.7. Approximately 150 were ordered. These aircraft are believed to have had Schmitt construction numbers 33 to 180. B86.1130.

Below: Paul Schmitt P.S.7 with a 250-hp Renault engine. B83.4916.

were attacked. Eleven days later, the 18 aircraft bombed German camps at Lanquetaille, and two machines of PS 125 bombed the Cambrai train station. An attempt to bomb targets in the vicinity of Pouilly sur Serre proved disastrous; enemy fighters caused heavy casualties and forced the escadrilles to miss the target. The PS units were then assigned to support the G.A.N. during the Second Battle of the Somme and l'Aisne. Various train stations along the front were attacked in September. October would be the last month the PS 7s would be used operationally, when Catele, Ramicourt, and the cement plant at Mont d'Origny were attacked. It was now obvious that the Paul Schmitt 7s were useless in either the day or night bombing role. PS 128 received Sopwith 1½ Strutters in June. PS 126 and PS 127 re-equipped with Breguet 14 B2s in November and PS 125 received Voisin 8s on 24 January.

Production delays had resulted in an aircraft that incorporated the technology of 1915 being introduced into combat in 1917. Given the rapid pace of aviation development in the intervening years, it is not surprising that the Paul Schmitt 7s proved to be a costly failure.

Paul Schmitt 9

Little is known about the Paul Schmitt 9, a two-seat bomber built in 1916 and powered by a 160-hp Canton Unné engine.

It should be noted, however, that the Paul Schmitt 3 had the same engine and was given c/n number 9. It is quite possible that the Paul Schmitt 9 some sources refer to is, in fact, the Paul Schmitt 3 prototype.

Paul Schmitt 10 B2

The Paul Schmitt 10 was built in 1917 and represented an attempt to produce an updated version of the obsolete PS 7 B2. It was given a more powerful 300-hp Renault 12Fe. Like the earlier PS 6 and 7, the PS 10 did not feature variable incidence wings. It had a three-bay wing, and, unlike Schmitt's previous designs, the wings were of unequal span, the lower being shorter.

The Paul Schmitt 7 had suffered heavy losses at least in part due to its inadequate defensive armament. Schmitt seemed determined to correct this deficiency by fitting the Type 10 with three machine guns. One was mounted on a revolving stand used by the rear gunner. A second gun, fired by the pilot, was synchronized to fire through the propeller. Finally, a third gun was mounted so that the gunner could fire through the floor of the aircraft. The plane carried 250 kg of fuel and 480 kg of bombs. The Paul Schmitt 10 B2 was tested in October 1917. It was recorded in *Janes All the Worlds Aircraft* for 1920 that "because of its remarkable efficiency, it was adopted by the Army as soon as it began production." In fact, no Paul Schmitt 10s were ever used by the Aviation Militaire, possibly because the superior Breguet 14 B2 was already entering service with the day bombing escadrilles. A contemporary report states that development of the P.S.10 had been abandoned by May 1918.

> **Paul Schmitt 10 Two-Seat Bomber with 300-hp Renault 12Fe**
> Span 14.650 m; length 9.105 m; height 3.585 m; wing area 48.910 sq. m
> Empty weight 1,150 kg; loaded weight 1,880 kg
> Maximum speed: 169 km/h at 5,000 m; climb to 2,000 m in 9 minutes 50 seconds; endurance 3 hours 30 minutes
> Armament: one synchronized machine gun fired by the pilot and two machine guns fired by the observer; 480 kg of bombs

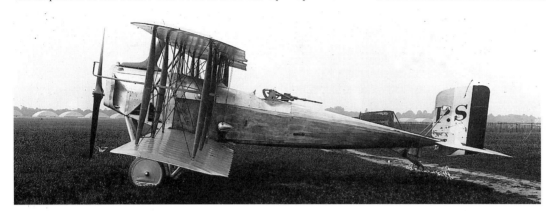

Paul Schmitt P.S.10. This aircraft was a refinement of the P.S.7 and had a more powerful 300-hp Renault 12 Fe engine and a three-bay wing. MA01013.

Paul Schmitt Floatplane

The Paul Schmitt floatplane was similar to the Paul Schmitt 10 B2 and may have been a development of that aircraft. Both were powered by a 300-hp Renault 12Fe. In place of the undercarriage there was a large centerline float flanked by two smaller floats. A crew of two, a pilot and observer/gunner, was carried. The armament was identical to the Schmitt 10: a gun synchronized through the propeller, a second gun fired by the observer, and a third gun in the belly of the airplane. There were shutters in the floor that could be opened by the pilot during the bomb run, enabling him to judge the best time to release the bombs. The aircraft could carry 285 kg of fuel and 295 kg of bombs. The floatplane was intended for the patrol/light bomber role. An unknown number were purchased by the Aviation Maritime. When the United States entered the war, the French supplied them with a number of Paul Schmitt floatplanes. At least one was at the U.S. naval aviation base at St. Trojan. It was subsequently sent to the U.S. where it was given the U.S. Navy serial number

> **Paul Schmitt Two-Seat Floatplane with a 300-hp Renault 12Fe**
> Span 15.650 m; length 10.450 m; height 4 m; wing area 52 sq. m
> Empty weight 1,425 kg; loaded weight 2,005 kg
> Maximum speed: 195 km/h at sea level; climb to 3,000 m in 17 minutes 5 seconds; endurance 3 hours 45 minutes
> Armament: one synchronized machine gun fired by the pilot and two machine guns fired by the observer; 295 kg of bombs

A-5636. Others were supplied to the U.S. naval aviation base at Moutich. In U.S. service the Paul Schmitt seaplane was used primarily as a trainer. At least one was fitted with a 370-hp Liberty engine.

Paul Schmitt floatplane with a Renault engine in 1916. The Paul Schmitt floatplane was similar to the Paul Schmitt 10 B2 and may have been a development of that aircraft. MA36444.

Paul Schmitt C2

The Schmitt firm had a two-seat fighter under construction in 1918, a biplane powered by a 400-hp Lorraine engine. The Paul Schmitt C2 was never selected for service with the Aviation Militaire; indeed, it is uncertain construction of the aircraft was ever completed.

> **Paul Schmitt Two-Seat Fighter with 400-hp Lorraine (Provisional Data)**
> Wing area 49 sq. m
> Empty weight 1,052 kg; loaded weight 1,633 kg
> Maximum speed (estimated): 130 mph at 4,000m; climb to 4,000 m in 14 minutes

Paul Schmitt BN3/4

In late 1917 Paul Schmitt designed a heavy bomber to meet the BN3/4 (night bomber with three or four crew members). It was to have been a biplane powered by four 200-hp Hispano-Suiza engines. The design remained a project only and further development had been abandoned by April 1918.

Paul Schmitt Three- Or Four-Seat Bomber with Four 200-hp Hispano-Suiza Engines (estimated data)
Wing area 412 sq. m
Empty weight 2,425 kg; loaded weight 4,310 kg; payload 1,000 kg
Maximum speed (estimated): 103 mph at 4,000 m, climb to 4,000 m in 23 minutes; endurance 5 hours

Schneider *Henri-Paul* S3

When the S class specification was formulated in 1918 by the STAé it was hoped that French manufacturers could produce a series of ground attack aircraft as effective as the Junkers J.I series. The S3 classification called for a three-seat attack aircraft heavily armored for low altitude attacks with a 75-mm cannon; it was also to be equipped with a radio and camera. Obviously, such an aircraft was going to be large and it appears that only two were built that could fulfill the STAé's requirements—the Voisin 12 S3 and the Schneider *Henri-Paul* S3.

The *Henri-Paul* was designed by Établissements Schneider but was built at least in part by Farman, as Schneider did not have a factory. It was a huge three-bay biplane with equal span wings whose outer panels were sharply swept back. The engines were four 370-hp Lorraines mounted back-to-back in twin nacelles on either side of the fuselage. There were twin wheels under each of the nacelles, as well as a small nose wheel. The angular fuselage terminated in a biplane tail with elevators. A large central fin and rudder, as well as two lateral fins, were fitted between the biplane tail. Construction was entirely of steel and light alloy. Photographs clearly show that the crew had an open cockpit on the prototype, but drawings show that an enclosed cabin was under consideration. Gunners were located in open cockpits in the nose and amidships. It was intended to fit 37-mm cannons in each of these positions. Consideration was also given to placing a 75-mm cannon in the nose inside a special balcony underneath the fuselage. However, it was never fitted.

The *Janes All the Worlds Aircraft* for 1924 shows that the aircraft was just being completed. The annual notes that the aircraft was now in the BN2 category, suggesting that it was no longer being considered for use in the ground attack role but rather was to been employed as a heavy bomber.

The aircraft was tested at Harfleur by Établissements Schneider in the early 1920s and subsequently the *Henri-Paul* was flown to Villacoublay for official tests. It was not selected for production.

Schneider *Henri-Paul* S3 Three-Seat Ground Attack Aircraft with Four 370-hp Lorraine Engines
Span 30 m; length l9.9 8m; height 6.1 m; wing area 220 sq. m
Loaded weight 10,000 kg
Maximum speed: 160 km/h; ceiling 5,000 m; range 750 km
Armament (planned, but never fitted): two 37-mm cannons and one 75-mm cannon
One built

Schneider *Henri Paul* S3. The SHAA archives identify this aircraft as a Farman type. In fact, the *Henri-Paul* was designed by Établissements Schneider, but at least in part was built at the Farman factory. B83.3718.

Schneider Henri-Paul S3

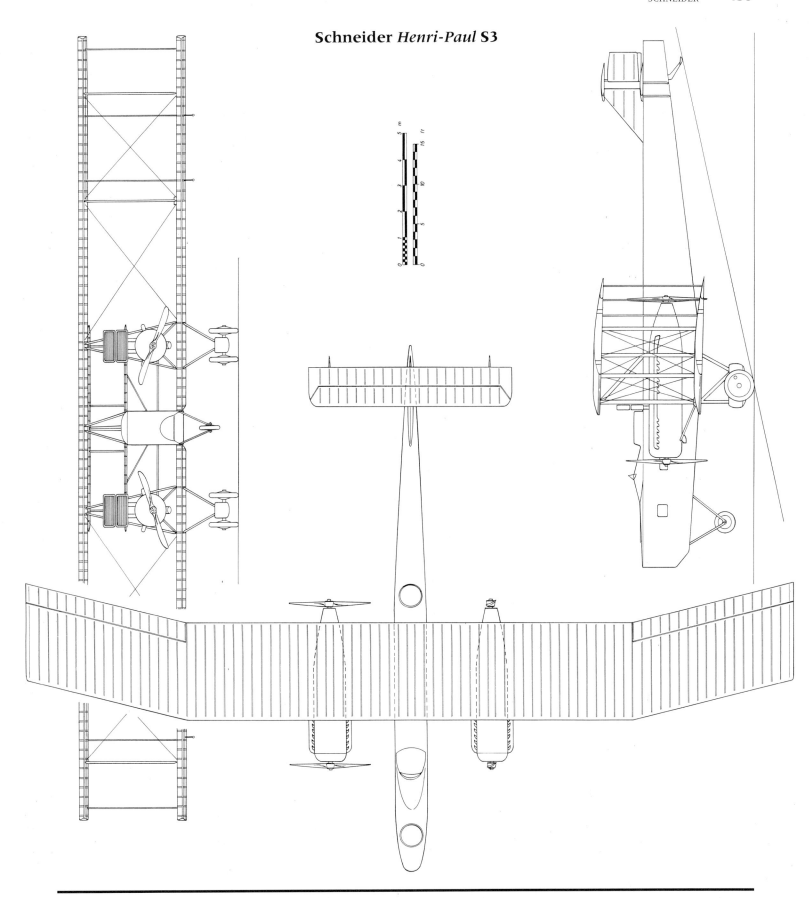

Société d'Études Aéronautiques (SEA)

SEA 1

The Sociètè d'Études Aéronautiques (SEA) was formed in 1916 by Marcel Bloch and Henry Potez at the former Antoinette factory at Suresnes. While producing SPAD 7s under license, Bloch and Potez initiated the design of a two-seat reconnaissance aircraft intended to replace the now obsolescent Sopwith 1½ Strutter.

Work on this new aircraft was done in the evening after the production of SPAD 7s had finished. Louis Coroller played a major role in producing the prototype. As the team's experience grew, the airframe gradually took shape. The aircraft was powered by a newly-designed 200-hp Clerget rotary engine and was a conventional tractor biplane with two-bay, negatively-staggered wings. It appears to have had least one machine gun synchronized to fire through the propeller. The Clerget engine proved unsatisfactory and this resulted in test flights of only a few minutes' duration. It was soon discovered that because of the unsatisfactory engine the SEA 1 was no better than the Sopwith 1½ Strutter it was intended to replace, and further development was abandoned. (See Addendum for drawing.)

Although of poor quality, this is one of the few photographs available of the little-known SEA 1. MA28764.

SEA 4

The SEA 2 and SEA 3 were projects only and were never built, but the SEA 4 was to prove more successful. Despite the unfortunate experience with the Clerget engine, the team decided to gamble on another new engine—the 370-hp Lorraine-Dietrich 12Da. The prototype was completed in 1918, intended to meet the C2 classification for a two-seat fighter. The aircraft featured a streamlined fuselage and even used piano wire for bracing because it was hoped that its thin cross-section would reduce drag. The two-bay wing had no sweepback. The ribs were closely spaced to maintain the airfoil shape.

The fuselage had a rectangular cross section with rounded angles. There were four longerons made of spruce, and the cowling was made of aluminum. The vertical fin was triangular in shape and made of wood. The landing gear consisted of V-shaped struts made of ash. These struts were fixed to the fuselage and carried a heavy steel axle. Bungee chords served as shock absorbers. The landing gear was reinforced by steel cables. The armament was a synchronized 7.7-mm Vickers machine gun and two T.O.3 ring-mounted 7.7-mm Lewis guns. A Winchester rifle could be fired from a trapdoor beneath the aircraft. There were special supports within the fuselage on which a camera or T.S.F. unit could be mounted.

The aircraft was tested in April 1918 at Villacoublay, Plessis-Belleville (where the machine was tested in combat conditions), and the training field at Perthe. One thousand were ordered by the STAé. To meet that demands a new firm, called the Anjou Aéronautique Company, was formed. However, it was far from Paris, resulting in production delays. This angered the Minister of Munitions, who demanded an increased production rate from both the SEA firm and its subcontractors. Potez was sent to Angers to supervise assembly and the first production aircraft was finally ready on November 11 1918—Armistice Day. The SEA 4 was also to be built at the Societè SCAF (Stè de Construction Aéronautiques Francaise) at Lavallois-Perret, Janoir, and SAIB.

A version of the SEA 4 was produced which was intended for long-range flights. This version, designated the P.M., had additional fuel tanks fitted and had an endurance of 6 hours.

With the end of the war the order for 1,000 aircraft was canceled, but the government accepted the first series of 115. Those built for the Aviation Militaire were used to meet the type Ap2 classification for a long-range reconnaissance aircraft. Many of these were assigned to the 34th RAO at Bourget.

To make use of surplus airframes, a number of planes were converted to airliners by the addition of an enclosed cabin for two passengers in the position previously occupied by the

SEA 4 PM. This version of the SEA 4 was intended for long-range flights. It had additional fuel tanks which gave it an endurance of six hours. MA11414.

SEA 4 Two-Seat Fighter with 370-hp Lorraine-Dietrich 12Da

Span 12.00 m; length 8.50 m; height 3.0 m; wing area 36.80 sq. m
Empty weight 1,040 kg; loaded weight 1,620 kg

Maximum speed:
	ground level	218 km/h
	1,000 m	215 km/h
	2,000 m	210 km/h
	3,000 m	206 km/h
	4,000 m	203 km/h
	5,000 m	197 km/h

Climb:
	1,000 m	3 minutes 1 seconds
	2,000 m	6 minutes 18 seconds
	3,000 m	10 minutes 45 seconds
	4,000 m	16 minutes
	5,000 m	23 minutes 29 seconds

Endurance: 2.25 hours
Armament: synchronized 7.7-mm Vickers machine gun and two T.O.3 ring-mounted 7.7-mm Lewis guns. A Winchester rifle could be fired through a trapdoor beneath the aircraft.
115 aircraft built

gunner. About 25 of these were used by the Franco-Roumaine de Navigation and operated as far as Warsaw and Constantinople. At least two passed to the French airline CIDNA in 1925.

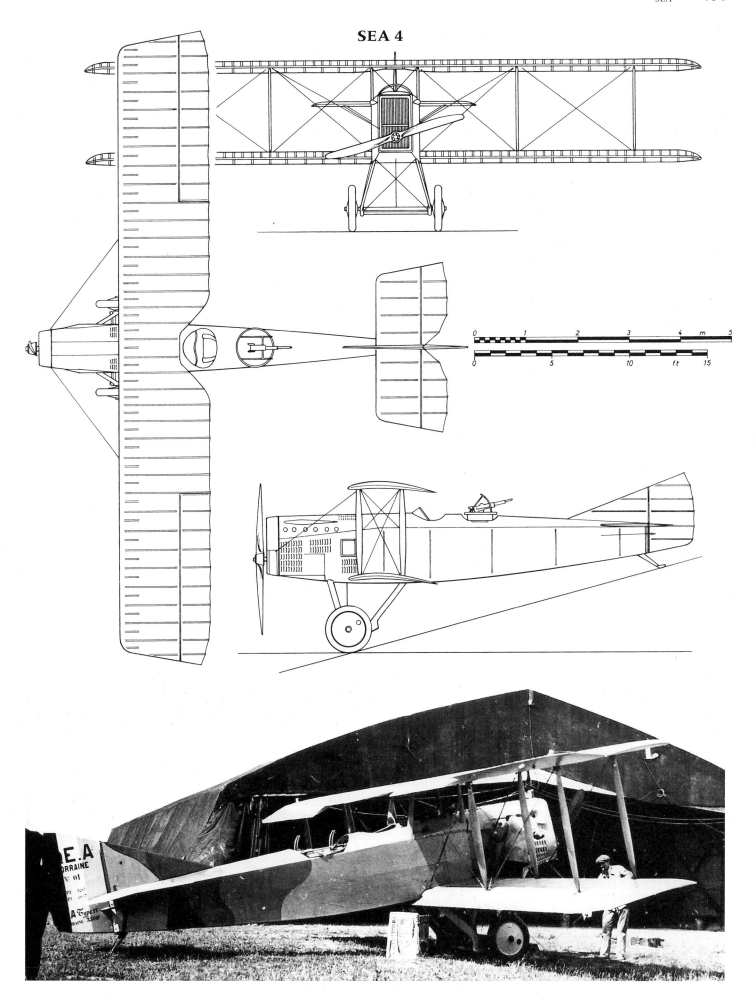

The prototype (serial 01) of the SEA 4 in August 1918. B78.1515.

SEA 4. The aircraft featured a streamlined fuselage and even used piano wire for bracing because it was hoped that its thin cross-section would reduce drag. The two-bay wing had no sweepback. MA25498.

SEA Floatplane

SEA had a floatplane under construction in mid-1918. It had twin floats and carried a crew of two. The engine was a 400-hp Lorraine. It is not known if construction was ever completed.

SEA Two-Seat Floatplane with 400-hp Lorraine
Wing area 35 sq. m
Payload 120 kg
Estimated Maximum speed: 200 km/h at 2,000 m; climb to 2,000 m in 10 minutes

Semenaud

Semenaud C1

The C 1 specification of 1918 called for a single-engine aircraft with twin machine guns or a single machine gun and a 37-mm cannon. The type was to have been suitable for further development as a photo reconnaissance or escort fighter.

A number of different engines were made available to manufacturers designing aircraft to meet the C1 requirement. Those using the 300-hp Hispano-Suiza 8F included the SPAD 18, 20, 21, 22, Nieuport-Delange 29, Hanriot HD.7, De Marcay 2, SAB 1, Moineau, Descamps Types 2 and 3, and the Semenaud fighter.

On 29 November 1918 the commission appointed to study new aircraft projects evaluated a design proposed by the Semenaud firm. An unusual feature was its forward-swept biplane wings.

The Semenaud fighter was built in 1918. It had a wing area of 17.50 square meters, an empty weight of 675 kg, a loaded weight of 900 kg, a maximum speed of 235 km/h at 4,000 m, could climb to 6,000 meters in 17 minutes, and had a ceiling of 8,200 meters.

It was not selected for production, the Nieuport Delange 29 with the same engine being preferred. One of the complaints about Semenaud's design was that the position of the engine under the upper wing would have made it difficult for ground crews to service it.

Semenaud Single-Engine Flying Boat

In 1918 Semenaud designed and built a flying boat probably intended for coastal patrol and "alerte" missions. It carried a crew of two separated by a 200-hp Hispano-Suiza engine powering a tractor airscrew. The pilot was seated ahead of the engine and could fire a Lewis machine gun fixed in the nose. The gunner, who also had a set of controls (which would enable him to relieve the pilot of flying duties on long missions) sat behind the engine and controlled a machine gun on a swivel mount. The wooden hull had a prominent keel, but there was no step on the underside of the hull. The airplane was not selected for use by the Aviation Maritime.

Semenaud Flying Boat with 200-hp Hispano-Suiza
Empty weight 750 kg; loaded weight 1,450 kg
Armament: two 7.7-mm Lewis machine guns
One built

Semenaud Twin-Engine Flying Boat

In 1918 Senemaud designed a twin-engine flying boat that would have been a conventional biplane (described as having a "round section hull") with two 300-hp Hispano-Suiza engines. A crew of three was to have been carried. It remained an un-built project.

Semenaud Twin-Engine Flying Boat with Two 300-hp Hispano-Suiza Engines (all data provisional)
Wing area 80 sq. m
Payload 450 kg
Maximum speed (estimated): 160 km/h at 2,000 m

Short Bomber

A single example of the RNAS Short Bomber was obtained by the STAé. On 31 July 1918 this plane (no.9311) was given the STAé designation Sho 1 B2. It was evaluated in the B2 day bomber role, intended to replace the outdated M.F.11s and Voisin 5s. However, the Sopwith 1½ Strutter was selected to fill the B2 requirement and no further Short Bombers were purchased.

S.I.A.

S.I.A. BN2

This bomber was designed by Henri Coanda to meet the 1917 specification for a heavy night bomber. The aircraft of this class were to be able to carry 1,200 kg of bombs. Other aircraft designed to meet the BN requirements included the Caudron C.23, Caproni 3 BN3, Letord 9, Delattre BN3, Bassan-Gue BN4, Voisin 12, Renault O 1, and the Sikorsky BN3.

The aircraft was built at Angers and made wide use of duralumin construction. Duralumin had been looked upon with some suspicion by the STAé, but Louis Breguet had employed it in his very successful Breguet 14 design. Its combination of strength and low weight was perfectly suited for use in a heavy bomber.

The aircraft was a large biplane powered by two 400-hp Liberty engines. The wings appear to have been of equal span with three bays of struts outboard of each engine. The engines were suspended on struts between the upper and lower wings near the fuselage. The short nose protruded just ahead of the engine nacelles, and there was a gunner's station in the nose position. There was a biplane tail with three vertical stabilizers mounted between the upper and lower horizontal stabilizers. Two rudders projected behind the horizontal stabilizers. The landing gear consisted of two pairs of wheels, each pair beneath one of the engines. Prominent bracing struts extended from the base of each engine nacelle. These were V-shaped and attached to the base of each wheel axle. The armament actually fitted to the prototype (if any) is not known but a model of the project shows several machine guns and a cannon were being considered.

Construction apparently was still under way at the time of the Armistice. The Coanda-S.I.A.bomber was subsequently evaluated in 1919 by the STAé and a favorable report was issued. However, probably as a result of the Armistice and the entry into service of the Caudron C.23, the S.I.A. bomber was canceled.

Above: The S.I.A. bomber was intended for use as a heavy bomber. It is shown here near the time of its completion on 17 April 1919. MA21086.

Right: The S.I.A. was powered by two 400-hp Liberty engines. MA36411.

S.I.A. BN2
(Provisional)

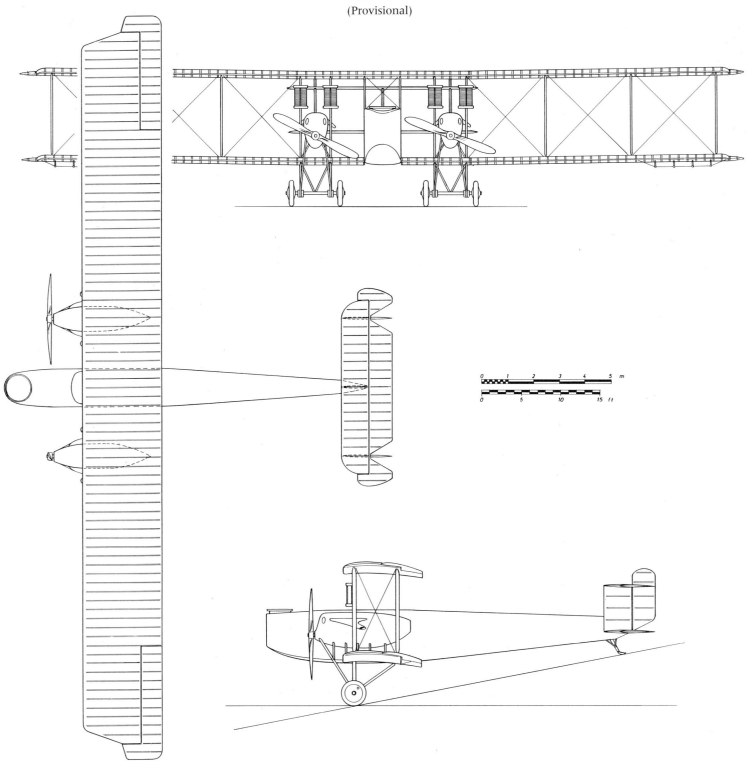

S.I.A.-Coanda Lorraine Bomber

Apparently this aircraft was developed in parallel with the Liberty-engined S.I.A. bomber described above. It had identical dimensions and performance data except for being powered by two 400-hp Lorraine engines. It is probable that two parallel designs were necessitated by the engine shortage at that time, bombers being given low priority for available engines. In any event, the Coanda-S.I.A. bomber actually built was fitted with Liberty engines and it seems likely that the Lorraine-powered version was never completed.

Sikorsky Bomber

Igor Sikorsky arrived in France in 1917 after the Russian Revolution had begun. Supplied with a letter of introduction from the chief of the French military mission to Russia, Sikorsky met with the representatives of STAé. He was asked to use his talent for producing large aircraft (such as the Ilya Muromets) to design one capable of carrying a 1,000-kg bomb load (BN class). It has been reported that Sikorsky examined parts of German bombers that had been shot down over Paris, searching for any technical advances over his earlier designs.

The final design was to have been powered by two 400-hp Liberty engines, although another source states that four 300-hp Hispano-Suizas were planned. Probably due to the unavailability of these engines, Sikorsky was ordered to use four 180-hp Hispano-Suizas.

Five aircraft were ordered but the Armistice ended the urgent need for a night bomber. The project was terminated before a single aircraft could be completed, and Sikorsky left for the United States a few months later.

Sopwith

Sopwith 1½ Strutter (Sop 1 A2, 1 B1, and 1 B2)

The Aviation Militaire faced a dual crisis in 1916. The army cooperation escadrilles were equipped with ineffective Caudron G.4s and Farman F.40s, while the bomber escadrilles were still using obsolescent M.F.11s and Voisin 5s. French-built replacements for both categories were planned, but these would not become available until mid-1917. What was needed was a proven type that would be immediately available for mass production and fill the gap until more modern types could enter service.

The French decided to build British Sopwith 1½ Strutters under license until more sophisticated types, such as the Breguet 14 A2s and B2s, could enter front-line service. While waiting for full-scale production to begin, the French acquired nine single-seat and two-seat examples of the Sopwiths from No.3 Wing RNAS. The first was a two-seater with serial number 9413. The other eight were single-seat bombers with serials 9651, 9655, 9657, 9661, 9664, 9666, 9720, and 9742.

French production was, yet again, to fall far behind the optimistic projections of the manufacturers. These delays may have had more to do with national pride than any major flaws in the aircraft or manufacturing process. The French claimed to have found numerous deficiencies in the design and insisted that it was structurally unsafe. The latter opinion was based on an incident in February 1917 when a bomber flown by a French pilot disintegrated in mid-air. The British felt that the machine had been over-stressed by the pilot; the French believed structural failure was to blame. Since the type was now being built by the French, it is not surprising that the modifications were introduced, causing further production delays. A GQG memo dated 22 November 1916 reported that tests of the French-built 1½ Strutters had been completed. The upper wings could withstand a weight of 5,680 kg with a load coefficient of 7.25. These results were "comparable to those attained by those aircraft built by the English." The pilots were warned that,

Sopwith 1½ Strutter 1B1 single-seat bomber. The winged serpent was the insignia of SOP 107. B 75.4292.

Sopwith 1A2 floatplane. Floatplane versions of the Sopwith 1½ Strutter were built by Hanriot. A total of 17 Sopwith 1½ Strutter floatplanes were delivered from January to September 1917. MA18249.

although these tolerances were acceptable for an army cooperation plane, they were unacceptable for fighters. The message was clearly given that the Sopwith pilots were not to treat their new machines as fighters.

The engines used were the 135-hp Clerget 9Ba or Bb, 145-hp Clerget 9Bc, or 135-hp Le Rhône 9Jby. Performance data is given below. The STAé designation was Sop. 1 A2 for the reconnaissance machine and Sop 1 B1 or B2 for the bomber.

A major modification to the 1½ Strutter was the fitting of a smaller empennage. This resulted in improved performance, as noted below. It is not known how many of the modified 1½ Strutters were built, but while the modified Sopwiths had a superior performance at low altitude, at high altitude they were less effective than the standard model.

A total of approximately 4,200 Sopwiths were built in France by Amiot, Bessoneau, Darracq, Lioré et Oliver, Hanriot, Sarazin, S.E.A., and R.E.P. Floatplane versions of the Sopwith 1½ Strutter were built by Hanriot. A total of 17 floatplane versions of the Sopwith 1½ Strutter were delivered between January and September 1917.

Operational Service
Army Cooperation

There seems to have been a tremendous amount of excitement on the part of the army cooperation escadrilles about receiving the new planes. First priority was given to reconnaissance units, then fighter escadrilles (to use as reconnaissance planes), followed by artillery aviation units. The fighter and artillery escadrilles were to receive either four Sopwith 1½ Strutters or Morane Saulnier Type Ps, and the Sopwiths were to be given to only the best pilots in each army cooperation escadrille. The 1½ Strutters seemed to have performed well, although a GQG memo dated 30 August 1917 called for the undercarriage axle to be strengthened.

The Sopwiths manufactured in France did not become available until the spring of 1917, by which time the type was obsolescent. By 1 August 1917 there were 142 Sopwith 1 B1 and B2 and 243 Sopwith 1 A2 aircraft in service. Eighteen reconnaissance escadrilles used the Sopwith 1½ 1A2s. They were:

SOP 5, formed from F 5 in September 1917, assigned to the 33rd C.A. and commanded by Capitaine de Peyronnet. It re-equipped with Salmson 2s in July 1918.

SOP 7, created from MF 7 in mid-1917, equipped with Sopwith 1½ Strutters and a few Caudron R.4s and assigned to the 6th C.A. Based in the 5th Armée sector and commanded by Capitaine Saqui-Sannes, SOP 7 was involved in the battle of Chemin des Dames. It subsequently moved to Plessis-Belleville where it re-equipped with Breguet 14 A2s during 1917.

SOP 9, created from C 9 with Caudron G.4s in July 1917. Assigned to the 30th C.A., it moved to Aisne, where it participated in the offensive at La Malmaison. It subsequently moved to Mont-Saint-Martin and the Moissy-Cramayel, where it re-equipped with Breguet 14 A2s in May 1918.

SOP 13, formed from C 13 with Caudron G.4s in September 1917, assigned to the 2nd Armée in the Verdun sector and commanded by Capitaine Pecquet. It re-equipped with Salmson 2s early in 1918.

SOP 17, formed from C 17 (with Caudron G.4s) in October 1917. It was assigned to the 1st C.A. until it re-equipped with Salmson 2s in June 1918.

SOP 24, formed from F 24 (with Farman F.40s) in 1917. It was assigned to the 5th Armée and re-equipped with Salmson 2s in March 1918.

SOP 28, formed from C 28 during 1917. It was assigned to the 2nd C.A. and beginning in 1918 was commanded by Lieutenant Seyer. It re-equipped with Salmson 2s early in 1918.

SOP 36, formed from F 36 in July 1917. It was assigned to the 31st C.A. and was sent to Italy in November as part of the 10th Armée (see introduction). In March 1918 it was assigned to the 12th C.A. in Italy. It returned to France in April 1918 and was re-equipped with SPAD 16s in June 1918.

SOP 39, formed from C 39 (with Caudron G.4s) in October 1917. Assigned to the 38th C.A., it was based at Rosnay when it re-equipped with Salmson 2s in February 1918.

SOP 43, formed from C 43 in February 1917. It was assigned to the 47th D.I., and was based at Bulainville in April 1917. It was under command of Capitaine Simon. SOP 43 converted to Breguet 14s in June 1918.

SOP 51, formed C 51 with Caudron G.4s in February 1917. Based at Frétoy and assigned to the 1st C.A.C., it was initially commanded by Capitaine Houdry (May to August 1917) and then by Capitaine Rocard (August to December 1917). It became SAL 51 in December 1917.

SOP 55, created from F 55 (with Farman F.40s) in late 1917. It was assigned to the 11th C.A. and based near Soissons. It converted to SPAD 16s in September 1918.

SOP 60, formed from F 60 in late 1917. It was assigned to the 14th C.A. and converted to SPAD 16s in June 1918.

SOP 61, formed from C 61 in September 1917. It was assigned to the 34th C.A. and converted to Salmson 2s in early 1918.

SOP 104, formed from in May 1917 from C 104. It was assigned to the 2nd Armée and later to the 31st C.A. at Montagne. It converted to Breguet 14 A2s in May 1917.

Paul Schmitt P.S.7

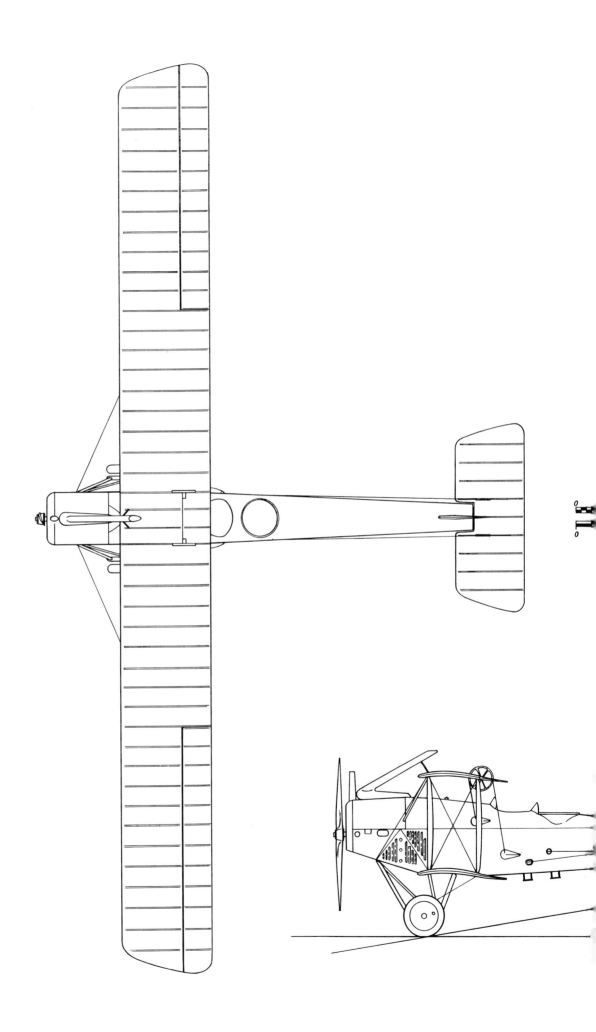

466B

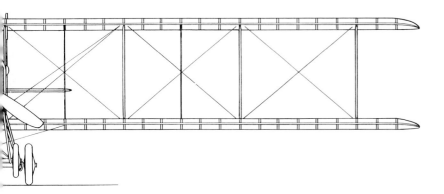

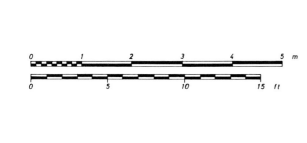

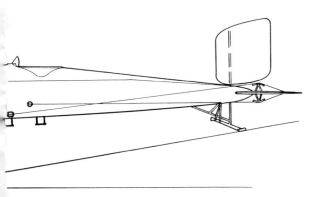

Paul Schmitt P.S.6

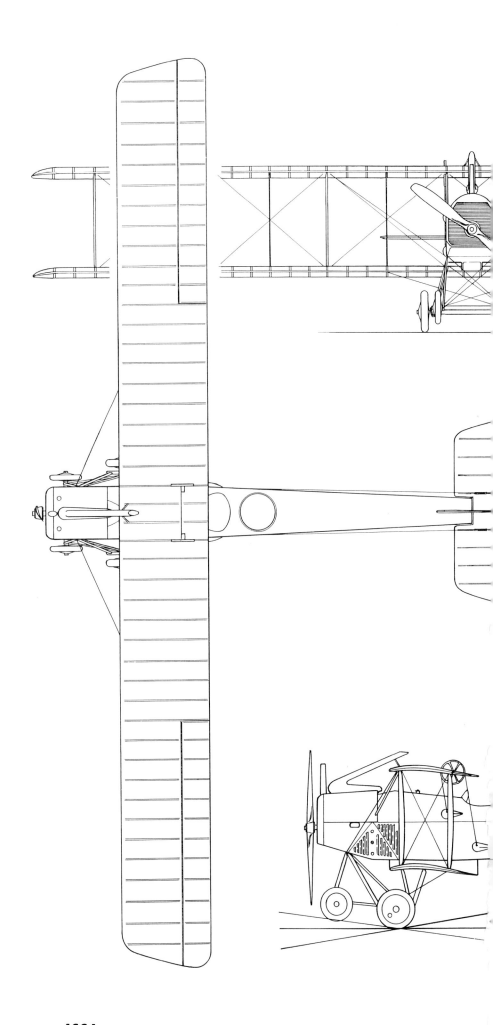

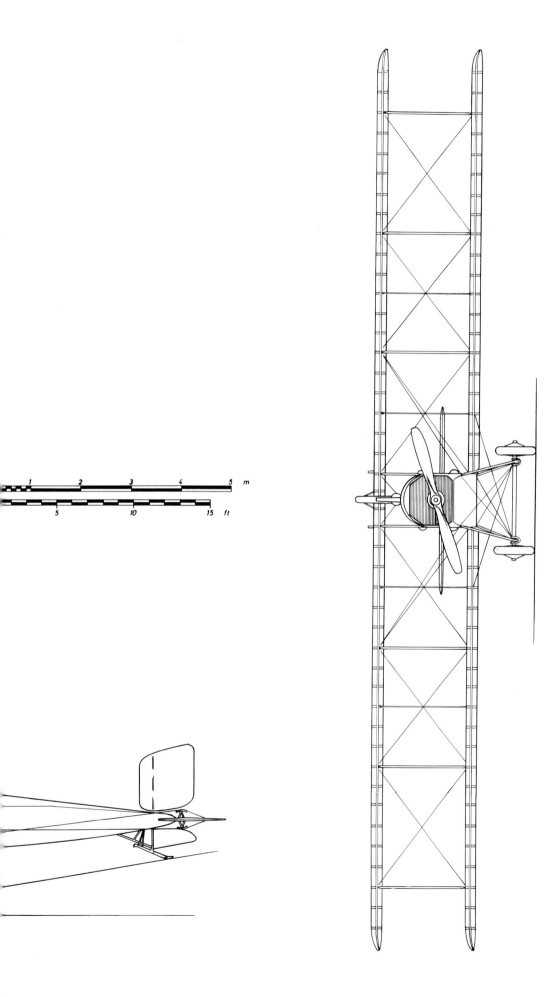

Sopwith 1A2 Strutter of SOP 43 undergoing gun testing. SOP 43 was assigned to the 47th D.I., and was based at Bulainville in April 1917. B79.3012.

SOP 105, created from VB 105 in October 1917. It was assigned to the 4th Armée and was based at Rosnay. Later it was assigned to the 5th C.A. and participated in the battles at Noyon (March and April 1918), after which it converted to Salmson 2s.

SOP 106, formed from C 106 in early 1917. It was assigned to the 21st C.A. and based at Corcieux in April 1917. It re-equipped with Salmson 2s in May 1918.

SOP 141, formed in December 1917. It was active in the Battle of the Somme and was based at Esquennoy. It was assigned to the 6th CA and moved to Lunéville in May 1918. In August 1918 it re-equipped with Breguet 14 A2s.

Artillery Cooperation Escadrilles

Forty-one artillery escadrilles used the Sopwith 1½ 1A2s. They were:

SOP 204, formed from F 204 in October 1917. It was assigned to the 6th C.A. and was serving in the 2nd Armée sector when it converted to Salmson 2s in February 1918.

SOP 206, created from F 206 in October 1917. It was assigned to the 8th Armée and was active over the Alsace front. It was also active with the 10th Armée and was based at Porto Nuova. SOP 206 converted to Breguet 14 A2s in May 1918.

SOP 207, formed from C 207 in October 1917 and was active in the 4th Armée sector. It transitioned to Breguet 14 A 2s in April 1918.

SOP 208, created from F 208 in late 1917. It converted to Breguet 14 A 2s in February 1918.

SOP 212, formed from F 212 in early 1917 and changed to Caudron R.4s in March 1917 to become C 212 (it also had Morane-Saulnier Type Ps on strength).

SOP 216, formed from F 216 in 1917. In March 1918 it was assigned to the 6th Armée and was based at Saint-Amand. It re-equipped with Breguet 14 A2s in April 1918.

SOP 217, formed from C 217 in 1917 and assigned to the 4th Armée. In early 1918 it was based at Alger and commanded by Lieutenant Parmentier. In April it moved to Poix and participated in the battles at Picardie. It became BR 217 in May 1918.

SOP 219, formed from C 219 in July 1917. It was assigned to the 1st Armée during the Flanders offensive. In November it moved to the 3rd Armée sector. SOP 219 transferred to the 6th Armée sector in early 1918. In February it became BR 219.

SOP 221, formed from F 221 in 1917. Assigned to the 2nd Armée initially, it was based at Étampes in March 1918. It also participated in the French mission to Italy with the 10th Armée, based at Ghedi. It became BR 221 in May 1918.

SOP 222, formed from C 222 in 1917. In April 1918 it was assigned to the 10th Armée and was based at May-en-Multien. It received Breguet 14 A2s in May 1918.

SOP 223, created from F 223 in July 1917. It was based at Lemmes in the 4th Armée sector. From January to Mach 1918 it was assigned to an American division and in April 1918 it was assigned to the 21st C.A. and based at Fontaine. SOP 223 was re-equipped with Breguet 14 A2s in May 1918.

SOP 226, formed from C 226 in 1917. It was based at Mairy in April 1918 and was assigned to the 36th C.A. It re-equipped with Breguet 14 A2s in April 1918.

SOP 229, created from C 229 in March 1917. It was based at Mairy in April 1917 and converted to Breguet 14 A2s in May 1918.

SOP 231, formed from elements of SOP 222 in December 1917. It was assigned to the 88th Artillery Lourde and was based at Sacy. It converted to Breguet 14 A2s in April 1918.

SOP 232, formed in January 1918. It was assigned to the 1st

Sopwith 1A2 Strutter of SOP 216. SOP 216 was assigned to the 6th Armée and was based at Saint-Amand. B93.2627.

Sopwith 1A2

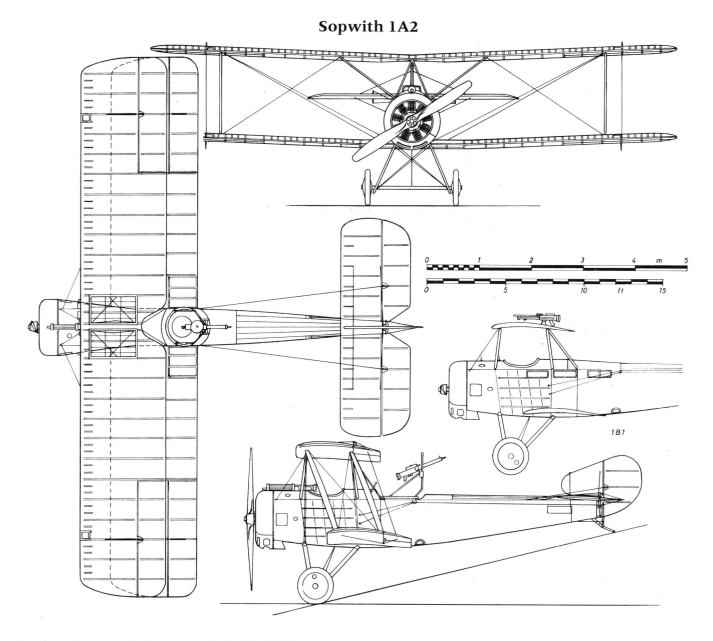

Armée and converted to Breguet 14 A2s in May 1918.

SOP 234, formed from SOP 229 in January 1918. It was assigned as to the 3rd Armée and based at Remy. It became BR 234 in May 1918.

SOP 235, formed from elements of SOP 207 in April 1918. However, as it re-equipped with Breguet 14 A2s in May 1918, it appears unlikely that this unit ever saw operational service with Sopwith 1½ Strutters.

SOP 236, formed from SOP 217 in March 1918. It was assigned to the 14th C.A. and was based at Fienvillers. It re-equipped with Breguet 14 A2s in May 1918.

SOP 237, formed from elements of SOP 223 in March 1918. It was assigned to the 9th C.A. and was based at Grandvillers. It re-equipped with Breguet 14 A2s in May 1918.

SOP 238, formed from elements of SOP 226 in March 1918. It was assigned to the 47th D.I. and was based at Rozoy-en-Brie. It became BR 238 in May 1918.

SOP 250, formed on 13 November 1917 at Étampes-Mondésir. Commanded by Lieutenant Sclafer, it moved to La Cense on 5 December. It was assigned to the 5th C.A. in the 6th Armée sector. It later moved to Courcelles-sous-Jouarre. In late March 1918 SOP 250 moved to Remy and supported the 9th and 10th D.I. before moving to support the 77th D.I. in April. It later moved to the 17th D.I. and was based at Moyencourt-lès-Poix. On 20 April it was placed under the control of the 18th D.I. In May SOP 250 moved to Luxeuil (May 27) and then Trécon (April 1) where it participated in Battle of Chemin des Dames. It re-equipped with Breguet 14 A2s in September 1918.

SOP 251, formed from elements of SOP 51 in November 1917. It was assigned to the 11th C.A. and was based at Mont-de-Soissons. It re-equipped with Salmson 2s in August 1918.

SOP 252, formed in November 1917. It was assigned to the 21st C.A. and was based at Dognerville. It re-equipped with Salmson 2s in May 1918.

SOP 255, formed in January 1918. It was assigned to the 1st C.A. and was based at La Cense. It became SPA-Bi 216 in June 1918.

SOP 260, formed from elements of SOP 51 in January 1918. It was assigned to the 1st C.A.C. and was based at Bouzy. It re-equipped with Breguet 14 A2s in June 1918.

SOP 263, formed from elements of SOP 60 in January 1918. It was assigned to the 14th C.A. and converted to Salmson 2s in May 1918.

SOP 269, formed from elements of SOP 28 in January 1918. It was assigned to the 2nd C.A. and was based at Ambrief. In July 1918 it became BR 269.

SOP 270, formed in February 1918 from elements of SOP 13. It was assigned to the 15th C.A. and was based at Villers-les-Nancy. It re-equipped with Salmson 2s in July 1918.

SOP 271, formed from elements of SOP 61 in February 1918. It

Sopwith 1A2. Armament was a fixed, synchronized 7.7-mm Vickers machine gun and a 7.7-mm Lewis machine gun on a swivel mount. Renaud.

Sopwith 1A2 landing at Venice as part of the French units assigned to the 10th Armée. MA27702.

was assigned to the 34th C.A. and was based at Remy. It became BR 271 in July 1918.

SOP 273, formed from SOP 39 in February 1918. It was assigned to the 38th C.A. and was based at Rosnay. It re-equipped with Salmson 2s in June 1918.

SOP 276, formed from elements of SOP 74 in February 1918. It was assigned to the 36th C.A. and was based near Bray-Dunes. It became SPA-Bi 276 in June 1918.

SOP 277, formed in February 1918. It was assigned to the 2nd C.A.C. and was based at Belrain. It re-equipped with Salmson 2s in May 1918.

SOP 278, formed from elements of SOP 18 in February 1918. It was assigned to the 30th C.A. and based at Mont-Saint-Martin. It became SPA-Bi 278 in June 1918.

SOP 279, formed from elements of SOP 24 in March 1918. It was assigned to the 2nd C.C. and based near Bray-Dunes. It re-equipped with Breguet 14 A2s in June 1918.

SOP 280, formed from elements of SOP 4 in March 1918. It was assigned to the 3rd C.A. and was based at La Cheppe. It became SAL 280 in June 1918.

SOP 281, formed from elements of SOP 56 in March 1918. It was assigned to the 17th C.A. and was based at Vadelaincourt. It re-equipped with Breguet 14 A2s in August 1918.

SOP 282, formed from elements of SOP 10 in February 1918. It was assigned to the 35th C.A. and based at Sacy-le-Grand. It became BR 282 in September 1918.

SOP 283, formed in March 1918. It was assigned to the 67th D.I.; and re-equipped with Breguet 14 A2s in June 1918.

SOP 284, formed in March 1918. It was assigned to the 30th C.A. and was based at Mont-de-Soissons. It was subsequently assigned to the American 3rd Army Corps and was based at Beauzee. SOP 284 re-equipped with SPAD 16s later in 1918.

SOP 285, formed from elements of SOP 9 in March 1918. It was assigned to the 34th C.A. and was based at Remy. It was later attached to the 62nd D.I. and re-equipped with SPAD 16s in July 1918.

SOP 287, formed from elements of SOP 104 in March 1918. It was assigned to the 68th D.I. and was at based at Beauzee. It was later reassigned to the 52nd D.I. and re-equipped with Breguet 14 A2s in June 1918.

SOP 582, formed in February 1917 and assigned to the Russian 7th Army. It began operations in May 1917, based at Boutchatch, Galicia. It was later transferred to Buzcacz and moved to the Romanian front, where it was assigned to the 3rd Corps. In February 1918 it was ordered back to Moscow and its planes given to the Russians. The French aircrew returned to France in 1918.

SOP 583 was based in Serbia.

By late 1917 the appearance of the Breguet 14, Salmson 2, and SPAD 11 permitted the units at last to receive more modern equipment.

Bombing

The Sopwith 1 B1s and 1 B2s were supplied to 11 bomber units in the spring of 1917. Two Farman units, F 29 and MF 123, received Sopwiths in September 1916. A Caudron escadrille, C 66, was re-equipped with 1½ Strutters in 1917. A single escadrille that used the deficient P.S.7 bombers, PS 128, finally received Sopwiths in July 1917. Three Voisin 8 units re-equipped with Sopwiths: VB 107, VB 108, and VC 111. Finally, four new escadrilles were formed on Sopwiths in mid-1917—SOP 129, 131, 132, and 134. These escadrilles were assigned these Groupes d'Bombardement:

GB 1: SOP 66 and SOP 111.
GB 3: SOP 29, SOP 107, SOP 108, SOP 128, and SOP 129.
GB 4: SOP 123, SOP 131, SOP 132, and SOP 134.

GB 1 was the first Groupe d'Bombardement to receive the Sopwith 1½ Strutters. GB 1 had two Voisin units and a single Caproni escadrille, which gave it a limited ability to undertake daylight bombing. The Sopwiths of SOP 66 and SOP 111 restored that capability. Both were equipped with both the single-seat (1 B1) and two-seat bomber versions (1 B2). The latter provided fighter escort for the single-seat bombers, which carried no defensive armament. In the 1 B1 variant the rear gunner's compartment was deleted to permit the internal storage of bombs. They were dropped through four bomb bay doors in the floor of the aircraft.

SOP 66 and 111 were active over the Aisne front in April 1917, bombing barracks, train stations, and marshaling yards. During these early raids it was common for several attacking

A lineup of Sopwith 1B1 Strutters of SOP 111 in July 1917. In the 1B1 variant the rear gunner's compartment was deleted to permit internal bomb storage. B79.3149.

aircraft to turn back, reportedly due to engine trouble. In May these attacks continued with escort provided by SPAD 7s. June was also a busy month for the units but, again, many aircraft returned due to mechanical difficulties; on 7th June 50 percent of the attacking force was forced to abort a bombing mission. Daytime attacks continued against train stations in July but now the Sopwiths were also being used for night missions. On 12 August SOP 66 and 111 were detached from GB 1 to serve on the Verdun front, assigned to support the troops who would be participating in that offensive. Aerial activity intensified as the date of the battle approached. The Sopwiths bombed train stations and camps along the front and encounters with German fighters occurred frequently. Night attacks were coordinated with the daylight raids so that German camps could be attacked around the clock. In September Sopwiths accompanied Voisin 8s on night missions. Attacks were launched against Frankfurt, but because the Sopwith bombers did not carry a second crew member to do the navigating, they often had difficulty finding the target, especially in bad weather. By 6 October both SOP 66 and 111 were re-equipped with Breguet 14 bombers.

GB 3, based at Cambrai, received two Sopwith escadrilles, SOP 107 and 108, in June/July 1917. From June through August the units usually had only between six and ten aircraft because of combat attrition and equipment failure. Frequently the single- and two-seat versions of the 1½ Strutters would fly in mixed formation with the two-seaters doing the navigating and providing air cover. The mechanics of GB 3 found the supply and performance of the engines were adequate; however, the airframes were too fragile. Furthermore, the bomb load was too small to do much damage. GB 3 moved from Cambrai to

Sopwith 1B1 showing the bomb rack, which was placed where the observer's cockpit was located in the 1A2. B88.1024.

Sopwith 1A2 serial 5047 belonging to Colonel Hamonic. MA7187.

Below: Sopwith 1B2 serial 3123. MA15430.

Sopwith 1A2 Strutter of the Japanese air service in 1920. In 1917 Japan purchased a number of Sopwith 1½ Strutters which had been built by the Lioré et Olivier firm. B85.1806.

Laon at the beginning of August and from this new base bombed train stations at Roules and targets near Houthulst. PS 128 retired its P.S.7 bombers and received Sopwiths by September. All three Sopwith escadrilles continued attacks on train stations in the vicinity of Cambrai. In October SOP 108 was detached to GB 1; the other two units continued attacks on train stations. SOP 128 switched to Breguet 14s in November, thus leaving SOP 127 the sole Sopwith 1½ Strutter unit attached to GB 3. SOP 127 continued to fly sorties until January 1918 when it, too, was equipped with the new Breguet 14s.

GB 4 received Sopwith 1½ Strutters in March 1917 when F 29 and 123 converted to the type. In April both escadrilles concentrated on targets at Colmar and in May also attacked Sieren and the airfield at Habsheim. GB 4 continued the practice of sending out mixed formations of single- and two-seat Sopwiths. Beginning on 11 July GB 4's units began a series of attacks on German airfields that continued during August. GB 4 was reinforced that month when SOP 129 joined the group.

Reinforced by SOP 129, both SOP 29 and 123 staged a large number of raids in September. French records claim that aircraft of GB 4 attacked German airship L.45 on 20 September; German sources, however, do not support this assertion. Raids on train stations continued for the remainder of the month.

November was to be an even busier month, with the units attempting to complete as many sorties as possible before winter weather brought an end to operations. In December depots and airfields were bombed. SOP 132 and SOP 134 joined GB 4 in January 1918, thus raising the number of Sopwith 1½ Strutter escadrilles to six. GB 4 moved to Luxeuil in mid-January. Gradually, however, the Sopwith units began to receive Breguet 14s and between January and April they frequently flew mixed formations of Sopwiths and Breguets. Targets attacked in January and February included Baden, Württemberg, the Mauser factory at Oberndorff, the gunpowder factory at Rottweil, the poison gas factory at Ludwigshafen, and airfields. By March the Sopwiths had been withdrawn from service and GB 4 became an all-Breguet 14 unit.

France's Aviation Maritime acquired several 1½ Strutters from the Aviation Militaire. Apparently four of them (probably with floats) were assigned to the Centre de Corfu and it is likely that others were assigned to naval bases. Floatplane versions of the Sopwith 1½ Strutter were built by Hanriot. A total of 17 floatplane versions of the Sopwith 1½ Strutter were delivered from January to September 1917. Postwar, some of these planes were initially assigned to the naval base at Saint Raphaël as part of escadrille AR 2. It was obvious that the machines were too obsolescent to form an operational unit but they proved to be useful for training. The Sopwiths were apparently easier to fly than the Salmson 2s and thus became the first aircraft to be assigned to the carrier *Bearn* in October 1920. The Sopwiths were replaced by Breguet 14s in 1922.

Foreign Service
Belgium
The Belgian air service purchased a number of Sopwith 1½ Strutters, at least three of them French-built machines. They were assigned to the 2nd, 3rd, and 4th Squadrons from 1916 through 1917. They were withdrawn from operational squadrons in October 1921.

Brazil
Three French-built Sopwith 1½ Strutters were sent to Brazil in 1919. They were used by the Escola de Aviacao Militar for liaison and army cooperation. They had serial numbers 3061, 3514, and 3633.

Japan
In 1917 Japan purchased a number of Sopwith 1½ Strutters built by the Lioré et Olivier firm. The army arsenals at Tokyo built 18 1½ Strutters under license.

Sopwith 1B1 single-seat bomber. MA4010.

Netherlands

A single French-built 1½ Strutter was interned on 22 April 1917. It had been built by Hanriot and carried serial number 115. It is believed that it was assigned to SOP 111. It was given the Dutch serial LA-45, changed to S-701 in 1918.

Russia/Soviet Union

Although most of the Sopwith 1½ Strutters used by the Russian air service and the RKKVF (Worker's and Peasant's Red Air Fleet) came from England, a few were license-built French versions. The Siberian Air Fleet of Admiral Kolchak had six 1½ Strutters on strength in 1919. The Siberian Air Fleet of the Far Eastern Republic had three 1½ Strutters in 1920; they had serial numbers 5, 6, and 7.

Sopwith 1½ Strutter 1B1 Single-Seat Bomber with 135-hp Clerget 9Ba or Bb

Span 10.2 m; length 7.7 m; height 3.125 m; wing area 32.25 sq. m
Empty weight 592 kg; loaded weight 975 kg
Maximum speed:
  1,000 m   160 km/h
  2,000 m   156 km/h
  3,000 m   147 km/h
  4,000 m   136 km/h
Climb:
  1,000 m   5 minutes 40 seconds
  2,000 m   15 minutes 20 seconds
  3,000 m   18 minutes 50 seconds
  4,000 m   31 minutes
Ceiling 4,725 m; endurance 3 hours 45 minutes
Armament: one fixed, synchronized 7.7-mm Vickers machine gun and 150 kg of bombs
Between 4,200 and 4,500 Sopwith 1½ Strutters of all types were built in France

Sopwith 1½ Strutter 1A2 Two-Seat Reconnaissance Aircraft with 135-hp Clerget 9Ba or Bb

Span 10.2 m; length 7.7 m; height 3.125 m; wing area 32.25 sq. m
Empty weight 597 kg; loaded weight 1,062 kg
Maximum speed: 158 km/h at 3,050 m; climb to 1,980 m in 12 minutes 40 seconds; ceiling 3,960 m
Armament: a fixed, synchronized 7.7-mm Vickers machine gun and a 7.7-mm machine gun on a swivel mount

Ukraine

A single Sopwith 1 A2 was obtained by the Ukrainians (presumably from Russia) in 1918. It carried serial number 1136.

United States

The Sopwith bomber had been the primary French day bomber for almost exactly one year, but as the Breguet 14s became available the French supplied the older Sopwiths to the American air service. The United States purchased 514 1½ Strutters in the spring of 1918. A total of 384 reconnaissance versions and 130 single-seat bombers were obtained. The reconnaissance (A2) machines had either 150-hp Le Rhône 9Jby

Sopwith 1½ Strutter 1B1 Single-Seat Bomber with 135-hp Le Rhône 9Jby

Span 10.2 m; length 7.7 m; height 3.125 m; wing area 32.25 sq. m
Empty weight 626 kg; loaded weight 926 kg
Maximum speed: 156 km/h at 3,000 m; climb to 2,000 m in 11 minutes; climb to 3,000 m in 18 minutes; ceiling 3,960 m
Armament: a fixed, synchronized 7.7-mm Vickers machine gun and 150 kg of bombs

Sopwith 1½ Strutter with Smaller Empennage and 135-hp Clerget 9Ba or Bb

Span 10.2 m; length 7.7 m; height 3.125 m; wing area 32.25 sq. m
Maximum speed:
  2,000 m   165 km/h
  3,000 m   157 km/h
  4,000 m   146 km/h
Climb:
  1,000 m   5 minutes
  2,000 m   12 minutes 20 seconds
  3,000 m   22 minutes 50 seconds
  4,000 m   38 minutes 10 seconds
Ceiling: 4,880 m

Sopwith 1½ Strutter 1A2 Two-Seat Reconnaissance Aircraft with 145-hp Clerget 9Bc

Span 10.2 m; length 7.7 m; height 3.125 m; wing area 32.25 sq. m
Empty weight 626 kg; loaded weight 926 kg
Maximum speed: 161 km/h at 3,000 m; 150 km/h at 4,000 m; climb to 3000 m in 24 minutes; climb to 4,000 m in 41 minutes; ceiling 3,960 m
Armament: one fixed, synchronized 7.7-mm Vickers machine gun and a 7.7-mm machine gun on a swivel mount

engines (236 examples) or the 130-hp Clerget 9Bc (148 examples). The single-seat bomber version had a 135-hp Clerget 9Ba (130 examples). Both versions were assigned to the 19th, 88th, and 99th Aero Squadrons, which were posted to the 4th, 3rd, and 5th Corps Observation Groups respectively. All the groups were assigned to the 1st Army. Apparently, none of the Sopwith bombers were assigned to the day bombardment groups. The Sopwiths were intensely disliked by the Americans, who felt they were obsolete and underpowered. They were quickly replaced by more modern types.

The U.S. Navy purchased at least four Sopwith 1½ Strutters directly from France. These had been built under license by the Hanriot firm. The only modifications to them was the addition of hydrovanes to the landing gear to make water landings safer. The Sopwiths were based at Moutchic and used for training. Two were subsequently sent to the base at Pauillac. One of these was sent to the United States, where it was given serial A 5660. After the war an additional 21 French-built Sopwiths were obtained from the U.S. Army. They were given serials A 5725–A 5728 and A 5734–A 5750 and used for observation. One was carried by the battleship USS *Texas* during 1919 maneuvers in the Caribbean.

Sopwith Dauphin

The French were pleased with the performance of the 300-hp Hispano-Suiza 8F engine and for the C1 specification of 1918 a number of aircraft were designed to use it (SPAD 18, 20, 21, and 22; SAB 1, Moineau, Semenaud, De Marcay 2, Descamps 27, Nieuport 29, and the Hanriot Dupont HD.7). The French had learned that the British were planning to place the engine in a Sopwith Dolphin and ordered an example of the Mark 2 Dolphin for evaluation. If the type proved satisfactory it was planned to produce it under license for use by the French and Americans.

The Dolphin (in French Dauphin) was a single-seat fighter with negatively staggered, two-bay wings. The Mark 1 was powered by a 200-hp Hispano-Suiza engine and an example (D 3615) was provided with the 300-hp Hispano-Suiza 8F in 1918. This aircraft was designated Dolphin 2 and featured a rounder cowling enclosing twin Vickers machine guns. Auxiliary mid-bay flying wires were fitted to the upper wing in the inboard bracing bay. An adjustable tailplane was also fitted.

The aircraft was test flown by the French and had excellent performance, but the war ended before the type could be combat tested. In any event, the Nieuport 29, which used the same engine, was selected for series production. At least one Sopwith Dolphin was built by the French firm SACA (Société Anonyme de Constructions Aéronautiques), presumably for evaluation, and, therefore, merits inclusion in this volume. The S.F.A. number of the French-built machine, 007, suggests that possibly as many as seven Dolphins were built by the French, but this cannot be confirmed. It had been anticipated that the SACA was to build as many as 2,194 Dolphin 2s for the American Expeditionary Force. However, the SACA failed to produce any machines before the Armistice and the American order was subsequently canceled.

A Sopwith Dauphin was possibly the first plane to fly with a 220-hp Hispano-Suiza engine fitted with a Rateau supercharger. It attained a speed of 210 km/h at 2,650 m and 192 km/h at sea level.

Sopwith Dauphin Single-Seat Fighter with 300-hp Hispano-Suiza 8F

Height 2.6 m; wing area 24.5 sq. m
Empty weight 710 kg; loaded weight 1,072 kg
Maximum speed: at 10,000 feet was 140 mph; climb to 10,000 feet in 8 minutes 20 seconds; ceiling 24,600 feet
Armament: two 7.7-mm Vickers machine guns
One example acquired by the French for evaluation; at least one was built by the SACA firm

---

Sopwith Dauphin (Dolphin) undergoing evaluation by the French. B81.2132.

Société Anonyme pour l'Aviation et ses Derives (SPAD)

In 1913 Armand Deperdussin was jailed on fraud charges and the existence of the Société Provisoire des Aéroplanes Deperdussin was endangered. The assets of the company were obtained by a syndicate headed by Louis Blériot. With original designs being produced by the firm's chief engineer, Louis Béchereau, a new company was formed. It retained the name SPAD; this time, however, SPAD stood for Société Anonyme pour l'Aviation et ses Derives. The Deperdussin factory built Caudron G.3s and G.4s under license while Béchereau began the manufacture of his own projects.

SPAD SA.1, SA.2, SA.3, and SA.4

Of all the problems that faced early aircraft designers, one of the most challenging was finding a way for combat aircraft to fire a fixed machine gun without damaging the propeller. The SPAD firm and Béchereau took a novel approach to this problem. The new design was intended to give the gunner an unobstructed field of fire ahead of the aircraft while maintaining the agility and safety (from rear attack) of the tractor configuration. This design was felt to be so promising that a French patent (No.498.338) was applied for on 27 February 1915. Originally, the plane was a tractor biplane with a gunner's nacelle suspended in front of the propeller by struts attached to the upper and lower wings. This made access to the engine too difficult and the design was modified (Addition No.22.088) to permit the nacelle to be hinged to the undercarriage struts.

The SPAD SA.1 was in many ways a conventional biplane. The fuselage was of wooden construction with four spruce longerons faired top and bottom with stringers. Later, lateral ducts ("cheeks") were added to either side of the engine compartment; these were later retrofitted to early versions of the A series when they were brought in for maintenance. It seems likely that these were to ensure adequate engine cooling. The fuel and oil tanks were mounted on top of the fuselage and in front of the pilot. The wings featured a novel design unique to the SPAD firm. At the intersection points of the landing and flying wires was a two-part articulated auxiliary strut. The genius of this design was that it permitted a biplane with large span wings to be braced as a single-bay structure. This gave the appearance of a two-bay aircraft, although the A series actually had only a single bay of struts. Otherwise, the wings were quite conventional, of equal span and chord with ailerons on both upper and lower wings. The engine was a 80-hp Le Rhône mounted in the nose and in front of this the gunner's nacelle (or "pulpit") was attached to the landing gear. This arrangement enabled the nacelle to be moved out of the way to start the engine (although it could be started with the nacelle in place) or for maintenance. The nacelle was built of ash longerons and covered with plywood. In addition to the struts to the landing gear, the nacelle was supported by a bearing attached to the rear bulkhead and seated on the propeller shaft. An additional attachment came in the form of two struts from the top wing to an L-shaped pylon built into the aft bulkhead. Armament consisted of a 7.7-mm Lewis machine gun mounted on a vertical track. The tail had wood framing and a steel tube elevator spar. The tail was of the same triangular outline used in later SPAD designs.

To facilitate communication between gunner and pilot there was a communication tube passing from the gunner's compartment through the propeller hub to the pilot's cockpit. This was of critical importance because the crewmen were separated from each other by a noisy engine and, while the gunner could fire the gun, it was the pilot who had to maneuver him into position to use it.

The SPAD SA.1 was evaluated in May 1915. Complaints about the type included inadequate cooling because of the location of the nacelle and excessive vibration in the gunner's nacelle. The excellent field of fire was praised. It was suggested that better results might by obtained if the gunner was replaced by four fixed machine guns. In fact this modification was carried out and became the Type G. (See below.)

The flight tests recorded the following data: span 9.10 m; length 7.30 m, height 2.60 m; area 21.35 sq. m, maximum speed 150 km/h; climb to 500 m in 2 min. 10 sec.; 1,000 m in 4 min. 33 sec.; 1,500 m in 5min 45 sec; 2,000 m in 7min. 40 sec.; 2,500 m in 11 min. 35 sec. ;and 3,000 m in 16 min. 5 sec.

Although the SPAD SA.1 was considered a successful design, it was not produced in large numbers (approximately ten were

SPAD SA.1, serial number 1, at Châlons in 1915. B85.406.

Above: SPAD SA.1 at Buc airfield. B85.576.

Above right: SPAD SA.1 at the R.G.A. The front cockpit could be lowered to service or start the engine. B79.3181.

Right: SPAD SA.2 in 1915. MA4016.

Below: SPAD SA.1 serial S.3. B88.2253.

built) because an improved version, the SA.2, became available.

The SA.2 was powered by a more powerful 110-hp Le Rhône 9J engine. The only other differences from the SA.1 consisted of minor changes to the nacelle attachments, gun mount, and horizontal tail surfaces (which now had parallel leading and trailing edges). The prototype flew on 21 May 1915. A total of 42 SA.2s were ordered by the Aviation Militaire; an additional 57 were purchased by the Imperial Russian Air Service. However, Shavrov identifies the Russian SA.2s as being fitted with 80-hp Le Rhône engines. The wings of some, if not all, Russian SA.2s and SA.4s had a center-section, unlike the French machines, which had the two upper wing panels joined in the center.

The SA.2s were never formed into a single escadrille by the French, but were instead divided among the reconnaissance units to provide fighter escort. It is clear that the SA.2 was not liked by the French crews. The gunners felt that the pulpit position was hazardous and meant certain death if an accident should occur during landing—the gunner would almost certainly be crushed by the engine. The occupant of the nacelle also was at risk of injury or death from the adjacent propeller. In addition, the pilots noted that the cockpit, which was located in the middle of the fuselage, meant that the view was very poor, especially during landing.

A SPAD SA.2 with the gunner's cockpit lowered to show the engine. Note the screen behind the gunner's cockpit and the guide for the machine gun. B85.1155.

SPAD SA.2 & SA.4

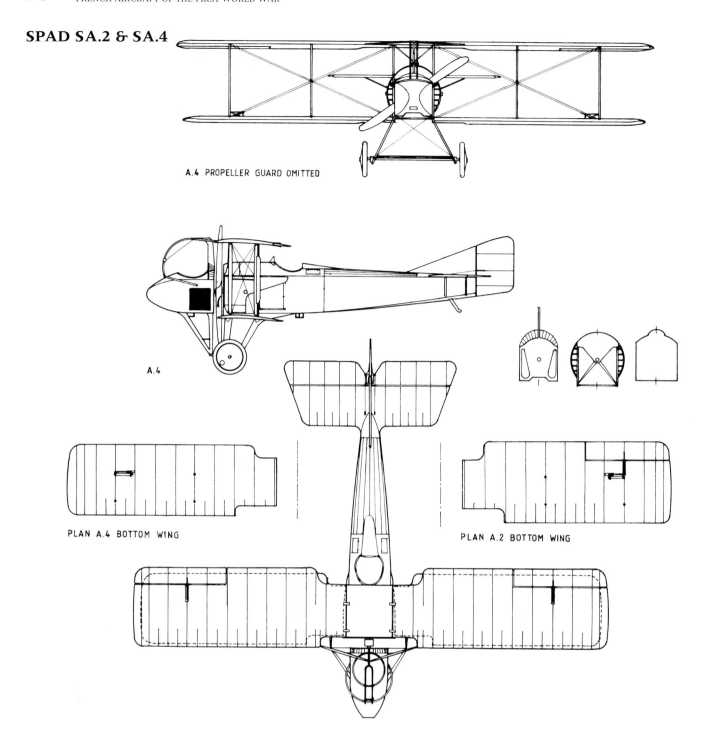

The SA.2s had a short life in French service. By 4 February 1916 there were only four still in front-line service with an additional five in service with training units. N 49 had only one SA.2 on strength in late March 1916; less than two months later this aircraft was no longer listed with N 49.

Although the SA.2 had only briefly served at the front, the SPAD firm continued a number of developments and modifications to its basic layout. The SA.3 was similar to the SA.2 but featured two gunners, one in the nose pulpit and another in a separate cockpit behind the wing. The rear gunner was in the position occupied by the pilot in the SA.1 and SA.2. Indeed, the only major change was the fitting of dual controls and providing a rearward-firing machine gun. This certainly represented a logical development of the A series, since one of the advantages of the tractor layout was that it permitted a rearward-firing gun for defense against stern attacks. The sole example of the SA.3 was given the serial number S.40. Although the SA.3 was not pursued further, the basic layout for a front and rear gunner was repeated in may of Béchereau's subsequent designs, including the SPAD C, D, E, and F. The idea of gunners with dual controls would also appear in the contemporary Ponnier twin-engine fighter (see above).

The SPAD SA.4 was an SA.2 airframe fitted with an 80-hp Le Rhône engine, possibly because of cooling problems with the 110-hp Le Rhône 9J. The other basic change was that ailerons were fitted to the top wing only. Also, the wings were positioned a few millimeters further aft than on the SA.1 and SA.2; this may have been intended to offset the tail-heaviness found on the earlier SPADs. The first flight of the SPAD SA.4 was on 22 February 1916. Only 11 were built. One was evaluated by the Aviation Militaire and the remaining ten were purchased by the Imperial Russian Air Service. However, Shavrov identifies the Russian SA.4s as being fitted with 110-hp Le Rhône 9J engines. It is possible that Shavrov had confused the SA.2 with the SA.4 by assigning the more powerful engine to the later type. Another possibility is that these designations were unique to the IRAS. In any event, this was the last variant of the SPAD SA series to see combat service. While the type was shunned by its

SPAD SA.2 serial S.19 named *Ma Jeanne*. The machine gun was mounted on a double-track mounting. B82.4228.

Above: SPAD SA.2 serial S.17 with pillar-type gun mount. Via Colin Owers.

SPAD SA.3 serial S.40. In this aircraft both the pilot and gunner had flight controls. B83.3460.

Above: SPAD SA.4 of the IRAS.

Left: Closeup of the SPAD SA.4 engine installation.

aircrews and rapidly withdrawn from service, many of the innovations of the series found their way into Béchereau's elegant and far more successful SPAD 7 and 13.

The STAé system of designations for SPAD aircraft are not known, but may have been as follows:
SPAD SA.1 = SPAD 1
SPAD SA.2 = SPAD 2
SPAD SA.3 = SPAD 3
SPAD SA.4 = SPAD 4

Foreign Service
Russia

In Russian service the SPAD SA.2 and SA.4 were known as SPADs "with cabin." Shavrov estimates that 50 SA.1s and six SA.4s were used by Russia, while other sources state that 57 SA.2s and ten SA.4s were obtained. There were victories credited to pilots who flew the SPAD SA.2s and 4s, A.A. Kozakov and observer-pilot Yu. A. Bratolyubov being one such successful team. It was reported that "insufficient construction" of the forward cockpit resulted in the pulpit falling off when the mounting had been damaged by bullets. This appears to be the main reason the type was disliked by Russian pilots. There was a report that one gunner's neck was broken when his scarf became caught in the propeller.

SPAD SA.1 Two-Seat Fighter with 80-hp Le Rhône 9C
Span 9.55 m (other sources state 9.10 m); length 7.29 m; height 2.60 m; wing area 25.36 sq. m
Empty weight 421 kg; loaded weight 708 kg
Maximum speed: 135 km/h at sea level; endurance 2.75 hours
Armament: one 7.7-mm Lewis machine gun mounted on a swivel in the gunner's pulpit
Approximately 10 built

SPAD SA.2 Two-Seat Fighter with 110-hp Le Rhône 9J
Dimensions same as SA.1 except length 7.85m (other sources state 7.30 m)
Empty weight 414 kg; loaded weight 674 kg
Maximum speed: 140 km/h at sea level; climb to 1,000m in 6 minutes 30 seconds; climb to 2,000 m in 12.5 minutes; climb to 3,000 m in 23 minutes 30 seconds; ceiling 4,000 m; endurance 3 hours
Armament: one 7.7-mm Lewis machine gun mounted on a swivel in the gunner's pulpit
A total of 52 were built for the Aviation Militaire

SPAD SA.2 Two-Seat Fighter with 80-hp Le Rhône 9C in Russian Service
Span 9.55 m (other sources state 9.10 m); length 7.3 m; height 2.60 m; wing area 25.3 sq. m
Empty weight 535 kg; loaded weight 815 kg
Maximum speed: 112 km/h; climb to 1,000 m in 8 minutes; climb to 2,000 m in 20 minutes; ceiling 3,000 m; endurance 2 hours
Armament: one 7.7-mm Lewis machine gun mounted on a swivel in the gunner's pulpit
A total of 47 were built for the IRAS

SPAD SA.3 Two-Seat Fighter with 110-hp Le Rhône 9J
Dimensions same as SPAD SA 2
Armament: two 7.7-mm Lewis machine guns mounted on swivels; one in the forward gunner's pulpit and the second in the rear gunner's cockpit
One built

SPAD SA.4 Two-Seat Fighter with 80-hp Le Rhône 9C
Span 9.55m; length 7.85 m (other sources state 7.35 or 7.290 m); height 2.60 m
Maximum speed: 154 km/h at sea level; climb to 1,000 m in 5 minutes; climb to 3,000 m in 23.5 minutes; endurance 2.5 hours
Armament: one 7.7-mm Lewis machine gun mounted on a swivel in the gunner's pulpit
One built for the Aviation Militaire

SPAD SA.4 Two-Seat Fighter with 110-hp Le Rhône 9J in Russian Service
Span 9.5 m; length 7.3 m; wing area 25.3 sq. m
Empty weight 565 kg; loaded weight 960 kg
Maximum speed: 135 km/h; climb to 1,000 m in 6 minutes; climb to 2,000 m in 12.3 minutes; climb to 3,000 m in 34 minutes; ceiling 3,500 m; endurance 2 hours
Armament: one 7.7-mm Lewis machine gun mounted on a swivel in the gunner's pulpit
A total of 11 were built for the IRAS

SPAD SA.4 of the IRAS fitted with skis. The SPAD SA.4 was an SA.2 airframe fitted with an 80-hp Le Rhône engine and had ailerons on the top wing only. B85.2601.

The No.1 Fighter Squadron and the 2nd Guard Air Squadron had SPAD SA.2s on strength. The latter unit had three SA.2s in June 1917; these had serials 95, 98, and 100. The 1st Turkastan Air Squadron of the 11th Air Division had two SA.2s and one SPAD SA.4 with serials 54, 78, and 68 (or 63).

As of April 1 1917 there was a total of 25 SPAD SA.2s and SA.4s still in service. This represented more than ten percent of the Russian fighter force at that time. It would appear that at least a few of the SA.2s and SA.4s saw service during the Russian civil war. As of 9 December 1917 a total of 19 SPAD fighters of all types, including SA.2s and SA.4s, were listed on strength.

SPAD SB

The SPAD SB designed by Béchereau was a monoplane with a crew of two. The construction was simplified to permit rapid series production. The fuselage was to be molded from stamped metal and the two halves joined. The engine, landing gear, fuel tank, seat, and wings were then to be attached to the fuselage. It was obviously intended as a means of providing the Aviation Militaire with a modern military aircraft that could be mass produced. It is not known if it was meant primarily for the reconnaissance or the fighter role. The SB was probably never built and may have remained a 1914 patent only. Unfortunately for the SPAD firm, Béchereau's design did not appear until 1915. The monoplane at that time had fallen into disfavor and only biplanes were being ordered in large numbers. Consequently the SPAD SB was not selected for production.

SPAD SC

The SPAD SC first appeared in August 1915 and seems to have been a more powerful version of the SPAD SA.3. It was powered by a 220-hp Renault engine (as opposed to the SA.3's 110-hp Le Rhône engine) and had a crew of three: a pilot in the center of the fuselage, a gunner in a pulpit in front of the propeller, and a second gunner in the rear fuselage behind the pilot. Although it is likely that the more powerful engine resulted in an improved performance over the SPAD SA.3, by this time the SA series was held in contempt by French aircrews because of the inherent danger of the pulpit design as well as the type's poor performance. The prototype of the excellent Nieuport 11 had also appeared in the summer of 1915 and the Aviation Militaire selected this more modern design to fill the fighter requirement.

SPAD SD

The SPAD SD was a bomber design built in March 1915. It retained the same basic layout as the SA series of fighters, which meant it was a tractor biplane with a gunner situated in a "pulpit" in front of the engine. It had a crew of four: a gunner in the nose, a gunner behind the cockpit, and two pilots seated side by side. The engine was a 250-hp Panhard. The SD was a large biplane and photos believed to be the SD show that it had a two-bay wing. It appears that a portion of the upper wing featured a rectangular frame extending over the engine housing but stopping short of the roll-bar at the rear portion of the gunner's nacelle. The aircraft was also fitted with a bomb rack. The type was intended for the 1915 concours puissant (for a bomber capable of reaching Essen) but was not selected for production. The prototype was flown by Maurice Prevost during the competition.

Although it did not win the 1915 concours, the SPAD firm decided to submit another bomber design for the 1916 competition. This design, the SPAD SE, was powered by two pusher engines, thus eliminating the unpopular pulpit position.

SPAD SD Four-Seat Bomber with 240-hp Panhard
One source gives the following dimensions: span 9.55 m, length 7.30 m, height 2.60 m. However, photographs suggest a considerably larger aircraft.
Armament would have been two machine guns: one in the front pulpit on a swivel mount and one on a swivel mount in the rear cockpit
One built

SPAD SD
(Provisional)

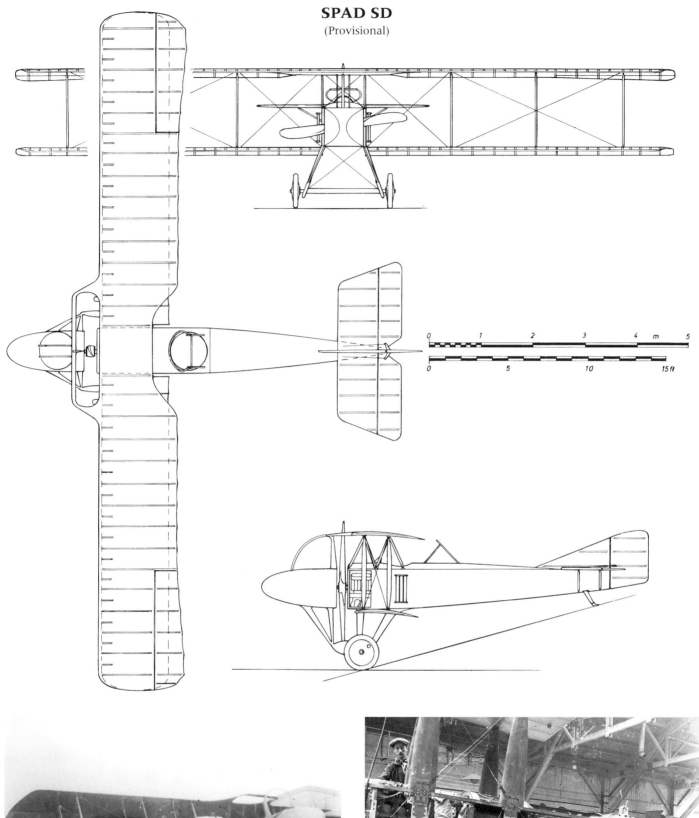

This aircraft is believed to be the SPAD SD. Blueprints available at the Musée de l'Air are very similar to the aircraft pictured.

SPAD SD undergoing construction. It was powered by a 240-hp Panhard engine. BOFFY #10.

SPAD SE

Although his most famous creations were to be fighters, Béchereau produced a credible bomber prototype in 1915. Despite the failure of the SPAD D, the SPAD firm decided to contribute a bomber design to the 1916 concours. However, Béchereau was given a very short amount of time to design and build it.

Béchereau's design was much larger than the preceding SPAD SD and was powered by two pusher engines. The front pulpit was eliminated, replaced with a more conventional machine gun position in the extreme nose. Another machine gun was located in the mid-fuselage area behind the wings. Two pilots were seated side by side in separate cockpits. The aircraft was powered by two 220-hp Renault engines, each mounted on one of the wings.

Both the Morane-Saulnier S and the SPAD SE were selected as winners of the competition, but only the Morane-Saulnier design was chosen for mass production. The SPAD SE was not developed further, possibly so the SPAD firm could concentrate on fighter designs.

> **SPAD SE Four-Seat Heavy Bomber with Two 220-hp Renault Engines**
> Span 24.30 m; length 13.0 m; height 3.45 m
> Payload 1,200 to 1,300 kg
> Maximum speed: 140 km/h
> Armament would have been one machine gun in the nose and one amidships on swivel mounts.
> One built

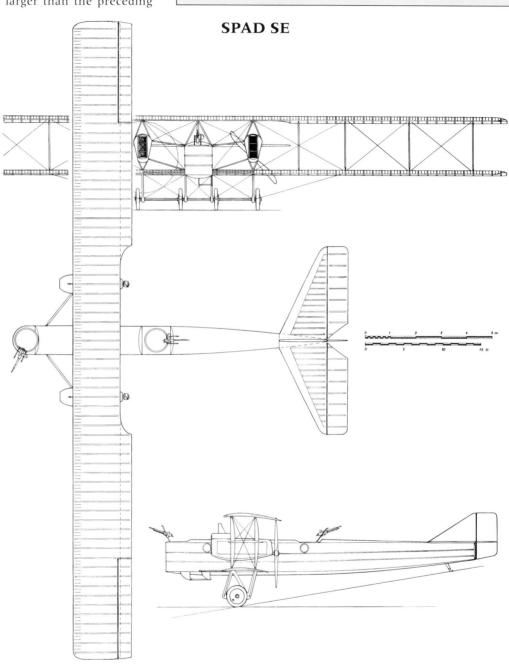

SPAD SE

The SPAD SE and the Morane-Saulnier S were the only aircraft able to meet the requirements of the bomber competition. BOFFY #7.

SPAD SE heavy bomber. The aircraft was powered by two 220-hp Renault engines mounted as pushers. MA92.

Below: The SPAD SE was entered in the 1916 concours puissant. BOFFY #1.

SPAD SF

The SPAD SF represented yet another attempt by the firm to design a fighter with a clear field of fire for the gunners. The SA series had used a pulpit design, which placed the gunner in a precarious location and partly for this reason had been only a limited success. The type SF, designed in February 1916, was probably intended to rectify this problem by moving the gunners as far away from the propeller as possible. The design featured a fuselage with two "tails," one in the front and one in the rear. This seems to have meant that it had a canard arrangement in the front and a second tail in the rear. It has been described in a contemporary publication as literally two biplanes joined "nose-to-nose" at their engines. There were machine gun positions in the nose and tail, which gave both gunners a clear field of fire. The two rotary engines were buried in the fuselage and drove two propellers located in the center of the fuselage (possibly in a similar arrangement as that used on the Dufaux fighter). The pilot's cockpit was also located in the fuselage. It is not known if this machine was ever completed or flown; the concept of a canard fighter with engine and propellers located in mid-fuselage was not further developed by the SPAD firm.

SPAD SG

The SPAD SG was a modification of the SA.2 series. The primary difference was that the SG had a fixed Hotchkiss infantry machine gun in the nose (although on the prototype four dummy machine gun barrels had been fitted). The SG also dispensed with the gunner, carrying only the pilot and a machine gun. The nacelle had been extensively modified to allow the Hotchkiss to be fitted. As there was no gunner, the gun was aimed by the pilot. The engine was a 110-hp Le Rhône 9J.

SPAD SG prototype modified from the prototype SA.1. The nacelle was extensively modified in order to allow a Hotchkiss to be fitted, being considerably shorter than that used on the SA series, and of circular cross section. The guns are mockups. MA13362.

SPAD SG Single-Seat Fighter with 110-hp Le Rhône 9J

Wing area 18.58 sq. m
Maximum speed: 161 km/h at 2,000 m; 154 km/h at 3,000 m; climb to 2,000 m in 7.25 min.; climb to 3,000 m in 11.5 min.
Armament: one fixed Hotchkiss machine gun with 1,000 rounds
One built

Above, left, and below: These three photographs, provided by courtesy of Tom Darcey and Alan Durkota, show an extensively modified SPAD SA fighter in IRAS service. Three machine guns were fitted in the gunner's position and were, presumably, fired by the pilot. The shape of the nacelle and the retained position of the front gunner suggests that this aircraft was not a true SPAD SG, but was a SPAD SA modified in the field.

Photographs of the prototype SPAD SG show it was otherwise quite similar to the SA series, although the pulpit/nacelle was considerably shorter and of circular cross-section. The engine was the same as the SA.1, an 80-hp Le Rhône. The prototype SG appears to have been a modification of a standard SA.1, although this has not been established.

The actual SG, which was significantly different from the prototype, was evaluated in April 1916. The deletion of the second crewman enabled a more powerful armament to be carried and the aircraft might have proved useful in the ground attack role; it was not likely to be very effective as an interceptor given the inferior performance of the SPAD SA series. It may have been felt that the Voisin 4 "canons" were adequate for the role the SG was intended to fill, as no orders were placed for the SPAD design. Some sources suggest that the Type SG may have been given the STAé designation SPAD 9.

At least one Imperial Russian Air Service SA.4 was modified in similar fashion to the proposed SPAD SG.

SPAD SH

The identity of the SPAD SH remains a mystery; it may have been the prototype for the SPAD 7 fighter. It is believed that the STAé designation "SPAD 5" was applied to this aircraft, which was built and tested in March 1916 powered by a 140-hp Hispano-Suiza engine. A development of the Type SH with an 18 square meter wing was built in April 1916. Neither type was selected for production. An official drawing of the Type SH was discovered by the authors at the Musée de L'Air and is the basis for the three-view.

A version of the SH may have been fitted with an 80-hp Le Rhône 9C in April 1916. It is not known if the aircraft was flown; one source suggests that this variant formed the basis for the SPAD 15.

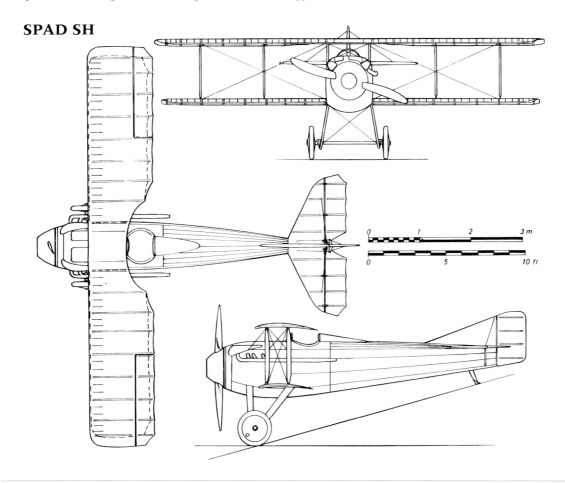

SPAD SH

SPAD SI

The SPAD SI was intended from the outset to use a new 130-hp rotary engine then undergoing development. The identity of this engine is not known. The Type I was a petite single-seat biplane with a fixed, synchronized machine gun. It was of monocoque construction. The rotary engine that had been intended for use by the SPAD SI never materialized and it would seem likely that the type was never flown.

SPAD SJ

The SPAD SJ was to be powered by a new 130-hp rotary engine then undergoing development (its exact identity is not known). The aircraft was a single-engine fighter with a synchronized machine gun. The fuselage was of monocoque construction. The most interesting feature of the Type SJ was that it was a monoplane—the first to be designed by Béchereau since the ill-fated Type SB. It is possible that the success of the Fokker monoplane had given Béchereau the impetus to try another monoplane design. The Type SJ may have been given the STAé designation SPAD 6. Since the rotary engine for which the Type SJ was intended never materialized it seems likely that the aircraft, which was constructed in April 1916, was never flown.

SPAD SK

The SPAD SK was a small twin-engine fighter intended to use the same rotary engine (the designation for which is not known) as the SPAD SI and the SPAD SJ. The engine was never completed so it seems likely that the Type SK was never flown. It has been reported that another SPAD design was also designated SK. This was a biplane with two 150-hp Hispano-Suiza engines, which suggests that the SK airframe may have been fitted with these engines because of the unavailability of the planned rotaries. This aircraft reportedly had a crew of three: a pilot and two gunners. It was undergoing flight testing in November 1916.

It is interesting to note that there were three separate and quite distinct designs intended to use the new rotary engine, which suggests that it held great promise. Although the failure to produce it may have seemed to be a setback for the SPAD firm, the company was to have its greatest success when it abandoned the rotary engine and turned to the fixed Hispano-Suiza V 8—the engine that would power the SPAD 7 fighters.

SPAD 5

The SPAD 7 was to become one of the most famous fighters of the First World War, yet its origins lay in the failure of three of Béchereau's earlier designs. The types SI, SJ, and SK all were designed to accept a new rotary engine that was never completed. Thus Béchereau, who had spent considerable effort to find a suitable airframe for the new engine, was left with three unusable airframes.

Fortunately for Béchereau a new engine was completing development—the 150-hp Hispano-Suiza 8Aa engine, lightweight and powerful. When it appeared in the late spring of 1915 the STAé quickly saw that it would be ideal as a power plant for fighter aircraft. In response to the request for a new fighter to use it, Béchereau designed a compact single-engine biplane incorporating a number of features found in his earlier aircraft, especially the SA and SH series.

The prototype may have been given the STAé designation SPAD 5, although this is not certain. Features of the aircraft that had been incorporated in the SPAD SA series included the patented two-part articulated auxiliary strut that gave the aircraft the appearance of a two-bay biplane. In addition, the basic airframe closely resembled that of the SA series. The undercarriage, control system, and tail were also based on the SA.2, SA.3, and SA.4. series. However, in the new SPAD the armament was a single synchronized Vickers machine gun, enabling Béchereau to dispense with the unwieldy gunner's nacelle. The airframe was extremely strong by virtue of its plywood engine bearers and main spars. On the prototype the propeller was covered by a large spinner and there was a circular radiator in an annular cowling. It may be that Béchereau's experience with the sleek nacelle used on the A series convinced him of the value of aerodynamic streamlining. Probably because of problems with engine cooling, the large spinner was deleted, although the annular cowling was retained.

Flight trials of the new SPAD took place during July 1916. It had an impressive performance; the top speed was reportedly 170 km/h and it could climb to 3,000 meters in 9 minutes.

The Aviation Militaire was enthusiastic about this aircraft and an order for 268 was placed on 10 May 1916. Despite the uncertainty over the SPAD 5 designation, it is known for certain that these production aircraft were designated SPAD 7.

SPAD 7

The production SPAD 5 was designated SPAD 7 (the STAé designation SPAD 6 possibly being applied to the SPAD SJ). The first came off the production lines in late 1916 and a total of 24 had been delivered by September of that year. From October through December 143 were built.

As the SPAD 7s began to reach operational units it was discovered there were some minor problems. The biggest was with engine cooling. It was found that the radiator system and cooling vents were inefficient, and the severe winter weather was taking a heavy toll on the new Hispano-Suiza engines. In addition to this problem the engines overheated in the summer. Early aircraft had a tight cowling and a small opening in the front for engine cooling. In the field, several (nine to ten) small holes were drilled into the top half of the cowling. As 1917 progressed the opening in the front cowling was greatly enlarged, thus obviating the need for the cooling holes. A GQG memo dated 1 October suggested that the spacers on the engine mounts be modified to give more space between the radiator and the front of the engine. It was hoped that this would improve cooling. The last production model of the SPAD 7 had the enlarged cowling opening and a system of nine vertical shutters to permit better control over the amount of air let into the engine.

Engine cooling was not the only problem. In a report on 11 November 1916 Capitaine le Reverand of the Ministere de la Guerre stated that a number of modifications to the SPAD 7 would be required. These changes were to the bracing of the fuselage (the wire bracing being simplified) and aluminum sockets were substituted for steel ones. The engine bearers were reinforced with steel plates, as it was found that there were frequent failures as a result of engine vibration.

An STAé memo noted that there had been criticism of Blériot's method of construction. To evaluate these complaints, a SPAD 7 built by Blériot was given a static test in which it supported 4,870 kg with a coefficient of 7.9. As the required coefficient for fighters was 6.0, this was more than enough to enable the Blériot-built SPAD 7s to enter service.

The SPAD 7 had a good performance and because it was well-constructed was able to absorb heavy punishment and dive rapidly. But these very qualities worked to the aircraft's disadvantage; it was so heavy that even with the 150-hp Hispano-Suiza engine it was less maneuverable than the aircraft it had been intended to replace—the Nieuport 17. This problem was solved by fitting the SPAD 7 with the new 180-hp Hispano-Suiza 8Ab. This made the SPAD 7 competitive with the newer German fighters. From the table given below it can be seen that the re-engined SPAD 7 was 16 km/h faster at 4,000 meters and could climb to 3,000 meters in 9 minutes and 50 seconds

SPAD 7 of SPA 124 (better known as the *Lafayette Escadrille*) at Ham in April 1917 with a Nieuport in the background. B83.3461.

SPAD 7 serial number S 1379 of SPA 65. MA6953.

compared with 11 minutes and 20 seconds for the earlier SPAD 7s. The service ceiling was increased from 6,000 meters to 6,553 meters. Endurance, however, was reduced by more than one hour. By early April only the newer engines were being fitted to SPAD 7s.

Variants

1. SPAD 7 with 150-hp Renault engine—Probably intended for further development if the Hispano-Suiza engine had proved to be a failure (Béchereau undoubtedly had remembered what delays in engine production could do to an aircraft design from his experience with the SPADs SI, SJ, and SK). The engine was of a different configuration from the Hispano-Suiza Ab as its cylinder banks were set at 60 degrees instead of 90 degrees. Considerable modifications had to be carried out to the nose in order to fit the Renault. Fortunately, the Hispano-Suiza had proved successful so the Renault was not needed.
2. SPAD 7 with reduced wing span—Tested by the STAé in November 1916, the aircraft had a span reduced to 7.315 meters, but increased area of 18 square meters. Its performance was inferior to the standard SPAD 7.
3. SPAD 7 with flat wings—Presumably an aircraft with a wing airfoil modified from the standard SPAD 7. It was tested in February 1917 but production did not ensue.
4. SPAD 7 with a 200-hp Hispano-Suiza 8B —At least two were tested with combat units but the type was never ordered into widespread production.
5. SPAD 7 with a 190-hp supercharged Hispano-Suiza engine—this aircraft was tested in 1917 and it has been reported that its performance was equivalent to the SPAD 7 with the Hispano-Suiza 8Ab.
6. SPAD 7 with "Blériot wing"—Possibly the same aircraft as the SPAD 7 with the flat wings.

Operational Service

By 25 February 1917 the number of SPAD fighters delivered was 268. The output, while far short of official expectations, improved when the aircraft was produced under license by Grémont, Janoir, Kellner, de Marcay, Régy, Sommer, and the SEA (Société d'Études Aéronautiques). By 1 August there were 495 SPAD 7s in service with 445 in the escadrilles and aircraft parks, 32 with the RGA, and 18 others. More than 50 escadrilles were using SPAD 7s at this time.

A GQG memo in April 1917 noted that planes were reaching front-line escadrilles covered with rust, with broken airframes, and damaged cowlings. It appears that the damage occurred in transit, however, and could not be directly traced to problems at the factory.

Delays in production of the SPAD 13 resulted in SPAD 7s remaining in service longer than had been anticipated. On 1 April 1918 there were 372 SPAD 7s and 290 SPAD 13s at the front, while the RGA had 99 SPAD 7s and 64 SPAD 13s. The aircraft parks had 36 SPAD 7s and 18 SPAD 13s. This gives a total of 507 SPAD 7s and 362 SPAD 13s. Between 1 April 1917 and 31 March 1918 a total of 1,200 SPAD 7s had been produced, just short of the anticipated number of 1,276. Few SPAD 7s were in service with the GDE units because so many were needed at the front. Many escadrilles still had examples on strength up to the end of the war. SPA 167, for example, was formed in September 1918 and equipped with a mix of 12 SPAD 13s and six SPAD 7s. Approximately 3,500 SPAD 7s were produced during the war.

The first aircraft to arrive at the front included a single example sent to N 49 on 7 November 1916 followed by a second machine sent to N 49 in December. It is interesting to note that there was no attempt to equip an entire escadrille with SPAD 7s. Rather, a few examples were sent to each unit and when, during 1917 enough machines were received, the escadrille would change its designation to SPA. It is known that the French fighter pilots were impatiently awaiting a replacement for their now obsolescent Nieuports. Therefore it is likely that the unit's best pilots (those with higher rank and scores) would have the first opportunity to fly the SPAD 7 in combat. This is supported by the fact that Georges Guynemer (then the highest-scoring French pilot), as well as the aces Paul Sauvage and Armand Pinsard, all were flying SPAD 7s before the end of 1916.

Army Cooperation Escadrilles

SPAD 7s were assigned to fighter units beginning in early 1917. SPAD 7 escadrilles assigned to individual Armées were:

SPA 23, formed from N 23 and assigned to the 2nd Armée.

SPA 38, formed from N 38 in late 1917 and assigned to the 4th Armée. It was commanded by Lieutenant Madon, then Capitaine Echard.

SPA 49, formed from N 49 in December 1917, assigned to the 7th Armée and commanded by Capitaine Charles Taver beginning in February 1918.

SPA 62, formed from N 62 in November 1917, assigned to the 6th Armée and based at Saint-Amond. In December Capitaine Blamauticr assumed command. SPA 62 was active over the Picardie front in march 1918.

SPA 68, formed from N 68 and assigned to the 8th Armée in the Lorraine sector at the end of 1917. It became part of GC 20 in February 1918.

SPAD 7 wearing the Griffin insignia of SPA 65. SPA 65 was assigned to GC 13. B83.4021.

SPA 69, formed from N 69 in October 1917 and assigned to the 10th Armée. At that time SPA 69 accompanied the 10th Armée to Italy and was commanded by Lieutenant Malavialle. It returned to France in March 1918.

SPA 76, formed from N 76 in late 1917 and assigned to the 5th Armée. It was commanded by Capitaine Perrins, then Capitaine Verdon, and finally Lieutenant Vitoux.

SPA 77, formed from N 77 in July 1917 at Manoncourt and assigned to the 8th Armée, joining GC 17 on 5 February 1918 and commanded by Capitaine Muranval.

SPA 79, formed from N 79 in January 1918, assigned to the 3rd Armée, and commanded by Capitaine Franc-Robert. It was based at Maisonneuve from January to March 1918.

SPA 82, formed from N 82 in March 1918, assigned to the 6th Armée and based at Vaubéron. It commanders were Lieutenant Echard until mid-1918, then Lieutenant Lecoq deKerland.

SPA 87, formed from N 87 in May 1918 (although it had SPAD 7s on strength long before then). It was assigned to the 8th Armée. The next month Lieutenant Joseph Jochaud de Plessis assumed command.

SPA 90, formed from N 90 in April 1918, assigned to the 10th Armée, and commanded by Lieutenant Weiss.

SPA 102 was how N 102 was re-designated in August 1918; however, it had SPAD 7s and 13s on strength long before then. It was assigned to the 1st and, later, the 6th Armée sectors. It was led by Lieutenant Derode until March 1918 when Capitaine Lemercier de Maisoncelle-Vertille de Richemont assumed command.

SPA 124, known better as the Lafayette Escadrille, received SPAD 7s in June 1917. It was assigned to the 6th Armée sector in the Aisne sector. In July it was assigned to the 1st Armée and in August moved to Verdun. In September it moved to the 6th Armée in the Aisne sector. In December 1917 N 124 was assigned to the 4th Armée. It disbanded in February 1918 to become the 103rd U.S. Aero Squadron.

SPA 150, formed from N 150 in December 1917 under command of Lieutenant L'Huillier. It was assigned to the 7th Armée and joined GC 16 in January 1918.

SPA 152, formed from N 152 in May 1917 and assigned to the 7th Armée.

SPA 153, formed from N 153 in November 1917 and assigned to the GC Provisoire de Bonneuil in the 3rd Armée sector. It joined GC 18 in January 1918.

SPA 156, formed from MS 156 in May 1918, assigned to the 4th Armée and under command of Lieutenant Jean Thobie and later Capitaine Paumier.

SPA 158, formed from MS 158 in May 1918. It was initially assigned to the 3rd Armée but joined GC 23 later in May.

SPA 160, formed from N 160 in May 1918 under command of Lieutenant Barès with both SPAD 7s and 13s. It was assigned to the 2nd Armée and joined GC 23 in August.

SPA 161, formed from MS 161 in May 1918 under command of Lieutenant Parfait and assigned to the 6th Armée at Maisonneuve. In June it moved to Champaubert, where it performed reconnaissance over the front.

SPA 313, formed from N 313 in June 1917 and initially equipped with SPAD 7s. It was commanded by Lieutenant Duron. This was a D.C.A. escadrille assigned to Dunkerque.

SPA 314, formed in July 1918 and assigned as a D.C.A. escadrille for Nancy, it was commanded by Capitaine Perrin.

SPA 315, formed from N 315 in July 1918 and assigned as a D.C.A. escadrille for Nancy, it was based at Chaux on the date of its conversion. It was commanded by Capitaine Etienne.

SPA 412, formed in June 1917 as the Escadrille de Tarcenay, which was the D.C.A. unit for Belfort.

SPA 441, formed as the Escadrille de Antilly and later designated the 304th D.C.A. The unit's assignment was to protect the town of Creusot. It became SPA 441 in July 1917.

SPA 442, formed in 1916 as the Escadrille de Meyzieu; the type of aircraft used at this time is not known. It was subsequently redesignated the 307th and assigned as a D.C.A. unit for Lyon. It became SPA 442 in July 1917.

SPA 461, assigned to the C.R.P., it became Escadrille 461 in July 1915 and then redesignated 95th D.C.A. It later became Escadrille 395, and, finally, SPA 461 in July 1917.

SPA 462, formed in March 1917 and initially designated Escadrille 350. It was assigned to the C.R.P. and became SPA 462 in July 1917.

SPA 463, formed in March 1917 and assigned to the C.R.P. Initially designated Escadrille 351, it became SPA 463 in July 1917.

Escadrille 464, formed in January 1916 as 93rd D.C.A. and assigned to the C.R.P. It was designated Escadrille 393 and became 464 in July 1917. It is believed that SPADs were assigned to the unit but this cannot be confirmed.

Escadrille 466, formed in March 1915 as 94th D.C.A. and assigned to the C.R.P. It was redesignated Escadrille 394 and became Escadrille 466 in July 1917. It is believed that SPADs were assigned to the unit but this cannot be confirmed.

Escadrille 467, formed in March 1917 as 96th D.C.A. and assigned to the C.R.P. It was designated Escadrille 396 and became 467 in July 1917. It is believed that SPADs were assigned to the unit but this cannot be confirmed.

Escadrille 469, formed in March 1915 as 96th D.C.A. and assigned to the C.R.P. It was designated Escadrille 397 and became Escadrille 467 in July 1917. It is believed that SPADs were assigned to the unit but this cannot be confirmed; it is known that unit carried out several night bombing missions.

Escadrille 470, formed in March 1917 and assigned to the C.R.P. It was designated Escadrille 352 and became 470 in July 1917. It is believed that SPADs were assigned to the unit but this cannot be confirmed.

SPAD 7 of N 561 *"Forse Che Si."* N 561 (later SPA 561) was assigned to the protection of Venice. B85.997.

Escadrille 471, formed in February 1918 as the 2nd Escadrille Américaine (2nd American Squadron). It became Escadrille 471 sometime later and was assigned to the C.R.P. It is believed that SPADs were assigned to the unit but this cannot be confirmed.

Escadrille 472, formed in 1918 as an escadrille postale. It became Escadrille 472 sometime later. It is believed that SPADs were assigned to the unit but this cannot be confirmed.

T.O.E. Units

SPA 506, formed from N 390 in June 1917 with SPAD 7s. It was assigned to the A.F.O.

SPA 507, formed from N 391 in June 1917 with SPAD 7s and was assigned to the A.F.O.

SPA 523, formed from N 523 in 1917/18 and attached to the Serbian Army.

SPA 531, formed in March 1918 with some SPADs on strength. It had both Greek and French pilots. It was disbanded in 1919.

SPA 561, formed in 1917/18 from N 561 and assigned to the protection of Venice.

SPA 581, formed from N 581 with eight Nieuport 17s and seven SPAD 7s. It was assigned to the Russian 7th Army and began operations in May 1917. Based at Boutchatch, Galicia, it served alongside Captain Kozakov's Nieuport fighter group. In response to the German offensive of 19 July on the 6th Army front, it was transferred to Buzcacz. In August it moved to the Romanian front and was assigned to the 3rd Corps. In February 1918 N 581 (which became SPA 581 because it had re-equipped entirely with SPAD 7s) was ordered back to Moscow. SPA 581 passed its SPAD 7s to what became the Czechoslovakian air service.

While the SPAD 7s are associated primarily with aerial combat and in particular with units assigned to the Groupes d'Combat; the escadrilles assigned directly to the armées employed the SPAD 7 for a wide variety of missions including ground attack and bombing. By mid-1917 SPAD 7s had been fitted with a bomb rack attached to each rear undercarriage leg. These racks each held a 10-kg Anilite bomb. Units equipped with SPAD 7s are known to have flown bombing missions. SPA 49 (with ten SPAD 13s and five SPAD 7s), for example, flew a night bombing mission on 2 October in which six bombs were dropped on stations, then the aircraft strafed trains.

Ground attack was also of critical importance and SPAD 7s and 13s were used extensively to strafe retreating German troops during the closing weeks of the war. SPA 93 alone expended 50,000 rounds of ammunition while attacking troops in March 1918. These units also performed long-range reconnaissance; some SPAD 7s were fitted with vertical cameras mounted behind the cockpit. These aircraft were marked with the word "PHOTO" on each side of the fuselage.

Groupes d'Combat

GC 11 was, by January 1918, an all-SPAD unit composed of SPA 12, SPA 31, SPA 57, and SPA 154. It was assigned to the 5th Armée. On 27 February GC 11 joined Groupement Féquant which also consisted of GC 13 and GC 17 and was assigned to the G.A.N. (the 1st and 3rd Armées). GC 11 was active in the Battle of Picardie in March and was based at Champagne. GC 11's escadrilles saw intense activity at this time; in addition to combat patrols the pilots flew escort missions for army cooperation units and attacked balloons. GC 11 also supported the 4th and 5th Armées.

GC 12 had three SPAD-equipped units (SPA 3, SPA 26, and SPA 73) and one Nieuport-equipped unit that soon converted to SPADs (SPA 103). GC 12 was assigned to the 10th Armée at Lorraine in January 1917 and took part in the Battle of Chemin des Dames in March. All GC 12's units saw considerable activity during the Battle of the Aisne. In December Escadrilles SPA 63 and SPA 102 briefly joined GC 12. GC 12 moved to Chadon on the Aisne front in support of the 6th Armée. The main mission of the fighter units on this relatively quiet front was to provide fighter escort for army cooperation units and prevent German reconnaissance aircraft from photographing the French front. In January 1918 GC 12 was assigned to the 2nd Armée on the

SPAD 7. The five color camouflage scheme can be discerned on the tail surfaces and wing. MA.

SPAD 7 serial 31905 and SPAD 13 43629. MA6250.

SPAD 7 of SPA 103 fitted with Le Prieur rocket rails on the interplane struts. B83.4994.

Argonne front; poor weather limited activity. By 7 March SPA 73 was detached to form the nucleus of GC 19. It was replaced by SPA 67. GC 12 was then assigned to the 1st Armée east of the Champagne front. The successful German offensive on the Somme resulted in GC 12's units being forced to transfer to new airfields. At this time GC 12's escadrilles were flying low-altitude reconnaissance in an attempt to determine the location of the front lines. As the aerial activity became even more intense over this front, the fighter units would fly patrols consisting of several escadrilles on a single mission.

GC 13 was equipped with SPAD fighters by July 1917. It consisted of SPA 15, SPA 65, SPA 84, N 124 (became SPA 124 in November 1917). In June GC 13 was assigned to the 1st Armée on the Flanders front. The units saw considerable action during the Third Battle of Ypres, where a number of trench-strafing missions were flown. In August GC 13 was assigned to the G.A.C. and provided fighter cover for the 2nd Armée. The missions of the fighter escadrilles remained unchanged: prevent German aerial reconnaissance and attack German troops with machine guns and bombs. After the battle, GC 13 moved briefly to an airfield at Sénard and was active over the left bank of the Meuse River. It moved on 26 September to the Aisne front in support of the 10th Armée. On 1 October SPA 88 joined GC 13. On 8 December 1917 GC 13 was assigned to the G.A.N. SPA 88 joined GC 13 early in 1918, while SPA 124 left the group. On 18 February GC 13 was assigned to Escadre de Combat 1, which was part of Groupement Féquant along with GC 11 and GC 17. GC 13 was active in the 6th Armée sector and, along with the Groupes de Combat assigned to Groupement Féquant, provided aerial cover during the Second Battle of the Somme from 21 March to 8 April.

GC 14 had escadrilles SPA 75, SPA 80, SPA 83, and N 86. By April 1917 SPA 75, SPA 80, and SPA 83 had completely re-equipped with SPAD 7s. Based at Marencourt, the escadrilles of GC 14 participated in the Battle of Chemin des Dames. In June 1917 GC 14 moved to Souilly, near Verdun, where it supported the 10th Armée. In July it briefly supported the 2nd Armée in the Aisne sector. It returned to the 6th Armée sector on 4 October; based at Vaubéron, GC 14 participated in the Battle of La Malmaison and was particularly active between Fére-en-Tardenois and Berry-au-Bac. Ten days later GC 14 was attached to the British army and in November participated in the Battle of Cambrai. In December it returned to the Chemin des Dames. By January 1918 GC 14 was based at Vauberon in the 7th Armée sector. GC 14 flew a number of bomber escort and reconnaissance missions during the Battle of Picardie, along with more unusual missions such as delivering spies behind enemy lines and dropping propaganda leaflets. GC 14 was assigned to the G.A.N. in early April.

GC 15 had six fighter escadrilles assigned to it in June 1917; two units (SPA 37 and SPA 81) used SPAD 7s, the other four (N 78, N 92, N 93, and N 112; later N 85) flew Nieuport 23s and 24s. GC 15 was based at Champagne until July when it moved to Vadelaincourt and was assigned to the 2nd Armée. From 20 August to 8 September GC 15's main task was to provide fighter escort for reconnaissance planes and bombers during the Second Battle of Verdun. By November N 95 joined GC 15 while N 78 and N 93 re-equipped with SPAD 7s and 13s. By the end of 1917 there were only two units assigned to GC 15 still flying Nieuports—N 85 and N 112. By January 1918 GC 15 was an all-SPAD unit with SPA 37, SPA 81, SPA 93, and SPA 97. It was based at Villeneuve-les-Vertus and was soon assigned to Escadre de Combat 1, which consisted of GC 15, GC 18, and GC 19. During this time GC 15 continued to provide support for the 2nd Armée. Formed on 18 February 1918, SPA 124 was briefly assigned to GC 15. However, difficulties in obtaining adequate equipment and personnel limited SPA 124 to reconnaissance missions until it was transferred to GC 21 on 25 February 1918. In February GC 15 moved to Plessis-Belleville and was active in the Battle of Picardie in March. On 27 March it moved to support the 6th Armée in response to the German offensive that began on 21 March and was also known as the Second Battle of the Somme. GC 15s played an important role in this battle by strafing enemy troops. On 9 April it was composed of SPA 37, SPA 81, SPA 93, SPA 97, and R 46. It was used primarily to provide fighter escort for the army cooperation and bombing escadrilles during the Battle of Lys. In May GC 15 had been incorporated into Groupement Ménard and was based at Montagne.

GC 16 was formed on 10 January 1918 and consisted of SPA 78, SPA 112, SPA 150, and SPA 151. GC 16 was based at Chaux in the 7th Armée sector. During 16 to 30 January it flew 169 sorties and engaged in 31 combats. On 1 April 1918 GC 16 moved: SPA 78 and SPA 112 to the 3rd Armée sector; SPA 150 to the 6th Armée sector; and SPA 151 to the 2nd Armée sector.

GC 17 was formed in February 1918 with three SPAD escadrilles (SPA 77, SPA 89, SPA 91) and one Nieuport escadrille (N 100, which became SPA 100 in March 1918). GC 17 was initially based in the 7th Armée sector as part of Escadre de Combat 1 (GC 11, GC 13, and GC 17) which, in turn, was assigned to Groupement Féquant. On 8 March GC 17 moved from Mannoncourt (where it had been formed) to Saponay on the Aisne front and was assigned to the 6th Armée. GC 17 was active from 21 March to 8 April during the Second Battle of the Somme.

GC 18 was formed in February 1918 with SPA 48, SPA 94, SPA 153, and SPA 155. It was assigned to Escadre de Combat 2 in support of the G.A.N. The German offensive began on the Oise front on 21 March. Eight days later GC 18 moved to the 6th Armée sector on the Somme front and was active in the Battles of Picardie (March), Matz (June), and Chemin des Dames (July).

GC 19 was formed in February 1918 and was composed of

SPAD 7 of S/Lt. Bayal of SPA 48 which was assigned to GC 18. B81.2133.

SPA 73, SPA 85, SPA 95, and SPA 96 (the latter unit having just converted to SPAD fighters). It was assigned to Escadre 1 along with GC 15 and GC 18. Based at Plessis-Belleville (27 March) and then Pierrefonds (7 April), GC 19 gave air cover to the 6th Armée trying to stem the German advance on Paris. A move to Airaines was necessitated by the start of the Second Battle of the Somme. Missions were flown between Arras and Anizy and in support of the 6th Armée.

GC 21 was formed in March 1918 with Escadrilles N 98, N 124, and SPA 157, and the American 103rd Aero Squadron. GC 21 was initially assigned to the 4th Armée. SPA 154 and the 103rd Aero Squadron were created by a division of N 124 (the famed Lafayette Escadrille) on 28 November 1917. At the time of its formation SPA 124 and the 103rd Aero Squadron were based at La Ferme de la Noblette. However, soon after it was formed the 103rd Aero Squadron was transferred to the 6th Armée sector and never played an active role with GC 21. SPA 124 was assigned to GC 21 on 25 February 1918. GC 21 was assigned to the 4th Armée and flew reconnaissance missions and combat patrols. SPA 163 was formed in April 1918 by a division of SPA 124. It was also attached to GC 21 as a replacement for the 103rd Aero Squadron.

By the spring of 1918 SPAD 13s began to replace the SPAD 7s in service with the army cooperation escadrilles and Groupes d'Combat (see the entry under SPAD 13).

Foreign Service

Argentina
Two SPAD 7s and two SPAD 13s were purchased by Argentina in 1919. They were assigned to the Escuela de Aviacion Militar, later re-designated Grupo de Aviacion 1.

SPAD 7 *Nemesis III* of SPA 48 piloted by Robert Bajac. B82.4349.

Belgium
The SPAD 7s ordered by Belgium were intended to replace Nieuport 17s. Fifteen were ordered in 1917 and given serials SP 1 through SP 15. They arrived in September 1917 and were assigned to the 5th Escadrille, (later re-designated the 10th Escadrille) and were based at Le Moeres in 1918. The first went to Belgian pilot Edmond Thieffy, who scored the first Belgian victory on the aircraft. Seven additional aircraft arrived later and were given serials SP 17 through SP 22.

Brazil
Although some sources suggest that Brazil received SPAD 7s, these were in fact SPAD 13s. (See SPAD 13.)

Chile
A single SPAD 7 was obtained by the Chilean air service in 1918.

Czechoslovakia
Fifty SPAD 7s and 13s were obtained by Czechoslovakia after the war. The SPAD 7s were designated B-13s and assigned in 1919 to 2 Letecka Setnia (at Nitra), the flying school (Cheb) and Letecke Dilny (Olomouc); in 1923 to 32 and 34 Stihaci Rota (Prahha-Kbely), 33 Stihaci Rota (Olomouc), and 7 Pozorovaci Rota. They were withdrawn in 1924.

Estonia
Two SPAD 7s were obtained by the Estonian air service in 1925; serial numbers were 78 and 79.

Finland
A single Dux-built SPAD 7 was captured by the Finns in 1918. It was given serial D 76/18 (later, ID445) and remained in service until 1923.

Greece
The Greek air service purchased 16 SPAD 7s. 531 Mira (Squadron) had a mixture of Nieuport 24bis and SPAD 7s and was based at Gorgupi. It disbanded in 1919 and many of its aircraft were sent to the 534 Mira, based at Salunica-Lebet.

Italy
Italy's fighter force had been equipped exclusively with modern French fighters. It is not surprising, then, that when the SPAD 7 entered service Italy would acquire some. It seems the type was not liked by most Italian pilots and the Nieuport 17s and Hanriot HD.1s remained the main fighter types. The SPAD 7s were found to be particularly useful as high-speed reconnaissance aircraft. However, the top Italian ace, Baracca, used the SPAD 7

because he was willing to accept its single machine gun armament as a trade-off for its improved speed and maneuverability. It appears that the type first entered service in March 1917 with the 77th and 91st Squadriglias. By August 1917 the following units had SPAD 7s on strength:

1 Gruppo (3rd Armata): 71a and 77a Squadriglias.
2 Gruppo (2nd Armata): 76a Squadriglia.
3 Gruppo (Supreme Command): 91a Squadriglia.
9 Gruppo (1st Armata): 75a Squadriglia.
10 Gruppo (Supreme Command): 78a Squadriglia.

All of these units were flying a combination of SPAD 7s and Nieuport 11 and 17s. By October 1917 the number of units had increased from six to eight:

1 Gruppo (3rd Armata): 77a, 80a, and 84a Squadriglias.
2 Gruppo (4th Armata): 76a Squadriglia.
3 Gruppo (1st Armata): 72a Squadriglia.
9 Gruppo (1st Armata): 71a, 75a Squadriglias.
10 Gruppo (Supreme Command): 91a Squadriglia.

In November 1917 Nos.76a, 80a, and 84a Squadriglias re-equipped with Hanriot HD.1s and gave up their SPAD 7s. This was in part due to the fact that the Hanriots were being built under license by the Macchi firm, unlike the SPAD 7s which had to be supplied by the overtaxed French aircraft industry. By June 1918 the SPAD 7s were no longer in front-line service and were relegated primarily to training units. The SPAD 7s did have some notable successes. 91a Squadriglia led by Baracca destroyed 14 Austro-Hungarian aircraft between 20 and 26 October 1918. As previously noted, the type was also used for reconnaissance; of particular importance was a mission by three SPAD 7s over Trieste and Istria. In these aircraft a camera was located behind the cockpit.

The remaining squadriglias replaced their SPAD 7s with SPAD 13s during 1918.

Japan
SPAD 7s were imported by the Japanese army air service in 1918 but the type was not selected for license production in Japan.

Netherlands
SPAD 7 (serial number 1832) of SPA 73 was interned in November 1917 when it landed in Holland. It was purchased from the French and was supplied as a pattern aircraft to the Goedkoop firm. It was intended to produce the type under license, but no SPAD 7s were ever built in the Netherlands.

Peru
Two SPAD 7s were purchased by the Peruvian air service in 1919 after samples had been brought to the country by a French military mission. Both were based at the Centro de Aviacion Militar at Maranga.

Poland
The 1st Polish Combat Aviation Unit was formed early in 1918 with several SPAD 7s.

The 2nd Polish Combat Aviation Unit was formed December 1917 with one SPAD 7 or 13. The unit was based at Kanev and the SPAD was soon captured by the Germans. The 7th Squadron, which was equipped with SPAD 7s, was based at Warsaw in 1919. Later, an additional 15 SPAD 7s were ordered from the French.

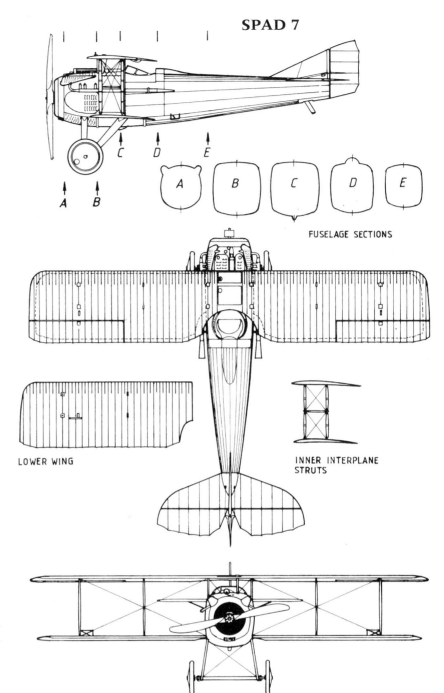

SPAD 7 in Italian markings. The Italians found the SPAD 7 to be particularly useful as a high speed reconnaissance aircraft. Via Colin Owers.

SPAD 7 with wing-mounted machine guns. The purpose of this field modification is not known, but it seems likely that the aircraft was to be used for ground attack. Renaud.

Portugal

SPA 124 had three Portuguese pilots assigned to the escadrille. After the war, the Portuguese air service obtained 12 SPAD 7s. These were assigned to the EMD in 1919 and in 1923 to the GIAC (Grupo Independente de Aviacao de Caca). Three years later the Portuguese air service had 17 SPAD 7s and Martinsyde F.4 Buzzards on strength. These were now assigned to the Esquadrilha Nol de Caca. In 1927 this unit became the Grupo Independente de Aviacao de Proteccao e Combate. The SPAD 7s were withdrawn in 1932.

Romania

Eight SPAD 7s were used by the Romanians in 1919. SPAD 7s and 13s were assigned to Grupl 3 at Galati.

Russia

Initially, the French sent SPAD 7s directly to Russia and by early 1917 43 had been delivered to Russia. The type was well-liked and plans were made to produce the aircraft in Russia. Approximately 100 SPAD 7s were built by the Aktsionyernoye Obshchestovo Dux plant at Moscow during 1917. However, the lack of availability of the requisite Hispano-Suiza engines resulted in the termination of production in 1918 before the order for 200 could be completed. The few SPAD 7s available were assigned to the premiere Russian unit—No.1 Fighter Group, which was composed of four squadrons. A Soviet history of the IRAS states that "It is necessary to emphasize the importance of the single-seater SPADs... These airplanes...had a radius of operation which made them suitable for escorting the Russian bombers in long-range raids." On June 1 1917 the disposition of 26 SPAD A.2s, A.4s, and 7s was: western front (4); southwestern/Romanian front (13); Caucasus front (9).

Relatively few SPAD 7s saw service during the Russian civil war. On 9 December 1917 only 579 aircraft of all types were available. Of these, 19 were SPADs (mostly SPAD 7s, but this figure probably includes SA.2s and SA.4s).

Postwar, 76 SPADs of all types were assigned to various RKKVF units. In 1921 they were in service with all the military districts, most of them were in the Ukraine. In 1922 the SPADs were with the Istrebitel'naya Eskadril'ya in Moscow, the 3rd Otryad of the Istrebitel'naya Eskadril'ya at Kiev, the 1st and 3rd naval Istrootryady at Leningrad and Sevastopol, the 2nd naval Istrootryady at Odessa, and the School of Military Pilots.

In 1923 the SPAD 7s were with the 4th Otdek'naya Istrebitel'naya Aviaeskadril'ya at Minsk (four aircraft), the 2nd Otdel'nyi Istrebitel'nyi Aviaotryad at Kharkov (four), and the 1st Morskoi Istrebitel'nyi Vozdukhchast (three). Others remained in service as trainers with the Training Eskadrill'ya and the 1st Higher School of Military Pilots. All had been withdrawn from service in 1925.

Serbia

By July 1918 the Serbian air service had some SPAD 7s on strength. These served along with Nieuport fighters in the newly-formed 1st and 2nd Serbian Escadrilles. The 1st was based at Ostrow and the 2nd at Verbano. Later they formed the 1st Pursuit Group. The 1st Army also had a pursuit escadrille at Verbano. These units supported the Serbian attack on the Bulgarians in September 1918. The aircraft flew from Vertekop and made a large number of ground attack sorties. The units later moved to Uskub, where they attacked retreating German troops. The SPAD 7s were phased out of service after the war.

Siam

Several examples of the SPAD 7 were acquired by the Siamese air service in 1920. Most were based at Don Muang.

Spain

Spain planned to produce the SPAD 7 under license. These planes, known in contemporary literature as the Barrón España, never entered service with the Servicio de Aeronautica Militar. The initial plans to build the SPAD 7 in Spain began after Coronel Rodriguez Mourelo, director of the Aeronautica Militar, visited Paris and reviewed plans for it. Capitan Eduardo Barrón directed the adaptation of the French design. The firm of Pujol, Comabella, and Cia built the plane, which unfortunately had numerous defects. In 1920 the British Martinsyde F.4 was selected for production instead.

Ukraine

Two SPADs were obtained by the Ukrainian air service in 1918. It is likely these were SPAD 7s captured from the Russians; one had serial 1516.

United Kingdom

The RFC placed orders for SPAD 7s in April 1916. The first, S 126, was sent from France in September 1916. and allocated to No.60 Squadron, where it stayed until October. Pleased with the new aircraft, the British placed orders for additional machines with Blériot (or possibly Janoir). The first, delivered to the RFC in late 1916, had French serial S.1002. The SPAD 7s were used by Nos.19 and 23 Squadrons; both began re-equipment in October 1916. As the intensity of the air battles over the front increased, the RNAS was asked to release its

SPAD 7 built by the de Marcay firm. MA13929.

SPAD 7 Single-Seat Fighter with 150-hp Hispano-Suiza 8Aa

Span 7.822 m; length 6.080 m; height 2.20 m; wing area 17.85 sq. m
Empty weight 500 kg; loaded weight 705 kg

Maximum speed:	sea level	193 km/h
	2,000 m	187 km/h
	3,000 m	180 km/h
	4,000 m	174 km/h
Climb:	2,000 m	6 minutes 40 seconds
	3,000 m	11 minutes 20 seconds

Ceiling 5,500 m; range 400 km; endurance 2.66 hours
Armament: one 7.7-mm Vickers machine gun and two 10-kg Anilite bombs
Approximately 3,500 SPAD 7s of all types were built in France during the War

SPAD 7 Single-Seat Fighter with 19 Square Meter Wing and 150-hp Hispano-Suiza 8Aa

Span 7.315 m; length 6.080 m; height 2.20 m; wing area 19.00 sq. m
Empty weight 510 kg; loaded weight 714 kg
Maximum speed: 187 km/h at sea level, 185 km/h at 2,000 m; 179 km/h at 3,000 m; climb to 2,000 m in 6 minutes 55 seconds; climb to 3,000 m in 11 minutes 50 seconds; ceiling 5,501 m; endurance 2 hours
Armament: a 7.7-mm Vickers machine gun and two 10-kg Anilite bombs

SPAD 7 Single-Seat Fighter with Flat Wing and 150-hp Hispano-Suiza 8Aa

Length 6.080 m; height 2.20 m
Empty weight 499 kg; loaded weight 704 kg
Maximum speed: 199 km/h at sea level; 194 km/h at 2,000 m; 188 km/h at 3,000 m; climb to 2,000 m in 5 minutes 50 seconds; climb to 3,000 m in 9 minutes 50 seconds; ceiling 6,004 m; endurance 2 hours
Armament: a 7.7-mm Vickers machine gun and two 10-kg Anilite bombs

SPAD 7 Single-Seat Fighter with 180-hp Hispano-Suiza 8Ab

Span 7.822 m; length 6.080 m; height 2.20 m; wing area 17.85 sq. m
Empty weight 500 kg; loaded weight 705 kg

Maximum speed:	2,000 m	212 km/h
	3,000 m	204 km/h
	4,000 m	200 km/h
	5,000 m	187 km/h
Climb:	2,000 m	4 minutes 40 seconds
	3,000 m	8 minutes 10 seconds
	4,000 m	12 minutes 49 seconds

Ceiling 6,553 m; endurance 1.5 hours
Armament: one 7.7-mm Vickers machine gun and two 10-kg Anilite bombs

SPAD 7 Single-Seat Fighter with 140/150-hp Hispano-Suiza License-Built in Russia by Aktsionyernoye Obshchestovo Dux

Span 7.8 m; length 6.1 m; height 2.20 m; wing area 18.0 sq. m
Empty weight 543 kg; loaded weight 795 kg
Maximum speed: 185 km/h (with 140-hp H-S); 195 km/h (with 150-hp H-S); climb to 1,000 m in 2 minutes 30 seconds; climb to 2,000 m in 5 minutes 30 seconds; climb to 3,000 m in 9 minutes; ceiling 6,000 m; endurance 2 hours
Armament: one 7.7-mm Vickers machine gun
Approximately 100 built

Hispano-Barron SPAD 7 Single-Seat Fighter with 180 Hispano-Suiza 8Ab (Later Fitted with 220-hp Hispano-Suiza 8Ba) License-Built in Spain by Pujol, Comabella, and Cia

Loaded weight 840 kg (H-S 8Ab)
Maximum speed: 188 km/h (H-S 8Ab); 206 km/h (H-S 8Ba); climb to 1,000 m in 2 minutes 40 seconds (H-S 8Ab); climb to 2,000 m in 6 minutes (H-S 8Ba)
No armament fitted
One built

SPAD 7 Single-Seat Fighter with 150-hp Hispano-Suiza 8Aa, 180-hp Hispano-Suiza 8Ab; 150-hp Wolseley W.4A Python I, or 180-hp Wolseley W.4A Python II Produced Under License in England by Mann, Egerton & Co. and Blériot (Aeronautics) of Brooklands

Span 7.822 m; length 6.08 m; height 2.2 m; wing area 17.85 sq. m
Empty weight 500 kg; loaded weight 705 kg

Maximum speed (150-hp H-S or Python I):		
	sea level	193 km/h
	2,000 m	187 km/h
	3,000 m	180 km/h
	4,000 m	174 km/h
Maximum speed (180-hp H-S or Python II):		
	2,000 m	212 km/h
	3,000 m	204 km/h
	4,000 m	200 km/h
Climb (150-hp H-S or Python I):		
	2,000 m	6 minutes 40 seconds
	3,000 m	11 minutes 20 seconds
Climb (180-hp H-S or Python II):		
	2,000 m	4 minutes 40 seconds
	3,000 m	8 minutes 10 seconds
	4,000 m	12 minutes 49 seconds

Ceiling: 5,500 m (150-hp H-S or Python I); 6,553 m (180-hp H-S or Python II)
Armament: one synchronized 0.303-in Vickers machine gun and two 25-lb Cooper bombs
Approximately 120 built

SPAD 7s to the RFC, and a large number were. It was hoped that SPAD 7s ordered by the RNAS from Mann, Egerton & Co. could be used to help replace the SPAD 7 losses at the front. However, these aircraft had structural defects requiring modifications and it seems that few of the SPAD 7s produced by this firm were ever sent to the front. An additional 100 were delivered to the RFC by Blériot (Aeronautics) Ltd. of Brooklands. These saw service with Nos.30 and 63 Squadrons in Mesopotamia and No.72 Squadron in Palestine. At the front, Nos.19 and 23 Squadrons suffered significant losses of aircraft and, because of the delays in the production of SPAD 7s by the Mann, Egerton and Co. in Britain and Kellner et ses Fils in France, it was feared that it would not be possible to keep the squadrons up to strength. Fortunately, enough aircraft were forthcoming to prevent this from happening. No.19 Squadron was active in the

Battle of Arras, Messines Ridge, and the Third Battle of Ypres. The RFC units discovered that the SPAD 7s were quite effective in the ground attack role. Both combat patrol and ground attack missions continued until January 1918, when the squadron re-equipped with Sopwith Dolphins. No.23 Squadron was also active in the same vicinity of the front as No.19 Squadron. In addition to patrol and ground attack missions, the SPAD 7s were used on wireless interruption flights. When the unit re-equipped with Sopwith Dolphins in 1918 it retained a number of SPAD 7s. These, when retired from service, were sent back to Britain for service with the training units. Training units that had SPAD 7s were 27, 31, 54, 56, and 60th Squadrons, 17th Wing Beaulieu, and 18th Wing Fighting School. In addition, No.81 Squadron had some SPAD 7s on strength.

SPAD 7 of SPA 88 piloted by Marcel Coadou; the escadrille was assigned to GC 18. B87.6270.

Below: SPAD 7. Via B. Millot and Colin Owers.

As mentioned above, the RNAS also had an interest in the SPAD 7 and evaluated S.211 in November 1916 at Dunkerque. Based on a favorable report by evaluation pilots, an order for 75 aircraft (N6210–N6284) was placed with Mann, Egerton & Co. and another order for 50 (N6030–N6079) was placed with British Nieuport. An additional 25 were ordered from Mann, Egerton & Co. in January 1917 (N6580–N6604). Because of the intense need for fighters at the front, the RNAS agreed to transfer all its forthcoming SPAD 7s to the RFC. Even the initial SPAD 7 (S.211/9611) was turned over to the RFC and given the serial B 388. Thus, the SPAD 7 never saw operational service with the RNAS.

United States

The Bolling Commission recommended on 30 July 1917 that the A.E.F. should purchase the SPAD 13 for its fighter squadrons. However, it was also decided to purchase SPAD 7s for use by training units. A total of 167 SPAD 7s were eventually obtained by the A.E.F. Many of these were built by the Mann, Egerton & Co. firm of Britain and these were decidedly inferior to those produced in France. Unfortunately for the Americans, French production estimates once again proved to be optimistic and this, combined with the pressing need of the French fighter escadrilles for SPAD 13s, resulted in a number of A.E.F. squadrons using SPAD 7s.

The 41st Aero Squadron used SPAD 7s during much of 1918.

Initially based in Scotland, it moved to Romorantin, France, in August. After the war it was based at Lay-St. Remy and was attached to the 5th Pursuit Group after the Armistice.

The 138th and 638th Squadrons were also formed as the war ended and were assigned to the 5th Pursuit Group. It seems that these new units spent most of their time training and it is unlikely that the SPAD 7s saw much combat.

After the war a number of SPAD 7s were sent to the United States for use as trainers. Some were still in service at Kelly Field, Texas, as late as 1926.

SPAD 11

The SPAD family of fighters had proved to be an outstanding success and so it was perhaps natural that Béchereau would have turned his attention to producing a two-seat fighter to meet the C2 specification. This aircraft, which received the STAé designation SPAD 11 (presumably the SPAD 6, 8, 9, and 10 designations had been assigned to the SPAD projects mentioned above), was developed in parallel with the SPAD 7.

The SPAD 11 appeared in many ways to be a scale-up of the basic SPAD 7 layout. It was a two-bay biplane with sweepback and positive stagger but without dihedral, and powered by a 220-hp Hispano-Suiza 8Bc. There were cutouts in the lower wings to improve the crew's downward view. Rib spacing was 190 mm in the upper wing and 175 mm in the lower. The wing trailing edge was made from metal wire. Ailerons were on the upper wing only. The fuselage had four longerons with a rectangular cross-section. A trap-door was located in the

SPAD 11 of SPA-Bi 34 which was assigned to the 7th C.A. B84.544.

Right: SPAD 11. University of Texas at Dallas.

Below right: SPAD 11. An evaluation of the SPAD 11 published 28 November 1917 stated that the aircraft's performance, described as comparable to the SPAD SA series, was inadequate for a two-seat fighter. B76.1175.

Below: SPAD 11. B92.1557.

floor behind the observer/gunner's cockpit. The tailplane was attached to the upper longeron at zero degrees incidence; both sides were cambered. The Hispano-Suiza engine was attached to the fuselage on engine bearers connected to the longerons by transverse wooden supports braced by aluminum fittings. The pressurized fuel tank held 140 liters and there was a gravity tank in the upper wing. The oil tank held 15 liters and was located on the fuselage floor behind the engine. The annular radiator in front of the engine was connected to a reservoir next to the front spar of the upper wing. The aircraft had dual controls, and the observers cockpit could hold up to six ammunition drums. The pilot had a forward-firing Vickers machine gun and the gunner had a single Lewis gun on a ring mounting. A bomb load of 70 kg could be carried.

Aircraft SPA 6049 with a loaded weight of 1,167.5 kg, a Hispano-Suiza 8Ee engine, and Ratmanoff H.R.1 propeller had the following performance during STAé testing on 25 October 1917: maximum speed 181 km/h at 2,000 m; 177 km/h at 3,000 m; 168 km/h at 4,000 m; climb to 500 m in 1 min. 35 sec.; 1,000 m in 3 min. 25 sec.; 2,000 m in 7 min. 35 sec.; 3,000 m in 12 min. 30 sec.; and 4,000 m in 17 min. 30 seconds.

An evaluation of the SPAD 11 published 28 November 1917 stated that the aircraft's performance was inadequate for a two-seat fighter and was described as comparable to the SPAD SA series. It was hoped that fitting the SPAD 11 with a Lorraine engine could salvage the program. The Hispano-Suiza engines were noted as being out of service two out of every three days.

Furthermore, it was decided that the SPAD 11's handling qualities were unacceptable for a fighter and therefore could not meet the requirements of the C2 category. At the time, the army

Above: SPAD 11 nightfighter fitted with a searchlight. MA6956.

Right: Closeup of the SPAD 11 converted to a nightfighter by the addition of a nose-mounted light. University of Texas at Dallas.

SPAD 11

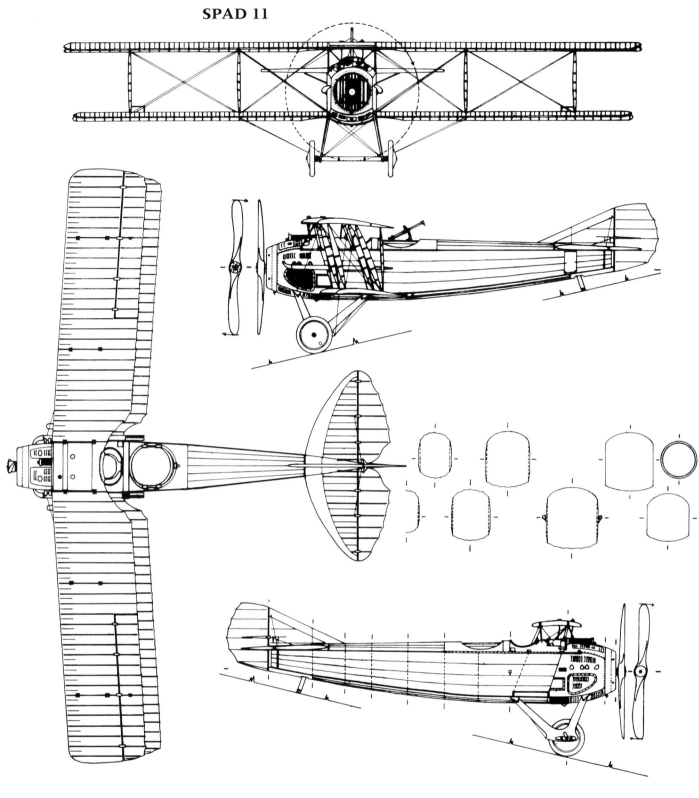

SPAD 11 serial S 6035. The prototype first flew in September 1916 and apparently was found to be satisfactory for the A2 category as the STAé demanded in August 1917 that as many of the SPAD 11s (as well as Salmson 2s and Breguet 14s) as possible be obtained. 78.1516.

SPAD 11 (S.619) of BR 272. This escadrille used Breguet 14 A2s as its primary equipment. B84.622.

Left: SPAD 11 serial 2045 of SPA 102. SPA 102 was a fighter unit and used the SPAD 11s for reconnaissance missions in support of the 2nd Armée. B88.2437.

Below: SPAD 11. B83.5713.

cooperation squadrons were beginning to realize that the A.R.1s and Sopwith 1½ 1A2 Strutters were obsolescent. The class A2 requirement was initiated by the STAé as a result and it was decided to procure the SPAD 11 as an army cooperation machine along with the Salmson 2 and Breguet 14.

Because of production difficulties with the Hispano-Suiza engines, as well as an explosion that heavily damaged the Bernard factory at Courneuve, production of the SPAD 11 was greatly delayed. In October 1917 complaints were being made about problems with the control system. Specifically, the aircraft was noted to be tiring to fly and difficult to control because it was tail heavy and lacked maneuverability. Worse, it stalled easily, often ending up in a spin. Despite these glaring problems, it was decided that the A.R.1s and Sopwith 1½ Strutters had to be replaced and the SPAD 11 had the virtue of being a more modern design that was readily available.

A night fighter version of the SPAD 11 was designed. It had a searchlight mounted in similar fashion to the pulpit of the SPAD A series. Designated the SPAD 11 Cn2, only a single example was built.

Difficulties with the Hispano-Suiza engine led the SPAD firm to fit a 220-hp Renault to a SPAD 11. It has been reported that the performance with this engine was inferior to the standard production versions. At least one SPAD 11 was fitted with a 310-hp Hispano-Suiza 8Fb with a Rateau supercharger. It was designated SPAD 11 CE.

Although it was inferior to the Breguet 14 A2 and Salmson 2 A2, 1,000 SPAD 11s were built. There were still 85 SPAD 11s at the front in October 1918.

Operational Service

Escadrilles re-equipped with SPAD 11s were:

SPA-Bi 2, formed from AR 2 in November 1917 when SPAD 11s were assigned to the escadrille at Poix. SPA-Bi 2 was attached to the 5th C.A. and was active over Noyon during the March German offensive. It was subsequently equipped with SPAD 16s.

SPA-Bi 20, formed from AR 20 in early 1918 at Fienvillers. Assigned to the 14th C.A., it re-equipped with SPAD 16s in August 1918.

A lineup of SPAD 11s of SPA-Bi 34. MA38563.

SPA-Bi 34, formed from C 34 in October 1917 and assigned to the 7th C.A. It was commanded by Capitaine Molinié and re-equipped with SPAD 16s in 1918.

SPA-Bi 42, formed from C 42 (with Caudron G.6s) in October 1917. It was assigned to 74th D.I. and in early 1918 placed under control of the 5th Armée as part of the G.A.R. In April it returned to the American 1st Division and was based at Boismont, Esquennoy, and Viefvillers. It was then reassigned to the 10th Armée sector. SPA-Bi 42 participated in the Battle of Cantigny in support of the American army. It re-equipped with SPAD 16s in October.

SPA-Bi 53, formed from C 53 (with Caudron G.4s) in December 1917 when the unit received seven SPAD 11s. It was assigned to the 1st C.A. at La Cense and later moved to la Ferte-sous-Jouarre. In January it moved to the vicinity of Fisme and later to Oise, where it participated in the Battle of Chemin des Dames. In July it moved to Soissons and participated in the Battle of Noyon. It re-equipped with SPAD 16s in September.

SPA-Bi 54, formed from F 54 in January 1918 when the escadrille received seven SPAD 11s. It was assigned to the 8th C.A. at Dancourt, but used the SPAD 11s only briefly until August 1918, when it re-equipped with SPAD 16s.

SPA-Bi 63, formed from in December 1917 when the unit received SPAD 11s. The escadrille was assigned to the 1st C.C. at Plessis-Beleville and took part in the Battle of Picardie in April. It re-equipped with SPAD 16s in October.

SPA-Bi 64, formed from C 64 in February 1918 when the unit received SPAD 11s and a few Letords. Assigned to the 13th C.A. at Vadelaincourt, the escadrille was active over the woods of Cheppy in March and over Champagne during the German offensive in mid-July. During these battles it suffered 30 casualties. SPA-Bi 64 re-equipped with SPAD 16s in August.

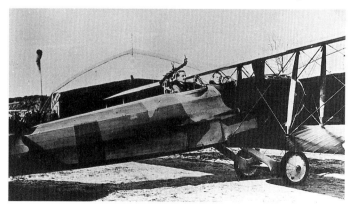

SPAD 11 in Belgian service. SPAD 11s were used by the 4th, 5th, and 6th Belgian Escadrilles d'Observation. Via Colin Owers.

SPA-Bi 140, formed from MS 140 (with Morane-Saulnier Type Ps) in late 1917 when the SPAD 11s replaced the troubled Morane-Saulniers. It was assigned to the 4th C.A. at Ferme D'Alger, and re-equipped with SPAD 16s in October.

SPA-Bi 212, formed from MS 140 (with Morane-Saulnier Type Ps) in January 1918. In April it was serving with the 1st C.A. at La Cense. It was assigned as a S.A.L. unit for the 4th Armée and was disbanded in March 1919.

SPA-Bi 228, formed from C 228 (with Caudron G.6s) in January 1918. It was with the 1st Armée at Esquennoy and was then assigned to the 87th D.I. It became BR 228 in July when it re-equipped with Breguet 14 A2s.

SPA-Bi 266, formed in January 1918 with SPAD 11s. It was assigned to the 30th C.A. under command of Capitaine Pertusier. It was based on the Champagne front and later was assigned to the 4th Armée, based at Bouy. It was later assigned to the 87th D.I and re-equipped with SPAD 16s in September.

Other units which received a small number of SPAD 11s included SPA 76 (5th Armée), and SOP 284 (3rd C.A.).

Problems with the SPAD 11 continued after it entered service. A GQG memo dated 9 January 1918 suggested that repairs of the Hispano-Suiza engines were frequently required and Adjudant Foidit visited the SPA-Bi units to ensure that the repairs were done correctly. A memo of 11 July 1918 noted that the SPAD 11's oil coolers were deficient and required repair or replacement. It was also reported that month that the tail skids were weak and needed to be reinforced. The modifications, suggested by Adjudant Girardon of SPA-Bi 266, enabled the skid to better withstand the shock of landing and also helped to keep the plane straight during takeoff and landing. A less serious problem was the gun mount. When the plane was flown as a single-seater pilots were warned to lock the turret in place or it would swivel violently in flight. A GQG memo dated 27 June 1918 noted that in the field some of the SPAD 11s required modifications to the rigging.

SPAD 11s were gradually withdrawn from service during the first months of 1918. They were replaced by SPAD 16s with 240-hp Lorraine engines or, in a few cases, Breguet 14 A2s.

Foreign Service
Belgium
SPAD 11s were acquired by the Belgian Air Service in 1918. These served with the 4th Escadrille d'Observation at Hondschoote, 5th Escadrille d'Observation at Houtem (along with Breguet 14 A2s), 6th Escadrille d'Observation at Houtem, and with the Escadrille Royale. The SPAD 11s were retired from service in 1921.

Japan
At least one SPAD 11 was purchased by the Japanese in April 1918. It was used as a trainer by the army aerial reconnaissance corps based at Shimoshizu.

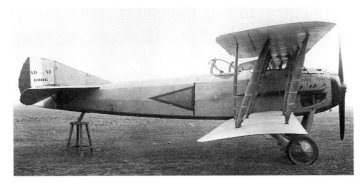

Above: Four views of SPAD 11 S 60066. University of Texas at Dallas.

Below: SPAD 11 front view. University of Texas at Dallas.

SPAD 11 Two-Seat Reconnaissance Aircraft with 220-hp Hispano-Suiza 8Bc

Wing span 11.24m; length 7.80m; height 2.50 m; wing area 30 sq. m
Empty weight 679 kg; loaded weight 1035 kg
Maximum speed: 185 km/h at sea level; 172 km/h at 4,000 m; climb to 2,000 m in 7.5 minutes; climb to 3,000 m in 12.6 minutes; climb to 4,000 m in 17 minutes; ceiling 6300 m; endurance 2.25 hours
Armament: one fixed, synchronized 7.7-mm Vickers machine gun fired by the pilot and one 7.7-mm Lewis gun mounted on a ring stand fired by the observer, and 70 kg of bombs
Approximately 1,000 built

Russia

At least one SPAD 11 was used by the RKKVF as late as 1923. It served with the 2nd Otdel'nyi Istrebitel'nyi Aviaotryad.

United States

The American Expeditionary Force acquired 35 SPAD 11s in 1918. They were supplied to the 1st and 90th Aero Squadrons.

The 1st Aero Squadron received 18 SPAD 11s in February 1918. They were reported to be troublesome to maintain and apparently most of the difficulties were attributed to the 220-hp Hispano-Suiza engines. On 1 May 1918 there were 14 SPAD 11s on strength with the 1st Aero Squadron. They were replaced beginning in June by the far more successful Salmson 2s.

The 90th Aero Squadron had a few SPAD 11s on strength but these were apparently replaced by Sopwith 1½ Strutters (!) by May 1918.

Uruguay

A few SPAD 11s were acquired by the Uruguayan air service after the First World War ended.

SPAD 11 in Belgian service. In 1921 the SPAD 11s were retired from Belgian service. MA5021.

SPAD 12

While the SPAD 7 was a definite improvement over the Nieuport series and more than a match for contemporary German fighters, it had one definite weakness—its armament. Unlike many of the German fighters in 1917, the SPAD 7 was armed with only a single machine gun. One of the pilots who had put the SPAD 7 to such good use was Georges Guynemer. He approached Béchereau with the idea of fitting a cannon mounted to fire through the hollow hub of a 200-hp Hispano-Suiza 8Cb engine. Béchereau designed the SPAD 12 to meet Guynemer's request, and the prototype first appeared in December 1916.

The cannon selected for the SPAD 12 was a 12-round, 37-mm S.A.M.C. In addition to this powerful weapon, a single Vickers machine gun was fitted to starboard. The only other obvious changes were the elimination of the tear-drop cylinder fairings and a more streamlined cowling. More subtle additions were the increase in wing area (necessitated by the weight of the cannon), positive stagger of the wings, and rounded wing tips. Subsequent modifications included fitting of a 220-hp Hispano-Suiza 8Cb engine and addition of pocket extensions to the wing tips. Initial examples of the SPAD 12 were tested in combat by Guynemer himself, and he achieved notable success with the type. However, it would seem that it took pilots of Guynemer's skill to use the SPAD 12 successfully in combat.

The SPAD 12 prototype was successful and a total of 300 were ordered, although it seems that not all were completed. These aircraft were supplied in limited numbers to some fighter units but it does not appear that they ever equipped an entire escadrille. In August 1917 only one was on strength. The mystery as to why the operational career of the SPAD 12 was so short may be found in a memo dated 28 November 1917. Here it was noted that there were only two SPAD 12s at the front several months after the type was accepted for operational service. Because of delays in production it was decided to abandon plans to supply more aircraft to front-line units. By April 1918 the number of SPAD 12s had risen to eight, and by 1 October there were still only eight at the front. The cannon-armed SPAD 12s remained available in small numbers and were intended for use by the best pilots in each escadrille. Some French aces scored numerous victories with the SPAD 12.

It was soon found that the weight of the cannon severely limited the aircraft's performance and that the cannon was at best tricky to aim and fire. The Vickers machine gun was apparently used to aim along the line of fire and, once this was determined, the pilot had to fire the cannon at the correct moment. The single-shot weapon had to be reloaded in flight by the pilot through the breech, which extended into the cockpit between the pilot's knees. This extension prevented a control stick from being fitted, so the aircraft used a control system similar to that developed by the Deperdussin firm. In addition to

> **SPAD 12 Single-Seat Fighter with 220-hp Hispano-Suiza 8Cb**
> Span 8.00 m; length 6.40 m; height 2.55 m; wing area 20.2 sq. m
> Empty weight 587 kg; loaded weight 883 kg
>
> | Maximum speed: | 2,000 m | 203 km/h |
> |---|---|---|
> | | 3,000 m | 198 km/h |
> | | 4,000 m | 190 km/h |
> | | 5,000 m | 177 km/h |
> | Climb: | 2,000 m | 6 minutes 3 seconds |
> | | 3,000 m | 10 minutes 2 seconds |
> | | 4,000 m | 15 minutes 42 seconds |
> | | 5,000 m | 23 minutes 13 seconds |
>
> Ceiling 6,850 m; endurance 1.75 hours
> Armament: a 12-round, 37-mm S.A.M.C. and a synchronized 7.7-mm Vickers machine gun
> A total of 300 were ordered but it is unlikely all were built.

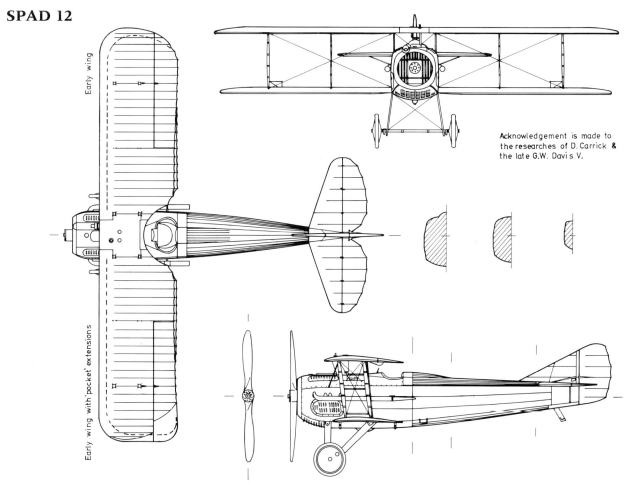

Acknowledgement is made to the researches of D. Carrick & the late G.W. Davis V.

SPAD 12; the cannon (a 12-round, 37-mm S.A.M.C.) fired through the center of the propeller hub. 02735.

Left: SPAD 12. The most obvious changes from the SPAD 7 were the elimination of the tear-drop cylinder fairings and a more streamlined cowling. B76.1875.

Below: Prototype SPAD 12 S.382 test flown by Guynemer—the underside was painted black. B84.854.

these difficulties, the engine vibration and gases from the cannon made the SPAD 12 extremely difficult to fly in combat. It has been recorded that for all these reasons the SPAD 12 was disliked by most pilots and saw little front-line service.

Foreign Service
Russia
At least six SPAD 12s were in service with the RKKVF in the early 1920s. These had serials 1545, 1581, 1586, 1600, 1604, and 1609. They were withdrawn from service in 1925. For details on SPAD aircraft in service with the IRAS see the entry under SPAD 7.

United Kingdom
A single SPAD 12 was acquired by the RFC and was tested in January 1918. It had serial S.449; it was given RFC serial B 6877. It was sent to the Candas site of the No.2 Aircraft Depot

and on 18 March was flown to Martlesham Heath for testing. It crashed on 4 April 1918 while in transit to the Isle of Grain.

United States
A single SPAD 12 was obtained by the A.E.F.

SPAD 13

The SPAD 7 was a successful fighter but was regarded as inadequately armed because it carried only a single machine gun. The better-armed SPAD 12 proved to be too difficult for the average fighter pilot and, clearly another solution was needed.

The answer appeared in the form of the SPAD 13, which appears to have been under development in early 1917. The prototype (S.392) flew in April 1917 at Buc. Extensive tests at Villacoublay seemed to have been completed without difficulty, the main complaint being the limited view from the cockpit.

The armament was two 7.65-mm Vickers machine guns. The added weight required a more powerful engine, and the one chosen was the 200-hp Hispano 8Ba, a version of which had powered the SPAD 12. In keeping with its new engine and increased armament, the SPAD 13 was larger than the SPAD 7. The SPAD 13 retained the characteristic SPAD single-bay bracing with intermediate struts. The cabane struts were staggered but the wings were not. To accommodate the new engine, the cowling was more tapered and teardrop cylinder fairings (which had been eliminated entirely in the SPAD 12) were more prominent than those on the SPAD 7. The fin, rudder, and tailplanes were more rounded on the SPAD 13 than on the SPAD 7. The ailerons were inversely tapered.

SPAD 13 (S.2179) of N 561 flown by Garros. This is an early Blériot-built machine with rounded wingtips. B85.937.

It was hoped that the new SPAD would be in widespread service with front-line escadrilles by mid-1917, and the firm had built 20 pre-production planes instead of a single prototype. Unfortunately, the SPAD 13's introduction to operational service was delayed by difficulties with the new engine, and it did not begin to replace the SPAD 7 until 1918. Although an initial batch of 250 was ordered by the STAé, by December 1917 only 131 had been delivered. It was hoped that 2,230 would be delivered by 31 March 1918; but only 764 were ready for service on that date.

There were many reasons why the SPAD firm was unable to meet the initial production goals, but the main difficulty seems to have stemmed from the Hispano 8B. The SPAD 11s had suffered from poor engine serviceability (which grounded the aircraft two-thirds of the time) and, similarly, the geared Hispano engine was causing difficulties with the SPAD 13s. Most of the problems seemed to be caused by the reduction gearing. These difficulties were secondary to excessive engine vibration and inadequate lubrication. Relatively "quick" fixes were used to solve these problems: the camshaft covers were stayed to decrease vibration, and the pipe from the oil tank to the oil pump was enlarged. Attempts were made to fit a SPAD 13 with a 200-hp Renault 8Gd engine but nothing seems to have come of this.

Other problems came from the main fuel tank, in which bulging occurred because of over-pressure. A change in the pressure relief and a gauge to monitor fuel pressure helped to correct this problem. Other changes included partitioning the fuel tank and giving each compartment its own fuel line. This helped to prevent the interruption of fuel flow when the aircraft was dived or climbed. The radiators used on the SPAD 13 were prone to leakage. The radiator was bolted to the front of the fuselage and, secondary to the previously mentioned problems with engine vibration, leaks developed at the attachment sites, tubes, and the central grill. The problem was eventually corrected by replacing all the radiators with new strengthened ones. Further, an STAé document dated 1 October 1918 noted that the radiator attachment sites were changed to allow more freedom of movement and thus prevent component failure secondary to stress. The replacements were made when the aircraft underwent servicing.

Other changes included revising the wing tips from the rounded configuration to a more squared-off layout. This improved handling by increasing the surface area of the wings and ailerons. It may have been because of the pressing need for putting the SPAD 13s into service (the formidable Fokker D.VII having begun to appear during the spring of 1918) that yet another "quick" fix was devised. Two pieces of three-ply in the shape of a triangle were slipped over the rounded wing tips and laced onto the wing. An RFC report noted that this modification should not be carried out on British aircraft because should the tips break off the aileron could be damaged or jammed. These tips were introduced in the winter of 1918. Several hundred SPADs with rounded tips were built before production changed to SPAD 13s with square tips. A GQG memo of 11 January 1918 warned all SPAD units to watch for damage to or failure of the wing tips. Most of this surveillance was to be carried out at the RGA.

Another modification to existing SPAD 13s was the fitting of a wooden fairing to the drag wire extending from the forward center-section strut to the upper longeron on each side. In addition to these changes, the cooling louvers were replaced by screens to facilitate cooling.

Production of the SPAD 13 was hampered by the problems already mentioned. Aircraft at the front numbered 17 in August 1917 and 764 in October 1918. Manufacturers and estimated production were: SPAD (1,141), A.C.M. de Colombes (361), Adolphe Bernard (1,750), Blériot (2,300), Borel (300), Kellner (1,280), Levasseur (340): Nieuport (700), and Société Anonyme Française de Construction Aéronautiques (300), giving a total of 8,472.

The Kellner aircraft were noted in a GQG memo dated 18 August 1918 as having errors in the rigging requiring modification in the field. Numerous failures of the rigging turnbuckles were also noted, requiring them to be replaced in the field. The rigging wires were said to be weak, and in April 1918 units were advised to replace them.

Variants

1. At least one SPAD 13 was fitted with the Rateau supercharger. This aircraft, S.706, had the supercharger placed behind the pilot's seat and cooling louvers were placed around the fuselage sides and rear to help dissipate the heat from it. It has been reported that the plane with this installation was actually inferior to the standard SPAD 13 and further development was not pursued.
2. In late 1917 another attempt was made to fit the SPAD 13 with a more powerful engine, in this case a 300-hp Hispano-Suiza 8Fb. A cannon-equipped version of the engine (designated as the Hispano-Suiza 8G) underwent testing in September 1918, but development was apparently abandoned in favor of the SPAD 20. Neither of the re-engined SPAD 13s were developed further. Béchereau later designed a fighter using the 300-hp Hispano-Suiza 8Fb which became the SPAD 17.

Operational Service
Army Cooperation

Although best remembered for its service with the Groupes de Combat, a large number of the SPAD 13 escadrilles were assigned directly to Armée units. These escadrilles were used for reconnaissance, light attack, and escort for reconnaissance and bomber units. Reconnaissance missions sometimes ranged as deep as 60 kilometers behind enemy lines. In addition to these duties, the destruction of enemy balloons still had high priority, as did attacking German aircraft and strafing enemy troop concentrations. The SPAD 13s also flew bombing missions. For

SPAD 13 of SPA 48. This machine was built by Blériot. Assigned to GC 18, SPA 48 flew bomber escort missions for Escadre de Bombardement 12 during the Battle of Picardie. B81.2138.

example, on 22 October 1918 SPA 49 (attached to the 7th Armée), bombed the train station at Freiburg. When fitted with cameras the SPAD 13 could be used for high-speed reconnaissance. SPA 38, for example, flew a large number of reconnaissance missions for the 4th Armée in the Champagne sector during 1918. SPA 62 (attached to the 6th Armée) was used to locate one of the huge guns which was shelling Paris. SPA 49 flew as far as 200 kilometers behind enemy lines and took 6,000 photographs during September 1918 alone.

SPAD 13s also saw service with the French escadrilles sent to Serbia, Greece, Italy, and possibly Russia (although the latter probably used only SPAD 7s). They were also employed by units assigned to the D.C.A. to protect French cities from German bombers. These units, numbered in the 300 series, also were employed in combat in the vicinity of the cities they were assigned to protect.

SPAD 13 units assigned to individual Armées from April 1918 were:

SPA 23, assigned to the 2nd Armée, commanded by Capitaine Pinsard and based at Souilly. It was disbanded in August 1919.

SPA 38, assigned to the 4th Armée. It received its SPAD 13s in the spring of 1918 and was under command of Lieutenant Georges Madon. It was assigned to GC Noblette (later GC 22) in 5 June 1918. It was disbanded in August 1919.

SPA 49, assigned to the 7th Armée and commanded by Capitaine Charles Taver beginning in February 1918. It was based at Chaux. On 8 November Lieutenant Roger Labauve assumed command. It became the 6th Escadrille of the 3rd Regiment d'Aviation de Chasse in January 1920.

SPA 62, assigned to the 6th Armée and based at Saint-Amond. It was active over the Picardie front in March 1918. It was commanded by Capitaine Coli until Capitaine Blamautier assumed command it late 1918. In May it was active over Chemin des Dames, where its operations included strafing German troops in the Ourcq valley. In June it participated in attacks at Champagne and the Marne. It moved to the Aisne briefly and then to the 6th Armée sector at Flanders and Brussles. It became the 9th Escadrille of the 1st Regiment d'Aviation de Chasse in January 1920.

SPA 69, assigned to the 10th Armée and based at Dugny and saw action in the Battle of Picardie in March 1918. It was commanded by Lieutenant Malavaille. In June it was active over Chemin des Dames. From 15 June to 15 July it was active during the counter-offensive near the Marne. SPA 69 later moved to Autrey. After the war, it was based Darmstadt in June 1919 and Gemersheim in July. It became the 4th Escadrille of the 3rd Regiment d'Aviation de Chasse in January 1920.

SPA 76, assigned to the 5th Armée and commanded by Lieutenant Vitoux until disbanded in 1919.

SPA 79, assigned to the 3rd Armée. It was commanded by Capitaine Franc-Robert and was based at Maisonneuve from January to March 1918. In March it moved to Ambriel, then Mensil-Saint-Georges. It moved to Sacy-le-Grand in April 1918 and to Auvillers in September 1918. Subsequently SPA 79 moved to Luithellier, Croutoy, and Chambry, then relocated to Lachelle in 1919. On 2 May it made a final move to Sommesous, where it disbanded.

SPA 82, assigned to the 6th Armée and based at Vaubéron. In April it was assigned to the 10th Armée and moved to Fienvillers. By mid-1918 SPA 82 was commanded by Lieutenant Lecoq de Kerland. It was briefly detached to the G.A.N. at Flanders. In May it was active over Chemin des Dames. As the war came to an end, it was assigned to GC 14.

SPA 87, assigned to the 8th Armée. Lieutenant Joseph Jochaud de Plessis assumed command in May 1918. It was disbanded in August 1919.

SPA 90, assigned to the 10th Armée, commanded by Capitaine Weiss, and disbanded in August 1918.

SPA 102, assigned to the 1st and, later in April, to the 6th Armée. It returned to the 1st Armée sector in September. On 5 September Capitaine Andre d'Humeries assumed command. SPA 102 was redesignated the 3rd Escadrille of the 3rd Regiment d'Aviation de Chasse in January 1920.

SPA 152, assigned to the 7th Armée. In June 1918 it joined GC de la Noblette, which became GC 22 in July. It was commanded by Capitaine Echard until August when Capitaine Bonne assumed command.

SPA 156, formed from MS 156 in May 1918. It was assigned to the 4th Armée under command of Lieutenant Jean Thobie and later Capitaine Paumier. It was disbanded in March 1919.

SPA 158, formed from MS 158 in May 1918 and commanded by Lieutenant Chaudron. It was initially assigned to the 3rd Armée but joined GC 23 later in May.

SPA 160, formed from N 160 in May 1918. SPA 160 was formed with both SPAD 7s and 13s. It was active near Verdun assigned to the 2nd Armée and joined GC 23 in August. It was commanded by Lieutenant Barès.

SPA 161, formed from MS 161 in May 1918 under command of Lieutenant Parfait. It was assigned to the 6th Armée at Maisonneuve. In June it moved to Champaubert, where it performed reconnaissance over the front. In September SPA 161 moved to the 6th Armée sector at Flanders. It joined GC 23 in August 1918.

SPA 171, formed in October 1918 and assigned to the 1st Armée under command of Lieutenant Louis Tron. Initially based at Mercières-aux-Bois, SPA 171 moved to Séraucourt-le-Grand the next month. It moved to Trécon in December 1918 and was disbanded in March 1919.

SPAD 13 with photo equipment for high-speed reconnaissance. Renaud.

SPA 173, formed in October 1918 and assigned to the 4th Armée under command of Lieutenant Allez. It was based at La Noblette and moved to Hauviné then Dugny in November. By the end of the war it was assigned to GC 12. Postwar SPA 173 moved to Revigny-sur Ornain, La Neuveville, Manoncourt, and Haguenau. SPA 173 moved to Neustadt-Speyerdorf as part of the occupation force, where it was disbanded on 9 April 1919.

SPA 313, a D.C.A. escadrille commanded by Lieutenant Duron and assigned to Dunkerque. It was disbanded in December 1918.

SPA 314, a D.C.A. escadrille assigned to Nancy. It was commanded by Capitaine Perrin and disbanded in January 1919.

SPA 315, a D.C.A. escadrille assigned to Nancy. It was based at Chaux, commanded by Capitaine Etienne, and disbanded in January 1919.

SPA 412, formed in June 1917 as the Escadrille de Tarcenay which was a D.C.A. unit for Belfort. It was disbanded in early 1919.

SPA 441, assigned to protect the town of Creusot. It became SPA 441 in July 1917 and was disbanded in 1919.

SPA 442, previously designated Escadrille 307 and assigned as a D.C.A. unit for Lyon. It became SPA 442 in July 1917 and was disbanded in 1919.

SPA 461, originally designated Escadrille 395 as part of the C.R.P., was redesignated SPA 461 in July 1917 as part of the C.R.P. It was disbanded in 1919.

SPA 462, formed in March 1917 and initially designated Escadrille 350. It became SPA 462 in July 1917 and was disbanded in 1919.

SPA 463, formed in March 1917 and assigned to the C.R.P. Initially designated Escadrille 351, it became SPA 463 in July 1917. It was disbanded in 1919.

Escadrille 464, formed from Escadrille 393 in July 1917 as part of the C.R.P. It is believed that SPADs were assigned to the unit but this cannot be confirmed. It was disbanded in 1919.

Escadrille 466, formed in March 1915 as 94th D.C.A. and assigned to the C.R.P. It was designated Escadrille 394 and became Escadrille 466 in July 1917. It is believed that SPADs were assigned to the unit but this cannot be confirmed. It disbanded in November 1918.

Escadrille 467, formed in March 1917 as 96th D.C.A. and assigned to the C.R.P. It was designated Escadrille 396 and became Escadrille 467 in July 1917. It is believed that SPADs were assigned to the unit but this cannot be confirmed. It was disbanded in November 1918.

Escadrille 469, formed in March 1915 as 96th D.C.A. and assigned to the C.R.P. It was later redesignated Escadrille 397 and became Escadrille 467 in July 1917. It is believed that SPADs were assigned to the unit but this cannot be confirmed; it is known that the unit carried out several night bombing missions. It was disbanded in November 1918.

Escadrille 470, formed in March 1917 and assigned to the C.R.P. It was designated Escadrille 352 and became Escadrille 470 in July 1917. It is believed that SPADs were assigned to the unit but this cannot be confirmed. It disbanded in November 1918.

Escadrille 471, formed in February 1918 as the 2nd Escadrille Américaine (2nd American Squadron). It became Escadrille 471 sometime later and was assigned to the C.R.P. It is believed that SPADs were assigned to the unit but this cannot be confirmed. It was disbanded in 1918.

Escadrille 472, formed in 1918 as an escadrille postale. It became Escadrille 472 sometime later. It is believed that SPADs were assigned to the unit but this cannot be confirmed. It disbanded before the Armistice.

T.O.E. Units

SPA 506, assigned to the A.F.O. and disbanded in 1919.

SPA 507, assigned to the A.F.O. It became the 3rd Regiment d'Aviation de Chasse in January 1920.

SPA 523, formed from N 523 in 1917/18. It was attached to the Serbian army and was disbanded in 1919.

SPA 531, formed in March 1918 with some SPADs on strength. It had both Greek and French pilots and was disbanded in 1919.

SPA 561, formed in 1917/18 from N 561. It was assigned to the protection of Venice. It was disbanded in December 1918.

Groupes de Combat

The SPAD 13 equipped 90 fighter escadrilles, or nearly every fighter unit in 1918 (except for the one unit with Hanriot HD.3s and those flying Caudron R.11 escort fighters). Most SPAD units served with the Groupes d'Combat, in which several escadrilles were joined as a highly mobile force to provide air support over selected areas of the front. The GC units were usually sent to areas where offensives were planned or under way. In addition

SPAD 13 of SPA 37 in April 1919. By 1920 SPA 37 had become part of 101e Escadrille assigned to 1er GC. B85.125.

to destroying German fighters and reconnaissance planes, the Groupes d'Combat were tasked with the destruction of enemy balloons. These were often more difficult to attack than aircraft because they were well protected by anti-aircraft artillery. Despite the danger, many SPAD units became proficient at destroying balloons. SPA 77 alone destroyed 22 balloons.

In the final months of the war the SPAD units flew extensive ground-attack missions. The units of GC 15, for example, were used for ground attack during the fighting around the Marne from 18 July to 4 August.

Of course, it is in its primary mission of air-to-air combat that the SPAD 13 will be best remembered. The aircraft was flown by such famous aces as Guynemer, Fonck, and Nungesser, to name but a few. As a fighter, the SPAD 13 was more than a match for the Albatros and Pfalz fighters in service during 1917. However, the appearance of the Fokker D.VII was the beginning of the end for the SPAD 13's mastery of the air. A British report dated 3 October 1918 notes that "the 200-hp Hispano-Suiza SPADs are becoming more outclassed every day. Their visibility is bad and their climbing powers are insufficient." Nevertheless, the SPAD 13s remained in service until 1923.

GC 11 (SPA 12, SPA 31, SPA 57, and SPA 154) was assigned to Escadre de Combat 2 early in 1918 as part of Groupement Féquant. In March it participated in the Battle of Picardie in the 1st and 3rd Armée sectors. On 1 June it moved to the 5th Armée sector. The units of GC 11 saw considerable activity during the summer months as they were sent to trouble spots in the 5th Armée sector. GC 11 was involved in the Battle of Chemin des Dames (May), Matz (June), and Champagne (July) On 15 July a new fighter escadrille, SPA 165, joined GC 11. In September GC 11 was in the 4th Armée sector. In early November SPA 172 joined GC 11. On 5 November a move was made to the 8th Armée sector.

On 15 April 1918 **GC 12** consisted of four SPAD escadrilles; their primary equipment was the SPAD 13. SPA 3, SPA 26, SPA 67, and SPA 103 were based at Hetomesnil. GC 12 was assigned to the 1st Armée, and from March to August the groupe was active in the Montdidier sector. May was to prove to be especially active as the Luftstreitkräfte was now using the formidable Fokker D.VII. At the beginning of September GC 12 moved to the 8th Armée sector. During September SPA 167 was assigned to GC 12 when the group moved to the 4th Armée sector; SPA 167 had six SPAD 7s and 12 SPAD 13s on strength. In October GC 12 was based in the Ardennes area. In November it was based at Hauviné in the 4th Armée sector, where bad weather inhibited operations until the end of the war. SPA 173 joined in November.

In April 1918 **GC 13** was assigned to Groupement Féquant and consisted of SPA 15, SPA 65, SPA 84, and SPA 88; all were based at Fouquerolles. SPA 77 joined GC 13 from 14 May to 10 June. GC 13 was active in the battles at the Aisne, Oise, and Marne from 27 May through 19 June. After participating in the Battle of Matz in June, GC 13 transferred to Villers-Saint George, where, during 14 to 24 July, it took part in the Battle of Champagne. In August it was active during the Battle of Picardie. The escadrilles assigned to GC 13 flew a large number of ground-attack missions against German troops, especially during the Battle of Saint-Mihiel in September.

On 15 April 1918 **GC 14** consisted of four SPAD escadrilles all equipped with SPAD 13s (along with some SPAD 7s). These were SPA 75, SPA 80, SPA 83, and SPA 86; all were based at Fienvillers. GC 14 was assigned to the 10th Armée in the reserve of the GQG. From 9 April to 1 May the escadrilles were active during the Third Battle of Flanders. During the German drive on Chemin des Dames and the breakthrough near the Marne, GC 14 supported the 10th and 3rd Armées in the Compiegne region. GC 14 flew contact patrols and attacked enemy troops and transportation. From 18 July to 9 August it was active during the Second Battle of the Marne, the Second Battle of Ourcq, and the Third Battle of Picardie. During the Battles of Champagne, Argonne, and St.-Quentin, GC 14 operated primarily in the army cooperation role, performing reconnaissance and ground attack. SPA 166 was assigned to GC 14 in September. During the Battle of Serre the units of GC 14, especially SPA 83, participated in a large number of aerial combats. As the war drew to a close GC 14 moved to the 1st Armée sector, where it participated in the Second Battle of Guise on 4 November.

GC 15 on 15 April 1918 consisted of four escadrilles with SPAD 13s (SPA 37, SPA 81, SPA 93, SPA 97) and one unit with Caudron R.11 long-range escort fighters. GC 15 was assigned to Groupement Ménard as part of the 1st Escadre. It provided protection for GB 5, GB 6, and GB 9. GC 15 moved Plessis-Belleville during the Third Battle of the Aisne. All the escadrilles saw action during the Battle of Lassigny (Metz) from 9 to 11 June. GC 15 later participated in the Allied counter-attacks at the Marne (Second Battle of the Marne) from 15 to 18 July and the Aisne-Marne operation 18 July to 4 August. During these battles the escadrilles performed a large number of trench-strafing missions. On 17 August the group moved to the 1st Armée sector, where GC 15 participated in the reduction of the Amiens salient. Beginning on 6 September, it participated in the St. Mihiel Offensive.

On 15 April 1918 **GC 16** was assigned to the 3rd Armée and consisted of four SPAD 13 escadrilles: SPA 78, SPA 112, SPA 150, and SPA 151. They were based in the 3rd Armée sector on 6 July. GC 16 was assigned to support the 1st Army, American Expeditionary Force, for the St. Mihiel Offensive on 2 September. At this time the group was under the administrative control of the 8th Armée. During the St. Mihiel Offensive GC 16 was to maintain a presence over the western sector of the front, which was being covered by the 1st Pursuit Wing. The group attempted to destroy enemy balloons, flew reconnaissance patrols, and provided protection for observation aircraft.

SPAD 13 wearing the wasp insignia of SPA 89; the escadrille was assigned to GC 17. B91.2284.

SPA 168 joined GC 16 on 24 September. On 26 September GC 16 moved to Trecon on the 4th and 5th Armée fronts. On 31 October SPA 176 joined the group.

On 15 April 1918 **GC 17** was assigned to Groupement Féquant and had four SPAD 13 escadrilles: SPA 77, SPA 89, SPA 91, and SPA 100. They were all based at Fouquerolles. GC 17 was active from 27 May to 3 June during the Battle of the Aisne. SPA 77, which had been briefly assigned to GC 13 from 14 May to 10 June, rejoined GC 17 in mid-June. Also in June SPA 100 was briefly detached to provide air support for the 3rd Armée during the Battle of Matz and the Battle of Montagne de Reims. In preparation for the forthcoming Battle of Matz, GC 17 moved to an airfield in the 3rd Armée sector on 10 June. By 23 July it had moved to Plessis-Belleville, where all the escadrilles were active during the Aisne-Marne operations. GC 17 moved to Rancourt on 6 September in preparation for the St. Mihiel Offensive, in which it would provide air support for the American armies. On 29 October another move was made to the 4th Armée sector, where GC 17 remained until the end of the war. By 11 November it was based at Auve and consisted of SPA 77, SPA 89, SPA 91, SPA 100, and HD 174. GC 17 was still part of Escadre de Combat 2 assigned to Groupement Féquant.

In April 1918 **GC 18** consisted of four SPAD escadrilles: SPA 48, SPA 94, SPA 153, and SPA 155. It was assigned to Escadre de Combat 1. During the Battle of Picardie (March to April) the fighter escadrilles flew a large number of bomber escort missions for Escadre de Bombardement 12 (GB 5, GB 6, GB 9, and R 46). GC 18 was based at Montagne in the 2nd Armée sector during the Third Battle of the Aisne and the Marne (May to June). It moved to the Champagne front on 18 July to support the 5th Armée. In August GC 18 moved to the Oise sector to provide air support for the 1st Armée. Later in August GC 18 moved to the 2nd Armée sector, where the escadrilles flew numerous ground-attack and reconnaissance sorties during the counter-offensive at the Marne, as well as during the Battle of Santerre, the St. Mihiel Offensive, and the Battle of Champagne. On 23 September, GC 18 returned to the 4th Armée sector. One of GC 18's last missions of the war was to support the attack on Philipeville and Rocroi by the bomber units of Groupement Ménard on 10 November.

GC 19 consisted of four SPAD 13 units: SPA 73, SPA 85, SPA 95, and SPA 96. Assigned to the 1st Escadre and initially based at Pierrefonds, the group moved to Airaines and participated in the Battle of the Somme. On 29 May a move was made to Verrines, where GC 19 was assigned to support the 5th, 6th and 10th Armées. This move was necessitated by the initial success of the German 1st and 7th Armies along the front from Berry-au-Bac to Anizy. On 2 June GC 19 moved to Plessis-Bellville and participated in operations west of the Ham-Noyon sector. During the German attack on Château Thierry and Argonne GC 19 operated in support of the 4th Armée. In July it was active in the Battle of Champagne and provided air cover for the 5th, 6th, 9th, and 10th Armées during the drive to recapture Château-Thierry. GC 19 later supported the 1st and 3rd Armées during the reduction of the Amiens salient from 8 to 15 August; it was based at Fouquerolles. GC 19 was placed at the disposal of Colonel William Mitchell during the reduction of the St. Mihiel salient from 12 to 15 September. GC 19 was

SPAD 13 of Le Cicognes. Renaud.

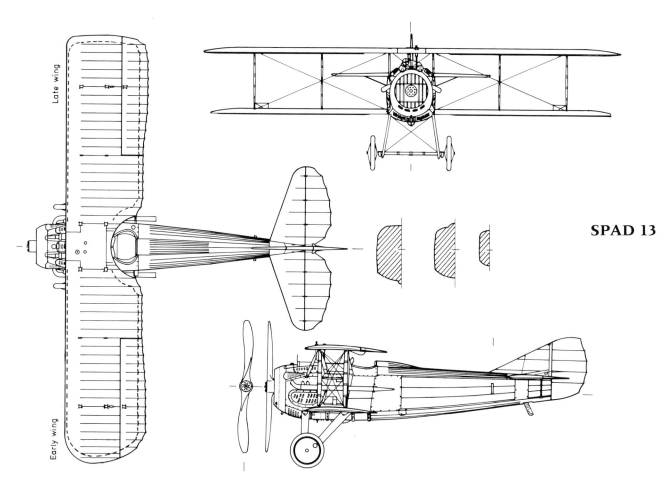

SPAD 13

initially based at Alger, but subsequently moved to airfields at Chalôns-sur-Marne and finally Gonderville.

GC 20 was formed in February 1918 with SPA 68, N 99, N 159, and N 162. By 15 April GC 20 was based at Mannoncourt and assigned to the G.A.E. (7th Armée). By May all the units had re-equipped with SPAD 7s and 13s, and GC 20 was transferred to the Aisne sector in response to a major German offensive in that sector. However, GC 20's pilots were up against Jagdgeschwader 1, and this powerful opposition extracted a heavy toll. In July GC 20 was transferred to the Champagne front and was assigned to Escadre 2. During the Somme offensive, it moved to Plessis-Belleville. By the time of the Armistice GC 20 was based at Tilley, still assigned to Escadre 2.

On 15 April **GC 21** had two SPAD escadrilles (SPA 124 and SPA 163) and two Nieuport escadrilles (N 98 and N 157). GC 21 was assigned to the 4th Armée and was stationed at la Noblette. On 28 May it was assigned to the 6th Armée sector. However, due to the rapid advances being made by the German armies between Noyon and Reims, GC 21 was sent instead to support the 1st Colonial Corps in the 5th Armée sector south of Reims. The units were active during the Battle of Chemin des Dames and performed fighter patrols and liaison missions with the ground units. During the five-day battle (25 May through 1 June) GC 21 flew 209 sorties and downed 12 aircraft and four balloons. On 2 June it moved to the 6th Armée sector, where it patrolled the area between Château-Thierry and Villers-Cotterets. GC 21 was assigned to the 4th Armée on 3 July 1918. GC 21 and GC 22 provided aerial coverage for the 4th and 21st Corps d'Armée. During August a new unit, SPA 164, was assigned to GC 21. In November SPA 124 was detached from GC 21 to serve with the Commandant de l'Aéronautique de la Region Parisienne (SPA 124 was reassigned to GC 21 after the war ended). Shortly before the Armistice SPA 175 joined GC 21.

GC 22 was formed in June 1918 with SPA 38, SPA 87, SPA 92, and SPA 152. By 15 July it was based at la Noblette and assigned to the 4th Armée. Its units were active in the Battle of Champagne, which began 18 July. The main duty of the fighter units at this time was to provide support for the advancing French armies. This took the form of fighter sweeps to clear the air over the battlefield of enemy fighter and reconnaissance aircraft, as well as balloons. In addition to these missions, the escadrilles provided escort for the bomber and army cooperation units and made ground-attack sorties against German troops. In late July, GC 22 moved to Bezu in support of the 6th Armée. At this time GC 21 and GC 22 were assigned to provide support for the 4th and 21st C.A. Throughout August GC 22 participated in the Second Battle of the Marne and the Battles of Soisonnais and l'Ourcq. Its escadrilles flew in mixed formations consisting of planes from several units for maximum striking power. At the end of August GC 22 left the 6th Armée sector and was based at Coincy in support of the 5th Armée. From this base it participated in the Battles of Champagne, Argonne, and Saint Thierry. GC 22 moved to the 4th Armée sector in mid-October. Also in October a new unit, SPA 169, joined GC 22. The escadrilles of GC 22 had been employed in providing direct support for the various French army units and the group was never assigned to any of the Escadres de Combat.

GC 23 was formed 24 August 1918 and was assigned to the 3rd Armée. It consisted of escadrilles SPA 82, SPA 158, SPA 160, and SPA 161. Shortly after its formation, GC 23 was reassigned to the 1st Armée. By 25 September it was based in the 6th Armée sector, and on 28 September it was joined by SPA 170. GC 23 participated in the attacks on the Hindenburg line and at Noyon in the 6th Armée sector. During the German retreat the fighter units were being utilized for ground attack; despite inclement weather the escadrilles of GC 23 continued their attacks. In late September GC 23 moved to Capelle, where it was active over the Belgian front. After participating in the Battle of Lys and Escaut, GC 23 moved to Rumbecke, where it ended the war. The unit had nine confirmed and three probable victories for the loss of ten pilots during its brief existence.

SPAD 13 with camera gun. MA8925.

SPAD 13 downed by Ballon Kompagnie No.3 on 3 September 1918. R. Stach via Colin Owers.

Postwar Service

Groupes d'Combat GC 11, GC 14, GC 16, GC 20, GC 21, GC 22, and GC 23 were disbanded in 1919. The remaining GC units were assigned to Groupement de Chasse 1; GC 15 (1er Groupe), GC 18 (2e Groupe), GC 19 (3e Groupe), and Groupement de Chasse 2; GC 12 (1er Groupe), GC 13 (2e Groupe), and GC 17 (3e Groupe).

By 1920 many of the SPA escadrilles had also been disbanded. Those that remained were reorganized. They were:
1er GC, 101e Escadrille (SPA 37, SPA 3, SPA 96); 102e Escadrille (SPA 81, SPA 26, SPA 97); 103e Escadrille (SPA 93, SPA 103, SPA 102).
2er GC, 104e Escadrille (SPA 31, SPA 15, SPA 69); 105e Escadrille (SPA 48, SPA 65, SPA 88); 106e Escadrille (SPA 94, SPA 84, SPA 49).
3er GC, 107e Escadrille (SPA 73, SPA 57, SPA 67); 108e Escadrille (SPA 95, SPA 77); 109e Escadrille (SPA 62, HD 174).

By January 1920 there were a total of 25 fighter escadrilles, all but one being SPAD units. In addition to those listed above there were four SPA escadrilles still in existence after the war. SPA 124 became the 10th Escadrille of the 2nd RAC on 1 January, 1920. The other SPA 124 (the Lafayette Escadrille or 103rd Aero Squadron) became the 7th Escadrille of the 35th RAO in 1920. SPA 162 became the 8th Escadrille of the 3rd RAC on 1 January, 1920. SPA 507 became the 9th Escadrille of the 3rd RAC on 1 January, 1920.

The SPAD 13 remained in service as late as 1923. However, as the new SPAD 20, Nieuport 29, and Gourdou-Leseurre C1 became available the SPAD 13s were replaced in front-line service.

Foreign Service

Argentina
Argentina purchased two SPAD 13s after the war. In 1919 these were assigned to the Escuela de Aviacion Militar, later redesignated Grupo de Aviacion 1.

Belgium
Thirty-seven SPAD 13s were purchased by Belgium from France in 1918. They were given serials S1 to S37 and arrived in March 1918. They were assigned to the 10th Escadrille to allow that unit to replace its SPAD 7s. Postwar, an additional order for a number of SPAD 13s was placed. These were assigned to the 3rd, 4th, and 10th Escadrilles. They were replaced in the early 1920s by Nieuport 29s.

Brazil
Brazil obtained SPAD 13s in 1920, given serial numbers 2952–2961 and 2971–2980. They were assigned to the Escola de Aviacao Militar at Campo dos Afonsos. Nine were assigned to the 1st Esquadrilha de Caca when that unit was established in 1922. By 1928 the SPAD 13s had been struck off charge.

Czechoslovakia
Fifty SPAD 7s and 13s were obtained by Czechoslovakia after the war. SPAD 13s were initially labeled SPAD IIs, but later were redesignated B-21s. SPAD 13s were assigned to in 1919 to Letecke Sklady (at Praha), and in 1923 to 32 and 34 Stihaci Rota (Praha-Kbely), 33 Stihaci Rota (Olomouc), and 31 Stihaci Rota (Nitra). They were withdrawn in 1924.

Greece
By the end of 1918 the Greek 531 Mira had received eight SPAD 13s. The unit was disbanded in 1919 and many of its aircraft were sent to the 534 Mira, which was based at Salunica-Lebet. It had been planned to send 534 Mira to support the White Russians. However, only one SPAD 13 was sent before the expedition was abandoned. It appears that 532 Mira also had some SPAD 13s on strength.

During the Greco-Turkish War only one SPAD-equipped unit saw active service—534 Mira, re-designated B Squadron. The other SPAD-equipped unit was 532 Mira, re-designated A Squadron. It, however, remained in Greece until 1921. By 1923 the SPAD 13s were barely air worthy and had been replaced by Gloster Nighthawks.

Italy
Although the SPAD 7 had not been a great success in Italian service, the Italians arranged to import several SPAD 13s. It may have been felt that the twin guns would offset the biggest complaint about the SPAD 7, that its armament was inadequate.

Among the first units to be equipped with the SPAD 13 were the only two squadriglias still using SPAD 7s: 77a and 91a.

In 1918 the following units had SPAD 13s on strength:
10 Gruppo (Supreme Command): 70a Squadriglia.
17 Gruppo: 71a, 77a, and 91a Squadriglias.

As of October 1918 there were 48 SPAD fighters on strength, including both SPAD 7s and 13s. By January 1923 only two units were still using SPAD 13s:
13 Gruppo: 78a Squadriglia.
23 Gruppo: 91a Squadriglia.

By the end of 1923 the following units had converted to SPAD 7/13s: 6 Gruppo: 72a and 76a Squadriglias. In 1925 the number of squadriglias equipped with SPAD 13s was at its peak with eight units:
13 Gruppo: 77a, 78a, and 82a Squadriglias.
17 Gruppo: 83a Squadriglia.
23 Gruppo: 74a, 75a, 76a, 91a Squadriglias.

The new Nieuport 29 now began to replace the SPAD 13s. In 1925 Squadriglias 74a, 75a, 76a, and 91a all changed from SPAD 13s to Nieuport 29s. Four units still had SPAD 13s on strength:
6 Gruppo: 81a Squadriglia.
13 Gruppo: 78a and 82a Squadriglias.
17 Gruppo: 83a Squadriglia.

By 1926 there were only two units using SPAD 13s; Nos. 78a

and 82a Squadriglias. Both were assigned to 13 Gruppo.

By 1927 both 78a and 82a Squadriglias had re-equipped with Fiat CR.1s. Although the last SPAD 13 had been retired from front-line service with the Regia Aeronautica, some would continue to serve with training units.

Japan

The SPAD 13 was selected by the Japanese army air service for front-line service. Originally, 40 were imported to Japan by the French military mission. The French instructed the Japanese pilots in advanced aerobatics. In December 1920 the SPAD 13 was officially adopted by the army and given the designation Hei I.

The SPAD 13s equipped the Japanese army fighter squadrons (which consisted of three squadrons assembled into one battalion) and several squadrons divided among four "mixed" battalions. The SPAD 13s were subsequently replaced by Nieuport 29s.

Katsunami Ishibashi, a famous civilian pilot, bought three SPAD 13s after the war, but they were destroyed in a hangar fire. He then built a racing aircraft from the salvaged parts and equipped it with a 180-hp Hispano-Suiza engine.

Poland

The fledgling Polish air service used a number of SPAD 13s. The 2nd Polish Combat Unit was formed in December 1917 and was equipped with either SPAD 7s or 13s. The unit was based at Kanev and was soon captured by the Germans.

In May 1919 the French sent SPAD 13s to the Poles when SPA 162 was redesignated the Polish 19th Squadron. An additional 40 SPADs were later ordered from the French.

The 19th Squadron had ten SPAD 13s in September 1920. The unit was based at Warsaw and was active on the northeast front from April to October 1920 during the Polish-Soviet War. In 1921 the 19th Squadron was incorporated into the 18th Fighter Squadron.

The 18th Fighter Squadron, using SPAD 13s absorbed from the 19th, was based at Deblin. The unit was assigned to the 1st Dyon based at Warsaw. Later, the 19th Fighter Squadron was attached to the 3rd Fighter Dyon. At the time of the cease-fire in the Polish-Soviet War, the unit was based at Grudziadz.

The 14th Fighter Squadron was equipped with ten SPAD 13s in July 1919. By March 1920 the unit had given them up for Fokker D.VIIs.

Russia

It seems that the French need for SPAD 13s and the beginning of the Russian revolution resulted in no SPAD 13s being used by the Imperial Russian Air Service. A single SPAD 13 was obtained for evaluation but is unclear if this was sent from France or was captured during the Polish-Soviet War.

Siam

Several SPAD 13s were purchased by Siam in 1920. They made a number of air mail flights on an experimental basis. In 1920 the SPAD 13s, along with Nieuport 29s, were front-line fighters for the Royal Siamese Aeronautical Service and equipped the 1st Pursuit Group. They appear to have remained in service as late as 1930.

Spain

A small number of SPAD 13s were obtained by Spain, probably in 1921. By 1922 they were based at Los Alcazares and used as advanced trainers for fighter pilots. Three participated in air races at Valencia.

Turkey

Several SPAD 13s were obtained by the Turkish air service after the war. They saw service in Asia Minor in the war against the Greeks.

United Kingdom

A SPAD 13 was examined by a representative of the RFC in late April 1917, resulting in a single example (S.498) being

SPAD 13 of the Japanese air service in 1919. A total of 40 were imported to Japan by the French Military Mission. B85.1863.

purchased and sent to No.2 Aircraft Depot at Candas in June 1917. Allotted the RFC serial B 3479, it was tested extensively and found to be markedly superior to other fighters in RFC service. The aircraft was subsequently sent to No.19 Squadron and four victories were recorded on it. It was later sent to No.23 Squadron, where it survived until March 1918, when it was struck off charge.

The aircraft's performance had proven sufficiently impressive for the RFC to issue an order for 60 SPAD 13s from the Kellner firm. The first were delivered in November 1917. Production delays caused by shortages of aluminum and fabric resulted in

SPAD 13 Single-Seat Fighter with 200-hp Hispano-Suiza 8Ba, 8Bb, or 8Bd

Span 8.25 m (later SPAD 13s had span of 8.08 m); length 6.25 m; height 2.60 m, wing area 21.11 sq. m (later SPAD 13s had an area of 20.2 sq. m)

Maximum speed:	1,000 m	211 km/h
	2,000 m	208.5 km/h
	3,000 m	205.5 km/h
	4,000 m	201 km/h
	5,000 m	190 km/h
Climb:	1,000 m	2 minutes 20 seconds
	2,000 m	5 minutes 17 seconds
	3,000 m	8 minutes 45 seconds
	4000 m	13 minutes 5 seconds
	5,000 m	20 minutes 10 seconds

Ceiling 6,800 m; endurance 2 hours
Armament: two 7.65-mm Vickers machine guns: in U.S. service two 0.303-inch Marlin machine guns; four 25-lb Cooper bombs
Production totals given for SPAD 13s range from 7,300 to 8,472

SPAD 13 Single-Seat Fighter with 220-hp Hispano-Suiza 8Bc or 8Be

Span 8.25 m (later SPAD 13s had span of 8.08); length 6.25 m; height 2.60 m, wing area 21.11 sq. m (later SPAD 13s had an area of 20.20 sq. m)
Empty weight 601.5 kg; loaded weight 856.5 kg
Maximum speed: 218 km/h at 2,000 m; climb to 2,000 m in 4.67 minutes; ceiling 6,800 m; endurance 1.67 hours
Armament: two 7.65-mm Vickers machine guns

SPAD 13 Single-Seat Fighter with Supercharged Hispano-Suiza

Span 8.20 m; length 6.30 m; height 2.30 m; wing area 20 sq. m
Empty weight 599 kg; loaded weight 856 kg
Maximum speed: 225 km/h; climb to 2,000 m in 4 minutes 40 seconds; ceiling 7,000 m
One built

SPAD 13 of the 22nd Aero Squadron flown by Lt. W. Watson. This machine was built by Kellner and has serial number 7697. The Kellner aircraft were noted to have defective rigging which required modification in the field. B76.615.

Below left: SPAD 13 of the USAS built by Levasseur, which produced a total of 340 SPAD 13s during the war. MA20990.

Below: SPAD 13 of the USAS. This aircraft was built by Levasseur. MA742.

the last not being delivered until late March 1918. These aircraft featured the rounded wing tips, although some were later modified to have the squared-off edges. It appears from RFC records that only No.23 Squadron used the SPAD 13 in any numbers. The squadron was not completely re-equipped with Sopwith Dolphins until 4 May 1918, when it finally gave up its last SPAD 13. The unit was active in the Battles of Arras, Messines Ridge, and the Third Battle of Ypres, its SPAD 13s being used for fighter, ground attack, and wireless interruption flights. No.19 Squadron may also have received some SPAD 13s to supplement its SPAD 7s.

United States

The Bolling Commission had recommended that the United States standardize its fighter production on the SPAD 13. However, all 893 SPAD 13s obtained during the war came from France or Britain. The Curtiss company had contracted to produce 2,000, but these were later canceled. Production delays resulted in the Americans using Nieuport 28s and even SPAD 7s. When the SPAD 13s did arrive there were many pilots who expressed a preference for the more agile Nieuports. According to the 22nd Aero Squadron's historian, the SPAD 13s were found to have leaky tanks, broken water pumps, clogged water systems, fouled carburetors, shorting magnetos, broken gun gears, and even inaccurate rigging. These problems were gradually overcome and the United States became the largest foreign user of the SPAD 13. In the field many of the U.S. SPAD 13s were fitted with Marlin machine guns. Most American fighter units began to receive the SPAD 13s in July 1918. They were:

1st Pursuit Group (1st Army): 24th Aero Squadron, 27th Aero Squadron, 94th Aero Squadron, 95th Aero Squadron, 147th Aero Squadron, 185th Aero Squadron (joined October 1918 as the only U.S. night fighter unit in World War I).
2nd Pursuit Group (1st Army): 13th Aero Squadron, 22nd Aero Squadron, 49th Aero Squadron (beginning August 1918), 103rd Aero Squadron (July 1918), 139th Aero Squadron.
3rd Pursuit Group (1st Army): 28th Aero Squadron, 49th Aero Squadron (until August 1918), 93rd Aero Squadron, 103rd Aero Squadron (from August 1918), 213th Aero Squadron 1st Army Observation Group (1st Army): 24th Aero Squadron (September 1918 to April 1919), 91st Aero Squadron.
4th Pursuit Group (2nd Army): 17th Aero Squadron, 141st Aero Squadron.

The SPAD 13 squadrons participated in these battles:
Champagne-Marne offensive (July to August 1918)—27th, 94th, 95th, and 147th Aero Squadrons.
Aisne-Marne (July to August 1918)—27th, 94th, 95th, and 147th Aero Squadrons.
Toul (September 1918)—49th, 91st, 93rd, 94th, 95th, 103rd, 139th, 147th, and 213th Aero Squadrons.
St. Mihiel Offensive (September 1918)—13th, 22nd, 24th, 27th, 28th, 49th, 91st, 93rd, 94th, 95th, 103rd, 139th, 147th, and 213th Aero Squadrons.
Meuse-Argonne Offensive (September to November 1918)— 13th, 22nd, 24th, 27th, 28th, 49th, 91st, 93rd, 94th, 95th, 103rd, 139th, 147th, 185th, and 213th Aero Squadrons.

Postwar, many of the SPAD 13s were sent to the United States. Relegated to training duties, many of them were equipped with 180-hp Wright-Hispano-Suiza Es. These aircraft, designated SPAD 13Es, had a revised radiator and nose contour.

The 1st Pursuit Group (27th, 94th, 95th, and 147th Pursuit Squadrons) was assigned to the 8th Corps Area at Kelly Field in 1920. It was equipped with the S.E.5a supplemented by SPAD 13s withdrawn from storage as the British aircraft were destroyed or damaged in accidents. The 1st Pursuit Group moved to Ellington Field in Texas on 1 July 1921 and then to Selfridge Field on 1 July 1922. Its squadrons used the SPAD 13s until they were replaced by Boeing-built MB-3As.

Uruguay

Eight SPAD 13s were purchased by Uruguay in 1917. They were based at Pazo de Mendoza. The first operational unit, the Escuadrilla de Caza, was formed with two SPAD 13s in January 1924.

Above: SPAD Serial S.8170 in 1918. M.T. Cottam via Colin Owers.
Top: SPAD 13 of the USAS in 1919. M.T. Cottam via Colin Owers.

SPAD 13. University of Texas at Dallas.

SPAD 14

The Aviation Maritime depended on land-based aircraft to provide protection for seaplanes on anti-submarine patrol. However, land-based fighters had too limited a range to permit them to accompany the seaplanes during these patrols. An attempt was made to use Voisin, Farman, and Letord long-range reconnaissance aircraft for this duty, but the performance of these planes was not adequate.

To remedy this situation, André Herbemont designed a floatplane version of the SPAD 12. Built by Pierre Levasseur, it featured a fuselage from a SPAD 12 fitted with two floats and larger wings and tail surfaces. The larger wings provided more lift to compensate for the weight and drag of the floats, while the enlarged tail surfaces improved lateral stability. Because of the larger wings, it was rigged as a true two-bay aircraft. The floats were designed by Maurice Payonne (who would become the head of F.B.A. in 1926). Those on the prototype were built by the Tellier firm. The engine was the same 220-hp Hispano-Suiza 8Cb used on the SPAD 13. Armament consisted of a 0.303-inch Vickers machine gun and the 12-round, 37-mm S.A.M.C. cannon.

The prototype first flew in November 1917. It was tested at Saint Raphaël by a pilot named Bequet. Apparently these trials were successful enough to warrant the Aviation Maritime to place an order for 39 aircraft. They were built by the SPAD firm and the floats by Levasseur.

SPAD 14 on its beaching gear. These aircraft were built by the SPAD firm and the floats were built by Levasseur. B87.5019.

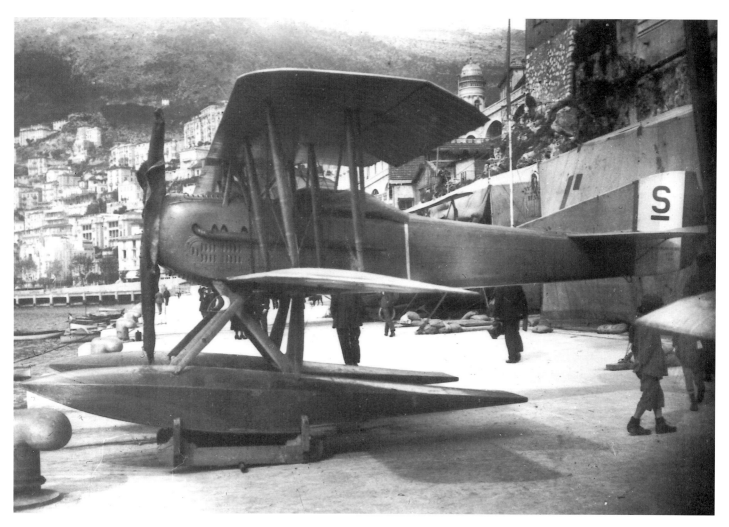

SPAD 14. These aircraft were assigned to the French naval air station at Dunkerque under the command of Lieutenant La Burthe. B87.30.

These aircraft were assigned to the French naval air station at Dunkerque under the command of Lieutenant La Burthe. The SPAD 14 remained in service until 1919. Postwar, one was entered in the 1920 Monaco seaplane race. A landplane version was developed as the SPAD 24. (See SPAD 24 section.)

SPAD 14 Single-Seat Floatplane Fighter with 220-hp Hispano-Suiza 8Cb (Some Sources State 8Bc)

Span 9.80 m; length 7.40 m; height 4.00 m; wing area 26.20 sq. m
Empty weight 770 kg; loaded weight 1,060 kg
Maximum speed: 205 km/h at sea level; climb to 2,000 m in 7 minutes and 18 seconds; ceiling 5,000 m
Armament: one 0.303 Vickers machine gun and a 12-round, 37-mm S.A.M.C.
Forty were built (including the prototype)

The prototype of the SPAD 14. Because of its enlarged wings, the aircraft was rigged as a true two-bay aircraft. B80.750.

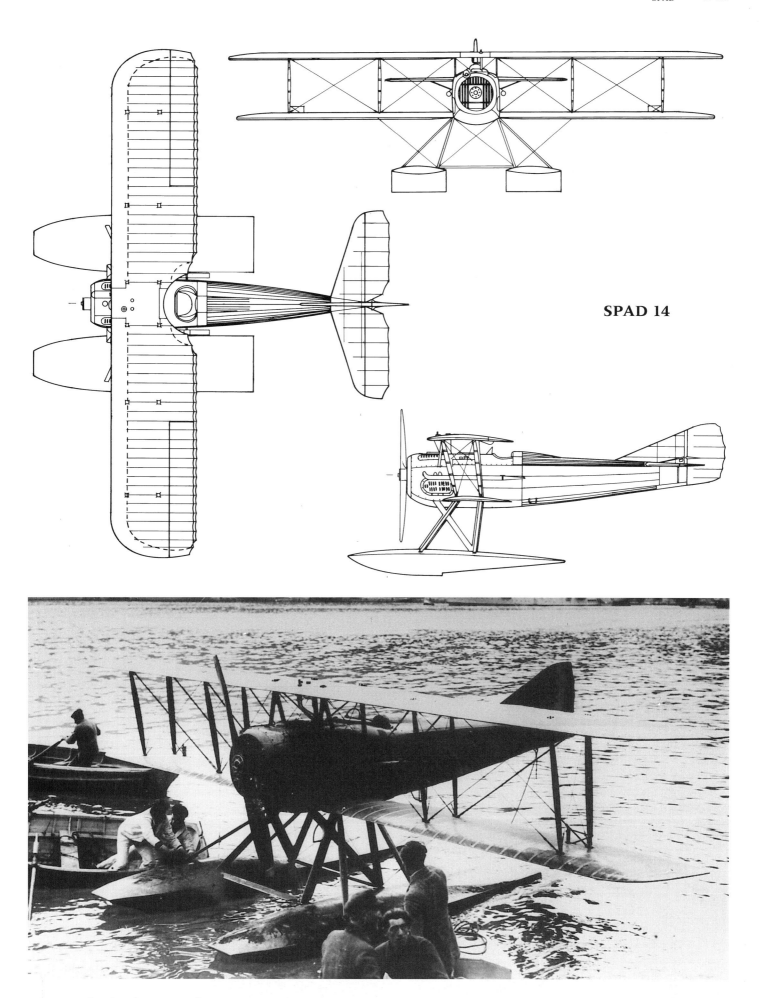

SPAD 14. University of Texas at Dallas.

SPAD 14 on its beaching gear. MA6895.

SPAD 15

The C1 category of 1918 called for a medium-altitude fighter with a 220-kg payload, a service ceiling of 6,500 m, a maximum altitude of 9,000 m, and a speed of 240 km/h. Among the engines selected for the category was the 160-hp Gnome 9Nc rotary. Five aircraft used this engine—the Morane-Saulnier 27 and 29, the Courtois-Suffit-Lescop, the Nieuport 28, and the SPAD 15.

The SPAD 15/1 was designed by André Herbemont and its molded monocoque fuselage of wood was reminiscent of the SPAD B. The SPAD 15 was apparently intended to be given to the French ace Nungesser. A single-valve, 160-hp Gnome 9Nc Monosoupape engine was fitted. The single-bay wings were unstaggered and had equal spans.

The aircraft first flew on 31 July 1917. Unfortunately, because of the low power of the engine its performance with two 0.303 Vickers machine guns fitted was no better than that of the SPAD 13 it was intended to replace. However, the molded fuselage proved to be quite lightweight (hence the performance comparable with the much more powerful SPAD 13) and the aircraft was remarkably maneuverable. Thus, while the SPAD 15/1 was not selected for production because of the engine's inadequacies, the concept of the molded wooden fuselage was considered to be quite successful and was widely used by Herbemont in his subsequent designs.

The SPAD 15/2 was built in August 1917, intended to be piloted by Madon. It featured a slightly larger wing span and a redesigned tail. As with the SPAD 15/1, troubles with the Gnome Monosoupape prevented further development.

SPAD 15/1 Single-Seat Fighter with 160-hp Gnome 9Nc Monosoupape

Span 7.10 m
Armament: two 0.303 Vickers machine guns
One built

SPAD 15/2 Single-Seat Fighter with 160-hp Gnome 9Nc Monosoupape

Span 7.10 m; length 5.35 m; height 2.30 m; wing area 17.50 sq. m
Empty weight 368 kg; loaded weight 625 kg
Maximum speed: 199 km/h at 2,000 m; climb to 2,000 m in 5.67 minutes; endurance 2.5 hours
One built

SPAD 15/3 Single-Seat Fighter with 160-hp Gnome 9Nc Monosoupape

Span 7.10 m; length 5.51 m; height 2.30 m; wing area 16 sq. m
Empty weight 350 kg; loaded weight 500 kg
Maximum speed: 170 km/h; ceiling 5,000 m
One built

SPAD 15/5 Sporting Aircraft with 80-hp Le Rhône

Span 7.10 m; length 5.51 m; height 2.30 m; wing area 16 sq. m
Empty weight 350 kg; loaded weight 500 kg
Maximum speed: 170 km/h
Two built

SPAD 15 fitted with a cône de pénétration flown by Charles Nungesser. This photograph was taken at Villacoublay in June 1919. B81.1543.

The SPAD 15/3 had a wing with reduced chord and lengthened fuselage; it was flown in 1918.

The SPAD 15/4 was to have been powered by a 170-hp Le Rhône 9K, but apparently it was never completed.

The SPAD 15/5 was a postwar development of the SPAD 15 series, designed at the request of Fonck and Nungesser, who predicted a need for a light sporting aircraft with performance comparable to combat aircraft. It was powered by an 80-hp Le Rhône engine with a huge cowling ("casserole"). It was test-flown on 18 May 1919. The aircraft built for Nungesser was painted with the insignia of the "Hussards de la Mort." The one built for Fonck (serial N 2944) was painted with the letters "RF" on the side and carried the registration F-ONCK.

None of the military SPAD 15s was used by the Aviation Militaire.

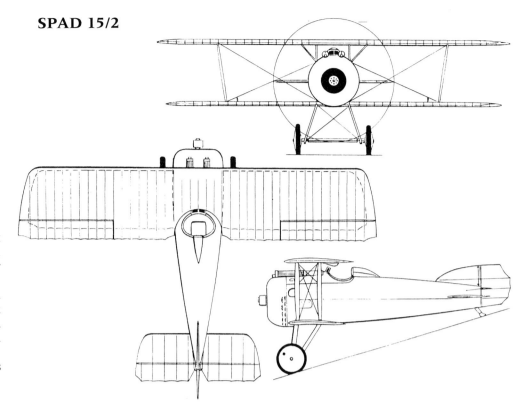

SPAD 15/2

SPAD 15. The SPAD 15/1 was designed by André Herbemont and employed a molded wooden monocoque fuselage. University of Texas at Dallas.

SPAD 16

An improved version of the SPAD 11, designated SPAD 16, was produced in 1917. It featured a 240-hp (some sources say 250-hp) Lorraine 8Bb engine. The more powerful engine improved the aircraft's performance, but did nothing to correct the design flaws that were present in the SPAD 11. Furthermore, the type was still slower than the Breguet 14 A2 and Salmson 2 A2.

The SPAD 16 was a two-bay biplane. There were ailerons on the top wing only; both wings had a slight sweepback of 0.40 degrees. There were two spruce spars per wing. The leading edge was made of spruce, while the trailing edge was made of steel wire. The ribs were made of plywood and the wing was fabric-covered and coated with a special enamel ("Emaillite") recommended by the SFA. The tail had a D-shaped rudder with a triangular, fixed fin. The rudder was 1.20m in height with a surface area of 0.840 sq. m; the fin had a surface area of 0.40 sq. m. The surface area of the horizontal stabilizer was 1.440 sq. m.

The fuselage had a quadrangular shape and was constructed of four spruce longerons held together by piano wire. The wire criss-crossed the entire fuselage frame to ensure its strength. The fuselage formers were also made of spruce. The undercarriage had V-shaped struts which were fixed to the fuselage spars, and bungee chords served as shock absorbers. The undercarriage also had diagonal struts to ensure rigidity. The articulated axle passed through openings in the landing gear struts; steel tubes on either side of the axle were used for added strength. The tail skid was made of ash and had a steel tip. As with the main undercarriage, the skid was mobile and bungee chords served as shock absorbers. The fuel tank held 140 liters and was located under the pilot's seat. The tank could be removed through an access panel in the fuselage. Panels in the fuel tank could be opened to immediately empty the tank in case of fire. There was an auxiliary fuel tank with a capacity of 25 liters located in the top wing. A cylindrical oil reservoir was located behind the motor and was mounted laterally. The radiator was octagonal in shape

A lineup of SPAD 16s of SPA-Bi 255. The aircraft nearest the camera has serial S.8133 and was built by Blériot. SPA-Bi 255 received its SPAD 16s in June 1918. B84.3168.

and had adjustable slats. Armament consisted of a synchronized gun mounted just to the left of mid-line fired by the pilot and a flexible gun fired by the observer.

The problems with the SPAD 16 were similar to those encountered with the SPAD 11. A GQG memo of 7 July 1918 noted that the SPAD 16's tail skids were weak and needed to be reinforced. Similarly, when the plane was flown as a single-seater, pilots were warned to lock the gun ring in place or it would swivel violently in flight.

Operational Service

By early 1918 there were 235 SPAD 16s in service, and 130 SPAD 16s were in service in October 1918. They were in service with these escadrilles:

SPA-Bi 2, re-equipped with SPAD 16s in early 1918. Assigned to the 5th C.A. in the vicinity of Noyon, the unit suffered significant losses during the early German offensives. On 2 May SPA-Bi 2 moved to Alsace and on the 27th to Chemin des Dames. During the latter part of 1918 the escadrille participated in the French attacks at the Meuse and Mézières.

SPA-Bi 20, converted from SPAD 11s to ten SPAD 16s in August 1918. It was assigned to the 14th C.A. (as part of the 4th Armée) and was based at Champagne. It was active along the front from Champagne to Argonne.

SPA-Bi 21, formed from AR 21 when it received SPAD 16s in March 1918. It was assigned to the 3rd Armée and later the 37th D.I., and was disbanded in February 1919.

SPA-Bi 34, formed from C 34 in October 1917. Assigned to the 7th C.A., it was under command of Capitaine Molinié. Postwar, it participated in the occupation of Germany, based on the left bank of the Rhine river. It was disbanded in April 1919.

SPA-Bi 36, formed from SOP 36 after it returned from Italy. Assigned to the 47th D.I., it was disbanded in March 1919.

SPA-Bi 42, formed from SPA-Bi 42 when it re-equipped with SPAD 16s in October. It was assigned to the 2nd D.M. Postwar, it was based at Bourguignon and was disbanded in July 1919.

SPAD 16 of SPA-Bi 21 piloted by Paul Menet, serial S 12101. B86.1947.

SPAD 16

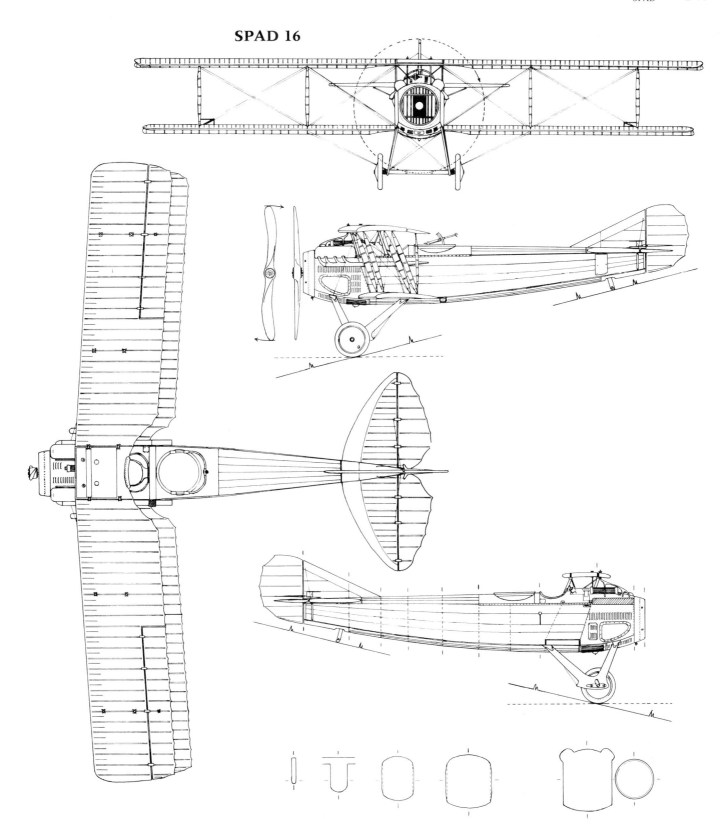

SPA-Bi 47, formed from Escadrille 47 (which used Salmson 2s for only a brief time) in early 1918. It was assigned to the 2nd C.A.C.

SPA-Bi 53, formed from SPA-Bi 53 when the unit received ten SPAD 16s in September 1918. Assigned to the 1st C.A., it moved to Chaux in early October. At the time of the Armistice it was based at Ruestenhard. As part of the German occupation force it was based at Gosenheim.

SPA-Bi 54, formed from SPA-Bi 54 in August 1918 when it received ten SPAD 16s. It remained assigned to the 8th C.A. and was disbanded in March 1919.

SPA-Bi 55, formed from SOP 55 in June 1918 when it received ten SPAD 16s. Based at Dugny and under command of Capitaine Astruc, it was assigned to the 11th C.A. The unit was especially active over the Ardennes during the Allied offensive late in the war. It came under the control of the 5th Armée after the war.

SPA-Bi 60, formed from SOP 60 in June 1918 when it received ten SPAD 16s. It was assigned to the 14th C.A. and was disbanded in March 1919.

SPA-Bi 63, re-equipped with ten SPAD 16s in October 1918. Still assigned to the 1st C.C., it was based postwar in succession at Pierre-Morins, Autreville, and Frescaty. SPA-Bi 63 was disbanded in December 1918.

SPA-Bi 64, re-equipped with ten SPAD 16s in August 1918. It

SPAD 16 of SPA-Bi 21 piloted by Paul Menet. B86.1911.

Below: SPAD 16 with a Breguet 14 A2 in the background. The unit is SPA-Bi 21. B76.962.

remained attached to the 13th C.A. near St. Mihiel and was active over Vesle and the Aisne. It was based (as part of the occupation force) at Mayence and was disbanded in early 1919.

SPA-Bi 140, re-equipped with SPAD 16s in October 1918 and assigned to the 40th C.A. It was disbanded in March 1919.

SPA-Bi 212, (assigned to the 4th Armée as an S.A.L. unit), probably also had some SPAD 16s on strength in addition to SPAD 11s.

SPA-Bi 255, equipped with SPAD 16s in June 1918; these replaced SOP 255's Sopwith 1½ Strutters. Assigned to the 1st C.A., the escadrille was disbanded in July 1919.

SPA-Bi 258, formed from AR 256 in July 1918 when the unit received ten SPAD 16s. Under command of Lieutenant Roger, SPA-Bi 258 was assigned to the 32nd C.A. Initially based at Toul, it moved to Essey-lès-Nancy in August and flew reconnaissance sorties from Bezange-la-Grande to Port-sur-Seille. It ended the war at Villers-lès-Nancy and was disbanded in April 1919.

SPA-Bi 261, formed from AR 261 in July 1918 when it received ten SPAD 16s. Assigned to the 10th C.A., it was disbanded in April 1919.

SPA-Bi 265, created in January 1918 from SPA-Bi 64. Assigned to the 48th D.I, it was disbanded in March 1919.

SPA-Bi 266, received ten SPAD 16s in September 1918. Known serial numbers were 9600, 9385, 9678, 9458, 9653, 9481, 9654, and 9675. It remained attached to the 87th D.I. The escadrille was active over Chavigny and Saint-Pierre-Aigle. It also participated in the Second Battle of the Marne in July. It was based at Wargemoulin-Hurlus at the time of the Armistice. SPA-Bi 266 was disbanded in January 1919.

SPA-Bi 268, formed from AR 268 in August 1918 when it received ten SPAD 16s. Assigned to the 18th C.A., it was disbanded in March 1919.

SPA-Bi 276, formed from SOP 276 in June 1918 when it received ten SPAD 16s. Assigned to the 36th D.I., it was disbanded in March 1919.

SPA-Bi 278, formed from SOP 278 in June 1918 with ten SPAD 16s. Assigned to the 30th C.A., it was disbanded in April 1919.

SPAD 16 of SPA-Bi 60 piloted by Lt. Georges Cuvelier. B91.6284.

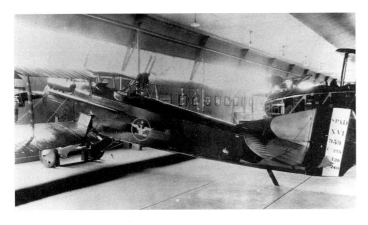

Preserved SPAD 16 S 959 in USAS markings. This is probably the example flown by Billy Mitchell. University of Texas at Dallas.

SPAD 16 Two-Seat Reconnaissance Aircraft with 240-hp Lorraine 8Bb

Span 11.210 m; length 7.840 m; height 2.840 m; wing area 30 sq. m
Loaded weight 1,140 kg

Maximum speed:	sea level	179.8 km/h
	1,000 m	178 km/h
	2,000 m	175 km/h
	3,000 m	169 km/h
Climb:	1,000 m	4 minutes 28 seconds
	2,000 m	9 minutes 15 seconds
	3,000 m	15 minutes 51 seconds
	4,000 m	29 minutes 27 seconds

Armament: one fixed, synchronized 7.7-mm Vickers machine gun, one or two 7.7-mm Lewis guns mounted on a ring stand, and 70 kg of bombs

Approximately 1,000 built

Above: SPAD 16. The engine was a Lorraine 8Bb which improved the aircraft's performance, but did nothing to correct the design flaws that were present in the SPAD 11. B87.4340.

Right: SPAD 16 at Neuhof near Strasbourg as part of the French Occupation force. It is likely that this aircraft was assigned to SPA-Bi 53. B89.3688.

Below: SPAD 16 of SPA-Bi 21. That escadrille's insignia was a woman (painted in white) leading the Marseilles with a vermilion background. MA38080.

Above: Close-up details of the observer's cockpit and ring mount. University of Texas at Dallas.

Above left: SPAD 16 of SPA-Bi 212. University of Texas at Dallas.

Left: SPAD 16 at Colmar on 22 June 1919. B83.1994.

Below: SPAD 16 flown by Col. "Billy" Mitchell in USAS service 1919. M. Cottam via Colin Owers.

SPA-Bi 284, formed from SOP 284 by 1918 with SPAD 16s. It was assigned to the 3rd C.A. and disbanded in January 1919.

SPA-Bi 285, formed from SOP 285 in July 1918 when ten SPAD 16s replaced its Sopwith 1½ Strutters. Assigned to the 62nd D.I., it was disbanded in January 1919.

SPA-Bi 286, formed from SPA-Bi 42 in March 1918. Assigned to the 40th C.A., it was disbanded in March 1919.

SPA-Bi 289, formed from AR 289 in June 1918. It had ten SPAD 16s and was assigned to the 66th D.I. It was disbanded in March 1919.

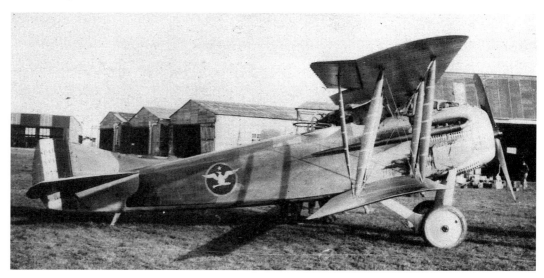

Postwar, the remaining units equipped with SPAD 16s were re-equipped with Salmson 2s and Breguet 14s. Most of the wartime units were disbanded in March or April 1919. These escadrilles remained active in 1919:

SPA-Bi 2, which became the 3rd Escadrille of the 2nd Group assigned to the 7th RAO at Pau. Later it was assigned as the 9th Escadrille of the 37th Regiment at Rabat, Morocco.

SPA-Bi 20, which became the 7th Escadrille of the 4th Group assigned to the 5th RAO (later the 35th RAO) at Bron.

SPA-Bi 47, which became the 7th Escadrille of the 4th Group assigned to the 3rd RAO at Beavais.

SPA-Bi 53, which became the 2nd Escadrille of the 1st Group assigned to the 5th RAO at Bron. It was attached to the forces occupying the Rhine as the 11th Escadrille of the 33rd RAO at Mayence.

SPA-Bi 55, which became the 1st Escadrille of the 1st Group assigned to the 7th RAO at Pau. Later it was assigned to the 37th Regiment at Rabat, Morocco.

One source estimates that 1,000 SPAD 16s were built, but this cannot be confirmed. By 1920 the SPAD 16s had been withdrawn from service.

Foreign Service
Belgium
A single SPAD 16 was purchased for evaluation.

United States
Six SPAD 16s were acquired by the A.E.F. in August 1918. One that was flown by Colonel William "Billy" Mitchell is preserved.

SPAD 17

The Aviation Militaire decided at the beginning of 1918 that a new fighter with a more powerful engine was needed to maintain air superiority. The C1 requirement called for a single-seater with two machine guns (7.7-mm or 11-mm) or a single machine gun and a 37-mm cannon. The aircraft of this category were also to be capable of carrying out high-speed reconnaissance missions and so had to have the option of being fitted with cameras. The engines available to the designers included the 300-hp Hispano-Suiza 8Fb and a 320-hp ABC Dragonfly. Both the Wibault 1 and the SPAD 17 were initially fitted with the 220-hp Hispano-Suiza 8Be because problems with the 300-hp 8Fb had delayed its availability.

The SPAD 17 was essentially a SPAD 13 redesigned to accept the new Hispano-Suiza 8Fb. The SPAD 13 airframe was modified and strengthened to accommodate the larger engine, and the fuselage was fully faired. The wing cellule was also strengthened to a load factor of 9 and the horizontal stabilizers were enlarged. The SPAD 17 was capable of carrying two cameras but in this configuration only a single machine gun could be fitted. Manufacturer trials began in April 1918, and operational trials were conducted by Capitaine de Slade in June.

The overall dimensions were little changed from the standard SPAD 13, and the new engine endowed the aircraft with a marginally higher performance. Twenty aircraft were produced, most of them assigned to the units with GC 12 ("Les Cigognes"), which were SPA 3, SPA 26, SPA 67, SPA 103, and SPA 167. No further aircraft besides the initial test batch were acquired, possibly because the STAé had decided to standardize production on the Nieuport 29.

> **SPAD 17 Single-Seat Fighter with 300-hp Hispano-Suiza 8Fb**
>
> Span 8.08 m; length 6.25 m; height 2.60 m; wing area 20.00 sq. m
> Empty weight 687 kg; loaded weight 942 kg
>
> Maximum speed:
> 2,000 m   217 km/h
> 3,000 m   214 km/h
> 4,000 m   211 km/h
> 5,000 m   201 km/h
>
> Climb:
> 2,000 m   5 minutes 24 seconds
> 3,000 m   8 minutes 20 seconds
> 4,000 m   12 minutes 32 seconds
> 5,000 m   17 minutes 21 seconds
>
> Ceiling 7,175 m; endurance 1.25 hours
> Armament: two 7.7-mm Vickers machine gun; only a single gun carried if the aircraft was equipped with two cameras
> A total of 20 were built

Below: SPAD 17 serial number 745. The SPAD 17 was essentially a SPAD 13 redesigned to accept the new 300-hp Hispano-Suiza 8Fb. B82.4341.

SPAD 17 of SPA 3. Twenty aircraft were produced and most were assigned to the units with GC 12 (Les Cicognes—The Storks) which were SPA 3, SPA 26, SPA 67, SPA 103, and SPA 167. B82.781.

FRENCH AIRCRAFT OF THE First World War

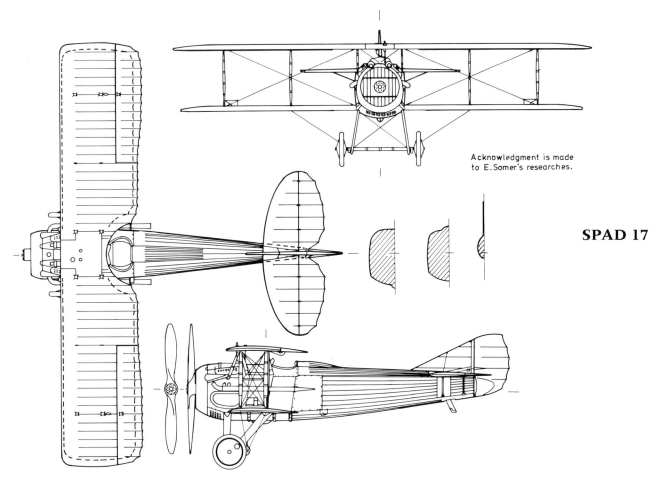

Acknowledgment is made to E. Somer's researches.

SPAD 17

Left and below: Two views of a SPAD 17. MA14066.

SPAD 18

The SPAD 18 was a large single-seat monocoque design with a 300-hp Hispano-Suiza 8G fitted with a 37-mm Putueax cannon firing through a hollow hub. However, it was found that because of excessive vibration the Hispano-Suiza 8G was unusable. The airframe of the SPAD 18 (which was apparently never equipped with the cannon) was fitted with a version of the 300-hp Hispano-Suiza 8Fb engine that did not have a cannon. This aircraft was designated the SPAD 20. The SPAD 18 had the option of a rear gunner, hence the Ca.1–2 designation.

> **SPAD 18 Single-Seat or Two-Seat Fighter with 300-hp Hispano-Suiza 8G (never fitted)**
> Span 9.70 m; length 7.30 m; height 2.80 m; wing area 30 sq. m
> Intended armament was a 37-mm Putueax cannon (never fitted)
> One built (but never flown with Hispano-Suiza 8G)

Below: Official drawing of the SPAD 18 single-seat fighter with an optional position for a rear gunner. Construction of the aircraft was never completed.

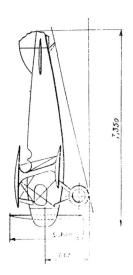
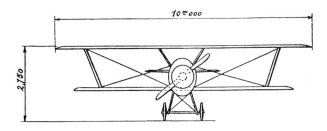
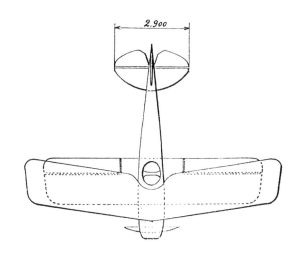

SPAD 19

Little is known about the SPAD 19, and only a single photograph of the type is believed to exist. It was a two-seater and appeared to have a larger wing span than the SPAD 11 or 16. The single central exhaust stack also suggests a different engine than the Lorraine-Dietrich used in the SPAD 16. However, one source states that the SPAD 19 was fitted with a 280-hp Lorraine engine. Finally, the wing bracing was significantly different from the SPAD 11 or 16. Historian Jean Deveaux believes this plane was intended as an army cooperation type. As far as can be determined, no SPAD 19s were used operationally.

The SHAA archives identify this aircraft as SPAD 19. No further details are available (see text). B86.4467.

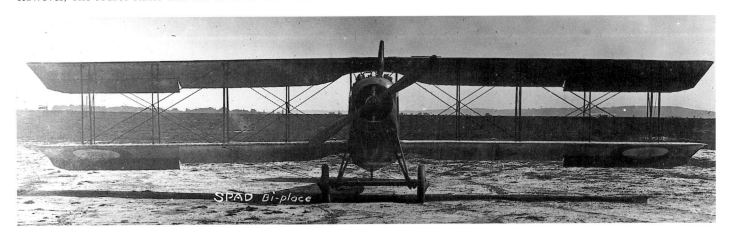

SPAD 20

The SPAD 20 was produced in response to the C2 requirement of 1918 calling for a crew of two, a useful load of 375 kg, a maximum ceiling of 8,000 m, a service ceiling of 5,000 m, and a maximum speed of 220 km/h. The aircraft offered in the C2 category included the SEA 4, the Hanriot-Dupont HD.3, and the SPAD 20.

The SPAD 20 was developed from the SPAD 18, which had failed because the Hispano-Suiza engine fitted with a cannon had proved unsuccessful. The SPAD 20 incorporated many of the features that Herbemont would include in his postwar designs. These included a monocoque fuselage of molded wood, an upper wing which had a slight sweepback, a straight lower wing fitted with ailerons, thick I-shaped interplane struts, and wooden construction with fabric covering. The tail surfaces were quite large and the tail section was molded in the same way as the fuselage. The engine was a 300-hp Hispano-Suiza 8Fb. A notable feature was that the engine attachments were designed so that it was easily accessible and could be removed quickly in the field. The fuel tanks were protected to prevent fire from breaking out should they be damaged.

Because single-seat aircraft were vulnerable to attack from the rear, the C2 requirement called for a rear gunner. On the SPAD 20 the gunner was provided with one 7.7-mm Vickers machine gun on a T.O.3 mounting, although on early examples a Lewis gun was used. The pilot had two synchronized 7.7-mm Vickers machine guns.

On 7 August 1918, the SPAD 20 was flown by test pilot Sadi Lecointe at Buc airfield. During testing Lecointe set a world altitude record of 8,900 meters. By the beginning of September the prototype was sent to the CIACB at Villacoublay for military trials. These revealed some shortcomings, and changes were mandated for the armament and the fuel tanks. After these modifications, the STAé ordered 300 SPAD 20s. However, after the war ended the order was reduced (on 2 October 1920) to 100, of which 95 were actually delivered.

The aircraft delivered in 1920 were armed with two synchronized Vickers machine guns and a single Lewis gun on a ring mount. The performance was comparable to the single-seat Nieuport 29s then in service.

One major variant was the SPAD 20bis, which had twin rear guns. The wing surface area was augmented by 1.50 square meters and the fin and rudder were enlarged slightly to compensate for the additional rear gun. The first flight took place on 7 July 1921. Two prototypes were constructed, but development was not pursued.

The SPAD 20s were flown before delivery at Bourget by four pilots drawn from the 1st and 2nd Aviation Regiments. They praised the machine's maneuverability and rapid climb. The aircraft served with the 2nd Air Regiment at Strasbourg, the 32nd Regiment Mixte at Dijon (5th and 6th Escadrilles, fighter units assigned to the Groupe d'Chasse), the 3rd Regiment d'Aviation de Chasse at Chateauroux, which consisted of the 1st Groupe d'Chasse (1st, 2nd, and 3rd Escadrilles), and the 2nd Groupe d'Chasse (5th, 6th, and 7th Escadrilles).

Foreign Service

Japan

A pair of SPAD 20bis were sold to the Japanese, who dissected them to learn the latest French manufacturing techniques. They were designated Hei 2.

Paraguay

A few SPAD 20s were reportedly obtained by Paraguay postwar. These were used during the 1922 civil war.

SPAD 20 Prototype with 300-hp Hispano-Suiza (Data Refers To Performance When Flown as a Single-Seater)

Span 9.7720 m; length 7.200 m; height 2.87 m; wing area 30 sq. m
Empty weight 850 kg; loaded weight 1,310 kg
Maximum speed: 237 km/h at sea level; 201 km/h at 6,000 m; 195 km/h at 7,000 m; 193 km/h at 7,500 m; climb to 5,000 m in 15 minutes 25 seconds; climb to 6,000 m in 21 minutes 24 seconds; ceiling 8,900 m

SPAD 20 Two-Seat Fighter with 300-hp Hispano-Suiza 8Fb

Span 9.80 m; length 7.34 m; height 2.87 m; wing area 29 sq. m
Empty weight 867 kg; loaded weight 1,106 kg
Maximum speed: 242 km/h at sea level; 229 km/h at 2,000 m; climb to 2,000 m in 4.6 minutes; ceiling 8,900 m
Armament: two synchronized 7.7-mm Vickers machine guns and a single 7.7-mm Lewis gun on a ring mount
A total of 95 were built

SPAD 20. University of Texas at Dallas.

SPAD 20

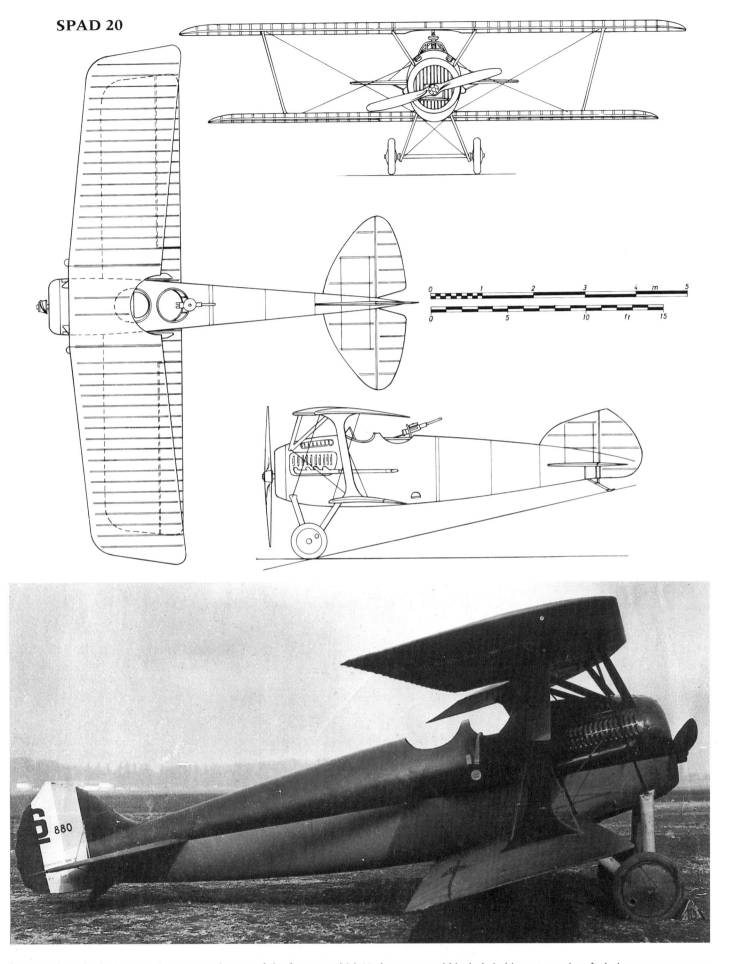

SPAD 20 (S.880). The SPAD 20 incorporated many of the features which Herbemont would include in his postwar aircraft designs: a monocoque fuselage of molded wood, an upper wing which had a slight sweepback, a straight lower wing, thick I-shaped interplane struts, and wooden construction with fabric covering. 87.6005.

The SPAD 20 prototype. The engine was a 300-hp Hispano-Suiza 8Fb. B83.3459.

Below and below left: SPAD 20. University of Texas at Dallas.

SPAD 21

The SPAD 21 was a development of the SPAD 17. The SPAD 17's performance was almost identical to the SPAD 13 and it was probably hoped that by improving the SPAD 17's maneuverability the STAé could be persuaded to order some of these aircraft.

Except for having upper and lower wings of equal span and chord and fitted with ailerons, the SPAD 21 was identical to the SPAD 17. The placement of ailerons on both wings was intended to improve maneuverability. However, the aircraft showed no significant improvement over the SPAD 17, and in fact its rate of climb was actually slower. Tests were undertaken in October 1918, but development was abandoned in November. Postwar, two aircraft were used to test fuel equipment.

SPAD 21 Single-Seat Fighter with 300-hp Hispano-Suiza 8Fb

Span 8.435 m; length 6.40 m; height 2.60 m; wing area 23.5 sq. m
Empty weight 750.5 kg; loaded weight 1,047.5 kg

Maximum speed:	2,000 m	221 km/h
	3,000 m	217 km/h
	4,000 m	214 km/h
	5,000 m	205 km/h
Climb:	2,000 m	5 minutes 40 seconds
	3,000 m	8 minutes 41 seconds
	4,000 m	13 minutes 3 seconds
	5,000 m	18 minutes 18 seconds

Ceiling 7,000 m; endurance 1 hour and 40 minutes
Armament: two 7.7-mm Vickers machine guns; only one gun was carried if the aircraft was equipped with two cameras

SPAD 21. The SPAD 21 was essentially a SPAD 17 modified to improve the aircraft's maneuverability. MA36412.

SPAD 21. Except for having upper and lower wings of equal span and chord both fitted with ailerons, the SPAD 21 was identical to the SPAD 17. MA36413.

SPAD 22

The SPAD 22 was a radically re-designed SPAD 17. Béchereau left the fuselage and engine unchanged, but the upper wing was strengthened by three main spars. The lower wing was still braced by the landing gear struts. The upper wing (as would appear on many of Herbemont's SPAD designs) had a pronounced sweepback. The lower wings were actually swept forward with an inverse taper toward the root. Because of the changes to the wing, the single-bay interplane bracing was extensively modified. Only the upper wing was fitted with ailerons. The elevators and tailplane were also modified, incorporating horn-balanced elevators and braced by inverted V-struts.

The reason for these changes is not totally clear; it may have been hoped that the new wings would provide the pilot with a better field of vision than the SPAD 13. The SPAD 22 was tested in 1919, but only a single example was built.

SPAD 22 Single-Seat Fighter with 300-hp Hispano-Suiza 8Fb

Span 8.08 m; length 6.25 m; wing area 20.20 sq. m
Armament: two 7.7-mm Vickers machine guns
One built

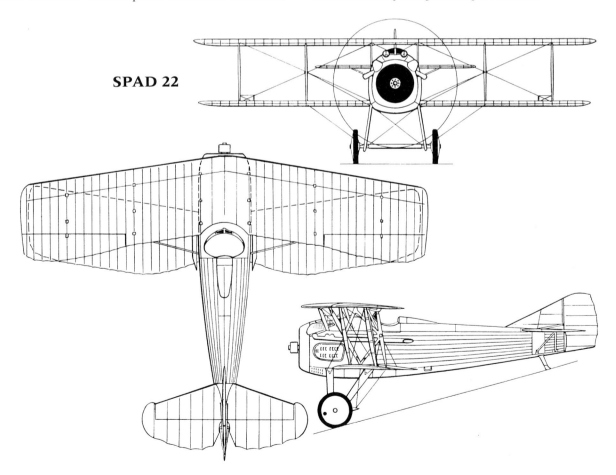

SPAD 22

The SPAD 22. The upper wings were swept back and the lower wings were swept forward with an inverse taper towards the root. MA11639.

SPAD 23

The SPAD 23 was a project for single-seat biplane that was to have been fitted with a Rateau turbocharger. It was ordered by the Aviation Militaire in December 1918 but it appears that it was never completed.

SPAD 24

As the war neared its end the Aviation Maritime began to evaluate the possibility of flying aircraft from battleships by using platforms mounted on the gun turrets. Several aircraft were tested in this role and the SPAD firm was asked to develop a version of the SPAD 14 floatplane with a conventional undercarriage for platform takeoff. This aircraft became the SPAD 24.

It flew on 5 November 1918, just six days before the war ended. It has been reported that the SPAD 24 was 5 km/h slower then the same airframe equipped with the huge floats and the same 220-hp Hispano-Suize 8Bc engine, although this seems unlikely. In any event, it appears that the navy preferred the Hanriot HD.1 and HD.2, both of which were test-flown from ships' turrets postwar. However, there is no evidence that the SPAD 24 was ever tested in this fashion and only the single example was built.

> **SPAD 24 Single-Seat Fighter with 220-hp Hispano-Suiza 8Cb (Some Sources State 8Bc)**
> Span 9.80 m; length 6.48 m; height 2.56 m; wing area 26.20 sq. m
> Empty weight 650 kg, loaded weight 1,000 kg (approximately)
> Maximum speed: 200 km/h; ceiling 5,000 meters
> Armament: a 0.303 Vickers machine gun and a 12-round, 37-mm S.A.M.C. cannon
> One built

SPAD 24 (S.09). The SPAD 24 was a SPAD 14 without the floats. The SPAD 14 was a modified SPAD 12. MA36405.

Hydravions Tellier

Alphonse Tellier built his first airplane in 1910. The T.1, as it was believed to have been designated, was an aircraft intended for racing and was powered by a 35-hp Panhard-Levasseur engine. It was built for M. Dubonnet in 1909. The T.1 was a monoplane with an exceptionally long, uncovered fuselage. A horizontal stabilizer was fixed to the top of the tail and a prominent rudder with a fixed fin was placed above the fuselage.

The Société des Chantiers Tellier was dissolved in January 1911. Tellier subsequently turned his attention to producing motor boats. In 1914, he built floats and hulls for other manufacturers of seaplanes such as Marcel Besson.

Tellier T.1 Racer with 35-hp Panhard-Levasseur
Span 11 m; length 11 m; wing area 24 sq. m
Loaded weight 500 kg
Maximum speed: 70 km/h at sea level; endurance 6 hours

Tellier T.2

In 1916 Tellier was given 25,000 francs by M. Salct to build a patrol flying boat. To reduce the time needed to build the prototype, Tellier built the wooden hull at l'Ile de la Jatte, while Voisin constructed the wings (with metal ribs) at his factory at Issy-les-Moulineaux. The aircraft was completed in May 1916 at Tellier's factory.

Designated the T.2, the aircraft had a 200-hp 8Ba Hispano-Suiza engine and carried a crew of two. The T.2 was initially flown from the Seine in June 1916 by an Aviation Militaire pilot (Sgt. Duyck) with Tellier as a passenger. Many successful flights were made, and the aircraft was found to be so stable that the pilot was able to release the controls for several minutes during the flight. Based on these tests no major modifications were planned. On one of the landings the hull was damaged, but did not sink due to its watertight compartments. Tellier took out a patent on his seaplane design (Brevet #493 411). The airplane crashed in June 1916 due to carburetor failure.

T.2 Two-Seat Patrol Flying Boat with 200-hp Hispano-Suiza 8Ba
Span 15.60 m; length 11.84 m; height 3.40 m; wing area 47 sq. m
Empty weight 1,150 kg; loaded weight 1,750 kg
Maximum speed: 145 km/h; climb to 1,000 m in 6 minutes 30 seconds; range 700 km
One built

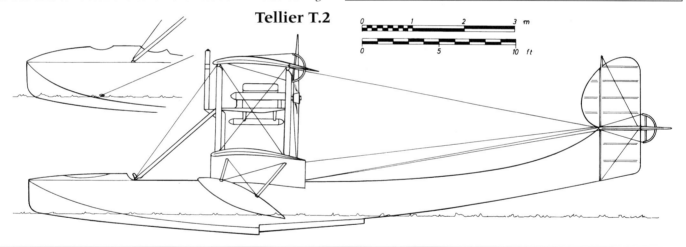

Tellier T.2

Tellier T.3

Despite the crash of his T.2 flying boat, Tellier was awarded a contract for two more patrol flying boats by Emile Dubonnet for whom he had built the T.1 for in 1909. Both were completed in August 1916. The first flight from the Seine, also by Sgt. Duyck, was in August.

The new design was a biplane with unequal span wings; the upper wing being longer. The top wing had a span of 15.60 m and no dihedral. There were ailerons on the upper wing only; they had an area of 4.26 sq. m. The upper wing was attached to the four motor-mount struts. The bottom wing, which was bolted to the fuselage, had prominent dihedral to lift the wing clear of the water and spanned 11.60 m. The wings had wooden spars and metal ribs and were supported by intertwined metal cables. There were two bays of struts mounted at oblique angles. The struts were made of extruded metal with metal attachment points; this eliminated the need to constantly adjust the wing ribbing. Two floats were located on the lower wing tips; each had a volume of 0.170 cubic meters. The engine, a 200-hp Hispano-Suiza 8Ac, was mounted just below the upper wing as a pusher. It was supported by four struts. Two inclined metal struts ran from the top decking of the fuselage to the motor base. The pilot sat directly ahead of the engine while the gunner was seated in the extreme nose. There was a third seat for a passenger next to the pilot. The fuel tank was mounted in the fuselage behind the pilot. Two oil reservoirs, which were cooled, were mounted side-by-side under the motor. The radiator was placed in front of the engine. The aircraft had a slim, wooden hull with an upswept tail that was supported by bracing wires to the upper and lower wings. The fuselage had several watertight compartments to prevent the aircraft from sinking should damage occur to the hull. There was a horn-balanced rudder with no fin.

The T.3 was selected for production after a single example was purchased by a Monsieur Salet and presented to the Aviation Maritime. Flight testing began 13 September and was halted after a total of 6 hours and 24 minutes because of engine failure. A new engine was fitted and testing resumed on the 3rd, 4th, and 5th of October for a total of 5 flights lasting 3 hours 55 minutes. Additional test flights following modifications to the fuel and oil reservoir system were conducted on 14, 17, and 18 October which completed testing. The T.3 had been flown for a total of 19 hours 30 minutes. The T.3 was evaluated at Saint-

Tellier T.3. The CEPA (Marine Commission) had favorable comments on the T.3's two-step hull which helped to dampen oscillations of the aircraft while it was on the water, and the fuselage construction which was similar to that of boat hulls. B88.3797.

Raphaël by the CEPA (Marine Commission) which offered the following comments about Teller's design:

1. The two step hull helped to dampen oscillations of the aircraft while it was on the water.
2. The construction of the fuselage was similar to that used on the hulls of boats.
3. The sesquiplane wings had a marked dihedral on the lower wings only (probably to keep them well above the water on takeoff and landing); the upper wings had no dihedral. The longerons were steel tubes crossed by steel rods and piano cord. The ribs were reinforced poplar. The covering was a strong linen made from flax. The airfoil was made of twelve frames constructed of wooden tubes covered in wood.
4. The airplane's center of gravity had been so finely adjusted that with the engine shut off the airplane would enter a stable glide path even when the pilot was not touching the controls. The control wires were made of steel cables.
5. The two-step hull facilitated the separation of the airplane from the water on takeoff and enabled the T.3 to maintain a stable line of flight upon leaving the water.
6. The mounting of the engine and fuel tanks was similar to that used on racing boats. The fuel tanks were mounted inside the hull, behind the pilot. they had a combined capacity of 310 liters of fuel. An air pump controlled the fuel flow.
7. There was a cooled oil tank, but its capacity was felt to be insufficient.
8. All struts could articulate freely, preventing structural deformation.
9. Static testing revealed a load factor 4.5.

It is clear from this report that the committee was most impressed by the features of the airplane based on Tellier's boat-building experience. These qualities meant that the T.3 could operate more effectively from water and, again according to the committee, served "to distinguish the airplane from any other in service in 1916...The airplane is a boat with wings." It was also noted to be as easy to land a T.3 on the sea as it was to land a land-based airplane on the ground. Additional modifications were suggested to the fuel and oil reservoir system and, once these changes had been affected, then the T.3 would be ready for series production.

Apparently the Navy was also impressed by the T.3 and placed an order for ten, followed by two additional orders on 21 March and 11 April for a total of 100 additional aircraft. The hulls were built by Tellier at Argenteuil, and the wing ribs, tail, and motor mounts were built by Cicles Alcyon (specialist in metal production), while Voisin assembled the wings. When Nieuport took over the Tellier firm, it built 47 of the aircraft. The first T.3s arrived at the Centre de Aviation Maritime at Corfu in late 1916.

Although the T.3s performed adequately, there were several problems with the design. The fuselage was weak, particularly in the vicinity of the second former. This lead to an accident on 4 July 1917 when a T.3 broke up in the air, killing its two crew members. It was noted that when newer pilots landed they tended to set the aircraft's tail down first, which resulted in additional stress on this area. A warning was circulated to the naval air stations that the aircraft should be landed with the fuselage in a horizontal position.

The wing was also prone to failure, particularly where the metal spars of the lower wing were attached to the wooden longerons of the upper fuselage. The bomb racks for the Corpet lance-bombs were at this site, which further contributed to stress. On 19 April 1918 the STIAé Marine recommended that the T.3s remaining in service be strengthened by metal struts attaching from the base of the hull to the spar of the lower wing.

The Hispano-Suiza engines also proved to be troublesome and suffered from fuel leaks and inadequate lubrication. Even the Regy propellers were prone to failure, resulting in numerous T.3s being grounded while awaiting new propellers.

The T.3 entered service in February 1917. Deliveries began in February 1917 and were as follows: February—one aircraft was delivered, March—three, April—five, May—two, June—15, July—20, August—12, and September—22; October—12. This gives a total of 92 Tellier 3s delivered by 1 November 1917. Between February 1917 and December 1918 approximately 245 T.3s and T.6s were built by Tellier, five by Fabre, and 11 at Cherbourg. Another source places the number at 283. The aircraft were built by Tellier, the State Arsenal at Cherbourg, Georges-Levy, Henri Fabre, S.A.C.A, S.F.A. Dubonnet, and Nieuport.

Tellier T.3s were used for anti-submarine patrols and saw extensive service from the French naval stations located on the English Channel and the Mediterranean. In the patrol mission the T.3 was armed with two 35-kg bombs and a single Hotchkiss 8-mm machine gun on an Éteve turret.

Postwar, three T.3s were converted to airliners capable of carrying two passengers in the gunner's position, now covered with a windshield. F-ACBO, F-ACBU, and F-ACFS were used by Compagnie Franco-Bilbaine de Transport Aeronautique in 1919 (along with a Farman F.40H and a Levy HB.2). The company went bankrupt by October 1921.

Tellier T.7

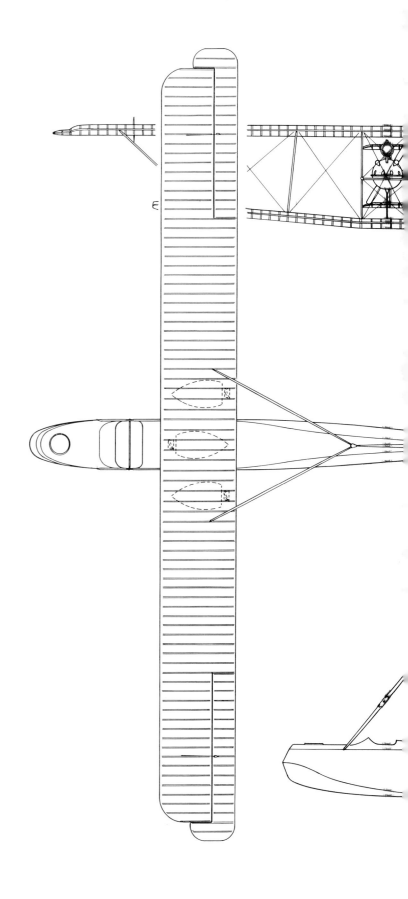

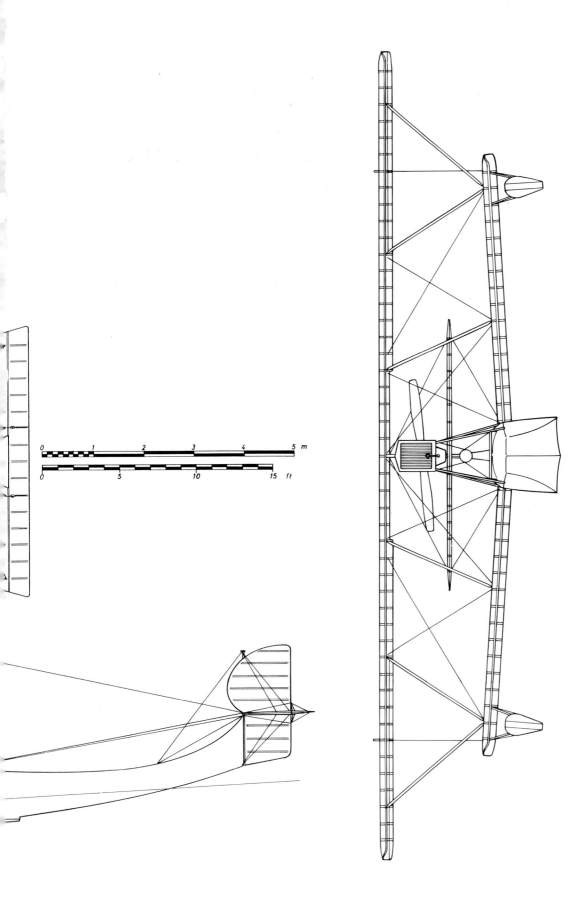

Tellier T.3

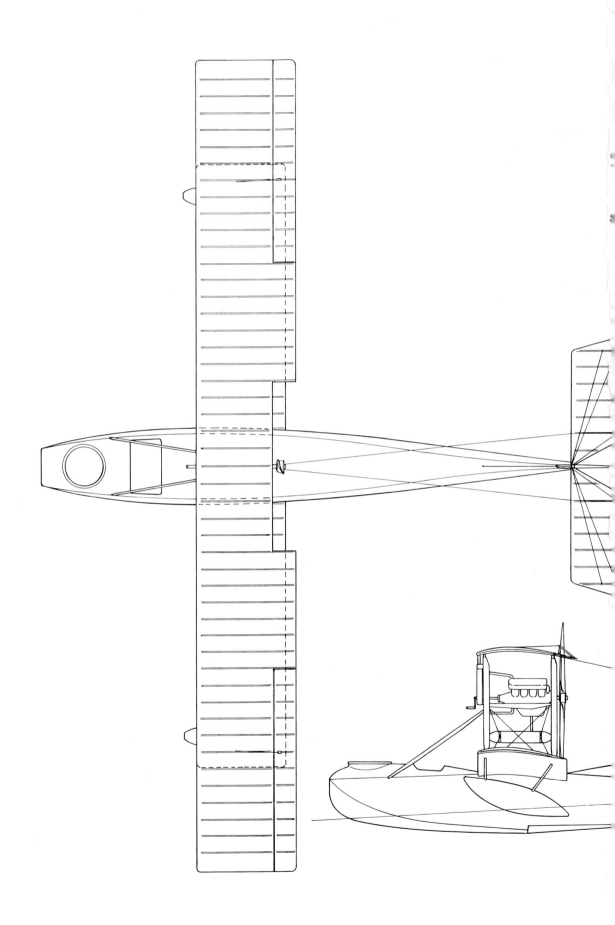

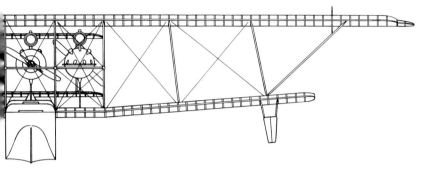
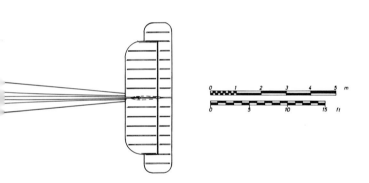
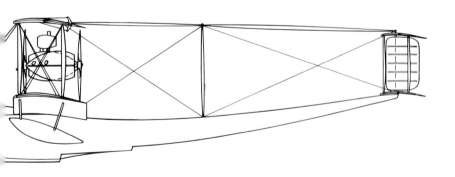

Foreign Service

Japan

The Japanese Navy purchased a single Tellier T.3 in 1918 with money donated by Kamesaburo Yamashita. The aircraft was evaluated by the Navy, but the trials were unsatisfactory, and no further aircraft were purchased. However, Borget reports that three T.3s, all built by Fabre, were acquired by the Japanese.

Portugal

Portugal purchased five T.3s in 1917 and 1920. The first two aircraft (given serials Numbers 1 and 5) arrived in 1917. An additional two or three were obtained from the French in August 1918. One airplane was given serial Number 12. The T.3 remained in service until 1928. These airplanes were assigned to the Centro de Aviacao Naval which had a base in Lisbon and a second base at Aveiro.

Russia

A single Tellier T.3 was purchased by the Russian government in mid-1916. It was sent to the Dux plant where it was planned to produce the aircraft under license in place of the indigenous M-9 flying boat. However, the Allies did not deliver the requisite 200-hp Hispano Suiza engines and production by the Dux firm, as well as by the Lebedev and Meltser firms, was canceled. In fact, only 20 hulls and sets of wings were ever built by the Russians. In 1920 these parts were used to assemble ten Tellier T.3s at GAZ No.3 at Leningrad (it is not known how the Russians obtained the Hispano-Suiza 8Ac engines). By the summer of 1921 one of the assembled Tellier T.3s had crashed and the rest were withdrawn from service. Michel Borget has noted that six Tellier T.3s (T.73, 82, 87, 89, 91, and 93) were ordered by the Russians but never delivered.

United Kingdom

The RNAS ordered two Tellier T.3s for evaluation in November 1917. They were assigned serials N84 and N85 and were delivered to Grain in 1918 where they were used for gun and camouflage trials.

United States

The Tellier T.3 series of flying boats were flown from the Le Croisic U.S. Naval aviation station and from the training school at Moutchic. These aircraft had long since been replaced in the French Aviation Maritime and it is reasonable to conclude that those aircraft supplied to the U.S. had already seen extensive service. The airplanes were delivered from S.F.A. Dubonnet. The U. S. Navy liked at least one point about the T.3s. Since they knew the Tellier firm had produced yachts, the naval pilots were not surprised that the T.3s had the best-built hull of any French flying boat.

T.3 Three-Seat Patrol Flying Boat with 200-hp Hispano-Suiza 8Ac

Span 15.60 m; length 11.83 m; height 3.60 m; wing area 47 sq. m
Empty weight 1,150 kg; loaded weight 1,796 kg; payload 560 kg
Maximum speed: 130 to 135 km/h; climb to 500 m in 2 minutes 45 seconds; climb to 1,000 m in 6 minutes 30 seconds; climb to 1,500 m in 11 minutes 30 seconds; climb to 2,000 m in 15 minutes 30 seconds; climb to 2,500 m in 24 minutes; endurance 4 hours 30 minutes
Armament: one machine gun in the nose and two 35-kg bombs
Approximately 245 T.3 and T.6s were built

T.3 Two-Seat Flying Boat Built at GAZ No.3 in Russia with 200-hp Hispano-Suiza 8Ac

Span 15.60 m; length 11.84 m; height 3.40 m; wing area 46.55 sq. m
Empty weight 1,134 kg; loaded weight 1,792 kg
Maximum speed: 125 km/h; climb to 1,000 m in 13 minutes 30 seconds; ceiling 3,500 m; endurance 4 hours
20 built

The Le Croisic station was the only U.S. Naval air station to use the Tellier T.3s operationally. The first mission was flown on 18 November 1917 with six T.3s. Two aircraft were kept at combat readiness at all times. The aircraft flew patrols looking for mines and submarines. Forced landings at sea became more frequent as the T.3s began to break down; often patrol boats had to tow the aircraft back to base. Later, a number of convoy escort missions were flown. Although convoys in the area were "ably protected," the official history does not record that any enemy submarines were destroyed. Serial numbers were 10, 11, 21, 24, 25, 40 ,56, 69, 70, 71, 72, 73, 74, 86, 87, 88, 89, 93, 105, 106, 111, 114, 116, 139, 140, 143, 144, 145, 146, 151, 152, and 153.

Tellier T.3 flying boat at Juvisy. B85.2009.

Postwar view of a Tellier T.3 of La Compagnie Franco-Bilbaine de Transports Aeronautiques. B87.4623.

Tellier T.4

The French Navy decided it needed a flying boat capable of carrying a heavier payload and issued a new specification calling for a payload of 1,000 kg while maintaining the same performance as the T.3 and HB.2 flying boats. Unfortunately for the seaplane manufacturers, the 450-hp and 500-hp engines manufactured in France were reserved for the SPAD series of fighters.

Tellier responded to this requirement by designing two different aircraft: the T.4, which was essentially a T.3 fitted with the more powerful 350-hp Sunbeam engine (a water cooled V-12 Maori), and the T.5, which had two Hispano-Suiza engines mounted in tandem.

The T.4 required few changes to the standard T.3 aside from strengthening the engine supports to carry the heavier engine. The T.3 had experienced structural failures, and Tellier attempted to remedy this fault by strengthening the fuselage behind the wings. Other changes included a redesign of the bow and revision of the wings and tail. The hull had two steps composed of three watertight compartments, each of which had a 10 cm inspection hatch. The hull was made of three layers of mahogany from the nose to the first step. From first step to the second step the fuselage was constructed of two thicknesses of plywood. The remainder of the fuselage was also made of plywood, but of decreasing thickness. There was a position for a gunner in the nose, and two seats in the main cockpit just ahead of the bottom wing. The fuel tanks, which carried 620 liters of fuel, were located just behind the pilot's cockpit. There was a Chausson radiator and a Letombe starter. The wing structure was almost identical to the preceding T.3 and T.6. The wing

> **T.4 Two-Seat Patrol Flying Boat with 350-hp Sunbeam**
> Span 23 m; length 14.750 m; height 4.10 m; wing area 84.50 sq. m
> Empty weight 2,100 kg; loaded weight 3,257 kg; payload up to 1,350 kg
> Performance (at a total weight of 3,100 kg); maximum speed: 80 knots; climb to 500 m in 4 minutes 15 seconds; climb to 1,000 m in 8 minutes 27 seconds; climb to 2,000 m in 19 minutes 13 seconds
> Performance (at a total weight of 3,732 kg): climb to 500 m in 5 minutes 45 seconds; climb to 1,000 m in 13 minutes 23 seconds; climb to 2,000 m in 35 minutes and 40 seconds; endurance 5 hours
> Armament: one 7.7-mm machine gun in the nose and four Corpet lance bombs

Tellier T.4

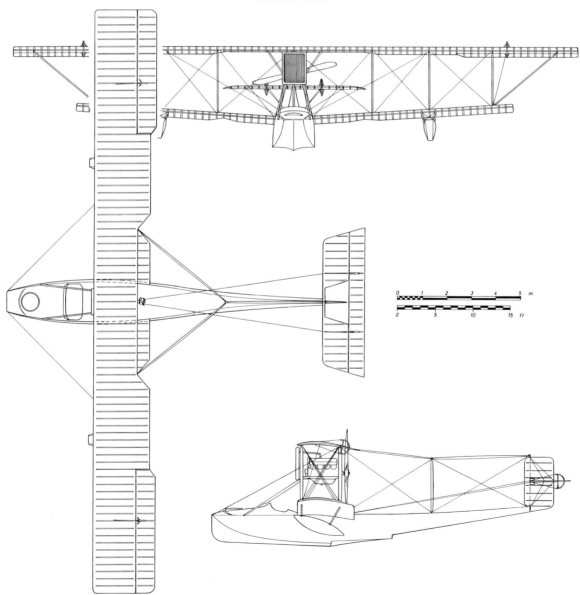

Tellier T.4 with the Sunbeam engine with the designation SU 16 on the rudder. This aircraft is undergoing testing by the CEPA at Saint Raphaël. B87.5008.

Below: Tellier. T.4. In order to support the tail unit an extensive array of struts connected the trailing edge of the upper wing to the upper part of the tail and a post connected the rear part of the hull to the struts connecting the upper wing and tail. B90.0738.

spars were metal tubes held in place by piano wire on which the ribs (made of ash reinforced with plywood) were placed. The wing structure was covered in linen with Emaillite coating. The upper wing was fitted with ailerons and the lower wing had a marked dihedral. The stabilization floats were attached to the lower wing. The wings were supported by 16 metal struts. A horizontal "Y" of extruded steel connected the trailing edge of the upper wing to the upper part of the tail. A post connected the rear part of the hull to the struts connecting the upper wing and tail; this served to make the Y-strut more rigid. This structure was, in turn, reinforced by numerous bracing wires. Tellier applied for a patent for this system.

The aircraft was completed by May 1917 and flight testing began in late December. George Duick made the first flight carrying a ton of ballast. In early 1918 the T.4 was dismantled and sent to Saint Raphaël for evaluation by the Aviation Maritime.

The Aviation Maritime requested that newer seaplane designs incorporate folding wings to make storage at its naval air stations easier. The prototype T.4 had fixed wings but was modified to meet this new requirement. The modifications included changing the wing struts which were canted inwards to a vertical orientation; the dihedral and incidence of both the upper and lower wings remained unchanged. The center struts were split and the wings were hinged to pivot around them. The span of the folded wings was 8.7 meters, and the wings could be folded rapidly without need to adjust the rigging.

The airplane was able to carry the specified 1,000 kg payload; when the Navy tested the machine they found it could carry a 1,504 kg payload. They were suitably impressed by this achievement, and the T.4 was ordered into production. However, the Navy made plans to equip T.4s with other engines, including the 340-hp Panhard 12C, the 400-hp Liberty 12A, and the 370-hp Lorraine 12Da. However, as far as can be determined these engines were never fitter to Tellier T.4s.

The aircraft were used from French naval bases on the Atlantic and Africa and superseded the T.3s then in service. They remained in front-line service until 1922.

Tellier T.4 at Saint Raphaël. B90.0736.

Tellier T.5

The Tellier T.5 was designed to meet the French navy's requirement for a flying boat capable of carrying a 47-mm cannon and 300 kg of bombs, a considerably heavier payload than the Tellier T.3. As noted above, it was designed to the same specification as the T.4 but used two Hispano-Suiza 8B engines mounted in tandem rather than the single Sunbeam engine of the T.4. It was intended to use these aircraft in the anti-submarine role where the cannon could disable a submarine and then bombs could destroy the vessel once it had been immobilized.

The T.5 was considerably larger than the previous Tellier flying boats. It was a biplane with unequal span wings and had the wing/hull/tail system of support struts that were used on the T 4. The hull had a position for a gunner in the nose and two seats in the main cockpit just ahead of the bottom wing. The fuel tanks, which carried 620 liters, were located just behind the pilot's compartment. The aircraft also had floats underneath the bottom wing. The T.5 was later modified to have folding wings (for further details see the entry for the T.4). The two 250-hp Hispano-Suiza engines were uncowled and mounted back-to-back beneath the upper wing. Two Chausson radiators were mounted between the forward engine and upper wing and Brizon starters were used. The aircraft was completed in May 1917 and its first flight was in early January 1918. Various payloads were tested including 800, 1,000, and 1,235 kg. Flight tests with one engine shut down revealed that the aircraft flew very well while carrying a 711 kg payload. In early 1918 the T.5 was dismantled and sent to Saint Raphaël for evaluation by the Aviation Maritime. Here it was test flown beginning 3 March 1918. It was capable of carrying a 1,500 kg payload. Advantages of the twin-engine design included stable flight on only a single engine, asymmetry being virtually eliminated by the tandem engines. Furthermore, with the Navy's decision to end production of the Tellier T.3 and T.6, there was a surplus of Hispano-Suiza engines available. The T.5 was also a very stable gun platform in flight.

Approximately 10 T.5s were ordered by the Aviation Maritime in May 1918. These were built at the State Arsenal at Cherbourg.

> **T.5 Two-Seat, Cannon-Armed Flying Boat with Two 250-hp Hispano-Suiza 8B Engines**
> Span 22.96 m; length 15.88 m; wing area 86.5 sq. m
> Empty weight 2,100 kg; loaded weight 3,300 kg
> Maximum speed: 135 km/h; endurance 4 hours 30 minutes
> Armament: a 47-mm Hotchkiss cannon and 300 kg of bombs
> Approximately ten built

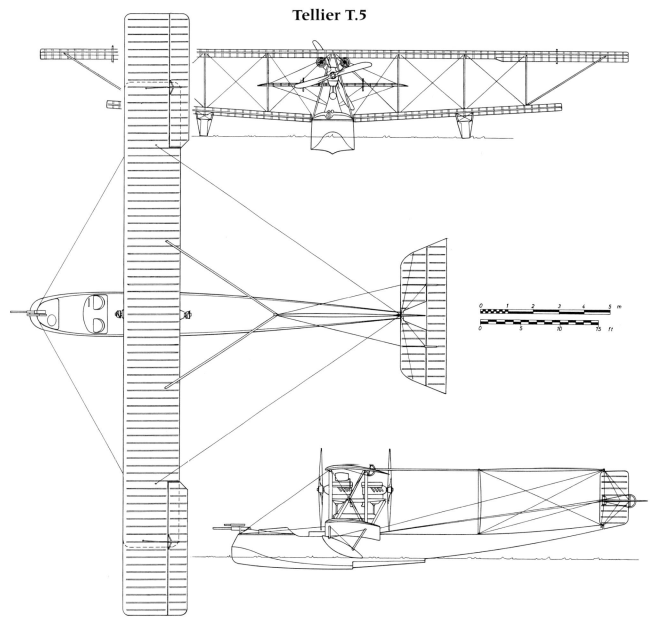

Tellier T.5

Tellier T.5. This aircraft had two 250-hp Hispano-Suiza engines mounted back-to-back. Armament was a 47-mm cannon on a traversable mount in the nose. The rear engine is running while the front engine has not been started. MA 30259.

Tellier T.6

Both the Aviation Maritime and Aviation Militaire displayed considerable interest in the use of cannon-armed aircraft. The Aviation Militaire used the cannon for air-to-air combat and for ground attack. However, they proved of limited use in either of these roles. The Aviation Maritime, on the other hand, had a very specific mission in mind for its cannon-armed aircraft. The bombsights of the time were very inaccurate and the bombs were so small that only a direct hit could disable an enemy vessel. Furthermore, the slim outline of a submarine was difficult to hit from the air. Naval aviators hoped that a cannon would be a more effective weapon against submarines because it was easier to aim and a single hit was more likely to destroy an enemy vessel.

The French modified several of their patrol flying boats to carry a cannon. Tellier modified its T.3 with a 47-mm Hotchkiss cannon (model 1885) placed on a traversal mount in the nose. The hull was strengthened and modified because the weight of the cannon and gunner in the extreme nose substantially altered the center of gravity. To compensate for this change, the fuselage was lengthened 0.87 m at the location of the second step on the hull. The aircraft was designated the T.6 and its cannon mount received Brevet #493411. The T.6 carried 20 to 30 cannon shells. These changes resulted in the T.6 being 60 kg heavier than the standard T.3. Except for the longer fuselage, the dimensions of the T.6 were identical to the T.3.

The T.6 underwent testing in the summer of 1917 at Saint-Raphaël. The tests were successful and the aircraft was ordered in limited numbers. However, the T.6 suffered from the same structural problems as the T.3 and in April 1918 the STIAé Marine added wooden reinforcements to the interior of the lower wing.

Only 55 were built before the Armistice caused the remainder of the order to be canceled. These aircraft were used in the Atlantic and Mediterranean. They were reported to be quite effective in the convoy escort role.

T.6 Two-Seat Patrol Flying Boat with 200-hp Hispano-Suiza 8Ac

Span 15.60 m; length 12.71 m; height 3.40 m; wing area 47 sq. m

Empty weight 1,210 kg; loaded weight 2,670 kg (including the 700 kg Hotchkiss cannon)

Maximum speed: 125 km/h; climb to 500 m in 7 minutes; climb to 1,000 m in 13 minutes; climb to 2,000 m in 29 minutes; endurance 3 hours

Armament: a 47-mm Hotchkiss cannon (model 1885) with up to 30 rounds

Approximately 245 T.3s and T.6s were built, at least 55 of which were T.6s.

Tellier T.6

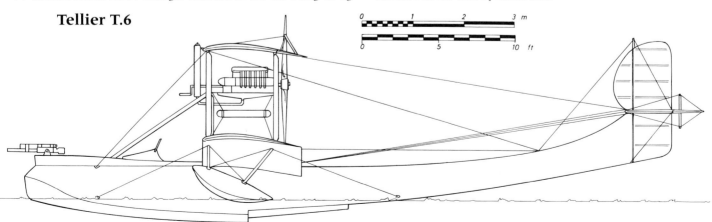

Above: Tellier T.6 flying boat (serial T.66) of the Ecole de Tir at Cazaux with cannon mount in the nose. B86.321.

Left: Tellier T.6. This was a T.3 modified to accept a 47-mm cannon on a traversable mount in the nose. B88.4844.

Tellier T.7

In 1917 the French navy issued a requirement for what it termed a "high seas" flying boat. It was intended that this aircraft would be the French equivalent of the British Felixstowe flying boats which had proved so successful. The specification called for a four-seat flying boat with a wireless radio, a 75-mm cannon with 35 rounds, and two machine guns. Aircraft were submitted to meet this requirement by the Donnet-Denheut, Farman, Levy, and Tellier firms.

The Tellier T.7 was much larger than any of the firm's other aircraft. The basic shape of the T.3 was retained, including the biplane wings of unequal span, the lower wing with the marked dihedral, the complex arrangement of struts connecting the upper wing/tail/hull, and the pair of smaller floats under the bottom wing. There were three 250-hp Hispano-Suiza engines which were mounted side by side and suspended between the upper and lower wings. The central engine was mounted as a tractor, while the two lateral engines were pushers. It had been intended to equip the T.7 with 350-hp Lorraine-Dietrich 12D engines but these were unavailable at the time the prototype was completed. Armament consisted of the 75-mm cannon as specified.

Construction began in April or May 1918, and the aircraft was completed before the Armistice. The first flight took place from the Seine in December 1918. The pilot on the initial flight was Corpet and the aircraft was flown, without armament, to an altitude of 500 meters. Military testing was hampered by the

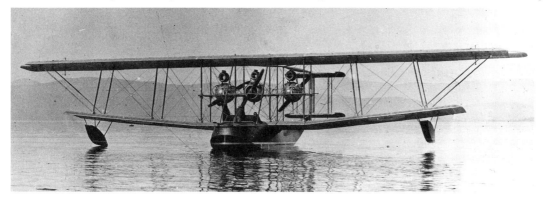

Tellier T.7 with three 350-hp Lorraine engines, one mounted as a tractor and the other two mounted as pushers. MA36446.

Tellier T.7 with three 250-hp Hispano-Suiza engines. B87.5017.

low-powered engines fitted to the prototype. Furthermore, probably in late January, the aircraft was damaged while being towed and sunk; it was several days before it could be raised. The aircraft was repaired but when flight testing resumed at Saint-Raphaël the aircraft would not take off because of the excessive weight of water which had been absorbed by airframe; as well as the weight of structural reinforcements added while the aircraft was undergoing repair. The aircraft was allowed to dry out and when the Lorraine 12D engines arrived in November 1919 they replaced the less powerful Hispano-Suiza 8Bs.

Tellier T.7 "High Seas" Flying Boat with Three 250-hp Hispano-Suiza 8B Engines
Span 30 m; length 21.35 m; height 5.89 m; wing area 156 sq. m
Empty weight 4,650 kg; loaded weight 7,150 kg
Maximum speed: 130 km/h
One built

Tellier T.8

The Tellier T.8 was intended to meet the same requirement for a "high seas" flying boat as the T.7. When the 350-hp Lorraine D12 engines finally arrived in November 1919, they replaced the 250-hp Hispano-Suiza 8Bs used on the T.7 airframe. The T.8, as the re-engined T.7 may have been designated, was able to complete its military trials in December—a year after flight testing with the T.7 had begun. The aircraft appears to have met the requirements of the "high seas" category of flying boat, and the only change requested to the design was reducing the size of the rudder. In early 1920 the 75-mm cannon was at last fitted and fired in flight. However, the end of the war eliminated the need for such an aircraft and the T.7 did not enter service.

The airplane was subsequently flown by Commandant de Labore, who flew the T.7, armed with its cannon and equipped with bombs, at the Monaco air race in April 1920. The airplane remained in service as a testbed for the 75-mm cannon until retired in 1922.

By the end of the war the Tellier firm had been completely absorbed by the Nieuport-Astra firm and, at least in part due to Tellier's poor health, the firm ceased further development of the T.7 and T.8 flying boats.

Tellier T.8 "High Seas" Flying Boat with Three 350-hp Lorraine 12D engines
Span 30 m; length 21.35 m, height 5.89 m; wing area 156 sq. m
Empty weight 4,250 kg; loaded 6,835 kg
Maximum speed 125–130 km/h; endurance six hours
One built (re-engined T.7)

Tellier Four-Engine Flying Boat

In July 1918, Tellier began the design of a four-engine flying boat. The aircraft was intended to be used to attempt an Atlantic crossing postwar. The aircraft, christened *Vonna*, was to have been far larger than any aircraft previously built by the Tellier firm. The basic design was a four-engine triplane with a loaded weight of 1,840 kg. A crew of three was planned and power was to be supplied by four 325-hp Panhard and Levassor V-12 engines. The engines were to be mounted in tandem on either

Tellier *Vonna* Flying Boat with Four Panhard and Levassor V-12 Engines (Provisional Data)
Span 40 m; length 21.50 m; height 9.50 m; wing area 285.30 sq. m
Empty weight 9,200 kg; loaded weight 15,000 kg
Maximum speed: 125 km/h; range 600 km

side of the fuselage and mounted on the middle wing.

It was planned to build two flying boats and to order ten Panhard and Levassor V-12 engines (two for use as spares). However, as construction was nearing completion in early 1919, Tellier left his position with Nieuport. Work on the project slowed with Tellier's departure, and was finally abandoned by Nieuport in early 1920. For further details see the section under Nieuport.

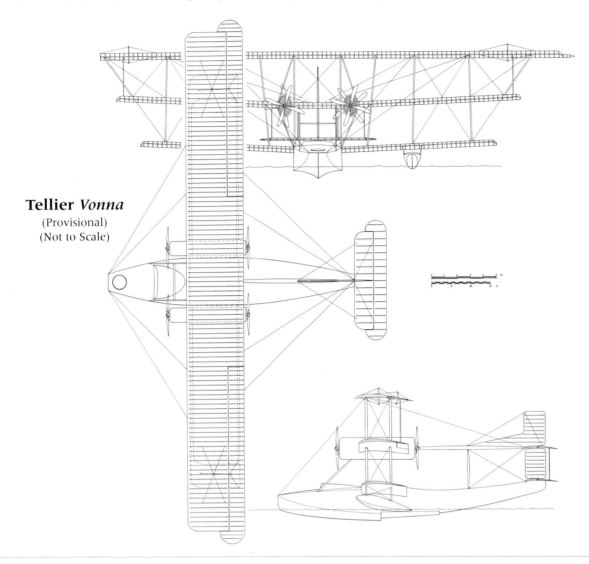

Tellier *Vonna*
(Provisional)
(Not to Scale)

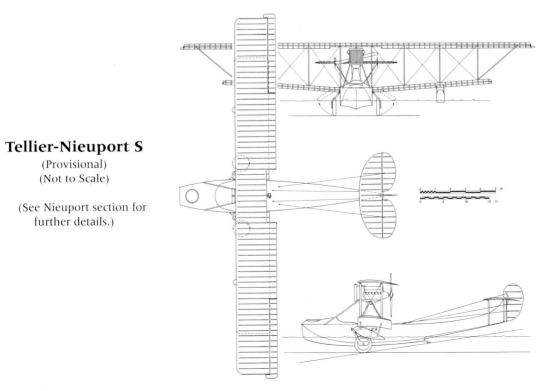

Tellier-Nieuport S
(Provisional)
(Not to Scale)

(See Nieuport section for further details.)

Voisin 4

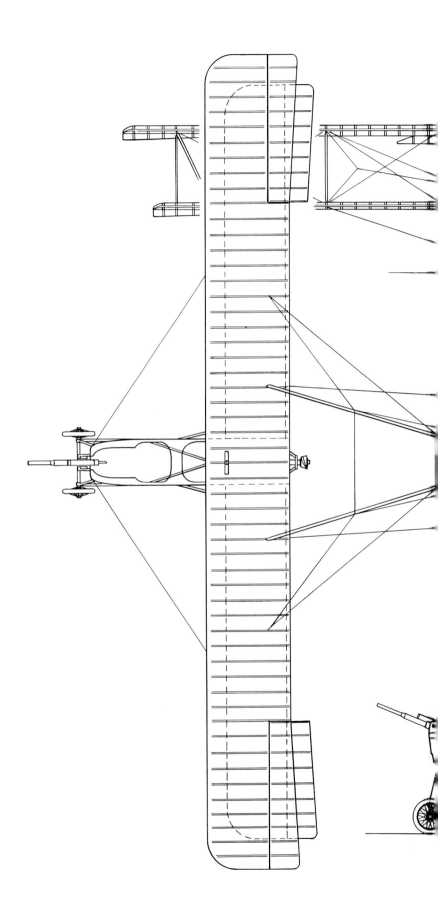

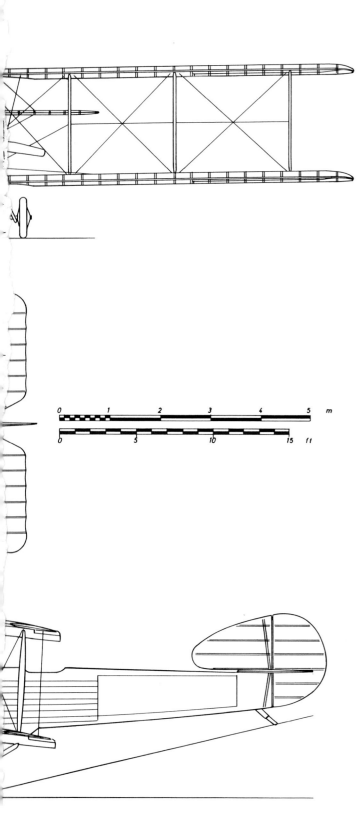

Salmson 4

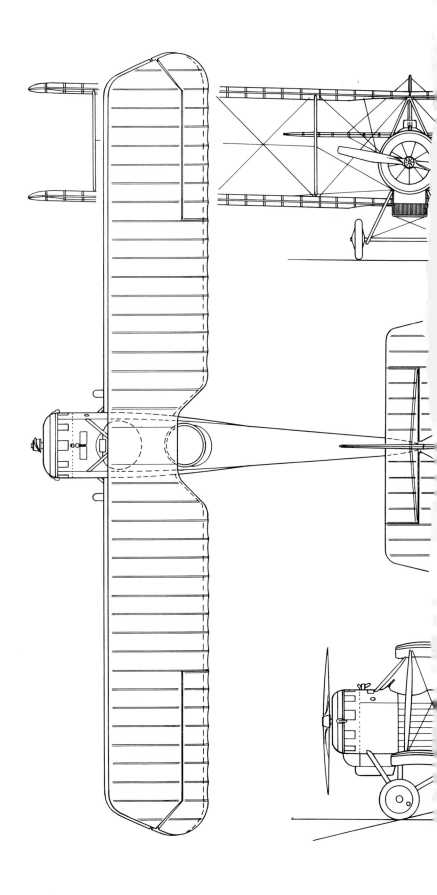

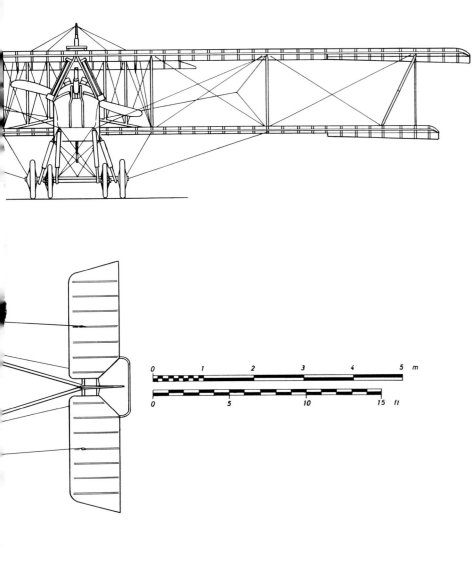
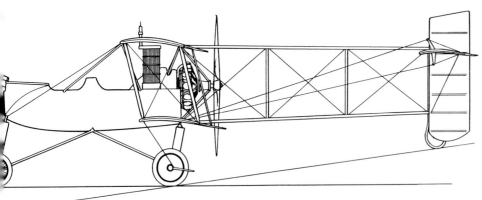

Van Den Born F5

The French F category called literally for an aircraft that would be a flying fortress. It was hoped that it would be able to destroy enemy aircraft and airships simply by the strength of its armament. The Van Den Born F5 was intended to meet this specification. It was a trimotor and carried a crew of five. It was originally intended to use three 230-hp Salmson radial engines. However, it was initially fitted with three 80-hp Le Rhônes, presumably due to a lack of availability of the Salmsons. The decrease in planned power by almost two-thirds cannot have done much to help the aircraft's performance, and after flight testing at Villacoublay in late 1917 it was rejected for use.

Vendôme

Vendôme Monoplane

Raul Vendôme produced several airplanes before the First World War. These were:
1. Vendôme 1.
2. Vendôme 2—biplane.
3. Vendôme 3—monoplane with 50-hp Anzani engine.
4. Raoul-Vendôme 2—monoplane with 50-hp Anzani engine.
5. Odier-Vendôme—biplane with Turcot-Mery engine.
6. Vendôme 4—monoplane with 50-hp Anzani engine.
7. Vendôme 5—same as 4 but with altered tail surfaces.
8. Vendôme 6—monoplane with 25-hp Anzani engine.
9. 1911 monoplane with 50-hp Viale.
10. 1911 military monoplane.

An armored version of the Vendôme was evaluated by the military in 1914. The aircraft had a good climb rate, good visibility, and could be disassembled for easy transport.

A Vendôme 1914 monoplane was in service with the Aviation Militaire in August 1915. It used wing warping for control and had a 60-hp Le Rhône rotary covered with a horseshoe cowling, a spindly undercarriage, and a thin fuselage with the extreme tail section uncovered as in the Blériot 11. The Vendôme was assigned to a front-line escadrille but had been withdrawn by February 1916.

Although the Vendôme monoplane saw only limited service with the Aviation Militaire, it was used by the Spanish air service from 1914 through 1917. The first was taken to Spain by Salvador Hedilla in 1914. The Aviacion Militar ordered six copies built by the Pujol, Comabella, and Cia firm. These machines are believed to have been sent to the Escuela Catalana in early 1917. Some had 60-hp Le Rhône engines; others may have been fitted with 80-hp Gnomes. They remained in service so long because of the difficulty in obtaining more modern equipment from the Allies during the first few years of the war.

An experimental Vendôme airplane of 1914. This diminutive airplane had a 60-hp engine and was described as being easy to assemble and disassemble. Photo 69413.

Vendôme 1914 military monoplane. A single example of a Vendôme 1914 military monoplane was in service with the Aviation Militaire in August 1915. Photo 12663.

Vendôme Monoplane with 60-hp Le Rhône
Span 9.47 m; length 5.85 m; height 2.64 m

Vendôme A3 Biplane

The Vendôme biplane was built in 1916, intended to meet the STAé A3 specification calling for a three-seat aircraft for army cooperation and light attack missions. It was similar to the Salmson-Moineau S.M.1, also designed to the A3 specification.

The Vendôme's two 120-hp Gnome Monosoupape rotary engines were buried inside the fuselage, each driving a propeller via a bevel gear. The engines rotated in opposite directions to counteract torque. As in the S.M.1, the pilot was seated in the center cockpit with gunners in the nose and aft. The slender rear fuselage had a large fin and a rudder with a crescent shaped trailing edge. It seems likely that the complex transmission system and the large number of struts would have reduced performance and created maintenance problems. The type was not selected for production.

Vendôme A3 biplane. Photo 0179.

Vendôme A3 biplane. Photo 0180.

Vendôme A3 biplane. This photograph clearly shows the unusual mounting for the two 120-hp Monosoupape Gnome rotary engines. Photo 181.

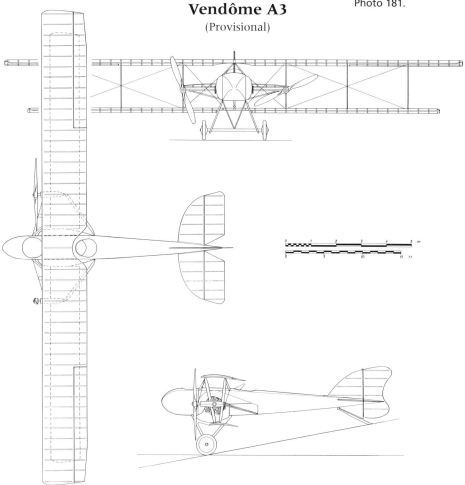
Vendôme A3 (Provisional)

Vendôme Ab2

In 1917 the STAé formulated requirement Ab2, calling for a two-seat reconnaissance aircraft for short-range missions. It was to have light armor and carry a radio and camera. In 1918 two aircraft were produced to meet this requirement: the Salmson 4 and the Vendôme, which was powered by a 300-hp Hispano-Suiza engine. No production was undertaken.

Vickers F.B.24G

The Vickers F.B.24 was a design of the British Vickers firm intended to serve in the fighter and armed reconnaissance role. Although designed to use the Hart engine, the failure of that engine resulted in the aircraft being fitted with a variety of others. One airframe was fitted with a 375-hp Lorraine-Dietrich engine by the French firm of S.A. Darracq and designated the F.B.24G. Its top wing was attached directly to the top of the fuselage. It was listed as a contender for the C2 category of 1918 but did not fly until after the war and was not selected for production.

Aéroplanes Voisin

Gabriel Voisin's first aircraft design, created in 1897, was an ornithopter to have been powered by an electric motor. He became involved with more practical machines from 1903 to 1905, building and testing gliders based on the Hargrave boxkite concept. In 1905 Voisin and his brother, Charles, established the Appareils d'Aviation Les Freres Voisin at Billancourt; this was the first commercial aircraft factory in Europe. Two Blériot floatplanes (Types 3 and 4) were built in 1907—neither of them capable of flying. The Voisins' first successful design was the Delagrange 1, which flew in 1907. It had a pusher configuration that would be featured in most of Voisin's wartime designs. The biplane wings used the Hargrave boxkite layout and an elevator was placed in the front of the aircraft. The fuselage consisted of an abbreviated nacelle with a 50-hp Antoinette engine at the rear. The longest flight of the Delagrange 1 covered only 500 meters. Subsequent attempts to convert it to a floatplane failed. The next Voisin aircraft, also built in 1907, featured significant changes: the wings had dihedral and the frontal elevator had one wing instead of two. These changes greatly improved the aircraft's stability, and it became the first European machine to stay aloft for a full minute. It was also the first European aircraft to complete a one-kilometer circuit. While other French aviation pioneers tried new designs, the Voisins stayed with their reliable boxkite layout. Initially they prospered with this configuration, as it was safe and reliable. However, it also resulted in inferior performance. After completing almost 20 aircraft, the Voisin firm began to build airplanes designed by other aviation pioneers, including Farman, Moore-Brabazon, Ambroise Goupy, and Wright. Drawing on their experience in building these, the Voisin brothers produced their first aircraft with a tractor layout. However, it was unsuccessful and its failure probably reinforced the brothers' conviction that the pusher configuration was superior to a tractor layout. The Voisins equipped one of their boxkite designs with a quick-firing gun in 1910. In 1911 a canard machine was built fitted with floats designed by Henri Fabré. A modified version of this aircraft was built in 1912 and was offered to the Marine Francaise, which purchased an unknown number of canards. The army's Aviation Militaire purchased 12 Voisin pushers with quadricycle landing gear. They were later designated the Voisin 1912 Type (it may have later been called Voisin Type 1). This aircraft featured the basic configuration that virtually all of Voisin's aircraft would use throughout the war. It was an equal-span biplane with no dihedral, a short nacelle with the crew in front and the engine in the rear, a cruciform tail attached to the wing by a set of booms, and a quadricycle landing gear. Although they were clearly obsolete by the start of the war, the sturdiness and reliability of the Voisin designs enabled them to form the backbone of the French night bomber force until late 1918.

Prewar Voisin Aircraft

1907

1. Kapferer—25-hp Buchet engine.
2. Delagrange No.1—50-hp Antoinette engine.
3. Delagrange No.2—50-hp Antoinette engine.
4. Farman No.1—50-hp Antoinette engine.

1908

5. Farman No.1bis—50-hp Antoinette engine.
6. ? —58 hp Renault engine.
7. Farman No.1bis triplane—50-hp Antoinette engine.
8. Moore-Brabazon No.1—50-hp Antoinette engine.
9. ? —50-hp Vivinus engine.
10. de Caters No.1 triplane—50-hp Vivinus engine.
11. Zipfel No.1—50-hp Antoinette engine.

1909

12. Moore-Brabazon No.2 triplane—52 hp Chenu engine.
13. ? —60-hp ENV Type F engine.
14. Moore-Brabazon No.3—60-hp ENV Type F engine.
15. Moore-Brabazon No.4—60-hp ENV Type F engine.
16. de Caters No.2 biplane—80-hp Gobron engine.
17. de Caters No.2—80-hp Gobron engine.
18. Vivinus No.1—50-hp Vivimus engine.
19. ? —52 hp Chenu engine.
20. Delagrange No.3—50-hp Antoinette engine.
21. Farman No.2—58 hp Renault engine.
22. Koch No.1—50-hp Vivinus engine.
23. Fournier No.1—50-hp Vivinus engine.
24. Simms No.1—50-hp Antoinette engine.
25. Hein No.1
26. Euler No.1—Adler engine.
27. Aero-Club d'Odessa—50-hp Antoinette engine.
28. Kaulbars No.1—50-hp Antoinette engine.
29. Fletcher—50-hp Itala engine.
30. Rougier—50-hp Antoinette engine.
31. ? —58 hp Renault engine.
32. Alsace—50-hp Antoinette engine.
33. Ile-de-France—50-hp Antoinette engine.
34. Daumont No.1—50-hp Antoinette engine.
35. Octavie No.3—50-hp Antoinette engine.
36. ? —38 hp Gnome engine.
37. Hansen No.1 —50-hp Antoinette engine.
38. Bolotoff triplane—100-hp Panhard engine.
39. Gaudart—50-hp Gnome Omega engine.
40. Gobron No.1— 80-hp Gobron engine.
41. Buneau-Varilla No.1—60-hp Wolseley engine.
42. Sanchez-Besa No.1—50-hp Antoinette engine.
43. Sanchez-Besa No.2—50-hp ENV Type F engine.
44. Da Zara—Rebus engine.
45. Swendsen—50-hp Gnome Omega engine.
46. Cockburn No.1—50-hp Gnome Omega engine.
47. De Baeder—Mutel engine.
48. Metrot-Marce No.1—58-hp ENV Type G engine.
49. de Laroche—?
50. Italian Aeronautical Society—58 hp ENV Type G engine.

1910

51. ? —50-hp Gnome Omega engine.
52. Champel —50-hp ENV Type F engine.
53. Pauwels No.1—50-hp Mutel engine.
54. Stoeckel No.1—50-hp Gnome Omega engine.
55. Mignot —50-hp Gnome Omega engine.
56. Ravetto—50-hp Gnome Omega engine.
57. Paul—50-hp ENV Type F engine.
58. Chailley—50-hp Gnome Omega engine.
59. Antelme—50-hp ENV Type F engine.
60. Allard—50-hp ENV Type F engine.
61. Poillot—50-hp Vivinus engine.
62. Adorjan—?
63. Croquet—50 p Gnome Omega engine.
64. De Montigny—50-hp Gnome Omega engine.
65. Haeffely—58 hp ENV Type G engine.
66. Cagno—50-hp ENV Type F engine.
67. Economo—58 hp ENV Type G engine.
68. Rigal—ENV engine.

69. "Militaire"—100-hp Gnome.
70. "Tourisme"—Dual-controls, 100-hp Gnome.

1911
71. "Militaire"—75-hp canard.
72. "Militaire"—140-hp Gnome canard.

73. "Militaire"—100-hp Gnome.
74. Astra Triplane—75-hp Renault.

1912
75. Sanchez-Besa—Renault.
76. Bathiat-Sanchez—75-hp or 100-hp Renault.

Voisin 1912 and 1913 Types (L)

Two types of Voisin airplanes were in service at the outbreak of the First World War. The earlier was built in 1912 and so was given the designation "1912 Type." It is possible it was later given the SFA designation Type 1. The Voisin factory designation was Type L. Initially, the 1912 Voisins were fitted with Gnome engines; later ones were given an 80-hp Le Rhône 9C. When the war began the 1912 Voisins were assigned to the training unit at Crotoy.

The second pre-war Voisin was similar to the 1912 version but powered by a 70-hp Gnome 7A engine. It was designed in 1913 and some sources refer to it as the 1913 Voisin. Others state that it carried the designation 13.50, which indicated the wing span. Some 1913 Voisins were in service with V 14 and V 21 at the beginning of the war. Both escadrilles were assigned to the 4th Armée and performed short-range reconnaissance. However, by 20 August 1914 both V 14 and V 21 were re-equipped with Voisin 3s. After being withdrawn from front-line escadrilles, the 1913 Voisins were assigned to training units.

The Voisin 1913 types, however, played an important part in the development of aircraft armament. Capitaine Faure of V 24 experimented with ways to arm his Voisin aircraft beginning in 1914. He took his plans to the War Ministry, which dismissed his ideas. However, Gabriel Voisin heard of Faure's work and decided to develop a way to fit a gun to an aircraft. Eventually a tripod was developed that could hold an 8-mm Hotchkiss gun. Work was completed on 15 August 1914. A short time later, Voisin arrived at escadrille V 24 (based at Mezières) with six Hotchkiss machine guns and mounts and a pass signed by Faure

1912 Voisin (Voisin 1) Two-Seat Reconnaissance Airplane with 80-hp Le Rhône 9C

Span 13.52 m; length 10.50 m; height 2.91 m; wing area 41.93 sq. m
Empty weight 825 kg; loaded weight 1,100 kg
Maximum speed: 94 km/h at 2,000 m; climb to 2,000 m in 28 minutes; endurance 4.3 hours

1913 Voisin (Voisin 2) Two-Seat Reconnaissance Airplane with 70-hp Gnome 7A

Dimensions: same as Voisin 1
Empty weight 820 kg; loaded weight 1,100 kg;
Maximum speed 94 km/h at 2,000 m; climb to 2000 m in 28 minutes
Approximately 70 Voisin 1s and 2s were built

Voisin L armed with a Hotchkiss machine gun. A spare propeller is mounted on the left side of the fuselage. B83.3423.

Voisin 1913 (Type L) with an 80-hp Le Rhône 9C engine at the 1913 military maneuvers. B83.3464.

The Voisin Canon of 1913. Changes from the Voisin L series included an increase in the number of wheels to six (to support the increased weight of the 37-mm Hotchkiss cannon) and a simplified tail unit consisting of a metal tube with three rudders. The engine was a 200-hp Canton-Unné (Salmson). The aircraft underwent evaluation by the military in May 1914, but the beginning of the war and the urgent

Voisin L floatplane of the Aviation Maritime. The Aviation Maritime had several Voisin Ls on strength at the Boulogne and Dunkirk bases. Reairche.

need for Voisin 3s (L/LA) prevented the type from being developed further. MA95486.

enabling him to gain entry to the base with his suspicious cargo. Six of V 24's Voisins were armed with the guns and were able to fly combat missions within a few hours of receiving their new weapons. It appears however, that none of the Voisin 1913 types saw combat before being replaced by the newer Voisin 3s, also fitted with Hotchkiss guns.

Several armored versions of the 1913 Voisin were built to meet General Bernard's 1913 specification for an armored two-seater for short-range reconnaissance. Three of these were in service with the C.R.P. on 4 November 1914.

The Aviation Maritime had several Voisin 1913s on strength at the Boulogne and Dunkerque bases, used as bombers.

It is believed that a total of 70 Voisins (Types 1912 and 1913) were built.

Russia also had a number of Voisin Ls, some probably built under license. A table in Shavrov's book shows the engine as a 140-hp Salmson, which suggests that the Voisins were at some point retrofitted with these (see entry under Voisin 3).

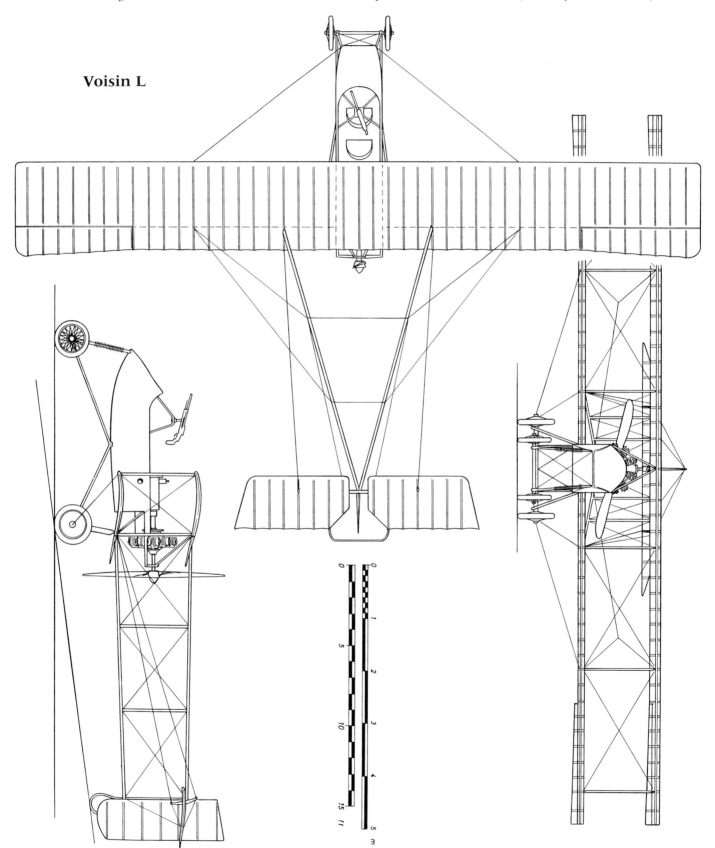

Voisin L

Voisin L piloted by Adjudant Possi. B83.4976.

Voisin 3 (LA/LAS)

It was clear to Gabriel Voisin that the 1912 and 1913 types were underpowered. Furthermore, while the metal structure of the series made the airplanes sturdy, it degraded performance significantly. A new engine would permit a stronger airframe without a reduction in performance. Voisin elected to use the 120-hp Salmson M9. In 1914 one was experimentally fitted to a Voisin 1913 type that was tested at Toulon by the Aviation Militaire. The improvement in performance was impressive enough to warrant series production. The new aircraft received the SFA designation Voisin 3; the factory designation was Type LA. It has been estimated that more than 1,000 Type 3s and Type 4s (a development of the Type 3) were built.

Most Voisin 3s were equipped with 120-hp Salmson M9 engines, and because the more powerful engine was fitted with a larger propeller the wings and nacelle had to be altered. A cutout was placed in the lower wing to permit adequate clearance for the propeller. The engine was attached to a platform, ensuring the propeller would not strike the ground, and the tail booms were spread farther apart and strengthened. Because of the increased horsepower of the M9, the Voisin 3 was the first Voisin design to feature an all-metal fuselage. The rear wheels were fitted with improved, and very powerful, brake drums to handle the increased weight. In addition to these brakes, the airplane could be brought to a rapid halt simply by forcing the nose down onto the front wheels while landing. The forward wheel struts could be adjusted to accommodate various loads. All four struts of the quadricycle landing gear were fitted with spring shock absorbers. The rear wheels were set directly beneath the engine and fuel tank in order to better support their weight. Windows were fitted to the floor of the nacelle to improve the crew's downward view; in some aircraft a machine gun was mounted to fire through these windows. Some Voisin 3s were fitted with lights on the front landing gear legs. These permitted the crew to observe objects on the ground during night missions from 150 meters in the air and also facilitated landing. An additional light was installed in the cockpit to help the pilot read his instruments. However, in escadrille service these lights were not used because they might have destroyed the pilot's night vision. Instead, the pilot would continually look outside the cockpit to preserve his night vision while the observer would call out the instrument readings. Some Voisin 3s were also fitted with mufflers for night bombing missions to reduce the risk of detection. When the airplane maintained an altitude of 800 meters it was virtually inaudible—possibly the first use of aerial "stealth" technology in wartime. Voisin 3s were often fitted with an 8-mm Hotchkiss machine gun, and the crew had a Winchester and a Chauchat gun.

On later Voisin 3s the Salmson M9 was mounted on an elevated platform; these planes were designated LAS, the S standing for surélevé, or raised. As can be seen from the accompanying tables, the increase in performance was modest. Some Voisin 3s had Salmson P9 or R9 engines.

An armored version of what is believed to have been a Voisin 3 was tested by V 14 in August 1914. Because of the added weight it could only carry the pilot. Performance include a climb to 1,000 m in 40 minutes. Two of these aircraft were built powered by a 130-hp Canton-Unné engine.

Operational Service
Army Cooperation

Beginning in August 1914 the Voisin 1913s in service were replaced by Voisin 3s which soon became the primary night bomber for the Aviation Militaire. The army cooperation escadrilles exchanged their Voisin 3s for Caudron G.3s and M.F.11s and passed their Voisin 3s to the newly-formed Groupes d'Bombardement (Bomb Groups).

Three escadrilles used Voisin 3s for army cooperation duties:

V 14, formed in 1913, commanded by Lieutenant Mouchardand and assigned to the 4th Armée. It became part of GB 1 in 1914.

V 21, formed in December 1913 and assigned to the 4th Armée. In addition to reconnaissance missions, V 21 flew bombing raids as far as 50 km behind enemy lines. In October it moved to Belgium and participated in the Battle of Flanders and later the Battle of Yser. On 7 June 1915 V 21 became part of GB 1.

V 24, formed at Reims in August 1914, commanded by Capitaine Faure and assigned to the 5th Armée in September. V 24 operated over the Champagne front and re-equipped with Farman F.40s in May 1916.

Although the Voisin 3s were successful as reconnaissance aircraft, they will always be best remembered as the airplanes that equipped the world's first dedicated bomber units.

Bombing

On 29 September 1914 Voisin Escadrilles V 14 and V 21 were designated as bomber units. Groupe d'Bombardement 1 (GB 1) was formed on 23 November 1914 with three escadrilles assigned—V 14 (now redesignated VB 1), VB 2 (which was formed from BR 17), and VB 3, an entirely new unit. All three were equipped with six Voisin 3s, well-suited to bombing

Voisin 3 (probably an LA) armed with lance bombs in the container mounted on the fuselage side. Windows were fitted in the floor of the nacelle to improve the crew's downward view. B87.304.

missions because of their inherent stability and because they could carry a bomb load of up to 150 kg.

GB 1 moved from its home base at Malzéville to the Somme front on 19 December. The next day the unit flew its first combat mission when 11 of its aircraft bombed train stations near Lens. Four aircraft were damaged, but none was lost during the attack. Although bad weather hindered operations, GB 1 was able to fly a large number of sorties throughout 1915. Most of the time train stations were attacked to prevent troops and supplies from reaching the front. Initially the Voisin 3s carried either flechettes or modified 40-kg artillery shells.

GB 2 was formed on 16 January 1915 with escadrilles VB 3, VB 4, and VB 5 assigned to it. GB 2 was initially based at Saint-Pol-sur-Mer and went into action on 3 February, when a single crew from VB 5 flew a reconnaissance mission over the front. Later in the month the naval facilities at Bruges and Ostende and the airfields at Laffinghe and Ghistelles were attacked.

By 15 March 1915 a new unit, GB 3, had been formed at Alsace. All Voisin bomber units had been given new designations on 4 March 1915, as bomber escadrilles were to have numbers in the 100 series. VB 107, VB 108, and VB 109 were assigned to GB 3. GB 1's units became VB 101, VB 102, and VB 103 and GB 2's became VB 104, VB 105, and VB 106.

During March GB 1 and 2 attacked train stations, artillery positions, and barracks; GB 2 also attacked the airfields at Gits on 20 March. GB 3 flew its first combat mission on 20 March, during which train stations at Altkirch and Mulheim were bombed.

In April GB 1 attacked train stations and barracks and concentrated on targets in the Metz and Moselle valleys. GB 2 continued to have difficulties; raids on 1 and 11 April failed because the crews were unable to reach the targets. However, a notable success occurred on 15 April, when ten Voisin 3s struck the harbor at Zeebrugge with 155-kg bombs. Significant damage was inflicted, but all the Voisins were hit by anti-aircraft fire. For the remainder of the month GB 2's crews attacked barracks, airfields, and supply convoys. GB 3 continued attacks on train stations along the Alsace front and also bombed the airfield at Habsheim.

One of the most important raids of the war took place 27 May, when units from GB 1, joined by V 21, bombed the poison gas factory at Ludwigshafen. The aircraft used in this mission had been modified by Voisin to have long-range fuel tanks and an improved ignition system. The factory was hit with 47 90-kg bombs and two 155-kg bombs. The plant was damaged and lead shields were breached, resulting in the release of toxic fumes.

In early May GB 2 flew infantry support missions, concentrating on German artillery positions and train stations. Some German airplanes were encountered but they refused to engage in combat.

GB 3 flew missions from Artois beginning on 15 May. Targets along that front were attacked throughout the spring and summer months. Artillery positions, balloons, and blast furnaces were bombed.

GB 4 was formed on 19 May. VB 110, VB 111, and VB 112 were assigned to the new group.

The next major raid was on 15 June, when GB 1 and V 21 bombed the city of Karlsruhe in retaliation for German attacks on French cities. It was hoped that by replying in kind the French could deter future such attacks on their cities. A total of 124 bombs were dropped. Two aircraft were lost, one to a night-fighter and one to anti-aircraft artillery. Civilian casualties numbered about 23, and the raid was effective enough to cause the German authorities to double the city's anti-aircraft artillery positions and to station 18 fighters near the town.

GB 2 continued attacks in support of the French troops from its base at Herlin le Sec, bombing artillery positions and airfields. However, increasingly accurate ground fire resulted in significant damage to several Voisins and the inaccurate bombsights resulted in very little damage being done by the aircraft.

On 23 June GB 2 joined GB 1 at Malzéville, followed by GB 3 on 19 July. All three groups prepared for a major series of strikes on the petroleum refineries at Pechelebron. On 30 July GB 1 and GB 2 sent 38 aircraft to bomb the refineries, while GB 3 attacked barracks in the vicinity of Pechelebron. The Germans retaliated by bombing the airfield at Malzéville. The Voisins were unable to intercept the intruders, and consequently Caudron G.4s and Nieuport 10s were sent to the base to provide fighter protection and escort. During August and September GB 1, 2, and 3 bombed targets in the Saar Valley, Saarbrucken, and the steelworks at Dillingen.

GB 3 moved to Artois on 3 September and concentrated its raids on transportation targets, particularly train stations. GB 2

Voisin 3 (LA) serial V.499 at the Ecole d'Aviation at Amberieu. B83.3017.

Voisin 3 (LAS) serial V.685 of the ecole d'Aviation at Amberieu; its elevated engine mount is visible. B83.2959.

Voisin 3 (LAS). Renaud.

Voisin 3 (LA) serial V.443. B78.1234.

moved to Matogues on 18 September and from this base attacked train stations and rail lines. GB 1, still at Malzéville, continued to attack railroad targets.

By mid-October GB 1 was rejoined at Malzéville by GB 2 and 3. GB 4, on the other hand, was disbanded. The decisions to deactivate GB 4 and to switch from strategic bombing missions to tactical support were probably the result of the relatively poor results the Groupes de Bombardement had achieved after almost a year-long campaign. Furthermore, the damage inflicted by anti-aircraft artillery and fighters was steadily increasing. For this reason C 66 and N 65 had been assigned to GB 2 to provide fighter protection. The bomber units were dispersed throughout September and October and were relegated to tactical support of ground troops.

Bad weather hindered operations during November and December. Because of increasing losses the bomber units now switched almost exclusively to night attacks. GB 3 was withdrawn from action in December and would spend four months retraining as a night bombardment unit. GB 1 continued night attacks during December.

When V 24 re-equipped with Voisin 3s, several aircraft were fitted with Voisin's tripod mount carrying an 8-mm Hotchkiss gun. The gun was not fired in anger until a Voisin 3 (serial V.84) was on patrol with a crew of two—Sergeant Frantz and his mechanic, Louis Quénault. They engaged a German Aviatik B.I heading for the German lines. Quénault began to fire but the gun jammed after only 47 rounds. Both aircraft continued to maneuver while Quénault furiously tried to clear the jammed weapon. Incredibly, just as the gun was repaired, the Aviatik flipped over and dived into the ground.

The victory, however tentative, over the German observation plane led to the widespread acceptance that heavy armament could be effectively used by military aircraft. However, the Hotchkiss was not a reliable weapon; eventually the Lewis machine gun would become preferred for French combat aircraft.

At the beginning of January 1916 the Voisin 3s were being rapidly replaced by the improved Voisin 5s. By March most of the Voisin 3s had been retired to training units.

The Aviation Maritime had at least two Voisin 3s based at Dunkerque in 1915 and these were used on numerous bombing raids. Additional Voisin 3s were sent to Dunkerque at the end of February 1915 (see section on the Aviation Maritime). Two Voisin 3 floatplanes were aboard the seaplane carrier *Foudre* in the Adriatic. They carried out reconnaissance of the Kotor section along the Balkan coast early in the war.

Voisin 3 of VB 103, which was assigned to GB 1. B87.1148.

Voisin 3 The ground crew is loading the aircraft with bombs. B83.3443.

Voisin 3 with machine gun mounted in the nose. B87.152.

Foreign Service
Belgium
Belgium acquired Voisin 3s in 1915. French units based in Belgium often carried Belgian observers during reconnaissance missions. Apparently the Belgian officers were impressed enough by the Voisin 3s to order seven for the Belgian Militar Vliegwezen (Belgian Air Service). By 5 January 1915 the first had been handed over but crashed on its initial flight.

Undaunted, the Belgians acquired additional (estimated to be between five and seven) Voisin 3s. They equipped Escadrilles 6 (based at Houthem) and 3 (at Koksijde). By 1917 the Voisin 3s had been replaced by Farman F.40s. The Voisin 3s were subsequently used for pilot training at Étampes.

Italy
After receiving several Voisin 3s in early 1915, the Italian Servizio Aeronautico (Air Service) decided to produce the type under license. More than 100 were built by the S.I.T. (Societe Italiana Transaera) firm. Their Salmson M9 engines were built under license by the Isotta-Franschini firm. Other engines used were the 190-hp Isotta-Franschini V.4, 100-hp Fiat A.10, and (on at least one Italian Voisin) a 120-hp Le Rhône with external reduction gear. Deliveries began in January 1916. On some Italian Voisins a second gun, usually a 9-mm Revelli machine pistol, was carried.

In 1915 Voisin 3s acquired from France were assigned to the 5a and 7a Squadriglias, both of which were attached to the 1st Gruppo Squadriglia Aviatori (assigned to the 3rd Armata).

By 1916 the 5a Squadriglia had been redesignated 25a Squadriglia. It was still assigned to the 3rd Armata. The 7a Squadriglia was redesignated 27a Squadriglia and was assigned to the 2nd Armata. A new Voisin unit, 35a squadriglia, was attached to the 2nd Gruppo Aeroplani Udine, which was also assigned to the 2nd Armata.

Some S.I.T.-built Voisin 3s were assigned to the 303rd (Novi Ligure) and the 305th (Cairo Montenotte) air defense units.

By 1917 only 25a Squadriglia had Voisin 3s as its primary equipment. It was assigned to the 1st Gruppo Aeroplani and was still attached to the 3rd Armata.

The Voisin 3s were withdrawn from front-line units in 1917 and assigned to reconnaissance training. Others were used to drop spies behind enemy lines. For these missions they were fitted with special mufflers to quiet engine noise.

Romania
Romania acquired eight Voisin 3s in 1915. They were assigned to Grupul 2. By 1916 there were two assigned to Grupul 1 and two serving with Grupul 3. The Voisin 3s were withdrawn from front-line units by the end of 1916.

Russia
After France, Russian was the main user of the Voisin 3. More than 800 were purchased by Russia during the war and an additional 400 Voisin 3s (of all types) were built in Russian factories. Voisin 3s were built by five Russian factories: Anatra, Breshnev-Moller, Dux, Lebedev, and Schetinin. The Lebedev plant alone produced 98 aircraft from 1917 through 1918. Engines were supplied by a subsidiary of Salmson which, at the insistence of the Russian government, had been established by the firm in Moscow. Shavrov states that the Voisin LAs were equipped with 140-hp Salmson engines and the LAS with 150-hp Salmson engines. This would seem unlikely unless the Voisins were re-equipped with these engines during the latter

Voisin 3 of the IRAS assigned to the 21st Escadrille. B86.21.

Voisin 3 LAS

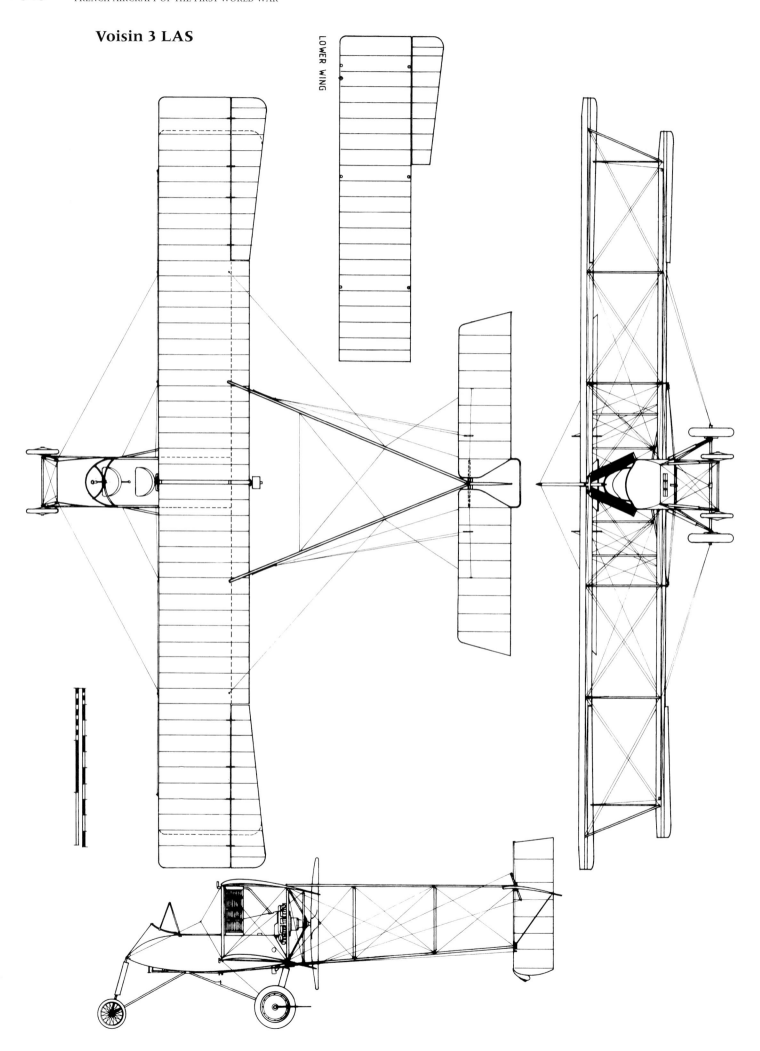

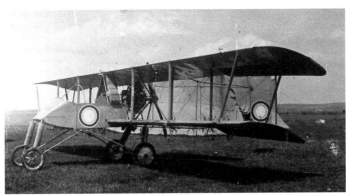

Above, above left, and left: Voisin 3 (LAS) in IRAS service. Via Colin Owers.

part of the war. Some Voisins built in Italy were fitted with water-cooled Isotta-Fraschini engines. Shavrov suggests that some of the Voisins with the Isotta-Fraschini were used by the Russians.

Most of the Voisins were used by units assigned to the Second Army, particularly to the Second Aviation Artillery Unit. The Voisins were used extensively in the army cooperation role. Artillery spotting was a common mission for them, but the most dangerous activities were "decoy" missions in which one aircraft would fly near German lines to draw fire while another performed reconnaissance. Other Voisin 3s were used as night bombers by the simple expedient of fitting them with electric lights so the crew could read the instruments in the dark

Voisin 3 (LA) Two-Seat Bomber with 140-hp Salmson Built by Anatra, Breshnev-Moller, Dux, Lebedev, and Schetinin in Russia

Span 14.74 m; length 9.50 m; height 3.60 m; wing area 42.00 sq. m
Empty weight 900 kg; loaded weight 1,250 k
Maximum speed: 100 km/h; climb to 1,000 m in 12 min.; climb to 2,000 m in 26 min.; ceiling 2,800 m; endurance 4 hours

Voisin 3 (LAS) Two-Seat Bomber with 150-hp Salmson Built by Anatra, Breshnev-Moller, Dux, Lebedev, and Schetinin in Russia

Span 14.74 m; length 9.50 m; height 3.60 m; wing area 42.00 sq. m.
Empty weight 900 kg; loaded weight 1,250 kg;
Maximum speed: 105 km/h; climb to 1,000 m in 11 min.; climb to 2,000 m in 23 min.; climb to 3,000 m in 55 min.; ceiling 3,000 m; endurance 4 hours

Voisin 3 (LA) Two-Seat Reconnaissance/Bomber with 120-hp Salmson M9 Built by Savages of King's Lynn in Britain

Span 14.74 m; length 9.50 m; height 2.95 m; wing area 49.65 sq. m
Empty weight 950 kg; loaded weight 1,350 kg
Maximum speed 105 km/h at sea level: climb to 1,000 m in 12 min.; climb to 2000 m in 30 min.; endurance 4.5 hours
Armament: an 0.303-in Lewis machine gun and up to 91 kg of bombs
Approximately 50 Voisin 3s were built

Voisin 3 (LAS) Two-Seat Reconnaissance/Bomber with 120-hp Salmson M9 Built by Savages of King's Lynn in Britain

Span 14.74 m; length 9.50 m; height 2.95 m; wing area 49.65 sq. m
Empty weight 950 kg; loaded weight 1,350 kg
Maximum speed 110 km/h at sea level: climb to 1,000 m in 10 min.; climb to 2000 min. 24.5 min.; endurance 4 hours
Armament: an 0.303-in Lewis machine gun and up to 91 kg of bombs

Voisin 3 (LA) Two-Seat Bomber with 120-hp Salmson M9

Span 14.74 m; length 9.50 m; height 2.95 m; wing area 49.65 sq. m (some sources report 54 sq. m)
Empty weight 950 kg (825 kg); loaded weight 1,350 kg (1,100 kg)
Maximum speed: 98 km/h at 2000 m; climb to 2000 min. 24.5minutes; range 200 km
Armament: an 8mm Hotchkiss machine gun, a Winchester and Chauchat gun; when used as bomber the Voisin 3 could carry up to 150 kg of bombs
Approximately 1,000 Voisin 3s and 4s were built

Voisin 3 (LAS) Two-Seat Bomber with 120-hp Salmson M9

Span 14.74 m; length 9.50 m; height 2.95 m; wing area 49.65 sq. m
Empty weight 994 kg; loaded weight 1,400 kg
Maximum speed: 100 km/h at 2000 m; climb to 2000 min. 24 minutes; range 200 km
Armament: an 8mm Hotchkiss machine gun, a Winchester and Chauchat gun; when used as bomber the Voisin 3 could carry up to 150 kg of bombs
Voisin 3 could carry up to 150 kg of bombs

Voisin 3 Two-Seat Floatplane with 120-hp Salmson M9

Span 15.00 m; length 8.00 m; height 3.71 m; wing area 49.65 sq. m
Payload 450 kg
Maximum speed: 96 to 110 km/h

Voisin 3 (LA) Two-Seat Bomber with 120-hp Salmson M9 built by S.I.T. in Italy

Span 14.74 m; length 9.50 m; height 3.60 m; wing area 53.60 sq. m
Empty weight 800 kg; loaded weight 1,200 kg
Maximum speed: 120 km/h; climb to 1,000 m in 7 minutes; endurance 3 hours 30 minutes
A total of 112 were built

(Voisins serials 385, 487, 672, 689, and 690 were so modified). When used as a bomber, these aircraft could carry up to 176 kg of fuel and 80 kg of bombs. The latter were usually 4-, 10-,16-, or 32-kg Oranovskiv bombs. The Voisins were also used for ground attack. Voisin 3s even had limited success when used as fighters. Voisins were assigned to the 2nd Fighter Group and Captain Kruten of that unit used his Voisin 3 to destroy a number of German reconnaissance airplanes.

By March the distribution of the 116 Voisin 3s at the front was: northern front (33), western front (26), south-western/Romanian front (45), Caucasus front (12). In April 1917 there were 50 serviceable Voisin 3s at the front. In June 1917 there were 73: northern front (19), western front (17), southwestern/Romanian front (19), Caucasus front (18).

Voisin 3s were also widely used by training units. The syllabus, based on the French method of instruction, called for students to transition from M.F.7 trainers to Voisin 3s.

A few Voisin 3s were equipped with floats for use by the Russian naval air service, and some Voisins were fitted with skis.

The Voisin 1 (L), Voisin 3 (LA/LAS), and Voisin 4 (LAS) are known to have been used in Russia after 1917. In 1920, 81 Voisins of all types were still in service with operational units; most of these were probably Voisin 3s. Additional Voisins were with the training units. Units identified as having Voisins in 1923 included the Higher School of Aerial Observers, the 1st Voennaya Shkola Letchikov, the Voennaya Shkola KVF, and the Strel'bom school. Most of the Voisins had been struck off charge by 1925.

Ukraine

Two Voisins of undisclosed types were obtained by the Ukrainians in 1918. Since these were almost certainly obtained from the Russians it is likely they were Voisin 3s. Approximately six Voisins were assigned to the Don Cossack squadron under the command of General P.N. Krasnov. Four Voisins (with known serials 624, 730, 737, and 745) were assigned to the Ukrainian air service.

United Kingdom

Britain acquired several Voisin 3s in early 1915 and a decision was made to manufacture 50 in England. They were assigned to Nos.4, 5, 7, 12, and 16 Squadrons. They were withdrawn from service in September 1915; none was ever sent to France. Subsequently the Voisin 3s served with No.1 and 8 Reserve Aeroplane Squadrons. Some were also used by No.4 Wing at Netheraron for training.

The Royal Naval Air Service acquired 36 Voisin 3s. These were 8501–8509, 8518–8523, 8700–8707, A154, and 3821–3832. They were assigned to Nos.1 and 3 Wings Eastbourne, Eastchurch, Force D RFC, No.3 Wing Aegean, Grain, White City, and No.8 Squadron in South Africa.

Voisin 4 (LB/LBS)

It had proved possible to equip Voisin 3s with Hotchkiss guns and, based on this success, Gabriel Voisin decided to modify a Voisin 3 airframe to accept a 37-mm cannon. The wings of the new design were shifted backward 40 cm to compensate for the weapon's recoil and weight. The cannon was mounted on a metal tripod designed by Voisin and built by Aircraft Park 101. The new aircraft had identical dimensions to the Voisin 3 except for a slightly longer fuselage nacelle. The position of the crew was reversed so the gunner was in the front, where he could load and fire the cannon. The aircraft retained the Voisin 3's 120-hp Salmson M9.

The SFA designated the new aircraft the Voisin 4, but it became far better known as the Voisin Canon. Only a small number were built, as they proved to be of limited operational usefulness. Production of the Type 4 continued until late 1916, when it was superseded by cannon-equipped Voisin 8s. The factory designation for the Voisin 4 was LB. A version with a raised motor mount was designated LBS (S for surélevé, or raised).

Operational Service

The Voisin 4 began to arrive in front-line escadrilles by April 1915. GB 3 received some of the first and used them to equip VB 113; the escadrille's designation was subsequently changed to VC 113 (for Voisin Canon) on 25 September 1915. GB 1 also received enough to convert VB 110 to VC 110 in April 1916. VC 116 was formed from Section D'avions Canons 1, 12, 13, and 14 in February 1916. It was assigned to GB 3. VC 111 was also formed from three Section D'avions Canons Nos.5, 8 and 9 in December 1915. Flying from Bar-le-Duc, the unit was assigned to the 2nd Armée and was especially active during the Battle of Verdun.

The Voisin 4s were soon discovered to be too slow and unwieldy to be useful as fighters. According to a note in the SHAA files, the Voisin 4s were tested by GB 3 in both the air-to-air and ground attack roles in May 1915. The aircraft were found to be of limited utility in attacking aircraft because the cannon was difficult to aim, and it was concluded that enemy planes could easily evade the slower Voisins. The Voisin 4s, however, were considerably more successful when used to

A lineup of Voisin 4s. The wings were shifted rearward 40 cm (compared to the Voisin 3) to compensate for the weapon's recoil and weight. The aircraft had identical dimensions to the Voisin 3 except for a slightly longer fuselage nacelle. MA73357.

Voisin 4 showing the mounting for the 37-mm cannon. B86.2385.

Below: Voisin 4 (Canon) of VC 116. B93.1633.

attack targets on the ground, especially if the targets were stationary (i.e., railway guns and battery emplacements). It was recommended that the airframe and cannon mounting on the Voisin 4s be lightened.

Indeed, the Voisins were more successful in the ground attack role. They used their cannons on searchlight positions, and these attacks forced the German searchlight crews to install anti-aircraft guns. In addition to escort missions, the VC 110 aircraft were sometimes used as conventional bombers. For example, on the night of 10/11 July three VC 110 aircraft bombed train stations while two other Voisins attacked searchlights that were attempting to locate the bombers. Finally, the VC escadrilles flew night-fighter missions in an attempt to bring down German night raiders.

VC 110 and VC 116 replaced their Voisin 4s in late 1916 with Voisin 8s. VC 111 re-equipped with Sopwith 1½ Strutters in March 1917. VC 113 was given Breguet 14s in May 1917.

V 83 was formed in December 1915 with Voisin 4s as Escadrille 501.

Foreign Service
Argentina
Argentina received a single Voisin 4 in 1917. It was apparently in poor condition as it has been described as being "worthless."

Netherlands
A single Voisin 4 landed in the Netherlands on 3 February 1915. It was a machine from GB 2 whose crew had become lost. A year later the Dutch government bought the machine from the French and gave it serial LA 22. It served with the Dutch air service until 1917.

Russia
Russia used a number of Voisin 4s (LBS), some of which may have been built under license (see Voisin 3).

Right: Voisin 4. The cannons were mounted on a metal tripod which had been designed by Voisin and built by Aircraft Park 101. B87.650.

Voisin 4 (LB) Two-Seat Cannon-Armed Fighter with 120-hp Salmson M9
Span 14.75 m; length 9.75 m; height 2.95 m; wing area 49.65 sq. m
Empty weight 1,050 kg; loaded weight 1,550 kg;
Maximum speed: 90 km/h at 2,000 m; climb to 2,000 m in 35 minutes; endurance 3.15 hours
Voisin 4 (LBS) Two-Seat Cannon-Armed Fighter with 120-hp Salmson M9
Span 14.75 m; length 9.75 m; height 2.95 m; wing area 49.65 sq. m
Empty weight 1,050 kg; loaded weight 1,550 kg;
Maximum speed: 90 km/h at 2,000 m; climb to 2,000 m in 35 minutes; endurance 3.15 hours

Voisin 4 armed with a 37-mm cannon. B76.1930.

Voisin Triplane 1915

In response to a government requirement for a heavy bomber capable of attacking the German industrial center of Essen, Voisin's team labored day and night for five weeks to produce a monstrous triplane. Testing of the aircraft began in 1915 with Joseph Frantz as test pilot. It was powered by four 270-hp Salmson radial engines mounted back to back in nacelles on the center wing. The fuselage had a partially enclosed cabin for the pilots. A huge rectangular beam extended from just aft of the cockpit through the front of the center and top wings. This structure contained the radiators for all four engines. A second triangular beam connected the top wing with the rudder. The landing gear consisted of three wheels in tandem below the lower wing. A massive nosewheel was semi-recessed in the nose in order to prevent the aircraft from rolling over on landing. The three wings were supported by a complex series of struts. Ailerons were located on the center and top wings and were controlled by two concentric wheels inside the cabin. The outer wheel controlled the ailerons on the top wing; the inner wheel controlled the ailerons on the center wing. The latter were used only to compensate for asymmetric flight as a result of engine failure.

The aircraft, named the *J. Benoist—L. Mijduant*, was the first four-engine aircraft to fly in France. It was presented to the army on 15 August, 1915. The military authorities saw the aircraft not as a heavy bomber capable of smashing Essen, but as a slow-moving behemoth that would present an easy target for anti-aircraft artillery. After examining the other entrants in the 1915 concours, the pilots and organizers agreed to another competition.

The efforts of the Voisin team had been wasted and the triplane was dismantled. However, Voisin did not yet abandon the triplane layout and used the wings from the 1915 bomber to produce his next heavy bomber, the 1916 E.28.

Voisin 1915 Triplane Bomber Powered by Four 270-hp Salmsons

Span 38 m; length 23.80 m; height 5.42 m; wing area 200 sq. m
Empty weight 4,500 kilograms; loaded weight 6,500 kilograms
Maximum speed: 140 km/h at 2,000 m; climb to 2,000 m in 20 min.; range 420 km; endurance 3 hours
One built

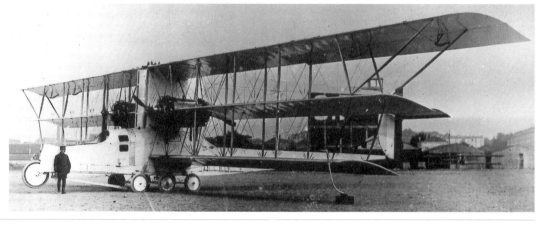

Voisin 1915 Triplane at Issy-les-Moulineaux. It was christened the *J. Benoist—L. Mijduant*. B77.950.

Voisin 5 (LAS)

The Voisin 3 had proved a successful bomber, but its payload was limited by the Salmson M9 engine, which produced only 120-hp. The Aviation Militaire wished to obtain a more powerful airplane but with the concours puissant (competition for a heavy bomber) not due to take place until mid-1915, it was decided to produce a Voisin 3 with a new engine. A Voisin 3 airframe was fitted with a 150-hp Salmson P9 engine, and the airframe was strengthened and the central nacelle streamlined. The new engine was placed on a raised platform to provide clearance for the propeller and was angled to provide downward thrust. The landing gear was strengthened and the wing chord was increased from the roots to the wing tips. The new aircraft was given the STAé designation Voisin 5, while the factory designation was LAS. The S stood for surélevé (raised) which indicated the raised engine mount. While the exhaust system on the Voisin 3 permitted fumes to escape freely, that of the Voisin 5 ejected the fumes upward through two exhaust pipes.

One Voisin 5 was transformed into a twin-engine aircraft in 1916. This was accomplished by adding a second Salmson in the front of the fuselage driving a tractor propeller. It is believed this was done to test a possible configuration for a new bomber planned by Voisin. The twin-engine Voisin 5 first flew in early 1916; apparently the type was not developed further.

The first Voisin 5 reached VB 101 in 1915 and soon replaced the Voisin 3 on the production lines. However, the Voisins 150-hp (as they were referred to at the front) were held in low regard by their crews. Despite the more powerful engine, the Voisin 5s' payload was only marginally better and the maximum speed was only 13 km/h faster. Approximately 300 Voisin 5s were built, and these served alongside the Voisin 3s in front-line escadrilles during 1915 and well into 1916.

Voisin 5.

Operational Service
Army Cooperation
A number of army cooperation escadrilles and artillery cooperation units used Voisin 5s. These included:

V 21, re-equipped with Voisin 5s in early 1915. It participated in a raid on Karlsruhe on 7 June 1915. It was later reassigned to the 4th Armée and flew reconnaissance, artillery registration, and bombing missions over the Saint-Mihiel salient. In 1916 the unit moved to Palennes and re-equipped with Caudron G.4s.

V 24, formed in 1914 at Reims, commanded by Capitaine Faure and assigned to the 5th Armee. It was active over Champagne until re-equipped with Farman F.40s in May 1916.

V 207, formed in May 1916 for artillery registration. It was assigned to the 13th C.A. in the 3rd Armée sector in October 1916. V 207 was commanded by Capitaine Devéze. It was active over Saint-Quentin and moved to Eppeville in April 1917. It re-equipped with Caudron G.6s and Sopwith 1½ Strutters in May 1917.

V 209, formed as an S.A.L. unit in January 1916. In October 1916 it was re-designated V 209 but it re-equipped with Farman F.40s two months later.

V 210, formed from elements of V 24 in January 1916. It was based at Rosnay in the 5th Armee sector and was under the command of Lieutenant Thébault. It participated in the battles at Bois de Buttes. In June it re-equipped with Farman F.40s.

V 212, formed in January 1916 as an S.A.L. unit. It re-equipped with Farman F.40s six months later.

V 216, formed in February 1916. It quickly became clear that the Voisins were unsuitable for this mission, and the unit re-equipped with Farman F.40s in June 1916.

V 220, formed as an S.A.L. unit in February 1916, based at Lemmes under the command of Capitaine Bastien. It re-equipped with Caudron G.4s in September 1916.

V 222, formed as an S.A.L. unit in April 1916. It re-equipped with Caudron G.4s six months latter.

V 223, formed in February 1916 under the command of Lieutenant Thévenot and assigned to the 102nd Regiment d'Artillerie Lourde of the D.A.L. It was based at Pont-à-Mousson. In May 1916 it was reassigned to the 1st Armée and in September to the 129th Division d'Infanterie. It subsequently re-equipped with Farman F.40s.

Voisin 5. This aircraft was a Voisin 3 airframe fitted with a 150-hp Salmson P9 engine, a streamlined nacelle, strength-ened landing gear, and a wing chord which increased from the roots to the wing tips. B75.624.

Experimental conversion of a Voisin produced in Russia. This may be the Ivanov Voisin described under the section on the Voisin 5. B83.4977.

T.O.E.

T.O.E. escadrilles included:

V 83, formed in December 1915 (including Voisin 4s) as Escadrille 501. It became V 83 in December 1915 and V 383 in June 1916. It re-equipped with Breguet 14s in July 1917.

V 84, formed as V 84 in December 1915 and assigned to the A.F.O. units in Serbia. It was assigned to the Serbian army on 30 May 1915. It re-equipped with M.F.11s in September 1916.

V 85, formed in January 1916 and assigned to the A.F.O. It became V 385 six months later. It was commanded by Capitaine Maire. V 85 re-equipped with Breguet 14s in July 1917.

V 86, formed in November 1915 and assigned to the A.F.O. It was re-designated V 386 in June 1916. It was commanded by Lieutenant Chaignon. V 386 re-equipped with Breguet 14s in June 1917.

V 88, formed in December 1915 and assigned to the A.F.O. In June 1916 it was re-designated V 388. A year later it received Breguet 14s.

V 89, formed in October 1915 as an A.F.O. unit. It became V 389 in June 1916 and a year later re-equipped with Breguet 14s.

Bombing

By early 1916 the Groupes de Bombardement were in disarray. GB 4 had been disbanded and two escadrilles of GB 1 (VB 102 and 103) had been withdrawn to retrain with Nieuport 11s. GB 3 was training to become a night bombardment unit. These changes were due to the relatively disappointing performance of the Voisin 3s. It had been hoped to replace these with more powerful types but the failure of the 1915 concours to produce a suitable bomber meant that the Voisin 3s and the only slightly better Voisin 5s would serve well into 1916. On 1 February 1916 there were 159 Voisin 3s and 5s at the front, with an additional

> **Voisin 5 (LB) Two-Seat Bomber with 150-hp Salmson P9**
> Span 14.74 m; length 10.28 m (some sources say 9.50); height 3.80 m; wing area 45 sq. m
> Empty weight 1,000 kg (970 kg); loaded weight 1,450 kg (1,370 kg); bomb load 180 kg;
> Maximum speed: 109 km/h at 2,000 m; climb to 2,000 m in 22 minutes; endurance 4 hours.
> Approximately 459 Voisin 5 and Voisin 6 were built
>
> **Voisin 5 (LBS) Two-Seat Reconnaissance Airplane with 160-hp Salmson Built in Russia**
> Span 15.70 m; length 9.5 m; height 2.95 m; wing area 47.00 sq. m
> Empty weight 975 kg; loaded weight 1,325 kg;
> Maximum speed: 105 km/h at 2,000 m; climb to 1,000 m in 10 minutes; climb to 2,000 m in 22 minutes; climb to 3,000 m in 40 minutes; ceiling 3,500 m; endurance 4 hours
>
> **Voisin 5 Two-Seat Cannon-Armed Fighter with 225-hp Salmson built in Russia in 1915—Possibly Based on Voisin 5**
> Span 18.80 m; length 11.0 m; height 2.95 m; wing area 63.00 sq. m
> Empty weight 1315 kg; loaded weight 1,865 kg;
> Maximum speed: 120 km/h at 2,000 m; climb to 1,000 m in 9 minutes; climb to 2,000 m in 20 minutes; climb to 3,000 m in 36 minutes; ceiling 4,000 m; endurance 2.8 hours

17 in foreign theaters, four in Paris, and 132 with the training units and the reserves. By March 1915 most of the Voisin bombers in service were Voisin 5s.

The GQG planned to concentrate its bomber attacks on targets in the Metz, Alsace-Lorraine, and Briey areas, all of which were within the 300-km radius considered to be acceptable for the Voisin 5s. The targets, in order of priority, were railroads, airfields, communication centers, and cities. The latter were to be attacked only in reprisal for similar German raids on French cities.

GB 1, consisting of VB 101, VB 114, and VC 110 (the latter with Voisin 4s), played a major part in the Battle of Verdun. Joined by fighter escorts, these units attacked railroad stations, camps, and factories during the first four months of 1916. On 24 April Voisin 4s and 5s of VB 101 and VB 114 moved from Malzéville to Autrecourt to support the Verdun offensive. For the next five days these units dropped 112 bombs on train stations, and during May the escadrilles based at Autrecourt dropped 430 bombs. Aircraft of VB 101, based at Lemmes, attacked targets near the Verdun front from 24 April through 13 May. Bad weather hampered operations in June (when only a single major mission was flown, on the night of 22/23 June).

Voisin 5 serial number V.1321.

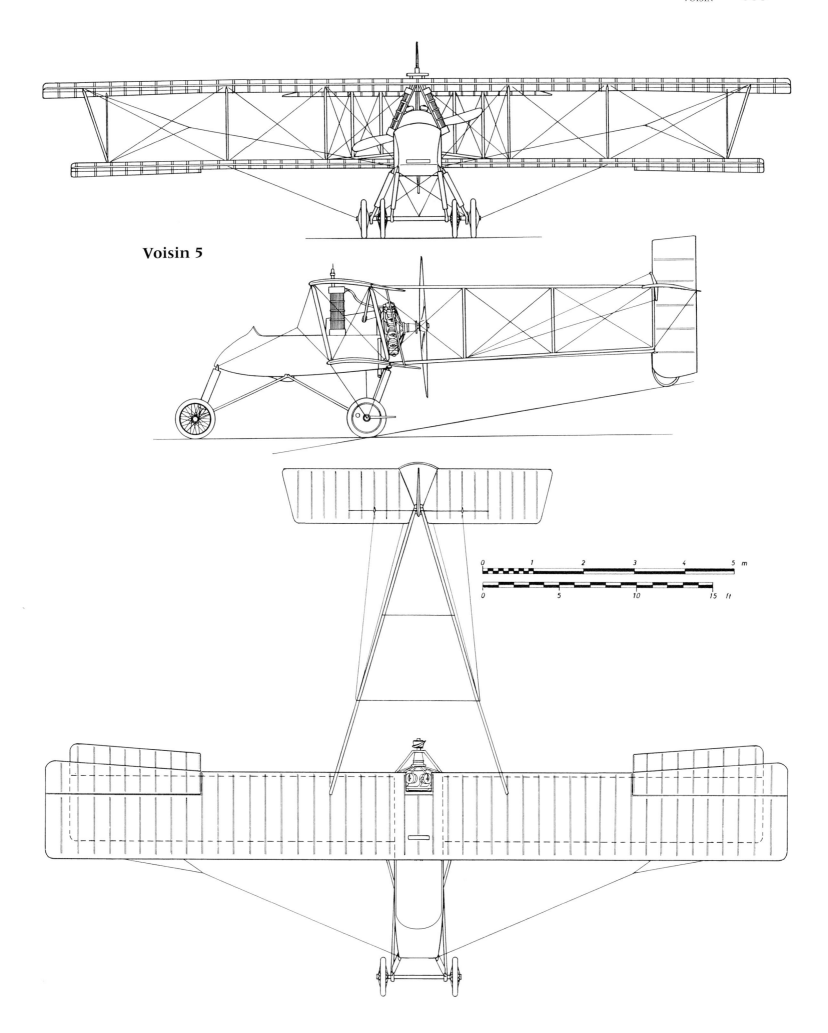
Voisin 5

Voisin 5 serial number V.1310.

Similarly there was little activity in July, when a total of only 295 bombs were dropped. There was considerably more activity in August and September, with numerous attacks on enemy troops and train stations. By the end of September VB 114 had returned to Malzéville. However, VB 101 had now been detached from GB 1 and was assigned directly to the 10th Armée. Numerous raids were made by GB 1's units on railroads, factories, and airfields during the fall of 1916 but bad weather seriously reduced the number of sorties. By November 1916 the first Voisin 8s began to arrive and soon replaced those Voisin 5s still in service.

GB 2 had only a brief existence in 1916. It consisted of VB 104, 105, and 106, all equipped with Voisin 5s. The unit was based at Malzéville and often joined GB 1 in attacks against various targets. However, the severe weather limited operations until 21 February. On that date GB 2 sent 11 crews to bomb targets at Pagny sur Moselle. GB 2 was reduced to two escadrilles when VB 106 was detached to re-equip with Caudron G.4s. With its two escadrilles, GB 2 was able to send only seven crews to bomb the Metz-Sablons station on 8 March. GB 2's last major mission was flown on 14 March, when seven crews dropped 23 bombs on the Brieulles station. Shortly after this the remaining units, VB 104 and VB 105, re-equipped with Caudron G.4s for army cooperation duties and GB 2 was disbanded.

GB 3 and VB 101 were active over the Somme in 1916. GB 3 completed its training as a night bomber unit in April and was based at Villers-Bretonneux. During April GB 3 attacked fortifications at Biaches, the munitions depot at Cremery, and targets in the vicinity of Roye. In June VB 101 joined GB 3. During June and July factories at Noyon and train stations along the Somme front were bombed. During August and September GB 3 struck targets at Ham, Nesle, Guiscard, and Roiseel with bomb, incendiary, and cannon attacks. GB 3 then returned to the airfield at Matigny and attacked enemy camps along the front. Meanwhile, VB 101 concentrated its assaults on train stations along the Somme front. On 23 November, VB 101 dropped 171 bombs on the German airbase at Grisolles; the raid was so successful that the Germans had to abandon the airfield.

In December VB 101 and GB 3 concentrated on tactical missions along the Somme front, including attacks on bivouacs, airfields, and lines of communication. As the first Voisin 8s began to arrive in late 1916, GB 3 bombed enemy airfields at Matigny and Flez. By the end of 1916 the Voisin 5s had been withdrawn from front-line escadrilles.

Foreign Service
Russia

At least one Voisin 5 was sent to the air service, a Type 5 being attached to the 26th Air Group. It was modified by a pilot named Petr Ivanov and became known as the Ivanov Voisin. These changes performed at the 6th Aviapark included a new plywood gondola, a machine gun mounted in the nose and fired by the observer, and an aluminum bulkhead to separate the fuel tank from the engine. The fuel tank was self-sealing. The wings and tail were made of steel tubes with wooden fairings. The aircraft was test flown on 6 April 1916 with excellent results; it was 20 km/h faster than the standard Voisin despite being powered by the same 150-hp Salmson P9 engine. The Anatra plant was ordered to stop building standard Voisins and instead produced 125 Ivanov Voisins. Production began in late 1916 and by 1917 no fewer than 150 airplanes had been built. These remained in service as late as 1922, and some saw service in the civil war.

It appears that a heavily-modified version of the Voisin 5 was produced in Russia. This was a two-seat, cannon-armed fighter with a 225-hp Salmson. It was considerably larger and heavier than the Voisin 5, but is listed in Shavrov as being based on the 1915 design. The Salmson engine suggests that this type was built late in the war, or possibly postwar.

Switzerland

The Swiss obtained a single Voisin 5 when one was interned in August 1915. It was given the registration number 32 and used for reconnaissance. Postwar it was returned to the French but the engine still resides in a Swiss museum at Lucerne.

Voisin 6

The Voisin 6 (factory designation LAS) was the same airframe as the Type 5, but was fitted with a 155-hp Salmson R9 engine. It is not known exactly how many Voisin 6s were built, but probably about 50. The specifications were the same as the Type 5. The maximum speed was slightly higher at 113 km/h and the climb to 2,000 m was reduced to 21 minutes and 30 seconds.

Voisin 6 (Canon) armed with a cannon. The sailor in the gunner's cockpit was not uncommon, it being believed that sailors would be better gunners because of their experience with these weapons on ships. B92.953.

Voisin 7 A2 (LC)

Gabriel Voisin decided that he need to produce a more powerful airplane to carry a heavier bomb load over longer distances. He began by designing a sturdier airframe to accept a larger engine. The new airplane was experimentally fitted with a Salmson P9 or R9 engine and a variety of 150- to 180-hp Renaults. Eventually two types were fitted to the new Voisins: the 180-hp Renault 8G and the 150-hp Salmson P9. However, only a few Voisin 7s received the Salmsons and only the Renault-engined aircraft were used operationally.

The new Voisin was intended for service with the reconnaissance and bombing escadrilles. It had a machine gun on a flexible mount in the nose that greatly increased the gunner's field of fire. Radiators for the Renault engine were attached to either side of the fuselage just ahead of the crew cabin. The wing span was enlarged. It was the first Voisin type to feature detachable fuel tanks attached to the wings. These could be jettisoned in an emergency to reduce the risk of losing the aircraft if fire broke out.

The Type 7, however, was felt to be too underpowered to carry an adequate bomb load; it was used only by army cooperation escadrilles. Only about 100 Type 7s were produced and were primarily assigned to T.O.E. units; for example, Voisin 7s served with V 83 and V 84, which were stationed with the A.F.O. (see the section on the Voisin 5 for further details).

Serbia acquired some Voisin 7s when the equipment of V 384 (previously V 84) was turned over to the Serbian air service in 1917. The new unit was redesignated Escadrille 522.

Voisin 7 Army Cooperation Airplane with 180-hp Renault 8G
Span 17.93 m; length 10.35 m; height 3.95 m; wing area 61.14 sq. m
Empty weight 1,250 kg; loaded weight 1,700 kg
Maximum speed: 112 km/h at 2,000 m; climb to 2,000 m in 18min 30 sec; endurance 2.3 hours
Armament: one machine gun on a flexible mount in the nose
Approximately 100 built

Voisin 7 Army Cooperation Airplane with 160-hp Salmson X9
Span 17.93 m; length 10.35 m; height 3.95 m; wing area 61.14 sq. m
Empty weight 1,250 kg
One built

A Voisin 7. The standing crewman is holding a Gros-Andeau bomb made from a 120-mm cannon shell. B81.1504.

Voisin E.28

The Voisin 1915 triplane had been a failure, but this did not prevent Voisin from developing the layout further. He elected to adopt it for his 1916 bomber.

In December 1916 he had opened study E.28 for a four-man triplane armed with two cannons. It retained the wing used on the 1915 triplane. The new aircraft, labeled Triplane No.2, was powered by four 220-hp Hispano-Suiza 8Bc engines. Reportedly, four 250-hp Salmson engines were initially planned, but were not fitted. The engines were mounted in tandem in nacelles for each pair in the middle wing. The undercarriage was redesigned and had four wheels beneath the lower wing and individual wheels beneath each of the engine nacelles. A single nosewheel was mounted on a prominent strut extending forward. The fuselage had a circular cross-section, and an enlarged boom extended from the top wing to the rudder. The rudder was mounted between the fuselage and top boom and the tailplane at the upper third of the rudder. The wing size decreased from top to bottom, and there were ailerons on the top wing only. There was a crescent-shaped rudder, two triangular sails (one ahead of the rudder and the other emerging from the upper fuselage) and a sweptback horizontal stabilizer. Unlike the 1915 design, the bottom wing was suspended below the fuselage. A Hotchkiss 37-millimeter cannon was to be mounted in the front cockpit. The main cabin had two pilots and was located between the two front engines. It was found that the pilot's field of vision was greatly reduced in this location. Another gunner with a Hotchkiss cannon was to be located behind the cockpit. Of most interest was the beam suspended above the fuselage and passing from the top wing to the tail surfaces. The space between the two fuselages was triangular and was quite rigid. This apparently was the reason Voisin chose this unique layout. A gunner was located in the upper fuselage, where he would have had an unimpeded field of fire.

> **E.28 Four-Seat Heavy Bomber with Four 220-hp Hispano-Suiza 8Bc Engines**
> Span 35.40 m (other sources say 36 and 37 m); length 21.80 m; height 5.80 m; wing area 200 sq. m
> Loaded weight 6,500 kg
> Maximum speed: 125 m/h at 2,000 m; climb to 2000 m in 25 minutes 30 seconds
> Armament (proposed): two 37-mm Hotchkiss cannons
> One built

Obsolete even by the standards of 1916, the E.28 was apparently not completed until 1919, when flight testing began. It was not selected for production.

There was an additional design based on the 1916 bomber. It was for a triplane intended to carry 12 passengers. It was to have carried 740 kilograms of fuel and to have had an endurance of three hours. It was never built.

Right: Voisin E.28. The aircraft retained the same wing used on the 1915 triplane. The E.28 was powered by four 220-hp Hispano 8Bc engines.

Above: Voisin E.28 Triplane at Issy-les-Moulineaux on 29 May 1919. B88.4872.

Above right: This view of the Voisin E.28 emphasizes the double fuselage; a design intended to increase the aircraft's structural strength. B80.309.

Right: Voisin E.28. B85.1166.

Voisin 8 BN2 (LAP) and Ca2 (LBP)

It had become clear to the Aviation Militaire that the failure of the 1915 and 1916 concours puissant to produce a suitable strategic bomber would mean that the Voisins would remain in service until late 1917. Gabriel Voisin was asked to develop an improved bomber with a significantly better performance and heavier bomb load than the Voisin 5s then in service.

To power his new bomber he chose the 300-hp Hispano engine. However, production difficulties resulted in the Hispano engines being in short supply. Furthermore, the STAé decided that those available should be reserved for fighters. This decision is not surprising, as the bomber had not brought about any meaningful improvement in the war situation aside from the propaganda value of long-distance raids. It was decided that a Peugeot engine would be substituted for the Hispano-Suiza. The 220-hp Peugeot 8Aa not only supplied less power than the anticipated 300-hp engine, but it was also bulky and heavy. Voisin had no choice but to design an enlarged and strengthened airframe to carry the new engine. To support this added weight the wing span had to be increased by more than three meters.

The new Voisins received the STAé designation Type 8; the factory designation was LAP for the BN2 bombers. Some were armed with cannons and these Ca2 class airplanes were given the factory designation LBP. The fuselage of the basic Voisin series was enlarged to permit the Type 8 to carry an increased bomb load. A huge propeller, four meters in diameter, was fitted. Other changes included an indentation at the base of the rudder and strengthened landing gear struts. Because trapped exhaust bubbles could cause trouble with the Peugeot in-line engine, Voisin used six vertical tubes to eject exhaust over the top of the wing. The undercarriage retained the quadricycle arrangement with brakes on the rear wheels. Because the fuselage was enlarged, the wheels were set farther apart than on previous aircraft in the Voisin series; however, this arrangement created new problems with the suspension system. As with other Voisin aircraft, the structure was entirely of steel and the fuselage was covered with aluminum sheets. The wings and tail surfaces were still covered with cloth.

The first Voisin 8 to be produced was armed with a 37-mm cannon in the front of the fuselage. This position could also be used by a conventional machine gun on a semi-circular mount. At least one aircraft was fitted with a 47-mm cannon for use against ground targets.

Operational Service

The Voisin 8s began to leave the factory in August 1916 and had entered service with the escadrilles by November. There were 51 Type 8s (LAP) and 33 Type 8s with cannons (LBP) at the front in August 1917, and ten LAPs and 26 LBPs with the general reserve. The Type 8 presented serious maintenance difficulties because of its Peugeot engine. Eventually, a 280-hp Renault replaced the troublesome Peugeot; these airplanes were designated Type 10. The Type 8s were phased out of front-line service beginning in September 1917.

Voisin 8s began to reach front-line escadrilles in November 1916; the first to receive the type was VB 114. By the beginning of 1917 Voisin 8s equipped a substantial portion of the bombing escadrilles and were used almost exclusively for night bombing. Despite their many improvements, the Voisin 8s were no match for German fighters.

Voisin 8. Because the fuselage was enlarged the wheels were set farther apart than on previous aircraft in the Voisin series, creating new problems with the suspension system. B77.1199.

On 20 February 1917 the French War Committee drew up a list of priority targets for night bombers, including factories, railroad stations, munitions depots, barracks, and arsenals. Cities still could be attacked only in reprisal and could be bombed only by direct order of the GQG.

Only GB 1 and GB 3 still had Voisin bombers on strength. In December 1917 the units assigned to GB 1 were VB 110 (which had converted from Voisin 4s to 8s) and VB 114. GB 3 had three escadrilles: VB 107, VB 108, and VB 109. VC 113 had re-equipped with Nieuport 17s in January 1917. VB 101 remained an independent escadrille.

Bad weather prevented GB 1 from flying many missions in January. Based at Malzéville, GB 1's escadrilles were more active in February, attacking the barracks and airfield at Dieuze as well as bivouacs. During March the Frescaty airfield and Arnaville station were bombed.

In April VB 110 and VB 114 moved to Villenueve. From there they attacked train stations in an attempt to disrupt the flow of German reinforcements.

By the end of the May 13,495 kg of bombs had been expended on the train stations along the front. Major attacks on them continued in June, 5,000 kg of bombs being dropped in the second half of that month. The better weather in July enabled GB 1 to stage raids on train stations, enemy camps, and ammunition dumps. By the end of that month 8,805 kg of bombs had been delivered by GB 1's escadrilles.

VB/VC 110 and VB 114 moved to Senard on 8 August to support a planned attack in that area. Fighter escort was provided by SPAD 13s of GC 13. To disrupt the enemy supply lines, attacks were concentrated on ammunition dumps and train stations. The airfield at Remonville was also bombed. September proved to be equally busy. GB 1 now concentrated its aircraft for mass attacks at night. Twenty Voisins and two Sopwiths attacked train stations, airfields, and camps with 3,575 kg of bombs and dropped 2,500 leaflets on the night of 4/5 September. A total of 2,980 kg of bombs were dropped on train stations on the nights of 24/25 and 25/26 September.

On 3/4 October VB 110 and VB 114 moved to GB 1's new base at Senard. The Voisins seem to have taken part in very few raids during November, possibly due to bad weather. However, raids on train stations continued throughout December. The last mission in 1917 to be flown by the Voisin 8s of GB 1 took place on the night of 23/24 December, when 16 airplanes attacked train stations with 4,330 kg of bombs. By early 1918 Voisin 10s replaced the Voisin 8s in service.

GB 3 started 1917 with a large number of Voisin 8s. However, two escadrilles switched to Sopwith 1½ Strutters;

Voisin 8 (Canon) fitted with a 220-hp Peugeot 8Aa. B76.1483.

VB 107 and VB 108 became SOP 107 and SOP 108 respectively. At the same time VC 113 joined GB 3. By the end of July only VB 109 and VC 113 were using Voisin 8s. VB 101, while remaining an independent unit, participated with GB 3 on several missions.

VB 101 and GB 3 were active over the Somme front. GB 3 covered the areas from Oise to Aisne and, along with VB 101, concentrated on tactical missions against German troops. Both units had some success in disrupting the German transportation system.

During the German counterattack in mid-March VB 101 was especially active. Enemy convoys and train stations were bombed. April was equally busy. VB 101 flew 104 sorties and dropped 24,130 kg of bombs that month, and GB 3 concentrated its attacks on train stations and other targets in the Serre Valley.

GB 3 and VB 101 flew a number of sorties against railroad targets in the vicinity of Laon during May and both units continued their attacks on train stations throughout June. VB 109 was despatched to Flanders in June, leaving VB 101 and the newly-arrived VC 113 as the only Voisin units on the Somme front. For the remainder of 1917 VC 113 was the only escadrille attached to GB 3 still equipped with Voisin 8s. Joined by the independent unit VB 101, VC 113 flew only occasional sorties during the last six months of 1917.

Meanwhile, VB 109, having been sent to the Flanders front, remained quite active in 1917. Joined by VC 116, an independent unit, it made attacks on train stations and airfields. Interrogations of German prisoners revealed that heavy damage

Voisin 8 (Canon) in 1916. In order to support the added weight of the Peugeot engine the wing span had to be increased by more than three meters. B81.1507.

Voisin 8. Because trapped gas bubbles could cause troubles with the Peugeot in-line engine, Voisin used six vertical tubes to eject the gas over the top of the wing. B88.899.

Voisin 8. The fuel in the tanks mounted on the upper wing could be jettisoned in flight. B88.901.

had been done by these attacks. For the remainder of 1917 both VB 109 and VC 116 concentrated attacks on train stations and airfields at Ghistelles and Thielt.

As noted earlier, the Voisin 8s had proved to be troublesome in front-line service because of their unreliable Peugeot engines; by early 1918 the Voisin 8s were replaced by the Renault-powered Voisin 10s.

Approximately 20 Voisin 8s were also used by the Aviation Maritime, based at Dunkerque. Some of these were lost in a raid on their base by German bombers. The pilots and mechanics of the Aviation Maritime also found the Peugeot engines troublesome, and because of this the aircraft were often unavailable for service. At least one Type 8 was fitted with floats.

Foreign Service
United Kingdom

The RNAS purchased two Voisin 8 Ca cannons with 220-hp Peugeot engines (although the original order had called for 200-hp Hispano Suizas). These were assigned serials N544 and N545 and based at Grain. N545 was fitted with a 37-mm Hotchkiss gun and used for anti-submarine trials.

United States

The United States acquired a few Voisin 8s, although none was assigned to operational squadrons. In February 1918 the Americans had decided to establish two day bombing squadrons and one night bomber squadron. It was decided to supply the night bomber squadron with Voisin 8s. This selection was based on the experience of U.S. crews who had served with Voisin-equipped escadrilles. Deliveries of the Voisin 8 were very slow and it took ten weeks for eight aircraft to arrive. These aircraft were formed into a training squadron based at Amanty.

Voisin 8 (LAP) Two-Seat Bomber with Peugeot 8Aa

Span 18.00 m; length 10.35 m; height 3.95 m; wing area 61.14 sq. m

Empty weight 1,310 kg; loaded weight 1,860 kg

Maximum speed 118 km/h at 2,000 m; climb to 2,000 m in 17 minutes; range 350 km

Armament: one machine gun and approximately 180 kg of bombs

Approximately 1,100 Voisin 8 LAP and LBP were built

Voisin 8 (LBP) Two-Seat, Cannon-Armed Airplane with Peugeot 8Aa

Span 18.00 m; length 10.35 m; height 3.95 m; wing area 61.14 sq. m (some sources say 63 sq. m)

Empty weight 1,310 kg; loaded weight 1,860 kg

Maximum speed: 118 km/h at 2,000 m; climb to 2,000 m in 17 minutes; range 350 km

Armament: one 37-mm Hotchkiss cannon

Voisin 9 A2 (LC)

The Voisin 9 (which carried the factory designation LC and the project number E.60) was a modified Type 8 lightened for reconnaissance missions. It appears that the major modifications were internal,, and it was fitted with the 160-hp Renault 8G engine. The aircraft was given the STAé designation Type 9 A2, confirming that it was intended for the two-seat army cooperation role. Only one was built in 1917, and the Breguet 14, Salmson 2, and SPAD 11 were selected for service with the reconnaissance escadrilles.

Voisin 9 Two-Seat Reconnaissance Airplane with 160-hp Renault 8G

Span 14.54 m; length 10.00 m; height 3.75 m; wing area 48 sq. m
Empty weight 950 kg; loaded weight 1450 kg
Maximum speed: 120 km/h at 2000 m; climb to 2000 m in 20 minutes 10 seconds; endurance 4 hours
One built

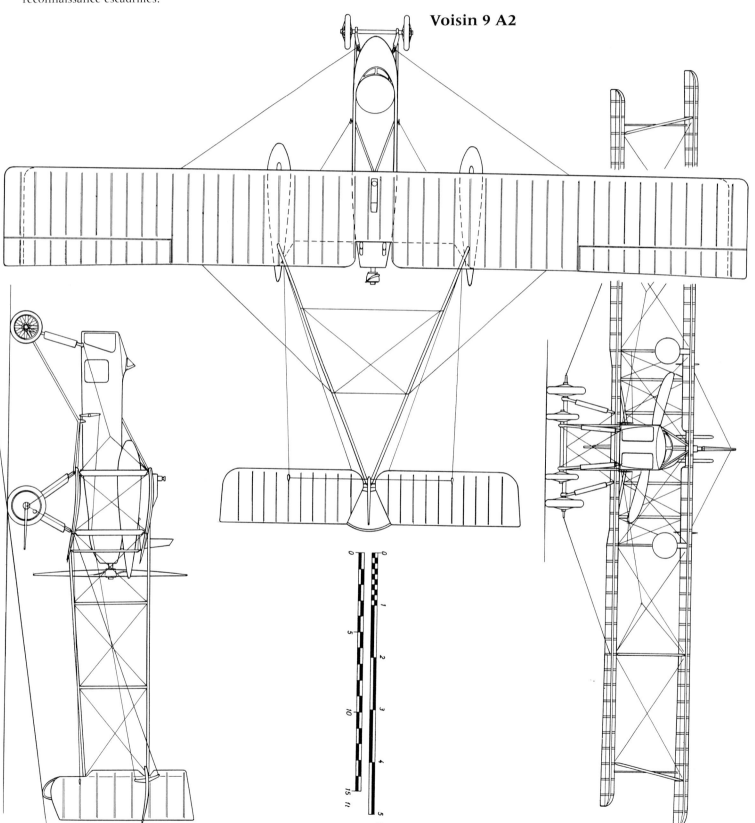

Voisin 9 A2

Voisin 10

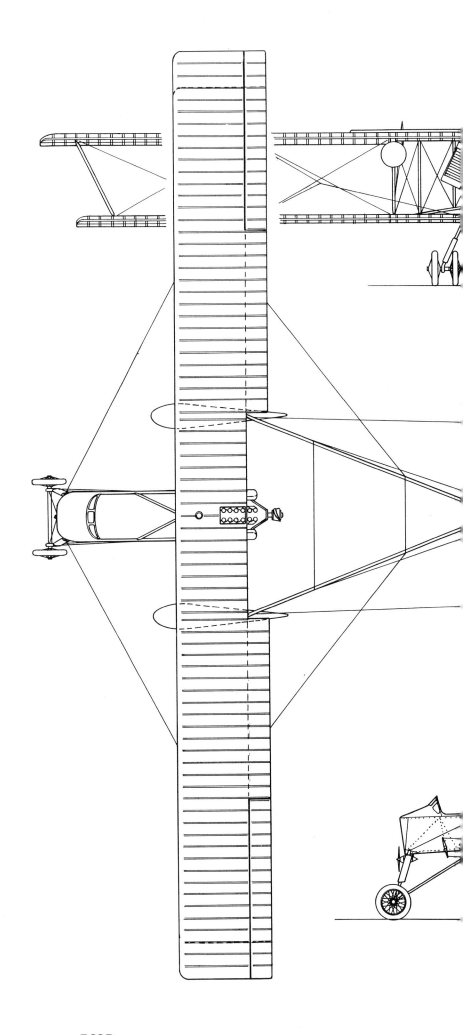

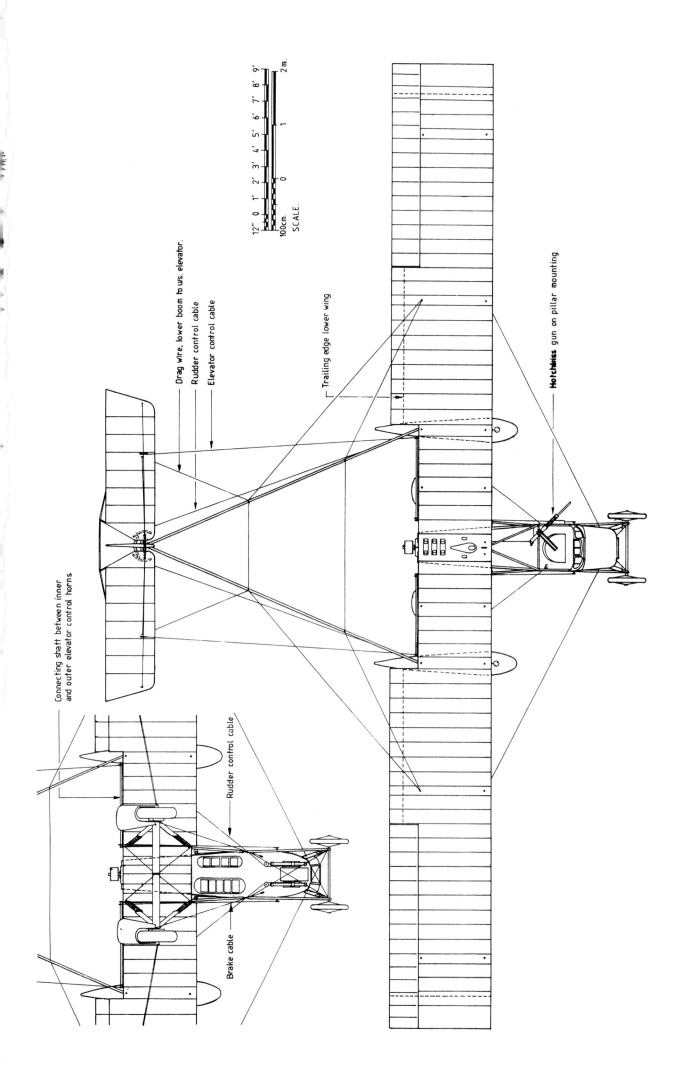

Voisin 8

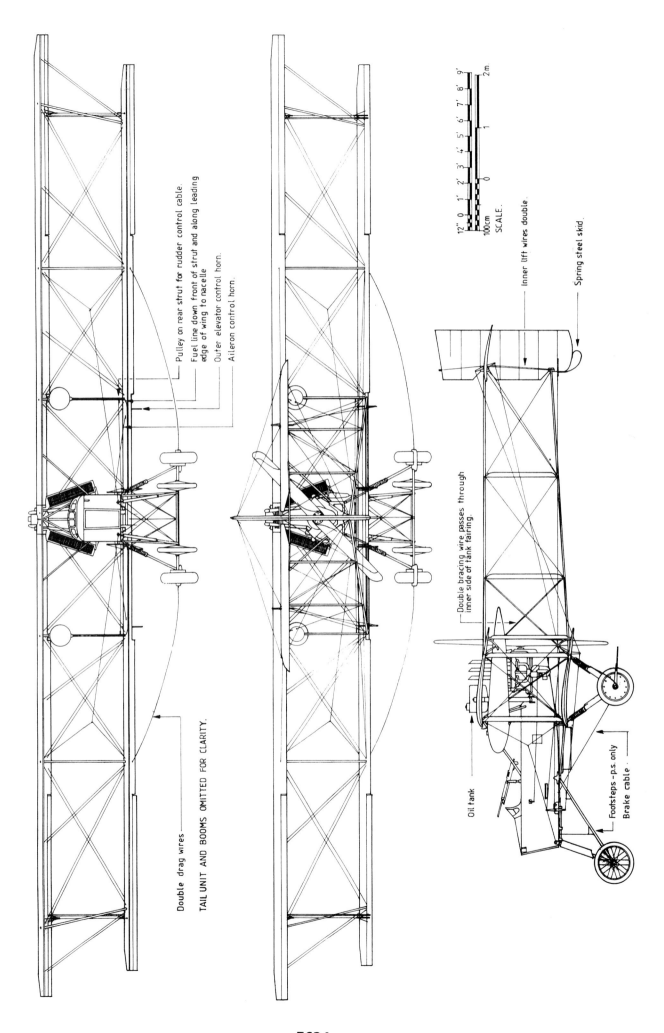

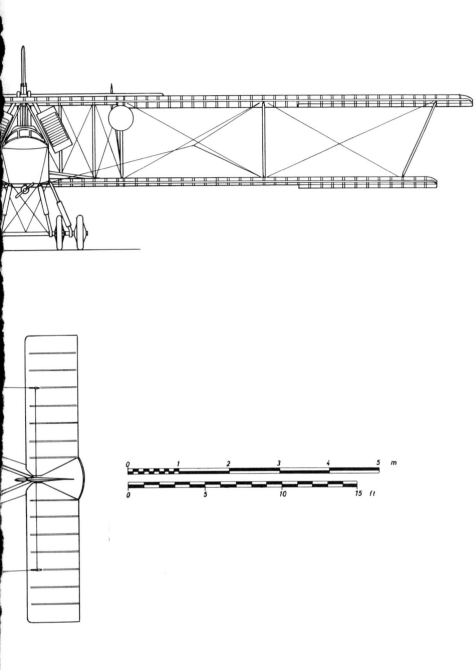
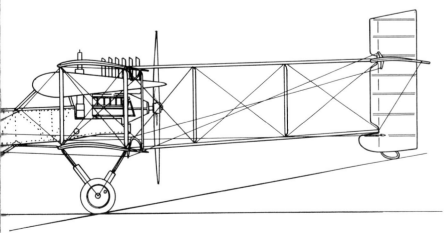

Voisin 10 (LAR/LBR)

As already noted, the Voisin 8's Peugeot engine had proved to be unreliable, seriously reducing the Type 8's serviceability. An attempt was made to salvage the Voisin 8 design which, aside from the engine, had proved to be quite successful. A new engine that would be compatible with the Type 8 airframe was chosen—the 280-hp Renault 12Fe. This engine was not only more powerful than the Peugeot but was also significantly lighter. A Lumière propeller was fitted and there were some minor changes made to the carburetor. The fuselage was enlarged so it could carry more fuel and a heavier bomb load—120 kg more than the Voisin 8. No major changes were made to the wing, but the rudder was enlarged and fitted with a small mass balance. The new Voisin carried the STAé designation Type 10 and had the Voisin designation E.54 and LAR.

Variants

As with most aircraft in the Voisin series, some Voisin 10s were fitted with cannons. The 37-mm Hotchkiss was usually carried. Those airplanes were designated the Voisin 10 Ca2 (Voisin designation LBR). The cannons were not widely used by the Voisin 10 escadrilles, however, and it appears that most Type 10 Ca2s had the weapons removed and were used solely as conventional bombers.

The Voisin 10 was put to more pacific use when one was converted by an engineer named Nemisrovsky and a physician named Tilmant into a flying ambulance. It was designated an "Aérochir," or ambulance. Its main function was to transport a surgeon, X-ray machine, and up to 360 kg of medical supplies to front-line hospitals. Stretcher cases could also be evacuated in what would have been the first aerial medivac operations in history. The ambulance plane was easily recognizable by a red cross on a white square painted on the nose. The "Aérochir" did not see operational service, as development was not completed until the end of the war.

A single Voisin 10 was converted to a radio-controlled drone in 1918. The first fully automatic flight took place in September 1918. The airplane, carrying a pilot who did not touch the controls, made a 100-km circuit at Étampes. The project was discontinued after the first flight, but testing was resumed in 1923.

Operational Service

The Voisin 10 entered service slowly. By 1 January 1918 there were to have been 300 at the front; in reality there were only 104. This shortfall caused the STAé to try to speed production. New night bombers were needed and the Voisin 10 was the only design readily available. An additional order for 300 aircraft was placed in May 1918 even though the initial 300 had not yet been completed. In time, production improved to the point that 11 aircraft were being manufactured every week. As delays continued to prevent the arrival of the more sophisticated Farman F.50s and Caudron C.23s, additional Voisin 10s were ordered. Eventually, a total of 900 were produced.

Reconnaissance

An unusual mission for the Voisins 10s was night reconnaissance to assess troop movements. Six night reconnaissance escadrilles were formed with Voisin 10s. They were:

VR 290, formed in September 1918 and assigned to 3rd Armée. It was redesignated the 8th Escadrille of the 4th RAO in January 1920 and based at Bourget.

Above right: The Voisin 10 Aérochir was adapted to carry a surgeon and medical equipment to front-line hospitals. B88.3401.

Voisin 10. A new engine which would be compatible with the Type 8 airframe was chosen; this was the 280-hp Renault 12Fe. B85.1898.

Voisin 10 of VB 125. Many of the Voisin 10 night bombers were painted black, although some retained camouflage on the upper wings. B76.1932.

VR 291, formed in November 1918 and assigned to the 4th Armée. It was disbanded in August 1919.

VR 292, formed in November 1918 and assigned to the 8th Armée. It was disbanded in May 1919.

VR 293, formed in November 1918 and assigned to the 1st Armée. It was disbanded in August 1919.

VR 294, formed in November 1918 and assigned to the 10th Armée. It was disbanded in August 1919.

VR 295 was never formed.

VR 296, formed in December 1918. It disbanded eight months later.

T.O.E.

Type 10s were also assigned to T.O.E. army units. These were:
VR 541, based in Tunisia.
VR 542, based in Tunisia.
VR 543, based in Tunisia.
VR 547, based in Algeria.
VR 551, based in Morocco.
VR 552, based in Morocco.
VR 553, based at Meknes.
VR 554, based at Marrakech.
VR 555, based at Taza.
VR 556, based at Rabat.
VR 557, based at Oujda.
VR 558, based at Tadla.
V 571, based in Indochina.

Bombing

The first Voisin 10s were supplied to VB 114. They were given a color scheme of solid black (except for camouflage on the upper wing), which was eventually applied to most Voisin night bombers. Voisin 10s were also used by escadrilles assigned to GB 1, 7, 8, 10, and 51 and were extremely active during the last year of the war.

The main mission of the Voisin 10 was night bombardment, and it was in this role that the airplane would become famous. The number of units equipped with Voisins actually increased in 1918. They were:
GB 1: VC 110, VB 114, VC 116, and V 25. VB 135, VB 136, and VB 137 were attached to GB 1 for a short time.
GB 3: VB 109, which served briefly with this group before being detached to GB 8.
GB 7: Formed in February 1918 with VR 118, VR 119, and VR 121.
GB 8: VB 109, VC 116, VB 125.
GB 9: VB 125 briefly served with this unit in 1918.
GB 10: Formed March 1918 with VB 101, VC 116, VB 133.
GB 51: Formed July 1918 with VB 135, VB 136, VB 137, and VR 293.

Thus during 1918 the number of Voisin escadrilles increased from five to 13. As the number increased the night bomber units were formed into Escadres. GB 8 and GB 10 were formed into Escadre 14 and GB 1, GB 7, and GB 51 were formed into Escadre 11.

The first three months of 1918 found the Voisin 10s continuing their attacks on enemy troop concentrations and supply lines. GB 1, now based at Rumont, used small numbers of Voisins to bomb railroad stations and airfields in January and February. On 25 February V 25 joined the group. In March an attempt was made to use a single Voisin as a pathfinder for the other airplanes. The lead Voisin would drop colored flares on the selected target. Unfortunately, clouds diminished the efficacy of this technique.

GB 3 had two Voisin escadrilles, VB 109 and VB 113, at the beginning of 1918. They were based at Champien in support of the 3rd Armée. In February they had moved to GB 8. Based at Creves, GB 8 had two Voisin escadrilles assigned to it in January, VC 116 and VB 125. By the time VB 109 and VB 113 joined GB 8 the entire group had combined with GB 10 to form Escadre 14.

Battles of Picardie and Flanders

The German offensive resulted in the Battles of Picardie and Flanders in March and April. The Voisin units were assigned to support the French troops involved in these battles. GB 1 was based at Passy-en-Valois and later at Arcy Sainte Restitue. GB 7 was initially based at May-en-Multien and later moved to Sapoonay. GB 8 was at Cramaille.

GB 1, joined now by VC 116, attacked troop columns and train stations in an attempt to reduce the movement of German troops to the front lines. GB 7, on the other hand, concentrated on enemy airfields and bivouacs. GB 8 attacked similar targets during March and April.

Battle of Aisne

Before the Battle of Aisne (27 May to 4 June) GB 1 flew a large number of sorties. GB 1 was based at Cernon during the Battle of Aisne and flew night reconnaissance and bombing missions.

During the Battle of Aisne GB 7 moved to Ferme des Greves and then to Mairy-sur-Marne It operated in support of the army units in Champagne and concentrated on train stations and

Voisin 10. A searchlight is fitted beneath the fuselage to illuminate the landing field during night raids. B85.1899.

> **Voisin 10 (LAR) Two-Seat Bomber with 280-hp Renault 12Fe**
> Span 17.90 m; length 10.35 m; height 3.95 m; wing area 61.14 sq. m
> Empty weight 1,400 kg; loaded weight 2,200 kg
> Maximum speed: 135 km/h at 2,000 m; climb to 2,000 m in 20 minutes; range 350 km; endurance 5 hours
> Armament: one machine gun and 300 kg of bombs
> Approximately 900 LAR and LBR Voisin 10s were built
>
> **Voisin 10 (LBR) Two-Seat, Cannon-Armed Airplane with 280-hp Renault 12Fe**
> Span 17.90 m; length 10.35 m; height 3.95 m; wing area 61.14 sq. m
> Empty weight 1,450 kg; loaded weight 2,200 kg
> Maximum speed: 130 km/h at 2,000 m; climb to 2,000 m in 14 minutes 40 seconds; range 350 km; endurance 5 hours
> Armament: one 37-mm Hotchkiss cannon

railroad yards in Reims.

GB 8 operated from Bettencourt and attacked a wide variety of targets. In early June GB 8 was joined by GB 10 to form Group Laurens, which later became Escadre 14. During the remainder of June and July Escadre 14 carried out a series of attacks along the Somme front.

GB 1 lost VB 116 when that unit was attached to GB 10 on 5 June. GB 1 continued attacks from its base at Cernon, while GB 7 continued attacks on railroad targets during June and July.

Battle of Ile-de-France

During the Battle of Ile-de-France (18 July to 14 August) GB 1 remained at Cernon and staged raids around Amifontaine, Saint-Gille, Jonchery, Claquedent, Fismes, and Bouffiquereux. At the end of July VB 110 received the new Farman F.50s and passed its Voisin 10s to the newly formed VB 133. Three days later VB 114 also re-equipped with Farman F.50 and its Voisin 10s were sent to VB 135, serving with the newly-formed GB 51. The Voisins now took part in fewer raids as the F.50s slowly became operational. However, because of engine difficulties with the Farmans, the Voisin 10s still formed the backbone of GB 1 through August and September. GB 1 and 7 (the latter still based at Mairy-sur-Marne) concentrated their attacks on targets in the Ardre and Suippe Valleys in support of the Armées in Champagne. GB 7 was assigned to attack stations along the Guignicourt-Laon railroad route. GB 8 and 10 were based at Chalôns and had a total of 30 Voisin 10s between them. Escadre 14 bombed munitions depots and train stations during the battle. GB 51 was formed with VB 135 and 136 at this time and, along with GB 1, was based at Cernon.

Voisin 10 serial V 2783 of VB 109. The unit's insignia was a winged devil holding a lighted bomb. Renaud.

Voisin 10 serial number V. 3237. B84.37.

Battle of Santerre
The Battle of Santerre lasted from the 8th to the 30th of August and all the Groupes d'Bombardement were active in supporting French troops during the battle. GB 1 and GB 51 remained at Cernon. The remaining Voisin 10s of GB 1 saw limited action and by 26 August the group lost another Voisin unit when V 25 re-equipped with Farman F.50s. Its Voisin 10s were sent to VB 137, the only escadrille attached to GB 1 still using Voisins.

GB 7 was based at Mairy-sur-Marne at the beginning of the battle; on 15 August it moved to Champagne and continued to concentrate its attacks on railroad stations, camps, and communications centers.

The escadrilles assigned to Escadre 14 were also active during the Battle of Santerre. GB 8 was stationed at Fourneuil and flew numerous strikes against train stations. GB 10 flew sorties from Champeaux to targets at Laon, Anizy-le-Château, Feres, Chauny, Tergnier, and Ham.

Battle of Saint Mihiel
During the Battle of Saint Mihiel (12 to 30 September) GB 1 and 51 were based at Cernon. GB 7 used the airfield at Mairy-sur-Marne. GB 8 was stationed initially at Fourneuil, then at Ferme des Greves, and finally at Coupru. GB 10 was initially at Château-Thierry but later joined with GB 8 at Coupru. All these Voisin units were active during the Battle of Saint Mihiel and in addition to the usual transportation targets a large number of raids were made on the blast furnaces in the Serre Valley.

Battles for Champagne and Argonne
The final major actions of the First World War were the Battles for Champagne and Argonne, which lasted from 25 September to 11 November. GB 1's affiliation with Voisin bombers ended ten days after the battle began when VB 137 left GB 1 and was assigned to GB 51.

GB 7 moved to La Cheppe during the battle and remained there until the war's end. GB 51 moved to Le Cheppe just before the Armistice.

Escadre 14 also provided support for French troops during this final battle. GB 8 moved back to Fermes des Greves while GB 10 remained at Corpu. In a two-week period from 25 September to 11 November, Escadre 14 flew 250 sorties and dropped 61,520 kg of bombs. Targets included the stations at Montecornet, Marles, Vervins, Hirson, and Chimay. The railroad lines between Guignicourt and Laon and the Serre depots were also frequently bombed.

Postwar Service
Postwar, the obsolete Voisin 10s were rapidly replaced. The Voisin escadrilles were redesignated in 1920 as follows:
Bomber units: VB 121, 119, 118, 109, 125, 101, and 113, assigned to the 2nd RB (Nuit) at Malzéville and redesignated Escadrilles 204, 205, 206, 207, 208, 209, and 210 respectively.
Reconnaissance units: VR 290, redesignated the 8th Escadrille of the 4th RAO at Bourget. VR 557 and 556 became the 6th and 8th Escadrilles of the 7th RAO at Pau.
T.O.E. units: VR 547, 543, 541, and 542 became Escadrilles 2, 3, 7, and 8 of the Regiment d'Algerie-Tunise. VR 551, 555, 552, 553, 554, and 558 became Escadrilles 1, 2, 3, 4, 5, and 6 of the Regiment du Maroc. V 571 was redesignated the 1st Escadrille d'Indochine in 1919.

VC 116, VB 133, VB 135, VB 136, VB 137, VR 291, VR 292, VR 293, VR 294, and VR 296 were disbanded in 1919.

Foreign Service
Czechoslovakia
Four Voisin 10s were obtained by Czechoslovakia after the war. Three were assigned to 4 Letecka Setnina at Cheb and one to Letecke Dilny at Cheb. They had been withdrawn from service by 1923.

United States
The United States obtained two Voisin 10s in 1919. They were used solely to train pilots for night bombing.

Voisin 11 Bn2
Near the end of the war Voisin designed a new aircraft he designated the type E.94. It was actually a Voisin 10 with a lighter wing and a new engine—a 350-hp Panhard 12Bc. Changes were made to the landing gear suspension system and its attachments to the fuselage. Also, the ailerons were modified. The wing span was slightly increased, the fuselage length slightly reduced, and the wing surface area was enlarged. Slight modifications were also made to the rudder. Performance was almost identical to the Voisin 10. A few Type 11s were built before the war's end but as there were no significant improvements provided by the new engine, large scale production did not ensue.

> **Voisin 11 Two-Seat Bomber with 350-hp Panhard 12Bc**
> Span 18.08 m; length 10.24 m; height 3.95 m; wing area 62.00 sq. m
> Empty weight 1,490 kg; loaded weight 2,050 kg
> Maximum speed: 125 km/h at 2,000 m
> Approximately 10 built

Voisin 11. The Voisin 11 was a Voisin 10 fitted with a 350-hp Panhard 12Bc engine. MA.

Voisin with 300-hp Fiat A12bis Engine

It appears that a development of the Voisin 8 appeared in 1918, powered by a 300-hp Fiat A12bis engine. It is possible that Voisin attempted this modification because of complaints that the Renault engines used on the Voisin 10 were prone to catching fire in flight. In any event, the Fiat-powered Voisin was not developed further.

Voisin Experimental Design with 300-hp Fiat A12bis
Span 17.93 m; length 10.95 m; wing area 61.14 sq. m
Empty weight 1,490 kg
Maximum speed 125 km/h at 2,000 m

Voisin with 400-hp Liberty Engine

A Voisin single-engine bomber was under construction in 1918, to have been fitted with a 400-hp Liberty engine. Known specifications include an empty weight of 1,300 kg, a loaded weight of 1,800 kg, and a military load of 400 kg. The surface area of the biplane wings was 146 sq. m. Maximum speed was estimated at 150 km/h at 4,000 m and climb to 4,000 m was to have taken 40 minutes. Endurance was estimated at four hours. As of May 1918, construction had been completed and the type was still awaiting delivery of the Liberty engine. It is not known if the engine ever arrived or what the results of testing revealed.

Voisin 12

In 1917 the STAé announced the BN2 specification for a new night bomber. Voisin tried a new approach to meet the specification. The airplane was a four-engine biplane with a conventional fuselage, horizontal stabilizer, and rudder. The only similarity with the Voisins 1 through 10 was the sloped nose and quadricycle landing gear. The bomber was powered by two pairs of 220-hp Hispano 8Bc engines mounted back to back in nacelles on either side of the fuselage and attached to the lower wings. The landing gear consisted of paired main wheels below each of the engine nacelles and the previously mentioned nose wheels. A tail skid was also located beneath the rudder. The crew consisted of a pilot and flight engineer/bombardier. The BN2 specification did not call for any gunners because the airplane was to fly only under the cover of darkness. Another feature of the design was that the top and bottom wings could be folded to permit hangar storage (a similar feature was found on the RAF's Handley Page bombers).

The performance of the new craft, designated Voisin 12, was impressive. It carried a payload of 20 120-kg bombs or 16 200-kg bombs. It also had a maximum speed 20 km/h faster than the Voisin 10.

During testing the only external change was the addition of

The Voisin 12 with the paired engines uncowled. MA13679.

Voisin 12 Two-Seat Heavy Bomber with Four 220-hp Hispano 8Bc Engines

Span 30.00 m; length 17.30 m; height 4.91 m; wing area 155.68 m

Empty weight 3,500 kg; loaded weight 5,700 kg

Maximum speed: 145 km/h at 2,000 m; climb to 2,000 m in 20 minutes; range 700 km; endurance 5 hours

Armament: 20 120-kg bombs or 16 200-kg bombs; a cannon was also experimentally fitted

One built

Voisin 12. During testing the only external change was the addition of two small auxiliary fins to either side of the horizontal stabilizer. B88.3421.

two small auxiliary fins to either side of the horizontal stabilizer. Construction of the Voisin 12 was completed by September 1918, too late to see service. At one point it was fitted with a cannon in an attempt to meet the S2/S3 specification for a ground-attack aircraft. However, the Armistice resulted in an end to the Aviation Militaire's interest in a new strategic bomber and the Voisin 12 was not selected for production.

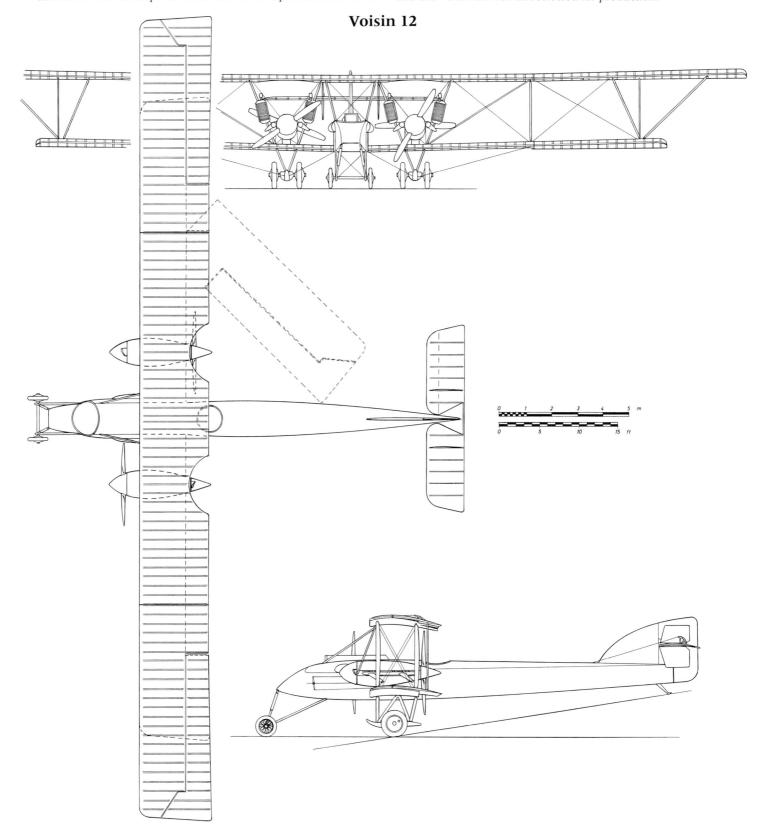

Voisin 12

Above and right: Voisin 12. University of Texas at Dallas.

Voisin E.50

The E.50 triplane was a project initiated in May 1917, to be powered by four 230-hp Hispano-Suiza engines. It was to be used as a heavy bomber and had five gun turrets (one of the first uses of covered gun positions in an aircraft). There was a single gun in the nose, a second beneath the forward fuselage, a third beneath the rear fuselage, a fourth behind the wings, and a fifth firing to the rear in the top fuselage. To reach the rear gun the crew had to climb a ladder three meters to the top fuselage. The overall configuration was similar to the E.28 triplane. However, unlike the E.28, the E.50 was never built and remained a project only.

E.50 Heavy Bomber Project with Four 230-hp Hispano-Suiza Engines
Empty weight 3507 kg; loaded weight 7,618 kg; payload 1,000 kg
Maximum speed: 131 km/h at 4000 m; climb to 4000 m in 30 minutes; endurance 7 hours

Voisin E.53

The Voisin E.53 was an attempt to adopt the unsuccessful E.28 triplane bomber to the ground attack (S) role. It was intended to fit it with one or more cannons, but the end of the war led to cancellation of the project. It is likely that the conversion was never completed and as far as can be determined, the E.28 never was fitted with cannons.

Voisin E.59

The Voisin E.59 was a study initiated in July 1917. It had the same wings as used on the Voisin 8 but with a one-piece fuselage of circular cross-section. The gunner in the nose controlled both a cannon and a machine gun, the latter in the lower part of the nose. A second machine gun position was in a separate position behind the pilot. The vertical tail was connected to the top wing by a thin beam of circular cross-section. The aircraft was to be powered by two 220-hp Hispano 8Bc engines, but was never built.

Voisin E.87

The Voisin E.87 was a four-engine bomber probably intended to meet the BN3/4 heavy bomber specification. It was to have been powered by four 400-hp Hispano engines. Empty weight was 3,300 kg, loaded weight was 5,613 kg, and payload was 1,000 kg. The maximum speed was estimated at 141 km/h at 4,000 m. Climb to 4,000 m was estimated to take 30 minutes. The airplane carried enough fuel for five hours of flight. This design was never built but was developed into the Voisin 12.

Voisin BN3/4 Bomber— Salmson Engines

Gabriel Voisin never lost his interest in the long-range heavy bomber. While his triplane designs of 1915 and 1916 had been failures, the Voisin firm produced several types to meet the BN3/4 requirements of 1918. One such design was for a large biplane powered by four 260-hp Salmson engines. Specifications included an empty weight of 3,300 kg, a loaded weight of 5,713 kg, and a payload of 1,000 kilograms. The wing surface area was 146 sq. m. Enough fuel was carried for seven hours of flight. Maximum speed was estimated at 157 km/h at 4,000 meters. Climb to 4,000 meters was estimated to take 30 minutes. By early April 1918 the design was still on paper and it appears that the aircraft was never built.

Voisin BN3/4 Bomber— Hispano-Suiza Engines

The Voisin firm submitted several designs to meet the BN3/4 heavy bomber specification of 1918. One called for a huge biplane with four 300-hp Hispano-Suiza engines. Empty weight was 3,407 kg, loaded weight 5,713 kg, and payload 1,500 kg. The combined wing surface area was 146 sq. m. The aircraft had an endurance of five hours, an estimated maximum speed of 146 km/h at 4,000 meters, and estimated climb to 4,000 meters within 30 minutes. It was under construction in April 1918 but it does not appear that construction was ever completed. This design may have been a development of the E.50 or E.87 projects.

Voisin Triplane Flying Boat

In 1917 it was reported that the Voisin firm was considering building a triplane flying boat to have four 200-hp Hispano-Suiza engines. It was almost certainly intended to fill the French "high seas" flying boat category. That specification called for a large seaplane with a crew of four, T.S.F. long-range wireless, a 75-mm cannon with 30 rounds, two machine guns, 120 kg of bombs, 140 km/h maximum speed, climb to 2,000 m in 25 minutes, and an endurance of eight hours.

The Voisin design was to have either a single hull or twin hulls similar to the Labourdette-Hallbron seaplane. As far as can be determined, the aircraft was never completed because of the end of the war.

Voisin Type M

In 1915 Gabriel Voisin decided to transform his Voisin 3 into an airplane with an underslung fuselage and landing gear in the center of the body. There were several disadvantages to this design. The fuselage was not able to carry the engine internally and the propeller would be very close to the ground. Voisin constructed a metal framework identical to the one used on his 1915 triplane. The fuselage enclosed the radiator of the engine and also served as an attachment point for the bracing of the flying surfaces. The front wheel was encased in the nose of the fuselage. A pair of wheels mounted in tandem supported the airplane along the midline and two skids at the end of the wings (span of 14.92 m) served to stabilize it during the landing. The engine was a 150-hp Salmson P9. The Voisin M, constructed in

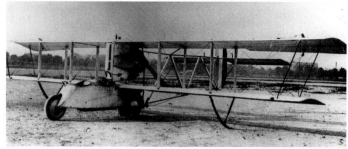

The Voisin Type M of 1916. The engine was a 150-hp Salmson P9. B87.6239.

July 1915, was not further developed.

Voisin Type O

The Type O was constructed at the same time as its predecessor, the type M. It was in fact a joining of two Voisin M types—two fuselages and two tails. One set of wings held the two fuselages together. It remained a prototype only.

The Voisin Type O was two Voisin Ms joined at the wings. Reairche.

The Creme De Menthe

Another project of Gabriel Voisin appeared in 1916—an armored biplane with a tractor Hispano-Suiza engine called *Creme de Menthe*. It had a tricycle landing gear and a lower wing with pronounced dihedral. The fuselage filled the space between the two wings. The crew of two sat in open positions on the top of the upper wing. A camera with a long focal length was installed below the engine block. The system of securing the wings was unusual, consisting of a pair of V-struts steeply canted inward toward the fuselage and reinforced by two additional struts inclined in the opposite direction. This gave the impression of an open W. The tail followed the classic Voisin layout with a rudder and square elevator. The *Creme de Menthe* also remained a prototype.

The unusual Voisin *Creme de Menthe*. A long focal-length camera was installed below the engine block. MA16270.

Wild Duck

Voisin *Wild Duck*. Note the deflated flotation bags underneath the lower wing. MA16266.

Voisin E.28

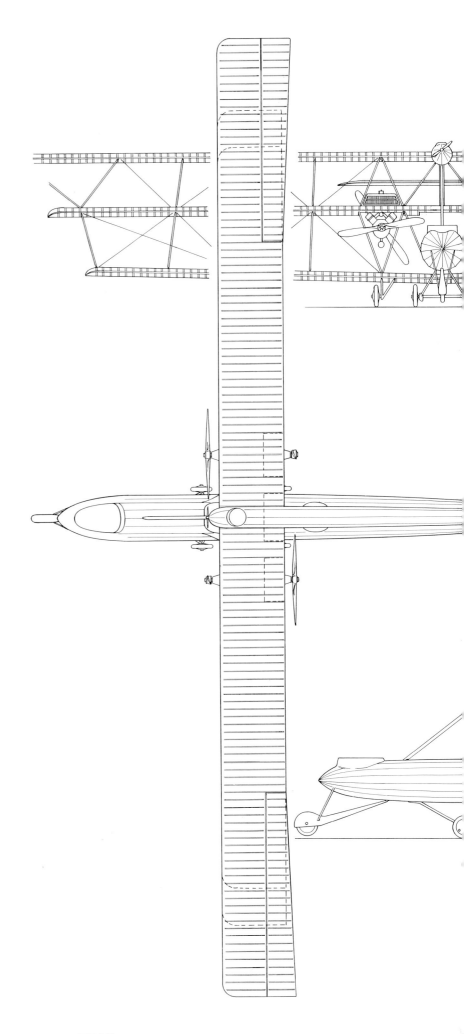

570B

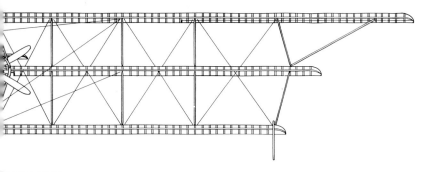
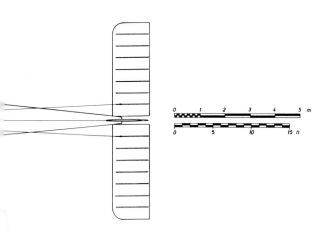
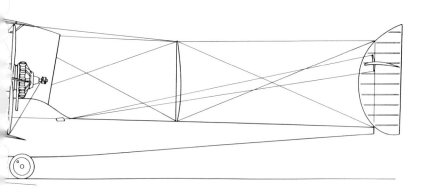

Voisin 1915 Triplane Bomber

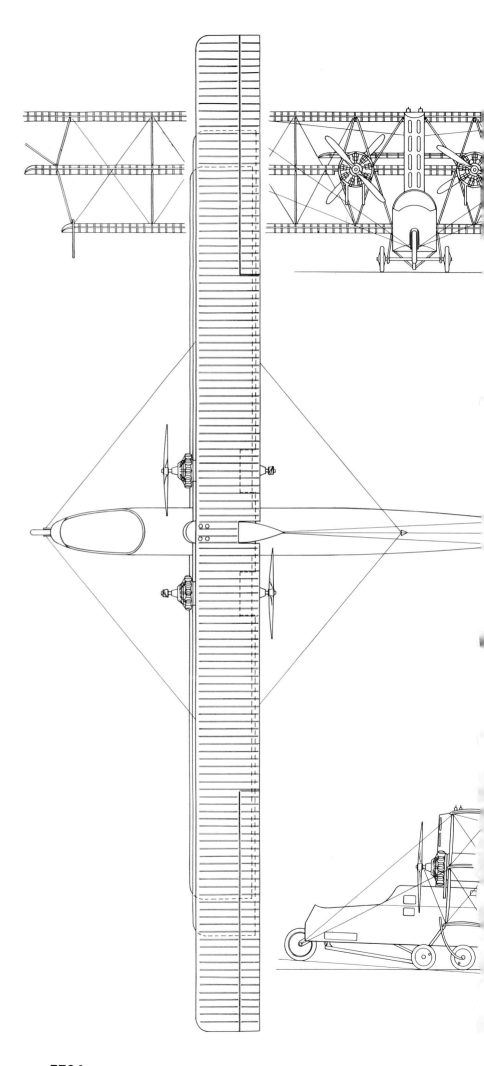

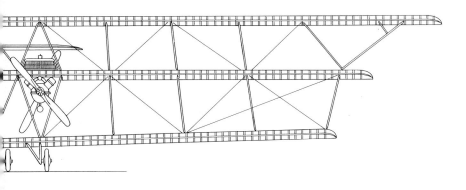

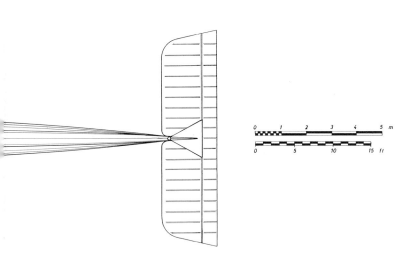

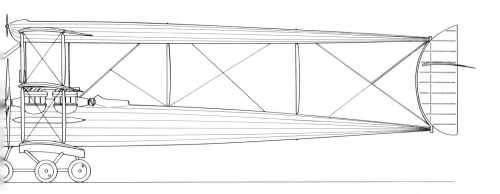

Weymann W.1

Charles Weymann designed an aircraft of unusual configuration in 1915. Designated the W.1, it was intended to meet the requirement for a single-seat fighter. It was built at Villacoublay by the Société de St. Chamond. Its 80-hp Clerget engine was located in the center of the fuselage and used a transmission to drive a propeller in the extreme tail. It had wings of equal span with ailerons on both wings. Wing area was 23 square meters. There was a cruciform tail and a tricycle landing gear. Armament was to consist of two fixed, forward-firing machine guns in the nose with the breeches extending into the pilot's cockpit. Yet another unusual feature was that the plane was made entirely of metal. It was stored at Villacoublay and it is believed that several test flights were carried out. Because of the location of the engine within the fuselage, engine cooling presented serious problems; flight testing is believed to have been abandoned in late 1915.

The Weymann C1 fighter. The aircraft's 80-hp Clerget engine was located in the center of the fuselage and used a transmission to drive a propeller located in the extreme tail. MA0141.

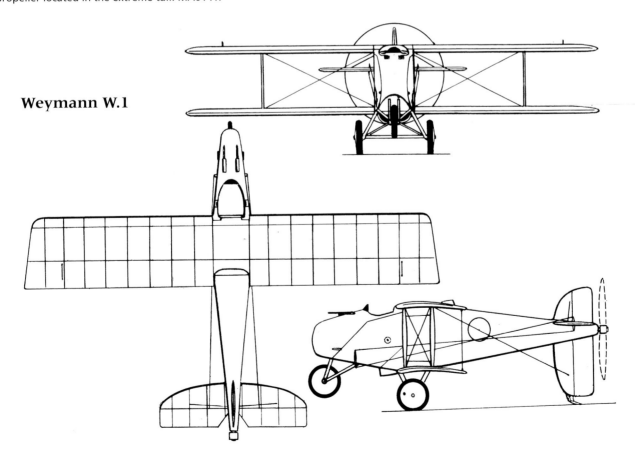

Weymann W.1

Wibault Wib.1

In 1917 Michel Wibault decided to initiate the construction of an airplane to meet the C1 category calling for a single-seat fighter with two machine guns. He intended to fit this aircraft with the new 280-hp Hispano Suiza 8F engine. However, this engine was reserved for the newer SPAD fighters, including the SPAD 17, and so Wibault was forced to use the older 220-hp Hispano Suiza 8Be.

The aircraft was of metal construction with single-bay, fabric-covered wings. It featured an abbreviated nose with equal-span staggered wings. Ailerons were on the lower wings only. The ailerons for the C1 category had been based on work done by a Dr. Margoulis, who was director of the Laboratoire Eiffel. Wibault, however, was not hesitant to redesign the ailerons to meet his own ideas, even though Margolis was critical of Wibault's modifications. Radiators were located on the bottom wing. Armament, as with most other aircraft of the C1 category, consisted of two synchronized 7.7-mm Vickers machine guns.

After completing testing with a model of the C1, Wibault initiated construction of the prototype (named the *Flandre*). The airplane was built by Niepce et Fetterer at Bolougne-Billancourt.

Construction was completed by October 1918 and the aircraft was sent to Villacoublay for tests by André Boillot, chief test pilot for the Blériot-SPAD firm. Flight tests, which began in November 1918, confirmed that the aircraft was both fast and maneuverable. At the official tests it attained a speed of 237 km/h, superior to the SPAD 13. However, the redesigned ailerons were found to be inadequate for sufficient control. They were subsequently modified but by then it was too late because the Nieuport 29, which had a superior climb rate and was more maneuverable, had been selected to fill the C1 requirement.

Wibault Wib.1 Single-Seat Fighter with 220-hp Hispano Suiza 8Be

Span 7.85 m; length 6.39 m; height 2.39 m; wing area 21.85 sq. m
Loaded weight 810 kg
Maximum speed: 244 km/h at sea level; 230 km/h at 3,000 m; 225 km/h at 5,000 m; ceiling 7,700 m; range 570 km
One built

Wibault C1 at Villacoublay where it was flown by André Boillot, chief test pilot for the Blériot-SPAD firm. CC.

Wibault C1. The aircraft was of metal construction with single-bay, fabric-covered wings. MA8841.

Wibault Wib.1

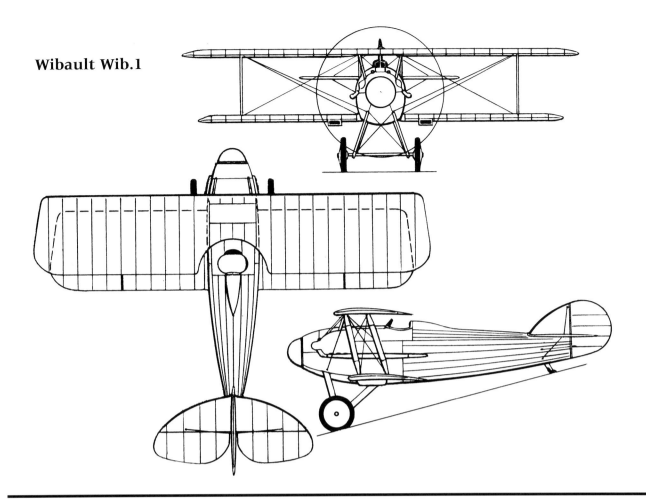

Appendix

Gliders

Several gliders were designed during the War. These include:
1. M-2—designed by Dr. Magnan in April 1914. It is uncertain how far construction proceeded after the original design was completed.
2. Laffont glider—designed in 1915 with the aid of Mouillard (wing design) and Dr. Cousin (the fuselage).
3. Nessler glider—designed in 1915. Designated the Nessler-I (N-1) *Aerovoilie,* it was tested at Douai. The biplane glider's lower wings were designed to have three settings adjustable in flight: climb, glide, and high speed. However, technical problems could not be overcome and the wings were installed in a fixed position. Nessler was wounded while serving with SPA 99 and was unable test the N-1 until 1921.

Glisseurs

Glisseurs were surface gliders which were comparable to the "penguins" used to train the pilots of the Aviation Militaire to taxi. The glisseurs, however, operated on the sea and were intend to train future flying boat pilots how to taxi their aeroplanes on the water; they were also used to practice machine-gun firing. In October 1915 a glisseur was constructed by engineer Fabre and enseigne de vaisseau Laperge; tests proved inconclusive and further testing was abandoned. Several other types were tested during the War and are shown below.

Fabre trainer in July 1914.
B87.4008.

Glisseur. B80.740.

Glisseur in operation on the North Sea. B83.3756.

Glisseur used as a machine-gun trainer at the Ecole de Tir at Cazaux in 1916. B86.301.

Color Section

Standard Colors

Methuen notations are given for known fabric samples. Pantone© equivalents (a printing color standard available from computer suppliers) are given in brackets. Actual colors varied from aircraft to aircraft based on weathering and paint batches; therefore, precise color values can only be approximated.

Color	Methuen	Pantone
Roundel Red	10C–D8	(1805C)
Roundel Blue	22–23D4	(3015C)
	23C–D8	
Clear-Doped Linen	2–3A3–4, 3B7	(127C, 110C)
	4A3, 4C3	(155C, 4525C)
Yellowish-Gray	4B2–3	(4545C)
Yellow	5C–D8, 5E4	(130C, 131C)
Camouflage Colors		
Chestnut Brown	6F 5–7	(497U)
Dark Green	29F3–6	(5605U)
Light Green	30D–E4–6	(3995U)
Beige	4–5D–E6	(132U, 139U)
Ecru	4–5C–D3–4	(451U, 4515U)

Aircraft Color Plates

1. Voisin 4 with 37-mm cannon of VB509. Finished overall in silver. Roundels appeared on wings in four positions. The unit insignia appeared on both sides of the nacelle.

2. Voisin 8 of VB109. Finished overall in silver, roundels appeared on the wings in four positions. This machine was armed with a Hotchkiss gun on a pillar mount. The unit insignia was applied to the port side of the nacelle; starboard side application is conjectural.

3. Voisin 10 of VB125. Finished overall in black, roundels appeared on the wings in four positions. It is believed the unit insignia appeared on both sides of the nacelle. For night operations, several landing lights were fitted to the front of the nacelle. See photo on page 564.

4. Morane-Saulnier H. Finished overall in clear-doped linen. Cowling, side panels, wheel covers, and all struts finished in black. Roundels appeared on wings in four positions.

5. Morane-Saulnier L, serial unknown, flown by Roland Garros, Dunkerque 1915. Finished overall in clear-doped linen. Cowling, wheel covers, and all struts finished in black. Roundels appeared on wing in four positions. An early production aircraft, it had the wide V-strut in its undercarriage and no central cut-out in the trailing edge of the wing. Armed with a Hotchkiss machine gun and bullet deflectors mounted on its airscrew, Garros obtained his first victory in this machine on 1 April 1915. He shot down two more German aircraft on April 15 and 18, but on the afternoon of the 18th was himself brought down by ground fire near Ingelmunster.

6. Morane-Saulnier L, serial number 39, flown by Corporal Georges Guynemer while with MS 3. Finished overall in clear-doped linen. Cowling, wheel covers, and all struts finished in black. Roundels appeared on wings in four positions. Soon after Guynemer mounted a Lewis gun to this aircraft, he and his gunner, Jean Guerder, shot down an enemy Aviatik in flames near Soissons (near the village of Septmonts) on 19 July 1915, for Guynemer's first victory.

7. Morane-Saulnier N A.178 of No.24 Squadron, RFC, July 1916. This machine was flown by Lt. T.P.H. Bayetto of No.3 Squadron on attachment to No.24 when he successfully forced down an enemy aircraft during the battle of the Somme, July 1916. Due to its similarity to the German Fokker E.III, the metal panels of the RFC type Ns were painted in red identity markings in July 1916. Roundels appeared on both sides of the fuselage and on the wings in four positions. An additional fuselage band for recognition was also added. A brass disc displaying the Morane-Saulnier trademark (letters "MS") was carried on each side of the engine cowling. Metal plates were fitted to the propeller blades to deflect bullets away from the un-synchronized Lewis gun.

8. Morane Saulnier P, serial number and unit unknown. Finished in clear-doped fabric with yellow painted cowling, metal panels, and struts. Roundels appeared on upper and lower surfaces of the wing in four positions. Cowling and wheel covers were painted red and white as shown. This machine carried two fuselage bands which crossed on the upper decking. Three stars—one red and two white—appeared on the port side of the fuselage; it is assumed this pattern was repeated on the starboard side. Armament consisted of a fixed, forward-firing Lewis gun attached to the center cabane strut and a movable gun in the gunner's position.

9. Morane Saulnier AI, serial number 1724, flown by Sgt. Rufus R. Rand, Jr., when with MS 158. On 4 March 1918 MS 158 was fully equipped with the Morane-Saulnier AI parasol. However, through a series of mishaps, the AI was considered structurally unsound and by 19 May 1918 all the AIs of MS 158 were replaced with SPAD 13 fighters. The specific French five-color camouflage pattern of chestnut brown, dark green, light green, beige, and black applied to this aircraft type is illustrated in the three-views. To minimize obstruction of the camouflage pattern, some markings are omitted in the smaller views. Under surfaces are illustrated as aluminum, but may have been ecru. Roundels appeared on wings in four positions. It is believed the white number "3" and black unit insignia appeared on both sides of fuselage.

10. Henri Farman H.F.20 flown by MF 44 near Verdun circa 1915. This H.F.20 is a typical machine with pale gray nacelle and all wings finished in clear-doped linen. The tailboom and its struts were varnished wood. Roundels appeared on the wings in four positions. The unit insignia for MF 44—black colored horse shoes—was applied to the upper tail surfaces.

11. Maurice Farman M.F.11 fitted with a forward 'visor' which could be lifted. Nacelle finished in pale gray and all wings finished in clear-doped linen overall. The tailboom and tailboom struts were varnished wood. Roundels appeared on wings in four positions. An image of the French Croix de Guerre award with one palm was painted on the port side of the nacelle, perhaps indicating the crew had received this award. The reference is the photo on page 255; the black and white disk is not shown since it may not be a permanent part of the aircraft.

12. Maurice Farman M.F.11 floatplane, Baltic Sea Fleet serial number 31, flown by Russian naval Ace Mikhail Safonov, circa March 1916. It was finished pale gray overall. Underside of the nacelle, wings struts, and tail boom were varnished wood. Rudders were white with the light blue Cross of Saint Andrew on each outer surface. Roundels appeared on wings in four positions. On each side of the nacelle, appeared a prefix of Cyrillic letters M (em) and ф (ef) to denote a Maurice Farman aircraft.

13. Caudron G.3, serial unknown, used as a trainer at Juvisy. It was finished in clear-doped linen overall with cowling in natural metal. The tailboom and tailboom struts were varnished wood. Roundels appeared on wings in four positions. An insignia of

two moons appears on the starboard side of the nacelle. Although illustrated as black, this color is conjectural.

14. Caudron G.4, serial number C1746, flown by Caporal Kenneth P. Littauer of C 74, circa winter 1916–17. Finished overall in clear-doped linen. The tailboom and tailboom struts were varnished wood. Roundels appeared in four positions on the upper wing only—its upper and lower surface. A white numeral "7" appeared on the red starboard cowling and a red numeral "7" appeared on the white port cowling. The forward observer's position was armed with a flexible Lewis machine gun firing forward; a second rearward-firing Lewis was mounted between the cockpits. The name of Littauer's wife, Helen, was stenciled on the port side of the nacelle, while the unit insignia—a white cat with red collar—was applied to the starboard side. The outer surfaces of the outer fins and rudders appeared in national markings and the inner surfaces were clear-doped linen and carried the serial number as illustrated.

15. Nieuport 11 serial number N557 from N 12. It was finished in clear-doped linen overall with natural metal cowling, side fairing plates, landing gear, and cabane struts. Roundels appeared in six positions on the wings. The numeral "12" superimposed on a blue and white flag was the unit insignia of N 12, the moon design being a personal marking of an unidentified pilot.

16. Nieuport 11, serial number unknown, flown by pilot Armand Galliot de Turrenne of N 48 in 1916. The fuselage was overpainted in national colors of blue, white, and red. Turrenne's family coat of arms in the form of a hunting horn with red and gold shield was painted on the white area of the fuselage. Wing and tail surfaces are clear-doped linen bordered in black. Roundels appeared on the wings in six positions. The cowling, side fairing plates, landing gear, and cabane struts were natural aluminum.

17. Nieuport 16, serial number N1454, flown by Sergeant Charles Johnson when with the Lafayette Escadrille (N 124). This machine had fuselage and upper wing surfaces finished in a two-tone camouflage scheme—solid bands of drab green and brown overall. Under surfaces of wings and tailplane were in plain linen. Edges of wings and tailplane had a narrow outline border of light blue. The cowling, side fairing plates, landing gear, and cabane struts were natural aluminum. Roundels appeared on the under-surfaces of the upper and lower wing only. Johnson's personal "white die" insignia was applied on each side of the fuselage and wheel, the fuselage "white die" being displayed in a different perspective on the starboard side of the fuselage. This aircraft is armed with a Lewis gun and strut-mounted Le Prieur rockets which Johnson used unsuccessfully to attack German observation balloons in July 1916.

18. Nieuport 12 serial number 1043 as flown by the 3rd Corps Detachment of the Imperial Russian Air Service. This machine had fuselage and upper wing surfaces finished in a two-tone camouflage scheme—solid bands of drab greens and browns overall. Under surfaces of wings and tailplane were plain linen. The edges of wings and tailplane had a narrow outline border of light blue. The cowling, side fairing plates, landing gear, and cabane struts were natural aluminum. The top wing center section was covered in clear cellulose. Roundels appeared on under-surfaces of upper and lower wings only.

19. Nieuport 17 serial number N1550 as flown by Sous Lieutenant George Guynemer when attached to N 3 in early 1917. This aircraft is armed with a Vickers gun and sported a large wrap-around windscreen. It was finished in aluminum overall with upper surfaces of its wings camouflaged in random patches of dark green and brown (pattern speculative) leaving a contrasting narrow aluminum outline. Roundels appeared in four positions on wing under surfaces, upper wing roundels being over-painted. This machine bears a large white numeral

"2" on the upper starboard wing. The fuselage sides carried the famous stork marking of N 3 and a numeral "2" (although shown in red, some historians suggest black—see detail a). In addition, a red (again, perhaps black) pennant was applied on the rear decking behind the cockpit. The aircraft's cône de pénétration was finished in red and white.

Detail (a) Nieuport 17, serial number N1331, as flown by ace Sous Lieutenant George Guynemer when attached to N 3 in 1917. This aircraft was finished in aluminum overall and armed with a single Vickers gun. Roundels appeared on wings in six positions. The fuselage sides carried a black stork with red beak, wing, and tail feathers, and a different style of numeral "2" which appeared in black. In addition, the interplane struts carried a black and white pennant with the famous phrase "Vieux Charles" (Old Charles). The aircraft's cône de pénétration was finished in red and white.

20. Nieuport 17 serial number N1490 as flown by ace Sous Lieutenant Charles Nungesser when attached to N 124 in July 1916. This aircraft is armed with both Vickers and Lewis guns. Upper surfaces are camouflaged in a random spray of dark green and brown (pattern speculative) leaving a contrasting narrow aluminum outline around flying and control surfaces. All under surfaces are aluminum. Roundels appeared on under-surfaces of upper and lower wings only, the upper wing roundels being over-painted. Noteworthy details include a natural metal cowling and a roundel blue cône de pénétration. Nungesser's personal marking was a macabre black heart insignia bearing a skull with cross bones, coffin, and candle sticks. Nungesser's personal insignia (detail b), appeared on each sides of the fuselage—with head of coffin facing tailplane on port side and facing cockpit on starboard side.

Detail (c) Top wing, port side panel, and insignia of Nieuport 17, serial number N1895, as flown by Charles Nungesser. This aircraft was armed with both Vickers and Lewis guns. It was finished overall in silver. Roundels appeared on wings in six positions. Color bands of red (outermost position), white, and blue appeared on all four wing panels—upper surface of top wing and lower surface of bottom wing, all angled so leading edges face toward cowling. Nungesser's personal insignia (detail c), appeared on both sides of the fuselage, with the head of the coffin facing the cockpit on each side.

Detail (d) Top wing, starboard side panel, and insignia of Nieuport 17 serial number N1574 as flown by Charles Nungesser. This aircraft was armed with both Vickers and Lewis guns and sported a silver cône de pénétration. It was finished overall in silver. Roundels appeared on wings in six positions. Color bands of Red (outermost position), white, and blue appeared on all four wing panels—upper surface of top wing and lower surface of bottom wing, all angled so trailing edges face towards tailplane. Nungesser's personal insignia (detail d), appeared on both sides of the fuselage, with the head of the coffin facing the cockpit on each side.

21. Nieuport 24 serial number N5340 flown by Sgt. Herschel Mckee of N 314 when he was shot down by German anti-aircraft fire and taken POW near Chateau-Salins on 8 February 1918. This machine was finished silver overall. Roundels appeared on wings in six positions. The unit insignia appeared on each side of the fuselage. Mckee individualized his machine by applying a white star on each blue painted wheel hub.

22. Nieuport 27 serial number N5246 flown by Sous Lieutenant Gilbert Discours (two victories) of N 87. The specific French five-color camouflage pattern of chestnut brown, dark green, light green, beige, and black applied to this aircraft type is displayed in the three-views. The undersurfaces were ecru. To minimize obstruction of the camouflage pattern, some markings are omitted in the smaller views. The unit's insignia—a cat—was applied to each side of the fuselage. It appeared in black on light

colored aircraft and in white on darker camouflaged machines.

23. Nieuport 28 serial N6169 flown by Major John Huffer (CO) and First Lieutenant Edward V. Rickenbacker of the 94th Aero Squadron. This was one of three Nieuport 28s flown by Rickenbacker during the war. This machine was normally used by then unit commander, Major Jon Huffer. The numeral "1" indicated the CO's machine. Rickenbacker, the leading American ace, reached ace status in this aircraft by obtaining victories 3–6 in it. His first two victories were obtained in Nieuport 28 serial number N6159. USAS roundels appeared on wings in four positions. The cowling was painted in USAS roundel colors. The squadron insignia—Uncle Sam's stovepipe hat being thrown in the ring—appeared on each side of the fuselage, as well as a white numeral "1" shadowed in red and blue. The unit insignia pointed forward on each side. The specific French five-color camouflage pattern of chestnut brown, dark green, light green, beige, and black applied to this aircraft type is displayed in the three-views. To minimize obstruction of the camouflage pattern, some markings are omitted in the top and starboard side views. The undersurfaces were ecru.

Nieuport 28 wing roundels were originally in French colors, with the upper wings having equal-size roundels on upper and lower surfaces. Roundels were overpainted in AEF colors by American squadrons. Fin and rudder colors were white-blue-red front to rear. From serial number N6201 onward Nieuport 28s were delivered in AEF markings, with fins and rudders in red-white-blue, front to rear, and the roundels on the lower surface of the upper wings deleted.

24. Nieuport 28, serial number unknown, flown by Lt. Zenos Miller of the 27th Aero Squadron, USAS. Miller used this machine to shoot down two German observation balloons on 16 July 1918 and three Fokker D.VIIs—one on 19 July and two on 20 July 1918. The machine bore the 27th's eagle insignia and a black numeral "14" outlined in white on its fuselage sides. The same numeral "14" also appeared on the top wing as illustrated. The aircraft was finished in the specific French five-color camouflage pattern as shown. Undersurfaces were ecru. USAS roundels appeared on the wings in four positions.

25. Hanriot HD.1 serial number 24 flown by Sous Lieutenant Willy Coppens of 9e Escadrille Belge, May 1918. Coppens flew several HD.1s during the war and achieved a final victory score of 37 (including 26 balloons), making him the leading Belgian ace. The specific French five-color camouflage pattern of chestnut brown, dark green, light green, beige, and black applied to this aircraft type is displayed in the three-views. To minimize obstruction of the camouflage pattern, some markings are omitted in the smaller side view. Upper surfaces of the horizontal tailplane were finished in alternating bands of white and blue. The aircraft also bears a white thistle on each side of the fuselage. Wheel covers, center W-shaped cabane struts, and all undersurfaces are shown aluminum. However, undersurfaces may have been ecru. Belgian roundels appeared on the wings in four positions. Rudder strips were black (leading), yellow, and red with white serial numbers outlined in red.

26. Hanriot HD.1 serial 6254 flown by ace Giorgio Michetti while attached to Squadriglia 76ª. The Hanriot HD.1 was accepted very readily by the Italians as the principle single-seat fighter to replace the Nieuports then in use. Although Italian production of the type was substantial (nearly 1,700 machines), some of the Hanriots used were built by the French firm. On this particular machine Michetti's personal marking was a white seahorse imposed inside a black disk This design appeared on each fuselage side and top of the tailplane. A red "76" appears at the center of the top wing and on the fuselage sides (below the cockpit). The word "ALZARE" ("LIFT") appears in small lettering near the top of fuselage sides (behind the cockade). Italian roundels appeared on the wings in four positions.

27. Hanriot HD.3 C2 of HD 174, serial unknown, The machine was finished in a five-color camouflage pattern—chestnut brown, light green, dark green, beige, and black. The complete pattern has not been totally documented. Undersurfaces and wheel covers were ecru. The "Mercury" unit insignia appeared in several versions. Roundels appeared on the wings in four positions.

28. A.R.1 serial number 366 of AR 14. It was finished in clear-doped linen overall with pale yellow metal and wood areas. Wing and landing gear struts were varnished wood. Roundels appeared in four positions on the wings. This unit's emblem—a winged griffin holding a shield bearing the cross of Lorraine and a black numeral "0"—appeared on each side of the fuselage.

29. A.R.1, serial number unknown, of AR 203. This aircraft was fitted with a 200 hp Renault engine. Finished with clear-doped linen overall with pale yellow metal and wood areas. Wing and landing gear struts were varnished wood. Roundels appeared in four positions on the wings. This unit's emblem and a white numeral "9" on a red disk appeared on each side of the fuselage. The unit's insignia, shown in detail, shows a clown holding a pair of binoculars inside a red disk.

30. A.R.2 serial number 717 of an unknown escadrille. The specific French five-color camouflage pattern of chestnut brown, dark green, light green, beige, and black applied to this aircraft type is displayed in the three-views. To minimize obstruction of the camouflage pattern, some markings are omitted in the smaller side view. Wheel covers and the wing undersurfaces were ecru. Roundels appeared on wings in four positions. A black numeral "17" appeared on each side of the fuselage.

31. Sopwith 1 A2 Strutter serial number unknown, of SOP 24, Toul, circa March 1918. Finished in clear-doped linen overall with aluminum cowling, cabane and landing gear struts. Outer wing struts and upper decking around cockpits varnished wood. Roundels appeared on wings in four positions. A red numeral "2" and the unit's insignia—a white star on a red disk, appeared on each side of the fuselage. A large tri-color fuselage band of red, white and blue appeared to wrap around the fuselage, however, it is not know if this continued on the underside. This machine is equipped with both Vickers and Lewis guns.

32. Sopwith 1 A2 Strutter, serial number and escadrille unknown. This machine had fuselage and upper wing surfaces camouflaged with solid bands of drab green over the plan linen. Undersurfaces of wings and tailplane remained plain linen. The metal cowling was painted over, but the landing gear and cabane struts retained their metal finish. Outer wing struts were varnished wood. Roundels appeared on the wings in four positions. A narrow fuselage band of black, white, and black appeared to wrap around the fuselage; however, it is not known if this continued on the underside. It is believed the black numeral "5" shadowed in white appeared on both sides of the fuselage. This machine was equipped with both Vickers and Lewis guns.

33. Sopwith 1 B1 Strutter, serial number unknown, of SOP 111. Finished in clear-doped linen overall with aluminum cowling, cabane and landing gear struts, and bomb rack access panel plates. Outer wing struts and upper decking around cockpit were varnished wood. Roundels appeared on wings in four positions. A red numeral "3" and the unit's insignia—a swan over a blue and red background—appeared on each side of the fuselage. This machine was equipped with single Vickers gun.

34. Sopwith 1 A2 Strutter, serial number 2572, of SOP 287. This machine had fuselage and upper wing surfaces camouflaged with solid bands of drab green over the plain linen. Undersurfaces of wings and tailplane remained plain linen. The metal cowling was painted over, but the landing gear and cabane struts retained their metal finish. Outer wing struts and

578 FRENCH AIRCRAFT OF THE FIRST WORLD WAR

upper decking around cockpits varnished wood. The tail fin was painted white. Roundels appeared on the wings in four positions. A narrow chevron of red and blue appeared centered on the top decking, with bands flaring out across the top deck as shown. This machine was equipped with both Vickers and Lewis guns. The unit's insignia—a buffoon or jester with hood and bells—appeared on both sides of the fuselage. The emblem was applied with a stencil and filled in with a variety of colors.

35. SPAD SA.2, serial number and unit unknown. Finished overall with fabric having a yellowish-gray color and all metal panels painted to closely match the fabric color. Roundels appeared on the wings in four positions. A large tri-color fuselage band appeared to wrap around the fuselage; however, it is not known if this continued on the underside.

36. SPAD 7, serial number 1215, flown by ace Jacques Rogues when with SPA 48, summer 1917. Clear-doped fabric overall with pale yellow metal and wood areas. Roundels appeared on the wings in four positions. The machine exhibits the "Tete de Cog" escadrille emblem and a black numeral "7" on each side of the fuselage. Rogues' personal markings included a blue cowling and the "Jolie Demoiselle" ("Pretty Maiden") which appeared on the starboard side below the cockpit. The port side application as shown is conjectural.

37. SPAD 7, serial unknown, as flown by ace Sergeant Pierre Cardon when with SPA 81. The machine was finished in a five-color camouflage pattern of chestnut brown, light green, dark green, beige and black—this pattern being one of several patterns applied to SPAD 7 aircraft. Undersurfaces and wheel covers were ecru. Markings omitted in smaller views so as not to obscure camouflage pattern. Roundels appeared on wings in four positions. The cowling was painted red. A white numeral "8" and the unit insignia—a white colored greyhound dog highlighted in black—appeared on each side of the fuselage. The greyhound insignia was reproduced from a stencil on all the airplanes of SPA 81.

38. SPAD 11 serial number 2043 of SPA-Bi 102. It was finished in clear-doped fabric overall with pale yellow metal and wood areas. Roundels appeared on the wings in four positions. Wing struts were varnished wood. The unit's insignia—the reddish-orange sun with yellow beams of light—appeared on both sides of the fuselage.

39. SPAD 11 serial number 6006 of SPA-Bi 49. It was finished in clear-doped fabric overall with pale yellow metal and wood areas. Roundels appeared on the wings in four positions. Wing struts were varnished wood. The unit's insignia—a yellow flag bordered in green—appeared on both sides of the fuselage. In addition, the cowling was painted green and a large band of green spanned the top wing as shown.

40. SPAD 11 (Bernard-built) serial number 6550 of SPA-Bi 34. It is finished in a five-color camouflage pattern typical of Bernard-built SPADs. The camouflage pattern used chestnut brown, light green, dark green, beige, and black. Undersurfaces and wheel covers were ecru. Roundels appeared on the wings in four positions. The unit's insignia—a white or gray fox—was applied on both sides of the fuselage. The fox insignia was reproduced from a stencil on all the airplanes of SPA-Bi 34. The complete Bernard five-color camouflage pattern is shown in three-views with the SPAD 16 illustrations.

41. SPAD 13 C1, serial unknown, as flown by ace Sgt. Frank L. Baylies when with SPA 73 in December 1917. This machine was an early model, non-camouflaged SPAD 13. It was clear-doped fabric overall with pale yellow metal and wood areas. The unit's insignia—a white stork with black wing tips and red beak and legs—appeared on both side of the fuselage over a blue and white band. The red numeral "13" also appeared on both sides of fuselage. Roundels were applied on the wings in four positions.

42. SPAD 13 C1 (Kellner-built) serial number S4523 as flown by ace First Lieutenant Edward V. Rickenbacker of the 94th Aero Squadron. Rickenbacker obtained 20 victories in this machine. It was finished in a five-color camouflage pattern typical of Kellner-built SPADs. The camouflage pattern used chestnut brown, light green, dark green, beige, and black. Undersurfaces and wheel covers were ecru. The complete camouflage pattern is illustrated in three views; however, some markings are omitted in the smaller side view to minimize obstruction of the pattern. The markings on this aircraft went through several changes over time due to battle damage and normal wear. The three white dots are battle damage patches on the fin and fuselage; each carried a small black cross. Three other patches were carried in other locations. A white numeral "1" and the squadron insignia—Uncle Sam's stovepipe hat being thrown in the ring—appeared on both sides of the fuselage, the hat pointing forward on each side. The outer wheel covers were blue and contained a white star with red dot. Additional markings included a red cowling and a red, white, and blue band which wrapped completely around the forward landing gear struts. The top wing carried a white and red band and a large numeral "1" outlined in red. USAS roundels appeared on the wings in four positions. The bottom of the lower wing carried a mirror image of the numberal "1" and the red and white band, but the "1" was under the left wing and the red and white stripe was under the right wing, with the red stripe forward. Unlike the top wing, the forward part of the stripe on the bottom wing was outboard.

43. SPAD 13 C1 (Blériot-built) serial number S2729 flown by Adjutant Reginald Sinclair of SPA 68, June 1918 (three victories). This machine is finished in a five-color camouflage pattern typical of Blériot-built SPADs. The camouflage pattern used chestnut brown, light green, dark green, beige, and black. Undersurfaces and wheel covers were ecru. The complete camouflage pattern is illustrated in three views; however, some markings are omitted in the smaller side view to minimize obstruction of the pattern. The unit's insignia—a white hunting horn—and a white numeral "14" appeared on each side of the fuselage and again on the top wing as illustrated.

44. SPAD 13 C1 (Blériot-built) serial number S2742 flown by ace First Lieutenant Gorman DeFreest Larner with the 103rd Aero Squadron (formerly Lafayette Escadrille, SPA 124). The machine is finished in a five-color camouflage pattern typical of Blériot-built SPADs. Undersurfaces and wheel covers were ecru. The fuselage sides were decorated with the unit's Indian head emblem and a yellow numeral "13" shadowed in black. The Indian head—a fierce Sioux warrior with brown face and a full red, white, and blue bonnet—was first applied to aircraft of SPA 124 in April 1917. The swastika was applied in both clockwise and counter-clockwise patterns and was patterned after a good-luck talisman which ace Raoul Lufbery had picked up in China. The same numeral "13" appeared again on the upper surface of the top wing partially covering the black footprint pattern on the starboard side of the top wing. A solid black numeral "13" of the same design appeared on the lower surface of the starboard wing. The cowling was painted white. USAS roundels appeared on the upper and lower wings in four positions.

45. SPAD 12 Ca.1 serial number S445 flown by ace Second Lieutenant Rene Fonck of SPA 103. Fonck, the leading allied ace, scored 11 of his 75 victories with the SPAD 12. The specific French five-color camouflage pattern of chestnut brown, dark green, light green, beige, and black applied to this aircraft type is displayed in the three-views. To minimize obstruction of the camouflage pattern, some markings are omitted in the smaller side view. Undersurfaces and wheel covers were ecru. Roundels appeared on the wings in four positions. The fuselage sides are decorated with the unit's stork emblem and a red Roman

numeral "VI." The Roman numeral is outlined in white. The stork emblem is black and white with reddish-orange beak and legs. The Roman numeral appeared again on the top wing with a five-pointed red star outlined in white.

46. SPAD 16 serial number S12102 flown by Paul Menet of SPA-Bi 21. This machine is finished in a five-color camouflage pattern typical of Bernard-built Spads. The camouflage pattern used chestnut brown, light green, dark green, beige, and black. Undersurfaces and wheel covers were ecru. Roundels appeared on the wings in four positions. The numeral "8" appeared on both sides of the fuselage with the unit's insignia—a woman with sword in hand, leading the charge on the Marsaille. The background circle is vermilion. The numeral "8" also appeared on the upper left wing; see photos on pages 516 and 518.

47. Salmson 2 A2 serial number 479 of SAL 58. The specific French five-color camouflage pattern of chestnut brown, dark green, light green, beige, and black applied to this aircraft type is displayed in the three-views. To minimize obstruction of the camouflage pattern, some markings are omitted in the smaller side view. Undersurfaces and wheel covers were ecru. Roundels appeared on the wings in four positions. Fuselage sides are decorated with the unit's rooster emblem positioned inside a flag. Although the flag is illustrated as gray with red outline, these colors are conjectural.

48. Breguet 14 B2 of BR 117 flown by American volunteer First Lieutenant Manderson Lehr. During a bombing mission on 15 July 1918 Lehr was mortally wounded and crash-landed soon after. His observer, Lieutenant Charles, was seriously injured but survived. They where probably the victims of Vizefeldwebel Eric Buder of Jasta 26. This Breguet had an overall finish of yellowish-gray, with all wood and metal areas painted to closely match the fabric color. Lehr personalized its port side with the words "HAM" and "NEBRASKA" (his home state). Roundels appeared on the wings in four positions. The radiator cowling was painted white. The unit's insignia—a red rooster holding a bomb in its claws—appeared on both sides of the fuselage. The insignia's background is shown as light blue, but may have been white.

49. Breguet 14 A2 serial number 63?? of BR 287. This Breguet is equipped with a Fiat engine and finished in a five-color camouflage pattern commonly referred to as the G-1 pattern, thought to indicate manufacture by Farman. The pattern used chestnut brown, light green, dark green, beige, and black. Undersurfaces were ecru. The complete camouflage pattern is illustrated with the three images on this page. Roundels appeared on the wings in four positions. The unit's insignia—a buffoon or jester with hood and bells—most likely appeared on both sides of the fuselage. The emblem was applied with a stencil and filled in with a variety of colors. The numeral "7" appeared on the fuselage, probably in red.

50. Breguet 14 A2 serial number 4644 of BR 234, circa autumn 1918. This Renault-built machine is finished in a five-color camouflage pattern commonly referred to as the G-2 pattern, thought to indicate manufacture by Breguet. The pattern used chestnut brown, light green, dark green, beige, and black. Undersurfaces were ecru. The complete camouflage pattern is illustrated with the three images on this page. Roundels appeared on wings in four positions. The unit's insignia—a white angelic figure carrying a torch with red flames, all stenciled on a medium blue disc—appeared below the observer's position on the port side. It is not known if the insignia was applied to the starboard side. A white numeral "3" appeared on the aft fuselage and is assumed to have been repeated on the aft position of the starboard side. A white "Cross of Lorraine" was applied to the port elevator (and most likely to the starboard as well) as illustrated in the top view. The engine cowling panel on the port side (radiator shell) appeared in a light color which

might have been blue. This aircraft's upper wing is equipped with a fitting for an additional Lewis gun on the center section—a common feature on many A.2 machines. This machine was wrecked at Nixeville on 26 September 1918.

51. Breguet 14 B2 serial number 1306 of BR 123. This machine is finished in the G-2 camouflage pattern. Undersurfaces were ecru. Roundels appeared on the wings in four positions. The unit's insignia—a light yellow swooping bird of prey with a gray bomb—was painted on the fin. A white numeral "6" appeared on the fuselage.

52. Breguet 14 B2, serial unknown, flown by Sous-Lieutenant Djibrail Nazare-Aga of BR 108, circa autumn 1918. This Michelin-built machine is finished in five-color camouflage pattern commonly referred to as the G-3 pattern, thought to indicate manufacture by Michelin (later pattern). The pattern used chestnut brown, light green, dark green, beige, and black. Undersurfaces were ecru. The complete camouflage pattern is illustrated by the three images on this page. Roundels appeared on the wings in four positions. The numeral "9" appeared in red on each side of the tail fin. The unit's insignia—a yellow Pegasus with a bomb in its mouth on a red background—appeared on each side of the fuselage. The Pegasus pointed forward on each side. Wing struts were gray.

53. Breguet 14 B2, probably serial number 4231, flown by ace pilot Capitaine Jean Jannekeyn (five victories) and ace observer Lieutenant Eugene Weismann (seven victories), both with BR 132. Prior to his transfer into aviation, Lt. Weismann was wounded in action four times in the infantry, one wound resulted in one of his feet being amputated in June 1916. Escadrille BR 132 produced a third ace—Sous Lieutenant Antoine Paillard (five victories). These three men were crew members in a four-plane formation from escadrille BR 132 which contributed to downing of four enemy planes on 14 September 1918. This Michelin-built machine is finished in the G-3 camouflage pattern. Undersurfaces were ecru. Roundels appeared on the wings in four positions. The unit's insignia—a yellow Chinese dragon on a black pennant with yellow fringe—appeared on the fin. A white outlined numeral "13" appeared on the fuselage side—its main color was probably black. It is assumed the unit insignia and numeral "13" appeared on both port and starboard sides.

54. Breguet 14 B2 serial number 4018 on July 29 1918. It was flown by Major Harry Brown in the 96th Aero Squadron's first mission over the lines on June 18, 1918. This Michelin-built machine was usually flown by First Lieutenant Arthur H. Alexander, with Second Lieutenant John C.E. McLennan as his observer. Flying 270 missions—100 of which were over the lines—this airplane took 110 bullet and shrapnel hits, had its tail section replaced three times, its lower wings six times, its upper wing twice, and its undercarriage twice. Although McLennan was wounded in it during the mission of 4 September 1918, no crewman was killed in this extraordinarily lucky aircraft. This machine is finished in the five-color camouflage pattern commonly referred to as the G-4 pattern, thought to indicate manufacture by Michelin (early pattern). The pattern used chestnut brown, light green, dark green, beige, and black. Undersurfaces were ecru. The complete camouflage pattern is illustrated with the three images. Roundels appeared on the wings in four positions. Aircraft of the 96th were also characterized by wing stripes of red-white-red. These stripes ran chordwise outboard of the port roundels on the upper and lower wings. In addition, the aircraft personal number, in this case numeral "18," appeared outbound of the starboard roundels on the upper and lower wings. The unit's insignia—the devil holding a bomb and gesturing with thumb on nose—appeared on both sides of the fuselage facing aft.

55. Breguet 14 B2 serial number 1333 of BR 117. This machine

is finished in the G-4 camouflage pattern. Undersurfaces were ecru. Roundels appeared on the wings in four positions. A red numeral "3" appeared on the fin and the unit's insignia—a red rooster holding a bomb in its claws—appeared on the fuselage. The insignia's background is shown white, but might have been light blue. The insignia also appeared on the port upper wing near the centerline. The numeral "3" appeared on the starboard upper wing near the centerline.

56. Blériot 11, serial unknown, finished in clear-doped linen overall with natural metal cowling, side fairing plates, and cabane struts. Wooden floats were varnished.

57. Hanriot HD.2, serial unknown, flown by Ensign George Clark Moseley of the USN fighter squadron at Dunkirk, early 1918. The machine was finished overall in a light matte gray with natural metal cowling, forward fuselage panels, and cabane struts. USN roundels appeared on upper and lower surfaces of wings. It is not known if the red heart appeared on both fuselage sides. It is believed the band of light blue wrapped around the fuselage.

58. F.B.A. Type C, serial unknown, flown by the French Aviation Maritime. The hull of this flying boat was varnished plywood and the wings were plain linen. The rudder was finished in national colors with the white section having a French naval insignia as shown. Roundels appeared on the wings in four positions. A white numeral "4" appeared on the hull. Armament was a single Lewis machine gun.

59. Nieuport 6H, serial unknown, flown by the French Aviation Maritime, circa autumn 1914, finished in clear-doped linen overall with natural metal cowling, side fairing plates, and cabane struts. Wooden floats were varnished. A black numeral "6" appeared on the rudder.

60. SPAD 14, serial unknown, flown by the French Aviation Maritime, circa autumn 1917. It was finished in clear-doped fabric overall with pale yellow metal and wood areas. Roundels appeared on the wings in four positions. The floats are shown as varnished wood.

61. F.B.A. Type H, serial number 3147, flown by naval pilots Guarnieri and Armaiolo G. Jannello, both with the Italian Naval Squadron 2A. The hull of this flying boat was varnished plywood. The plain linen wings had undersurfaces finished with the port side painted red and starboard side painted green, the center section remaining plain linen. Roundels appeared in two positions on the upper surfaces of the top wing, and in three positions on the hull—both sides and under the forward section. The unit's insignia—a light colored star—appeared on each side of the hull. The French flag located forward of the "F.B.A." is feathered at the aft end with an Italian flag in its center. The personal logo—a man with cap—most likely appeared on both sides of the hull. The cockpit of this aircraft is equipped with a Fiat Revelli machine gun. Reference photo is on page 262.

Paintings

1. U.S. Naval Aviators Of WWI Down At Sea After Engine Failure

May 6, 1918: Two U.S. naval aviation air crews left their base at Dunkerque, France, on anti-submarine patrol in Donnet-Denhaut flying boats, 814DD and 817DD. Twenty miles offshore the 200-hp Hispano-Suiza engine quit on 814DD and it crashed into the North Sea, killing forward observer QM-1c(A) Edward Smith. Enlisted pilot QM-1c(A) Herbert Lasher[1], trapped with a broken leg, was rescued by second observer QM-1c(A) Thomas Holliday before 814DD sank. "Duke"[2] set 817DD down and took survivors aboard but sea damage prevented takeoff. Homing pigeons were dispatched to Dunkerque requesting aid. Base commander Lt. Godfrey deC. Chevalier (N.A.#7) ordered a rescue mission by chief pilot Lt. Artemus Gates (N.A.#65) with MM-1c(A) Irving Sheely in 825DD, Ens. David Ingalls in an HD.2 scout, and MM-2c Alonzo Hildreth in a motor launch. Lt. Gates set 825DD down at the scene, took Lasher aboard, and returned him to Dunkerque.[3] Orbiting overhead, Ens. Ingalls guided Hildreth to disabled 817DD, where he took it in tow back to the base.[4]

Aug. 11, 1918: Ens. Edward de Cernea (N.A.#132) on patrol in 817DD, bombed and damaged a German submarine.[5]

Aug. 13, 1918: Ens. Julian Carson (N.A. #765) on patrol in 825DD bombed and sank the German submarine UB30.[5, 6]

(1) Ensign and N.A.#763, June 10, 1918.
(2) Lasher's diary, May 6, 1918.
(3) Sheely's letter home, May 11, 1918.
(4) Hildreth's diary, May 6, 1918.
(5) Hildreth's diary, Aug.11/13, 1918. Naval Aviation in WW1, Van Wyen, 1969.
(6) Simsadus London, J. Leighton, 1920.

© 1996 Lawrence D. Sheely

2. The Obendorf Raid

Two Breguet-Michelin 4s of BM 120 under attack by German fighters during the raid on the Mauser factory at Oberndorf. The Nieuport 17 is from N 124 which was assigned to accompany BM 120 as far as possible into enemy territory.

MF 29 (six F.40s), MF 123 (five F.40s), and BM 120 (seven Breguet-Michelin 4s and seven Breguet 5s) participated in the raid. In addition to N 124, support was supplied by other Nieuport escadrilles which staged raids on the German airfields at Habsheim, Colmar, and Fribourg. C 61 (Caudron G.4s) staged a diversionary raid on Lorrach.

The Oberndorf raid on 12 October 1916 showed the futility of attempting bombing missions during the day without adequate fighter escort. Twelve French aeroplanes were destroyed by the German defences. As a result of these appalling losses, the French GQG decided to abandon strategic bombing operations during the day and, instead, switched the day-bombing units to tactical targets near the front lines where adequate protection could be provided by French fighters.

© 1996 Jack Tyson

Photographs

1. Voisin 3 (Musée de l'Air)
2. Caudron G.3 and G.4 (Musée de l'Air)
3. Guynemer's SPAD 7 (Musée de l'Air)
4. Breguet 14 A2 (Musée de l'Air)
5. SPAD 13 (National Air & Space Museum)
6. Voisin 8 (National Air & Space Museum)
7. Voisin 8 (National Air & Space Museum)
8. Voisin 8 (National Air & Space Museum)

COLOR SECTION 581

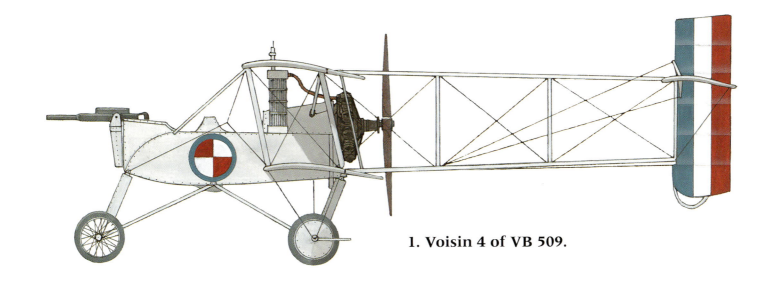

1. Voisin 4 of VB 509.

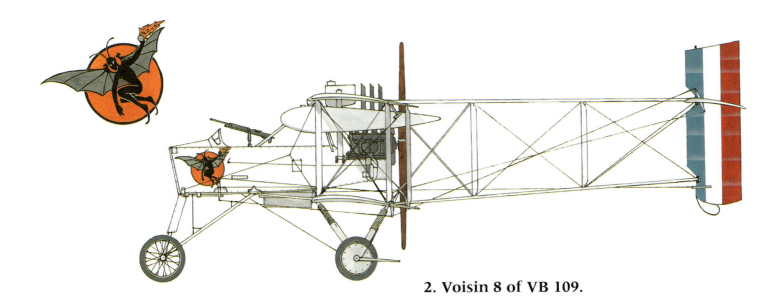

2. Voisin 8 of VB 109.

3. Voisin 10 of VB 125.

© Alan Durkota

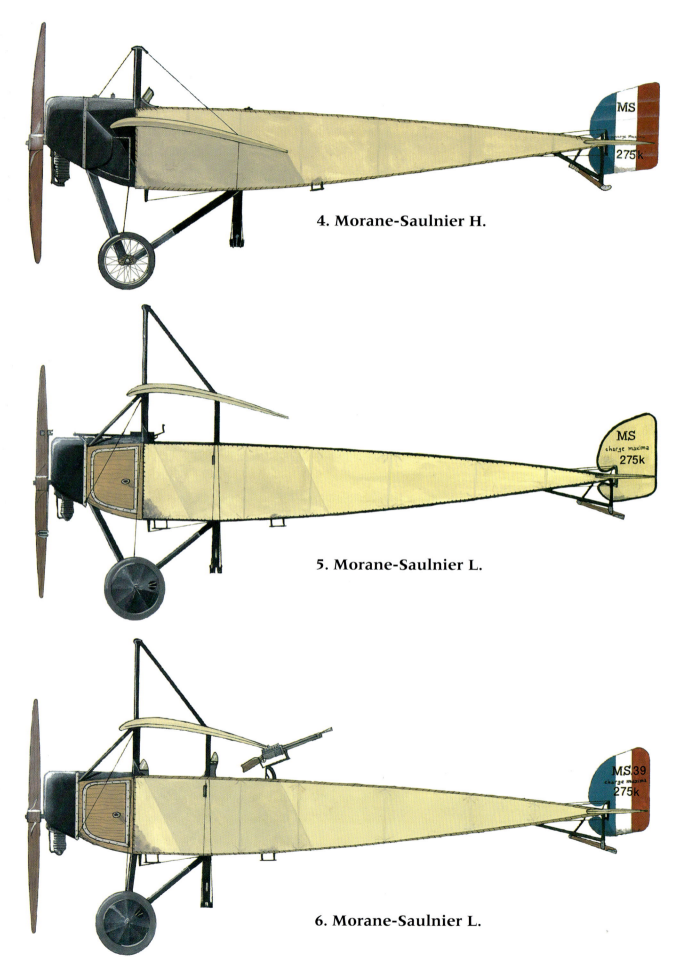

4. Morane-Saulnier H.

5. Morane-Saulnier L.

6. Morane-Saulnier L.

© Alan Durkota

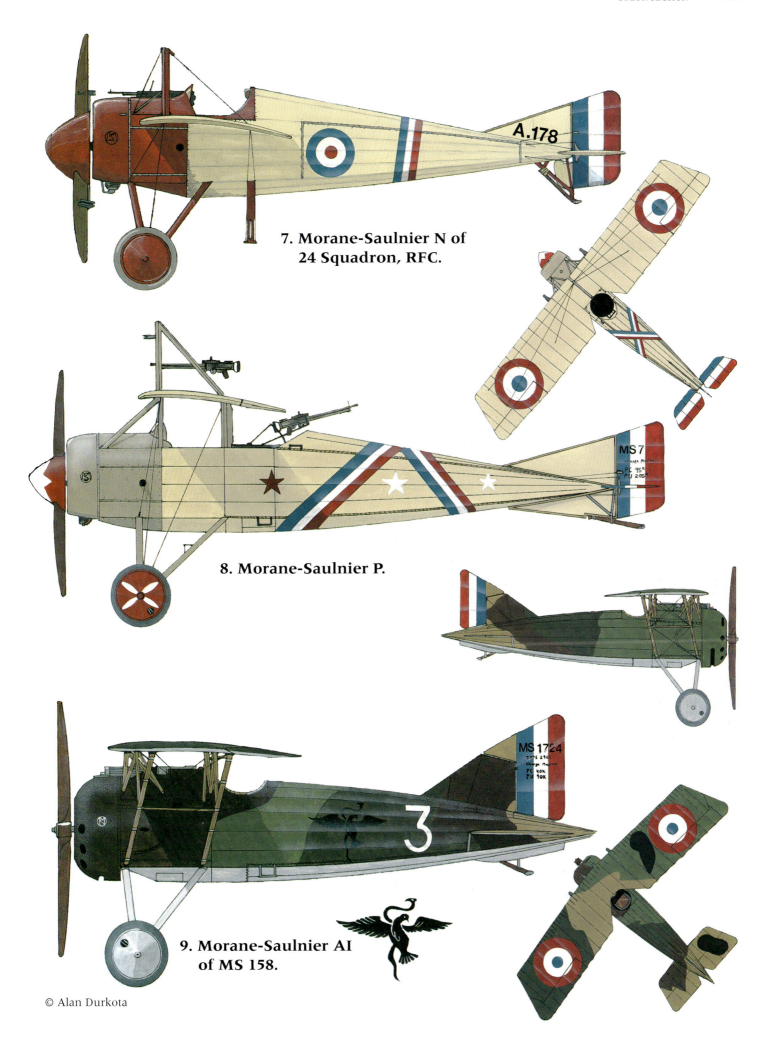

7. Morane-Saulnier N of 24 Squadron, RFC.

8. Morane-Saulnier P.

9. Morane-Saulnier AI of MS 158.

© Alan Durkota

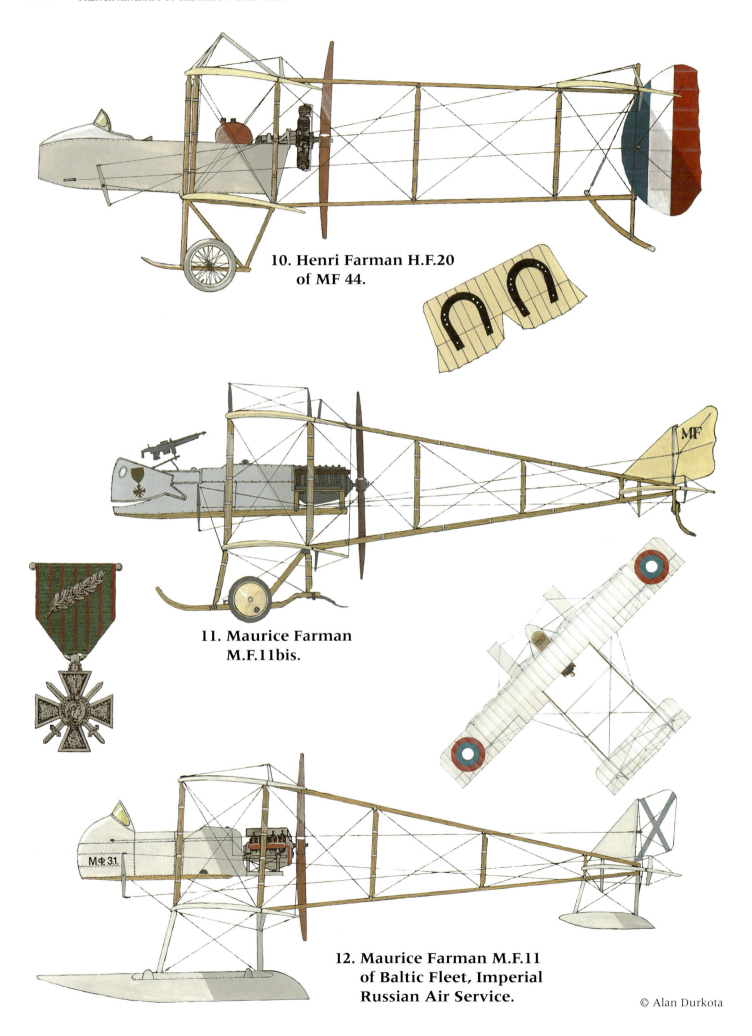

10. Henri Farman H.F.20 of MF 44.

11. Maurice Farman M.F.11bis.

12. Maurice Farman M.F.11 of Baltic Fleet, Imperial Russian Air Service.

© Alan Durkota

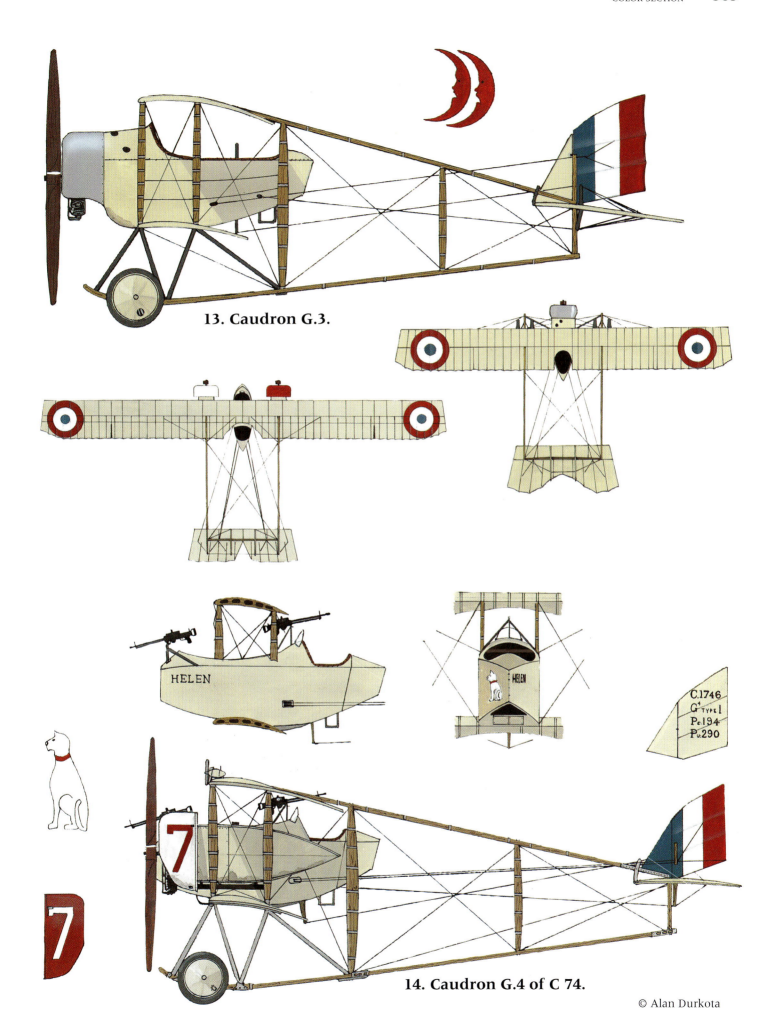
13. Caudron G.3.

14. Caudron G.4 of C 74.

© Alan Durkota

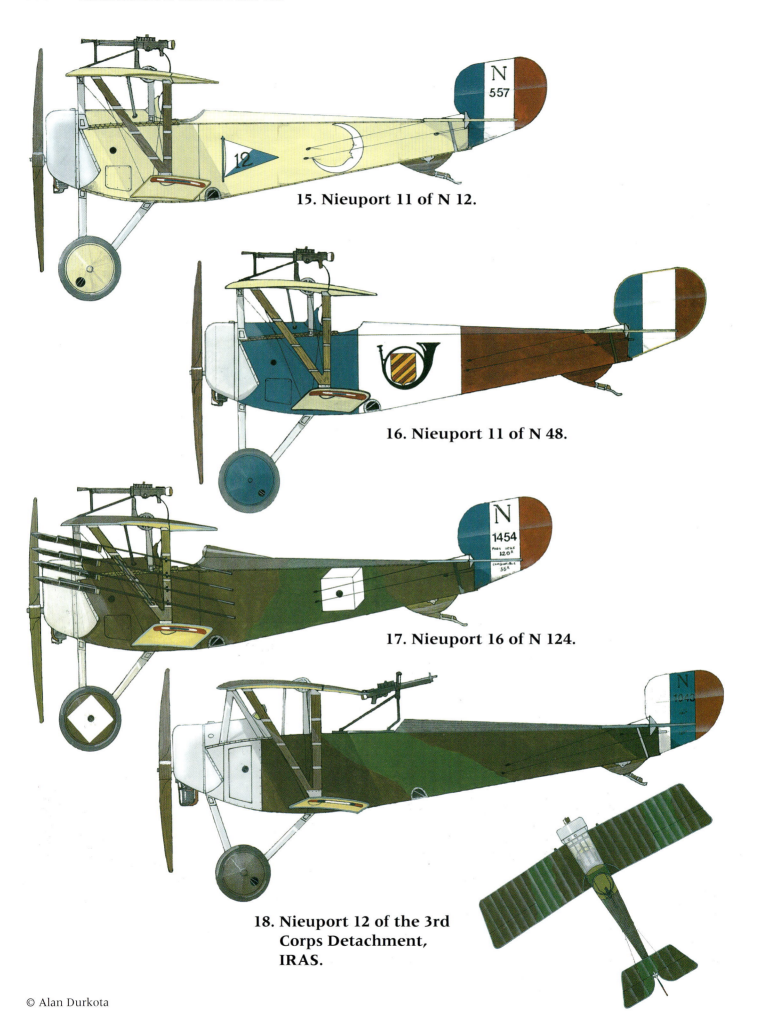

15. Nieuport 11 of N 12.

16. Nieuport 11 of N 48.

17. Nieuport 16 of N 124.

18. Nieuport 12 of the 3rd Corps Detachment, IRAS.

© Alan Durkota

21. Nieuport 24 of N 314.

22. Nieuport 27 of N 87.

© Alan Durkota

23. Nieuport 28 of the 94th Aero Squadron.

24. Nieuport 28 of the 27th Aero Squadron.

© Alan Durkota

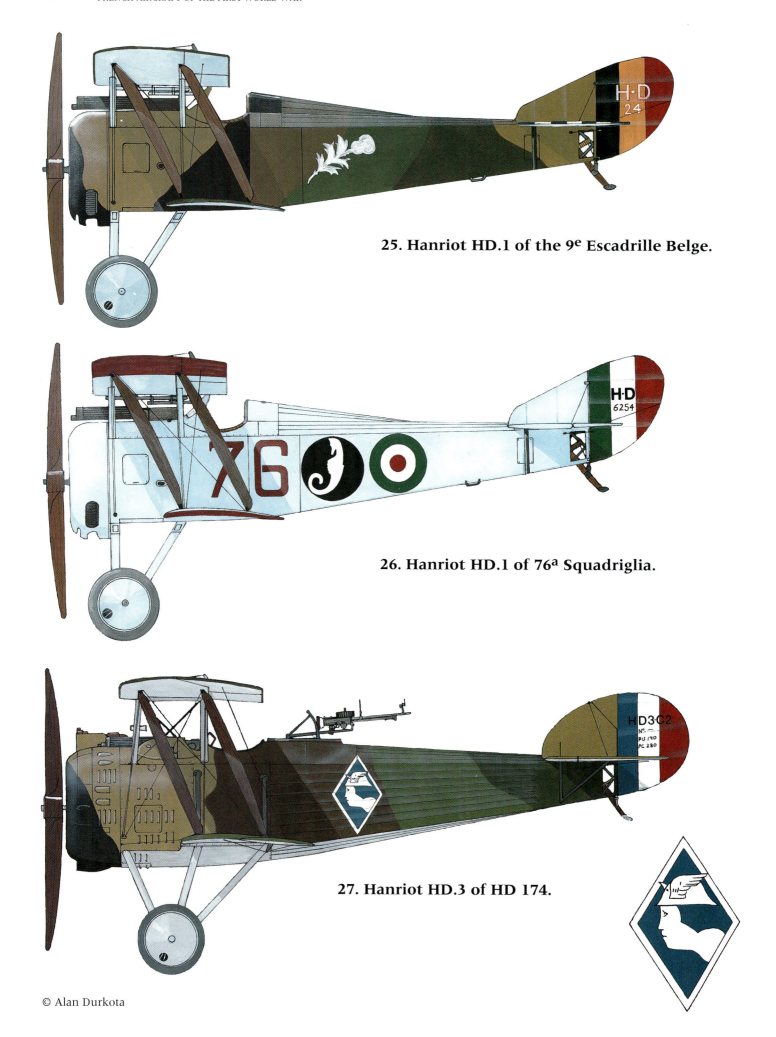

25. Hanriot HD.1 of the 9ᵉ Escadrille Belge.

26. Hanriot HD.1 of 76ª Squadriglia.

27. Hanriot HD.3 of HD 174.

© Alan Durkota

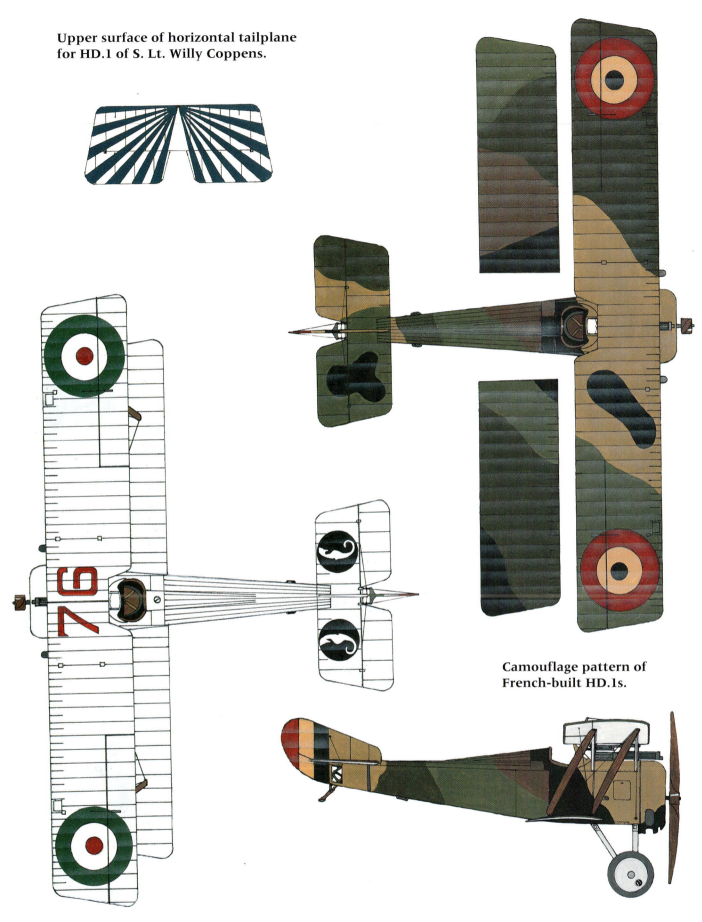

Upper surface of horizontal tailplane for HD.1 of S. Lt. Willy Coppens.

Camouflage pattern of French-built HD.1s.

Upper surfaces of HD.1 of 76ª Squadriglia.

© Alan Durkota

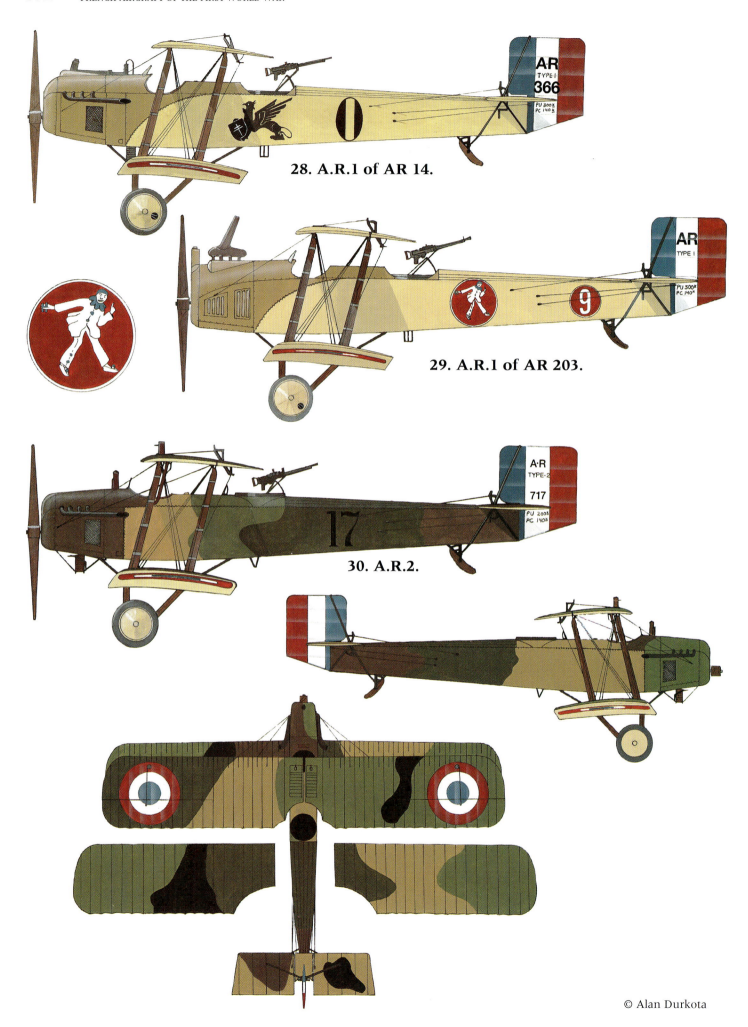

28. A.R.1 of AR 14.

29. A.R.1 of AR 203.

30. A.R.2.

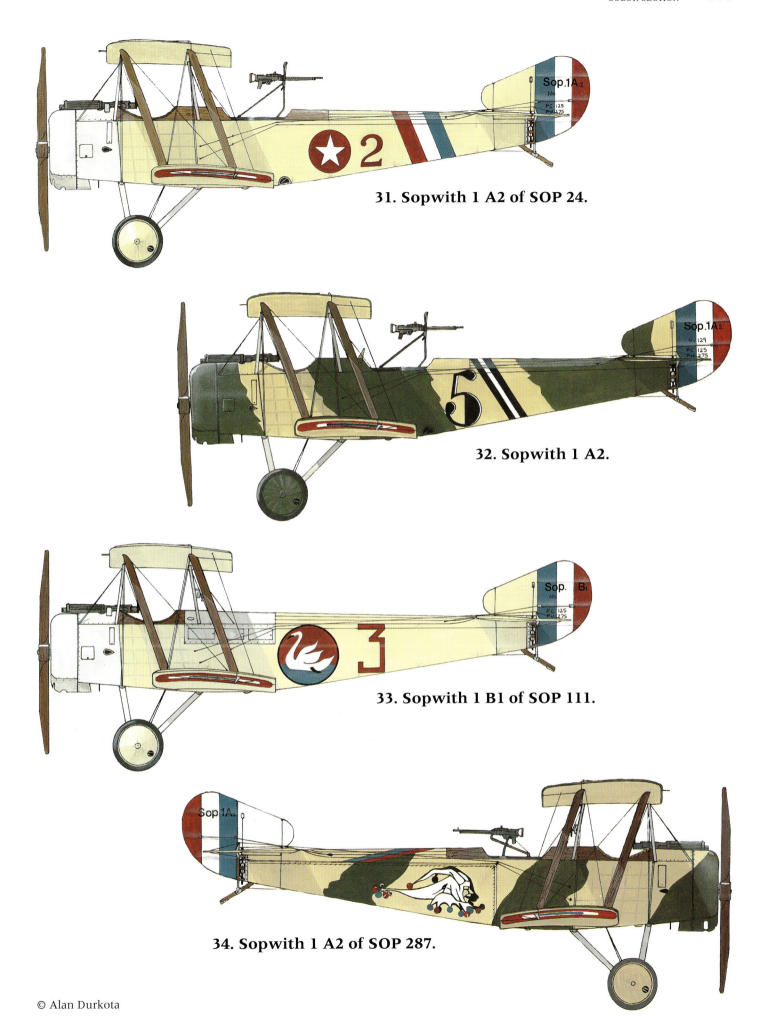

31. Sopwith 1 A2 of SOP 24.

32. Sopwith 1 A2.

33. Sopwith 1 B1 of SOP 111.

34. Sopwith 1 A2 of SOP 287.

© Alan Durkota

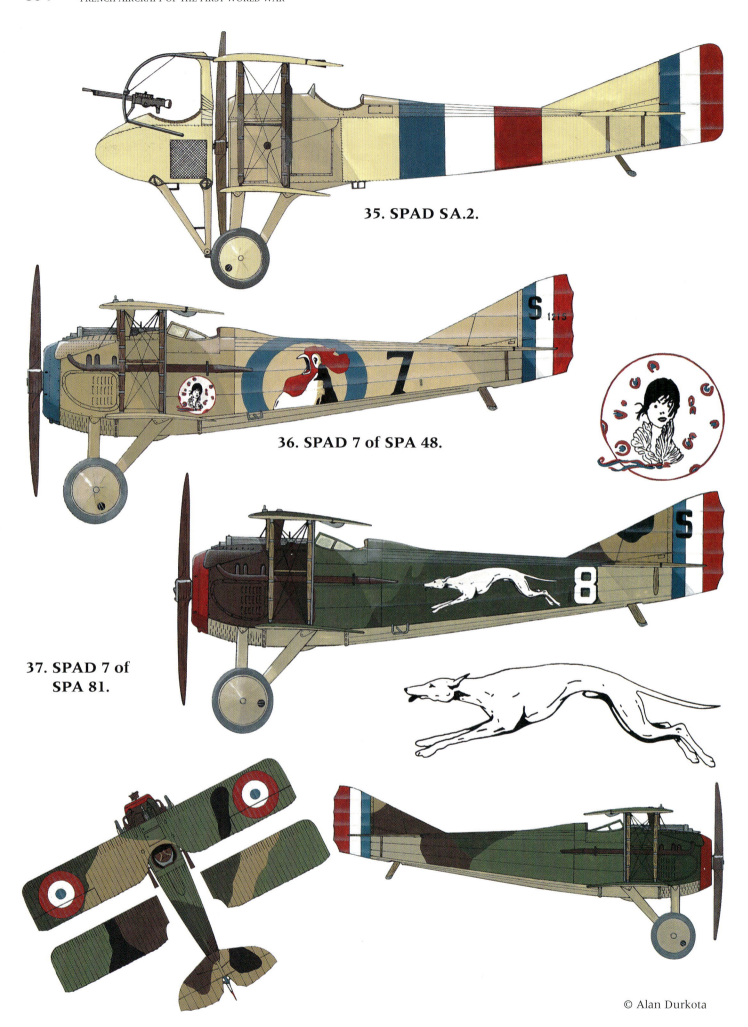

35. SPAD SA.2.

36. SPAD 7 of SPA 48.

37. SPAD 7 of SPA 81.

© Alan Durkota

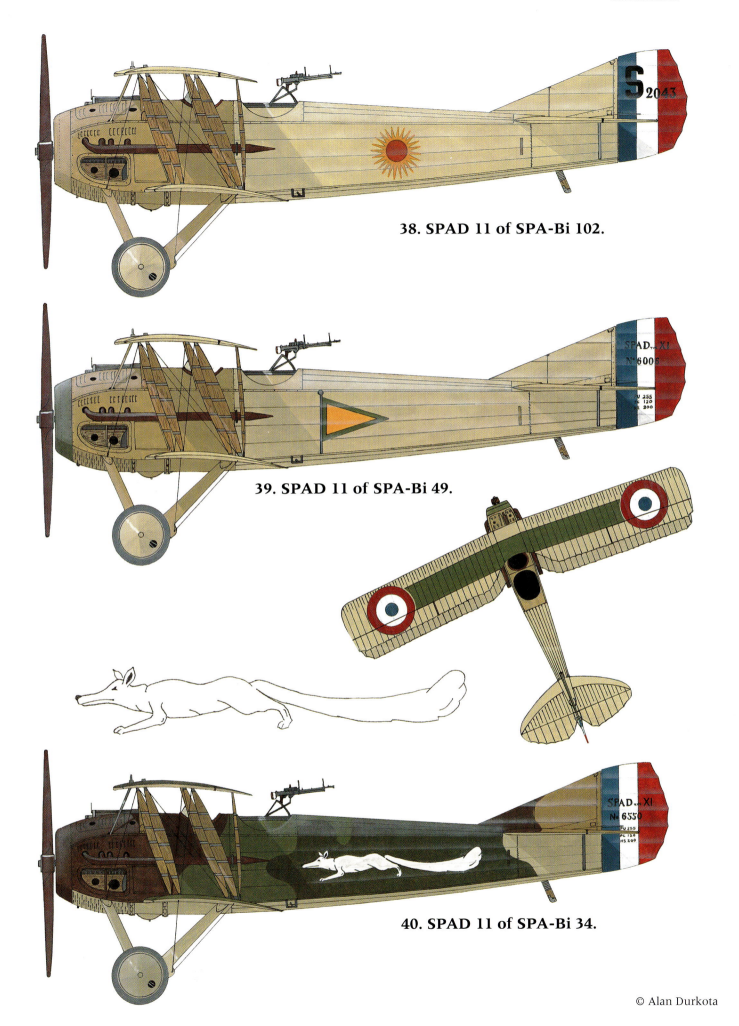

38. SPAD 11 of SPA-Bi 102.

39. SPAD 11 of SPA-Bi 49.

40. SPAD 11 of SPA-Bi 34.

© Alan Durkota

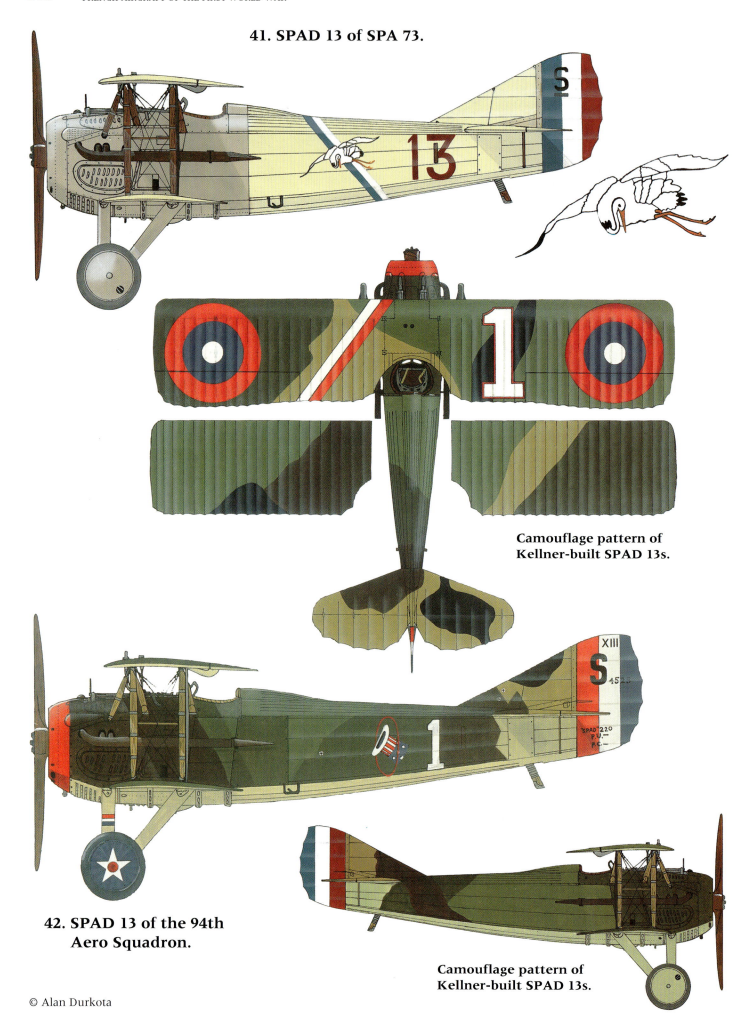

41. SPAD 13 of SPA 73.

Camouflage pattern of Kellner-built SPAD 13s.

42. SPAD 13 of the 94th Aero Squadron.

Camouflage pattern of Kellner-built SPAD 13s.

© Alan Durkota

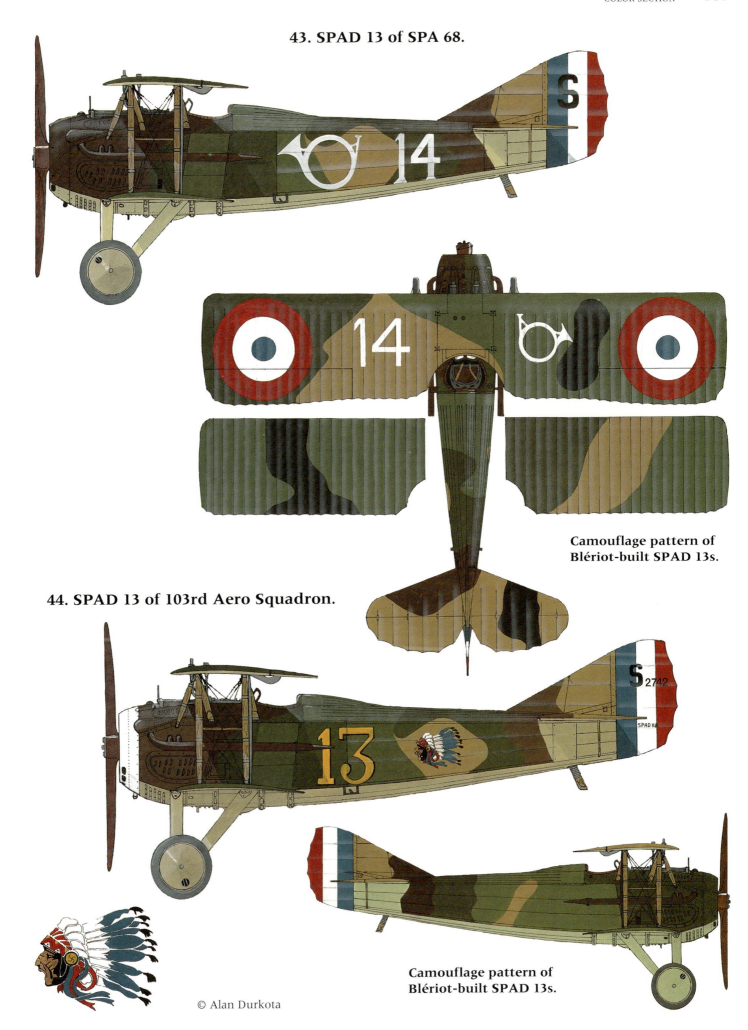

43. SPAD 13 of SPA 68.

Camouflage pattern of Blériot-built SPAD 13s.

44. SPAD 13 of 103rd Aero Squadron.

Camouflage pattern of Blériot-built SPAD 13s.

© Alan Durkota

45. SPAD 12 of SPA 103.

46. SPAD 16 of SPA-Bi 21.

© Alan Durkota

47. Salmson 2 A2 of SAL 58.

© Alan Durkota

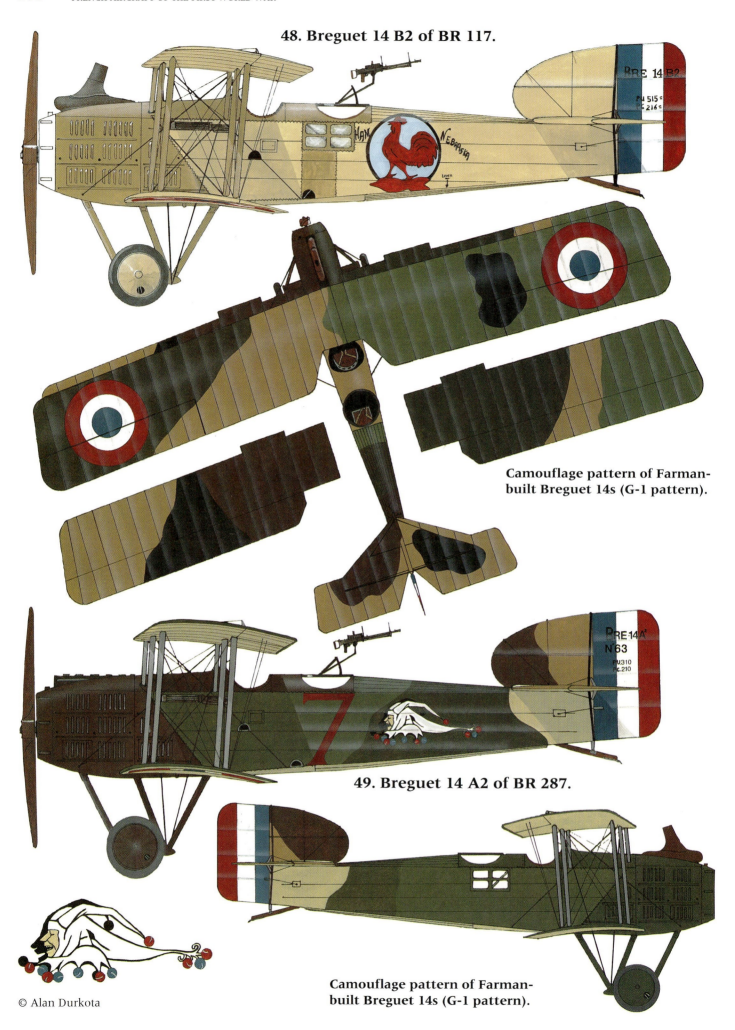

48. Breguet 14 B2 of BR 117.

Camouflage pattern of Farman-built Breguet 14s (G-1 pattern).

49. Breguet 14 A2 of BR 287.

Camouflage pattern of Farman-built Breguet 14s (G-1 pattern).

© Alan Durkota

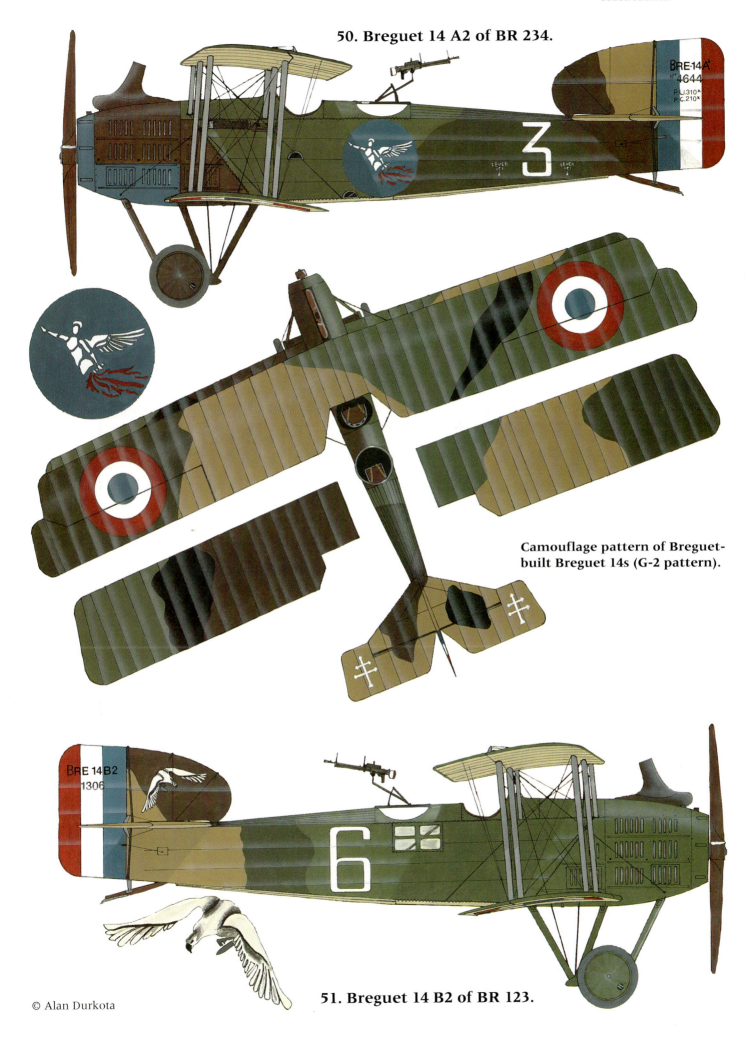

50. Breguet 14 A2 of BR 234.

Camouflage pattern of Breguet-built Breguet 14s (G-2 pattern).

51. Breguet 14 B2 of BR 123.

© Alan Durkota

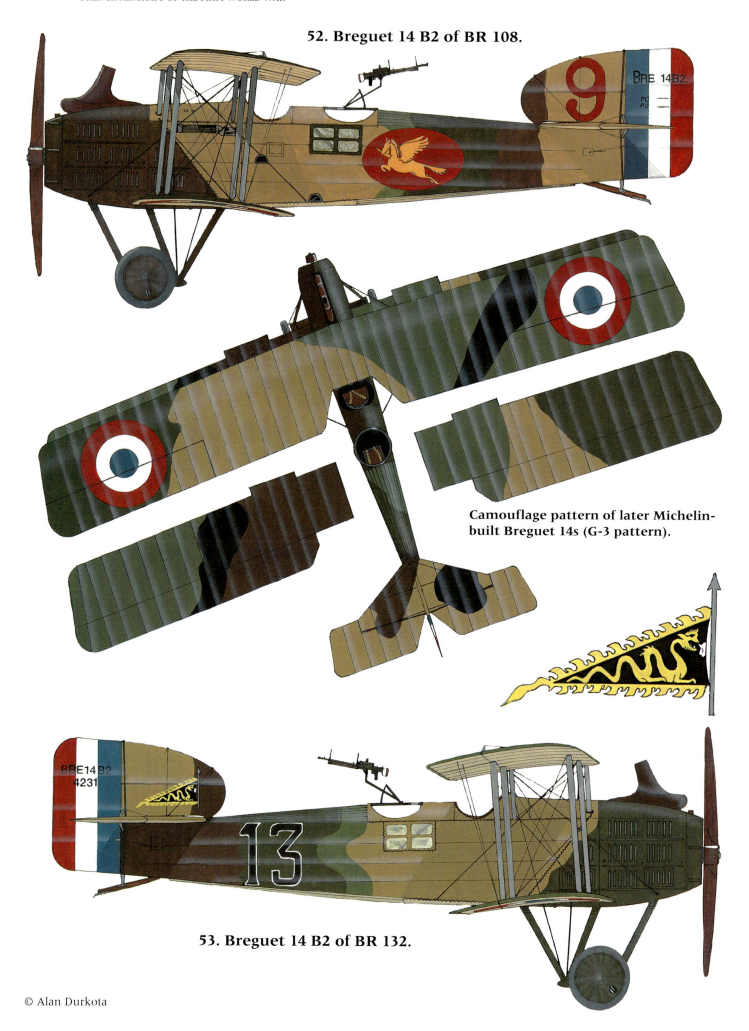

52. Breguet 14 B2 of BR 108.

Camouflage pattern of later Michelin-built Breguet 14s (G-3 pattern).

53. Breguet 14 B2 of BR 132.

© Alan Durkota

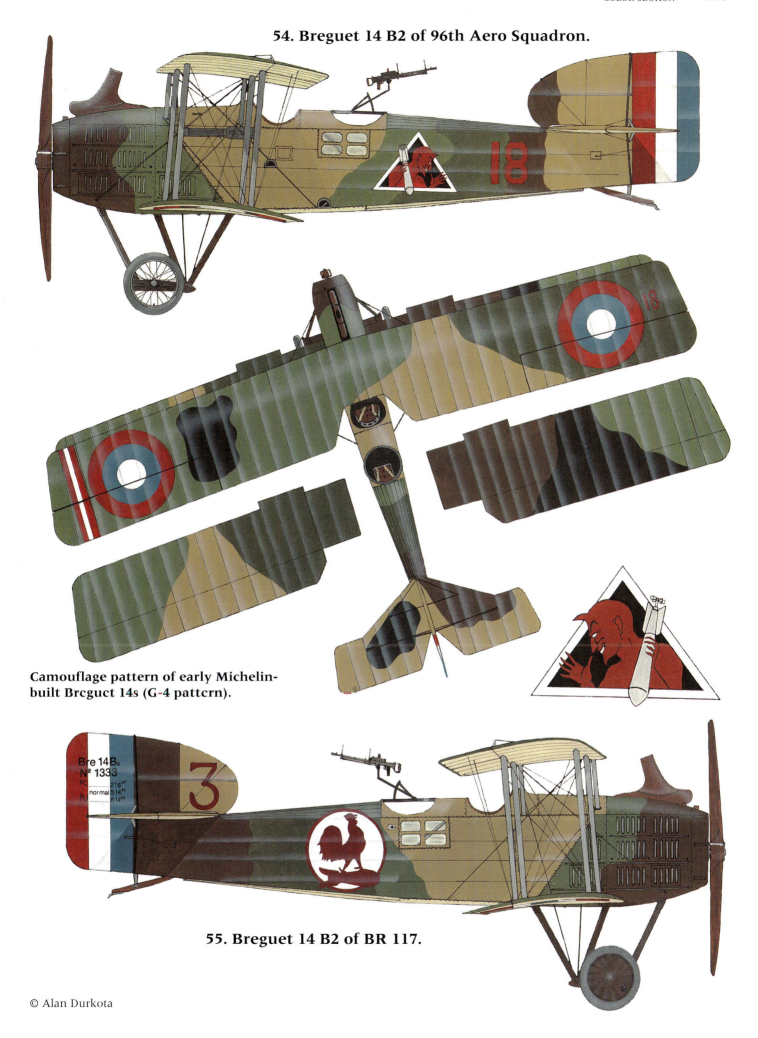

54. Breguet 14 B2 of 96th Aero Squadron.

Camouflage pattern of early Michelin-built Breguet 14s (G-4 pattern).

55. Breguet 14 B2 of BR 117.

© Alan Durkota

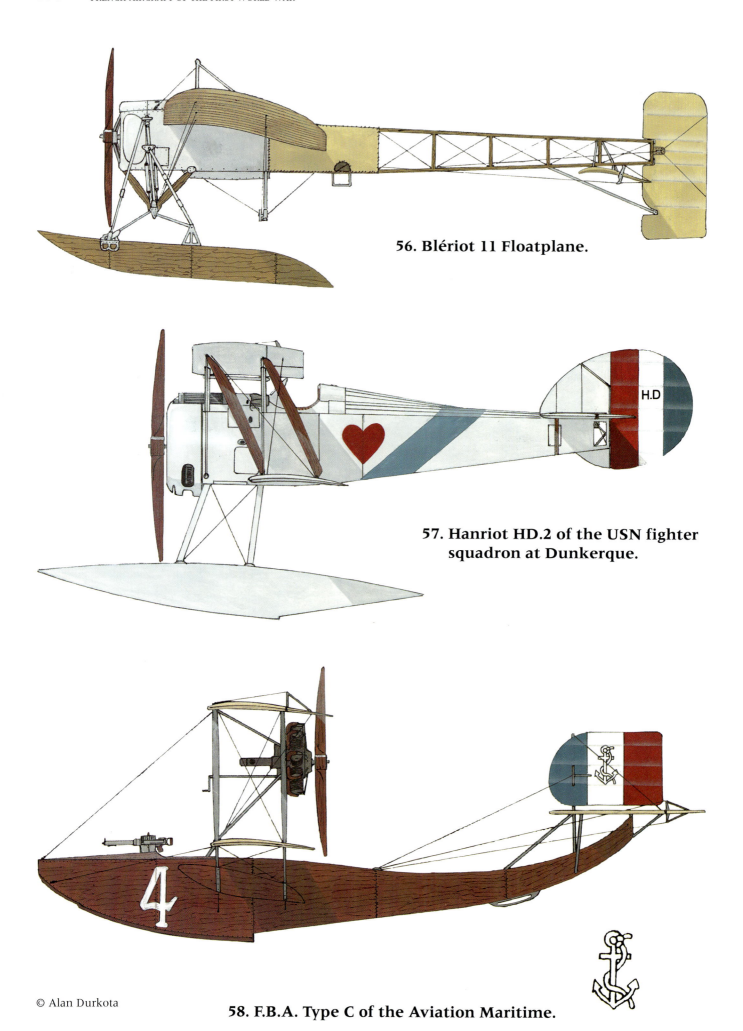

56. Blériot 11 Floatplane.

57. Hanriot HD.2 of the USN fighter squadron at Dunkerque.

58. F.B.A. Type C of the Aviation Maritime.

COLOR SECTION   **605**

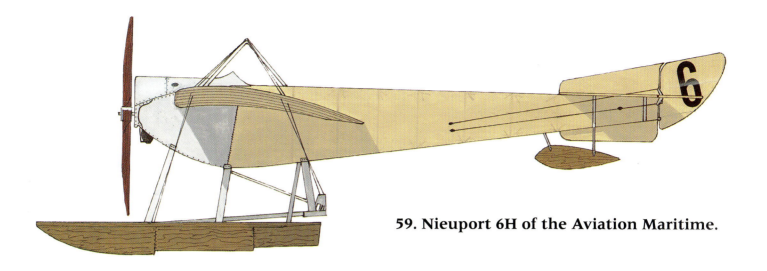

59. Nieuport 6H of the Aviation Maritime.

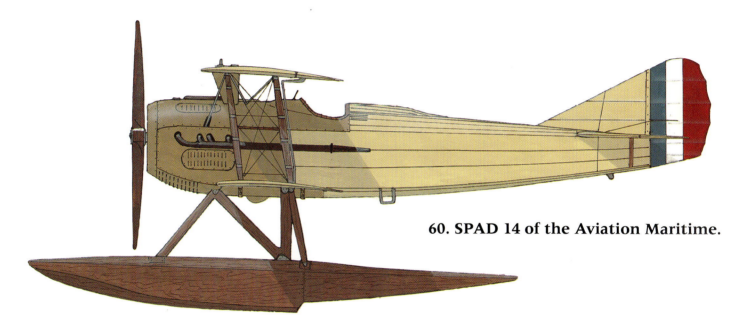

60. SPAD 14 of the Aviation Maritime.

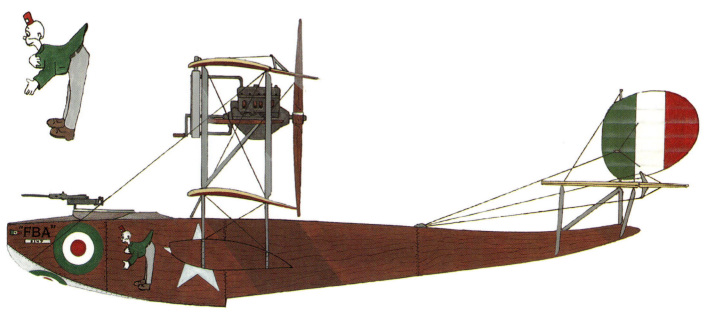

© Alan Durkota

61. F.B.A. Type H of the Italian Naval Squadron 2ª.

U.S. Naval Aviators Of WWI Down At Sea After Engine Failure

Bombing Raid on Obendorf, 1916

Voisin 3

Caudron G.4

Caudron G.3

Guynemer's SPAD 7

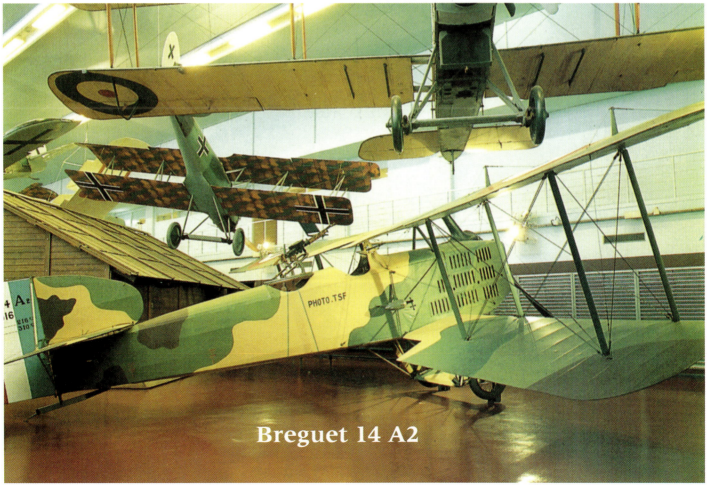
Breguet 14 A2

SPAD 13

Voisin 8

Voisin 8

Voisin 8

Addendum

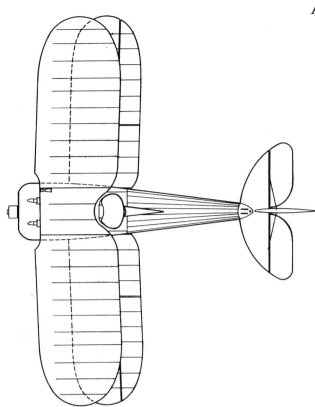

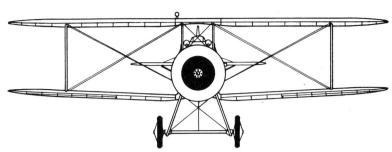

Nieuport 200-hp Clerget Experimental Fighter

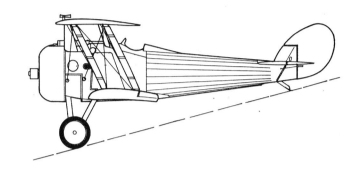

Breguet 14
The Italians evaluated a Fiat-engined Breguet 14 A2 in October 1918. As a result of these tests, 100 aircraft were ordered. Only a small number of Breguet 14s (designated BRE by the Italians) saw operational service during the 1920s. A report that FIAT manufactured some Breguet 14s in 1921 cannot be verified.

Farman H.F.30
Two used by the People's Revolutionary Army of the Far Eastern Republic in 1920.

Nieuport 17 C1
One used by Don Republic in 1918.

Salmson 2
Four used by the People's Revolutionary Army of the Far Eastern Republic in 1920.

Sopwith 1½ Strutter
Three used by the People's Revolutionary Army of the Far Eastern Republic in 1920.

SEA 2
The SEA 2 was to have been a single engine biplane with a crew of two. Intended to meet the A2 or C2 categories, the lack of a suitable engine resulted in the SEA's decision not to build a prototype.

SEA 3
This was to have been a three-engine aircraft intended to meet the same A3 specification as the Letord 4 and 5 and Caudron R.11 and R.12. As with the SEA 2, no prototype was constructed due to the lack of a suitable engine.

Paul Schmitt
Additional Paul Schmitt projects were:
1. Paul Schmitt 260-hp Salmson – Top wing level with engine cowling. In order to enhance pilotís vision, the upper wing was below the pilotís eve level, while the lower wing had an exceptionally narrow chord (0.6 meters)
2. Paul Schmitt armored aircraft – 250-hp Clerget engine with an all-steel fuselage and armored cockpit.
3. Paul Schmitt 9 – 230-hp Salmson engine; possibly the aircraft referred to above.
4. Paul Schmitt 13 – two-seat bomber with two 260-hp Fiat A12 engines.
5. Paul Schmitt 14 Bn2 – designation for heavy bomber with four Hispano-Suiza 8Be engines. See page 458.

Voisin
Six Voisins of unknown type used by the Don Republic in 1918.

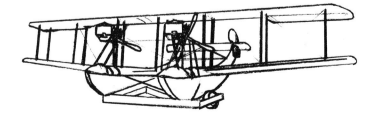

Above: LAF *Desmon* artist's impression.

Left: SEA 1 artist's impression.

Farman M.F.11 of MF 99 based in Albania. B87.2557.

Right: Hanriot HD.1 serial 6254 flown by Italian ace Giorgio Michetti while attached to Squadriglia 76ª. This aircraft is the subject of a color profile. Via Alan Durkota.

Below right: Closeup of Italian Nieuport 11 serial 1615. Via Alan Durkota.

Below: Hanriot HD.2. Via Alan Durkota.

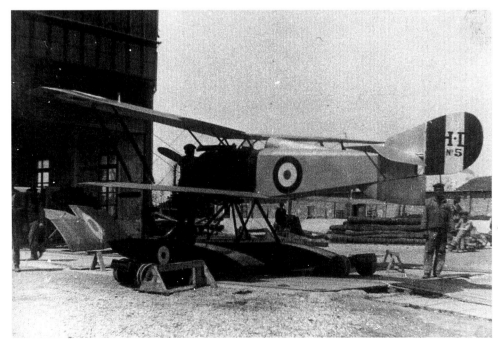

Thulin D (Swedish license-built Morane-Saulnier L). Via Lennart Andersson.

License-built copy of Nieuport 4G designated M-II. It was built by Svenska Aeroplans-konsortiets in 1913. Via Lennart Andersson.

Nieuport 10. Via Colin Owers.

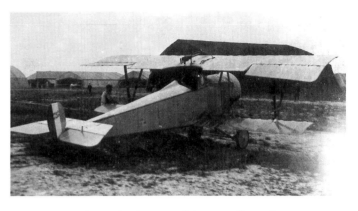

Nieuport 10 *Vieux Charles III* serial N 328 flown by Guynemer. B.86.1155.

SW 10 (Swedish license-built Farman H.F.22) serial number 8. See page 213 for details. Via Lennart Andersson.

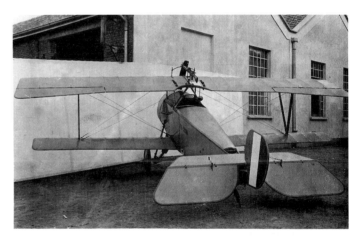

Italian-built Nieuport 11. Via Alan Durkota.

SVA number 6764 with Italian Nieuport 27s serial numbers 5345 and 5313 in the background. Via Alan Durkota.

Glossary

A.E.F. – American Expeditionary Forces
A.F.O. – Army Forces Oriental – Army Units in Eastern and Southern Europe
C.A. – Corps De Armée – Army Corps
C.R.P. – Camp Retranche Paris
D.A.L. – Division Armée Lorraine – Army Division At Lorraine
D.A.V. – Division Armée Verdun – Army Division At Verdun
D.C. – Divisione Cavalrie – Cavalry Division
D.C.A – Defense Contre Aviation – Air Defense
G.A.C. – Groupe Armée De Centre – Army Group Center
G.A.E. – Groupe Armée De East – Army Group East
G.A.N. – Groupe Armée De Nord – Army Group North
G.A.R. – Groupe Armée De Reserve – Army Group Reserve
GB – Groupe De Bombardement – Bomber Group
GC – Groupe De Chasse – Fighter Group
GQG – Grand Quartier Géneraél – Supreme Headquarters
IRAS – Imperial Russian Air Service
P.C. – Poste De Combat – Combat Post
P.R. – Poste De Relache – Rest Post
R.G.A. – Replacement Group Aerienne – Replacement Air Group
RFC – Royal Flying Corps
RNAS – Royal Naval Air Service
S.A.L. – Section Artiller Lourde – Heavy Artillery Section
SFA – Service Fabrication Aeronautique – Aeronautics Production Service
SI – Section Industriel – Industrial Section
STAé – Section Technical Aeronautique – Aeronautics Technical Section
T.S.F. – Transmission Sans Fibre – Wireless Transmission
USAS – United States Air Service

Escadrille Type Designations

AR – A.R.1 and A.R.2
BL – Blériot 11
BLC – Blériot Cavalerie
BM – Breguet Michelin
BR – Breguet
C – Caudron
CAP – Caproni
CEP – Caproni License-Built by R.E.P.
CM – Caudron Monoplace (G.2)
D – Deperdussin
DM – Deperdussin Monocoque
DO – Dorand DO.1
F – Farman (F.40 or F.50)
G – Caudron (G.6)
HD – Hanriot Dupont (H.D.3)
HF – Henri Farman
LET – Letord
MF – Maurice Farman
MS – Morane-Saulnier
MSP – Morane-Saulnier Parasol (Type AI)
N – Nieuport
PS – Paul Schmitt Type 6 or Type 7
R – Caudron R.4 or R.11
REP – Robert Esnault Pelterie – REP Type N
SAL – Salmson 2
SM – Salmson-Moineau S.M.1
SOP – Sopwith 1½ Strutter
SPA – SPAD 7 or 13
SPA-Bi – SPAD 11 or 16
V – Voisin
VC – Voisin Canon (Type 4)
VP – Voisin Peugeot (Type 8)
VR – Voisin Renault (Type 10)

Select Bibliography

Primary Sources

SHAA: A 21, A 22, A 23, A24, A 25; A 30, A 31, A 32, A 33, A 91, A 92, A 93, A 103, A 109, A 113, A 114, A 118, A 119, A 120, A 121, A 122, A 123, A 124, A 125, A 164, A 165, A 166, A 167, A 175, A 187, A 188, A 192, and A 193

SHAT: 18 N 171–174, 18 N 278–279; 18 N 368–369; 18 N 370–372; 18 N 412, 18 N 578, 18 N 580, 1 N 213–227, 19 N 225–226, 19 N 479, 19 N 483, 19 N 488, 19 N 490–492, 19 N 606, 19 N 608, 19 N 610–623, 19 N 774, 19 N 793–794; 19 N 944–947, 19 N 1117–1127, 19 N 1379, 19 N 1524, 19 N 1525, 1 N 1527, 19 N 1550–1501, 19 N 1701–1705, 20 N 32, 20 N 101–102, 20 N 163–165, 20 N 382, 20 N 383–385, 20 N 392, 20 N 398, 20 N 464, 20 N 577, 20 N 579, 20 N 644, 26 N 42, and 26 N 43

Secondary Sources

L'Aeronautique Militaire Francaise, Volumes 1 and 2; Icare numbers 85 and 88.
Ali Italiane 1908–22. Rizzoli Editore, 1978.
Andersson, L.; *Svenska Militara Flygplan 1911–1939*. 1986.
Andersson, L.; *Soviet Aircraft and Aviation 1917–41* Putnam Aeronautical Books, 1994.
Andersson, L. *Little Known Air Forces of the 1920s*. The Small Air Forces Observer; various issues.
Andrade, J. *Latin American Military Aviation*. Midlands Counties Publications; Leicester, 1982.
Apostolo, G. and Abate, R. *Caproni Nella Prima Guerra Mondiale.*

Europress; 1970.
Bleriot, L. *Bleriot XX Siecle;* 1994.
Blume, August G. *History of the Serbian Air Force*. Cross and Cockade Society; Volume 8, numbers 2, 3, and 4.
Bonte, Louis. *L'Historie des Essais en Vol. Docavia No.3 Editions;* Lariviere; Paris; 1974.
Borget, Michel. *Alphonse Tellier, Constructeur d'Hydro-Aeroplanes.* Le Fana de l'Aviation May, June, July, and August 1996 (#318–321).
Bruce, J.M. *Aeroplanes of the Royal Flying Corps (Military Wing)*. Putnam and Co., Ltd.; 1982.
Bruce J.M. *Breguet 14. Aircraft in Profile No. 161;* Profile Publications; London; 1967.
Bruce J.M. *Fighters. Volumes 4 and 5;* MacDonald and Co.; London; 1972.
Brunoff, M. *L'Aeronautique Militaire, Maritime, Colonial, and Marchande*. Paris; 1930.
Brunoff, M. *L'Aeronautique Pendant La Guerra Mondiale*. Paris; 1919.
Camurati, G. *Aerei Italiani 1914–1918*. Rivista Aeronautica.
Chadeau, E. *L'Industrie Aeronautique en France 1900–1950*. Librairie Artheme Fayard; 1987.
Christienne, C. and Lissarrague, P. *History of French Military Aviation*. Smithsonian Institute Press; Washington, DC; 1986.
Cuich, Myrone N. *De L'Aeronautique Militaire 1912 a la Armée de l'Air 1976.* 16 Rue de Rome, 59200 Tourcoiny; 1976.
Cuich, Myrone N. *Les Insignes de L'Aeronautique Militaire 1912–86.* 2 Volumes; 16 Rue de Rome, 59200 Tourcoiny; 1986.
Cuny, J. and Daniel, R. *L'Aviation de Chasse Francaise, 1918–40.* Docavia No.2; Editions Lariviere; Paris.

Cuny, J. and Daniel, R. *L'Aviation de Bombardement et de Renseignement 1918–40*. Docavia No.12; Editions Lariviere; Paris.

Cuny, J. *Latecoere:les Avions et Hydravions*. Docavia Editions Lariviere,n.d.

Cynk, J. *History of the Polish Air Force*. Osprey Publications, Ltd.; Berkshire; 1972.

Delear, F. *Igor Sikorsky: his Three Careers in Aviation* . Dodd, Mead, and Company; New York; 1969.

DOC-AIR-ESPACE Special Issue; May, 1968. (Review of all French aircraft 1909–68.)

Encyclopedia of Aircraft. Issues 1 through 216; Orbis Publishing Ltd.; London; 1985.

L'Encyclopdie Illustree de l'Aviation, Volume 15. Editions Atlas Paris, 1986.

English, A. *The Armed Forces of Latin America*. Janes Publishing Co.; London, 1984.

Eteve, A. *La Victories Des Cocades*. Robert Laffont; Paris; 1970.

Facon, P. *L'Aviation Francaise au "Chemin des Dames"*. Aviation Magazine No. 805, 1981.

Finken, L. and Holm, E. *Complete Civil Aircraft Register of Denmark*. Air Britain Publicatons; Kent; 1987.

Frank, N.and Bailey, F. *Over the Front*. Grub Street Publications, 1992.

Gomez, 1. *Breve Historia de ta Aviacion en Mexico*. Mexico; 1971.

Green, W. and Swanborough. *"Oil Well Top Cover."* Air Enthusiast. Volume 1, No. l; Fine Scroll Limited; London; 1971.

Green, W. and Swanborough. *"Wings of Hellas"*. Flying review International Hamilton; London; 1968.

Green, W.and Swanborough, G. *The Complete Book of Fighters*. Salamander Books Ltd., 1994.

Hagedorn, D. *Central American and Caribbean Air Forces*. Air Britain 1993.

van Haute, A. *Pictorial History of the French Air Force. Volume 1*. Ian Allan; London; 1974.

Historia de Forca Aerea Brasileira. Second edition. 1975.

Hudson, J. *Hostile Skies*. Syracuse University Press; New York; 1968.

Huisman, G. *Dans les Coulisses de l'Aviation 1914–1918*. La Renaissance du Livre, Paris, n.d.

Istoria Aviatiei Romane, Editura Stiintifica Si Enciclopedica, Bucharest, 1984.

Jackson, P. *Belgian Military Aviation 1945–77*. Midland Counties Publications; Leicestershire; 1977.

Janes All the World's Aircraft. 1909, 1910/11, 1912, 1913, 1914, 1916, 1917, 1918, 1919, 1920, 1922, 1923, and 1924.

Kandilakis, G.C., et al; *Hellenic Aeroplanes from 1912 until Today*. IPMS Greece, 1992.

Kreskinen, K.; Stenman, K.; and Niska, K. *Suomen llamauoimien Lentokoneet 1918–38*. Tietoteos Publishers; 1976.

Kilduff, P. *"History of Groupes de Bombardements 4 and 9."* Cross and Cockade. Volume 15, No. 1. Spring; 1974.

King, H.F. *Sopwith Aircraft 1912–20*. Putnam and Co., Ltd.; 1980.

Kohr, K.; Komon, L.; and Naito, I. *Airview's Fifty Years of Japanese Aviation 1910–60*. Kantosha Co., Ltd.; Tokyo; 1961.

Lamberton, et al. *Reconnaissance and Bomber Aircraft of the 1914–18 War*. Harleyford Publishers; Herts; 1962.

L'Industrie Aeronautique et Spatiale Francaise.Volume 1. Edite Par Le Groupment Des Industries Francaise Aeronautiques et Spatiales; 1984.

Liron, J. *Les Avions Bernard*. Docavia Editions Lariviere, n.d.

Liron, J. *Avions Farman*. Docavia No. 21. Editions Lariviere; Paris.

Liron, J. *Avions Latécoère*. Docavia No. 34. Editions Lariviere; Paris.

Liron, J. *Bechereau, Besson, Bleriot, Breguet, Donnet-Denhaut, F.B.A., Gourdou-Lesseurre, Hanriot, Liore et Olivier, Salmson, Tellier, Weymann, and Wibault*. Series of multipart articles on French manufacturers which appeared in Aviation Magazine.

Lomholdt, N. and Warren, W. *Aviation in Thailand*. Everhest Printing Co.; Hong Kong; 1987.

Mangin, J.; Charnpagne, J.; and Van Den Rul, M. *Sous Nos Ailes*. G. Everling; Arlon.

Martel, R. *L'Aviation Franciase en Bombardement*. Paul Hartmann Editeur; 1939.

Maurer, M. *U.S. Air Service World War I. 4 Volumes*. Office of Air Force History; U.S. Governmnent Printing Office; Washington, DC; 1971.

McKay, D. *The Morane-Saulnier Story*. Air Pictorial.

Mikesh, R. and Abe, S. *Japanese Aircraft 1910–41*, Putnam Aeronautical Books, 1990.

Moreau-Berillon, E. *Escadrilles de l'Aviation Francaise 1914–40*. 43 rue Boissy d'Anglas; Paris; 1968.

Munson, K. *Aircraft of World War I*. Doubleday; New York; 1968.

Netto, F. *Aviacao Militar Brasileira 1916–84*. Revista Aeronautica. 1984.

Nicolle, David. *L'Aviation "Française" Du Sultan: Les Unités Aériennes Ottomanes 1912–1915*. Avions Number 17.

Nilsson, T. *"Groupe de Bombardement 1: A History."* Over the Front. Volume 1 No.1.

Nilsson, T. *"History of the Original Groupe de Bombardement 2."* Over the Front. Volume 4 No.3.

Nowara, H. *Die Entwicklung der Flugzeuge 1914–18*. J.F. Lehmanns Verlag; Munich; 1959.

Nozawa, T. *Encyclopedia of Japanese Aircraft, Volume 6: Imported Aircraft*. Shuppan-Kyodo Publishers; Tokyo.

Pearce, M. *Aeronautique Militaire*. Cross & Cockade Great Britain Vol 22, Nos.3 and 4.

Pernet, A. *"Gabriel Voisin"*. A series of articles in Aviation Magazine.

Poirier, J. *Les Bombardements de Paris 1914–18*. Payot; Paris, 1930.

Porro, A. *La Guerra Nell'Aria*. 5th Edition. Corbaccio, Milan; 1940.

Robertson, Bruce. *Sopwith: The Man and His Aircraft*. Harleyford Publishers; Herts; 1970.

Schoenmaker, W. and Postma T. *KLU Vliegtuigen*. Vitgeverij De Alk B.U.; 1987.

Schroder,H. *The Royal Danish Air Force*. Tojhusmuseet. 1993.

Sekigawa, E. *Pictorial History of Japanese Military Aviation*. Ian Allan,1974.

Shavrov, B.V. *History of Aircraft Design in the U.S.S.R. Volume 1*. Translated by W. Schoemaker and appearing in the Bulletin of Russian Aviation, the journal of the Russian Research Group of Air Britain.

Shores, C. *Finnish Air Force 1918–60*. Arco-Aircam No. 14. Arco Publishing Co.; New York; 1969.

Stroud, J. *European Transport Aircraft Since 1910*. Putnams and Co., Ltd.; 1966.

Sturtivant, Ray and Page, G. *Royal Navy Aircraft Serials and Units 1911–1919*. Air Britain: 1992.

Tarnstrom, R. *Handbook of Armed Forces: France Volume 1* and *The Balkans Parts 1 and 2*. Privately published; 1981 and 1982.

Tarnstrom, R. *The Wars of Japan*. Privately published; 1992.

Thayer, L. *America's First Eagles*. R. James Bender Publisher; San Jose, CA.; 1983.

Thetford, O. *Aircraft of the Royal Air Force Since 1918*. 7th Edition. Putnams and Co., Ltd.; 1979.

Thetford, O. *British Naval Aircraft Since 1912*. 3rd Edition. Putnams and Co., Ltd.; 1971.

Titz, Z. *Czechoslovakian Air Force*. Arco-Aircam No. 30. Arco Publishing Co.; New York; 1971.

Vingt-Cinq Ans d'Aeronautique Francaise. Volumes 1 and 2. Edite par la Chambre Syndicale des Industries Aeronatiques; 1934.

Warleta, J.; Larrazabal, J.; and San Emeterio, C. *Aviones Militares Espanoles 1911–1986*. Instituto de Hisotria y Cultura Aerea; Madrid; 1986.

Aircraft Index

A
A.R.1: 6, 7, 9, 10, 17, 18, 34, 35, 37–46, 101, 106, 116, 151, 155, 159, 224–226, 234–237, 241, 245, 247, 249, 291, 439, 440, 441, 446, 497
A.R.2: 10, 37–46, 246, 439
Astoux-Vedrines: 46
Astra Bomber: 47
Astra-Paulhan Flying Boat: 48
Audenis E.P.2, C2: 48, 49
B
B.A.J. C2: 49
Bassan-Gué: 50
Bernard A.B. 1: 51, 171
Bernard A.B. 2: 51
Bernard A.B. 3: 51
Bernard A.B. 4: 51
Bernard S.A.B. C1: 53
Besson H-1: 299
Besson 150-hp Flying Boat: 300
Bille S.A.C.A.N.A.: 53
Blériot 11: 54–61, 63, 186, 208, 309, 316, 360, 431, 539
Blériot 36: 62, 63
Blériot 39: 62
Blériot 43: 63
Blériot 44: 63, 64
Blériot 45: 64
Blériot 53: 64
Blériot 65: 66
Blériot 67: 66, 68, 69, 74
Blériot 71: 69, 71, 105
Blériot 73: 71, 74
Blériot 74: 74, 75
Blériot 75: 74, 75
Blériot 76: 74, 75
Blériot 77: 75
Blériot 7L: 75
Blériot Monoplane: 65
Blériot Four-Engine Bomber: 65
Blériot La Vache: 62
Blériot Twin-Fuselage Monoplane: 64
BM 4: 91, 92, 94, 96, 101, 227
Borel C1: 79
Borel Floatplane: 76
Borel-Odier B.O.2: 77,78
Borel High Seas Flying Boat: 81
Borel Twin-Engine Floatplane: 81
Borel-Boccaccio Type 3000: 79, 80
Breguet BU 3/BUC/BLC/BC: 1, 3, 56,88, 91, 94, 356
Breguet 1 ï88
Breguet 2: 88
Breguet 3: 88
Breguet 4: 4, 6, 83, 93, 468
Breguet 5: 6, 91, 92, 94–96, 99, 101, 110, 264
Breguet 6: 96
Breguet 11: 99, 203, 379
Breguet 12: 96, 97
Breguet 13: 86, 102
Breguet 14: 10, 12, 15, 17, 35, 36, 39, 44, 45, 46, 69, 92, 101–126, 129, 132, 139, 150, 152, 155, 159, 160, 162, 165, 167–169, 237, 238, 241, 277, 286, 296, 305, 416, 439, 444, 450, 451, 456, 457, 463, 465–472, 497, 498, 515, 520, 551, 554, 562
Breguet 15: 126
Breguet 16: 51, 74, 126, 127, 171, 295
Breguet 17: 102, 104, 129, 249, 275, 276, 277, 341, 343
Breguet AG 4: 3
Breguet SN 3: 89, 453
Breguet Twin-Engine Bomber: 99
Breguet U2: 85
Breguet-Michelin 4: 27, 167
Breguet-Michelin BUM: 88
Brun-Cottan Flying Boat: 132
Brun-Cottan H.B. 2: 132
BUC: 88, 91, 94
C
Canton S2: 132, 133, 284, 306
Caproni C.E.P. 1: 133, 137
Caproni C.E.P. 2: 135, 137
Caproni C.E.P. 3: 137
Carroll A2: 139, 305
Caudron C.20: 170
Caudron C.21: 170, 171
Caudron C.22: 171
Caudron C.23: 69, 74, 173, 251, 298, 463, 563
Caudron CRB: 175
Caudron G.3: 360
Caudron G.2: 141, 142, 144
Caudron G.3: 3, 4, 6, 33, 56, 76, 142–149, 155, 184–186, 207, 208, 315, 362, 416, 421, 431, 474, 544
Caudron G.4: 3, 4, 5, 6, 7, 25, 56, 91, 144, 146, 149–156, 227, 228, 244, 284, 356, 357, 361, 465, 466, 498, 545, 553, 556
Caudron G.5: 156, 157, 163
Caudron G.6: 7, 9, 152, 158, 160, 161, 237, 290, 328, 440, 442, 498, 553
Caudron Heavy Bomber: 175
Caudron J: 141, 142
Caudron O2: 170, 175
Caudron R and R.3: 163
Caudron R.4: 4, 6, 39, 151, 159, 163–166, 236, 237, 290, 291, 466, 467
Caudron R.5: 166, 167
Caudron R.9: 167
Caudron R.10: 167
Caudron R.11: 15, 17, 107, 112; 167–170, 291
Caudron R.12: 14, 170
Caudron R.14: 170, 289
Clément-Bayard: 36, 176, 177, 198, 226
Coanda-Delaunay-Belleville: 177
Courtois-Suffit Lescop Fighter with Clerget Engine: 179
Coutant: 77, 179
D
D.D. 1: 77, 179, 192, 196, 198
D.D. 2: 24, 192, 194, 263
D.D. 8: 24, 194, 195, 196, 261
D.D. 9: 194, 195, 196
D.D. 10: 77, 179, 196, 198
D.N.F.: 6, 34, 203, 204
Donnet-Lévêque Type A: 190–192
Donnet-Lévêque Type B: 190–192
Donnet-Lévêque Type C: 190–192
De Bruyère Canard: 180
De Marcay 1 C1: 181
De Marcay 2 C1: 181
De Monge: 182
Deconde: 181
Delattre: 14, 173, 184, 463
Deperdussin Monocoque: 184, 185
Deperdussin TT: 185, 187
Descamps Type 27: 188, 189
Donnet-Denhaut P.10 and P.15: 198
Dorand Armored Interceptor: 198
Dorand BU: 201
Dorand DO.1: 198
Dorand Flying Boat: 201
Dormay: 201
Doutre: 201
Dufaux C1: 202
Dufaux Twin-Engine Fighter: 203
E
E.G.A.: 205
Farman
F.1,40: 242
F.1,40bis: 242
F.1,41: 242
F.1,41bis: 242
F.1,46: 244
F.2,40: 242
F.30 C2: 245
F.31: 249, 250, 275
F.40: 4, 6, 7, 10, 26, 27, 35, 37, 38, 44, 45, 91, 146, 153, 224–228, 230, 231, 233–238, 240–242, 244, 245, 247, 249, 284, 291, 324, 465, 466, 530, 544, 547, 553
F.41: 10, 238, 241
F.41 H: 241
F.41bis: 241
F.41bis H: 241
F.43: 10, 244
F.44: 245
F.45: 230, 247–249
F.47: 247, 248, 249
F.48: 247, 249
F.49: 247, 249
F.50: 15, 35, 36, 51, 69, 74, 170, 171, 206, 249, 251–254, 295, 563, 565, 566
F.51: 241, 254
H.F.20: 207–215, 219–221, 233
H.F.21: 212
H.F.22: 212–215
H.F.22 Floatplane: 213
H.F.23: 213, 214
H.F.24: 214
H.F.26: 215
H.F.27: 212, 213, 215–217, 221
H.F.30: 217, 218, 233, 240, 245
H.F.33: 233
H.F.35: 4, 233, 244
H.F.36: 244
M.F.7: 56, 185, 186, 207, 208, 217, 219–222, 224, 228–231, 233, 247, 316, 351, 550
M.F.11: 3, 35, 56, 58, 91, 144, 150, 199, 207, 209, 216, 217, 219–234, 237, 238, 240, 241, 291, 316, 357, 362, 416, 463, 465, 544, 554
M.F.12: 232
Farman BN2 with 400-hp Lorraine Engines: 254
Farman High Seas Flying Boat: 255
Farman Renault Flying Boat: 255
Farman Unknowns: 256
F
F.B.A. High Seas Flying Boat: 264
F.B.A. Triplane Flying Boat: 265
F.B.A. Type A: 257, 258
F.B.A. Type B: 258–260
F.B.A. Type C: 26, 27, 259, 301
F.B.A. Type H: 26, 259, 260, 261
F.B.A. Type S: 263
FLO L, La: 257
G
Galvin Floatplane Fighter: 265
Georges Levy 40 HB2: 303
Glisseurs: 573
Gliders: 573
Goupy: 1, 190, 266, 541
Gourdou-Leseurre 1C1,2C1: 266-269
H
Hanriot Twin-Engine Aircraft: 284
HD.1: 270–273, 278, 282, 284, 384, 403, 490, 491, 528
HD.2: 270, 273–275, 278, 284, 337, 528
HD.3: 49, 79, 129, 181, 275–279, 281, 282, 337, 412, 414, 473, 504, 524
HD.4: 277, 278, 284
HD.5: 79, 278, 279, 341
HD.6: 278–281
HD.7: 281, 282, 412, 462, 473
HD.8: 282
HD.9: 282, 283, 446
HD.12: 284
Henri Farman BN2: 254
Hochart: 14, 132, 284, 306
J
Janoir: 35, 36, 284, 285, 460, 486, 492
L
L.A.F.: 287, 611
L.D.: 249, 289
Laboratory Eiffel: 14, 75, 285, 286
Labourdette-Halbronn H.T.1, H.T.2: 287
Larnaudi: 287, 288, 302
Latécoère: 35, 36, 106, 288, 440, 444
Latham: 196, 198, 263, 264, 288
Levy-Besson 'Alerte' Flying Boat: 300, 301
Levy-Besson 450-hp Flying Boat: 301
Levy-Besson 300-hp Flying Boat: 301
Levy-Besson 500-hp Flying Boat: 301,302
Levy-Besson 'High Seas' Three-Engine Flying Boat: 303
Levy, George 40 HB2: 303ï304
LeO 3: 305
LeO 4: 305
LeO 4/1: 305
LeO 5: 14, 284, 306, 308
Letord 1: 10, 290, 292, 294
Letord 2: 290–292, 294
Letord 3: 295–297
Letord 4: 14, 291, 294
Letord 5: 291, 294
Letord 6: 289, 295, 296
Letord 7: 297, 298
Letord 9: 14, 74, 173, 298, 463
Levasseur: 299
Levy-Besson 300-hp Hispano-Suiza Flying Boat: 302
Levy-Besson 500-hp Bugatti Flying Boat: 302
Lioré et Olivier 3: 305
Lioré et Olivier 4: 305
Lioré et Olivier 5: 306
Lioré et Olivier S2: 308
M
Moncassin: 308
Morane-Saulnier Projects: 348
Morane Saulnier L: 355
Morane Saulnier P: 466
Morane-Saulnier TRK: 329
Morane-Saulnier AC: 332, 337
Morane-Saulnier AE: 335
Morane-Saulnier AF: 335, 337
Morane Saulnier AFH: 337
Morane-Saulnier AI: 337, 395, 402
Morane-Saulnier AN: 288, 345
Morane-Saulnier ANL: 345
Morane-Saulnier ANR: 345
Morane-Saulnier ANS: 345
Morane-Saulnier BB: 345
Morane-Saulnier BH: 348
Morane-Saulnier G: 2, 311, 314
Morane-Saulnier I: 313, 322, 331
Morane-Saulnier L: 56, 309, 312, 317, 318, 320, 355, 431
Morane-Saulnier LA: 318
Morane-Saulnier M: 320
Morane-Saulnier N: 321, 323
Morane-Saulnier O: 323
Morane-Saulnier P: 323, 326, 466, 467, 498
Morane-Saulnier S: 34, 167, 327
Morane-Saulnier T: 156, 328, 329
Morane-Saulnier U: 331
Morane-Saulnier V: 331
Morane-Saulnier WR: 309
Morane-Saulnier X: 332
Morane-Saulnier Y: 332
N
Nieuport 4R: 419
Nieuport 6H: 352, 354, 355
Nieuport 6M: 22, 208, 351, 352, 355
Nieuport 10: 3, 4, 6, 316–318, 322, 323, 349, 355, 357–360, 362, 364, 365, 368–370, 374, 388, 414, 416, 418, 545
Nieuport 11: 4, 6–10, 91, 203, 227, 238, 313, 316, 322, 331, 357–365, 368, 377–379, 384, 385, 388, 390, 416, 417, 423, 479, 491, 554
Nieuport 12: 10, 357, 359, 362, 365, 368, 369, 383, 387, 417, 418
Nieuport 13: 368, 374
Nieuport 14: 10, 374, 375, 418, 425
Nieuport 15: 376, 377, 379, 418
Nieuport 16: 7, 10, 270, 362, 377–379
Nieuport 17: 7, 9, 13, 17, 18, 270, 360, 363–365, 370, 378–386, 388, 390–392, 395, 397, 403, 404, 485, 488, 490, 560
Nieuport 18: 374, 379, 386
Nieuport 19: 386, 400
Nieuport 1916 Biplane with 150-hp Clerget: 400
Nieuport 20: 369, 387
Nieuport 21: 388–391
Nieuport 23: 338, 383, 390–392, 396, 405, 489
Nieuport 24: 10, 338, 386, 390, 392, 393, 396–400, 403, 415, 418, 490
Nieuport 24bis: 392, 393, 397, 398, 400, 403, 490
Nieuport 25: 399, 400
Nieuport 27: 272, 384, 392, 399, 400, 403–405
Nieuport 28: 14, 178, 338, 400, 403, 405–408, 510, 514
Nieuport 29: 14, 181, 182, 189, 281, 397, 400, 407, 411–415, 419, 473, 508, 509, 521, 524, 572
Nieuport 30: 14, 416
Nieuport 31: 410, 419
Nieuport 80: 416–418
Nieuport 81: 417
Nieuport 82: 375, 418
Nieuport 83: 417, 418
Nieuport Fighter with 150-hp Hispano-Suiza: 390
Nieuport Fighter with 230-hp Le Rhône 9L: 399
Nieuport Fighter with 250-hp Clerget: 399
Nieuport Fighter with 300-hp Hispano-Suiza 8Fb: 411
Nieuport Monoplane: 410
Nieuport Reconnaissance 150-hp Le RhÛne Engine: 369, 373
Nieuport Reconnaissance Aircraft with 150-hp Hispano-Suiza: 369, 373
Nieuport Reconnaissance Aircraft with 180-hp Lorraine-Dietrich: 373
Nieuport Reconnaissance Aircraft with 200-hp Hispano-Suiza: 373
Nieuport S: 409–411, 419
Nieuport-Tellier Designs: 419
Nieuport Triplanes: 369-371
Nieuport Twin-Fuselage Aircraft: 379
Nieuport Two-Seat Fighter with 370-hp Lorraine Dietrich 12Bd: 408
Nieuport Fighter with 275-hp Lorraine-Dietrich 8Bd: 409
P
Paul Schmitt 3: 451, 456
Paul Schmitt 7: 13, 110, 454–456
Paul Schmitt 9: 456
Paul Schmitt 10: 456, 457
Paul-Schmitt 13: Appendex
Paul-Schmitt 14: Appendex
Paul-Schmitt 260-hp Salmson: Appendex
Paul-Schmitt Armored Aircraft: Appendex
Paul-Schmitt BN 3/4: 458
Paul Schmitt C2: 457
Paul Schmitt Floatplane: 457
Paul Schmitt S.B.R.: 4, 453
Paul Schmitt Type 6: 453
Ponnier L.1: 422
Ponnier Armored Reconnaissance Aircraft: 423
Ponnier M.1: 423, 425
Ponnier M.2: 374, 425
Ponnier P.I: 426
Ponnier Pusher: 428
R
R.E.P. C1: 433-434
R.E.P. N: 208, 316, 430–433
R.E.P. Parasol: 430- 431
Rausser: 428
Renault Fighter: 430
Renault O1: 428
S
S.I.A. BN2: 463
S.I.A.-Coanda Lorraine Bomber: 464
S.M.1: 9, 156, 290, 308, 328, 435, 436, 437, 438, 539
S.M.2: 308, 438
Salmson 2: 12, 14, 15, 17, 35, 36, 38, 39, 44–46, 150–152, 155, 158–160, 162, 225, 235, 270, 275, 288, 291, 439–447, 449, 450, 466–469, 471, 497, 499, 515, 517, 520, 562
Salmson 3: 446
Salmson 4: 14, 447, 448, 540
Salmson 5: 449, 450
Salmson 6: 450
Salmson 7: 14, 450
Schneider Henri-Paul S3: 458
SEA 1: 459, 460, 611
SEA 2: 460, 611
SEA 3: 460, 611
SEA 4: 14, 275, 276, 341, 343, 408, 450, 460, 524
SEA Floatplane: 462
Semenaud C1: 181, 412, 462
Semenaud Single-Engine Flying Boat: 462
Semenaud Twin-Engine Flying Boat: 462
Short Bomber: 156, 463
Sikorsky Bomber: 465
SN 3: 89, 91, 453
Sopwith Dauphin: 473
Sopwith 11/2 Strutter 465-473
SPAD 5: 484, 485
SPAD 7: 8, 10, 12, 13, 17, 18, 36, 181, 270, 272, 299, 332, 335, 360, 361, 365, 380, 382–384, 390, 392, 393, 395–397, 400, 401, 403, 407, 408, 412, 442, 459, 460, 470, 478, 484–494, 500–503, 505, 507–510
SPAD 11: 10, 14, 15, 45, 46, 102, 151, 152, 159, 160, 291, 400, 442, 469, 494, 495, 497–499, 502, 515, 516, 518, 523, 562
SPAD 12: 10, 203, 500, 501, 511
SPAD 13: 10, 12, 15, 17, 36, 51, 112, 175, 188, 339, 397, 400, 403, 406, 408, 415, 433, 446, 486, 488, 490, 491, 494, 501–506, 508–511, 514, 521, 526, 527, 560, 572
SPAD 14: 299, 511, 512, 528
SPAD 15/1: 514
SPAD 15/2: 514
SPAD 15/3: 514, 515
SPAD 15/4: 515
SPAD 15/5: 514, 515
SPAD 16: 36, 39, 160, 162, 226, 235, 236, 324, 446, 466, 469, 497, 498, 515–520, 523
SPAD 17: 14, 502, 521, 526, 527, 572
SPAD 18: 14, 181, 188, 281, 412, 462, 473, 523, 524
SPAD 19: 523
SPAD 20: 14, 79, 80, 188, 281, 412, 502, 508, 523, 524
SPAD 21: 14, 188, 281, 412, 526
SPAD 22: 14, 281, 412, 527
SPAD 23: 528
SPAD 24: 512, 528
SPAD SA.1: 474, 478
SPAD SA.2: 478, 479
SPAD SA.3: 66, 167, 233, 478, 479
SPAD SA.4: 476, 478, 479
SPAD SB: 479
SPAD SC: 479
SPAD SD: 479, 481
SPAD SE: 479, 481
SPAD SF: 482
SPAD SG: 482, 483
SPAD SH: 484
SPAD SI: 484
SPAD SJ: 484, 485
SPAD SK: 484

T
Tellier Four-Engine Flying Boat: 537
Tellier T.1: 529
Tellier T.2: 24, 529
Tellier T.3: 194, 529–531, 534
Tellier T.4: 77, 179, 532, 533
Tellier T.5: 419, 534
Tellier T.6: 26, 419, 535
Tellier T.7: 536, 537
Tellier T.8: 419, 537
Tellier-Nieuport S: 409–411, 419
V
Van Den Born F5: 379, 539
Vendôme A3 Biplane: 539
Vendôme Ab2: 540
Vendôme Monoplane: 539
Vickers F.B.24G: 540
Voisin 1: 541–544, 550, 552
Voisin 1912: 541, 542
Voisin 1913: 2, 542–544
Voisin 2: 542
Voisin 3: 4, 6, 22, 27, 33, 85, 133, 150, 153, 199, 219, 315, 355, 362, 542–552, 554, 569
Voisin 4: 6, 264, 483, 550, 551, 554, 560
Voisin 5: 7, 10, 216, 463, 465, 546, 552–557, 559
Voisin 6: 554, 556
Voisin 7: 557
Voisin 8: 9, 10, 113, 126, 238, 456, 469, 470, 550, 551, 556, 559–561, 563, 567, 569
Voisin 9: 562
Voisin 10: 15, 36, 92, 105, 110, 560, 561, 563–567
Voisin 11: 14, 51, 171, 566
Voisin 12: 14, 74, 170, 173, 284, 458, 463, 567–569
Voisin BN3/4 Bomber - Hispano-Suiza Engines: 569
Volsln BN3/4 Bomber - Salmson Engines: 569
Voisin E.28: 557
Voisin E.50: 569
Voisin E.53: 569
Voisin E.59: 569
Voisin E.87: 569
Voisin L: 542,543
Voisin Triplane 1915: 552
Voisin Triplane Flying Boat: 569
Voisin Type M: 569
Voisin Type O: 570
Voisin with 400-hp Liberty Engine: 567
Voisin Creme de Menthe: 570
Voisin Wild Duck: 570
W
Weymann W.1: 571
Wibault Wib.1: 572

Unit Index
A
AR 1: 13, 17, 18, 38, 44, 49, 224, 234, 441
AR 2: 13, 16, 17, 18, 38, 39, 44, 106, 107, 109, 110, 236, 436, 437, 441, 443, 471, 497, 516, 518, 520
AR 8: 440
AR 14: 13, 17, 18
AR 19: 38, 44, 441
AR 20: 38, 44, 106, 107, 236, 497
AR 21: 39, 44, 107, 516
AR 22: 13, 17, 18, 39, 44, 441
AR 26: 16, 44, 110, 443, 518
AR 32: 39, 44, 441
AR 33: 39, 44, 441
AR 35: 39, 106, 225, 235
AR 40: 44, 441
AR 41: 17, 44
AR 44: 13, 16, 18, 44, 106
AR 45: 44, 106
AR 50: 44, 442
AR 52: 17, 44, 45, 442
AR 58: 17, 44, 291, 442
AR 59: 44, 291, 442
AR 70: 17, 44, 442
AR 71: 44, 442
AR 72: 17, 44, 442
AR 201: 44, 106, 236
AR 203: 44
AR 205: 44, 107
AR 211: 44, 107
AR 230: 44, 443
AR 233: 44, 109
AR 253: 17, 44, 443
AR 254: 17, 18, 44, 443
AR 256: 16, 44, 443, 518
AR 257: 16, 44, 110
AR 258: 17, 44
AR 259: 17, 44, 444
AR 261: 44, 518
AR 262: 44, 443

AR 264: 44, 443
AR 268: 16, 44, 518
AR 272: 16, 44, 110
AR 274: 17, 44, 110
AR 275: 16, 18, 44, 110
AR 288: 16, 44, 443
AR 289: 17, 436, 437, 520
AR 521: 17, 45
AR 533: 113
AR 556: 17
B
BL 3: 3, 5, 55, 56, 144, 316
BL 9: 3, 56, 150
BL 10: 3, 56, 144
BL 18: 3, 56, 144
BL 30: 56, 144
BLC 4: 3, 56
BLC 5: 56
BM 117: 91, 92, 110
BM 118: 91
BM 119: 91, 92
BM 120: 91, 92, 101, 110, 238
BM 121: 16, 91, 92
BR 7: 16, 20, 106, 114
BR 9: 20, 106, 114
BR 29: 19, 114
BR 35: 17, 19, 106, 114
BR 44: 19, 106, 114
BR 45: 16, 106
BR 66: 19, 110, 114, 117, 169
BR 104: 106, 114
BR 107: 16, 19, 110, 114, 169
BR 108: 19, 110, 114
BR 111: 19, 110, 114, 169
BR 113: 19
BR 117: 19, 110, 114, 115, 169
BR 120: 19, 110, 114, 115
BR 123: 19, 110, 114
BR 126: 16, 19, 110, 114
BR 127: 19, 110, 114
BR 128: 16, 19, 110, 114
BR 129: 19, 110, 114
BR 131: 16, 19, 110, 114, 169
BR 132: 16, 19, 110, 114
BR 134: 16, 19, 110, 114
BR 141: 20, 106
BR 201: 16, 19, 106, 114
BR 202: 19, 107
BR 206: 20, 107
BR 207: 17, 20, 107, 114
BR 208: 16, 20, 107
BR 209: 16, 19, 107
BR 210: 16, 20, 107
BR 211: 17, 20, 107, 114
BR 213: 20, 107
BR 214: 20, 107
BR 216: 107
BR 217: 19, 107, 467
BR 220: 17, 107, 114
BR 221: 20, 467
BR 223: 20, 108
BR 224: 108, 114
BR 226: 19, 108, 114
BR 227: 108, 114
BR 228: 20, 108, 114, 498
BR 229: 108
BR 231: 109
BR 232: 20, 109
BR 233: 17, 109
BR 234: 19, 20, 109, 114, 468
BR 235: 109
BR 236: 20, 109
BR 237: 20, 109
BR 238: 109, 468
BR 243: 20, 109, 114
BR 244: 17, 20, 109, 114
BR 245: 20
BR 250: 110
BR 257: 20, 110
BR 260: 110
BR 269: 20, 110
BR 271: 110, 469
BR 272: 110
BR 275: 19, 110
BR 279: 19, 110
BR 281: 20, 110
BR 282: 19, 110, 469
BR 283: 19
BR 501: 17, 113
BR 503: 17, 113
BR 504: 17, 113, 114
BR 505: 17, 113, 114
BR 508: 113, 114
BR 509: 17, 110, 113, 114
BR 509 NE: 110
BR 510: 17, 113, 114
BR 522: 17, 113, 117, 118
BR 524: 17, 113, 117, 118
BR 525: 17, 113, 117, 118
BR 532: 17, 113
BR 533: 113

BR 534: 113
C
C 4: 5, 6, 7, 8, 10, 11, 19, 144, 150, 151, 158, 159, 164, 165, 185, 291, 328, 396, 441, 466, 498, 506
C 6: 5, 6, 7, 8, 10, 11, 91, 144, 149, 150, 152, 153, 158, 160, 185, 227, 291, 356, 357, 361, 440, 469, 498, 546
C 9: 5, 6, 8, 11, 150, 158, 291, 466
C 11: 3, 5, 6, 9, 11, 106, 142, 143, 150, 159, 328
C 13: 5, 6, 8, 10, 144, 150, 159, 466
C 17: 5, 6, 8, 11, 19, 142, 144, 150, 159, 466
C 18: 5, 6, 8, 10, 144, 151, 159
C 21: 8, 39, 107, 151, 152, 159, 436, 466
C 27: 5, 6, 7, 11, 144, 151, 157, 159, 431
C 28: 5, 6, 8, 10, 144, 151, 159, 466
C 30: 5, 6, 8, 11, 142, 144, 151, 159, 291, 328, 441
C 34: 5, 6, 8, 11, 91, 144, 151, 153, 159, 516
C 39: 5, 6, 8, 11, 142, 144, 151, 159, 291, 328, 466
C 42: 6, 8, 11, 144, 151, 159, 498
C 43: 6, 8, 144, 151, 159, 466
C 46: 6, 7, 8, 19, 144, 151, 159, 165, 396, 506
C 47: 6, 11, 151, 159, 164, 328
C 51: 6, 8, 144, 151, 159, 466
C 53: 6, 7, 8, 11, 144, 152, 159, 291, 498
C 56: 6, 8, 144, 152, 160, 442
C 61: 8, 11, 91, 152, 153, 160, 227, 291
C 64: 6, 8, 149, 152, 160, 498
C 66: 6, 7, 10, 149, 153, 357, 469, 546
C 74: 152, 160, 442
C 106: 8, 149, 152, 153, 160, 467
C 122: 10, 152, 153, 160
C 202: 11, 107, 152, 160
C 207: 8, 152, 160, 324, 467
C 217: 467
C 219: 152, 436, 467
C 220: 8, 10, 107, 152, 160
C 224: 8, 10, 108, 152, 160
C 225: 152, 160
C 226: 152, 160, 467
C 227: 11, 108, 160
C 228: 8, 160, 498
C 543: 17, 153, 161
C 544: 17, 153, 162
C 545: 153, 162
C 546: 153, 162
C 547: 153, 162
C 548: 153, 162
C 549: 153, 162
C 575: 162
CAP 115: 20, 135, 137, 138, 173
CAP 130: 20, 135, 138
CEP 115: 11, 135
CM 39: 142, 144
D
D 4: 3, 185, 270
D 6: 3, 144, 185, 358, 391
D 36: 473
DO 14: 199
DO 22: 199, 224
F
F 16: 10, 38
F 22: 8, 10, 11, 39, 160, 324, 436, 467
F 25: 8, 10
F 29: 91, 469, 471
F 44: 8, 10, 44
F 58: 8, 11, 44, 436
F 70: 8, 11, 44
F 71: 8, 44, 436
F 72: 8, 11, 12, 44, 291, 436
F 201: 44
F 203: 44, 324
F 204: 8, 467
F 205: 44
F 206: 11, 324, 467
F 208: 8, 467
F 210: 8, 165
F 211: 44
F 215: 8, 324
F 216: 8, 10, 467
F 218: 107, 324
F 221: 8, 10, 467
F 223: 8, 11, 436, 467
F 553: 17
F 554: 17
G
G 488: 26, 153, 161
G 489: 153, 161
G 490: 153, 161
H
H 387: 9, 403
HF 1: 3, 144

HF 7: 3, 219
HF 13: 3, 144
HF 19: 3
HF 28: 144
HF 32: 5
HD 174: 277, 506, 508
L
LET 46: 291
LET 487: 28
M
MF 1: 3, 5, 6, 7, 9, 48, 49, 86, 91, 316, 469
MF 2: 3, 5, 6, 7, 9, 12, 91, 153, 199, 291
MF 5: 3, 5, 6, 7
MF 7: 5, 6, 466
MF 8: 3, 5, 6, 9
MF 14: 5
MF 16: 3, 5, 6, 86, 316
MF 20: 3, 5, 6
MF 22: 5, 6, 199
MF 25: 5, 6, 7, 9, 12
MF 29: 9, 91, 153
MF 32: 5
MF 33: 5, 6, 39
MF 35: 5, 6, 7, 39
MF 36: 5, 12, 185
MF 40: 6
MF 41: 6
MF 44: 6
MF 45: 6
MF 50: 6
MF 55: 7
MF 58: 6
MF 59: 6
MF 60: 6
MF 62: 6, 381
MF 63: 6
MF 70: 6
MF 82: 9
MF 98: 6, 9
MF 99: 6, 9
MF 123: 9, 91, 469
MF 382: 365
MF 384: 113
MF 398: 113
MF 399: 113
MS 3: 5, 56, 315, 316, 317, 321, 356, 360
MS 15: 16, 316, 338, 339, 356, 395, 402, 431, 487, 503
MS 12: 5, 315, 321, 351
MS 23: 5, 309, 315, 356
MS 26: 5, 309, 315, 355, 356
MS 31: 5, 315, 356
MS 37: 5, 315, 317, 356
MS 38: 5, 315, 356
MS 48: 316, 356
MS 49: 316, 356
MS 140: 324, 498
MS 161: 17, 339, 487, 503
N
N 3: 6, 7, 8, 10, 11, 12, 17, 62, 236, 317, 356, 360, 361, 362, 365, 366, 368, 377, 378, 381, 382, 383, 385, 390, 392, 393, 395, 396, 401, 403, 486, 487, 488
N 12: 3, 6, 7, 11, 12, 315, 324, 351, 360, 361, 365, 368, 378, 380, 382, 383, 390, 393, 395, 396, 401, 487, 489, 490
N 15: 6, 8, 11, 12, 16, 17, 232, 338, 356, 360, 361, 365, 378, 380, 383, 385, 386, 390, 395, 396, 402, 403, 487, 507
N 23: 6, 7, 8, 10, 356, 360, 361, 364, 365, 378, 380, 390, 393, 486
N 26: 7, 11, 12, 356, 360, 365, 378, 381, 383, 390, 396
N 31: 6, 11, 12, 17, 356, 360, 361, 368, 378, 381, 383, 385, 390, 395, 396, 403, 487
N 37: 6, 7, 11, 12, 236, 317, 356, 361, 366, 378, 381, 383, 390, 392
N 38: 6, 8, 10, 12, 356, 360, 361, 362, 365, 366, 368, 377, 378, 381, 390, 393, 486
N 48: 6, 7, 11, 12, 356, 357, 361, 366, 378, 381, 383, 390, 396
N 49: 6, 8, 11, 227, 356, 360, 361, 362, 366, 368, 377, 378, 381, 383, 390, 393, 476, 486
N 57: 6, 7, 11, 222, 316, 356, 361, 366, 378, 383, 390, 396, 403
N 62: 8, 11, 12, 13, 236, 366, 378, 381, 390, 393, 486
N 65: 6, 7, 11, 12, 153, 236, 356, 357, 361, 366, 378, 383, 385, 390, 396, 546
N 67: 6, 7, 8, 11, 324, 356, 361, 366, 378, 383, 390
N 68: 6, 8, 11, 356, 361, 366, 378,

381, 390, 393
N 69: 7, 8, 11, 12, 13, 361, 366, 378, 381, 390, 393, 487
N 75: 8, 11, 12, 368, 378, 381, 383, 390, 393, 396
N 76: 8, 11, 12, 381, 390, 393, 401, 487
N 77: 8, 11, 368, 378, 381, 390, 394, 487
N 78: 10, 368, 381, 383, 390, 396, 489
N 79: 10, 324, 368, 381, 390, 394, 401, 487
N 80: 11, 12, 368, 383, 390, 396
N 81: 11, 12, 368, 382, 383, 390, 394, 396, 401, 487
N 82: 11, 13, 17, 18, 382, 383, 390, 394, 396, 401, 487
N 83: 11, 12, 382, 383, 390, 396
N 84: 383, 390, 393, 396
N 85: 383, 390, 394, 395, 396, 401, 489
N 86: 383, 390, 396, 403, 489
N 87: 12, 17, 382, 390, 394, 395, 401, 487
N 88: 382, 390, 394, 401
N 89: 394, 403
N 90: 6, 9, 17, 382, 383, 390, 394, 395, 400, 401, 402, 487
N 91: 6, 9, 382, 390, 394, 401, 403
N 92: 17, 361, 382, 394, 395, 396, 401, 403, 489
N 93: 382, 383, 390, 394, 396, 489
N 94: 383, 390, 394, 396, 401
N 95: 394, 396, 401, 489
N 96: 395, 401
N 97: 383, 390, 395, 401
N 98: 16, 395, 401, 403, 490, 507
N 99: 395, 401, 403, 507
N 100: 403, 489
N 102: 8, 10, 12, 361, 368, 378, 382, 383, 395, 487
N 103: 11, 12, 361, 368, 378, 383, 390, 396
N 112: 8, 10, 12, 361, 378, 383, 390, 396, 489
N 124: 11, 324, 361, 368, 378, 382, 383, 390, 395, 396, 487, 489, 490
N 150: 395, 487
N 151: 395
N 153: 232, 390, 396, 487
N 154: 396
N 155: 385, 386, 395, 402
N 156: 338, 385, 386, 395, 402
N 157: 16, 395, 402, 403, 507
N 158: 338, 395, 402
N 159: 395, 402, 403, 507
N 160: 395, 402, 487, 503
N 161: 395, 402
N 162: 395, 403, 507
N 313: 395, 403, 487
N 314: 17, 395, 403
N 315: 17, 395, 403, 487
N 390: 488
N 391: 361, 365, 488
N 392: 361, 382, 395, 401, 403
N 523: 403, 488, 504
N 531: 17, 403
N 561: 17, 18, 395, 403, 488, 504
N 562: 403
P
PS 125: 455, 456
PS 126: 110, 455
PS 127: 110, 455
PS 128: 455, 469, 471
R
R 46: 15, 165, 168, 169, 291, 489
R 209: 291
R 210: 11, 237
R 241: 20
R 242: 20
R 246: 19
REP 15: 3, 5, 86, 316, 431
REP 27: 144, 431
S
SAL 1: 16, 17, 19, 20, 38, 291, 440, 441, 442, 443, 444
SAL 4: 17, 20, 291, 440, 441, 444
SAL 5: 14, 16, 17, 19, 20, 44, 440, 442, 444, 446, 466
SAL 6: 16, 19, 440, 442, 443
SAL 8: 20, 440, 443, 444
SAL 10: 16, 440, 442, 443
SAL 13: 17
SAL 14: 441
SAL 16: 16, 19, 441, 444
SAL 17: 20, 441, 443
SAL 19: 19, 38, 441, 443
SAL 22: 16, 20, 441
SAL 24: 17, 19, 441, 444
SAL 27: 17, 20, 441, 443
SAL 28: 17, 20, 441, 443, 469
SAL 30: 17, 441
SAL 32: 20, 39, 441, 443
SAL 33: 16, 20, 441
SAL 39: 20, 441, 443
SAL 40: 291, 441, 444

SAL 47: 17, 20
SAL 50: 17, 19, 44, 442
SAL 51: 442, 466
SAL 52: 16, 20, 442
SAL 56: 20, 442
SAL 58: 14, 20, 44, 442, 444, 446
SAL 59: 20, 44, 442
SAL 61: 16, 442
SAL 70: 19, 442
SAL 71: 19, 44, 442, 443
SAL 74: 17, 19, 442
SAL 105: 442
SAL 106: 442
SAL 122: 17, 291, 440, 442
SAL 203: 19, 44
SAL 204: 16, 19
SAL 230: 16, 44, 443
SAL 251: 20, 443
SAL 252: 443
SAL 253: 19, 443
SAL 254: 443
SAL 259: 443
SAL 262: 19, 443
SAL 264: 443
SAL 273: 20, 443
SAL 277: 443
SAL 280: 443, 469
SAL 288: 443
SAL 580: 446
SAL 581: 14, 446
SAL 582: 14, 446
SOP 5: 14, 16, 440, 442, 466, 468, 469, 517
SOP 7: 106, 466, 469
SOP 9: 106, 466, 469
SOP 13: 466, 469, 471
SOP 17: 441, 466
SOP 24: 441, 466, 469
SOP 28: 16, 110, 441, 443, 466, 468, 469, 498, 520
SOP 29: 238, 469, 471
SOP 36: 13, 17, 18, 466, 516
SOP 39: 441, 466, 469
SOP 43: 106, 466
SOP 51: 442, 466, 468
SOP 55: 466, 517
SOP 60: 466, 468, 517
SOP 61: 442
SOP 66: 469, 470
SOP 104: 16, 106, 469
SOP 105: 16, 442, 467

SOP 106: 17, 442, 467
SOP 107: 469, 470, 560
SOP 108: 12, 469, 471, 560
SOP 111: 110, 469, 472
SOP 123: 469
SOP 128: 469, 471
SOP 129: 469, 471
SOP 131: 469
SOP 132: 469, 471
SOP 134: 469, 471
SOP 141: 16, 106, 467
SOP 204: 467
SOP 206: 13, 17, 18, 107, 467
SOP 207: 107, 165, 467, 468
SOP 208: 107, 467
SOP 212: 467
SOP 216: 107, 467
SOP 217: 16, 107, 165, 467, 468
SOP 219: 107, 291, 467
SOP 221: 13, 17, 18, 291, 467
SOP 223: 17, 291, 467, 468
SOP 229: 17, 468
SOP 231: 16, 109, 291
SOP 234: 16, 109, 468
SOP 235: 109, 468
SOP 236: 17, 109, 468
SOP 237: 16, 109, 291, 468
SOP 238: 17, 109, 468
SOP 250: 16, 468
SOP 251: 443
SOP 252: 17, 443, 468
SOP 255: 468, 518
SOP 260: 110, 468
SOP 263: 443, 468
SOP 269: 110, 468
SOP 270: 17, 443
SOP 271: 16, 110
SOP 273: 443, 469
SOP 277: 17, 443, 469
SOP 278: 469, 520
SOP 279: 17, 110, 469
SOP 280: 443, 469
SOP 281: 110, 469
SOP 283: 469
SOP 284: 469, 498, 520
SOP 285: 16, 469, 520
SOP 287: 110, 469
SOP 583: 469
SPA 3: 15, 16, 17, 19, 20, 395, 396, 403, 486, 487, 488, 489, 503, 504, 505, 507, 508, 521

SPA 12: 16, 20, 396, 488, 489, 490, 492, 505, 507, 508
SPA 15: 15, 16, 19, 20, 339, 395, 402, 403, 487, 488, 489, 490, 503, 505, 506, 507, 508
SPA 23: 20, 381, 393, 486, 503
SPA 26: 16, 17, 19, 20, 488, 505, 508, 521
SPA 31: 16, 19, 20, 395, 403, 487, 488, 504, 505, 508
SPA 37: 15, 19, 396, 489, 505, 508
SPA 38: 486, 503, 507
SPA 48: 15, 19, 506, 508
SPA 49: 17, 393, 486, 488, 503, 508
SPA 57: 16, 20, 403, 488, 505, 508
SPA 62: 19, 393, 486, 503, 508
SPA 65: 16, 19, 489, 505, 508
SPA 67: 16, 489, 505, 508, 521
SPA 68: 19, 361, 366, 381, 393, 403, 507
SPA 69: 13, 17, 18, 393, 487, 503, 508
SPA 73: 15, 19, 488, 490, 491, 506, 508
SPA 75: 17, 19, 393, 489, 505
SPA 76: 17, 487, 498, 503
SPA 77: 16, 19, 487, 489, 505, 506, 508
SPA 78: 16, 489, 505
SPA 79: 16, 19, 394, 401, 487, 503
SPA 80: 17, 19, 489, 505
SPA 81: 15, 19, 396, 489, 505, 508
SPA 82: 19, 487, 503, 507
SPA 83: 17, 19, 489, 505
SPA 84: 16, 19, 489, 505, 508
SPA 85: 15, 19, 490, 506
SPA 86: 17, 19, 505
SPA 87: 394, 401, 487, 503, 507
SPA 88: 16, 19, 382, 394, 401, 489, 505, 508
SPA 89: 16, 19, 394, 489, 506
SPA 90: 20, 382, 394, 401, 487, 503
SPA 91: 16, 19, 382, 394, 401, 489, 506
SPA 92: 507
SPA 93: 15, 19, 488, 489, 505, 508
SPA 94: 15, 19, 506, 508
SPA 95: 15, 19, 490, 506, 508
SPA 96: 15, 19, 490, 506, 508
SPA 97: 15, 19, 395, 401, 489, 505, 508
SPA 99: 19, 403
SPA 100: 16, 19, 489, 506
SPA 102: 16, 19, 395, 487, 488, 503, 508
SPA 103: 16, 488, 505, 508, 521
SPA 112: 16, 489, 505
SPA 124: 16, 396, 489, 490, 492, 507, 508

SPA 150: 16, 19, 395, 487, 489, 505
SPA 151: 16, 395, 489, 505
SPA 152: 487, 503, 507
SPA 153: 15, 19, 487, 506
SPA 154: 16, 20, 488, 490, 505
SPA 155: 15, 19, 395, 402, 506
SPA 156: 20, 339, 487, 503
SPA 157: 490
SPA 158: 19, 487, 503, 507
SPA 160: 19, 487, 503, 507
SPA 161: 19, 339, 487, 503, 507
SPA 162: 19, 403, 508, 509
SPA 163: 490, 507
SPA 164: 507
SPA 165: 20, 505
SPA 166: 19, 505
SPA 167: 486, 505, 521
SPA 168: 506
SPA 169: 19, 507
SPA 170: 19, 507
SPA 173: 504
SPA 175: 507
SPA 176: 506
SPA 313: 19, 403, 487, 504
SPA 314: 20, 395, 403, 487, 504
SPA 315: 487, 504
SPA 412: 487, 504
SPA 441: 487, 504
SPA 442: 487, 504
SPA 461: 487, 504
SPA 462: 487, 504
SPA 463: 487, 504
SPA 506: 17, 488, 504
SPA 507: 17, 488, 504, 508
SPA 531: 488, 504
SPA 561: 403, 488, 504
V
V 14: 3, 199, 542, 544
V 21: 3, 5, 6, 236, 237, 542, 544, 545, 553
V 24: 5, 6, 542, 543, 544, 546, 553
V 25: 238, 252, 564, 566
V 29: 5
V 83: 9, 551, 554, 557
V 84: 6, 9, 554, 557
V 85: 6, 9, 554
V 88: 9, 554
V 89: 9, 554
V 207: 152, 165, 553
V 209: 236, 553

507, 508
V 216: 237, 553
V 223: 237, 553
V 383: 113, 554
V 386: 113, 554
V 388: 113, 554
V 389: 113, 554
V 571: 17, 564, 566
VB 1: 6, 16, 19, 110, 152, 252, 361, 368, 374, 467, 469, 544, 545, 550, 552, 554, 556, 559, 560, 561, 564, 565, 566
VB 2: 85, 544
VB 4: 545
VB 5: 219, 545
VB 101: 545, 552, 554, 556, 560, 564
VB 103: 361, 368, 545
VB 104: 152, 545, 556
VB 105: 152, 467, 545, 556
VB 107: 469, 545, 560
VB 108: 469, 545, 560
VB 110: 16, 252, 545, 550, 560, 565
VB 111: 545
VB 112: 361, 374, 545
VB 113: 16, 110, 550, 560, 564
VB 114: 6, 16, 252, 554, 556, 559, 560, 564, 565
VB 125: 16, 19, 564
VB 133: 19, 564, 565, 566
VB 135: 564, 565, 566
VB 136: 564, 566
VB 137: 252, 564, 566
VC 110: 11, 550, 551, 554, 560, 564
VC 111: 469, 550, 551
VC 113: 550, 551, 560
VC 116: 12, 550, 551, 561, 564, 566
VR 118: 16, 564
VR 119: 16, 564
VR 121: 564
VR 291: 20, 564, 566
VR 292: 20, 564, 566
VR 293: 19, 564, 566
VR 294: 564, 566
VR 296: 564, 566
VR 541: 17, 564
VR 542: 17, 564
VR 551: 17, 564, 566
VR 557: 564, 566
VR 558: 17, 564

Related Books from Flying Machines Press

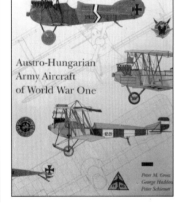

Air Aces of the Austro-Hungarian Empire 1914–1918
(ISBN 0-9637110-1-6) by Martin O'Connor. The standard reference on these courageous aces. ($49.95)
- Biographies of all 49 Austro-Hungarian Empire air aces, including photos, victory details, aircraft flown, and victories by aircrew role (fighter pilot, two-seater pilot, or gunner).
- 240 rare photos.
- 16 color pages illustrating 50 aircraft.
- 336 pages, 8.5" by 11" format.

Austro-Hungarian Army Aircraft of World War One
(ISBN 0-9637110-0-8) by Peter M. Grosz, George Haddow, & Peter Schiemer. The standard reference on these exotic aircraft. ($84.95)
- 903 rare photos and 25 sketches.
- 102 three-view aircraft drawings to standard 1/48 and 1/72 scales.
- 26 pages of all new color illustrations (56 profiles, 16 top views) complementing the color art in *Air Aces of the Austro-Hungarian Empire 1914–1918*
- 570 oversize 10" by 12" pages.

High in the Empty Blue
(ISBN 0-9637110-3-2)
Alex Revell's superb history of 56 Squadron RFC/RAF, the most famous and successful British fighter squadron in WWI. Full of personal anecdotes and air combat. ($49.95)
- 225,000 words of text.
- 430 photos and 20 maps, sketches, and documents.
- 8 pages of color illustrations showing 17 aircraft, including Voss's Fokker triplane in its true colors for the first time.
- 21 appendices.
- 450 pages, 8.5" by 11" format.

The Imperial Russian Air Service (ISBN 0-9637110-2-4)
By Durkota, Darcey, and Kulikov. The first comprehensive coverage of the Russian Air Service in WWI. ($79.95)
- 615 rare photos.
- Scale drawings of 40 aircraft (1/48 or 1/72 scale).
- 48 pages of color illustrations.
- Detailed biography of each Russian ace and each foreign ace in Russia.
- Sikorsky designs.
- Key Russian manufacturers.
- 560 pages, 9" by 12" format.

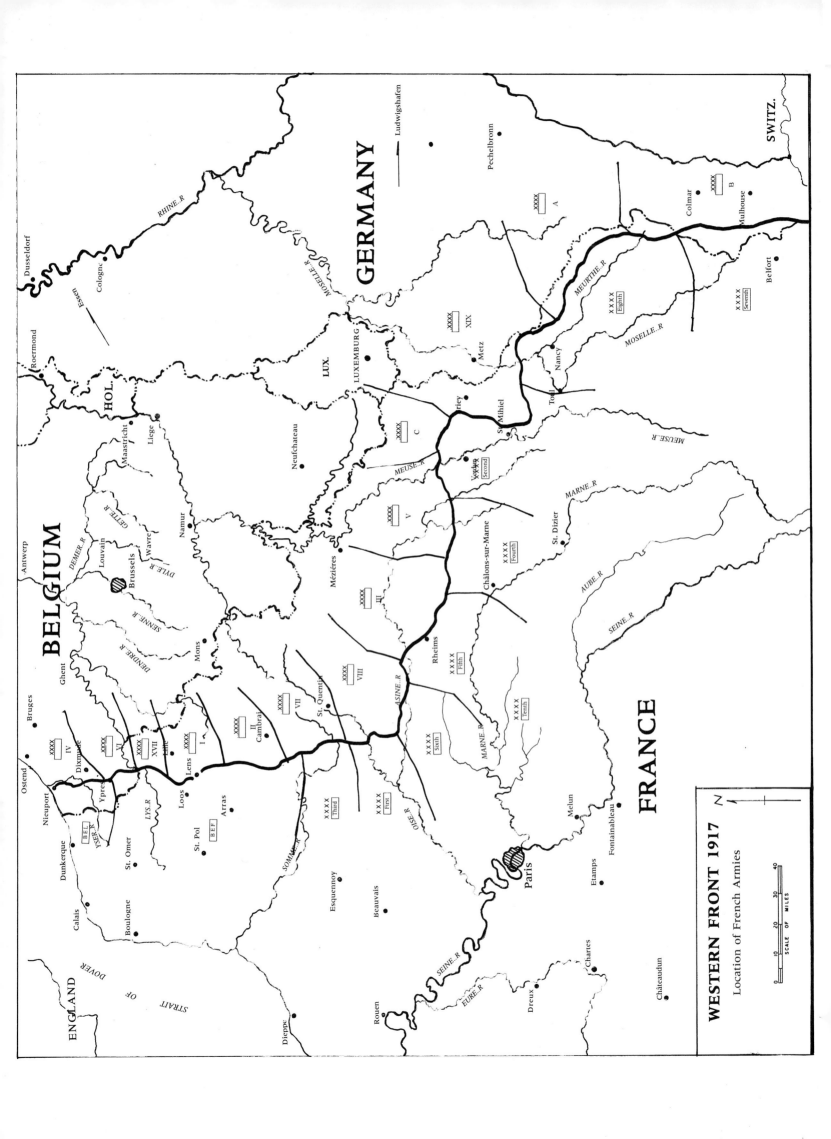